general mathematics: revision and practice

D.Rayner

Head of Mathematics
Richard Hale School, Hertford

Oxford University Press

Oxford University Press, Walton Street, Oxford OX2 6DP

Oxford New York Toronto
Delhi Bombay Calcutta Madras Karachi
Petaling Jaya Singapore Hong Kong Tokyo
Nairobi Dar es Salaam Cape Town
Melbourne Auckland

and associated companies in
Beirut Berlin Ibadan Nicosia

Oxford is a trade mark of Oxford University Press

ISBN 0 19 914242 4

First published 1984
Reprinted 1984, 1985 (twice)

Set by Hope Services, Abingdon
Printed in Great Britain
by R.J. Acford, Chichester

PREFACE

This book is for G.C.E. 'O' level and 16+ candidates in mathematics: it covers the requirements of all the major examination boards. The book contains teaching notes, worked examples and carefully graded exercises. These can be used selectively for classwork, for homework and for later revision.

The work is divided into short topics to suit any required order of treatment. The wide choice of questions provides plenty of practice in the basic skills and leads on to work of a more demanding nature. Each major part of the book concludes with two exercises: one of short revision questions and another of actual past examination questions. All numerical answers are given at the end of the book.

The author is indebted to the many pupils and colleagues who have assisted him in this work. He is particularly grateful to Reg Moxom, Philip Cutts, Michael Day and Micheline Dubois for their invaluable work of correction and checking and to Robert Melville for his illustrations. Thanks are also due to the following examination boards for kindly allowing the use of questions from their past mathematics papers.

Associated Examining Board
East Anglian Examination Board
Joint Matriculation Board
Northern Examining Association for Joint Examinations
Oxford and Cambridge Schools Examination Board
Oxford Delegacy for Local Examinations
Southern Universities' Joint Board
University of Cambridge Local Examinations Syndicate
University of London School Examinations Department
Welsh Joint Education Committee

CONTENTS

Part 7 Graphs

Part 8 Matrices and transformations

Part 9 Sets, vectors and functions

Part 10 Calculus

Part 11 Statistics and probability

ABBREVIATIONS

The following abbreviations are used against questions taken from past examination papers.

[AEB]	Associated Examining Board 'O' level
[C]	Cambridge Examination Syndicate 'O' level
[C16+]	Cambridge and East Anglian Board 16+
[JMB]	Joint Matriculation Board 'O' level
[L]	University of London 'O' level
[NEA]	Northern Examining Association
[O]	Oxford Delegacy 'O' level
[O & C]	Oxford and Cambridge Board 'O' level
[S]	Southern Universities' Board 'O' level
[SMP]	School Mathematics Project 'O' level
[W]	Welsh Joint Education Board 'O' level

1 Number

Karl Friedrich Gauss (1777–1855) was the son of a German labourer and is thought by many to have been the greatest all-round mathematician of all time. He considered that his finest discovery was the method for constructing a regular seventeen-sided polygon. This was not of the slightest use outside the world of mathematics, but was a great achievement of the human mind. Gauss would not have understood the modern view held by many that mathematics must somehow be 'useful' to be worthy of study.

1.1 DECIMALS

Example 1

Evaluate: (a) $7{\cdot}6 + 19$
(b) $3{\cdot}4 - 0{\cdot}24$
(c) $7{\cdot}2 \times 0{\cdot}21$
(d) $0{\cdot}84 \div 0{\cdot}2$
(e) $3{\cdot}6 \div 0{\cdot}004$

(a)
$$\begin{array}{r} 7{\cdot}6 \\ +\ 19{\cdot}0 \\ \hline 26{\cdot}6 \\ \hline \end{array}$$

(b)
$$\begin{array}{r} {}^{3\,1} \\ 3{\cdot}40 \\ -\ 0{\cdot}24 \\ \hline 3{\cdot}16 \\ \hline \end{array}$$

(c)
$$\begin{array}{r} 7{\cdot}2 \\ \times\ 0{\cdot}21 \\ \hline 72 \\ 1440 \\ \hline 1{\cdot}512 \\ \hline \end{array}$$

No decimal points in the working, '3 figures after the points in the question *and* in the answer'.

(d) $0{\cdot}84 \div 0{\cdot}2 = 8{\cdot}4 \div 2$

$$2\overline{)8{\cdot}4}\quad = 4{\cdot}2$$

Multiply both numbers by 10 so that we can divide by a whole number

(e) $3{\cdot}6 \div 0{\cdot}004 = 3600 \div 4$
$= 900$

Exercise 1

Evaluate the following:

1. $7{\cdot}6 + 0{\cdot}31$
2. $0{\cdot}074 + 0{\cdot}14$
3. $15 + 7{\cdot}22$
4. $112 + 0{\cdot}02$
5. $7{\cdot}004 + 0{\cdot}368$
6. $7 + 8{\cdot}1 + 150$
7. $0{\cdot}06 + 0{\cdot}006$
8. $25 + 0{\cdot}84$
9. $4{\cdot}2 + 42 + 420$
10. $0{\cdot}888 + 9{\cdot}78$
11. $3{\cdot}84 - 2{\cdot}62$
12. $5{\cdot}6 - 2{\cdot}42$
13. $11{\cdot}4 - 9{\cdot}73$
14. $0{\cdot}07 - 0{\cdot}062$
15. $4{\cdot}61 - 3$
16. $11 - 8{\cdot}2$
17. $17 - 0{\cdot}37$
18. $200 - 0{\cdot}76$

19. $18{\cdot}6 - 1{\cdot}741$
20. $13{\cdot}62 - 12{\cdot}764$
21. $8{\cdot}7 + 19{\cdot}2 - 3{\cdot}8$
22. $19 + 7{\cdot}61 - 0{\cdot}87$
23. $25 - 7{\cdot}8 + 9{\cdot}5$
24. $0{\cdot}62 - 0{\cdot}824 + 1{\cdot}5$
25. $3{\cdot}6 - 8{\cdot}74 + 9$
26. $25 - 0{\cdot}073 + 5{\cdot}72$
27. $17 - 25{\cdot}24 + 30$
28. $0{\cdot}007 - 0{\cdot}000\,777$
29. $20{\cdot}4 - 20{\cdot}399$
30. 15 000 + 6 million

31. $2{\cdot}6 \times 0{\cdot}6$
32. $8{\cdot}42 \times 0{\cdot}2$
33. $0{\cdot}72 \times 0{\cdot}04$
34. $7{\cdot}2 \times 0{\cdot}9$
35. $27{\cdot}2 \times 0{\cdot}08$
36. $0{\cdot}07 \times 0{\cdot}06$
37. $0{\cdot}1 \times 0{\cdot}2$
38. $(0{\cdot}2)^2$
39. $(0{\cdot}01)^2$
40. $(0{\cdot}3)^3$

41. $2{\cdot}1 \times 3{\cdot}6$
42. $0{\cdot}72 \times 2{\cdot}5$
43. $2{\cdot}31 \times 0{\cdot}34$
44. $0{\cdot}374 \times 0{\cdot}27$
45. $2{\cdot}61 \times 24{\cdot}3$
46. $7{\cdot}2 \times 1000$
47. $0{\cdot}36 \times 1000$
48. $0{\cdot}72 \times 0{\cdot}072$
49. $0{\cdot}34 \times 100\,000$
50. $(2{\cdot}41)^2$

51. $3{\cdot}6 \div 0{\cdot}2$
52. $3{\cdot}71 \div 0{\cdot}7$
53. $0{\cdot}592 \div 0{\cdot}8$
54. $4{\cdot}44 \div 0{\cdot}04$
55. $0{\cdot}1404 \div 0{\cdot}06$
56. $0{\cdot}447\,34 \div 0{\cdot}01$
57. $3{\cdot}24 \div 0{\cdot}002$
58. $0{\cdot}178 \div 0{\cdot}05$
59. $0{\cdot}968 \div 0{\cdot}11$
60. $0{\cdot}0002\,52 \div 0{\cdot}007$
61. $3{\cdot}38 \div 1{\cdot}3$
62. $0{\cdot}465 \times 0{\cdot}15$
63. $4{\cdot}08 \div 1{\cdot}7$
64. $24 \div 0{\cdot}04$
65. $600 \div 0{\cdot}5$
66. $0{\cdot}01 \div 0{\cdot}001$
67. $0{\cdot}007 \div 4$
68. $1500 \div 40$

69. $2640 \div 200$
70. $47{\cdot}2 \div 50$
71. $1100 \div 5{\cdot}5$
72. $(3{\cdot}6 - 0{\cdot}7) \div 0{\cdot}01$
73. $(11 + 2{\cdot}4) \times 0{\cdot}06$
74. $(0{\cdot}2)^2 \div 0{\cdot}4$
75. $(0{\cdot}4)^2 \div 0{\cdot}2$
76. $27 \div (0{\cdot}01)^2$
77. $3{\cdot}6 + (0{\cdot}71 \times 13)$
78. $10{\cdot}27 \div 100$
79. $77 \div 1000$
80. $0{\cdot}24 \div 100$

81. $(11 - 2{\cdot}72) \times 10\,000$
82. $(17 - 8{\cdot}1) \div 100$
83. $(0{\cdot}3)^2 \div 100$
84. $(0{\cdot}2)^2 \div (0{\cdot}1)^2$
85. $(0{\cdot}02)^3 \times 100$
86. $0{\cdot}008 \div (0{\cdot}1)^3$
87. $(2{\cdot}4 + 0{\cdot}14) \times 60$
88. $(0{\cdot}1 - 0{\cdot}01) \times 200$
89. $(0{\cdot}1)^4 \div 0{\cdot}01$
90. $1000 \div 0{\cdot}0001$

91. $\dfrac{92 \times 4{\cdot}6}{2{\cdot}3}$
92. $\dfrac{3{\cdot}3 \times 3 \times 5}{15}$
93. $\dfrac{180 \times 4}{36}$
94. $\dfrac{6{\cdot}6 \times 1{\cdot}8}{5{\cdot}4}$
95. $\dfrac{0{\cdot}55 \times 0{\cdot}81}{4{\cdot}5}$
96. $\dfrac{550 \times 3{\cdot}2}{2{\cdot}4 \times 11}$
97. $\dfrac{(1{\cdot}8 + 6{\cdot}12)}{2{\cdot}4}$
98. $\dfrac{(18{\cdot}61 - 9{\cdot}16)}{3{\cdot}5}$
99. $\dfrac{63 \times 600 \times 0{\cdot}2}{360 \times 7}$
100. $\dfrac{(7 + 3{\cdot}32)}{(9{\cdot}91 - 5{\cdot}11)}$

1.2 FRACTIONS

Common fractions are added or subtracted from one another directly only when they have a common denominator.

Example 1

Evaluate: (a) $\frac{3}{4} + \frac{2}{5}$ (b) $2\frac{3}{8} - 1\frac{5}{12}$
(c) $\frac{2}{5} \times \frac{6}{7}$ (d) $2\frac{2}{5} \div 6$

(a) $\frac{3}{4} + \frac{2}{5}$
$= \frac{15}{20} + \frac{8}{20}$
$= \frac{23}{20}$
$= 1\frac{3}{20}$

(b) $2\frac{3}{8} - 1\frac{5}{12}$
$= \frac{19}{8} - \frac{17}{12}$
$= \frac{57}{24} - \frac{34}{24}$
$= \frac{23}{24}$

(c) $\frac{2}{5} \times \frac{6}{7}$
$= \frac{12}{35}$

(d) $2\frac{2}{5} \div 6$
$= \frac{12}{5} \div \frac{6}{1}$
$= \frac{\overset{2}{\cancel{12}}}{5} \times \frac{1}{\underset{1}{\cancel{6}}} = \frac{2}{5}$

Exercise 2

Evaluate and simplify your answer.

1. $\frac{3}{4} + \frac{4}{5}$
2. $\frac{1}{3} + \frac{1}{8}$
3. $\frac{5}{6} + \frac{6}{9}$
4. $\frac{3}{4} - \frac{1}{3}$
5. $\frac{3}{5} - \frac{1}{3}$
6. $\frac{1}{2} - \frac{2}{5}$
7. $\frac{2}{3} \times \frac{4}{5}$
8. $\frac{1}{7} \times \frac{5}{6}$
9. $\frac{5}{8} \times \frac{12}{13}$
10. $\frac{1}{3} \div \frac{4}{5}$
11. $\frac{3}{4} \div \frac{1}{6}$
12. $\frac{5}{6} \div \frac{1}{2}$
13. $\frac{3}{4} + \frac{2}{3}$
14. $\frac{3}{4} - \frac{2}{3}$
15. $\frac{3}{4} \times \frac{2}{3}$
16. $\frac{3}{4} \div \frac{2}{3}$
17. $\frac{6}{7} + \frac{1}{8}$
18. $\frac{6}{7} - \frac{1}{8}$
19. $\frac{6}{7} \times \frac{1}{8}$
20. $\frac{6}{7} \div \frac{1}{8}$
21. $\frac{3}{8} + \frac{1}{5}$
22. $\frac{3}{8} - \frac{1}{5}$
23. $\frac{3}{8} \times \frac{1}{5}$
24. $\frac{3}{8} \div \frac{1}{5}$
25. $1\frac{3}{4} + \frac{2}{3}$
26. $1\frac{3}{4} - \frac{2}{3}$

27. $1\frac{3}{4} \times \frac{2}{3}$ **28.** $1\frac{3}{4} \div \frac{2}{3}$

29. $2\frac{2}{3} + 1\frac{1}{5}$ **30.** $2\frac{2}{3} - 1\frac{1}{5}$

31. $2\frac{2}{3} \times 1\frac{1}{5}$ **32.** $2\frac{2}{3} \div 1\frac{1}{5}$

33. $3\frac{1}{2} + 2\frac{3}{5}$ **34.** $3\frac{1}{2} - 2\frac{3}{5}$

35. $3\frac{1}{2} \times 2\frac{3}{5}$ **36.** $3\frac{1}{2} \div 2\frac{3}{5}$

37. $(\frac{3}{4} - \frac{2}{3}) \div \frac{3}{4}$ **38.** $(\frac{3}{5} + \frac{1}{3}) \times \frac{5}{7}$

39. $(3\frac{2}{3} - 2\frac{1}{2}) \times 1\frac{1}{7}$ **40.** $\dfrac{\frac{3}{8} - \frac{1}{5}}{\frac{7}{10} - \frac{2}{3}}$

41. $\dfrac{2\frac{2}{5} \times 1\frac{1}{4}}{3\frac{3}{4} \div 2\frac{1}{2}}$ **42.** $\dfrac{2\frac{3}{5} + 1\frac{7}{8}}{2\frac{3}{5} - 1\frac{7}{8}}$

43. $(2\frac{1}{2} - 1\frac{3}{5}) \times \frac{4}{5}$ **44.** $\dfrac{\frac{2}{3} + \frac{1}{5}}{\frac{3}{4} - \frac{1}{3}}$

45. $\dfrac{(2\frac{3}{4} \div \frac{5}{11})}{3\frac{1}{2}}$ **46.** $\dfrac{\frac{1}{3} \times \frac{1}{4}}{\frac{1}{3} - \frac{1}{4}}$

47. Arrange the fractions in order of size:

(a) $\dfrac{7}{12}, \dfrac{1}{2}, \dfrac{2}{3}$

(b) $\dfrac{3}{4}, \dfrac{2}{3}, \dfrac{5}{6}$

(c) $\dfrac{1}{3}, \dfrac{17}{24}, \dfrac{5}{8}, \dfrac{3}{4}$

(d) $\dfrac{5}{6}, \dfrac{8}{9}, \dfrac{11}{12}$

48. Find the fraction which is mid-way between the two fractions given:

(a) $\dfrac{2}{5}, \dfrac{3}{5}$ (b) $\dfrac{5}{8}, \dfrac{7}{8}$

(c) $\dfrac{2}{3}, \dfrac{3}{4}$ (d) $\dfrac{1}{3}, \dfrac{4}{9}$

(e) $\dfrac{4}{15}, \dfrac{1}{3}$ (f) $\dfrac{3}{8}, \dfrac{11}{24}$

Fractions and decimals

A decimal fraction is simply a fraction expressed in tenths, hundredths etc.

Example 2

(a) Change $\frac{7}{8}$ to a decimal fraction.
(b) Change 0·35 to a vulgar fraction.
(c) Change $\frac{1}{3}$ to a decimal fraction.

(a) $\frac{7}{8}$, divide 8 into 7

$$\frac{7}{8} = 0{\cdot}875 \qquad 8\,\overline{)7{\cdot}000}\;\;0{\cdot}875$$

(b) $0{\cdot}35 = \frac{35}{100} = \frac{7}{20}$

(c) $\frac{1}{3}$, divide 3 into 1

$$3\,\overline{)1{\cdot}0000}\;\;0{\cdot}3333\ldots$$

$$\frac{1}{3} = 0{\cdot}\dot{3} \text{ (0·3 recurring)}$$

Exercise 3

In questions **1** to **24**, change the fractions to decimals.

1. $\frac{1}{4}$ **2.** $\frac{2}{5}$ **3.** $\frac{4}{5}$ **4.** $\frac{3}{4}$

5. $\frac{1}{2}$ **6.** $\frac{3}{8}$ **7.** $\frac{9}{10}$ **8.** $\frac{5}{8}$

9. $\frac{5}{12}$ **10.** $\frac{1}{6}$ **11.** $\frac{2}{3}$ **12.** $\frac{5}{6}$

13. $\frac{2}{7}$ **14.** $\frac{3}{7}$ **15.** $\frac{4}{9}$ **16.** $\frac{5}{11}$

17. $1\frac{1}{5}$ **18.** $2\frac{5}{8}$ **19.** $2\frac{1}{3}$ **20.** $1\frac{7}{10}$

21. $2\frac{3}{16}$ **22.** $2\frac{2}{7}$ **23.** $2\frac{6}{7}$ **24.** $3\frac{19}{100}$

In questions **25** to **40**, change the decimals to vulgar fractions and simplify.

25. 0·2 **26.** 0·7 **27.** 0·25 **28.** 0·45
29. 0·36 **30.** 0·52 **31.** 0·125 **32.** 0·625
33. 0·84 **34.** 2·35 **35.** 3·95 **36.** 1·05
37. 3·2 **38.** 0·27 **39.** 0·007 **40.** 0·000 11

Evaluate, giving the answer to 2 decimal places:

41. $\frac{1}{4} + \frac{1}{3}$ **42.** $\frac{2}{3} + 0{\cdot}75$

43. $\frac{8}{9} - 0{\cdot}24$ **44.** $\frac{7}{8} + \frac{5}{9} + \frac{2}{11}$

45. $\frac{1}{3} \times 0{\cdot}2$ **46.** $\frac{5}{8} \times \frac{1}{4}$

47. $\frac{8}{11} \div 0{\cdot}2$ **48.** $(\frac{4}{7} - \frac{1}{3}) \div 0{\cdot}4$

Arrange the numbers in order of size (smallest first)

49. $\frac{1}{3}, 0{\cdot}33, \frac{4}{15}$ **50.** $\frac{2}{7}, 0{\cdot}3, \frac{4}{9}$

51. $0{\cdot}71, \frac{7}{11}, 0{\cdot}705$ **52.** $\frac{4}{13}, 0{\cdot}3, \frac{5}{18}$

1.3 APPROXIMATIONS

Example 1

(a) 7·8126 = 8 to the nearest whole number
↑ This figure is '5 or more'.

(b) 7·8126 = 7·81 to three significant figures
↑ This figure is not '5 or more'.

(c) 7·8126 = 7·813 to three decimal places
↑ This figure is '5 or more'.

Exercise 4

Write the following numbers correct to
(a) the nearest whole number
(b) three significant figures
(c) two decimal places

1. 8·174	**2.** 19·617	**3.** 20·041
4. 0·814 52	**5.** 311·14	**6.** 0·275
7. 0·007 47	**8.** 15·62	**9.** 900·12
10. 3·555	**11.** 5·454	**12.** 20·961
13. 0·0851	**14.** 0·5151	**15.** 3·071

Write the following numbers correct to one decimal place.

16. 5·71	**17.** 0·7614	**18.** 11·241
19. 0·0614	**20.** 0·0081	**21.** 11·12
22. 5·55	**23.** 9·99	**24.** 7·95

Write the following numbers correct to the nearest 100.

25. 2240	**26.** 3550	**27.** 360
28. 41 780	**29.** 247·2	**30.** 55 555
31. 10 020	**32.** 516·3	**33.** 0·765

34. If $a = 3{\cdot}1$ and $b = 7{\cdot}3$ correct to 1 decimal place, find the largest possible value of
(i) $a + b$ (ii) $b - a$

35. If $x = 5$ and $y = 7$ to one significant figure, find the largest and smallest possible values of
(i) $x + y$ (ii) $y - x$ (iii) $\frac{x}{y}$

36. In the diagram, ABCD and EFGH are rectangles with AB = 10 cm, BC = 7 cm, EF = 7 cm and FG = 4 cm, all figures accurate to 1 significant figure.
Find (a) the largest and (b) the smallest possible values of the shaded area.

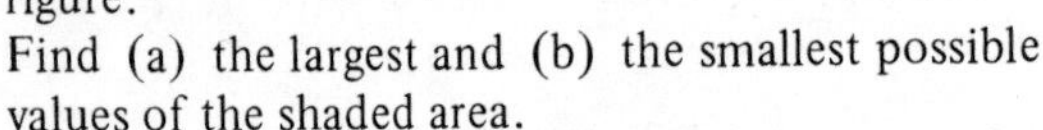

37. The velocity v of a body is calculated from the formula $v = \frac{2s}{t} - u$ where u, s and t are measured correct to 1 decimal place. Find the largest possible value for v when $u = 2{\cdot}1$, $s = 5{\cdot}7$ and $t = 2{\cdot}2$. Find also the smallest possible value for v consistent with these figures.

Estimation

It is always sensible to check that the answer to a calculation is 'about the right size'.

Example 2
Estimate the value of $\frac{57{\cdot}2 \times 110}{2{\cdot}146 \times 46{\cdot}9}$, correct to one significant figure.
We have approximately, $\frac{\not{50} \times 100}{2 \times \not{50}} \approx 50$
On a calculator the value is 62·52 (to 4 significant figures).

Exercise 5

Estimate the value of the following, giving the answer to one significant figure.

1. $31{\cdot}2 \times 97{\cdot}9$	**2.** $101{\cdot}2 \times 95{\cdot}6$
3. $\frac{507{\cdot}9}{49{\cdot}72}$	**4.** $2047 \times 96{\cdot}5$
5. $\frac{49{\cdot}6}{8{\cdot}94}$	**6.** $\frac{700 \times 39{\cdot}6}{378}$
7. $\frac{934 \times 51{\cdot}2}{62{\cdot}7 \times 9817}$	**8.** $\frac{20{\cdot}7 \times 48{\cdot}6}{50{\cdot}7 \times 41{\cdot}2}$
9. $\frac{1070}{0{\cdot}0112}$	**10.** $\frac{5074}{0{\cdot}02014}$
11. $\frac{0{\cdot}0997}{107{\cdot}4}$	**12.** $\frac{0{\cdot}0107}{985}$

In questions **13** to **20**, each calculation is followed by three numbers. Select the number which is nearest to the correct value.

13. $4110 \times 97{\cdot}84$	(4 000 000; 400 000; 40 000)
14. $\frac{79{\cdot}14}{0{\cdot}0896}$	(1000; 80; 8000)
15. $\frac{214{\cdot}7 \times 51{\cdot}7}{48{\cdot}2 \times 1970}$	(10; 1; 0·1)
16. $\frac{212}{2 \text{ million}}$	(10 000; 0·0001; 0·001)

17. $\dfrac{39{\cdot}4 \times 1097}{19{\cdot}2 \times 496}$ (40; 4; 0·4)

18. $\dfrac{97{\cdot}3 \times 5{\cdot}111}{0{\cdot}1999}$ (5000; 2500; 500)

19. $\dfrac{49{\cdot}8 - 0{\cdot}72}{0{\cdot}00111}$ (50 000; 10 000; 5000)

20. $\dfrac{110 - 60{\cdot}74}{4984}$ (0·1; 0·01; 0·001)

1.4 STANDARD FORM

When dealing with either very large or very small numbers, it is not convenient to write them out in full in the normal way. It is better to use standard form. Most calculators represent large and small numbers in this way.

The number $a \times 10^n$ is in standard form when $1 \leqslant a < 10$ and n is a positive or negative integer.

Example 1

Write the following numbers in standard form:
(a) 2000
(b) 150
(c) 0·0004

(a) $2000 = 2 \times 1000 = 2 \times 10^3$
(b) $150 = 1{\cdot}5 \times 100 = 1{\cdot}5 \times 10^2$

(c) $0{\cdot}0004 = 4 \times \dfrac{1}{10\,000} = 4 \times 10^{-4}$

Exercise 6

Write the following numbers in standard form:

1. 4000 2. 500 3. 70 000
4. 60 5. 2400 6. 380
7. 46 000 8. 46 9. 900 000
10. 2560 11. 0·007 12. 0·0004
13. 0·0035 14. 0·421 15. 0·000 055
16. 0·01 17. 564 000 18. 19 million
19. 0·000 001 23 20. 25×10^6
21. 383×10^2 22. $19{\cdot}6 \times 10^{-4}$
23. $0{\cdot}7 \times 10^{-3}$ 24. $0{\cdot}062 \times 10^{-2}$
25. 2000×10^{-12} 26. 550×10^7
27. $0{\cdot}8 \times 10$ 28. 700×10^{-2}

Calculations in standard form

The rules of indices are dealt with in detail on page 129 and we need here to use the laws of multiplication, division and raising to a power.

Example 2

Simplify (a) $10^2 \times 10^3$
(b) $10^3 \times 10^{-4}$
(c) $10^6 \div 10^2$
(d) $10^{-2} \div 10^3$
(e) $(10^3)^2$

For multiplication, the indices are added.
(a) $10^2 \times 10^3 = 10^5$
(b) $10^3 \times 10^{-4} = 10^{-1}$

For division the indices are subtracted.
(c) $10^6 \div 10^2 = 10^4$
(d) $10^{-2} \div 10^3 = 10^{-5}$

Raising to a power, the indices are multiplied together.
(e) $(10^3)^2 = 10^6$

Example 3

Evaluate the following and give the answer in standard form.

(a) $(2 \times 10^2) \times (3 \times 10^5)$
(b) $(8 \times 10^3) \div (2 \times 10^{-3})$
(c) $1500 \times 8\,000\,000$
(d) $400 \div 0{\cdot}0008$

Multiply (or divide) the numbers and the powers of 10 separately.

(a) $(2 \times 10^2) \times (3 \times 10^5) = 6 \times 10^7$
(b) $(8 \times 10^3) \div (2 \times 10^{-3}) = 4 \times 10^6$

(c) $1\,500 \times 8\,000\,000 = (1{\cdot}5 \times 10^3) \times (8 \times 10^6)$
$= 12 \times 10^9$
$= 1{\cdot}2 \times 10^{10}$

(d) $400 \div 0{\cdot}000\,8 = (4 \times 10^2) \div (8 \times 10^{-4})$
$= 0{\cdot}5 \times 10^6$
$= 5 \times 10^5$

Exercise 7

Simplify the following:

1. $10^3 \times 10^4$
2. $10^6 \times 10^2$
3. $10^7 \times 10^{-2}$
4. $10^{-3} \times 10^9$
5. $10^{-2} \times 10^{-4}$
6. $10^{-7} \times 10^{-10}$
7. $10^3 \div 10^2$
8. $10^4 \div 10^2$
9. $10^{-2} \div 10^3$
10. $10^3 \div 10^{-7}$
11. $10^{-3} \div 10^{-8}$
12. $10^6 \div 10^8$
13. $10^{-6} \div 10^{-12}$
14. $(10^4)^2$
15. $(10^{-3})^2$
16. $(10^5)^2$
17. $(10^2)^3$
18. $(10^{-2})^3$

Evaluate the following, giving the answer in standard form:

19. $(2 \times 10^2) \times (3 \times 10^6)$
20. $(4{\cdot}5 \times 10^4) \times (2 \times 10^7)$
21. $(2{\cdot}2 \times 10^6) \times (4 \times 10^{-2})$
22. $(3{\cdot}1 \times 10^{-3}) \times (3 \times 10^{-7})$
23. $(1{\cdot}2 \times 10^{-8}) \times (4 \times 10^{10})$
24. $(8 \times 10^6) \div (4 \times 10^2)$
25. $(9{\cdot}3 \times 10^8) \div (3 \times 10^{-2})$
26. $(8{\cdot}6 \times 10^{-4}) \div (2 \times 10^{-7})$
27. $(4{\cdot}2 \times 10^{-6}) \div (2{\cdot}1 \times 10^5)$
28. $(800\,000) \div (400)$
29. 5000×3000
30. $60\,000 \times 5000$
31. $0{\cdot}000\,07 \times 400$
32. $0{\cdot}0007 \times 0{\cdot}000\,01$
33. $8000 \div 0{\cdot}004$
34. $(0{\cdot}002)^2$
35. $150 \times 0{\cdot}0006$
36. $0{\cdot}000\,033 \div 500$
37. $(1{\cdot}8 \times 10^{-2}) + (2 \times 10^{-3})$
38. $(3{\cdot}4 \times 10^{-4}) + (4 \times 10^{-6})$
39. $(5{\cdot}9 \times 10^{-2}) - (6 \times 10^{-3})$
40. $(2 \times 10^{-3}) - (1{\cdot}4 \times 10^{-4})$
41. $0{\cdot}007 \div 20\,000$
42. $(0{\cdot}0001)^4$
43. $(2000)^3$
44. $0{\cdot}005\,92 \div 8000$
45. $4500 \div 0{\cdot}0027$
46. $22\,500 \div 250$
47. $\dfrac{(8 \times 10^3) + (4 \times 10^2)}{1{\cdot}2 \times 10^{-4}}$
48. $\dfrac{(5 \times 10^5) - (5 \times 10^3)}{0{\cdot}00011}$
49. $\dfrac{(2 \times 10^{-4}) + (4 \times 10^{-5})}{8000}$
50. $\left(\dfrac{4 \times 10^3}{3 \times 10^5}\right) \div \left(\dfrac{2 \times 10^{-5}}{6 \times 10^{-3}}\right)$
51. If $a = 512 \times 10^2$
    $b = 0{\cdot}478 \times 10^6$
    $c = 0{\cdot}0049 \times 10^7$
    arrange a, b and c in order of size (smallest first).
52. If the number $2{\cdot}74 \times 10^{15}$ is written out in full, how many zeros follow the 4?
53. If the number $7{\cdot}31 \times 10^{-17}$ is written out in full, how many zeros would there be between the decimal point and the first significant figure?
54. If $x = 2 \times 10^5$ and $y = 3 \times 10^{-3}$ correct to one significant figure, find the greatest and least possible values of
    (i) xy (ii) $\dfrac{x}{y}$
55. (a) The number 10 to the power 100 (10 000 sexdecillion) is called a 'Googol'! If it takes $\frac{1}{5}$ second to write a zero and $\frac{1}{10}$ second to write a 'one', how long would it take to write the number 100 'Googols' in full?
    (b) The number 10 to the power of a 'Googol' is called a 'Googolplex'. Using the same speed of writing, how long in years would it take to write 1 'Googolplex' in full? You may assume that your pen has enough ink.

1.5 RATIO AND PROPORTION

The word 'ratio' is used to describe a fraction. If the *ratio* of a boy's height to his father's height is 4 : 5, then he is $\frac{4}{5}$ as tall as his father.

Example 1

Change the ratio 2 : 5 into the form
(a) $1 : n$ (b) $m : 1$

(a) $2 : 5 = 1 : \frac{5}{2}$
$\quad\;\; = 1 : 2{\cdot}5$

(b) $2 : 5 = \frac{2}{5} : 1$
$\quad\;\; = 0{\cdot}4 : 1$

Example 2

Divide £60 between two people A and B in the ratio 5 : 7.

Consider £60 as 12 equal parts (i.e. 5 + 7). Then A receives 5 parts and B receives 7 parts.

∴ A receives $\frac{5}{12}$ of £60 = £25
B receives $\frac{7}{12}$ of £60 = £35

Example 3

Divide 200 kg in the ratio 1:3:4.

The parts are $\frac{1}{8}$, $\frac{3}{8}$ and $\frac{4}{8}$ (of 200 kg).
i.e. 25 kg, 75 kg and 100 kg.

Exercise 8

Express the following ratios in the form $1:n$:

1. 2:6
2. 5:30
3. 2:100
4. 5:8
5. 4:3
6. 8:3
7. 22:550
8. 45:360

Express the following ratios in the form $n:1$:

9. 12:5
10. 5:2
11. 4:5
12. 2:100

In questions **13** to **18** divide the quantity in the ratio given.

13. £40; (3:5)
14. £120; (3:7)
15. 250 m; (14:11)
16. £117; (2:3:8)
17. 180 kg; (1:5:6)
18. 184 minutes; (2:3:3)
19. When £143 is divided in the ratio 2:4:5, what is the difference between the largest share and the smallest share?
20. Divide 180 kg in the ratio 1:2:3:4.
21. Divide £4000 in the ratio 2:5:5:8.
22. If $\frac{5}{8}$ of the children in a school are boys, what is the ratio of boys to girls?
23. A man and a woman share a bingo prize of £1000 between them in the ratio 1:4. The woman shares her part between herself, her mother and her daughter in the ratio 2:1:1. How much does her daughter receive?
24. A man and his wife share a sum of money in the ratio 3:2. If the sum of money is doubled, in what ratio should they divide it so that the man still receives the same amount?
25. In a herd of x cattle, the ratio of the number of bulls to cows is 1:6. Find the number of bulls in the herd in terms of x.
26. If $x:3 = 12:x$, calculate the positive value of x.
27. If $y:18 = 8:y$, calculate the positive value of y.
28. £400 is divided between Ann, Brian and Carol so that Ann has twice as much as Brian and Brian has three times as much as Carol. How much does Brian receive?
29. A cake weighing 550 g has three ingredients: flour, sugar and raisins. There is twice as much flour as sugar and one and a half times as much sugar as raisins. How much flour is there?
30. A brother and sister share out their collection of 5000 stamps in the ratio 5:3. The brother then shares his stamps with two friends in the ratio 3:1:1, keeping most for himself. How many stamps do each of his friends receive?

Proportion

The majority of problems where proportion is involved are usually solved by finding the value of a unit quantity.

Example 4

If a wire of length 2 metres costs £10, find the cost of a wire of length 35 cm.

200 cm costs 1000 pence

$\therefore$ 1 cm costs $\frac{1000}{200}$ pence
= 5 pence

$\therefore$ 35 cm costs 5×35 pence
= 175 pence
= £1.75

Example 5

Eight men can dig a trench in 4 hours. How long will it take five men to dig the same size trench?

8 men take 4 hours
1 man would take 32 hours
5 men would take $\frac{32}{5}$ hours
= 6 hours 24 minutes.

Exercise 9

1. Five cans of beer cost £1.20. Find the cost of seven cans.
2. A man earns £140 in a 5-day week. What is his pay for 3 days?
3. Three men build a wall in 10 days. How long would it take five men?
4. Nine milk bottles contain $4\frac{1}{2}$ litres of milk between them. How much do five bottles hold?
5. A car uses 10 litres of petrol in 75 km. How far will it go on 8 litres?
6. A wire 11 cm long has a mass of 187 g. What is the mass of 7 cm of this wire?

7. A shopkeeper can buy 36 toys for £20.52. What will he pay for 120 toys?
8. A ship has sufficient food to supply 600 passengers for 3 weeks. How long would the food last for 800 people?
9. The cost of a phone call lasting 3 minutes 30 seconds was 52·5 p. At this rate, what was the cost of a call lasting 5 minutes 20 seconds?
10. 80 machines can produce 4800 identical pens in 5 hours. At this rate
    (a) how many pens would one machine produce in one hour?
    (b) how many pens would 25 machines produce in 7 hours?
11. Three men can build a wall in 10 hours. How many men would be needed to build the wall in $7\frac{1}{2}$ hours?
12. If it takes 6 men 4 days to dig a hole 3 feet deep, how long will it take 10 men to dig a hole 7 feet deep?
13. Find the cost of 1 km of pipe at 7p for every 40 cm.
14. A wheel turns through 90 revolutions per minute. How many degrees does it turn through in 1 second?
15. Find the cost of 20 grams of lead at £60 per kilogram.

Foreign exchange

Money is changed from one currency into another using the method of proportion.

Exchange rates April 1982.

Country	*Rate of exchange*	
Belgium (franc)	BF 87·0	= £1
France (franc)	Fr 10·9	= £1
Germany (mark)	DM 4·2	= £1
Italy (lire)	lire 2280	= £1
Spain (peseta)	Ptas 182	= £1
United States (dollar)	\$1·74	= £1

Example 6

If a bottle of wine costs 8 francs in France, what is the cost in British money?

$$10{\cdot}9 \text{ francs} = £1$$

$$\therefore \quad 1 \text{ franc} = £\,\frac{1}{10{\cdot}9}$$

$$8 \text{ francs} = £\,\frac{1}{10{\cdot}9} \times 8 = £0{\cdot}73$$

(to the nearest penny)

The bottle costs approximately 73p in British money.

Example 7

Convert \$500 into German marks.

$$\$1{\cdot}74 = £1$$
$$\text{DM } 4{\cdot}2 = £1$$
$$\therefore \quad \$1{\cdot}74 = \text{DM } 4{\cdot}2$$
$$\$1 = \text{DM}\left(\frac{4{\cdot}2}{1{\cdot}74}\right)$$
$$\therefore \quad \$500 = \text{DM}\left(\frac{4{\cdot}2}{1{\cdot}74} \times 500\right)$$
$$= \text{DM } 1206$$

Exercise 10

Give your answers correct to two decimal places. Use the exchange rates given in the table.

1. Change the amount of British money into the foreign currency stated.
   (a) £20 [French francs]
   (b) £70 [dollars]
   (c) £200 [pesetas]
   (d) £1·50 [marks]
   (e) £2·30 [lire]
   (f) 90p [dollars]
2. Change the amount of foreign currency into British money.
   (a) Fr. 500 (b) \$2500
   (c) DM 7·5 (d) BF 900
   (e) Lire 500,000 (f) Ptas 950
3. An L.P. costs £4·50 in Britain and \$4·70 in the United States. How much cheaper, in British money, is the record when bought in the USA?
4. A bottle of Cointreau costs 582 pesetas in Spain and Fr 48 in France. Which is the cheaper in British money, and by how much?
5. The EEC 'Butter Mountain' was estimated in 1982 to be costing Fr 218 000 per day to maintain the storage facilities. How much is this in pounds?
6. A Jaguar XJS is sold in several countries at the prices given below.
   Britain £15,000
   Belgium BF 1,496,400

France	Fr 194, 020
Germany	DM 52,080
USA	$24 882

Write out in order a list of the prices converted into pounds.

7. A traveller in Switzerland exchanges 1300 Swiss francs for £400. What is the exchange rate?
8. An Irish gentleman on holiday in Germany finds that his wallet contains $700. If he changes the money at a bank how many marks will he receive?
9. An English soccer fan is arrested in France and has to pay a fine of 2 000 francs. He has 10 000 German marks in his wallet. How much has he left in British money, after paying the fine?
10. In Britain, a pint of beer costs 65p. In France a third of a litre of the same beer costs 4 francs. If 1 pint is approximately 0·568 litre, calculate the cost in pence of a pint of the beer bought in France.

Map scales

When a map is drawn to a scale of 1 : 20 000, 1 cm on the map represents 20 000 cm on the land.

Example 8

A map is drawn to a scale of 1 to 50 000. Calculate:
(a) the length of a road which appears as 3 cm long on the map.
(b) the length on the map of a lake which is 10 km long.

(a) 1 cm on the map is equivalent to 50 000 cm on the Earth

$$\therefore \quad 3\text{ cm} \equiv 3 \times 50\,000 = 150\,000\text{ cm} = 1500\text{ m} = 1{\cdot}5\text{ km}$$

The road is 1·5 km long.

(b) 50 000 cm ≡ 1 cm on the map

$$1\text{ cm} \equiv \frac{1}{50\,000}\text{ cm on the map}$$

$$10\text{ km} = 1000\,000\text{ cm}$$

$$\therefore \quad 10\text{ km} \equiv \left(1000\,000 \times \frac{1}{50\,000}\right)\text{ cm on the map}$$

$$\equiv 20\text{ cm on the map}$$

The lake appears 20 cm long on the map.

Exercise 11

1. Find the actual length represented on a drawing by
(a) 14 cm (b) 3·2 cm
(c) 0·71 cm (d) 21·7 cm
when the scale is 1 cm to 5 m.
2. Find the length on a drawing that represents
(a) 50 m (b) 35 m
(c) 7·2 m (d) 28·6 m
when the scale is 1 cm to 10 m.
3. If the scale is 1 : 10 000, what length will 45 cm on the map represent:
(a) in cm; (b) in m; (c) in km?
4. On a map of scale 1 : 100 000, the distance between Tower Bridge and Hammersmith Bridge is 12·3 cm. What is the actual distance in km?
5. On a map of scale 1 : 15 000, the distance between Buckingham Palace and Brixton Underground Station is 31·4 cm. What is the actual distance in km?
6. If the scale of a map is 1 : 10 000, what will be the length on this map of a road which is 5 km long?
7. The distance from Hertford to St Albans is 32 km. How far apart will they be on a map of scale 1 : 5000?
8. The 17th hole at the famous St Andrews golf course is 420 m in length. How long will it appear on a plan of the course of scale 1 : 8000?

An area involves two dimensions multiplied together and hence the scale is multiplied *twice*.

For example, if the linear scale is $\frac{1}{100}$, then the area scale is $\frac{1}{100} \times \frac{1}{100} = \frac{1}{10\,000}$.

Exercise 12

1. The scale of a map is 1 : 1000. What are the actual dimensions of a rectangle which appears as 4 cm by 3 cm on the map? What is the area on the map in cm^2? What is the actual area in m^2?
2. The scale of a map is 1 : 100. What area does 1 cm^2 on the map represent? What area does 6 cm^2 represent?
3. The scale of a map is 1 : 20 000. What area does 8 cm^2 represent?
4. The scale of a map is 1 : 1000. What is the area on the map of a lake of area 5 km^2?
5. The scale of a map is 1 cm to 5 km. A farm is represented by a rectangle measuring 1·5 cm by 4 cm. What is the actual area of the farm?

6. On a map of scale 1 cm to 2 km the area of a car park is 3 cm^2. What is the actual area of the car park in hectares? (1 hectare = 10 000 m^2)
7. The area of the playing surface at Wembley Stadium is $\frac{3}{5}$ of a hectare. What area will it occupy on a plan drawn to a scale of 1:500?
8. The scale of a house plan is such that $\frac{1}{2}$ cm represents 4 m. What is the actual area of a room which has an area of 3 cm^2 on the plan?

1.6 PERCENTAGES

Percentages are simply a convenient way of expressing fractions or decimals. '50% of £60' means $\frac{50}{100}$ of £60, or more simply $\frac{1}{2}$ of £60. Percentages are used very frequently in everyday life and are misunderstood by a large number of people. What are the implications if 'inflation falls from 10% to 8%'? Does this mean prices will fall?

Example 1

(a) change 80% to a fraction
(b) change $\frac{3}{8}$ to a percentage
(c) change 8% to a decimal.

(a) $80\% = \frac{80}{100} = \frac{4}{5}$

(b) $\frac{3}{8} = \left(\frac{3}{8} \times \frac{100}{1}\right)\% = 37\frac{1}{2}\%$

(c) $8\% = \frac{8}{100} = 0{\cdot}08$

Exercise 13

1. Change to fractions
   (a) 60% (b) 24%
   (c) 35% (d) 2%
2. Change to percentages
   (a) $\frac{1}{4}$ (b) $\frac{1}{10}$
   (c) $\frac{7}{8}$ (d) $\frac{1}{3}$
   (e) 0·72 (f) 0·31
   (g) 0·575 (h) 0·075
3. Change to decimals
   (a) 36% (b) 28%
   (c) 7% (d) 13·4%
   (e) $\frac{3}{5}$ (f) $\frac{7}{8}$
   (g) $\frac{1}{100}$ (h) $\frac{2}{3}$
4. Change to fractions
   (a) 120% (b) 250%
   (c) 99% (d) $12\frac{1}{2}\%$
   (e) $2\frac{1}{2}\%$ (f) $62\frac{1}{2}\%$
   (g) $3\frac{1}{3}\%$ (h) $4\frac{1}{4}\%$
5. Change to percentages
   (a) 0·07 (b) 0·007
   (c) 0·0007 (d) $3\frac{1}{2}$
   (e) $2\frac{1}{4}$ (f) $\frac{5}{9}$
   (g) 3·2 (h) 9
6. Change to decimals
   (a) $7\frac{1}{2}\%$ (b) 301%
   (c) $\frac{1}{2}\%$ (d) $1\frac{1}{4}\%$
   (e) $\frac{9}{1000}$ (f) 9%
   (g) $\frac{9}{24}$ (h) 9·6%
7. Arrange in order of size (smallest first)
   (a) $\frac{1}{2}$; 45%; 0·6
   (b) 0·38; $\frac{6}{16}$; 4%
   (c) 0·111; 11%; $\frac{1}{9}$
   (d) 32%; 0·3; $\frac{1}{3}$
   (e) $(\frac{1}{2})^2$; ($\frac{1}{2}$ of 0·4); (15% of 2)
   (f) 0·075%; 8×10^{-3}; 8×10^{-4}
   (g) $(0{\cdot}2)^2$; $\frac{1}{20}$; 6%
   (h) 4×10^{-2}; 5×10^{-3}; 1·3%
   (i) (0·4 of 80%); $\frac{4}{12}$; $(0{\cdot}3)^2$
   (j) 10^{-3}; 0·2%; $\frac{1}{200}$
8. The following are marks obtained in various tests. Convert them to percentages.
   (a) 17 out of 20
   (b) 31 out of 40
   (c) 19 out of 80
   (d) 112 out of 200
   (e) $2\frac{1}{2}$ out of 25
   (f) $7\frac{1}{2}$ out of 20

Example 2

A car costing £2400 is reduced in price by 10%. Find the new price.

$$10\% \text{ of } £2400 = \frac{10}{100} \times \frac{2400}{1} = £240$$

$$\text{New price of car} = £(2400 - 240) = £2160$$

Example 3

After a price increase of 10% a television set costs £286. What was the price before the increase?

The price before the increase is 100%

$\therefore$ 110% of old price = £286

$\therefore$ 1% of old price = £ $\frac{286}{110}$

$\therefore$ 100% of old price = £ $\frac{286}{110} \times \frac{100}{1}$

Old price of TV = £260

Exercise 14

1. Calculate
   (a) 30% of £50 (b) 45% of 2000 kg
   (c) 4% of \$70 (d) 2·5% of 5000 people
2. In a sale, a jacket costing £40 is reduced by 20%. What is the sale price?
3. The charge for a telephone call costing 12p is increased by 10%. What is the new charge?
4. In peeling potatoes 4% of the mass of the potatoes is lost as 'peel'. How much is *left* for use from a bag containing 55 kg?
5. Johnny thinks his goldfish got chickenpox. He lost 70% of his collection of goldfish. If he has 60 survivors, how many did he have originally?
6. The average attendance at Everton football club fell by 7% in 1982. If 2030 fewer people went to matches in 1982, how many went in 1981?
7. When heated an iron bar expands by 0·2%. If the increase in length is 1 cm, what is the original length of the bar?
8. In the last two weeks of a sale, prices are reduced first by 30% and then by a *further* 40% of the new price. What is the final sale price of a shirt which originally cost £15?
9. During a Grand Prix car race, the tyres on a car are reduced in weight by 3%. If they weigh 388 kg at the end of the race, how much did they weigh at the start?
10. Over a period of 6 months, a colony of rabbits increases in number by 25% and then by a further 30%. If there were originally 200 rabbits in the colony how many were there at the end?

In the next exercise use the formula:

$$\text{Percentage profit} = \frac{\text{Actual profit}}{\text{Original price}} \times \frac{100}{1}$$

$$\text{Percentage loss} = \frac{\text{Actual loss}}{\text{Original price}} \times \frac{100}{1}$$

Example 4

A radio is bought for £16 and sold for £20. What is the percentage profit?

Actual profit = £4

$\therefore$ Percentage profit = $\frac{4}{16} \times \frac{100}{1}$ = 25%

The radio is sold at a 25% profit.

Example 5

A car is sold for £2280, at a loss of 5% on the cost price. Find the cost price.

Do *not* calculate 5% of £2280!
The loss is 5% of the cost price.

$\therefore$ 95% of cost price = £2280

1% of cost price = £ $\frac{2280}{95}$

$\therefore$ 100% of cost price = £ $\frac{2280}{95} \times \frac{100}{1}$

Cost price = £2400.

Exercise 15

1. The first figure is the cost price and the second figure is the selling price. Calculate the percentage profit or loss in each case.
   (a) £20, £25 (b) £400, £500
   (c) £60, £54 (d) \$9000, \$10800
   (e) £460, £598 (f) £512, £550·40
   (g) £45, £39·60 (h) 50p, 23p
2. A car dealer buys a car for £500, gives it a clean, and then sells it for £640. What is the percentage profit?
3. A damaged carpet which cost £180 when new, is sold for £100. What is the percentage loss?
4. During the first four weeks of her life, a baby girl increases her weight from 3·2 kg to 4·7 kg. What percentage increase does this represent? (give your answer to 3 sig. fig.)
5. When V.A.T. is added to the cost of a car tyre, its price increases from £16·50 to £18·48. What is the rate at which V.A.T. is charged?
6. In order to increase sales, the price of a Concorde airliner is reduced from £30 000 000 to £28 400 000. What percentage reduction is this?

7. Find the *cost* price in the following:
   (a) selling price £55, profit 10%
   (b) selling price £558, profit 24%
   (c) selling price £680, loss 15%
   (d) selling price £11·78, loss 5%
8. An oven is sold for £600, thereby making a profit of 20%, on the cost price. What was the cost price?
9. A pair of jeans is sold for £15, thereby making a profit of 25% on the cost price. What was the cost price?
10. A book is sold for £5·40, at a profit of 8% on the cost price. What was the cost price?
11. An obsolete can of worms is sold for 48p, incurring a loss of 20%. What was the cost price?
12. A car, which failed its MOT test, was sold for £143, thereby making a loss of 35% on the cost price. What was the cost price?
13. If an employer reduces the working week from 40 hours to 35 hours, with no loss of weekly pay, calculate the percentage increase in the hourly rate of pay.
14. The rental for a television set changed from £80 per year to £8 per month. What is the percentage increase in the yearly rental?
15. A greengrocer sells a cabbage at a profit of $37\frac{1}{2}$% on the price he pays for it. What is the ratio of the cost price to the selling price?

Income tax

The tax which an employee pays on his income depends on (a) how much he is paid
(b) his allowances
(c) the rate of taxation.

Tax is paid only on the 'taxable income'.

Total income = Allowances + Taxable income.

Example 6

Calculate the tax paid each month by a man whose annual salary is £8500, with allowances of £2100, where the rate of taxation is 30% of the first £10 000 of taxable income.

Total income = Allowances + Taxable income

∴ Taxable income = 8500 − 2100
= £6400

Tax paid in one year = 30% of £6400
= £1920

Tax paid per month = 1920 ÷ 12
= £160

Example 7

Calculate the annual salary of a man who has allowances of £2350 and who pays £1404 in tax. The rate of taxation is 30%.

Let his taxable income be £x

$$\frac{30}{100} \times x = 1404$$

$$\therefore \quad x = 1404 \times \frac{100}{30}$$

$$= 4680$$

∴ annual salary = taxable income + allowances
= £4680 + £2350
= £7030

Exercise 16

In questions **1** to **3**, the rates of taxation are as follows:

taxable income	*rate*
£1 → 11 250	30%
£11 251 → 13 250	40%
£13 251 → 16 750	45%
over £16 751	50%

1. Calculate the tax paid by a man with income and allowances as follows:
   (a) income £6000, allowances £1400
   (b) income £2750, allowances £1375
   (c) income £12 500, allowances £1375
   (d) income £14 500, allowances £2050
   (e) income £17 600, allowances £2760
   (f) income £35 000, allowances £2000
2. Calculate the income of a person whose amount of tax paid and allowances are as follows:
   (a) tax paid £2400, allowances £2000
   (b) tax paid £960, allowances £2000
   (c) tax paid £3360, allowances £1950
   (d) tax paid £4200, allowances £2660
   (e) tax paid £4000, allowances £3400
   (f) tax paid £5600, allowances £4055
3. In the budget the Chancellor raises allowances by 8% and reduces the rate of taxation from 32% to 30%.
   (a) How much less tax does a man pay on an income of £8800 with original allowances of £1800?

(b) What is the saving for a man with an income of £13 000 and allowances of £2000?

4. A man with an income of £7500 and allowances of £1800, paid £1824 in tax. What is the rate of taxation?

5. A man with an income of £10 850 and allowances of £2140, paid £2525·90 in tax. What is the rate of taxation?

1.7 SPEED, DISTANCE AND TIME

Calculations involving these three quantities are simpler when the speed is *constant*. The formulae connecting the quantities are as follows:

(a) distance = speed × time

(b) speed $= \dfrac{\text{distance}}{\text{time}}$

(c) time $= \dfrac{\text{distance}}{\text{speed}}$

A helpful way of remembering these formulae is to write the letters D, S and T in a triangle, thus:

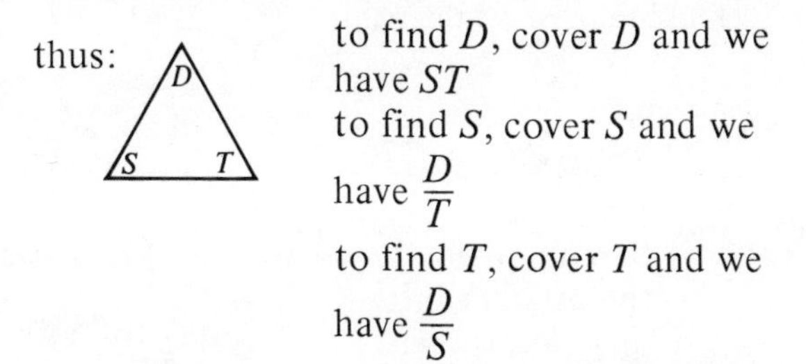

to find D, cover D and we have ST

to find S, cover S and we have $\dfrac{D}{T}$

to find T, cover T and we have $\dfrac{D}{S}$

Great care must be taken with the units in these questions.

Example 1

A man is running at a speed of 8 km/h for a distance of 5200 metres. Find the time taken in minutes.

$$\begin{aligned} 5200 \text{ metres} &= 5{\cdot}2 \text{ km} \\ \text{time taken in hours} &= \left(\frac{D}{S}\right) = \frac{5{\cdot}2}{8} \\ &= 0{\cdot}65 \text{ hours} \\ \text{time taken in minutes} &= 0{\cdot}65 \times 60 \\ &= 39 \text{ minutes.} \end{aligned}$$

Example 2

Change the units of a speed of 54 km/h into metres per second.

$$\begin{aligned} 54 \text{ km/hour} &= 54\,000 \text{ metres/hour} \\ &= \frac{54\,000}{60} \text{ metres/minute} \\ &= \frac{54\,000}{60 \times 60} \text{ metres/second} \\ &= 15 \text{ m/s.} \end{aligned}$$

Exercise 17

1. Find the time taken for the following journeys:
   (a) 100 km at a speed of 40 km/h
   (b) 250 miles at a speed of 80 miles per hour
   (c) 15 metres at a speed of 20 cm/s. (answer in seconds)
   (d) 10^4 metres at a speed of 2·5 km/h
2. Change the units of the following speeds as indicated:
   (a) 72 km/h into m/s (b) 108 km/h into m/s
   (c) 300 km/h into m/s (d) 30 m/s into km/h
   (e) 22 m/s into km/h
   (f) 0·012 m/s into cm/s
   (g) 9000 cm/s into m/s
   (h) 600 miles/day into miles per hour
   (i) 2592 miles/day into miles per second
3. Find the speeds of the bodies which move as follows:
   (a) a distance of 600 km in 8 hours
   (b) a distance of 31·64 km in 7 hours
   (c) a distance of 136·8 m in 18 seconds
   (d) a distance of 4×10^4 m in 10^{-2} seconds
   (e) a distance of 5×10^5 cm in 2×10^{-3} seconds
   (f) a distance of 10^8 mm in 30 minutes (in km/h)
   (g) a distance of 500 m in 10 minutes (in km/h)
4. Find the distance travelled (in metres) in the following:
   (a) at a speed of 55 km/h for 2 hours
   (b) at a speed of 40 km/h for $\frac{1}{4}$ hour
   (c) at a speed of 338·4 km/h for 10 minutes
   (d) at a speed of 15 m/s for 5 minutes
   (e) at a speed of 14 m/s for 1 hour
   (f) at a speed of 4×10^3 m/s for 2×10^{-2} seconds
   (g) at a speed of 8×10^5 cm/s for 2 minutes

5. A car travels 60 km at 30 km/h and then a further 180 km at 160 km/h. Find
   (a) the total time taken
   (b) the average speed for the whole journey
6. A cyclist travels 25 kilometres at 20 km/h and then a further 80 kilometres at 25 km/h. Find
   (a) the total time taken
   (b) the average speed for the whole journey
7. A swallow flies at a speed of 50 km/h for 3 hours and then at a speed of 40 km/h for a further 2 hours. Find the average speed for the whole journey.
8. Sebastian Coe ran two laps around a 400 m track. He completed the first lap in 50 seconds and then decreased his speed by 5% for the second lap. Find
   (a) his speed on the first lap
   (b) his speed on the second lap
   (c) his total time for the two laps
   (d) his average speed for the two laps.
9. The airliner Concorde flies 2000 km at a speed of 1600 km/h and then returns due to bad weather at a speed of 1000 km/h. Find the average speed for the whole trip.
10. A train travels from A to B, a distance of 100 km, at a speed of 20 km/h. If it had gone two and a half times as fast, how much earlier would it have arrived at B?
11. Two men running towards each other at 4 m/s and 6 m/s respectively are one kilometre apart. How long will it take before they meet?
12. A car travelling at 90 km/h is 500 m behind another car travelling at 70 km/h in the same direction. How long will it take the first car to catch the second?
13. How long is a train which passes a signal in twenty seconds at a speed of 108 km/h?
14. A train of length 180 m approaches a tunnel of length 620 m. How long will it take the train to pass completely through the tunnel at a speed of 54 km/h?
15. An earthworm of length 15 cm is crawling along at 2 cm/s. An ant overtakes the worm in 5 seconds. How fast is the ant walking?
16. A train of length 100 m is moving at a speed of 50 km/h. A horse is running alongside the train at a speed of 56 km/h. How long will it take the horse to overtake the train?

1.8 SQUARES AND SQUARE ROOTS

If you need the square or square root of a number when the batteries of your calculator are 'dead', it will unfortunately be necessary to refer to tables of these quantities. It is important to be able to get an approximate value for the answer *without* using tables because the tables will not tell you where to put the decimal point.

Example 1

Use tables to find

(a) 231^2 (b) $0{\cdot}0196^2$
(c) $\sqrt{47\,000}$ (d) $\sqrt{0{\cdot}000\,414}$

(a) We know that $200^2 = 40\,000$
From tables, the number for 231^2 is 5336.
But the answer must be about 40 000

$\therefore\ 231^2 = 53\,360$ (to four sig. fig.)

(b) Again, without tables, $0{\cdot}02^2 = 0{\cdot}0004$
From tables, the number for 196^2 is 3842

$\therefore\ 0{\cdot}0196^2 = 0{\cdot}000\,3842$ (to four sig. fig.)

(c) (i) Starting from the decimal point, divide the number into 'pairs' as shown. $\sqrt{04\ 70\ 00}$
(ii) Write down the approximate square root of each 'pair'. 2 8 0
So $\sqrt{47\,000}$ is approximately 280.
(iii) From square root tables for $\sqrt{47}$ we see either 2168 or 6856.
(iv) We choose 2168 because it is nearer to our approximation of 280.
(v) $\therefore\ \sqrt{47\,000} = 216{\cdot}8$ (to four sig. fig.)

(d) Use the same method as for (c)

$$\sqrt{0{\cdot}00\ 04\ 14}$$
$$\approx\ {\cdot}0\ \ 2\ \ 3$$

From square root tables for $\sqrt{414}$ we see either 2035 or 6434.
We choose 2035.

$\therefore\ \sqrt{0{\cdot}000\,414} = 0{\cdot}020\,35$ (to four sig. fig.)

Exercise 18

Estimate the values of the following correct to one significant figure.

1. $10{\cdot}72^2$ 2. $2{\cdot}94^2$ 3. $31{\cdot}4^2$
4. 78^2 5. 310^2 6. 700^2
7. 1111^2 8. $7{\cdot}024^2$ 9. 9020^2
10. $0{\cdot}0296^2$ 11. $0{\cdot}913^2$ 12. $0{\cdot}0073^2$
13. $0{\cdot}217^2$ 14. $\sqrt{40{\cdot}70}$ 15. $\sqrt{40}$
16. $\sqrt{40\,100}$ 17. $\sqrt{8005}$ 18. $\sqrt{800{\cdot}5}$
19. $\sqrt{80\,050}$ 20. $\sqrt{0{\cdot}0904}$ 21. $\sqrt{0{\cdot}009\,04}$
22. $\sqrt{0{\cdot}904}$ 23. $\sqrt{0{\cdot}000\,51}$ 24. $\sqrt{750\,000}$
25. $\sqrt{0{\cdot}628}$ 26. $\sqrt{4 \times 10^4}$ 27. $\sqrt{0{\cdot}006\,51}$

In questions **28** to **54**, use tables to find the values of the following:

28. $5{\cdot}63^2$ 29. $7{\cdot}09^2$ 30. $1{\cdot}415^2$
31. $11{\cdot}8^2$ 32. $23{\cdot}6^2$ 33. $35{\cdot}7^2$
34. $70{\cdot}6^2$ 35. $8{\cdot}614^2$ 36. 214^2
37. 523^2 38. $0{\cdot}723^2$ 39. $0{\cdot}217^2$
40. $0{\cdot}0415^2$ 41. $0{\cdot}01714^2$ 42. $0{\cdot}136^2$
43. $0{\cdot}0081^2$ 44. 271^2 45. $85{\cdot}6^2$
46. 609^2 47. 1110^2 48. $(5{\cdot}1 \times 10^3)^2$
49. $(2{\cdot}2 \times 10^{-3})^2$ 50. $(3{\cdot}7 \times 10^2)^2$
51. $(0{\cdot}23 \times 10^{-3})^2$ 52. $(1{\cdot}7 \times 10^5)^2$
53. $(1{\cdot}05 \times 10^{-4})^2$ 54. $0{\cdot}00314^2$

Exercise 19

Use tables to find the square roots of the following:

1. 8·97 2. 26·4 3. 85·06
4. 1·174 5. 79·9 6. 647
7. 6470 8. 214 9. 2140
10. 472·8 11. 179·3 12. 5700
13. 63 000 14. 0·782 15. 0·372
16. 0·143 17. 0·0836 18. 0·0567
19. 0·111 20. 0·009 41 21. 0·000 416
22. 0·009 23. 640 000 24. 10 million
25. 100 million 26. 0·00 001

Find the following without tables:

27. 300^2 28. $0{\cdot}2^2$ 29. $\sqrt{0{\cdot}04}$
30. $\sqrt{90\,000}$ 31. $\sqrt{0{\cdot}0016}$ 32. $(3 \times 10^3)^2$
33. $(7 \times 10^{-2})^2$ 34. $\sqrt{0{\cdot}000\,049}$ 35. $\sqrt{640\,000}$
36. $\sqrt{12\,100}$ 37. $(8 \times 10^4)^2$ 38 $\sqrt{10^{-4}}$
39. $\dfrac{1}{\sqrt{100}}$ 40. $\dfrac{1}{(0{\cdot}01)^2}$ 41. $\dfrac{1}{\sqrt{2500}}$
42. $\dfrac{1}{\sqrt{0{\cdot}0001}}$ 43. $\sqrt{\frac{1}{9}}$ 44. $\sqrt{\frac{1}{25}}$
45. $\sqrt{2\frac{1}{4}}$ 46. $\sqrt{6\frac{1}{4}}$ 47. $\sqrt{5\frac{4}{9}}$
48. $\dfrac{1}{0{\cdot}01^2}$ 49. $\dfrac{100}{0{\cdot}001^2}$ 50. $\dfrac{1}{\sqrt{7\frac{1}{9}}}$

1.9 NUMBER BASES

Consider the number 475. It means

$$(4 \times 100) + (7 \times 10) + (5 \times 1).$$

This is a number written in the ordinary decimal (or denary) system.

It is possible to use a number system which works on the powers of any number.
A common system used with computers is the binary system.

Example 1

Convert 2131_4 to base 10.

2131_4
$= (2 \times 4^3) + (1 \times 4^2) + (3 \times 4^1) + (1 \times 1)$
$= 128 + 16 + 12 + 1$
$= 157_{10}$

(2131_4 means 2131 in base 4)

Example 2

Convert 73_{10} into base 5.

divide 73 by 5 5 | 73
divide 14 by 5 5 | 14 r 3
divide 2 by 5 5 | 2 r 4
0 r 2 ↑

The answer is 243_5.

Example 3

Calculate $1010_2 + 111_2$

$$\begin{array}{r} 1010 \\ 111 \\ \hline 10001 \\ \hline {\scriptstyle 1\,1} \end{array}$$

The answer is 10001_2

Example 4

Calculate in base 8, $364_8 - 137_8$

```
 58
 364
-137
 ---
 225,  answer = 225_8
```

Exercise 20

Write the following numbers in base ten:

1. 312_4 2. 2113_4 3. 1012_4
4. 121_3 5. 2011_3 6. 112_5
7. 324_5 8. 1012_6 9. 514_6
10. 2014_8 11. 1012_8 12. 374_{11}
13. 1011_2 14. 11000_2 15. 1111_2
16. 10111_2

17. Write in order of size (smallest first):
(a) $101_4, 33_5, 1111_2$ (b) $21_4, 21_3, 1011_2$
(c) $131_4, 34_{10}, 51_6$ (d) $55_{12}, 102_8, 66_9$
(e) $100001_2, 40_8, 1011_3$

In questions **18** to **33**, change each base ten number into the base given:

18. 314 to base 4 19. 194 to base 4
20. 270 to base 5 21. 176 to base 6
22. 25 to base 2 23. 37 to base 2
24. 67 to base 2 25. 47 to base 2
26. 414 to base 8 27. 375 to base 9
28. 78 to base 12 29. 135 to base 12

In base 12, use T for 10 and E for eleven.

30. 165 to base 12 31. 227 to base 12
32. 375 to base 12 33. 600 to base 12

Calculate in base 2:

34. 101 + 101 35. 1110 + 101
36. 1001 + 111 37. 11001 + 1011
38. 110 − 10 39. 1001 − 11
40. 1110 − 111 41. 1011 − 101

Calculate in base 8:

42. 314 + 21 43. 373 + 232
44. 456 + 17 45. 1056 + 233
46. 416 − 252 47. 316 − 277
48. 2404 − 375 49. 4010 − 275

Calculate in base 2:

50. 101 × 11 51. 110 × 101
52. 1011 × 110 53. 1101 × 101
54. 111 × 11

55. If a number in base ten has two digits, what is the greatest number of digits it can have:
(a) in base 2? (b) in base 5?
56. (a) Write $1 + 4 + 4^2 + 4^3$ as a number in base 4
(b) Write $1 + 6 + 6^2 + 6^3 + 6^4$ as a number in base 6
(c) Write $1 + n + n^2 + n^3$ as a number in base n
(d) Write $n^3 + 2n^2 + 3n + 1$ as a number in base n
(e) Write $5n^2 + 7n + 2$ as a number in base n
(f) Write $3n^3 + 2n + 4$ as a number in base n
(g) Write $n^4 + 5$ as a number in base n
57. In the following additions, find
(a) the number base
(b) the missing digit.

```
(i)   2*6        (ii)  3*5
      743              334
     ----             ----
     1221             1032
```

58. (a) Write $1 + 3 + 3^2 + 3^3$ as a number in base 3
(b) Write 3^4 as a number in base 3
(c) Evaluate $3^4 - 1$ as a number in base 3
(d) Find k so that $3^4 - 1 = k(1 + 3 + 3^2 + 3^3)$
(e) If $1 + 3 + 3^2 + 3^3 + \ldots + 3^{10} = \frac{3^n - 1}{2}$, find the value of n.
59. (a) Write $1 + 5 + 5^2 + 5^3$ as a number in base 5
(b) Write 5^4 as a number in base 5
(c) Evaluate $5^4 - 1$ as a number in base 5
(d) Find k so that $5^4 - 1 = k(1 + 5 + 5^2 + 5^3)$
(e) If $1 + 5 + 5^2 + 5^3 + \ldots 5^{21} = \frac{5^n - 1}{c}$, find the values of n and c
60. In the following simultaneous equations, all the numbers are written in base 4.

$$2x + 3y = 21$$
$$x + y = 10$$

Find x and y, showing all the working in base 4.

1.10 CALCULATOR

In this book, the keys are described thus:

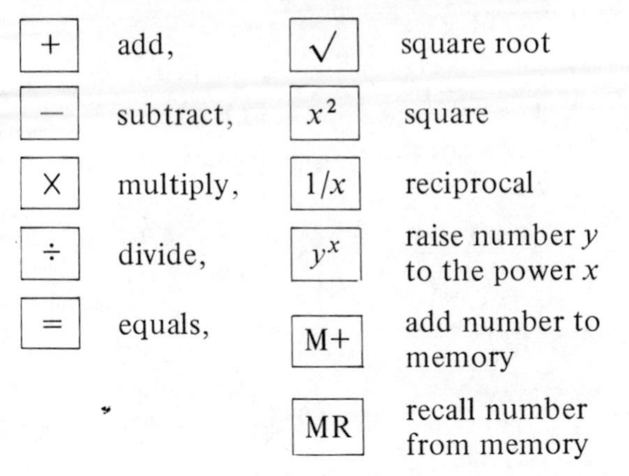

Example 1

Evaluate the following to 4 significant figures:

(a) $\dfrac{2\cdot3}{4\cdot7+3\cdot61}$ (b) $\dfrac{1\cdot74+0\cdot141}{0\cdot72-0\cdot508}$

(c) $\left(\dfrac{1}{0\cdot084}\right)^4$ (d) $\sqrt{(2\cdot3^2+6\cdot74^2)}$

(e) $\sqrt[3]{[3\cdot2\times(1\cdot7-1\cdot64)]}$

(a) $\dfrac{2\cdot3}{4\cdot7+3\cdot61}$ Find the denominator first

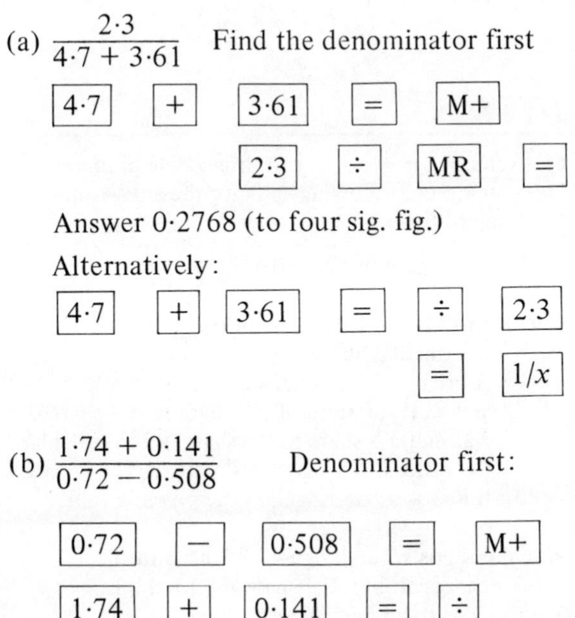

(b) $\dfrac{1\cdot74+0\cdot141}{0\cdot72-0\cdot508}$ Denominator first:

0·72 − 0·508 = M+

1·74 + 0·141 = ÷

MR =

Answer 8·873 (to four sig. fig.)

(c) $\left(\dfrac{1}{0\cdot084}\right)^4$

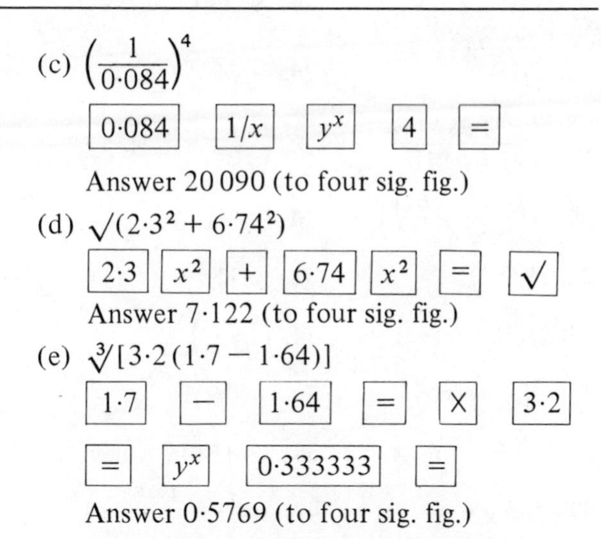

Note: To find a cube root, raise to the power $\frac{1}{3}$, or as a decimal 0·333 . . .

Exercise 21

Use a calculator to evaluate the following, giving the answers to 4 significant figures:

1. $\dfrac{7\cdot351\times0\cdot764}{1\cdot847}$
2. $\dfrac{0\cdot0741\times14700}{0\cdot746}$
3. $\dfrac{0\cdot0741\times9\cdot61}{23\cdot1}$
4. $\dfrac{417\cdot8\times0\cdot00841}{0\cdot07324}$
5. $\dfrac{8\cdot41}{7\cdot601\times0\cdot00847}$
6. $\dfrac{4\cdot22}{1\cdot701\times5\cdot2}$
7. $\dfrac{9\cdot61}{17\cdot4\times1\cdot51}$
8. $\dfrac{8\cdot71\times3\cdot62}{0\cdot84}$
9. $\dfrac{0\cdot76}{0\cdot412-0\cdot317}$
10. $\dfrac{81\cdot4}{72\cdot6+51\cdot92}$
11. $\dfrac{111}{27\cdot4+2960}$
12. $\dfrac{27\cdot4+11\cdot61}{5\cdot9-4\cdot763}$
13. $\dfrac{6\cdot51-0\cdot1114}{7\cdot24+1\cdot653}$
14. $\dfrac{5\cdot71+6\cdot093}{9\cdot05-5\cdot77}$
15. $\dfrac{0\cdot943-0\cdot788}{1\cdot4-0\cdot766}$
16. $\dfrac{2\cdot6}{1\cdot7}+\dfrac{1\cdot9}{3\cdot7}$
17. $\dfrac{8\cdot06}{5\cdot91}-\dfrac{1\cdot594}{1\cdot62}$
18. $\dfrac{4\cdot7}{11\cdot4-3\cdot61}+\dfrac{1\cdot6}{9\cdot7}$
19. $\dfrac{3\cdot74}{1\cdot6\times2\cdot89}-\dfrac{1}{0\cdot741}$
20. $\dfrac{1}{7\cdot2}-\dfrac{1}{14\cdot6}$
21. $\dfrac{1}{0\cdot961}\times\dfrac{1}{0\cdot412}$
22. $\dfrac{1}{7}+\dfrac{1}{13}-\dfrac{1}{8}$

23. $4{\cdot}2\left(\frac{1}{5{\cdot}5}-\frac{1}{7{\cdot}6}\right)$

24. $\sqrt{(9{\cdot}61+0{\cdot}1412)}$

25. $\sqrt{\left(\frac{8{\cdot}007}{1{\cdot}61}\right)}$

26. $(1{\cdot}74+9{\cdot}611)^2$

27. $\left(\frac{1{\cdot}63}{1{\cdot}7-0{\cdot}911}\right)^2$

28. $\left(\frac{9{\cdot}6}{2{\cdot}4}-\frac{1{\cdot}5}{0{\cdot}74}\right)^2$

29. $\sqrt{\left(\frac{4{\cdot}2\times 1{\cdot}611}{9{\cdot}83\times 1{\cdot}74}\right)}$

30. $(0{\cdot}741)^3$

31. $(1{\cdot}562)^5$

32. $(0{\cdot}32)^3+(0{\cdot}511)^4$

33. $(1{\cdot}71-0{\cdot}863)^6$

34. $\left(\frac{1}{0{\cdot}971}\right)^4$

35. $\sqrt[3]{(4{\cdot}714)}$

36. $\sqrt[3]{(0{\cdot}9316)}$

37. $\sqrt[3]{\left(\frac{4{\cdot}114}{7{\cdot}93}\right)}$

38. $\sqrt[4]{(0{\cdot}8145-0{\cdot}799)}$

39. $\sqrt[5]{(8{\cdot}6\times 9{\cdot}71)}$

40. $\sqrt[3]{\left(\frac{1{\cdot}91}{4{\cdot}2-3{\cdot}766}\right)}$

41. $\left(\frac{1}{7{\cdot}6}-\frac{1}{18{\cdot}5}\right)^3$

42. $\frac{\sqrt{(4{\cdot}79)}+1{\cdot}6}{9{\cdot}63}$

43. $\frac{(0{\cdot}761)^2-\sqrt{(4{\cdot}22)}}{1{\cdot}96}$

44. $\sqrt[3]{\left(\frac{1{\cdot}74\times 0{\cdot}761}{0{\cdot}0896}\right)}$

45. $\left(\frac{8{\cdot}6\times 1{\cdot}71}{0{\cdot}43}\right)^3$

46. $\frac{9{\cdot}61-\sqrt{(9{\cdot}61)}}{9{\cdot}61^2}$

47. $\frac{9{\cdot}6\times 10^4\times 3{\cdot}75\times 10^7}{8{\cdot}88\times 10^6}$

48. $\frac{8{\cdot}06\times 10^{-4}}{1{\cdot}71\times 10^{-6}}$

49. $\frac{3{\cdot}92\times 10^{-7}}{1{\cdot}884\times 10^{-11}}$

50. $\left(\frac{1{\cdot}31\times 2{\cdot}71\times 10^5}{1{\cdot}91\times 10^4}\right)^5$

51. $\left(\frac{1}{9{\cdot}6}-\frac{1}{9{\cdot}99}\right)^{10}$

52. $\frac{\sqrt[3]{(86{\cdot}6)}}{\sqrt[4]{(4{\cdot}71)}}$

53. $\frac{23{\cdot}7\times 0{\cdot}0042}{12{\cdot}48-9{\cdot}7}$

54. $\frac{0{\cdot}482+1{\cdot}6}{0{\cdot}024\times 1{\cdot}83}$

55. $\frac{8{\cdot}52-1{\cdot}004}{0{\cdot}004-0{\cdot}0083}$

56. $\frac{1{\cdot}6-0{\cdot}476}{2{\cdot}398\times 41{\cdot}2}$

57. $\left(\frac{2{\cdot}3}{0{\cdot}791}\right)^7$

58. $\left(\frac{8{\cdot}4}{28{\cdot}7-0{\cdot}47}\right)^3$

59. $\left(\frac{5{\cdot}114}{7{\cdot}332}\right)^5$

60. $\left(\frac{4{\cdot}2}{2{\cdot}3}+\frac{8{\cdot}2}{0{\cdot}52}\right)^3$

61. $\frac{1}{8{\cdot}2^2}-\frac{3}{19^2}$

62. $\frac{100}{11^3}+\frac{100}{12^3}$

63. $\frac{7{\cdot}3-4{\cdot}291}{2{\cdot}6^2}$

64. $\frac{9{\cdot}001-8{\cdot}97}{0{\cdot}95^3}$

65. $\frac{10{\cdot}1^2+9{\cdot}4^2}{9{\cdot}8}$

66. $(3{\cdot}6\times 10^{-8})^2$

67. $(8{\cdot}24\times 10^4)^3$

68. $(2{\cdot}17\times 10^{-3})^3$

69. $(7{\cdot}095\times 10^{-6})^{\frac{1}{3}}$

70. $\sqrt[3]{\left(\frac{4{\cdot}7}{2{\cdot}3^2}\right)}$

REVISION EXERCISE 1A

1. Evaluate, without a calculator:
   (a) $148\div 0{\cdot}8$ (b) $0{\cdot}024\div 0{\cdot}00016$
   (c) $(0{\cdot}2)^2\div(0{\cdot}1)^3$ (d) $2-\frac{1}{2}-\frac{1}{3}-\frac{1}{4}$
   (e) $1\frac{3}{4}\times 1\frac{3}{5}$ (f) $\frac{1\frac{1}{6}}{1\frac{2}{3}+1\frac{1}{4}}$
2. On each bounce, a ball rises to $\frac{4}{5}$ of its previous height. To what height will it rise after the third bounce, if dropped from a height of 250 cm?
3. A man spends $\frac{1}{3}$ of his salary on accommodation and $\frac{2}{5}$ of the remainder on food. What fraction is left for other purposes?
4. $a=\frac{1}{2}$, $b=\frac{1}{4}$. Which one of the following has the greatest value?
   (i) ab (ii) $a+b$ (iii) $\frac{a}{b}$ (iv) $\frac{b}{a}$ (v) $(ab)^2$
5. Express $0{\cdot}05473$
   (a) correct to three significant figures
   (b) correct to three decimal places
   (c) in standard form.
6. Evaluate $\frac{2}{3}+\frac{4}{7}$, correct to three decimal places.
7. Evaluate the following and give the answer in standard form:
   (a) $3600\div 0{\cdot}00012$ (b) $\frac{3{\cdot}33\times 10^4}{9\times 10^{-1}}$
   (c) $(30\,000)^3$
8. (a) £143 is divided in the ratio 2:3:6; calculate the smallest share.
   (b) A prize is divided between three people X, Y and Z. If the ratio of X's share to Y's share is 3:1 and Y's share to Z's share is 2:5, calculate the ratio of X's share to Z's share.
   (c) If $a:3=12:a$, calculate the possible values of a.
9. Labour costs, totalling £47·25, account for 63% of a car repair bill. Calculate the total bill.
10. (a) Convert to percentages
    (i) $0{\cdot}572$ (ii) $\frac{7}{8}$
    (b) Express 2·6 kg as a percentage of 6·5 kg.
    (c) In selling a red herring for 92p, a fishmonger makes a profit of 15%. Find the cost price of the fish.

11. The length of a rectangle is decreased by 25% and the breadth is increased by 40%. Calculate the percentage change in the area of the rectangle.

12. (a) What sum of money, invested at 9% interest per year, is needed to provide an income of £45 per year?
(b) A particle increases its speed from 8×10^5 m/s to $1{\cdot}1 \times 10^6$ m/s. What is the percentage increase?

13. A family on holiday in France exchanged £450 for francs when the exchange rate was 10·70 francs to the pound. They spent 3700 francs and then changed the rest back into pounds, by which time the exchange rate had become 11·15 francs to the pound. How much did the holiday cost? (Answer in pounds.)

14. A welder has an income of £8400 and allowances which total £1950. How much tax does he pay each month if the rate of taxation is 28% of taxable income? He receives a pay rise and subsequently pays £162·40 in tax each month. Assuming the same allowances and rate of taxation, calculate his new salary.

15. A map is drawn to a scale of 1 : 10 000. Find:
(a) the distance between two railway stations which appear on the map 24 cm apart.
(b) the area, in square kilometres, of a lake which has an area of 100 cm² on the map.

16. A map is drawn to a scale of 1 : 2000. Find:
(a) the actual distance between two points, which appear 15 cm apart on the map.
(b) the length on the map of a road, which is 1·2 km in length.
(c) the area on the map of a field, with an actual area of 60 000 m².

17. (a) On a map, the distance between two points is 16 cm. Calculate the scale of the map if the actual distance between the points is 8 km.
(b) On another map, two points appear 1·5 cm apart and are in fact 60 km apart. Calculate the scale of the map.

18. (a) A house is bought for £20 000 and sold for £24 400. What is the percentage profit?
(b) A piece of meat, initially weighing 2·4 kg, is cooked and subsequently weighs 1·9 kg. What is the percentage loss in weight?
(c) An article is sold at a 6% loss for £225·60. What was the cost price?

19. (a) Convert into metres per second:
(i) 700 cm/s (ii) 720 km/h
(iii) 18 km/h
(b) Convert into kilometres per hour:
(i) 40 m/s (ii) 0·6 m/s

20. (a) Calculate the speed (in metres per second) of a slug which moves a distance of 30 cm in 1 minute.
(b) Calculate the time taken for a bullet to travel 8 km at a speed of 5000 m/s.
(c) Calculate the distance flown, in a time of four hours, by a pigeon which flies at a speed of 12 m/s.

21. A motorist travelled 200 miles in five hours. His average speed for the first 100 miles was 50 m.p.h. What was his average speed for the second 100 miles?

22. (a) Without using tables, write down the value of:
(i) $\sqrt{0{\cdot}0025}$ (ii) $\sqrt{2\frac{1}{4}}$
(b) Given $\sqrt{15} = 3{\cdot}87$ and $\sqrt{1{\cdot}5} = 1{\cdot}22$, write down:
(i) $\sqrt{150}$ (ii) $\sqrt{1500}$ (iii) $\sqrt{0{\cdot}0015}$
(c) Given $\sqrt{19} = 4{\cdot}36$ and $\sqrt{1{\cdot}9} = 1{\cdot}38$, write down:
(i) $\sqrt{19\,000}$ (ii) $\sqrt{190}$ (iii) $\sqrt{0{\cdot}000\,19}$

23. Write the following in order of size (smallest first)
(a) $\sqrt{0{\cdot}04}$, $(0{\cdot}3)^2$, 5%, $\frac{1}{11}$.
(b) 11%, $\sqrt{0{\cdot}01}$, $\frac{1}{9}$, 91×10^{-3}.
(c) $\sqrt{(90\,000)}$, $2{\cdot}96 \times 10^3$, 310%, $21{\cdot}7^2$.

24. Express the following as binary numbers:
(a) 27_{10} (b) $2^4 + 2$
(c) $1001_2 + 110_2$ (d) $1000_2 - 1_2$
(e) 129_{10}

25. Express the following in base eight:
(a) 67_{10} (b) $8^4 + 3$
(c) 1010001_2 (d) $325_8 + 147_8$
(e) $357_8 - 273_8$

26. (a) If $120_n = 33_7$, find the value of n.
(b) In a certain base, $5 \times 8 = 44$. What is the base?
(c) Write the number $3y^2 + 2y + 4$ in base y. $(y > 4)$

27. Evaluate the following using either a calculator or tables: (answers to 4 sig. fig.)
(a) $\dfrac{0{\cdot}74}{0{\cdot}81 \times 1{\cdot}631}$ (b) $\sqrt{\left(\dfrac{9{\cdot}61}{8{\cdot}34 - 7{\cdot}41}\right)}$
(c) $\left(\dfrac{0{\cdot}741}{0{\cdot}8364}\right)^4$ (d) $\dfrac{8{\cdot}4 - 7{\cdot}642}{3{\cdot}333 - 1{\cdot}735}$

28. Evaluate the following and give the answers to 3 significant figures:
(a) $\sqrt[3]{(9{\cdot}61 \times 0{\cdot}0041)}$ (b) $\left(\dfrac{1}{9{\cdot}5} - \dfrac{1}{11{\cdot}2}\right)^3$
(c) $\dfrac{15{\cdot}6 \times 0{\cdot}714}{0{\cdot}0143 \times 12}$ (d) $\sqrt[4]{\left(\dfrac{1}{5 \times 10^3}\right)}$

EXAMINATION EXERCISE 1B

1. A man is trying to decide whether to buy or to rent a new television set.
   The model he wants costs £400 and the dealer charges an additional $3\frac{1}{2}\%$ of this cost to install it. During the first year no charge will be made for repairs. After this the man estimates that repairs will cost £20 for each of the next four years, and then £35 for each of the following three years. At the end of these eight years he expects to receive a trade-in value of £20 for the set when he buys a new one.
   Calculate (i) the installation charge,
   (ii) the total estimated repair cost,
   (iii) the estimated net cost of the set over the eight years (that is, the total he expects to pay less the trade-in value).
   The cost to rent the same set is £8·40 per month during the first year but $7\frac{1}{2}\%$ discount is allowed if the year's rental is paid in advance. Calculate the rental for this year if it is paid in advance.
   For the second and subsequent years the rental is reduced to £7·60 per month but no discount is allowed. Calculate the rental for the second year. Hence calculate the total rental if the set is kept for eight years, the first year's rental being paid in advance. [C]

2. The outfitter for the Mahos Comprehensive School was asked to stock three different styles of school jumper. She decided to purchase
   40 V-neck jumpers,
   30 polo-neck jumpers,
   and 30 crew-neck jumpers.
   (a) She bought the V-neck jumpers for £5·50 each and sold them at a profit of 40 per cent on the cost price. Calculate the selling price of each V-neck jumper.
   (b) She bought the polo-neck jumpers at £6·30 each and sold them at £8·40 each. Express the profit as a percentage of the cost price.
   (c) She sold the crew-neck jumpers at £7·20 each, thereby losing 10 per cent on the cost price. Calculate the cost price of each crew-neck jumper.
   (d) Calculate the total amount of money she received for the 100 jumpers.
   (e) Calculate the total profit she made on the sale of all the jumpers and express this profit as a percentage of her initial outlay, giving your answer correct to one decimal place.
   [C 16+]

3. At the beginning of 1976 Mrs Smart bought a secondhand car for £1500. During 1976 she ran the car for 20 000 miles for business purposes as representative of a cosmetics firm. The road tax for the year was £40 and insurance costs were £85. Service charges were £6·50 for each 5000 miles of running. On average the car consumed 1 gallon of petrol for every 25 miles and 1 gallon of oil for every 4000 miles. Petrol costs were 73p per gallon and oil costs £2·20 per gallon. Calculate the total expenditure incurred by Mrs. Smart in purchasing and running her car during 1976.
   At the end of 1976 Mrs. Smart sold the car for 80% of the price she paid for it. If she was allowed 7·2p per mile by the firm for running the car during the year, calculate the gain she showed in purchasing, running and selling the car during the year. [L]

4. A County Council surveyor calculating the costs of road construction determines the costs from the following:
(a) preparation, etc. £9 000 per km,
(b) excavation £300 000 per km,
(c) materials: hardcore £310 per m^3,
base sealer £150 per m^3,
final surface £170 per m^3.
Construction regulations require the hardcore to be 20 cm thick, the base sealer to be 8 cm thick and the final surface to be 6 cm thick.
Find the total cost involved in constructing each km of a new road 8 m wide.
For re-surfacing roads of the same width, no other costs apart from the final surfacing costs are incurred. If, in a year, the County Council decides to spend £42 432 000 on this task, calculate the total length of roads of this width which may be resurfaced. [L]

5. In 1980 a farmer planted potatoes in a field of area 10 hectares. [1 hectare = $10^4\,m^2$.] He incurred the following expenses:
(i) the rent of the field was £2000,
(ii) he required 1 kg of seed potatoes, which cost 20p per kg, to plant an area of 4 m^2,
(iii) he applied fertilizer, costing 15p per kg, to the soil at the rate of 100 g to 1 m^2,
(iv) the planting and harvesting required 4 men to work for 5 days, each man being paid £15 per day.
Find the total cost to the farmer.
At harvest time the total yield was 30 tonnes per hectare and the farmer sold the whole crop to a merchant at 4p per kg. Find the farmer's total receipts. [1 tonne = 10^3 kg.]
Hence find the farmer's profit and express this, correct to 1 decimal place, as a percentage of the total cost. [L]

6. In 1978 the basic rate for a workman's job was £58·80 for a 42 hour week. Calculate the basic hourly rate of pay. For overtime the workman is paid one and a half times the basic hourly rate. Calculate the man's total wage in a week in which he worked 56 hours.
The man's union made a claim on his behalf asking for £63 for a 36 hour week. Calculate:
(i) the hourly rate of pay represented by this claim,
(ii) the percentage increase in the hourly rate represented by this claim.
The award gave a $12\frac{1}{2}$% increase on the basic wage of £58·80, and at the same time reduced the basic working week from 42 hours to 40 hours. Under this award calculate the wage that the man could expect to receive for a week in which he worked 56 hours, given that overtime is now paid at one and a half times the new basic hourly rate. [AEB]

7. In a certain income tax year, the first £985 of income was exempt from tax. The next £750 of the income was taxed at the rate of 25%, and the remainder of the income was taxed at the rate of 30%.
(i) Calculate the total tax paid on an income of £4690 for the year, and express this tax as a percentage of the income,
(ii) Calculate the income on which a total tax of £963 was paid during the year. [O & C]

8. In May, the price of a chair in a certain shop was £25 and 360 of these chairs were sold. Calculate the income received from the sale of these chairs.
In June, the price of a chair was increased by 18 per cent and the number of chairs sold decreased by 10 per cent. Calculate the percentage change in the income received.
In July, the price of a chair was 25 per cent more than its price in May and the number of chairs sold was x per cent less than the number sold in May.
Calculate the value of x for which
   (i) the incomes received in July and May were equal,
   (ii) the income received in July was one-ninth greater than that received in May. [JMB]

9. (a) A journey of 23 km was completed in two stages by a train which took 30 minutes altogether. For the first stage the distance travelled was 8 km and the average speed was 40 km/h. Find the times taken for the two stages of the journey.
Calculate, in kilometres per hour, the average speed for the second stage of the journey.
   (b) An alloy contains 54% by weight of copper and 46% by weight of zinc. Find the number of grams of copper which must be combined with 138 g of zinc to give an alloy with this composition. [O & C]

10. (a) The population of a town in 1970 was 20 800, which was an increase of 4 per cent on that in 1965. The population in 1975 was an increase of 5 per cent on that in 1970.
What was the population
      (i) in 1965,
      (ii) in 1975?
   (b) The population of a village in 1970 was 330, which was an increase of 20 per cent on that in 1965. The population in 1975 was an increase of x per cent on that in 1970. Given that the population in 1975 was 88 greater than that in 1965, find x. [JMB]

11. In an election, the electors (the persons who were qualified to vote) were each entitled to give one vote to just one of the three candidates A, B and C. Some of the electors did not vote.
   (i) 30% of the electors voted for candidate A and he received 24 600 votes. Calculate the total number of electors.
   (ii) Candidate B received 20 500 votes. Calculate what percentage of the electors voted for him.
   (iii) Candidate C received 18% of all the votes recorded. Calculate the number of votes he received.
   (iv) Calculate what percentage of the electors actually voted in the election. [C]

2 Algebra 1

Isaac Newton (1642–1727) is thought by many to have been one of the greatest intellects of all time. He went to Trinity College Cambridge in 1661 and by the age of 23 he had made three major discoveries: the nature of colours, the calculus and the law of gravitation. He used his version of the calculus to give the first satisfactory explanation of the motion of the Sun, the Moon and the stars. Because he was extremely sensitive to criticism, Newton was always very secretive, but he was eventually persuaded to publish his discoveries in 1687.

2.1 DIRECTED NUMBERS

To add two directed numbers with the same sign, find the sum of the numbers and give the answer the same sign.

e.g. $+3+(+5) = +3+5 = +8$

$-7+(-3) = -7-3 = -10$

$-9{\cdot}1+(-3{\cdot}1) = -9{\cdot}1-3{\cdot}1 = -12{\cdot}2$

$-2+(-1)+(-5) = (-2-1)-5$

$= -3-5$

$= -8$

To add two directed numbers with different signs, find the difference between the numbers and give the answer the sign of the larger number.

e.g. $+7+(-3) = +7-3 = +4$

$+9+(-12) = +9-12 = -3$

$-8+(+4) = -8+4 = -4$

To subtract a directed number, change its sign and add.

e.g. $+7-(+5) = +7-5 = +2$

$+7-(-5) = +7+5 = +12$

$-8-(+4) = -8-4 = -12$

$-9-(-11) = -9+11 = +2$

Exercise 1

1. $+7+(+6)$	**2.** $+11+(+200)$
3. $-3+(-9)$	**4.** $-7+(-24)$
5. $-5+(-61)$	**6.** $+0{\cdot}2+(+5{\cdot}9)$
7. $+5+(+4{\cdot}1)$	**8.** $-8+(-27)$
9. $+17+(+1{\cdot}7)$	**10.** $-2+(-3)+(-4)$
11. $-7+(+4)$	**12.** $+7+(-4)$
13. $-9+(+7)$	**14.** $+16+(-30)$
15. $+14+(-21)$	**16.** $-7+(+10)$
17. $-19+(+200)$	**18.** $+7{\cdot}6+(-9{\cdot}8)$

19. $-1{\cdot}8+(+10)$
20. $-7+(+24)$
21. $+7-(+5)$
22. $+9-(+15)$
23. $-6-(+9)$
24. $-9-(+5)$
25. $+8-(+10)$
26. $-19-(-7)$
27. $-10-(+70)$
28. $-5{\cdot}1-(+8)$
29. $-0{\cdot}2-(+4)$
30. $+5{\cdot}2-(-7{\cdot}2)$
31. $-4+(-3)$
32. $+6-(-2)$
33. $+8+(-4)$
34. $-4-(+6)$
35. $+7-(-4)$
36. $+6+(-2)$
37. $+10-(+30)$
38. $+19-(+11)$
39. $+4+(-7)+(-2)$
40. $-3-(+2)+(-5)$
41. $-17-(-1)+(-10)$
42. $-5+(-7)-(+9)$
43. $+9+(-7)-(-6)$
44. $-7-(-8)$
45. $-10{\cdot}1+(-10{\cdot}1)$
46. $-75-(-25)$
47. $-204-(+304)$
48. $-7+(-11)-(+11)$
49. $+17-(+17)$
50. $-6+(-7)-(+8)$
51. $+7+(-7{\cdot}1)$
52. $-11-(-4)+(+3)$
53. $-2-(-8{\cdot}7)$
54. $+7+(-11)+(+5)$
55. $-610+(-240)$
56. $-7-(-3)-(-8)$
57. $+9-(-6)+(-9)$
58. $-1-(-5)+(-8)$
59. $-2{\cdot}1+(-9{\cdot}9)$
60. $-47-(-16)$

When two directed numbers with the same sign are multiplied together, the answer is positive.

(a) $+7\times(+3) = +21$
(b) $-6\times(-4) = +24$

When two directed numbers with different signs are multiplied together, the answer is negative.

(a) $-8\times(+4) = -32$
(b) $+7\times(-5) = -35$
(c) $-3\times(+2)\times(+5) = -6\times(+5) = -30$

When dividing directed numbers, the rules are the same as in multiplication.

(a) $-70\div(-2) = +35$
(b) $+12\div(-3) = -4$
(c) $-20\div(+4) = -5$

Exercise 2

1. $+2\times(-4)$
2. $+7\times(+4)$
3. $-4\times(-3)$
4. $-6\times(-4)$
5. $-6\times(-3)$
6. $+5\times(-7)$
7. $-7\times(-7)$
8. $-4\times(+3)$
9. $+0{\cdot}5\times(-4)$
10. $-1\frac{1}{2}\times(-6)$
11. $-8\div(+2)$
12. $+12\div(+3)$
13. $+36\div(-9)$
14. $-40\div(-5)$
15. $-70\div(-1)$
16. $-56\div(+8)$
17. $-\frac{1}{2}\div(-2)$
18. $-3\div(+5)$
19. $+0{\cdot}1\div(-10)$
20. $-0{\cdot}02\div(-100)$
21. $-11\times(-11)$
22. $-6\times(-1)$
23. $+12\times(-50)$
24. $-\frac{1}{2}\div(+\frac{1}{2})$
25. $-600\div(+30)$
26. $-5{\cdot}2\div(+2)$
27. $+7\times(-100)$
28. $-6\div(-\frac{1}{3})$
29. $100\div(-0{\cdot}1)$
30. -8×-80
31. $-3\times(-2)\times(-1)$
32. $+3\times(-7)\times(+2)$
33. $+0{\cdot}4\div(-1)$
34. $-16\div(+40)$
35. $+0{\cdot}2\times(-1000)$
36. $-7\times(-5)\times(-1)$
37. $-14\div(+7)$
38. $-7\div(-14)$
39. $+1\frac{1}{4}\div(-5)$
40. $-6\times(-\frac{1}{2})\times(-30)$

Exercise 3

1. $-7+(-3)$
2. $-6-(-7)$
3. $-4\times(-3)$
4. $-4\times(+7)$
5. $4-(+6)$
6. $-4\times(-4)$
7. $+6\div(-2)$
8. $+8-(-6)$
9. $-7\times(+4)$
10. $-8\div(-2)$
11. $+10\div(-60)$
12. $(-3)^2$
13. $40-(+70)$
14. $-6\times(-4)$
15. $(-1)^5$
16. $-8\div(+4)$
17. $+10\times(-3)$
18. $-7\times(-1)$
19. $+10+(-7)$
20. $+12-(-4)$
21. $+100+(-7)$
22. $-60\times(-40)$
23. $-20\div(-2)$
24. $(-1)^{20}$
25. $6-(+10)$
26. $-6\times(+4)\times(-2)$
27. $+8\div(-8)$
28. $0\times(-6)$
29. $(-2)^3$
30. $+100-(-70)$
31. $+18\div(-6)$
32. $(-1)^{12}$
33. $-6-(-7)$
34. $(-2)^2+(-4)$
35. $+8-(-7)$
36. $+7+(-2)$
37. $-6\times(+0{\cdot}4)$
38. $-3\times(-6)\times(-10)$
39. $(-2)^2+(+1)$
40. $+6-(+1000)$
41. $(-3)^2-7$
42. $-12\div\frac{1}{4}$
43. $-30\div-\frac{1}{2}$
44. $5-(+7)+(-0{\cdot}5)$
45. $(-2)^5$
46. $0\div(-\frac{1}{5})$
47. $(-0{\cdot}1)^2\times(-10)$
48. $3-(+19)$
49. $2{\cdot}1+(-6{\cdot}4)$
50. $(-\frac{1}{2})^2\div(-4)$

2.2 SUBSTITUTION INTO FORMULAE

Example 1

When $a = 3, b = -2, c = 5$, find the value of:
(a) $3a + b$ (b) $ac + b^2$
(c) $\frac{a+c}{b}$ (d) $a(c-b)$

(a) $3a + b$
$= (3 \times 3) + (-2)$
$= 9 - 2$
$= 7$

(b) $ac + b^2$
$= (3 \times 5) + (-2)^2$
$= 15 + 4$
$= 19$

(c) $\frac{a+c}{b}$
$= \frac{3+5}{-2}$
$= \frac{8}{-2}$
$= -4$

(d) $a(c-b)$
$= 3[5-(-2)]$
$= 3[7]$
$= 21$

Notice that working *down* the page is often easier to follow.

Exercise 4

Evaluate the following:
For questions **1** to **12** $a = 3, c = 2, e = 5$.

1. $3a - 2$ **2.** $4c + e$ **3.** $2c + 3a$
4. $5e - a$ **5.** $e - 2c$ **6.** $e - 2a$
7. $4c + 2e$ **8.** $7a - 5e$ **9.** $c - e$
10. $10a + c + e$ **11.** $a + c - e$ **12.** $a - c - e$

For questions **13** to **24** $h = 3, m = -2, t = -3$.

13. $2m - 3$ **14.** $4t + 10$ **15.** $3h - 12$
16. $6m + 4$ **17.** $9t - 3$ **18.** $4h + 4$
19. $2m - 6$ **20.** $m + 2$ **21.** $3h + m$
22. $t - h$ **23.** $4m + 2h$ **24.** $3t - m$

For questions **25** to **36**, $x = -2, y = -1, k = 0$

25. $3x + 1$ **26.** $2y + 5$ **27.** $6k + 4$
28. $3x + 2y$ **29.** $2k + x$ **30.** xy
31. xk **32.** $2xy$ **33.** $2(x + k)$
34. $3(k + y)$ **35.** $5x - y$ **36.** $3k - 2x$

$2x^2$ means $2(x^2)$
$(2x)^2$ means 'work out $2x$ and *then* square it'
$-7x$ means $-7(x)$
$-x^2$ means $-1(x^2)$

Example 2

When $x = -2$, find the value of
(a) $2x^2 - 5x$ (b) $(3x)^2 - x^2$

(a) $2x^2 - 5x = 2(-2)^2 - 5(-2)$
$= 2(4) + 10$
$= 18$

(b) $(3x)^2 - x^2 = (3 \times -2)^2 - 1(-2)^2$
$= (-6)^2 - 1(4)$
$= 36 - 4$
$= 32.$

Exercise 5

If $x = -3$ and $y = 2$, evaluate the following:

1. x^2 **2.** $3x^2$ **3.** y^2
4. $4y^2$ **5.** $(2x)^2$ **6.** $2x^2$
7. $10 - x^2$ **8.** $10 - y^2$ **9.** $20 - 2x^2$
10. $20 - 3y^2$ **11.** $5 + 4x$ **12.** $x^2 - 2x$
13. $y^2 - 3x^2$ **14.** $x^2 - 3y$ **15.** $(2x)^2 - y^2$
16. $4x^2$ **17.** $(4x)^2$ **18.** $1 - x^2$
19. $y - x^2$ **20.** $x^2 + y^2$ **21.** $x^2 - y^2$
22. $2 - 2x^2$ **23.** $(3x)^2 + 3$ **24.** $11 - xy$
25. $12 + xy$ **26.** $(2x)^2 - (3y)^2$ **27.** $2 - 3x^2$
28. $y^2 - x^2$ **29.** $x^2 + y^3$ **30.** $\frac{x}{y}$
31. $10 - 3x$ **32.** $2y^2$ **33.** $25 - 3y$
34. $(2y)^2$ **35.** $-7 + 3x$ **36.** $-8 + 10y$
37. $(xy)^2$ **38.** xy^2 **39.** $-7 + x^2$
40. $17 + xy$ **41.** $-5 - 2x^2$ **42.** $10 - (2x)^2$
43. $x^2 + 3x + 5$ **44.** $2x^2 - 4x + 1$ **45.** $\frac{x^2}{y}$

Example 3

When $a = -2, b = 3, c = -3$, evaluate
(a) $\frac{2a(b^2 - a)}{c}$ (b) $\sqrt{(a^2 + b^2)}$

(a) $(b^2 - a) = 9 - (-2)$
$= 11$

$\therefore \frac{2a(b^2 - a)}{c} = \frac{2 \times (-2) \times (11)}{-3}$
$= 14\frac{2}{3}$

(b) $a^2 + b^2 = (-2)^2 + (3)^2$
$= 4 + 9$
$= 13$

$\therefore \sqrt{(a^2 + b^2)} = \sqrt{13}$

Exercise 6

Evaluate the following:

In questions **1** to **16**, $a = 4, b = -2, c = -3$

1. $a(b + c)$
2. $a^2(b - c)$
3. $2c(a - c)$
4. $b^2(2a + 3c)$
5. $c^2(b - 2a)$
6. $2a^2(b + c)$
7. $2(a + b + c)$
8. $3c(a - b - c)$
9. $b^2 + 2b + a$
10. $c^2 - 3c + a$
11. $2b^2 - 3b$
12. $\sqrt{(a^2 + c^2)}$
13. $\sqrt{(ab + c^2)}$
14. $\sqrt{(c^2 - b^2)}$
15. $\frac{b^2}{a} + \frac{2c}{b}$
16. $\frac{c^2}{b} + \frac{4b}{a}$

In questions **17** to **32**, $k = -3, m = 1, n = -4$.

17. $k^2(2m - n)$
18. $5m\sqrt{(k^2 + n^2)}$
19. $\sqrt{(kn + 4m)}$
20. $kmn(k^2 + m^2 + n^2)$
21. $k^2m^2(m - n)$
22. $k^2 - 3k + 4$
23. $m^3 + m^2 + n^2 + n$
24. $k^3 + 3k$
25. $m(k^2 - n^2)$
26. $m\sqrt{(k - n)}$
27. $100k^2 + m$
28. $m^2(2k^2 - 3n^2)$
29. $\frac{2k + m}{k - n}$
30. $\frac{kn - k}{2m}$
31. $\frac{3k + 2m}{2n - 3k}$
32. $\frac{k + m + n}{k^2 + m^2 + n^2}$

In questions **33** to **48**, $w = -2, x = 3, y = 0, z = -\frac{1}{2}$

33. $\frac{w}{z} + x$
34. $\frac{w + x}{z}$
35. $y\left(\frac{x + z}{w}\right)$
36. $x^2(z + wy)$
37. $x\sqrt{(x + wz)}$
38. $w^2\sqrt{(z^2 + y^2)}$
39. $2(w^2 + x^2 + y^2)$
40. $2x(w - z)$
41. $\frac{z}{w} + x$
42. $\frac{z + w}{x}$
43. $\frac{x + w}{z^2}$
44. $\frac{y^2 - w^2}{xz}$
45. $z^2 + 4z + 5$
46. $\frac{1}{w} + \frac{1}{z} + \frac{1}{x}$
47. $\frac{4}{z} + \frac{10}{w}$
48. $\frac{yz - xw}{xz - w}$
49. Find $K = \sqrt{\left(\frac{a^2 + b^2 + c^2 - 2c}{a^2 + b^2 + 4c}\right)}$ if $a = 3, b = -2, c = -1$.
50. Find $W = \frac{kmn(k + m + n)}{(k + m)(k + n)}$ if $k = \frac{1}{2}, m = -\frac{1}{3}, n = \frac{1}{4}$.

2.3 BRACKETS AND SIMPLIFYING

A term outside a bracket multiplies each of the terms inside the bracket. This is the *distributive law*.

Example 1

$3(x - 2y) = 3x - 6y$

Example 2

$2x(x - 2y + z) = 2x^2 - 4xy + 2xz$

Example 3

$7y - 4(2x - 3) = 7y - 8x + 12$

In general,
numbers can be added to numbers
x's can be added to x's
y's can be added to y's
x^2's can be added to x^2's

but they must not be mixed.

Example 4

$2x + 3y + 3x^2 + 2y - x = x + 5y + 3x^2$

Example 5

$$\begin{aligned} 7x + 3x(2x - 3) &= 7x + 6x^2 - 9x \\ &= 6x^2 - 2x \end{aligned}$$

Example 6

$\frac{5}{x} - 3x + \frac{2}{x} + x = \frac{7}{x} - 2x$

Exercise 7

Simplify as far as possible:

1. $3x + 4y + 7y$
2. $4a + 7b - 2a + b$
3. $3x - 2y + 4y$
4. $2x + 3x + 5$
5. $7 - 3x + 2 + 4x$
6. $5 - 3y - 6y - 2$
7. $5x + 2y - 4y - x^2$
8. $2x^2 + 3x + 5$
9. $2x - 7y - 2x - 3y$
10. $4a + 3a^2 - 2a$
11. $7a - 7a^2 + 7$
12. $x^2 + 3x^2 - 4x^2 + 5x$

13. $\frac{3}{a} + b + \frac{7}{a} - 2b$
14. $\frac{4}{x} - \frac{7}{y} + \frac{1}{x} + \frac{2}{y}$
15. $\frac{m}{x} + \frac{2m}{x}$
16. $\frac{5}{x} - \frac{7}{x} + \frac{1}{2}$
17. $\frac{3}{a} + b + \frac{2}{a} + 2b$
18. $\frac{n}{4} - \frac{m}{3} - \frac{n}{2} + \frac{m}{3}$
19. $x^3 + 7x^2 - 2x^3$
20. $(2x)^2 - 2x^2$
21. $(3y)^2 + x^2 - (2y)^2$
22. $(2x)^2 - (2y)^2 - (4x)^2$
23. $5x - 7x^2 - (2x)^2$
24. $\frac{3}{x^2} + \frac{5}{x^2}$

Remove the brackets and collect like terms:

25. $3x + 2(x + 1)$
26. $5x + 7(x - 1)$
27. $7 + 3(x - 1)$
28. $9 - 2(3x - 1)$
29. $3x - 4(2x + 5)$
30. $5x - 2x(x - 1)$
31. $7x + 3x(x - 4)$
32. $4(x - 1) - 3x$
33. $5x(x + 2) + 4x$
34. $3x(x - 1) - 7x^2$
35. $3a + 2(a + 4)$
36. $4a - 3(a - 3)$
37. $3ab - 2a(b - 2)$
38. $3y - y(2 - y)$
39. $3x - (x + 2)$
40. $7x - (x - 3)$
41. $5x - 2(2x + 2)$
42. $3(x - y) + 4(x + 2y)$
43. $x(x - 2) + 3x(x - 3)$
44. $3x(x + 4) - x(x - 2)$
45. $y(3y - 1) - (3y - 1)$
46. $7(2x + 2) - (2x + 2)$
47. $7b(a + 2) - a(3b + 3)$
48. $3(x - 2) - (x - 2)$

Two brackets

Example 7

$$\begin{aligned}(x + 5)(x + 3) &= x(x + 3) + 5(x + 3)\\ &= x^2 + 3x + 5x + 15\\ &= x^2 + 8x + 15\end{aligned}$$

Example 8

$$\begin{aligned}(2x - 3)(4y + 3) &= 2x(4y + 3) - 3(4y + 3)\\ &= 8xy + 6x - 12y - 9\end{aligned}$$

Example 9

$$\begin{aligned}3(x + 1)(x - 2) &= 3[x(x - 2) + 1(x - 2)]\\ &= 3[x^2 - 2x + x - 2]\\ &= 3x^2 - 3x - 6\end{aligned}$$

Exercise 8

Remove the brackets and simplify.

1. $(x + 1)(x + 3)$
2. $(x + 3)(x + 2)$
3. $(y + 4)(y + 5)$
4. $(x - 3)(x + 4)$
5. $(x + 5)(x - 2)$
6. $(x - 3)(x - 2)$
7. $(a - 7)(a + 5)$
8. $(z + 9)(z - 2)$
9. $(x - 3)(x + 3)$
10. $(k - 11)(k + 11)$
11. $(2x + 1)(x - 3)$
12. $(3x + 4)(x - 2)$
13. $(2y - 3)(y + 1)$
14. $(7y - 1)(7y + 1)$
15. $(3x - 2)(3x + 2)$
16. $(3a + b)(2a + b)$
17. $(3x + y)(x + 2y)$
18. $(2b + c)(3b - c)$
19. $(5x - y)(3y - x)$
20. $(3b - a)(2a + 5b)$
21. $2(x - 1)(x + 2)$
22. $3(x - 1)(2x + 3)$
23. $4(2y - 1)(3y + 2)$
24. $2(3x + 1)(x - 2)$
25. $4(a + 2b)(a - 2b)$
26. $x(x - 1)(x - 2)$
27. $2x(2x - 1)(2x + 1)$
28. $3y(y - 2)(y + 3)$
29. $x(x + y)(x + z)$
30. $3z(a + 2m)(a - m)$

Be careful with an expression like $(x - 3)^2$. It is not $x^2 - 9$ or even $x^2 + 9$

$$\begin{aligned}(x - 3)^2 &= (x - 3)(x - 3)\\ &= x(x - 3) - 3(x - 3)\\ &= x^2 - 6x + 9\end{aligned}$$

Another common mistake occurs with an expression like $4 - (x - 1)^2$

$$\begin{aligned}4 - (x - 1)^2 &= 4 - 1(x - 1)(x - 1)\\ &= 4 - 1(x^2 - 2x + 1)\\ &= 4 - x^2 + 2x - 1\\ &= 3 + 2x - x^2\end{aligned}$$

Exercise 9

Remove the brackets and simplify:

1. $(x + 4)^2$
2. $(x + 2)^2$
3. $(x - 2)^2$
4. $(2x + 1)^2$
5. $(y - 5)^2$
6. $(3y + 1)^2$
7. $(x + y)^2$
8. $(2x + y)^2$
9. $(a - b)^2$
10. $(2a - 3b)^2$
11. $3(x + 2)^2$
12. $(3 - x)^2$
13. $(3x + 2)^2$
14. $(a - 2b)^2$
15. $(x + 1)^2 + (x + 2)^2$
16. $(x - 2)^2 + (x + 3)^2$
17. $(x + 2)^2 + (2x + 1)^2$
18. $(y - 3)^2 + (y - 4)^2$
19. $(x + 2)^2 - (x - 3)^2$
20. $(x - 3)^2 - (x + 1)^2$
21. $(y - 3)^2 - (y + 2)^2$
22. $(2x + 1)^2 - (x + 3)^2$
23. $3(x + 2)^2 - (x + 4)^2$
24. $2(x - 3)^2 - 3(x + 1)^2$

The same method is used even if there are more than two terms inside the brackets.

Example 10

$$\begin{aligned}&(x + y + 2)(2x + y - 3)\\ &= x(2x + y - 3) + y(2x + y - 3) + 2(2x + y - 3)\\ &= 2x^2 + xy - 3x + 2xy + y^2 - 3y + 4x + 2y - 6\\ &= 2x^2 + y^2 + 3xy + x - y - 6\end{aligned}$$

Expressions involving fractions are dealt with in a similar way to ordinary numerical fractions.

Example 11

$$\left(x+\frac{1}{x}\right)^2 = \left(x+\frac{1}{x}\right)\left(x+\frac{1}{x}\right)$$
$$= x\left(x+\frac{1}{x}\right)+\frac{1}{x}\left(x+\frac{1}{x}\right)$$
$$= x^2 + x\cdot\frac{1}{x}+\frac{1}{x}\cdot x+\frac{1}{x}\cdot\frac{1}{x}$$
$$= x^2+1+1+\frac{1}{x^2}$$
$$= x^2+2+\frac{1}{x^2}$$

Exercise 10

Remove the brackets and simplify.

1. $(2x+y+3)(x+y+4)$
2. $(x+2y-3)(2x-3y-2)$
3. $(x+y+2)^2$
4. $(a+b+2c)(a-b-c)$
5. $\left(x+\frac{1}{x}\right)\left(2x+\frac{3}{x}\right)$
6. $\left(x-\frac{1}{x}\right)\left(x+\frac{2}{x}\right)$
7. $\left(x+\frac{2}{x}\right)^2$
8. $\left(y+\frac{1}{y}\right)^2$
9. $\left(2x+\frac{3}{x}\right)^2$
10. $\left(x-\frac{1}{x}\right)^2$
11. $\left(2x+\frac{5}{x}\right)\left(x-\frac{1}{x}\right)$
12. $(x^2+1)\left(x+\frac{1}{x^2}\right)$
13. $\left(x^2-\frac{2}{x}\right)^2$
14. $\left(2x+\frac{1}{x^2}\right)^2$
15. $\left(x+1+\frac{1}{x}\right)\left(2x+1+\frac{1}{x}\right)$
16. $\left(x-2+\frac{3}{x}\right)\left(2x+1+\frac{1}{x}\right)$
17. $(x+1)^2-3(x-1)$
18. $(2x+1)^2-(x-3)$
19. $(x-3)^2-(x+2)^2$
20. $\left(x+\frac{3}{x}\right)^2+\left(x-\frac{2}{x}\right)^2$
21. $\frac{1}{x}(x+1)^2+\left(x+\frac{1}{x}\right)^2$
22. $\left(2x+\frac{4}{x}\right)^2+\left(x+\frac{1}{x}\right)^2$
23. $\left(2x-3+\frac{4}{x}\right)^2$
24. $\left(x+\frac{1}{y}\right)^2-\left(x-\frac{1}{y}\right)^2$

2.4 LINEAR EQUATIONS

(a) If the x term is negative, take it to the other side, where it becomes positive.

Example 1

$$4-3x = 2$$
$$4 = 2+3x$$
$$2 = 3x$$
$$\frac{2}{3} = x$$

(b) If there are x terms on both sides, collect them on one side.

Example 2

$$2x-7 = 5-3x$$
$$2x+3x = 5+7$$
$$5x = 12$$
$$x = \frac{12}{5} = 2\tfrac{2}{5}$$

(c) If there is a fraction in the x term, multiply out to simplify the equation.

Example 3

$$\frac{2x}{3} = 10$$
$$2x = 30$$
$$x = \frac{30}{2} = 15$$

Exercise 11

Solve the following equations:

1. $2x-5=11$
2. $3x-7=20$
3. $2x+6=20$
4. $5x+10=60$
5. $8=7+3x$
6. $12=2x-8$
7. $-7=2x-10$
8. $3x-7=-10$
9. $12=15+2x$
10. $5+6x=7$
11. $\frac{x}{5}=7$
12. $\frac{x}{10}=13$

13. $7 = \frac{x}{2}$

14. $\frac{x}{2} = \frac{1}{3}$

15. $\frac{3x}{2} = 5$

16. $\frac{4x}{5} = -2$

17. $7 = \frac{7x}{3}$

18. $\frac{3}{4} = \frac{2x}{3}$

19. $\frac{5x}{6} = \frac{1}{4}$

20. $-\frac{3}{4} = \frac{3x}{5}$

21. $\frac{x}{2} + 7 = 12$

22. $\frac{x}{3} - 7 = 2$

23. $\frac{x}{5} - 6 = -2$

24. $4 = \frac{x}{2} - 5$

25. $10 = 3 + \frac{x}{4}$

26. $\frac{a}{5} - 1 = -4$

27. $100x - 1 = 98$

28. $7 = 7 + 7x$

29. $\frac{x}{100} + 10 = 20$

30. $1000x - 5 = -6$

31. $-4 = -7 + 3x$

32. $2x + 4 = x - 3$

33. $x - 3 = 3x + 7$

34. $5x - 4 = 3 - x$

35. $4 - 3x = 1$

36. $5 - 4x = -3$

37. $7 = 2 - x$

38. $3 - 2x = x + 12$

39. $6 + 2a = 3$

40. $a - 3 = 3a - 7$

41. $2y - 1 = 4 - 3y$

42. $7 - 2x = 2x - 7$

43. $7 - 3x = 5 - 2x$

44. $8 - 2y = 5 - 5y$

45. $x - 16 = 16 - 2x$

46. $x + 2 = 3{\cdot}1$

47. $-x - 4 = -3$

48. $-3 - x = -5$

49. $-\frac{x}{2} + 1 = -\frac{1}{4}$

50. $-\frac{3}{5} + \frac{x}{10} = -\frac{1}{5} - \frac{x}{5}$

Example 4

$$\begin{aligned} x - 2(x - 1) &= 1 - 4(x + 1) \\ x - 2x + 2 &= 1 - 4x - 4 \\ x - 2x + 4x &= 1 - 4 - 2 \\ 3x &= -5 \\ x &= -\frac{5}{3} \end{aligned}$$

Exercise 12

Solve the following equations:

1. $x + 3(x + 1) = 2x$
2. $1 + 3(x - 1) = 4$
3. $2x - 2(x + 1) = 5x$
4. $2(3x - 1) = 3(x - 1)$
5. $4(x - 1) = 2(3 - x)$
6. $4(x - 1) - 2 = 3x$
7. $4(1 - 2x) = 3(2 - x)$
8. $3 - 2(2x + 1) = x + 17$
9. $4x = x - (x - 2)$
10. $7x = 3x - (x + 20)$

11. $5x - 3(x - 1) = 39$
12. $3x + 2(x - 5) = 15$
13. $7 - (x + 1) = 9 - (2x - 1)$
14. $10x - (2x + 3) = 21$
15. $3(2x + 1) + 2(x - 1) = 23$
16. $5(1 - 2x) - 3(4 + 4x) = 0$
17. $7x - (2 - x) = 0$
18. $3(x + 1) = 4 - (x - 3)$
19. $3y + 7 + 3(y - 1) = 2(2y + 6)$
20. $4(y - 1) + 3(y + 2) = 5(y - 4)$

21. $4x - 2(x + 1) = 5(x + 3) + 5$
22. $7 - 2(x - 1) = 3(2x - 1) + 2$
23. $10(2x + 3) - 8(3x - 5) + 5(2x - 8) = 0$
24. $2(x + 4) + 3(x - 10) = 8$
25. $7(2x - 4) + 3(5 - 3x) = 2$
26. $10(x + 4) - 9(x - 3) - 1 = 8(x + 3)$
27. $5(2x - 1) - 2(x - 2) = 7 + 4x$
28. $6(3x - 4) - 10(x - 3) = 10(2x - 3)$
29. $3(x - 3) - 7(2x - 8) - (x - 1) = 0$
30. $5 + 2(x + 5) = 10 - (4 - 5x)$

31. $6x + 30(x - 12) = 2(x - 1\frac{1}{2})$
32. $3(2x - \frac{2}{3}) - 7(x - 1) = 0$
33. $5(x - 1) + 17(x - 2) = 2x + 1$
34. $6(2x - 1) + 9(x + 1) = 8(x - 1\frac{1}{4})$
35. $7(x + 4) - 5(x + 3) + (4 - x) = 0$
36. $0 = 9(3x + 7) - 5(x + 2) - (2x - 5)$
37. $10(2{\cdot}3 - x) - 0{\cdot}1(5x - 30) = 0$
38. $8(2\frac{1}{2}x - \frac{3}{4}) - \frac{1}{4}(1 - x) = \frac{1}{2}$
39. $(6 - x) - (x - 5) - (4 - x) = -\frac{x}{2}$
40. $10\left(1 - \frac{x}{10}\right) - (10 - x) - \frac{1}{100}(10 - x) = 0{\cdot}05$

Example 5

$$\begin{aligned}(x+3)^2 &= (x+2)^2+3^2\\(x+3)(x+3) &= (x+2)(x+2)+9\\x^2+6x+9 &= x^2+4x+4+9\\6x+9 &= 4x+13\\2x &= 4\\x &= 2\end{aligned}$$

Exercise 13

Solve the following equations:

1. $x^2+4=(x+1)(x+3)$
2. $x^2+3x=(x+3)(x+1)$
3. $(x+3)(x-1)=x^2+5$
4. $(x+1)(x+4)=(x-7)(x+6)$
5. $(x-2)(x+3)=(x-7)(x+7)$
6. $(x-5)(x+4)=(x+7)(x-6)$
7. $2x^2+3x=(2x-1)(x+1)$
8. $(2x-1)(x-3)=(2x-3)(x-1)$
9. $x^2+(x+1)^2=(2x-1)(x+4)$
10. $x(2x+6)=2(x^2-5)$
11. $(x+1)(x-3)+(x+1)^2=2x(x-4)$
12. $(2x+1)(x-4)+(x-2)^2=3x(x+2)$
13. $(x+2)^2-(x-3)^2=3x-11$
14. $x(x-1)=2(x-1)(x+5)-(x-4)^2$
15. $(2x+1)^2-4(x-3)^2=5x+10$
16. $2(x+1)^2-(x-2)^2=x(x-3)$

In questions **17** to **22**, form an equation in x by means of Pythagoras' Theorem, and hence find the length of each side of the triangle. (All the lengths are in cm.)

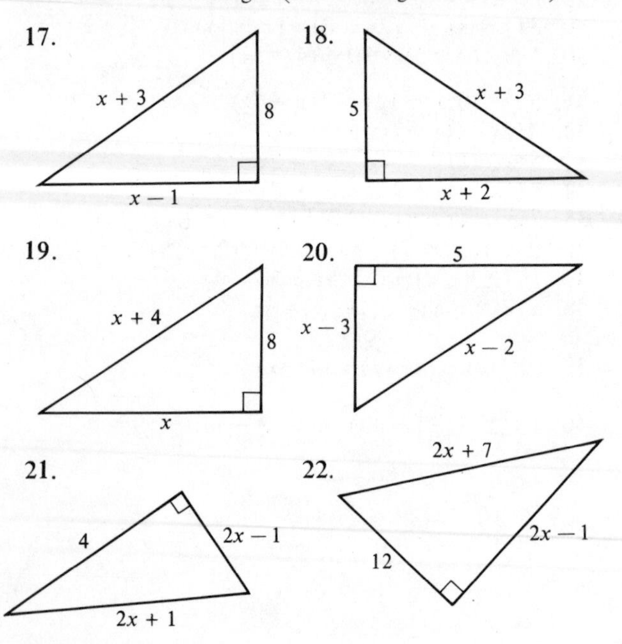

23. The area of the rectangle shown exceeds the area of the square by 2 cm². Find x.

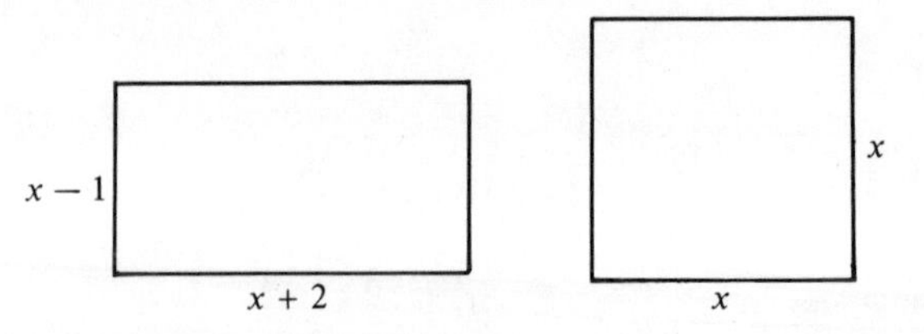

24. The area of the square exceeds the area of the rectangle by 13 m². Find y.

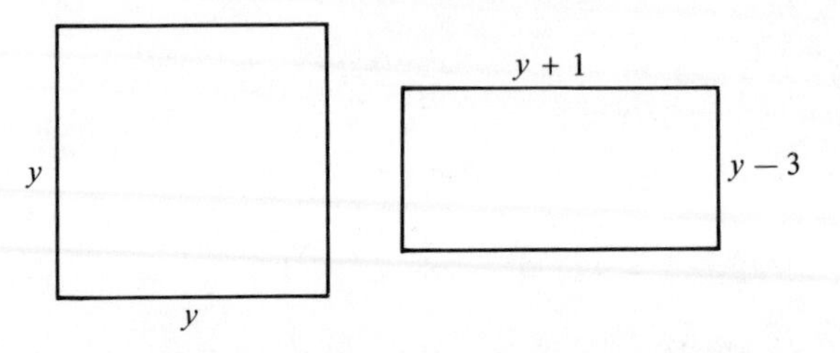

25. The area of the square is half the area of the rectangle. Find x.

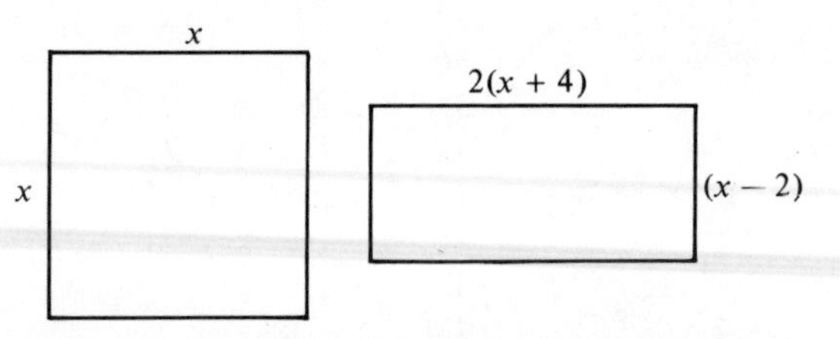

When solving equations involving fractions, multiply both sides of the equation by a suitable number to eliminate the fractions.

Example 6

$$\frac{5}{x} = 2$$

$5 = 2x$ (multiply both sides by x)

$$\frac{5}{2} = x$$

Example 7

$$\frac{x+3}{4} = \frac{2x-1}{3} \quad \ldots (A)$$

$$12\,\frac{(x+3)}{4} = 12\,\frac{(2x-1)}{3}$$

(multiply both sides by 12)

$$\therefore \quad 3(x+3) = 4(2x-1) \quad \ldots (B)$$
$$3x + 9 = 8x - 4$$
$$13 = 5x$$
$$\frac{13}{5} = x$$
$$x = 2\tfrac{3}{5}$$

Note: It is possible to go straight from line (A) to line (B) by 'cross-multiplying'.

Example 8

$$\frac{5}{(x-1)} + 2 = 12$$
$$\frac{5}{(x-1)} = 10$$
$$5 = 10(x-1)$$
$$5 = 10x - 10$$
$$15 = 10x$$
$$\frac{15}{10} = x$$
$$x = 1\tfrac{1}{2}$$

Exercise 14

Solve the following equations:

1. $\frac{7}{x} = 21$
2. $30 = \frac{6}{x}$
3. $\frac{5}{x} = 3$
4. $\frac{9}{x} = -3$
5. $11 = \frac{5}{x}$
6. $-2 = \frac{4}{x}$
7. $\frac{x}{4} = \frac{3}{2}$
8. $\frac{x}{3} = 1\frac{1}{4}$
9. $\frac{x+1}{3} = \frac{x-1}{4}$
10. $\frac{x+3}{2} = \frac{x-4}{5}$
11. $\frac{2x-1}{3} = \frac{x}{2}$
12. $\frac{3x+1}{5} = \frac{2x}{3}$
13. $\frac{8-x}{2} = \frac{2x+2}{5}$
14. $\frac{x+2}{7} = \frac{3x+6}{5}$
15. $\frac{1-x}{2} = \frac{3-x}{3}$
16. $\frac{2}{x-1} = 1$
17. $\frac{x}{3} + \frac{x}{4} = 1$
18. $\frac{x}{3} + \frac{x}{2} = 4$
19. $\frac{x}{2} - \frac{x}{5} = 3$
20. $\frac{x}{3} = 2 + \frac{x}{4}$
21. $\frac{5}{x-1} = \frac{10}{x}$
22. $\frac{12}{2x-3} = 4$
23. $2 = \frac{18}{x+4}$
24. $\frac{5}{x+5} = \frac{15}{x+7}$
25. $\frac{9}{x} = \frac{5}{x-3}$
26. $\frac{4}{x-1} = \frac{10}{3x-1}$
27. $\frac{-7}{x-1} = \frac{14}{5x+2}$
28. $\frac{4}{x+1} = \frac{7}{3x-2}$
29. $\frac{x+1}{2} + \frac{x-1}{3} = \frac{1}{6}$
30. $\frac{1}{3}(x+2) = \frac{1}{5}(3x+2)$
31. $\frac{1}{2}(x-1) - \frac{1}{6}(x+1) = 0$
32. $\frac{1}{4}(x+5) - \frac{2x}{3} = 0$
33. $\frac{4}{x} + 2 = 3$
34. $\frac{6}{x} - 3 = 7$
35. $\frac{9}{x} - 7 = 1$
36. $-2 = 1 + \frac{3}{x}$
37. $4 - \frac{4}{x} = 0$
38. $5 - \frac{6}{x} = -1$
39. $7 - \frac{3}{2x} = 1$
40. $4 + \frac{5}{3x} = -1$
41. $\frac{9}{2x} - 5 = 0$
42. $\frac{x-1}{5} - \frac{x-1}{3} = 0$
43. $\frac{x-1}{4} - \frac{2x-3}{5} = \frac{1}{20}$
44. $\frac{4}{1-x} = \frac{3}{1+x}$
45. $\frac{x+1}{4} - \frac{x}{3} = \frac{1}{12}$
46. $\frac{2x+1}{8} - \frac{x-1}{3} = \frac{5}{24}$
47. $\frac{x+1}{x-1} = \frac{x+5}{x+1}$
48. $\frac{x-1}{x+2} = \frac{x-1}{x-4}$
49. $\frac{x+4}{x} = \frac{x+1}{x-1}$
50. $\frac{2x+3}{x+4} = \frac{2x-1}{x}$
51. $\frac{x+3}{3x+7} = \frac{x-4}{3x}$
52. $\frac{1}{x+1} - \frac{4}{x+2} = 0$
53. $\frac{3}{x+2} - \frac{2}{x-1} = 0$
54. $\frac{2}{2x-1} - \frac{3}{x+1} = 0$

2.5 PROBLEMS SOLVED BY LINEAR EQUATIONS

(a) Let the unknown quantity be x (or any other letter) and state the units (where appropriate).
(b) Express the given statement in the form of an equation.
(c) Solve the equation for x and give the answer in *words*. (Do not finish by writing '$x = 3$'.)
(d) Check your solution using the problem (not your equation).

Example 1

The sum of three consecutive whole numbers is 78. Find the numbers.

(a) Let the smallest number be x; then the other numbers are $(x + 1)$ and $(x + 2)$.

(b) Form an equation:
$x + (x + 1) + (x + 2) = 78$

(c) Solve. $3x = 75$
$x = 25$

In words:
The three numbers are 25, 26 and 27.

(d) Check. $25 + 26 + 27 = 78$ ✓

Example 2

The length of a rectangle is three times the width. If the perimeter is 36 cm, find the width.

(a) Let the width of the rectangle be x cm. Then the length of the rectangle is $3x$ cm.

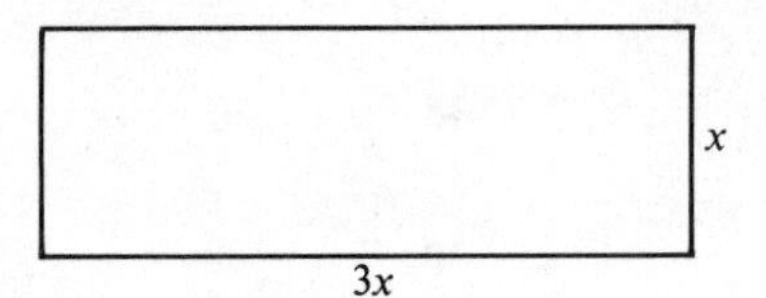

(b) Form an equation.

$x + 3x + x + 3x = 36$

(c) Solve $8x = 36$

$$x = \frac{36}{8}$$

$$x = 4{\cdot}5$$

In words:
The width of the rectangle is 4·5 cm.

(d) Check. If width $= 4{\cdot}5$ cm
length $= 13{\cdot}5$ cm
perimeter $= 2(4{\cdot}5 + 13{\cdot}5)$
$= 36$ cm ✓

Exercise 15

Solve each problem by forming an equation. The first questions are easy but should still be solved using an equation, in order to practise the method.

1. The sum of three consecutive numbers is 276. Find the numbers.
2. The sum of four consecutive numbers is 90. Find the numbers.
3. The sum of three consecutive odd numbers is 177. Find the numbers.
4. Find three consecutive even numbers which add up to 1524.
5. When a number is doubled and then added to 13, the result is 38. Find the number.
6. When a number is doubled and then added to 24, the result is 49. Find the number.
7. When 7 is subtracted from three times a certain number, the result is 28. What is the number?
8. The sum of two numbers is 50. The second number is five times the first. Find the numbers.
9. Two numbers are in the ratio 1:11 and their sum is 15. Find the numbers.
10. The length of a rectangle is twice the width. If the perimeter is 20 cm, find the width.
11. The width of a rectangle is one third of the length. If the perimeter is 96 cm, find the width.
12. If AB is a straight line, find x.

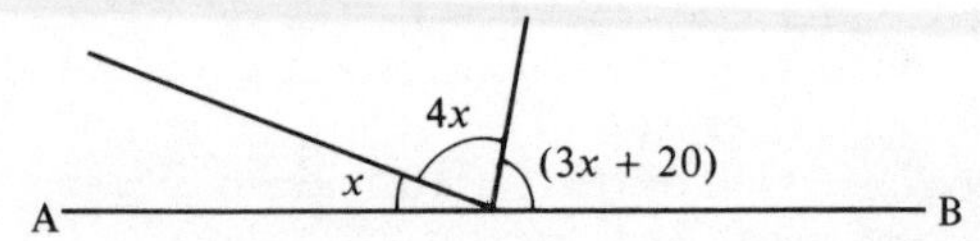

13. If the perimeter of the triangle is 22 cm, find the length of the shortest side.

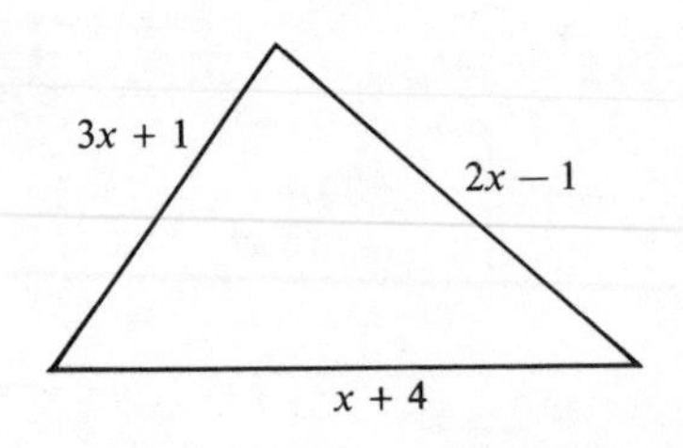

14. If the perimeter of the rectangle is 34 cm, find x.

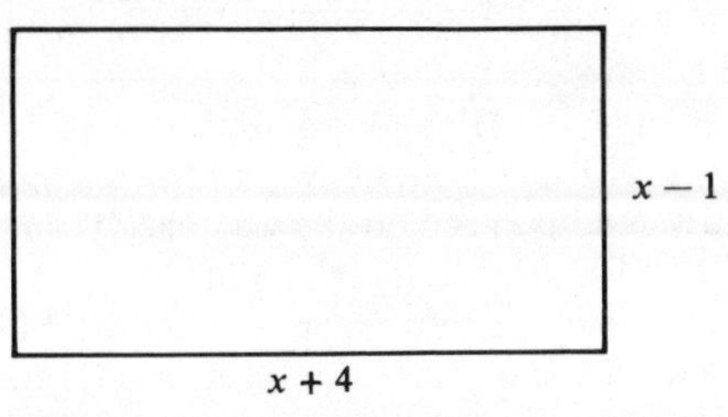

15. The difference between two numbers is 9. Find the numbers, if their sum is 46.
16. The three angles in a triangle are in the ratio 1:3:5. Find them.
17. The three angles in a triangle are in the ratio 3:4:5. Find them.
18. The product of two consecutive odd numbers is 10 more than the square of the smaller number. Find the smaller number.
19. The product of two consecutive even numbers is 12 more than the square of the smaller number. Find the numbers.
20. The sum of three numbers is 66. The second number is twice the first and six less than the third. Find the numbers.
21. The sum of three numbers is 28. The second number is three times the first and the third is 7 less than the second. What are the numbers?
22. David weighs 5 kg less than John, who in turn is 8 kg lighter than Paul. If their total weight is 197 kg, how heavy is each person?
23. Brian is 2 years older than Bob who is 7 years older than Mark. If their combined age is 61 years, find the age of each person.
24. Richard has four times as many marbles as John. If Richard gave 18 to John they would have the same number. How many marbles has each?
25. Stella has five times as many books as Tina. If Stella gave 16 books to Tina, they would each have the same number. How many books did each girl have?
26. The result of trebling a number is the same as adding 12 to it. What is the number?
27. Find the area of the rectangle if the perimeter is 52 cm.

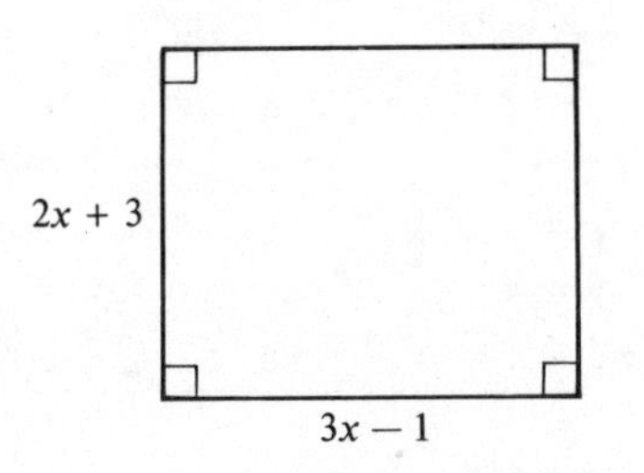

28. The result of trebling a number and subtracting 5 is the same as doubling the number and adding 9. What is the number?
29. Two girls have 76p between them. If the first gave the second 7p they would each have the same amount of money. How much did each girl have?
30. A tennis racket costs £12 more than a hockey stick. If the price of the two is £31, find the cost of the tennis racket.

Example 3

A man goes out at 16.42 h and arrives at a pillar box, 6 km away, at 17.30 h. He walked part of the way at 5 km/h and then, realising the time, he ran the rest of the way at 10 km/h. How far did he have to run?

(a) Let the distance he ran be x km. Then the distance he walked $= (6-x)$ km.

(b) Time taken to walk $(6-x)$ km at 5 km/h

$$= \frac{(6-x)}{5} \text{ hours.}$$

Time taken to run x km at 10 km/h

$$= \frac{x}{10} \text{ hours.}$$

$$\text{Total time taken} = 48 \text{ minutes} = \frac{4}{5} \text{ hour}$$

$$\therefore \quad \frac{(6-x)}{5} + \frac{x}{10} = \frac{4}{5}$$

(c) Multiply by 10:

$$\begin{aligned} 2(6-x) + x &= 8 \\ 12 - 2x + x &= 8 \\ 4 &= x \end{aligned}$$

He ran a distance of 4 km.

(d) Check:

$$\text{Time to run 4 km} = \frac{4}{10} = \frac{2}{5} \text{ hour}$$

$$\text{Time to walk 2 km} = \frac{2}{5} \text{ hour.}$$

$$\text{Total time taken} = \left(\frac{2}{5} + \frac{2}{5}\right) \text{h} = \frac{4}{5} \text{h} \checkmark$$

Exercise 16

1. Every year a man is paid £500 more than the previous year. If he receives £17 800 over four years, what was he paid in the first year?
2. A man buys x cans of beer at 30p each and $(x + 4)$ cans of lager at 35p each. The total cost was £3·35. Find x.
3. The sides of a rectangle measure 6 cm and x cm. If the diagonal is of length $(x + 2)$ cm, find x.
4. The length of a straight line ABC is 5 m. If AB : BC = 2 : 5, find the length of AB.
5. The opposite angles of a cyclic quadrilateral are $(3x + 10)^\circ$ and $(2x + 20)^\circ$. Find the angles.
6. The interior angles of a hexagon are in the ratio 1 : 2 : 3 : 4 : 5 : 9. Find the angles. This is an example of a concave hexagon. Try to sketch the hexagon.
7. A man is 32 years older than his son. Ten years ago he was three times as old as his son was then. Find the present age of each.
8. A man runs to a telephone and back in 15 minutes. His speed on the way to the telephone is 5 m/s and his speed on the way back is 4 m/s. Find the distance to the telephone.
9. A car completes a journey in 10 minutes. For the first half of the distance the speed was 60 km/h and for the second half the speed was 40 km/h. How far is the journey?
10. A lemming runs from a point A to a cliff at 4 m/s, jumps over the edge at B and falls to C at an average speed of 25 m/s. If the total distance from A to C is 500 m and the time taken for the journey is 41 seconds, find the height BC of the cliff.

2.6 SIMULTANEOUS EQUATIONS

To find the value of two unknowns in a problem, *two* different equations must be given that relate the unknowns to each other. These two equations are called *simultaneous* equations.

(a) Substitution method
This method is used when one equation contains a unit quantity of one of the unknowns, as in equation [2] of Example 1 below.

Example 1

$$3x - 2y = 0 \qquad \ldots [1]$$
$$2x + y = 7 \qquad \ldots [2]$$

(a) Label the equations so that the working is made clear.
(b) In *this* case, write y in terms of x from equation [2].
(c) Substitute this expression for y in equation [1] and solve to find x.
(d) Find y from equation [2] using this value of x.

$$2x + y = 7 \qquad \ldots [2]$$
$$y = 7 - 2x$$

Substituting in [1]

$$3x - 2(7 - 2x) = 0$$
$$3x - 14 + 4x = 0$$
$$7x = 14$$
$$x = 2$$

Substituting in [2]

$$2 \times 2 + y = 7$$
$$y = 3$$

The solutions are $x = 2, y = 3$.

These values of x and y are the only pair which simultaneously satisfy *both* equations.

Exercise 17

Use the substitution method to solve the following:

1. $2x + y = 5$, $x + 3y = 5$
2. $x + 2y = 8$, $2x + 3y = 14$
3. $3x + y = 10$, $x - y = 2$
4. $2x + y = -3$, $x - y = -3$
5. $4x + y = 14$, $x + 5y = 13$
6. $x + 2y = 1$, $2x + 3y = 4$
7. $2x + y = 5$, $3x - 2y = 4$
8. $2x + y = 13$, $5x - 4y = 13$
9. $7x + 2y = 19$, $x - y = 4$
10. $b - a = -5$, $a + b = -1$

11. $a + 4b = 6$
$8b - a = -3$

12. $a + b = 4$
$2a + b = 5$

13. $3m = 2n - 6\frac{1}{2}$
$4m + n = 6$

14. $2w + 3x - 13 = 0$
$x + 5w - 13 = 0$

15. $x + 2(y - 6) = 0$
$3x + 4y = 30$

16. $2x = 4 + z$
$6x - 5z = 18$

17. $3m - n = 5$
$2m + 5n = 7$

18. $5c - d - 11 = 0$
$4d + 3c = -5$

It is useful, at this point to revise the operations of addition and subtraction with negative numbers.

Example 2

Simplify:
(a) $-7 + (-4)$ (b) $-3x + (-4x)$
(c) $4y - (-3y)$ (d) $3a + (-3a)$

(a) $-7 + -4 = -7 - 4 = -11$
(b) $-3x + (-4x) = -3x - 4x = -7x$
(c) $4y - (-3y) = 4y + 3y = 7y$
(d) $3a + (-3a) = 3a - 3a = 0$

Exercise 18

Evaluate

1. $7 + (-6)$ 2. $8 + (-11)$ 3. $5 - (+7)$
4. $6 - (-9)$ 5. $-8 + (-4)$ 6. $-7 - (-4)$
7. $10 + (-12)$ 8. $-7 - (+4)$ 9. $-10 - (+11)$
10. $-3 - (-4)$ 11. $4 - (+4)$ 12. $8 - (-7)$
13. $-5 - (+5)$ 14. $-7 - (-10)$ 15. $16 - (+10)$
16. $-7 - (+4)$ 17. $-6 - (-8)$ 18. $10 - (+5)$
19. $-12 + (-7)$ 20. $7 + (-11)$

Simplify

21. $3x + (-2x)$ 22. $4x + (-7x)$
23. $6x - (+2x)$ 24. $10y - (+6y)$
25. $6y - (-3y)$ 26. $7x + (-4x)$
27. $-5x + (-3x)$ 28. $-3x - (-7x)$
29. $5x - (+3x)$ 30. $-7y - (-10y)$
31. $4a - (-6a)$ 32. $7a + (-5a)$
33. $-3x + (-5x)$ 34. $-5y - (-10y)$
35. $4x - (-5x)$ 36. $3k + (-2k)$
37. $2x - (+5x)$ 38. $-5y + (-6y)$
39. $2x - (-7x)$ 40. $-3k - (-7k)$

(b) Elimination method

Use this method when the first method is unsuitable (some prefer to use it for every question).

Example 2

$x + 2y = 8$ ... [1]
$2x + 3y = 14$ ... [2]

(a) Label the equations so that the working is made clear.
(b) Choose an unknown in one of the equations and multiply the equations by a factor or factors so that this unknown has the same coefficient in both equations.
(c) Eliminate this unknown from the two equations by subtracting them, then solve for the remaining unknown.
(d) Substitute in the first equation and solve for the eliminated unknown.

$x + 2y = 8$ ... [1]
[1] × 2 $2x + 4y = 16$ ... [3]
$2x + 3y = 14$ ... [2]

Subtract [2] from [3]

$y = 2$

Substituting in [1]

$x + 2 \times 2 = 8$
$x = 8 - 4$
$x = 4$

The solutions are $x = 4, y = 2$.

Example 3

$2x + 3y = 5$ ... [1]
$5x - 2y = -16$ ... [2]

[1] × 5 $10x + 15y = 25$ ... [3]
[2] × 2 $10x - 4y = -32$ ... [4]

[3] − [4] $15y - (-4y) = 25 - (-32)$
$19y = 57$
$y = 3$

Substitute in [1]

$2x + 3 \times 3 = 5$
$2x = 5 - 9 = -4$
$x = -2$

The solutions are $x = -2, y = 3$.

Exercise 19

Use the elimination method to solve the following:

1. $2x + 5y = 24$, $4x + 3y = 20$
2. $5x + 2y = 13$, $2x + 6y = 26$
3. $3x + y = 11$, $9x + 2y = 28$
4. $x + 2y = 17$, $8x + 3y = 45$
5. $3x + 2y = 19$, $x + 8y = 21$
6. $2a + 3b = 9$, $4a + b = 13$
7. $2x + 3y = 11$, $3x + 4y = 15$
8. $3x + 8y = 27$, $4x + 3y = 13$
9. $2x + 7y = 17$, $5x + 3y = -1$
10. $5x + 3y = 23$, $2x + 4y = 12$
11. $7x + 5y = 32$, $3x + 4y = 23$
12. $3x + 2y = 4$, $4x + 5y = 10$
13. $3x + 2y = 11$, $2x - y = -3$
14. $3x + 2y = 7$, $2x - 3y = -4$
15. $x - 2y = -4$, $3x + y = 9$
16. $5x - 7y = 27$, $3x - 4y = 16$
17. $3x - 2y = 7$, $4x + y = 13$
18. $x - y = -1$, $2x - y = 0$
19. $y - x = -1$, $3x - y = 5$
20. $x - 3y = -5$, $2y + 3x + 4 = 0$
21. $x + 3y - 7 = 0$, $2y - x - 3 = 0$
22. $3a - b = 9$, $2a + 2b = 14$
23. $3x - y = 9$, $4x - y = -14$
24. $x + 2y = 4$, $3x + y = 9\frac{1}{2}$
25. $2x - y = 5$, $\frac{x}{4} + \frac{y}{3} = 2$
26. $3x - y = 17$, $\frac{x}{5} + \frac{y}{2} = 0$
27. $3x - 2y = 5$, $\frac{2x}{3} + \frac{y}{2} = -\frac{7}{9}$
28. $2x = 11 - y$, $\frac{x}{5} - \frac{y}{4} = 1$
29. $4x - 0{\cdot}5y = 12{\cdot}5$, $3x + 0{\cdot}8y = 8{\cdot}2$
30. $0{\cdot}4x + 3y = 2{\cdot}6$, $x - 2y = 4{\cdot}6$

2.7 PROBLEMS SOLVED BY SIMULTANEOUS EQUATIONS

Adopt the procedure described on page 32, but now introduce two symbols for two unknown quantities.

Example 1

A motorist buys 24 litres of petrol and 5 litres of oil for £10·70, while another motorist buys 18 litres of petrol and 10 litres of oil for £12·40. Find the cost of 1 litre of petrol and 1 litre of oil at this garage.

Let cost of 1 litre of petrol be x pence.
Let cost of 1 litre of oil be y pence.

We have,
$$24x + 5y = 1070 \quad \ldots [1]$$
$$18x + 10y = 1240 \quad \ldots [2]$$

(a) Multiply [1] by 2,
$$48x + 10y = 2140 \quad \ldots [3]$$

(b) Subtract [2] from [3],
$$30x = 900$$
$$x = 30$$

(c) Substitute $x = 30$ into equation [2]
$$18(30) + 10y = 1240$$
$$10y = 1240 - 540$$
$$10y = 700$$
$$y = 70$$

1 litre of petrol costs 30 pence.
1 litre of oil costs 70 pence.

Exercise 20

Solve each problem by forming a pair of simultaneous equations.

1. Find two numbers with a sum of 15 and a difference of 4.
2. Twice one number added to three times another gives 21. Find the numbers, if the difference between them is 3.
3. The average of two numbers is 7, and three times the difference between them is 18. Find the numbers.
4. The line, with equation $y + ax = c$, passes through the points $(1, 5)$ and $(3, 1)$. Find a and c. Hint: For the point $(1, 5)$ put $x = 1$ and $y = 5$ into $y + ax = c$, etc.

5. The line $y = mx + c$ passes through $(2,5)$ and $(4,13)$. Find m and c.
6. The curve $y = ax^2 + bx$ passes through $(2,0)$ and $(4,8)$. Find a and b.
7. A fishing enthusiast buys fifty maggots and twenty worms for £1·10 and his mother buys thirty maggots and forty worms for £1·50. Find the cost of one maggot and one worm.
8. A television addict can buy either two televisions and three video-recorders for £1750 or four televisions and one video-recorder for £1250. Find the cost of one of each.
9. Half the difference between two numbers is 2. The sum of the greater number and twice the smaller number is 13. Find the numbers.
10. A pigeon can lay either white or brown eggs. Three white eggs and two brown eggs weigh 13 ounces, while five white eggs and four brown eggs weigh 24 ounces. Find the weight of a brown egg and of a white egg.
11. A tortoise makes a journey in two parts; it can either walk at 4 m/s or crawl at 3 m/s. If the tortoise walks the first part and crawls the second, it takes 110 seconds. If it crawls the first part and walks the second, it takes 100 seconds. Find the lengths of the two parts of the journey.
12. A cyclist completes a journey of 500 m in 22 seconds, part of the way at 10 m/s and the remainder at 50 m/s. How far does she travel at each speed?
13. A bag contains forty coins, all of them either 2p or 5p coins. If the value of the money in the bag is £1·55, find the number of each kind.
14. A slot machine takes only 10p and 50p coins and contains a total of twenty-one coins altogether. If the value of the coins is £4·90, find the number of coins of each value.
15. Thirty tickets were sold for a concert, some at 60p and the rest at £1. If the total raised was £22, how many had the cheaper tickets?
16. The wage bill for five men and six women workers is £670, while the bill for eight men and three women is £610. Find the wage for a man and a woman.
17. A kipper can swim at 14 m/s with the current and at 6 m/s against it. Find the speed of the current and the speed of the kipper in still water.
18. If the numerator and denominator of a fraction are both decreased by one the fraction becomes $\frac{2}{3}$. If the numerator and denominator are both increased by one the fraction becomes $\frac{3}{4}$. Find the original fraction.
19. The denominator of a fraction is 2 more than the numerator. If both denominator and numerator are increased by 1 the fraction becomes $\frac{2}{3}$. Find the original fraction.
20. In three years time a pet mouse will be as old as his owner was four years ago. Their present ages total 13 years. Find the age of each now.
21. Find two numbers where three times the smaller number exceeds the larger by 5 and the sum of the numbers is 11.
22. A straight line passes through the points $(2,4)$ and $(-1,-5)$. Find its equation.
23. A spider can walk at a certain speed and run at another speed. If she walks for 10 seconds and runs for 9 seconds she travels 85 m. If she walks for 30 seconds and runs for 2 seconds she travels 130 m. Find her speeds of walking and running.
24. A wallet containing £40 has three times as many £1 notes as £5 notes. Find the number of each kind.
25. At the present time a man is four times as old as his son. Six years ago he was 10 times as old. Find their present ages.
26. A submarine can travel at 25 knots with the wind and at 16 knots against it. Find the speed of the wind and the speed of the submarine in still air.
27. The curve $y = ax^2 + bx + c$ passes through the points $(1,8)$, $(0,5)$ and $(3,20)$. Find the values of a, b and c and hence the equation of the curve.
28. The curve $y = ax^2 + bx + c$ passes through the points $(1,4)$, $(-2,19)$ and $(0,5)$. Find the equation of the curve.
29. The curve $y = ax^2 + bx + c$ passes through $(1,8)$, $(-1,2)$ and $(2,14)$. Find the equation of the curve.
30. The curve $y = ax^2 + bx + c$ passes through $(2,5)$, $(3,12)$ and $(-1,-4)$. Find the equation of the curve.

2.8 FACTORISING

Earlier in this section we expanded expressions such as $x(3x-1)$ to give $3x^2-x$.
The reverse of this process is called *factorising.*

Example 1

Factorise: (a) x^2+7x (b) $3y^2-12y$
(c) $6a^2b-10ab^2$

(a) x is common to x^2 and $7x$.
$\therefore \quad x^2+7x = x(x+7)$.
The factors are x and $(x+7)$.

(b) $3y$ is common
$\therefore \quad 3y^2-12y = 3y(y-4)$

(c) $2ab$ is common
$\therefore \quad 6a^2b-10ab^2 = 2ab(3a-5b)$

Exercise 21

Factorise the following expressions completely.

1. x^2+5x
2. x^2-6x
3. $7x-x^2$
4. y^2+8y
5. $2y^2+3y$
6. $6y^2-4y$
7. $3x^2-21x$
8. $16a-2a^2$
9. $6c^2-21c$
10. $15x-9x^2$
11. $56y-21y^2$
12. $ax+bx+2cx$
13. $x^2+xy+3xz$
14. $x^2y+y^3+z^2y$
15. $3a^2b+2ab^2$
16. x^2y+xy^2
17. $6a^2+4ab+2ac$
18. $ma+2bm+m^2$
19. $2kx+6ky+4kz$
20. $ax^2+ay+2ab$
21. x^2k+xk^2
22. a^3b+2ab^2
23. $abc-3b^2c$
24. $2a^2e-5ae^2$
25. a^3b+ab^3
26. $x^3y+x^2y^2$
27. $6xy^2-4x^2y$
28. $3ab^3-3a^3b$
29. $2a^3b+5a^2b^2$
30. ax^2y-2ax^2z
31. $2abx+2ab^2+2a^2b$
32. $ayx+yx^3-2y^2x^2$

Example 2

Factorise $ah+ak+bh+bk$.

(a) Divide into pairs, $ah+ak \mid +bh+bk$.

(b) a is common to the first pair
b is common to the second pair
$a(h+k)+b(h+k)$

(c) $(h+k)$ is common to both terms.

Thus we have $(h+k)(a+b)$

Check this result by removing the brackets.

Example 3

Factorise $6mx-3nx+2my-ny$.

(a) $6mx-3nx \mid +2my-ny$

(b) $= 3x(2m-n)+y(2m-n)$

(c) $= (2m-n)(3x+y)$

Exercise 22

Factorise the following expressions:

1. $ax+ay+bx+by$
2. $ay+az+by+bz$
3. $xb+xc+yb+yc$
4. $xh+xk+yh+yk$
5. $xm+xn+my+ny$
6. $ah-ak+bh-bk$
7. $ax-ay+bx-by$
8. $am-bm+an-bn$
9. $hs+ht+ks+kt$
10. $xs-xt+ys-yt$
11. $ax-ay-bx+by$
12. $xs-xt-ys+yt$
13. $as-ay-xs+xy$
14. $hx-hy-bx+by$
15. $am-bm-an+bn$
16. $xk-xm-kz+mz$
17. $2ax+6ay+bx+3by$
18. $2ax+2ay+bx+by$
19. $2mh-2mk+nh-nk$
20. $2mh+3mk-2nh-3nk$
21. $6ax+2bx+3ay+by$
22. $2ax-2ay-bx+by$
23. $x^2a+x^2b+ya+yb$
24. $ms+2mt^2-ns-2nt^2$

Example 4

Factorise x^2+6x+8

(a) Find two numbers which multiply to give 8 and add up to 6.
In this case the numbers are 4 and 2.

(b) Put these numbers into brackets.
So $x^2+6x+8 = (x+4)(x+2)$.

Example 5

Factorise (a) $x^2+2x-15$
(b) x^2-6x+8

(a) Two numbers which multiply to give -15 and add up to $+2$ are -3 and 5.
$\therefore \quad x^2+2x-15 = (x-3)(x+5)$.

(b) Two numbers which multiply to give $+8$ and add up to -6 are -2 and -4.
$\therefore \quad x^2-6x+8 = (x-2)(x-4)$.

Exercise 23

Factorise the following:

1. $x^2 + 7x + 10$
2. $x^2 + 7x + 12$
3. $x^2 + 8x + 15$
4. $x^2 + 10x + 21$
5. $x^2 + 8x + 12$
6. $y^2 + 12y + 35$
7. $y^2 + 11y + 24$
8. $y^2 + 10y + 25$
9. $y^2 + 15y + 36$
10. $a^2 - 3a - 10$
11. $a^2 - a - 12$
12. $z^2 + z - 6$
13. $x^2 - 2x - 35$
14. $x^2 - 5x - 24$
15. $x^2 - 6x + 8$
16. $y^2 - 5y + 6$
17. $x^2 - 8x + 15$
18. $a^2 - a - 6$
19. $a^2 + 14a + 45$
20. $b^2 - 4b - 21$
21. $x^2 - 8x + 16$
22. $y^2 + 2y + 1$
23. $y^2 - 3y - 28$
24. $x^2 - x - 20$
25. $x^2 - 8x - 240$
26. $x^2 - 26x + 165$
27. $y^2 + 3y - 108$
28. $x^2 - 49$
29. $x^2 - 9$
30. $x^2 - 16$

Example 6

Factorise $3x^2 + 13x + 4$

(a) Find two numbers which multiply to give 12 and add up to 13.
In this case the numbers are 1 and 12.

(b) Split the '$13x$' term,
$3x^2 + x + 12x + 4$

(c) Factorise in pairs,
$x(3x + 1) + 4(3x + 1)$

(d) $(3x + 1)$ is common,
$(3x + 1)(x + 4)$

Example 7

Factorise $6x^2 + 5x - 6$

(a) Find two numbers which multiply to give -36 and add up to $+5$.
In this case the numbers are $+9$ and -4.

(b) $6x^2 + 5x - 6 = 6x^2 + 9x - 4x - 6$
(c) $\quad = 3x(2x + 3) - 2(2x + 3)$
(d) $\quad = (2x + 3)(3x - 2)$

Exercise 24

Factorise the following:

1. $2x^2 + 5x + 3$
2. $2x^2 + 7x + 3$
3. $3x^2 + 7x + 2$
4. $2x^2 + 11x + 12$
5. $3x^2 + 8x + 4$
6. $2x^2 + 7x + 5$
7. $3x^2 - 5x - 2$
8. $2x^2 - x - 15$
9. $2x^2 + x - 21$
10. $3x^2 - 17x - 28$
11. $6x^2 + 7x + 2$
12. $12x^2 + 23x + 10$
13. $3x^2 - 11x + 6$
14. $3y^2 - 11y + 10$
15. $4y^2 - 23y + 15$
16. $6y^2 + 7y - 3$
17. $6x^2 - 27x + 30$
18. $10x^2 + 9x + 2$
19. $6x^2 - 19x + 3$
20. $8x^2 - 10x - 3$
21. $12x^2 + 4x - 5$
22. $16x^2 + 19x + 3$
23. $4a^2 - 4a + 1$
24. $12x^2 + 17x - 14$
25. $15x^2 + 44x - 3$
26. $48x^2 + 46x + 5$
27. $64y^2 + 4y - 3$
28. $120x^2 + 67x - 5$
29. $9x^2 - 1$
30. $4a^2 - 9$

The difference of two squares

$$x^2 - y^2 = (x - y)(x + y)$$

Remember this result.

Example 8

Factorise (a) $4a^2 - b^2$
(b) $3x^2 - 27y^2$

(a) $4a^2 - b^2 = (2a)^2 - b^2$
$\quad = (2a - b)(2a + b)$

(b) $3x^2 - 27y^2 = 3(x^2 - 9y^2)$
$\quad = 3[x^2 - (3y)^2]$
$\quad = 3(x - 3y)(x + 3y)$.

Example 9

Factorise, and hence evaluate, $251^2 - 250^2$.

$251^2 - 250^2 = (251 - 250)(251 + 250)$
$\quad = 1 \times 501$
$\quad = 501$

Notice that this method involves less working than squaring 251, then squaring 250 and finally subtracting.

Exercise 25

Factorise the following:

1. $y^2 - a^2$
2. $m^2 - n^2$
3. $x^2 - t^2$
4. $y^2 - 1$
5. $x^2 - 9$
6. $a^2 - 25$
7. $x^2 - \frac{1}{4}$
8. $x^2 - \frac{1}{9}$

9. $4x^2 - y^2$
10. $a^2 - 4b^2$
11. $25x^2 - 4y^2$
12. $9x^2 - 16y^2$
13. $x^2 - \frac{y^2}{4}$
14. $9m^2 - \frac{4}{9}n^2$
15. $16t^2 - \frac{4}{25}s^2$
16. $4x^2 - \frac{z^2}{100}$
17. $x^3 - x$
18. $a^3 - ab^2$
19. $4x^3 - x$
20. $8x^3 - 2xy^2$
21. $12x^3 - 3xy^2$
22. $18m^3 - 8mn^2$
23. $5x^2 - 1\frac{1}{4}$
24. $50a^3 - 18ab^2$
25. $12x^2y - 3yz^2$
26. $36a^3b - 4ab^3$
27. $50a^5 - 8a^3b^2$
28. $36x^3y - 225xy^3$

Evaluate the following:

29. $81^2 - 80^2$
30. $102^2 - 100^2$
31. $225^2 - 215^2$
32. $1211^2 - 1210^2$
33. $723^2 - 720^2$
34. $3{\cdot}8^2 - 3{\cdot}7^2$
35. $5{\cdot}24^2 - 4{\cdot}76^2$
36. $1234^2 - 1235^2$
37. $3{\cdot}81^2 - 3{\cdot}8^2$
38. $540^2 - 550^2$
39. $7{\cdot}68^2 - 2{\cdot}32^2$
40. $0{\cdot}003^2 - 0{\cdot}002^2$

2.9 QUADRATIC EQUATIONS

So far, we have met linear equations which have one solution only.
Quadratic equations always have an x^2 term, and often an x term and a number term, and generally have two different solutions.

(a) Solution by factors

Consider the equation $a \times b = 0$, where a and b are numbers. The product $a \times b$ can only be zero if either a or b (or both) is equal to zero.
Can you think of other possible pairs of numbers which multiply together to give zero?

Example 1

Solve the equation $x^2 + x - 12 = 0$

Factorising, $(x-3)(x+4) = 0$

either $x - 3 = 0$ or $x + 4 = 0$

$x = 3$ $\quad x = -4$

Example 2

Solve the equation $6x^2 + x - 2 = 0$

Factorising, $(2x-1)(3x+2) = 0$

either $2x - 1 = 0$ or $3x + 2 = 0$

$2x = 1$ $\quad 3x = -2$

$x = \frac{1}{2}$ $\quad x = -\frac{2}{3}$

Exercise 26

Solve the following equations:

1. $x^2 + 7x + 12 = 0$
2. $x^2 + 7x + 10 = 0$
3. $x^2 + 2x - 15 = 0$
4. $x^2 + x - 6 = 0$
5. $x^2 - 8x + 12 = 0$
6. $x^2 + 10x + 21 = 0$
7. $x^2 - 5x + 6 = 0$
8. $x^2 - 4x - 5 = 0$
9. $x^2 + 5x - 14 = 0$
10. $2x^2 - 3x - 2 = 0$
11. $3x^2 + 10x - 8 = 0$
12. $2x^2 + 7x - 15 = 0$
13. $6x^2 - 13x + 6 = 0$
14. $4x^2 - 29x + 7 = 0$
15. $10x^2 - x - 3 = 0$
16. $y^2 - 15y + 56 = 0$
17. $12y^2 - 16y + 5 = 0$
18. $y^2 + 2y - 63 = 0$
19. $x^2 + 2x + 1 = 0$
20. $x^2 - 6x + 9 = 0$
21. $x^2 + 10x + 25 = 0$
22. $x^2 - 14x + 49 = 0$
23. $6a^2 - a - 1 = 0$
24. $4a^2 - 3a - 10 = 0$
25. $z^2 - 8z - 65 = 0$
26. $6x^2 + 17x - 3 = 0$
27. $10k^2 + 19k - 2 = 0$
28. $y^2 - 2y + 1 = 0$
29. $36x^2 + x - 2 = 0$
30. $20x^2 - 7x - 3 = 0$

Example 3

Solve the equation $x^2 - 7x = 0$

Factorising, $x(x-7) = 0$

either $x = 0$ or $x - 7 = 0$

$x = 7$

The solutions are $x = 0$ and $x = 7$.

Example 4

Solve the equation $4x^2 - 9 = 0$

(a) Factorising, $(2x-3)(2x+3) = 0$

either $2x - 3 = 0$ or $2x + 3 = 0$

$2x = 3$ $\quad 2x = -3$

$x = \frac{3}{2}$ $\quad x = -\frac{3}{2}$

(b) Alternative method

$4x^2 - 9 = 0$

$4x^2 = 9$

$x^2 = \frac{9}{4}$

$x = +\frac{3}{2}$ or $-\frac{3}{2}$.

Exercise 27

1. $x^2 - 3x = 0$
2. $x^2 + 7x = 0$
3. $2x^2 - 2x = 0$
4. $3x^2 - x = 0$
5. $x^2 - 16 = 0$
6. $x^2 - 49 = 0$
7. $4x^2 - 1 = 0$
8. $9x^2 - 4 = 0$
9. $6y^2 + 9y = 0$
10. $6a^2 - 9a = 0$
11. $10x^2 - 55x = 0$
12. $16x^2 - 1 = 0$
13. $y^2 - \frac{1}{4} = 0$
14. $56x^2 - 35x = 0$
15. $36x^2 - 3x = 0$
16. $x^2 = 6x$
17. $x^2 = 11x$
18. $2x^2 = 3x$
19. $x^2 = x$
20. $4x = x^2$
21. $3x - x^2 = 0$
22. $4x^2 = 1$
23. $9x^2 = 16$
24. $x^2 = 9$
25. $12x = 5x^2$
26. $1 - 9x^2 = 0$
27. $x^2 = \frac{x}{4}$
28. $2x^2 = \frac{x}{3}$
29. $4x^2 = \frac{1}{4}$
30. $\frac{x}{5} - x^2 = 0$

(b) Solution by formula

The solutions of the quadratic equation
$ax^2 + bx + c = 0$
are given by the formula

$$x = \frac{-b \pm \sqrt{(b^2 - 4ac)}}{2a}$$

Use this formula only after trying (and failing) to factorise.

Example 5

Solve the equation $2x^2 - 3x - 4 = 0$.
In this case $a = 2$, $b = -3$, $c = -4$.

$$x = \frac{-(-3) \pm \sqrt{[(-3)^2 - (4 \times 2 \times -4)]}}{2 \times 2}$$

$$x = \frac{3 \pm \sqrt{[9 + 32]}}{4} = \frac{3 \pm \sqrt{41}}{4}$$

$$x = \frac{3 \pm 6{\cdot}403}{4}$$

either $x = \frac{3 + 6{\cdot}403}{4} = 2{\cdot}35$ (2 decimal places)

or $x = \frac{3 - 6{\cdot}403}{4} = \frac{-3{\cdot}403}{4}$

$= -0{\cdot}85$ (2 decimal places).

Exercise 28

Solve the following, giving answers to two decimal places where necessary.

1. $2x^2 + 11x + 5 = 0$
2. $3x^2 + 11x + 6 = 0$
3. $6x^2 + 7x + 2 = 0$
4. $3x^2 - 10x + 3 = 0$
5. $5x^2 - 7x + 2 = 0$
6. $6x^2 - 11x + 3 = 0$
7. $2x^2 + 6x + 3 = 0$
8. $x^2 + 4x + 1 = 0$
9. $5x^2 - 5x + 1 = 0$
10. $x^2 - 7x + 2 = 0$
11. $2x^2 + 5x - 1 = 0$
12. $3x^2 + x - 3 = 0$
13. $3x^2 + 8x - 6 = 0$
14. $3x^2 - 7x - 20 = 0$
15. $2x^2 - 7x - 15 = 0$
16. $x^2 - 3x - 2 = 0$
17. $2x^2 + 6x - 1 = 0$
18. $6x^2 - 11x - 7 = 0$
19. $3x^2 + 25x + 8 = 0$
20. $3y^2 - 2y - 5 = 0$
21. $2y^2 - 5y + 1 = 0$
22. $\frac{1}{2}y^2 + 3y + 1 = 0$
23. $2 - x - 6x^2 = 0$
24. $3 + 4x - 2x^2 = 0$
25. $1 - 5x - 2x^2 = 0$
26. $3x^2 - 1 + 4x = 0$
27. $5x - x^2 + 2 = 0$
28. $24x^2 - 22x - 35 = 0$
29. $36x^2 - 17x - 35 = 0$
30. $20x^2 + 17x - 63 = 0$
31. $x^2 + 2{\cdot}5x - 6 = 0$
32. $0{\cdot}3y^2 + 0{\cdot}4y - 1{\cdot}5 = 0$
33. $10 - x - 3x^2 = 0$
34. $x^2 + 3{\cdot}3x - 0{\cdot}7 = 0$
35. $12 - 5x^2 - 11x = 0$
36. $5x - 2x^2 + 187 = 0$

The solution to a problem can involve an equation which does not at first appear to be quadratic. The terms in the equation may need to be rearranged as shown below.

Example 6

Solve $2x(x - 1) = (x + 1)^2 - 5$

$$2x^2 - 2x = x^2 + 2x + 1 - 5$$
$$2x^2 - 2x - x^2 - 2x - 1 + 5 = 0$$
$$x^2 - 4x + 4 = 0$$
$$(x - 2)(x - 2) = 0$$
$$x = 2$$

In this example the quadratic has a repeated root of $x = 2$.

Example 7

Solve $x + 2 = \frac{3}{x}$

Multiply each term by x

$$x^2 + 2x = 3$$
$$x^2 + 2x - 3 = 0$$
$$(x + 3)(x - 1) = 0$$
$$x = -3 \quad \text{or} \quad x = 1$$

Example 8

Solve $\dfrac{1}{x-1}+\dfrac{4}{x}=3$

Multiply each term by $(x-1)$

$$\frac{(x-1)\,1}{(x-1)}+\frac{4(x-1)}{x} = 3(x-1)$$

$$1+\frac{4(x-1)}{x} = 3(x-1)$$

Multiply each term by x

$$x+\frac{4(x-1)x}{x} = 3(x-1)x$$

$$x+4x-4 = 3x^2-3x$$

$$0 = 3x^2-3x-5x+4$$

$$3x^2-8x+4 = 0$$

$$(x-2)(3x-2) = 0$$

$$x = 2 \text{ or } x = \tfrac{2}{3}$$

Exercise 29

Solve the following equations, giving answers to two decimal places where necessary.

1. $x^2=6-x$
2. $x(x+10)=-21$
3. $3x+2=2x^2$
4. $x^2+4=5x$
5. $6x(x+1)=5-x$
6. $(2x)^2=x(x-14)-5$
7. $(x-3)^2=10$
8. $(x+1)^2-10=2x(x-2)$
9. $(2x-1)^2=(x-1)^2+8$
10. $3x(x+2)-x(x-2)+6=0$
11. $2x(x+5)+(x+1)^2=x^2+13x+11$
12. $2x(x-5)-x(x-6)=21$
13. $(2x-1)^2-(x-7)^2=-40$
14. $x(4-x)+2x(x-3)=7$
15. $(2x-3)^2=2x(x-3)+10$
16. $(2x-1)^2-(x-1)^2=3x+2$
17. $2x(4-x)+(x-3)^2=3x(1-x)+9$
18. $4(x+1)^2-(x-2)^2=0$
19. $2(x+3)^2+3(x-1)^2=20$
20. $3(x-2)^2-2(2x-1)^2=1$
21. $x=\dfrac{15}{x}-22$
22. $x+5=\dfrac{14}{x}$
23. $4x+\dfrac{7}{x}=29$
24. $10x=1+\dfrac{3}{x}$
25. $2x^2=7x$
26. $16=\dfrac{1}{x^2}$
27. $2x+2=\dfrac{7}{x}-1$
28. $\dfrac{2}{x}+\dfrac{2}{x+1}=3$
29. $\dfrac{3}{x-1}+\dfrac{3}{x+1}=4$
30. $\dfrac{2}{x-2}+\dfrac{4}{x+1}=3$
31. $\dfrac{1}{x-1}+\dfrac{3}{x+2}=\dfrac{5}{2}$
32. $\dfrac{1}{x+1}+\dfrac{1}{x+2}=\dfrac{5}{x+5}$
33. $\dfrac{1}{x-1}+\dfrac{1}{x+1}=\dfrac{8}{x+4}$
34. $\dfrac{1}{x-1}-\dfrac{1}{x+2}=\dfrac{6}{x+4}$
35. $\dfrac{2}{x-1}-\dfrac{1}{x+2}=\dfrac{-4}{x+3}$
36. $\dfrac{1}{x-2}-\dfrac{1}{x+3}=\dfrac{-5}{x+6}$
37. $\dfrac{1}{2x-1}-\dfrac{1}{x+1}=\dfrac{3}{x+5}$
38. $\dfrac{1}{x+2}-\dfrac{1}{x}=2$
39. $\dfrac{1}{x}+\dfrac{1}{x-3}=1\tfrac{1}{4}$
40. $20x=43-\dfrac{21}{x}$

2.10 PROBLEMS SOLVED BY QUADRATIC EQUATIONS

Example 1

The perimeter of a rectangle is 42 cm. If the diagonal is 15 cm, find the width of the rectangle.

Let the width of the rectangle be x cm.

Since the perimeter is 42 cm, the sum of the length and the width is 21 cm.

$\therefore$ length of rectangle $= (21 - x)$ cm

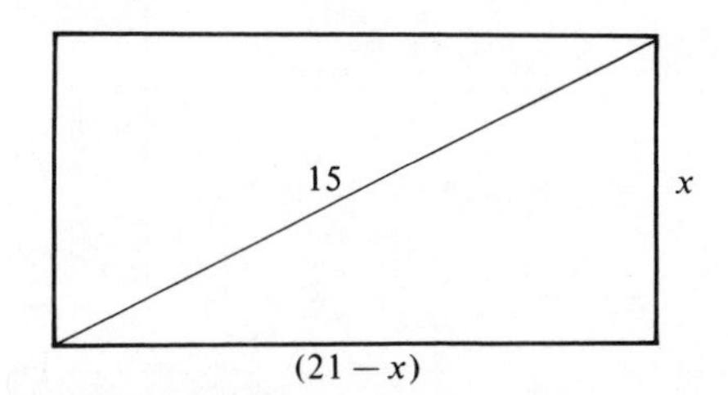

By Pythagoras' theorem

$$\begin{aligned} x^2 + (21 - x)^2 &= 152 \\ x^2 + (21 - x)(21 - x) &= 152 \\ x^2 + 441 - 42x + x^2 &= 225 \\ 2x^2 - 42x + 216 &= 0 \\ x^2 - 21x + 108 &= 0 \\ (x - 12)(x - 9) &= 0 \\ x &= 12 \\ \text{or} \quad x &= 9 \end{aligned}$$

Note that the dimensions of the rectangle are 9 cm and 12 cm, whichever value of x is taken.

$\therefore$ The width of rectangle is 9 cm.

Example 2

A man bought a certain number of golf balls for £20. If each ball had cost 20p less, he could have bought five more for the same money. How many golf balls did he buy?

Let the number of balls bought be x.

Cost of each ball $= \dfrac{2000}{x}$ pence

If five more balls had been bought

Cost of each ball now $= \dfrac{2000}{(x+5)}$ pence

The new price is 20p less than the original price.

$$\therefore \quad \frac{2000}{x} - \frac{2000}{(x+5)} = 20$$

(multiply by x)

$$x \cdot \frac{2000}{x} - x \cdot \frac{2000}{(x+5)} = 20x$$

(multiply by $(x + 5)$)

$$2000(x+5) - x\frac{2000}{(x+5)}(x+5) = 20x(x+5)$$

$$\begin{aligned} 2000x + 10000 - 2000x &= 20x^2 + 100x \\ 20x^2 + 100x - 10000 &= 0 \\ x^2 + 5x - 500 &= 0 \\ (x - 20)(x + 25) &= 0 \end{aligned}$$

$$\therefore \quad x = 20 \quad \text{or} \quad x = -25$$

We discard $x = -25$ as meaningless.
The number of balls bought = 20.

Exercise 30

Solve by forming a quadratic equation.

1. Two numbers, which differ by 3, have a product of 88. Find them.
2. The product of two consecutive odd numbers is 143. Find the numbers. (Hint: If the first odd number is x, what is the next odd number?)
3. The length of a rectangle exceeds the width by 7 cm. If the area is 60 cm^2, find the length of the rectangle.
4. The length of a rectangle exceeds the width by 2 cm. If the diagonal is 10 cm long, find the width of the rectangle.
5. The area of the rectangle exceeds the area of the square by 24 m^2. Find x.

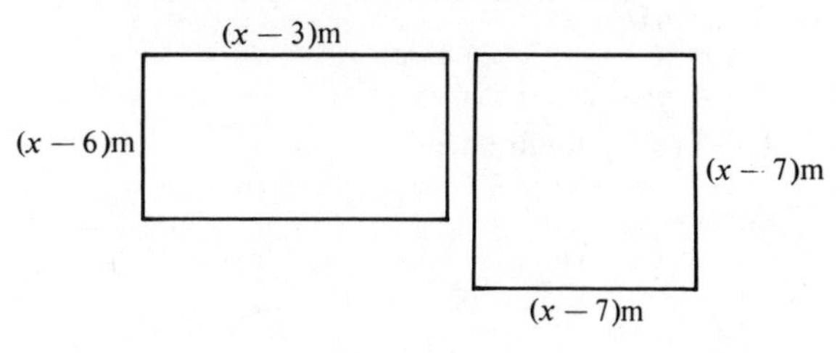

6. The perimeter of a rectangle is 68 cm. If the diagonal is 26 cm, find the dimensions of the rectangle.
7. A man walks a certain distance due North and then the same distance plus a further 7 km due East. If the final distance from the starting point is 17 km, find the distances he walks North and East.
8. A hen makes a profit of x pence on each of the $(x + 5)$ eggs she lays. If her total profit was 84 pence, find the number of eggs she lays.
9. A boy buys x eggs at $(x - 8)$ pence each and $(x - 2)$ rashers of bacon at $(x - 3)$ pence each. If the total bill is £1·75, how many eggs does he buy?
10. A number exceeds four times its reciprocal by 3. Find the number.
11. Two numbers differ by 3. The sum of their reciprocals is $\frac{7}{10}$; find the numbers.
12. A cyclist travels 40 km at a speed x km/h. Find the time taken in terms of x. Find the time taken when his speed is reduced by 2 km/h. If the difference between the times is 1 hour, find his original speed x.
13. An increase of speed of 4 km/h on a journey of 32 km reduces the time taken by 4 hours. Find the original speed.
14. A train normally travels 60 miles at a certain speed. One day, due to bad weather, the train's speed is reduced by 10 mph so that the journey takes 3 hours longer. Find the normal speed.
15. The speed of a sparrow is x mph in still air. When the wind is blowing at 1 mph, the sparrow takes 5 hours to fly 12 miles to its nest and 12 miles back again. He goes out directly into the wind and returns with the wind behind him. Find his speed in still air.
16. An aircraft flies a certain distance on a bearing of 135° and then twice the distance on a bearing of 225°. Its distance from the starting point is then 350 km. Find the length of the first part of the journey.
17. In Fig. 1, ABCD is a rectangle with AB = 12 cm and BC = 7 cm. AK = BL = CM = DN = x cm. If the area of KLMN is 54 cm² find x.

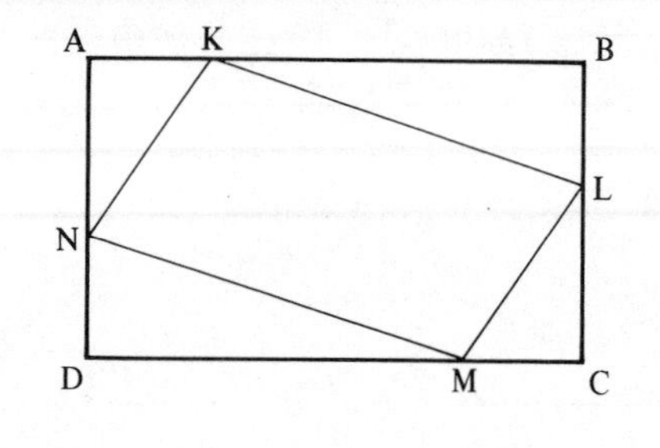

Fig. 1

18. In Fig. 1, AB = 14 cm, BC = 11 cm and AK = BL = CM = DN = x cm. If the area of KLMN is now 97 cm², find x.
19. The numerator of a fraction is 1 less than the denominator. When both numerator and denominator are increased by 2, the fraction is increased by $\frac{1}{12}$. Find the original fraction.
20. The perimeters of a square and a rectangle are equal. The length of the rectangle is 11 cm and the area of the square is 4 cm² more than the area of the rectangle. Find the side of the square.

REVISION EXERCISE 2A

1. Solve the equations
   (a) $x + 4 = 3x + 9$ (b) $9 - 3a = 1$
   (c) $y^2 + 5y = 0$ (d) $x^2 - 4 = 0$
   (e) $3x^2 + 7x - 40 = 0$
2. Given $a = 3, b = 4$ and $c = -2$, evaluate
   (a) $2a^2 - b$ (b) $a(b - c)$
   (c) $2b^2 - c^2$
3. Factorise completely
   (a) $4x^2 - y^2$ (b) $2x^2 + 8x + 6$
   (c) $6m + 4n - 9km - 6kn$
   (d) $2x^2 - 5x - 3$
4. Solve the simultaneous equations
   (a) $3x + 2y = 5$, $2x - y = 8$ (b) $2m - n = 6$, $2m + 3n = -6$
   (c) $3x - 4y = 19$, $x + 6y = 10$ (d) $3x - 7y = 11$, $2x - 3y = 4$
5. Given that $x = 4, y = 3, z = -2$, evaluate
   (a) $2x(y + z)$ (b) $(xy)^2 - z^2$
   (c) $x^2 + y^2 + z^2$ (d) $(x + y)(x - z)$
   (e) $\sqrt{[x(1 - 4z)]}$ (f) $\frac{xy}{z}$
6. (a) Simplify $3(2x - 5) - 2(2x + 3)$
   (b) Factorise $2a - 3b - 4xa + 6xb$
   (c) Solve the equation $\frac{x - 11}{2} - \frac{x - 3}{5} = 2$.
7. Solve the equations
   (a) $5 - 7x = 4 - 6x$ (b) $\frac{7}{x} = \frac{2}{3}$
   (c) $2x^2 - 7x = 0$ (d) $x^2 + 5x + 6 = 0$
   (e) $\frac{1}{x} + \frac{1}{4} = \frac{1}{3}$

8. Factorise completely
(a) $z^3 - 16z$ (b) $x^2y^2 + x^2 + y^2 + 1$
(c) $2x^2 + 11x + 12$

9. Find the value of $\frac{2x - 3y}{5x + 2y}$ when $x = 2a$ and $y = -a$.

10. Solve the simultaneous equations
(a) $7c + 3d = 29$, $5c - 4d = 33$ (b) $2x - 3y = 7$, $2y - 3x = -8$
(c) $5x = 3(1 - y)$, $3x + 2y + 1 = 0$ (d) $5s + 3t = 16$, $11s + 7t = 34$

11. Solve the equations,
(a) $4(y + 1) = \frac{3}{1 - y}$
(b) $4(2x - 1) - 3(1 - x) = 0$
(c) $\frac{x + 3}{x} = 2$ (d) $x^2 = 5x$

12. Solve the following, giving your answers correct to 2 decimal places.
(a) $2x^2 - 3x - 1 = 0$ (b) $x^2 - x - 1 = 0$
(c) $3x^2 + 2x - 4 = 0$ (d) $x + 3 = \frac{7}{x}$

13. Find x by forming a suitable equation.

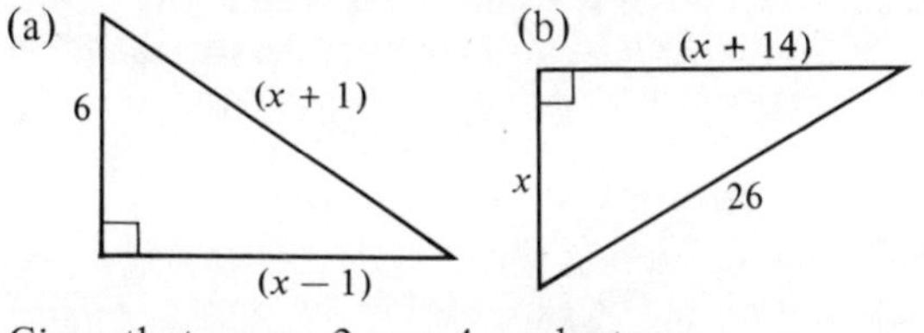

14. Given that $m = -2$, $n = 4$, evaluate
(a) $5m + 3n$ (b) $5 + 2m - m^2$
(c) $m^2 + 2n^2$ (d) $(2m + n)(2m - n)$
(e) $(n - m)^2$ (f) $n - mn - 2m^2$

15. A car travels for x hours at a speed of $(x + 2)$ miles per hour. If the distance travelled is 15 miles, write down an equation for x and solve it to find the speed of the car.

16. ABCD is a rectangle, where AB $= x$ cm and BC is 1·5 cm less than AB.

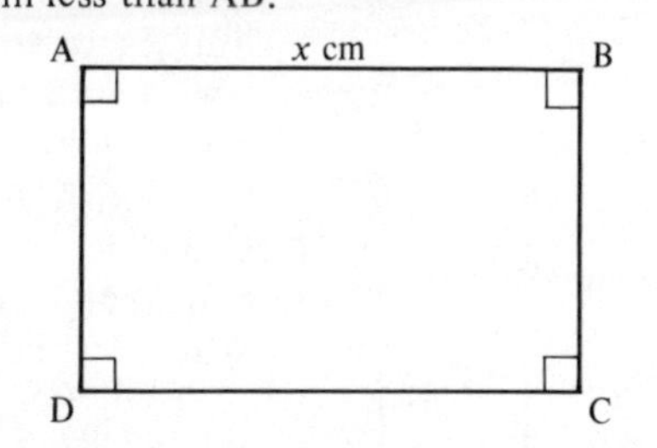

If the area of the rectangle is 52 cm^2, form an equation in x and solve it to find the dimensions of the rectangle.

17. Solve the equations
(a) $(2x + 1)^2 = (x + 5)^2$
(b) $\frac{x + 2}{2} - \frac{x - 1}{3} = \frac{x}{4}$
(c) $x^2 - 7x + 5 = 0$, giving the answers correct to two decimal places.

18. Solve the equation
$$\frac{x}{x + 1} - \frac{x + 1}{3x - 1} = \frac{1}{4}$$

19. Given that $a + b = 2$ and that $a^2 + b^2 = 6$, prove that $2ab = -2$. Find also the value of $(a - b)^2$.

20. The sides of a right-angled triangle have lengths $(x - 3)$ cm, $(x + 11)$ cm and $2x$ cm, where $2x$ is the hypotenuse. Find x.

EXAMINATION EXERCISE 2B

1. A student has a total of 126 marks in x tests. In the next two tests he has 9 marks and 8 marks respectively. Find, in terms of x, his average number of marks per test for
(i) the first x tests,
(ii) the $(x + 2)$ tests.

If his average for the first x tests was one greater than his average for the $(x + 2)$ tests, use the results of (i) and (ii) to form an equation and, hence, find the value of x.

Another student has an average of 13·5 marks for the first $(x + 1)$ tests, but his mark on the last test gave him a final average of 14 marks for the $(x + 2)$ tests. What was his mark on the last test? [AEB]

2. A man bought 3 boxes of Dutch cigars at x pence per box and 2 boxes of Havana cigars at y pence per box.
   (i) Calculate, in terms of x and y, the total cost of the 5 boxes.
   (ii) Hence, calculate the average cost per box in terms of x and y.
   He later sold the 3 boxes of Dutch cigars at a profit of $33\frac{1}{3}\%$ of their cost price but gave away the other 2 boxes as presents.
   (iii) Calculate the amount he received for the sale of the cigars.
   (iv) Find an expression, in terms of x and y, for his percentage profit on the whole transaction. [NEA]

3. In the rectangle ABCD, AB $= x$ cm and BC $= 1$ cm. The line LM is drawn so that ALMD is a square.
   Write down, in terms of x,
   (a) the length LB, (b) the ratios $\frac{AB}{BC}$ and $\frac{BC}{LB}$.
   (c) If $\frac{AB}{BC} = \frac{BC}{LB}$, obtain a quadratic equation in x. Hence find x correct to two decimal places. [L]

4. A stone is thrown into the air and its height, h metres above the ground, is given by the equation

$$h = pt - qt^2,$$

where p and q are constants and t seconds is the time it has been in the air.
   Given that $h = 40$ when $t = 2$ and that $h = 45$ when $t = 3$, show that

$$p - 2q = 20$$

and

$$p - 3q = 15.$$

Use these equations to calculate the values of p and q.
   Hence show that the equation for h can be expressed in the form

$$5t^2 - 30t + h = 0.$$

Use this equation to find the values of t when $h = 17$, giving your answers correct to two places of decimals. [AEB]

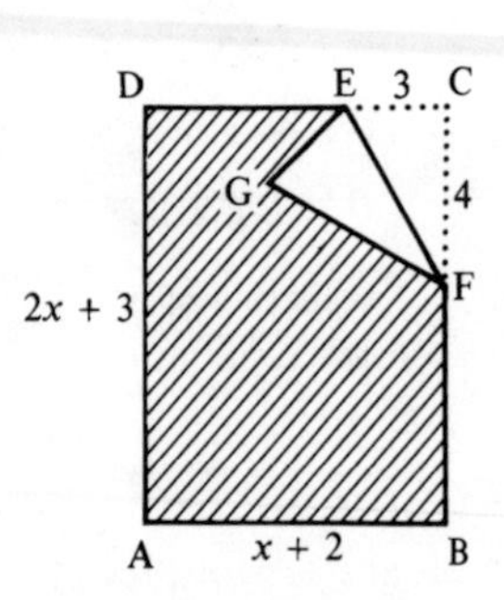

5. (a) Given that $(2x - y)(m + 5) = m(x - 1)$ and that $x = 3$ when $y = -1$, find the value of m.
   (b) The diagram represents a rectangular piece of paper ABCD which has been folded along EF so that C has moved to G.
   Given that EC $= 3$ cm, FC $= 4$ cm, AB $= (x + 2)$ cm and AD $= (2x + 3)$ cm,
   (i) calculate the area of ΔECF,
   (ii) find an expression for the shaded area ABFGED in terms of x.
   Given that the shaded area is 34 cm^2, show that

$$2x^2 + 7x - 40 = 0.$$

By solving this equation find the length of AB, giving your answer correct to two decimal places. [C]

6. A motorist makes a journey of 135 km at an average speed of x km/h. Write down an expression for the number of hours taken for the journey.
   Owing to road repairs, on a certain day his average speed for the same journey is reduced by 15 km/h. Write down an expression for the number of hours taken on that day.
   If the second journey takes 45 minutes longer than the first, form an equation in x and solve it. Hence find his average speed for each journey. [L]

7. Given that $y = \dfrac{a}{x} + bx$ and that $y = 6$ when $x = 2$ and $y = 10{\cdot}5$ when $x = 1$.
   Show that $a = 10$ and $b = 0{\cdot}5$. Calculate
   (i) the value of y when $x = 20$,
   (ii) the values of x when $y = 4{\cdot}5$,
   (iii) the values of x when $y = 5$, giving these values correct to 3 significant figures, where necessary. [JMB]

8. The diagram shows a rectangle ABCD with points E and F on AB and BC respectively; AB = 8 cm, BC = 4 cm, AE = x cm and BF = 3 cm. Write down expressions in terms of x for DE^2 and EF^2.
   Given that the angle DEF is a right angle, obtain an equation in x and solve it to find two possible values of x.
   Hence find the two possible values of the area of the triangle DEF.

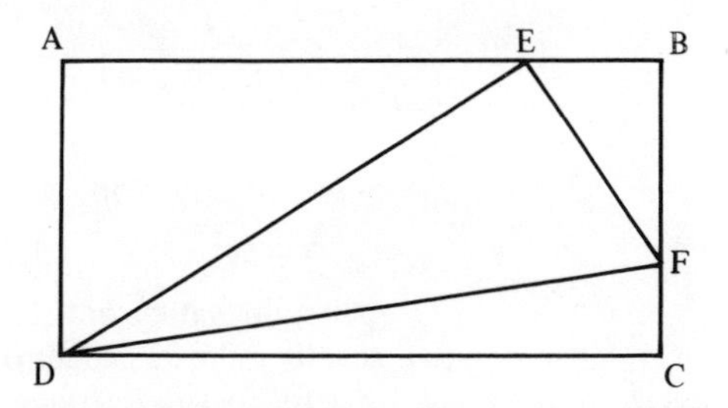

[O & C]

9. A racing cyclist completes the uphill section of a mountainous course of 75 km at an average speed of V km/h. Write down, in terms of V, the time taken for the journey.
   He then returns downhill along the same route at an average speed of $(V + 20)$ km/h. Write down the time taken for the return journey.
   Given that the difference between the times is one hour, form an equation in V and show that it reduces to
   $$V^2 + 20V - 1500 = 0.$$
   Solve this equation and calculate the time taken to complete the uphill section of the course.
   Calculate also the cyclist's average speed over the 150 km. [C]

10. (i) Solve the equation $\dfrac{1}{x-3} - \dfrac{1}{x} = \dfrac{3}{x+5}$.
    (ii) Given that $x^4 + 2x^2 + 9 = (x^2 + a)^2 - bx^2$ for all values of x, find, by putting $x = 0$ or otherwise, two possible values of a. Find also the corresponding values of b. Use one of these pairs of values of a and b to express $x^4 + 2x^2 + 9$ as a product of two factors. [O]

11. In the figure, AOB and COD are two straight lines intersecting at right angles at the point O. The area of the triangle AOD is 15 cm^2, that of the triangle COB is 75 cm^2, and the length of the line AOB is 20 cm. Taking the length of AO as x cm, show that the length of OD is $\frac{30}{x}$ cm and that the length of OC is $\frac{150}{20-x}$ cm.

Given also that the length of COD is 16 cm, obtain an equation in x and solve it to find two possible values of x.

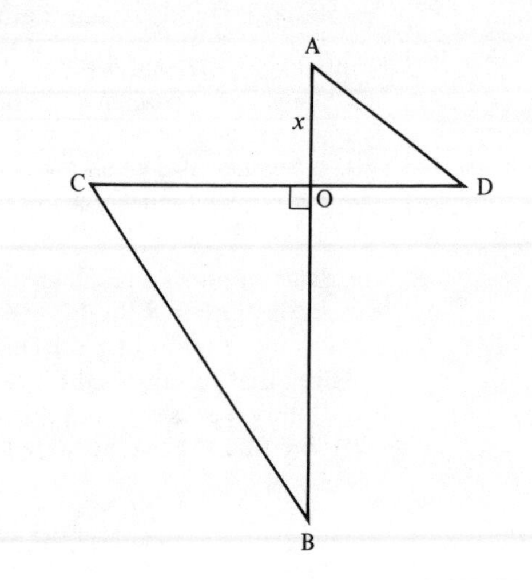

[O & C]

12. (i) Solve the equation

$$2x^2 - 5x - 13 = 0$$

giving the values of x correct to one place of decimals.

(ii) A solid cube has edges of length x centimetres. Write down an expression in x for its total surface area.

A solid rectangular block has its length 1 cm greater than that of the cube, its breadth 2 cm less than that of the cube and its height 50 per cent greater than that of the cube. Write down expressions in x for the dimensions of the rectangular block. Write down and simplify an expression in x for the total surface area of the rectangular block.

Given that the total surface area of the rectangular block is greater than that of the cube by 9 cm^2, prove that x satisfies the equation

$$2x^2 - 5x - 13 = 0.$$

By using your work in (i) find the volume of the cube. [AEB]

13. (a) Given that $y = 2x$ and $x^2 + 3xy - 5x = 18$, find the possible values of x.

(b) A rectangle has sides of length $(2x + 1)$ cm and $(x + 4)$ cm. Write down simplified expressions in x for

(i) the perimeter,

(ii) the area.

Given that the area of this rectangle is 63 cm^2, form an equation in x and solve it to find the positive value of x. Hence find the perimeter of the rectangle to the nearest mm. [AEB]

3 Mensuration

Archimedes of Samos (287–212 B.C.) studied at Alexandria as a young man. One of the first to apply scientific thinking to everyday problems, he was a practical man of common sense. He gave proofs for finding the area, the volume and the centre of gravity of circles, spheres, conics and spirals. By drawing polygons with many sides, he arrived at a value of π between $3\frac{10}{71}$ and $3\frac{10}{70}$. He was killed in the siege of Syracuse at the age of 75.

3.1 AREA

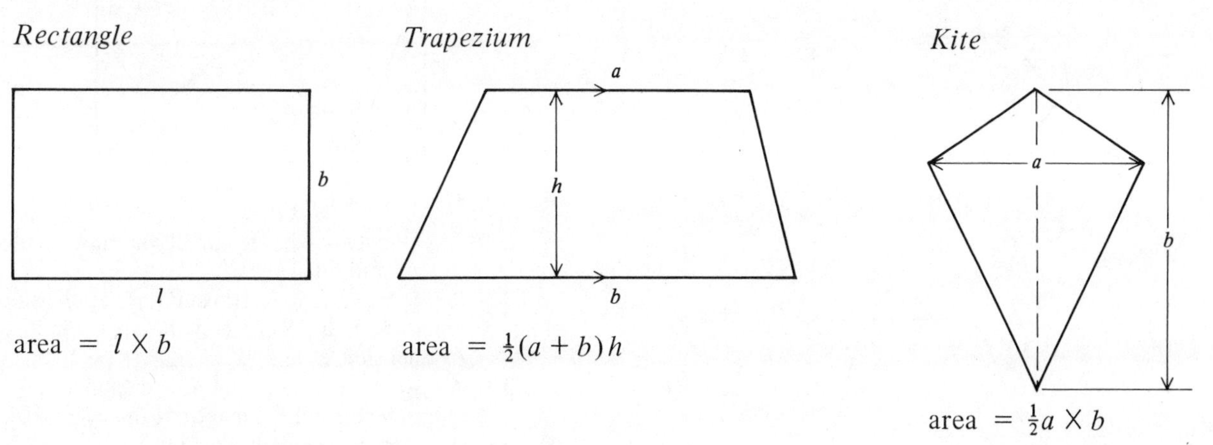

Rectangle: area $= l \times b$

Trapezium: area $= \frac{1}{2}(a + b)h$

Kite: area $= \frac{1}{2}a \times b$

$= \frac{1}{2} \times$ (product of diagonals)

Exercise 1

For questions **1** to **8**, find the area of each shape. Decide which information to use: you may not need all of it.

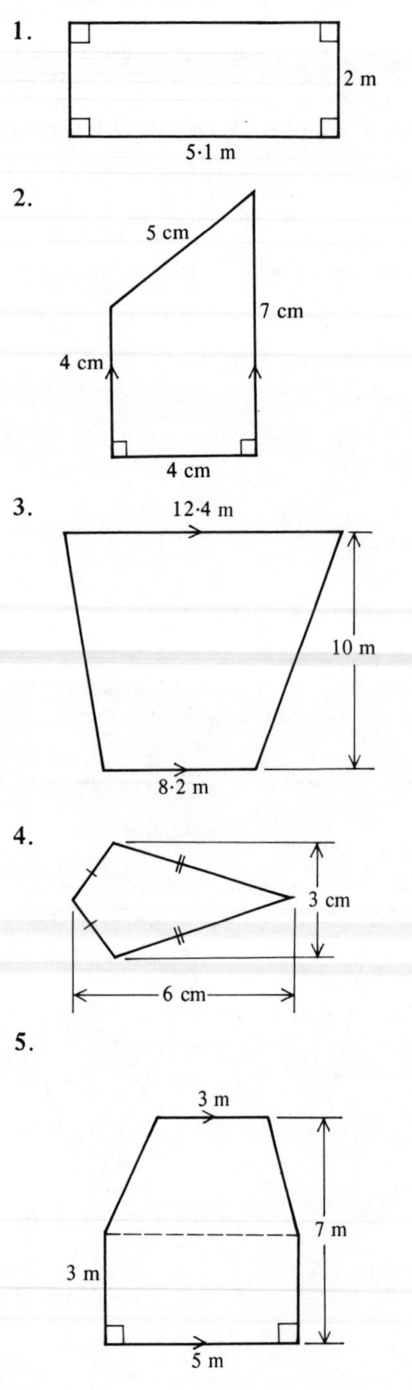

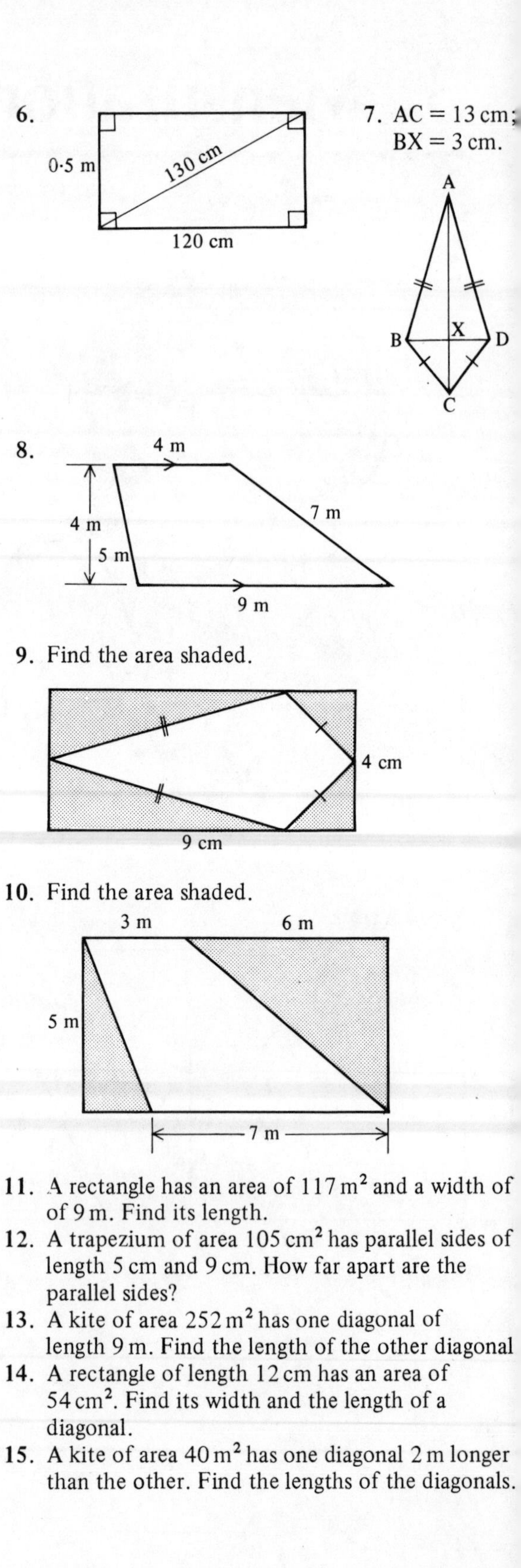

11. A rectangle has an area of $117\,m^2$ and a width of of 9 m. Find its length.
12. A trapezium of area $105\,cm^2$ has parallel sides of length 5 cm and 9 cm. How far apart are the parallel sides?
13. A kite of area $252\,m^2$ has one diagonal of length 9 m. Find the length of the other diagonal
14. A rectangle of length 12 cm has an area of $54\,cm^2$. Find its width and the length of a diagonal.
15. A kite of area $40\,m^2$ has one diagonal 2 m longer than the other. Find the lengths of the diagonals.

16. A trapezium of area 140 cm^2 has parallel sides 10 cm apart and one of these sides is 16 cm long. Find the length of the other parallel side.
17. A kite of area $11\frac{3}{8} \text{ cm}^2$ has one diagonal 3 cm longer than the other. Find the lengths of the diagonals.
18. A floor 5 m by 20 m is covered by square tiles of side 20 cm. How many tiles are needed?

Triangle

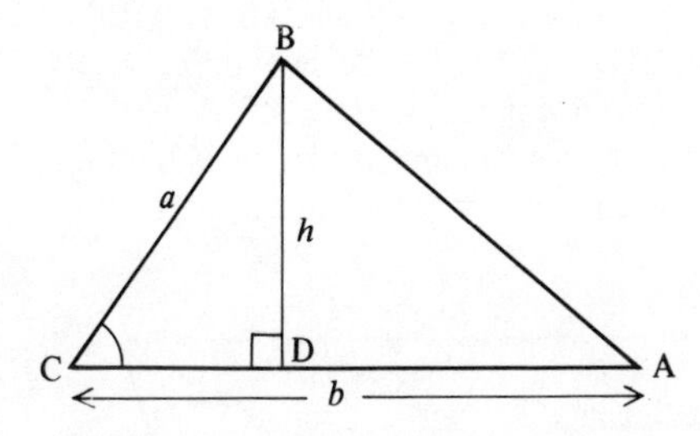

area $= \frac{1}{2} \times b \times h.$

In triangle BCD, $\sin C = \frac{h}{a}$

$\therefore \quad h = a \sin C$

$\therefore$ area of triangle $= \frac{1}{2} \times b \times a \sin C.$

This formula is useful when *two sides* and the *included* angle are known.

Example 1

Find the area of the triangle shown.

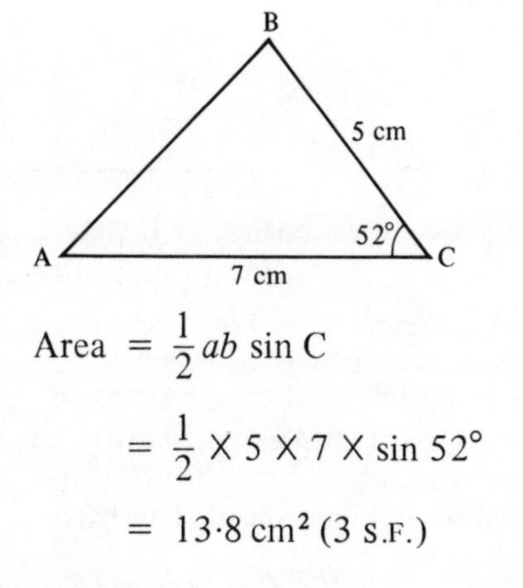

Area $= \frac{1}{2} ab \sin C$

$= \frac{1}{2} \times 5 \times 7 \times \sin 52°$

$= 13{\cdot}8 \text{ cm}^2$ (3 S.F.)

Parallelogram

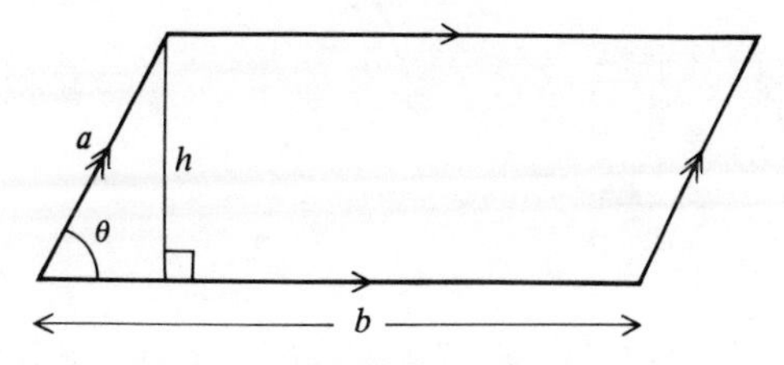

area $= b \times h$
area $= ba \sin \theta$

Example 2

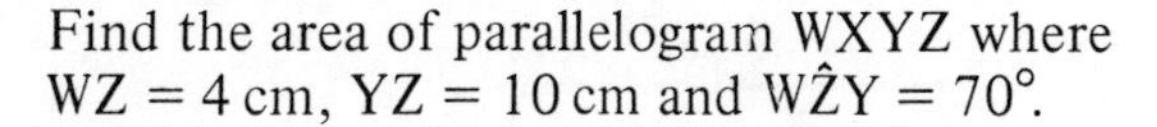

Find the area of parallelogram WXYZ where WZ = 4 cm, YZ = 10 cm and $W\hat{Z}Y = 70°$.

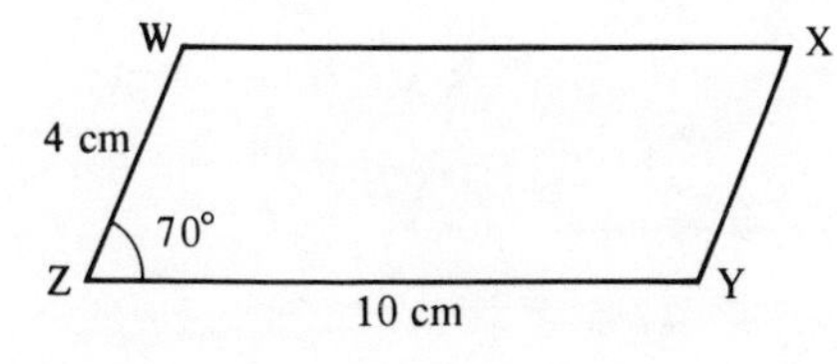

area $= 4 \times 10 \times \sin 70°$
$= 37{\cdot}6 \text{ cm}^2$ (3 S.F.)

Exercise 2

In questions **1** to **12** find the area of $\triangle ABC$ where $AB = c$, $AC = b$ and $BC = a$. (Sketch the triangle in each case.)

1. $a = 7$ cm, $b = 14$ cm, $\hat{C} = 80°$.
2. $b = 11$ cm, $a = 9$ cm, $\hat{C} = 35°$.
3. $c = 12$ m, $b = 12$ m, $\hat{A} = 67{\cdot}2°$.
4. $a = 5$ cm, $c = 6$ cm, $\hat{B} = 11{\cdot}8°$.
5. $b = 4{\cdot}2$ cm, $a = 10$ cm, $\hat{C} = 120°$.
6. $a = 5$ cm, $c = 8$ cm, $\hat{B} = 142°$.
7. $b = 3{\cdot}2$ cm, $c = 1{\cdot}8$ cm, $\hat{B} = 10°$, $\hat{C} = 65°$.
8. $a = 7$ m, $b = 14$ m, $\hat{A} = 32°$, $\hat{B} = 100°$.
9. $a = b = c = 12$ m.
10. $a = c = 8$ m, $\hat{B} = 72°$.
11. $b = c = 10$ cm, $\hat{B} = 32°$.
12. $a = b = c = 0{\cdot}8$ m.

In questions **13** to **20**, find the area of each shape.

13.

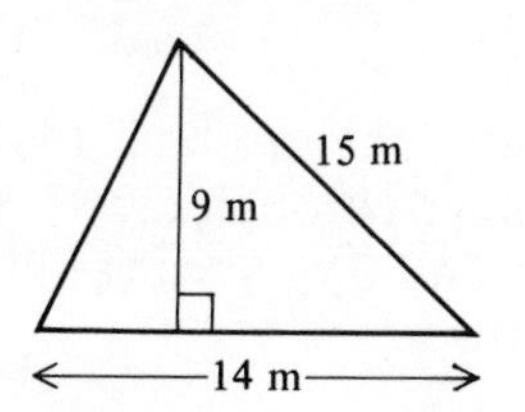

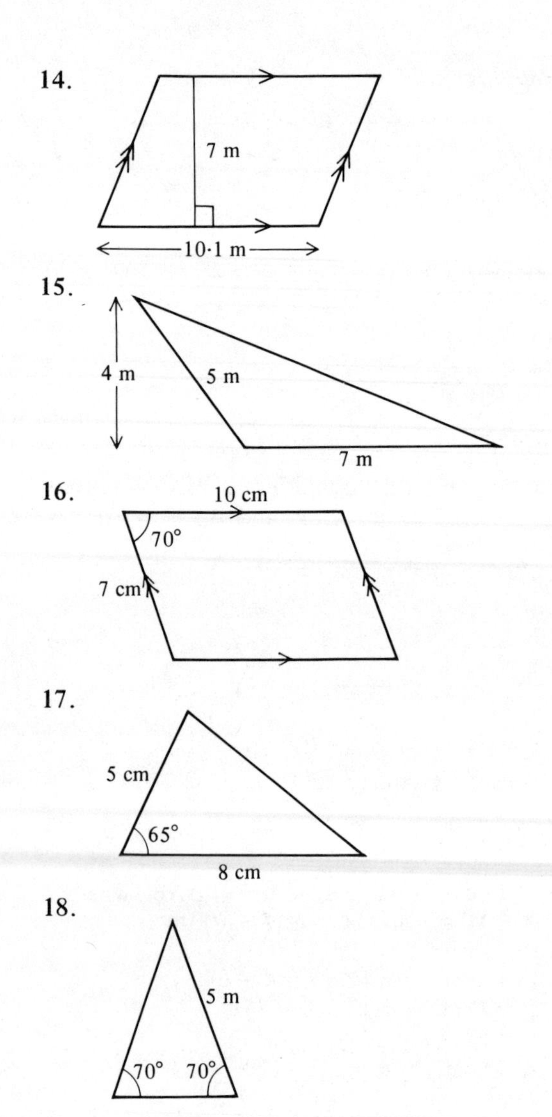

19. Find the area shaded.

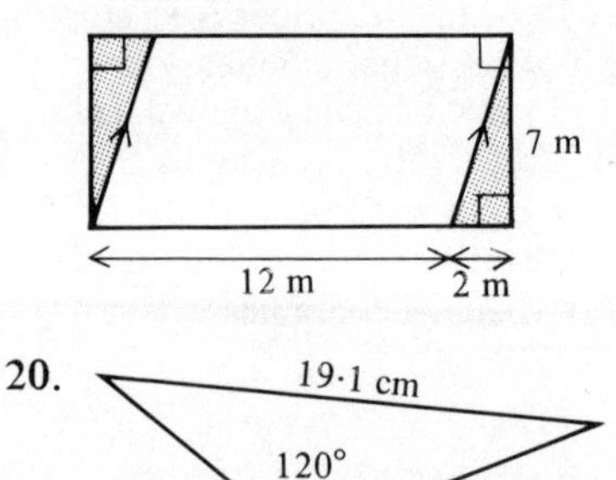

20.

19·1 cm
120°
10 cm
12 cm

21. Find the area of a parallelogram ABCD with AB = 7 m, AD = 20 m and $B\hat{A}D = 62°$.

22. Find the area of a parallelogram ABCD with AD = 7 m, CD = 11 m and $B\hat{A}D = 65°$.

23. In the diagram if $AE = \frac{1}{3}AB$, find the area shaded.

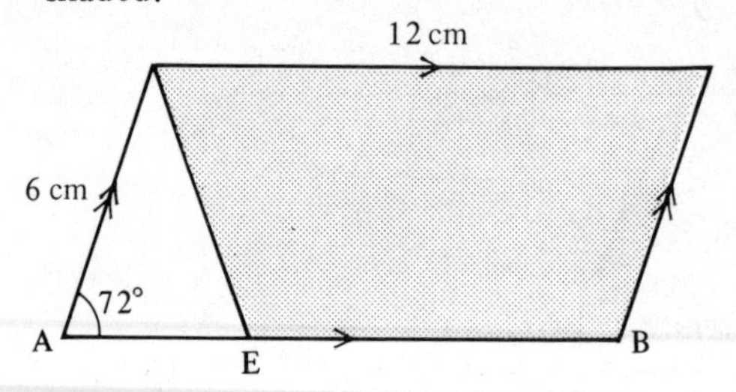

24. The area of an equilateral triangle ABC is 50 cm². Find AB.

25. The area of a triangle ABC is 64 cm². Given AB = 11 cm and BC = 15 cm, find $A\hat{B}C$.

26. The area of a triangle XYZ is 11 m². Given YZ = 7 m and $X\hat{Y}Z = 130°$, find XY.

27. Find the length of a side of an equilateral triangle of area 10·2 m².

28. A rhombus has an area of 40 cm² and adjacent angles of 50° and 130°. Find the length of a side of the rhombus.

3.2 THE CIRCLE

For any circle, the ratio $\left(\frac{\text{circumference}}{\text{diameter}}\right)$ is equal to π.

The value of π is usually taken to be 3·14, but this is not an exact value. Through the centuries, mathematicians have been trying to obtain a better value for π.

For example, in the third century A.D., the Chinese mathematician Liu Hui obtained the value 3·14159 by considering a regular polygon having 3072 sides! Ludolph van Ceulen (1540–1610) worked even harder to produce a value correct to 35 significant figures. He was so proud of his work that he had this value of π engraved on his tombstone.

Electronic computers are now able to calculate the value of π to many thousands of figures, but its value is still not exact. It was shown in 1761 that π is an *irrational number* which, like $\sqrt{2}$ or $\sqrt{3}$ cannot be expressed exactly as a fraction.

The first fifteen significant figures of π can be remembered from the number of letters in each word of the following sentence.

How I need a drink, cherryade of course, after the silly lectures involving Italian kangaroos.

There remain a lot of unanswered questions concerning π, and many mathematicians today are still working on them.

The following formulae should be memorised.

$$\text{circumference} = \pi d$$
$$= 2\pi r$$
$$\text{area} = \pi r^2$$

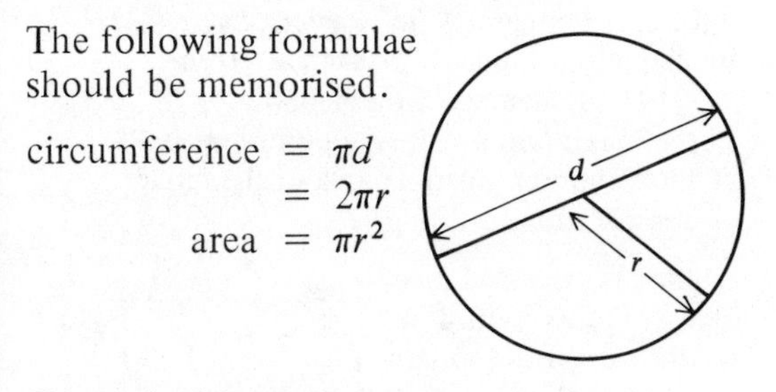

Example 1

Find the circumference and area of a circle of diameter 8 cm. (Take $\pi = 3{\cdot}14$.)

$$\text{Circumference} = \pi d$$
$$= 3{\cdot}14 \times 8$$
$$= 25{\cdot}1 \text{ cm (3 S.F.)}$$
$$\text{Area} = \pi r^2$$
$$= 3{\cdot}14 \times 4^2$$
$$= 50{\cdot}2 \text{ cm}^2 \text{ (3 S.F.)}$$

Example 2

A circle has a circumference of 20 m. Find the radius of the circle. (Take $\pi = 3{\cdot}14$.)

Let the radius of the circle be r m.

$$\text{Circumference} = 2\pi r$$
$$\therefore \quad 2\pi r = 20$$
$$\therefore \quad r = \frac{20}{2\pi}$$
$$r = 3{\cdot}18 \text{ (3 S.F.)}.$$

The radius of the circle is 3·18 m

Example 3

A circle has an area of 45 cm². Find the radius of the circle. (Take $\pi = 3{\cdot}14$.)

Let the radius of the circle be r cm.

$$\pi r^2 = 45$$
$$r^2 = \frac{45}{\pi}$$
$$r = \sqrt{\left(\frac{45}{\pi}\right)} = 3{\cdot}78 \text{ (3 S.F.)}$$

The radius of the circle is 3·78 cm.

Exercise 3

Take $\pi = 3{\cdot}14$ unless otherwise stated.

In questions **1** to **9**, find the circumference of the circle given the radius r or diameter d.

1. $r = 5$ cm
2. $r = 2$ m
3. $r = 0{\cdot}05$ cm
4. $d = 100$ m
5. $d = 5{\cdot}7$ m
6. $r = 5{\cdot}3 \times 10^4$ m
7. $d = 8{\cdot}1 \times 10^{-3}$ m
8. $d = 3{\cdot}14$ km
9. $r = 500\,000$ km

In questions **10** to **18**, find the area of the circle given the radius r or diameter d.

10. $r = 10$ cm
11. $r = 2$ m
12. $r = 0{\cdot}1$ m
13. $d = 200$ m
14. $d = 3 \times 10^4$ m
15. $r = 10^{-3}$ m
16. $d = \frac{1}{2}$ cm
17. $d = 6{\cdot}28$ cm
18. $r = 0{\cdot}02$ cm

In questions **19** to **22**, find the circumference and area of the circle. Take $\pi = \frac{22}{7}$.

19. $r = 7$ cm
20. $d = 28$ cm
21. $r = 35$ m
22. $d = 0{\cdot}7$ cm

In questions **23** to **28**, find the radius of the circle given the circumference c.

23. $c = 62{\cdot}8$ cm
24. $c = 25{\cdot}12$ m
25. $c = 4{\cdot}396$ m
26. $c = 0{\cdot}628$ km
27. $c = 1256$ m
28. $c = 2{\cdot}512 \times 10^4$ m

In questions **29** to **36**, find the radius of the circle given the area A.

29. $A = 12{\cdot}56$ cm²
30. $A = 314$ m²
31. $A = 28{\cdot}26$ m²
32. $A = 50$ cm²
33. $A = 750$ km²
34. $A = 0{\cdot}073$ cm²
35. $A = 8{\cdot}72 \times 10^6$ m²
36. $A = 1$ m²

Exercise 4

Take $\pi = 3{\cdot}14$.

1. Copy and complete the table.

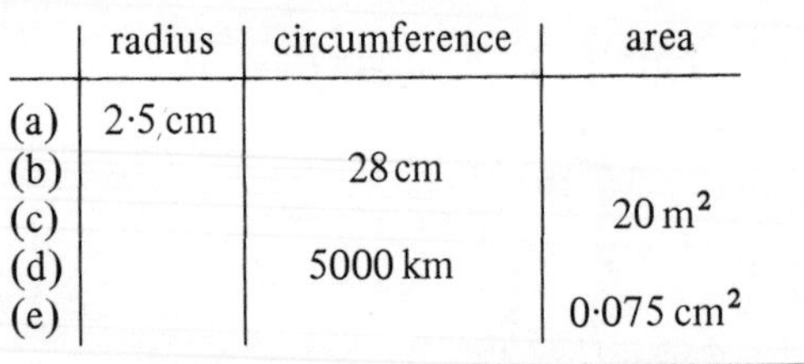

	radius	circumference	area
(a)	2·5 cm		
(b)		28 cm	
(c)			$20\ m^2$
(d)		5000 km	
(e)			$0{\cdot}075\ cm^2$

2. A circle of radius 5 cm is inscribed inside a square as shown. Find the area shaded.

5 cm

3. A circular pond of radius 6 m is surrounded by a path of width 1 m.
   (a) Find the area of the path.
   (b) The path is resurfaced with astroturf which is bought in packs each containing enough to cover an area of $7\ m^2$. How many containers are required?

4. Discs of radius 4 cm are cut from a rectangular plastic sheet of length 84 cm and width 24 cm. How many complete discs can be cut out? Find
   (a) the total area of the discs cut
   (b) the area of the sheet wasted.

5. The tyre of a car wheel has an outer diameter of 30 cm. How many times will the wheel rotate on a journey of 5 km?

6. A golf ball of diameter 1·68 inches rolls a distance of 4 m in a straight line. How many times does the ball rotate completely?
   (1 inch = 2·54 cm)

7. 100 yards of cotton is wound without stretching onto a reel of diameter 3 cm. How many times does the reel rotate?
   (1 yard = 0·914 m. Ignore the thickness of the cotton)

8. A rectangular metal plate has a length of 65 cm and a width of 35 cm. It is melted down and recast into circular discs of the same thickness. How many complete discs can be formed if
   (a) the radius of each disc is 3 cm?
   (b) the radius of each disc is 10 cm?

9. Calculate the radius of a circle whose area is equal to the sum of the areas of three circles of radii 2 cm, 3 cm and 4 cm respectively.

10. The diameter of a circle is given as 10 cm, correct to the nearest cm. Calculate
    (a) the maximum possible circumference
    (b) the minimum possible area of the circle consistent with this data.

11. A square is inscribed in a circle of radius 7 cm. Find
    (a) the area of the square
    (b) the area shaded.

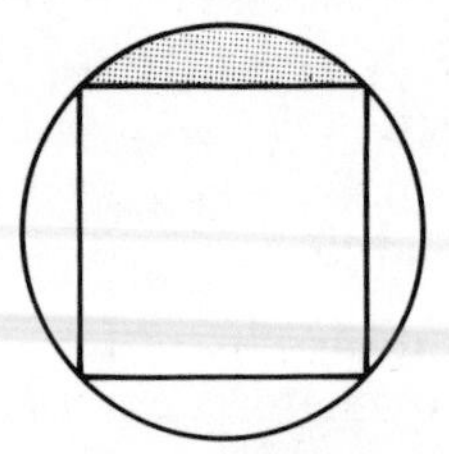

12. An archery target has three concentric regions. The diameter of the regions are in the ratio 1:2:3. Find the ratio of their areas.

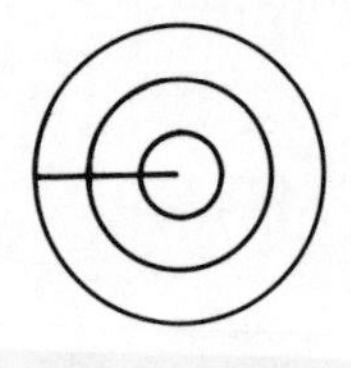

3.3 ARC LENGTH AND SECTOR AREA

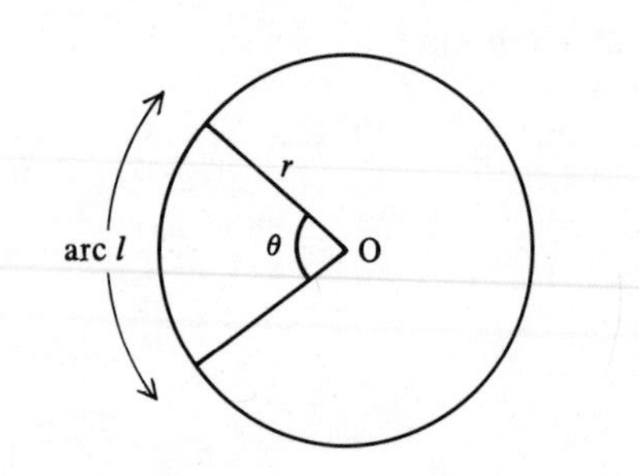

Arc length, $l = \dfrac{\theta}{360} \times 2\pi r$

We take a fraction of the whole circumference depending on the angle at the centre of the circle.

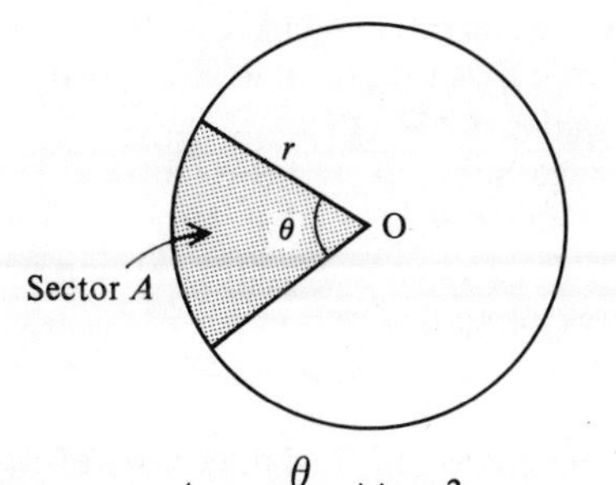

Sector area, $A = \frac{\theta}{360} \times \pi r^2$

We take a fraction of the whole area depending on the angle at the centre of the circle.

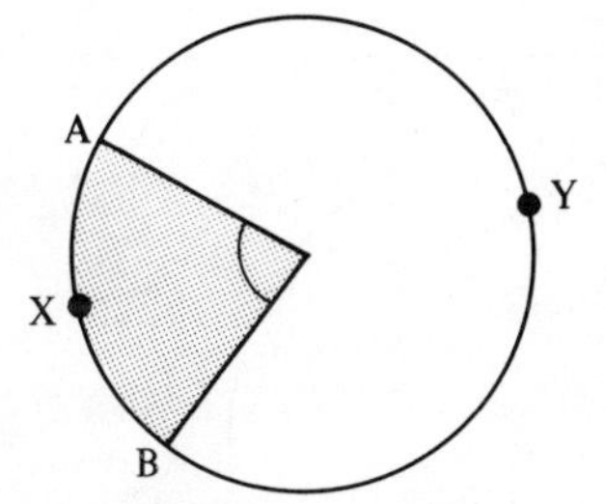

AXB is the *minor* arc.
AYB is the *major* arc.
The minor sector is shaded and the major sector is unshaded.

Example 1

Find the length of an arc which subtends to an angle of 140° at the centre of a circle of radius 12 cm. (Take $\pi = \frac{22}{7}$)

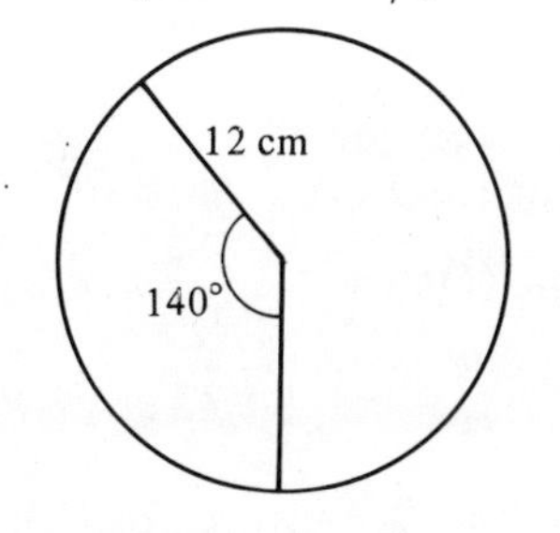

$$\text{Arc length} = \frac{140}{360} \times 2 \times \frac{22}{7} \times 12$$
$$= \frac{88}{3}$$
$$= 29\tfrac{1}{3}\text{ cm.}$$

Example 2

A sector of a circle of radius 10 cm has an area of 25 cm². Find the angle at the centre of the circle. (Take $\pi = 3{\cdot}14$)

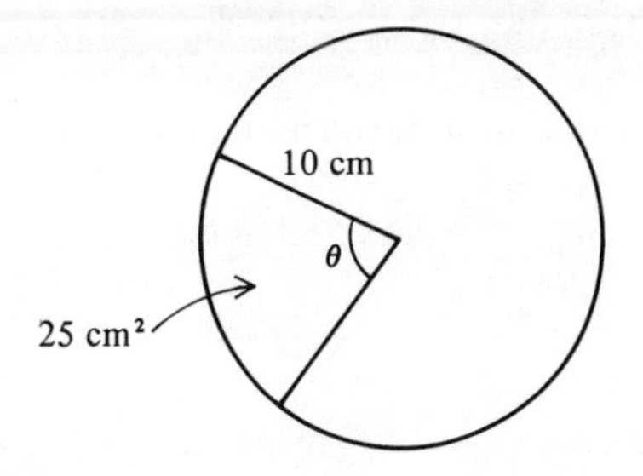

Let the angle at the centre of the circle be $\theta°$.

$$\frac{\theta}{360} \times \pi \times 10^2 = 25$$

$$\therefore \quad \theta = \frac{25 \times 360}{3{\cdot}14 \times 100} = 28{\cdot}7 \text{ (3 S.F.)}$$

The angle at the centre of the circle is 28·7°.

Exercise 5

For questions **1** to **17**, copy and complete the table. ($\pi = 3$)

	radius	angle at centre, θ	arc length	sector area
1.	4 cm	30°	7·5 cm	
2.	10 cm	45°		
3.	2 m	235°		
4.	0·3 m	300°		
5.	20 m	18°		
6.	5 cm		7·5 cm	
7.	14 m		70 cm	
8.	5·2 cm		11 cm	
9.	2 m			2 m²
10.	10 m			75 m²
11.	6 cm			95 cm²
12.		90°	6 cm	
13.		240°	48 cm	
14.		55°	100 m	
15.		72°		15 cm²
16.		135°		162 m²
17.		108°		27 cm²

18. The length of the minor arc AB of a circle, centre O, is 2π cm and the length of the major arc is 22π cm. Find (a) the radius of the circle, (b) the acute angle AOB.

19. The lengths of the minor and major arcs of a circle are 5·2 cm and 19·8 respectively. Find
 (a) the radius of the circle
 (b) the angle subtended at the centre by the minor arc. ($\pi = 3{\cdot}14$)

20. A wheel of radius 10 cm is turning at a rate of 5 revolutions per minute. Calculate
 (a) the angle through which the wheel turns in 1 second
 (b) the distance moved by a point on the rim in 2 seconds. ($\pi = 3{\cdot}14$)

21. The length of an arc of a circle is 12 cm. The corresponding sector area is 108 cm^2. Find
 (a) the radius of the circle
 (b) the angle subtended at the centre of the circle by the arc. (Take $\pi = 3$)

22. The length of an arc of a circle is 7·5 cm. The corresponding sector area is 37·5 cm^2. Find
 (a) the radius of the circle
 (b) the angle subtended at the centre of the circle by the arc. (Take $\pi = 3$)

3.4 CHORD OF A CIRCLE

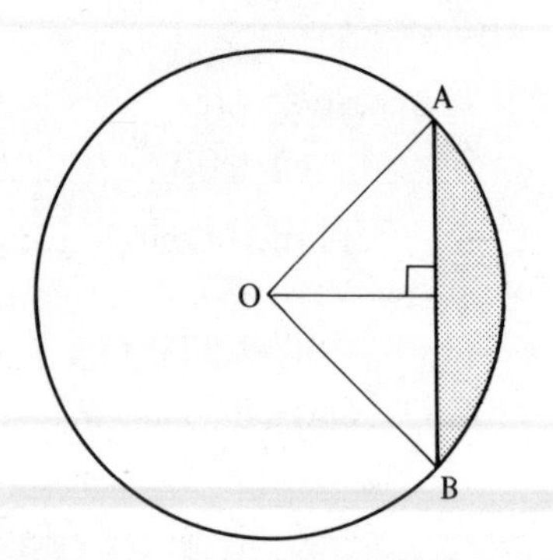

The line AB is a chord. The area of a circle cut off by a chord is called a *segment*. In the diagram the *minor* segment is shaded and the *major* segment is unshaded.

(a) The line from the centre of a circle to the mid-point of a chord *bisects* the chord at *right angles*.
(b) The line from the centre of a circle to the mid-point of a chord bisects the angle subtended by the chord at the centre of the circle.

Example 1

XY is a chord of length 12 cm of a circle of radius 10 cm, centre O. Calculate
(a) the angle XOY
(b) the area of the minor segment cut off by the chord XY.

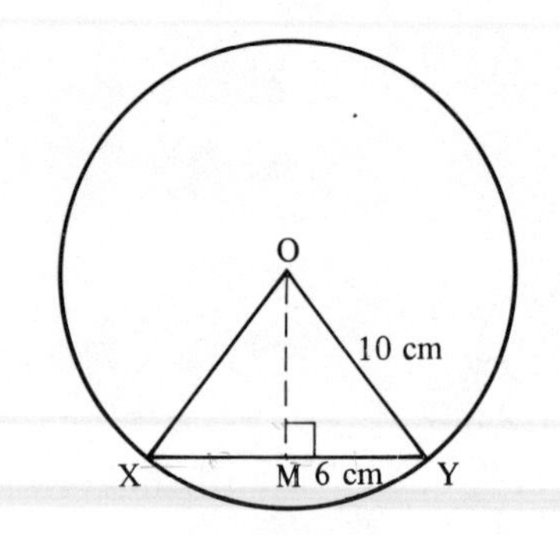

Let the mid-point of XY be M

$\therefore \quad MY = 6\text{ cm}$

$\sin M\hat{O}Y = \dfrac{6}{10}$

$\therefore \quad M\hat{O}Y = 36{\cdot}87°$

$\therefore \quad X\hat{O}Y = 2 \times 36{\cdot}87 = 73{\cdot}74°$

area of minor segment = area of sector XOY − area of ΔXOY

area of sector XOY $= \dfrac{73{\cdot}74}{360} \times \pi \times 10^2 = 64{\cdot}32\text{ cm}^2.$

area of $\Delta XOY = \dfrac{1}{2} \times 10 \times 10 \times \sin 73{\cdot}74° = 48{\cdot}00\text{ cm}^2$

$\therefore$ Area of minor segment $= 64{\cdot}32 - 48{\cdot}00 = 16{\cdot}3\text{ cm}^2$ (3 S.F.)

Example 2

The chord of a circle subtends an angle of 120° at the centre of the circle and cuts off a minor segment of area 60 cm². Find the radius of the circle. (Take $\pi = 3{\cdot}14$)

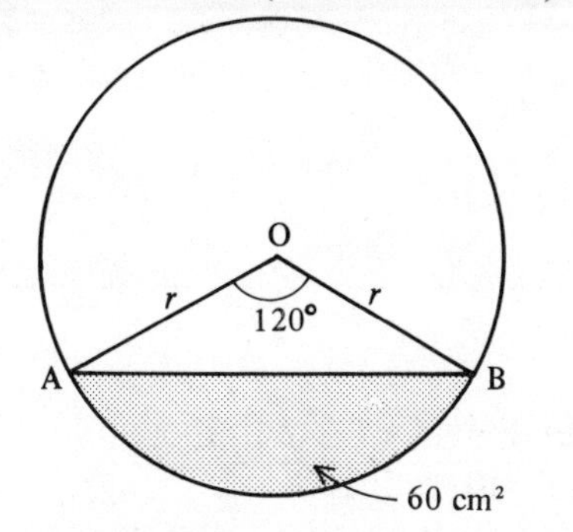

Let the radius of the circle be r cm.

area of minor segment

= area of sector AOB − area of △AOB

$$\text{area of sector AOB} = \frac{120}{360} \times \pi \times r^2$$

$$\text{area of } \triangle\text{AOB} = \frac{1}{2} \times r \times r \times \sin 120$$

$$\therefore \quad \frac{120}{360} \times \pi r^2 - \frac{1}{2} r^2 \sin 120 = 60$$

$$r^2\left(\frac{\pi}{3} - \frac{\sin 120}{2}\right) = 60$$

$$r^2(0{\cdot}6137) = 60$$

$$r^2 = \frac{60}{0{\cdot}6137}$$

$$r = 9{\cdot}89 \text{ (3 S.F.)}$$

The radius of the circle is 9·89 cm.

Exercise 6

Take $\pi = 3{\cdot}14$.

1. The chord AB subtends an angle of 130° at the centre O. The radius of the circle is 8 cm. Find
   (a) the length of AB,
   (b) the area of sector OAB,
   (c) the area of triangle OAB,
   (d) the area of the minor segment (shown shaded).

2. Copy the table and find the quantities marked * (take $\pi = 3{\cdot}14$).

	radius	angle at centre	chord length	arc length	area of minor segment
(a)	6 cm	70°	*	*	
(b)	14 m	104°	*	*	
(c)	5 cm	80°		*	*
(d)	8 cm	105°	*		*
(e)	10 cm	*	10 cm	*	*
(f)	5 cm	*	8 cm		*
(g)	6 cm	*	9 cm	*	*
(h)	*	100°	10 cm		*
(i)	*	63·2°	8 m	*	
(j)	15 cm	118°	*	*	
(k)	100 cm	*	173·2 cm		*
(l)	*	74·8°	12 cm	*	
(m)	*	90°			20 cm²
(n)	*	30°			35 cm²
(o)	*	150°	*		114 cm²

3. How far is a chord of length 8 cm from the centre of a circle of radius 5 cm?

4. How far is a chord of length 9 cm from the centre of a circle of radius 6 cm?

5. The diagram shows the cross section of a cylindrical pipe with water lying in the bottom.
   (a) If the maximum depth of the water is 2 cm and the radius of the pipe is 7 cm, find the area shaded.
   (b) What is the *volume* of water in a length of 30 cm?

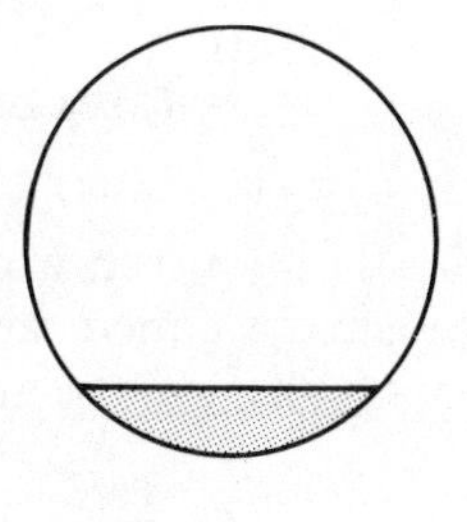

6. An equilateral triangle is inscribed in a circle of radius 10 cm. Find
   (a) the area of the triangle.
   (b) the area shaded.

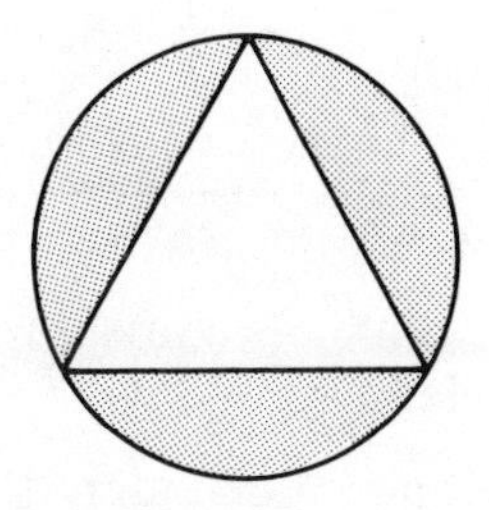

7. An equilateral triangle is inscribed in a circle of radius 18·8 cm. Find
   (a) the area of the triangle.
   (b) the area of the three segments surrounding the triangle.

8. A regular hexagon is circumscribed by a circle of radius 6 cm. Find the area shaded.

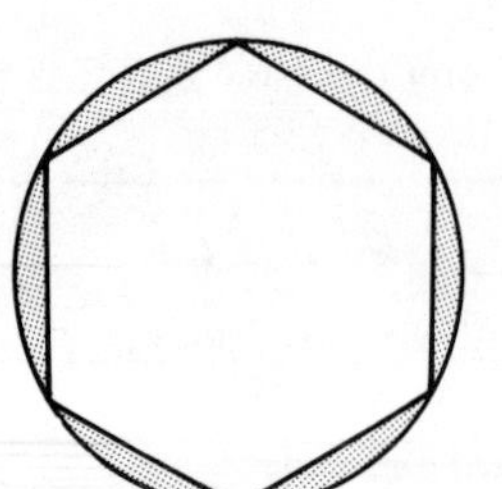

9. A regular octagon is circumscribed by a circle of radius r cm. Find the area enclosed between the circle and the octagon. (Give the answer in terms of r.)

3.5 VOLUME

Prism

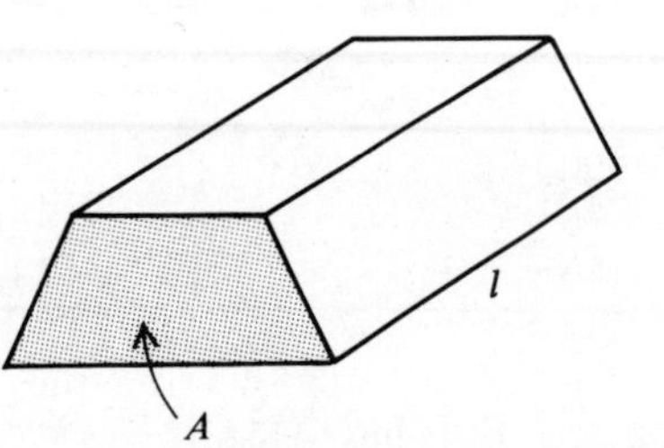

A prism is an object with the same cross section throughout its length.

Volume of prism

$$= \text{(area of cross section)} \times \text{length}$$
$$= A \times l.$$

A *cuboid* is a prism whose six faces are all rectangles. A cube is a special case of a cuboid in which all six faces are squares.

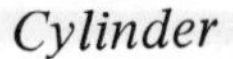

Cylinder

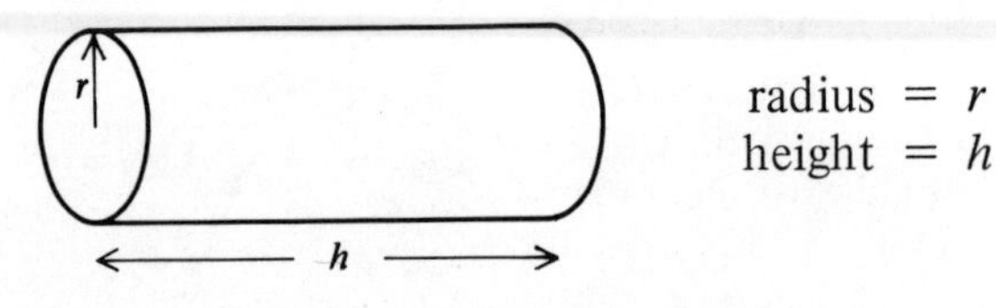

A cylinder is a prism whose cross section is a circle.

Volume of cylinder

$$= \text{(area of cross section)} \times \text{length}$$

$$\text{Volume} = \pi r^2 h$$

Example 1

Calculate the height of a cylinder of volume 500 cm^3 and base radius 8 cm.
Let the height of the cylinder be h cm.

$$\pi r^2 h = 500$$
$$3{\cdot}14 \times 8^2 \times h = 500$$
$$h = \frac{500}{3{\cdot}14 \times 64} = 2{\cdot}49 \text{ (3 S.F.)}$$

The height of the cylinder is 2·49 cm.

Exercise 7

1. Calculate the volume of the following prisms:

	cross section	length
(a)	rectangle 4 cm × 5 cm	8 cm
(b)	triangle: base = 12 cm, height = 5 cm	10 cm
(c)	equilateral triangle of side 7 cm	12 cm
(d)	trapezium: parallel sides 5 cm and 3 cm, height 6 cm	10 cm
(e)	kite: diagonals 9 cm and 12 cm	1 m
(f)	semicircle of radius 15 cm	2 m

2. Calculate the volume of the following cylinders (take $\pi = 3{\cdot}14$).
   (a) $r = 4$ cm, $h = 10$ cm
   (b) $r = 11$ m, $h = 2$ m
   (c) $r = 2{\cdot}1$ cm, $h = 0{\cdot}9$ cm
   (d) $r = 0{\cdot}01$ m, $h = 2$ cm
3. Find the height of a cylinder of volume 200 cm^3 and radius 4 cm.
4. Find the length of a cylinder of volume 2 litres and radius 10 cm.
5. Find the radius of a cylinder of volume 45 cm^3 and length 4 cm.
6. A prism has volume 100 cm^3 and length 8 cm. If the cross section is an equilateral triangle, find the length of a side of the triangle.

7. When 3 litres of oil is removed from an upright cylindrical can the level falls by 10 cm. Find the radius of the can.
8. A solid cylinder of radius 4 cm and length 8 cm is melted down and recast into a solid cube. Find the side of the cube.
9. A solid rectangular block of copper 5 cm by 4 cm by 2 cm is drawn out to make a cylindrical wire of diameter 2 mm. Calculate the length of the wire.
10. Water flows through a circular pipe of internal diameter 3 cm at a speed of 10 cm/s. If the pipe is full, how much water issues from the pipe in one minute? (answer in litres)
11. Water issues from a hose-pipe of internal diameter 1 cm at a rate of 5 litres per minute. At what speed is the water flowing through the pipe?
12. A cylindrical metal pipe has external diameter of 6 cm and internal diameter of 4 cm. Calculate the volume of metal in a pipe of length 1 m. If 1 cm^3 of the metal weighs 8 g, find the weight of the pipe.
13. For two cylinders A and B, the ratio of lengths is 3 : 1 and the ratio of diameters is 1 : 2. Calculate the ratio of their volumes.
14. A well-trained hen can lay eggs which are either perfect cylinders of diameter and length 4 cm, or perfect cubes of side 5 cm. Which eggs have the greater volume, and by how much? (Take $\pi = 3$)
15. Mr Gibson decided to build a garage and began by calculating the number of bricks required. The garage was to be 6 m by 4 m and 2·5 m in height. Each brick measures 22 cm by 10 cm by 7 cm. Mr Gibson estimated that he would need about 40 000 bricks. Is this a reasonable estimate?
16. A cylindrical can of internal radius 20 cm stands upright on a flat surface. It contains water to a depth of 20 cm. Calculate the rise in the level of the water when a brick of volume 1·5 litres is immersed in the water.
17. A cylindrical tin of height 15 cm and radius 4 cm is filled with sand from a rectangular box. How many times can the tin be filled if the dimensions of the box are 50 cm by 40 cm by 20 cm?
18. Rain which falls onto a flat rectangular surface of length 6 m and width 4 m is collected in a cylinder of internal radius 20 cm. What is the depth of water in the cylinder after a storm in which 1 cm of rain fell?

Pyramid

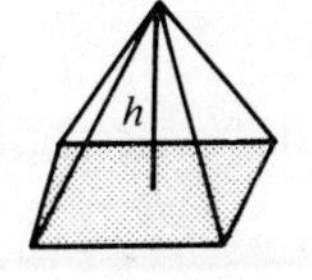

$$\text{Volume} = \frac{1}{3}\,(\text{base area}) \times \text{height}.$$

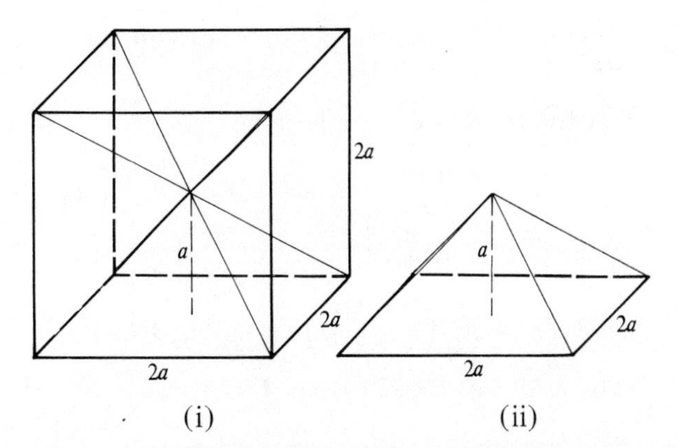

Figure (i) shows a cube of side $2a$ broken down in down into six pyramids of height a as shown in figure (ii).

If the volume of each pyramid is V,

then $6V = 2a \times 2a \times 2a$

$V = \frac{1}{6} \times (2a)^2 \times 2a$

so $V = \frac{1}{3} \times (2a)^2 \times a$

$V = \frac{1}{3}\,(\text{base area}) \times \text{height}.$

Cone

Volume $= \frac{1}{3}\pi r^2 h$

(note the similarity with the pyramid)

h

r

Sphere

Volume $= \frac{4}{3}\pi r^3$

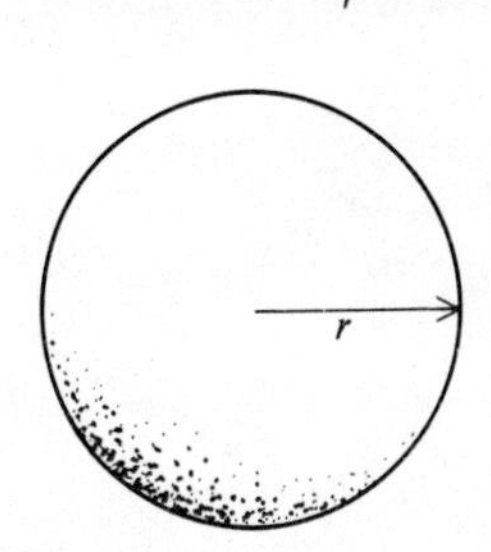

Example 2

A pyramid has a square base of side 5 m and vertical height 4 m. Find its volume.

Volume of pyramid $= \frac{1}{3}(5 \times 5) \times 4 = 33\frac{1}{3}$ m^3.

Example 3

Calculate the radius of a sphere of volume 500 cm^3. ($\pi = 3{\cdot}14$)

Let the radius of the sphere be r cm

$$\frac{4}{3}\pi r^3 = 500$$

$$r^3 = \frac{3 \times 500}{4\pi}$$

$$r = \sqrt[3]{\left(\frac{3 \times 500}{4\pi}\right)} = 4{\cdot}92 \text{ (3 S.F.)}$$

The radius of the sphere is 4·92 cm.

Exercise 8

(Take $\pi = 3{\cdot}14$ unless otherwise stated.)
Find the volumes of the following objects:

1. cone: height = 5 cm, radius = 2 cm
2. sphere: radius = 5 cm
3. sphere: radius = 10 cm
4. cone: height = 6 cm, radius = 4 cm
5. sphere: diameter = 8 cm
6. cone: height = x cm, radius = $2x$ cm
7. sphere: radius = 0·1 m
8. cone: height = $\frac{1}{\pi}$ cm, radius = 3 cm
9. pyramid: rectangular base 7 cm by 8 cm; height = 5 cm
10. pyramid: square base of side 4 m, height = 9 m
11. pyramid: equilateral triangular base of side = 8 cm, height = 10 cm
12. Find the volume of a hemisphere of radius 5 cm.
13. A cone is attached to a hemisphere of radius 4 cm. If the total height of the object is 10 cm, find its volume.

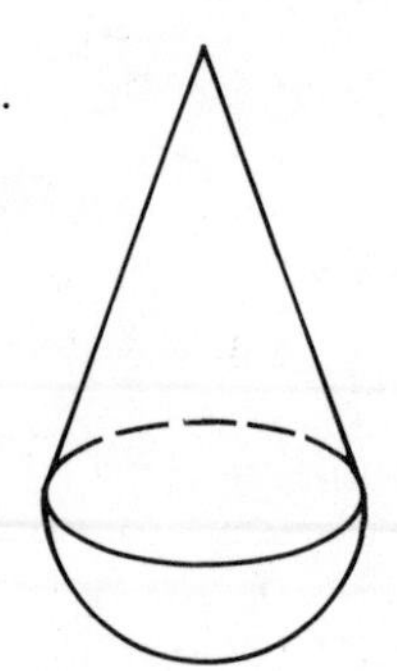

14. A toy consists of a cylinder of diameter 6 cm 'sandwiched' between a hemisphere and a cone of the same diameter. If the cone is of height 8 cm and the cylinder is of height 10 cm, find the total volume of the toy.

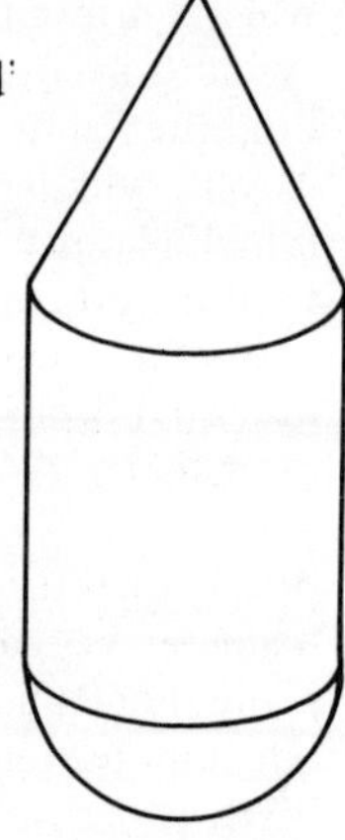

15. Find the height of a pyramid of volume 20 m^3 and base area 12 m^2.
16. Find the radius of a sphere of volume 60 cm^3.
17. Find the height of a cone of volume 2·5 litre and radius 10 cm.
18. Six square-based pyramids fit exactly onto the six faces of a cube of side 4 m. If the volume of the object formed is 256 cm^3, find the height of each of the pyramids.
19. A solid metal cube of side 6 cm is recast into a solid sphere. Find the radius of the sphere.
20. A hollow spherical vessel has internal and external radii of 6 cm and 6·4 cm respectively. Calculate the weight of the vessel if it is made of metal of density 10 g/cm^3.
21. Water is flowing into an inverted cone, of diameter and height 30 cm, at a rate of 4 litres per minute. How long, in seconds, will it take to fill the cone?
22. A solid metal sphere is recast into many smaller spheres. Calculate the number of the smaller spheres if the initial and final radii are as follows:
    (a) initial radius = 10 cm, final radius = 2 cm
    (b) initial radius = 7 cm, final radius = $\frac{1}{2}$ cm
    (c) initial radius = 1 m, final radius = $\frac{1}{3}$cm.
23. Spherical balls are immersed in water contained in vertical cylinders of various radii. Assuming the water covers the balls, calculate the rise in the water level in the following cases:
    (a) One sphere radius 3 cm, cylinder radius 10 cm
    (b) One sphere radius 2 cm, cylinder radius 5 cm
    (c) Ten spheres radius 3 cm, cylinder radius 10 cm.

24. Spherical balls are immersed in water contained in vertical cylinders. The rise in water level is measured in order to calculate the radii of the spherical balls. Calculate the radii of the balls in the following cases:
    (a) cylinder of radius 10 cm, water level rises 4 cm (one ball)
    (b) cylinder of radius 100 cm, water level rises 8 cm (one ball)
    (c) cylinder of radius 4 cm, water level rises 5 cm (three balls).

25. One corner of a solid cube of side 8 cm is removed by cutting through the mid-points of three adjacent sides. Calculate the volume of the piece removed.

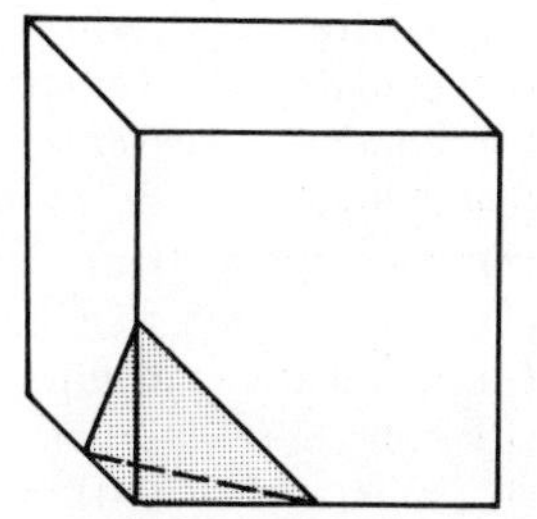

26. The cylindrical end of a pencil is sharpened to produce a perfect cone at the end with no overall loss of length. If the diameter of the pencil is 1 cm, and the cone is of length 2 cm, calculate the volume of the shavings.

27. Metal spheres of radius 2 cm are packed into a rectangular box of internal dimensions 16 cm × 8 cm × 8 cm. When 16 spheres are packed the box is filled with a preservative liquid. Find the volume of this liquid.

28. The diagram shows the cross section of an inverted cone of height MC = 12 cm. If AB = 6 cm and XY = 2 cm, use similar triangles to find the length NC. Hence find the volume of the cone of height NC.

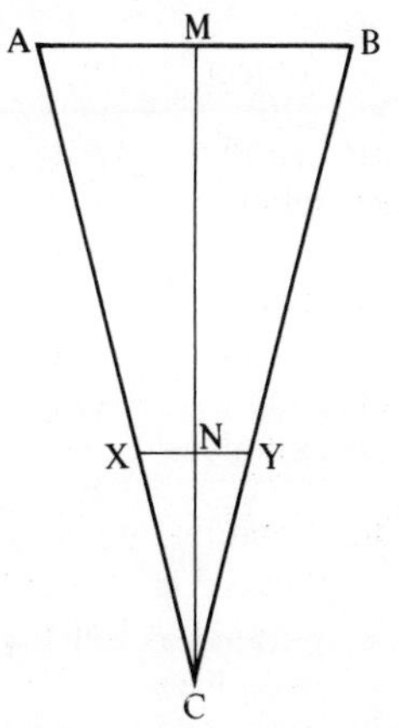

29. An inverted cone of height 10 cm and base radius 6·4 cm contains water to a depth of 5 cm, measured from the vertex. Calculate the volume of water in the cone.

30. An inverted cone of height 15 cm and base radius 4 cm contains water to a depth of 10 cm. Calculate the volume of water in the cone.

31. An inverted cone of height 12 cm and base radius 6 cm contains 20 cm^3 of water. Calculate the depth of water in the cone, measured from the vertex.

32. A frustrum is a cone with 'the end chopped off'. A bucket in the shape of a frustrum as shown has diameters of 10 cm and 4 cm at its ends and a depth of 3 cm. Calculate the volume of the bucket.

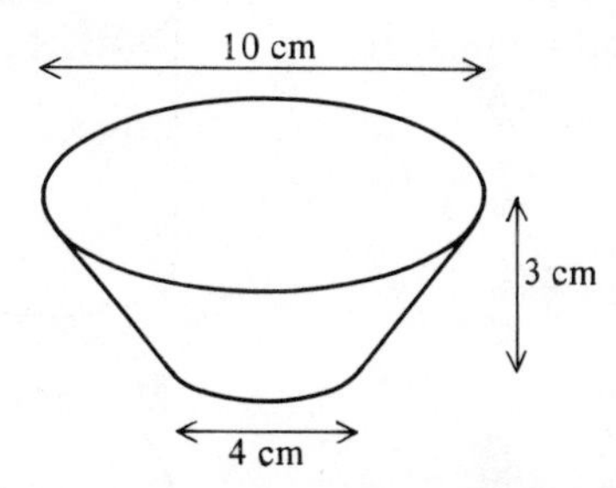

33. Find the volume of a frustrum with end diameters of 60 cm and 20 cm and a depth of 40 cm.

34. Find the volume of a regular tetrahedron of side 20 cm.

35. Find the volume of a regular tetrahedron of side 35 cm.

36. Day's formula for the volume of *any* tetrahedron of sides a, b, c, d, e, f is given below.

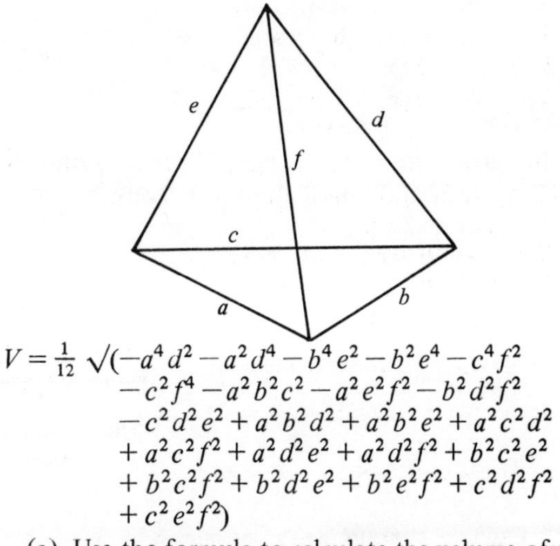

$$V = \tfrac{1}{12}\sqrt{(-a^4d^2 - a^2d^4 - b^4e^2 - b^2e^4 - c^4f^2 - c^2f^4 - a^2b^2c^2 - a^2e^2f^2 - b^2d^2f^2 - c^2d^2e^2 + a^2b^2d^2 + a^2b^2e^2 + a^2c^2d^2 + a^2c^2f^2 + a^2d^2e^2 + a^2d^2f^2 + b^2c^2e^2 + b^2c^2f^2 + b^2d^2e^2 + b^2e^2f^2 + c^2d^2f^2 + c^2e^2f^2)}$$

(a) Use the formula to calculate the volume of a regular tetrahedron of side 1 cm.

(b) Confirm that your result is $\dfrac{1}{(20)^3}$ of the result for the regular tetrahedron of side 20 cm.

(c) Use the formula to calculate the volume of a tetrahedron of sides 3 cm, 4 cm, 5 cm, 4 cm, 5 cm, $\sqrt{32}$ cm.

3.6 SURFACE AREA

We are concerned here with the surface areas of the *curved* parts of cylinders, spheres and cones. The areas of the plane faces are easier to find.

(a) Cylinder
Curved surface area
$= 2\pi rh$.

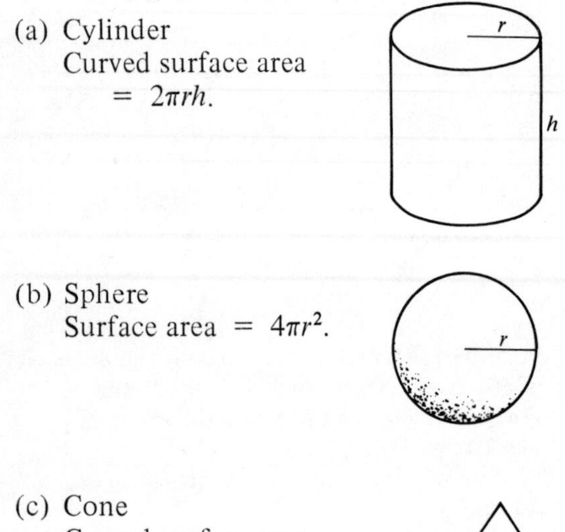

(b) Sphere
Surface area $= 4\pi r^2$.

(c) Cone
Curved surface area
$= \pi rl$
where l is the slant height.

Example 1

Find the *total* surface area of a solid cone of radius 4 cm and vertical height 3 cm.

Let the slant height of the cone be l cm.

$$l^2 = 3^2 + 4^2$$
$$l = 5$$

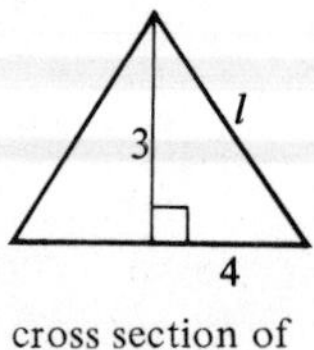

cross section of cone

Curved surface area $= \pi \times 4 \times 5$
$= 20\pi \text{ cm}^2$

Area of end face $= \pi \times 4^2 = 16\pi \text{ cm}^2$

$\therefore$ Total surface area $= 20\pi + 16\pi$
$= 36\pi \text{ cm}^2$
$= 113 \text{ cm}^2$ to 3 S.F.

Exercise 9

(Take $\pi = 3{\cdot}14$ unless otherwise instructed.)

1. Copy the table and find the quantities marked $*$. (Leave π in your answers.)

	solid object	radius	vertical height	curved surface area	total surface area
(a)	sphere	3 cm		*	
(b)	cylinder	4 cm	5 cm	*	*
(c)	cone	6 cm	8 cm	*	
(d)	cylinder	0·7 m	1 m	*	*
(e)	sphere	10 m		*	
(f)	cone	5 cm	12 cm	*	
(g)	cylinder	6 mm	10 mm		*
(h)	cone	2·1 cm	4·4 cm	*	
(i)	sphere	0·01 m		*	
(j)	hemisphere	7 cm		*	*

2. Find the radius of a sphere of surface area 34 cm^2.
3. Find the slant height of a cone of curved surface area 20 cm^2 and radius 3 cm.
4. Find the height of a solid cylinder of radius 1 cm and *total* surface area 28 cm^2.
5. Copy the table and find the quantities marked $*$. (Take $\pi = 3$.)

	object	radius	vertical height	curved surface area	total surface area
(a)	cylinder	4 cm	*	72 cm^2	
(b)	sphere	*		192 cm^2	
(c)	cone	4 cm	*	60 cm^2	
(d)	sphere	*		0·48 m^2	
(e)	cylinder	5 cm	*		330 cm^2
(f)	cone	6 cm	*		225 cm^2
(g)	cylinder	2 m	*		108 m^2

6. A solid wooden cylinder of height 8 cm and radius 3 cm is cut in two along a vertical axis of symmetry. Calculate the total surface area of the two pieces.
7. A cone of radius 3 cm and slant height 6 cm is cut into four identical pieces. Calculate the total surface area of the four pieces.
8. A tin of paint covers a surface area of 60 m^2 and costs £4·50. Find the cost of painting the outside surface of a hemispherical dome of radius 50 m. (Just the curved part.)

9. A solid cylinder of height 10 cm and radius 4 cm is to be plated with material costing £11 per cm^2. Find the cost of the plating.
10. Find the volume of a sphere of surface area 100 cm^2.
11. Find the surface area of a sphere of volume 28 cm^3.
12. Calculate the total surface area of the combined cone/cylinder/hemisphere.

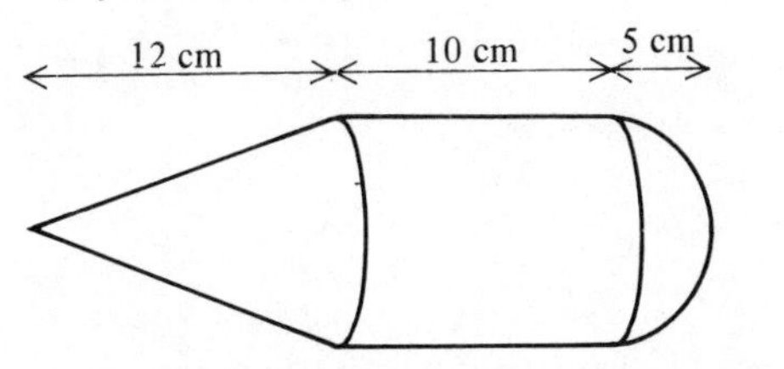

13. A man is determined to spray the entire surface of the Earth (including the oceans) with a revolutionary new weed killer. If it takes him 10 seconds to spray 1 m^2, how long will it take to spray the whole world? (radius of the Earth = 6370 km; ignore leap years.)
14. An inverted cone of vertical height 12 cm and base radius 9 cm contains water to a depth of 4 cm. Find the area of the interior surface of the cone not in contact with the water.
15. A circular paper of radius 20 cm is cut in half and each half is made into a hollow cone by joining the straight edges. Find the slant height and base radius of each cone.
16. A golf ball has a diameter of 4·1 cm and the surface has 150 dimples of radius 2 mm. Calculate the total surface area which is exposed to the surroundings. (Assume the 'dimples' are hemispherical.)

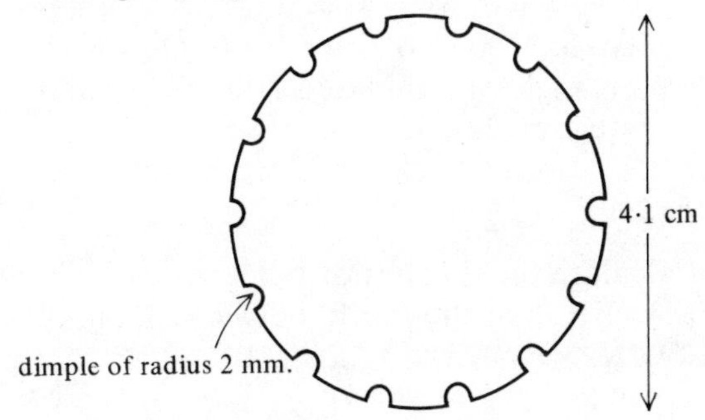

3.7 LATITUDE AND LONGITUDE

The position of a point on the Earth's surface is given by its latitude and longitude.

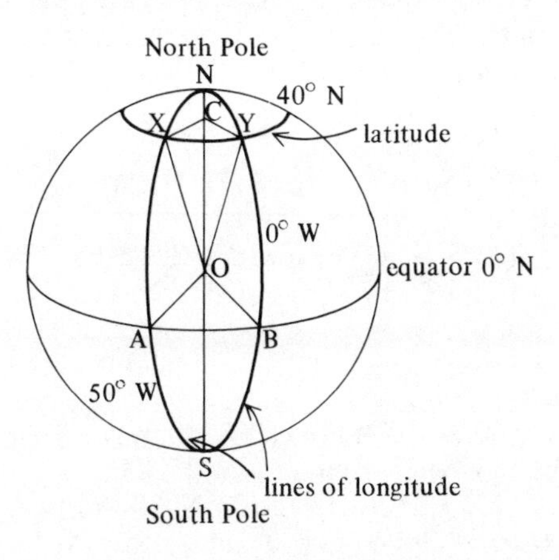

(a) 'Lines' of longitude are *great* circles which pass through the North and South poles.

The line of longitude through N, Y, B and S is the 0° line of longitude and is the reference line for all other lines of longitude.

The longitude of point A is given by the angle AOB. The longitude of point X is given by angle X$\hat{C}$Y. (A$\hat{O}$B = X$\hat{C}$Y since A and X are on the same line of longitude.) If A and X have longitude 50°W, then A$\hat{O}$B = X$\hat{C}$Y = 50°.

(b) The latitude of point Y is given by the angle YOB. If X and Y are on the 40°N parallel of latitude, then Y$\hat{O}$B = X$\hat{O}$A = 40°.

(c) A *great* circle is a circle with radius equal to that of the Earth. Thus, all lines of longitude and the equator are great circles.

(d) Nautical miles.

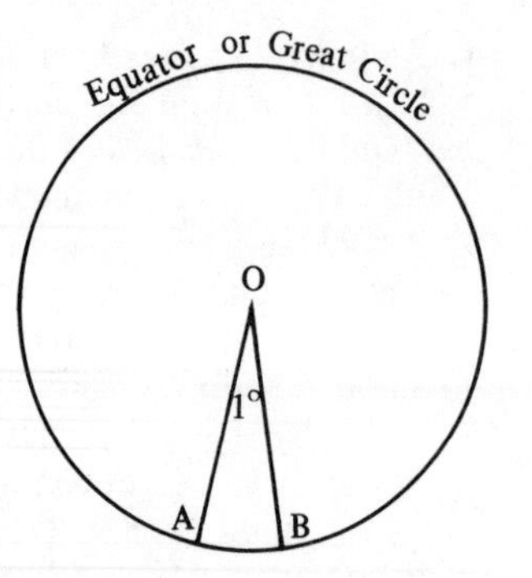

There are 60 nautical miles along the arc AB, where $A\hat{O}B = 1°$.
An arc length of 1 nautical mile subtends an angle of 1 minute at the centre of a great circle.

Example 1

Calculate the shortest distance, measured over the surface of the Earth, between the points A (50°N, 10°W) and B (20°S, 10°W)
(a) in km
(b) in nautical miles.
(Radius of the Earth = 6370 km.)

The shortest distance is the distance measured along a line of longitude from North to South.

(a) Consider a 'side view' of the Earth.

$A\hat{O}B = 70°$

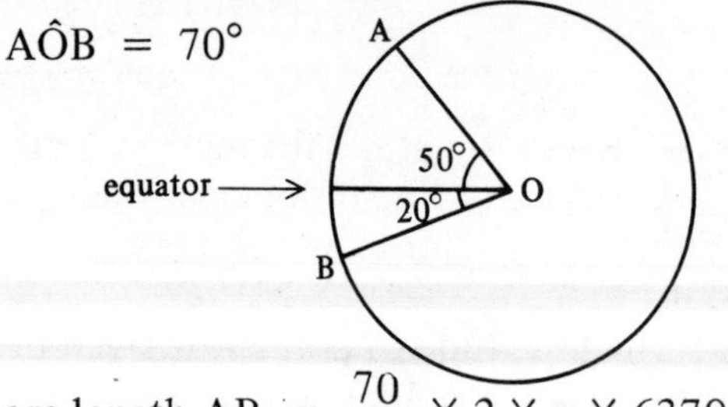

$$\text{arc length AB} = \frac{70}{360} \times 2 \times \pi \times 6370$$

$$AB = 7780\text{ km}.$$

The distance from A to B is 7780 km.

(b) $A\hat{O}B = 70°$.
$= 70 \times 60 \text{ minutes} = 4200'$

A 'line' of longitude is a great circle.

$\therefore$ arc AB = 4200 nautical miles

The distance from A to B is 4200 nautical miles.

Example 2

Find the shortest distance in km over the North pole from point A (65°N, 30°W) to point B (50°N, 150°E).

Since the longitudes of A and B differ by 180°, the points A and B are on the *same* line of longitude.

Take a side view of the Earth:

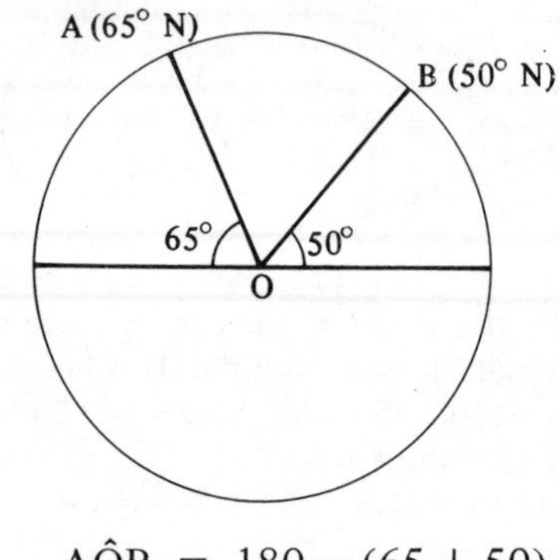

$$A\hat{O}B = 180 - (65 + 50)$$
$$A\hat{O}B = 65°$$

$$\text{Arc length AB} = \frac{65}{360} \times 2 \times \pi \times 6370$$
$$= 7220\text{ km (3 S.F.)}.$$

The shortest distance between A and B is 7220 km.

Exercise 10

1. Find the distance in km between each pair of points. (Take the radius of the Earth to be 6370 km and $\pi = \frac{22}{7}$.)
   (a) (30°N, 10°W) to (10°S, 10°W)
   (b) (75°N, 15°E) to (5°N, 15°E)
   (c) (27°S, 160°W) to (84°S, 160°W)
   (d) (17°S, 19°E) to (58°N, 19°E)
   (e) (25·5°N, 17°E) to (12·1°S, 17°E)
   (f) (58·3°N, 17·1°E) to (31·7°N, 17·1°E).

2. Find the distance in nautical miles between each pair of points.
   (a) (60°N, 30°W) to (10°S, 30°W)
   (b) (55°N, 25°E) to (15°N, 25°E)
   (c) (17° 32′N, 20°W) to (0°N, 20°W)
   (d) (0°N, 35°W) to (10°25′S, 35°W)
   (e) (14·5°N, 160°E) to (20°S, 160°E)
   (f) (72·6°N, 100°N) to (61·1°N, 100°W)

3. Find the shortest distance in km over the North or South pole between each pair of points (radius of Earth 6370 km, $\pi = 3{\cdot}14$).
   (a) (70°N, 50°W) to (60°N, 130°E)
   (b) (50°N, 10°W) to (66°N, 170°E)
   (c) (84°S, 17°E) to (75°S, 163°W)
   (d) (10°S, 119°W) to (0°, 61°E)
   (e) (17° 20′N, 97°W) to (60°N, 83°E)
   (f) (54° 10′S, 10°E) to (67°50′S, 170°W)
4. Repeat question **3**, finding the distance in nautical miles.
5. Find the distance in km between the following pairs of points on the equator:
   (a) (0°, 65°W) to (0°, 87°W)
   (b) (0°, 11°W) to (0°, 25°E)
   (c) (0°, 117°E) to (0°, 94° 30′E)
   (d) (0°, 17° 25′W) to (0°, 25° 5′E)
6. An aircraft flies due North or South from each of the following points. Find its new position.
   (a) 600 n. miles South from (70°N, 10°W)
   (b) 900 n. miles North from (10°N, 17°W)
   (c) 1620 n. miles North from (11°S, 111°W)
   (d) 3240 n. miles South from (5°30′S, 19°E)
   (e) 1050 n. miles South from (84°N, 11° 10′W)
   (f) 40 n. miles North from (17°N, 10°W)
7. A ship sails due West from (0°, 17°W) at a speed of 20 knots. Find its position after
   (a) 3 hours (b) 15 hours.
8. An aircraft flies due South from (65°N, 10°W) at a speed of 500 km/h. Find its position after
   (a) 1 hour (b) 3 hours (c) 7 hours.
9. A frigate sails due North from (19°S, 112°W) at a speed of 30 knots. Find its position after
   (a) 3 hours (b) 2 days.
10. An aircraft flies West from (0°, 10°E) to arrive at (0°, 17°W) in 4 hours. Find its speed in km/h.

Distances along lines of latitude

Apart from the equator, all circles of latitude have a radius smaller than that of the Earth. In the figure, point A has latitude x°N.

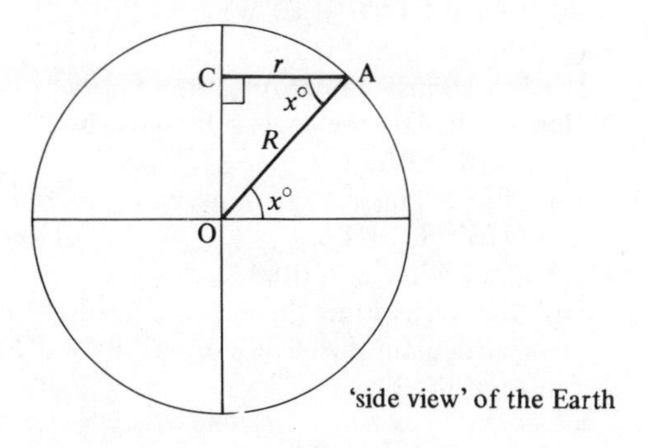

'side view' of the Earth

In triangle AOC

$$\cos x° = \frac{r}{R} \qquad \therefore \qquad r = R\cos x.$$

The radius of the circle of latitude x°N (or S) is $R\cos x$ (where the radius of the Earth is R). It is useful to remember this result.

Example 3

Find the distance in km along a line of latitude between the points A (55°N, 20°W) and B (55°N, 40°E).
($R = 6370$ km, $\pi = 3{\cdot}14$.)

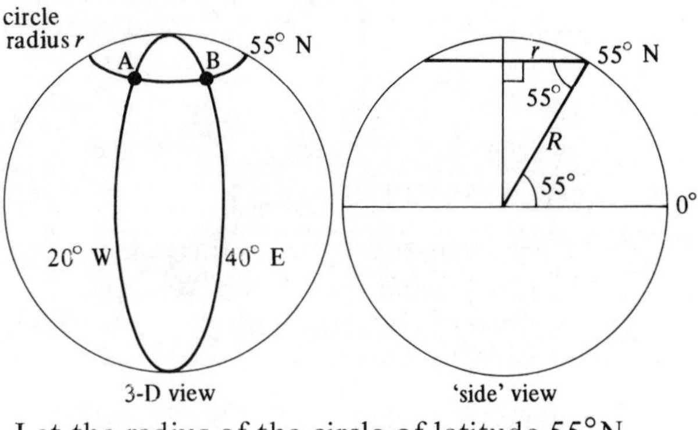

Let the radius of the circle of latitude 55°N be r km

$$\frac{r}{R} = \cos 55° \qquad \therefore \qquad r = R\cos 55°$$

Now draw a 'plan' view looking down on the North pole.

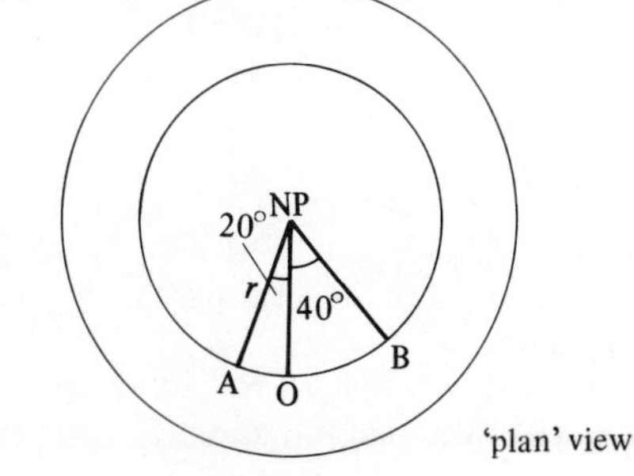

'plan' view

$$\text{Arc length AB} = \frac{60}{360} \times 2\pi r$$

$$= \frac{60}{360} \times 2 \times \pi \times 6370\cos 55°$$

The distance between A and B = 3820 km (3 S.F.)

Example 4

Find the distance in nautical miles along the line of latitude from X (40°N, 100°W) to Y (40°N, 65°W).

Take a 'plan' view looking down on the North pole.

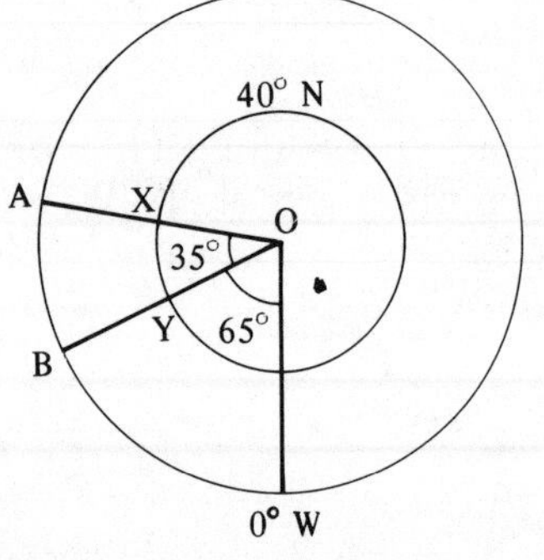

$X\hat{O}Y = 35°$ [100° − 65°]

Consider the distance between the points A (0°N, 100°W) and B (0°N, 65°W) (i.e. points on the equator with the same longitude as X and Y).

The distance between A and B = 35 × 60
= 2100 n.m.

Now the radius of the circle of latitude 40°N is smaller than the radius of the equator by a factor cos 40°.

∴ The distance between X and Y
= 2100 × cos 40
= 1610 n.m. (3 S.F.)

Exercise 11

(Take $R = 6370$ km, $\pi = 3{\cdot}14$.)

1. Calculate the radius of the following circles of latitude:
   (a) 60°N (b) 31°N (c) 78·2°S
   (d) 10° 15′N (e) 66° 6′ S (f) 25° 40′N
2. Find the shortest distance in km along a line of latitude between each pair of points.
   (a) (40°N, 75°W) to (40°N, 15°W)
   (b) (25°N, 102°E) to (25°N, 17°E)
   (c) (17°S, 11°W) to (17°S, 25°E)
   (d) (63° 30′S, 90°W) to (63° 30′S, 0°)
   (e) (87°N, 89°W) to (87°N, 85°E)
   (f) (54° 15′S, 10° 10′W) to (54°15′S, 28°20′E)
3. (a) Find the distance in nautical miles along the equator between A (0°, 65°W) and B (0°, 10°W).
   (b) Find the distance in nautical miles along the line of latitude between C (30°N, 65°W) and D (30°N, 10°W).
   [Hint: Use the first part and remember that the radius of the 30°N circle of latitude is R cos 30°.]
4. (a) Find the distance in nautical miles along the equator between W (0°, 24°W) and X (0°, 36°E).
   (b) Find the distance in nautical miles along the line of latitude between Y (56°N, 24°W) and Z (56°N, 36°E).
5. Find the speed in km/h of an aircraft which travels between the following points in the time given:
   (a) (59°N, 10°W) due East to (59°N, 31°E) in 2 hours
   (b) (17°S, 12° 10′W) due West to (17°S, 47° 25′W) in 3 hours
   (c) (0°N, 170°W) due East to (0°N, 0°W) in 8 hours.
6. Find the shortest distance in nautical miles between each pair of points:
   (a) (10°N, 41°W) to (27°N, 41°W)
   (b) (11°S, 165°E) to (85°S, 165°E)
   (c) (80°N, 100°W) to (70°N, 80°E)
   (d) (25°N, 84°W) to (25°N, 60°W)
   (e) (62°S, 100°W) to (62°S, 117°W)
   (f) (12° 15′N, 17°W) to (12° 15′N, 25°E)
7. A is 450 nautical miles due South of B (54°N, 10°W). Find the position of A.
8. P is 780 km due East of Q (25°N, 100°W). Find the position of P.
9. An aircraft can fly from X (65°N, 100°W) to Y (65°N, 80° E) along two different routes. Find the distance in km
   (a) over the North Pole
   (b) along the line of latitude.
10. Find the distance in km between E (40°N, 80°W) and F (40°N, 10°W)
    (a) along a line of latitude
    (b) in a straight line through the Earth.
    Hence calculate the angle subtended by EF at the centre of the Earth.
11. Find the distance in km between P (69°S, 10°W) and Q (69°S, 74°E)
    (a) along a line of latitude
    (b) in a straight line through the Earth.
    Hence calculate the angle subtended by PQ at the centre of the Earth.

12. Aircraft A flies due South from (72°N, 10°W) at the same time as aircraft B flies due North from (12°N, 10°W). If the speed of A is twice the speed of B, find where they meet. If it takes 4 hours for them to meet, what is the speed of A in knots?
13. Aircraft A and B start journeys at noon from point X (10°N, 56°W); A flies due North at 750 knots and B flies due South at 880 knots. At what time are they at diametrically opposite points on the surface of the Earth?
14. An aircraft flies due West from A (25°N, 10°W) at a speed of 500 km/h. Find the position of the aircraft after $2\frac{1}{2}$ hours.
15. An aircraft flies due East from M (66°S, 100°W) at a speed of 850 km/h. Find the position of the aircraft after 45 minutes.

REVISION EXERCISE 3A

(Take $\pi = 3{\cdot}14$ unless otherwise stated.)

1. Find the area of the following shapes:

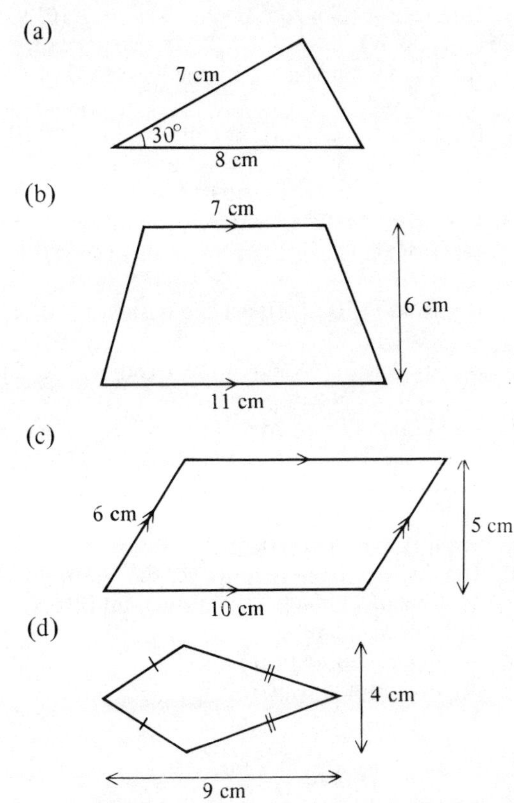

2. (a) A circle has radius 9 m. Find its circumference and area.
   (b) A circle has circumference 34 cm. Find its diameter.
   (c) A circle has area 50 cm². Find its radius.
3. A target consists of concentric circles of radii 3 cm and 9 cm.

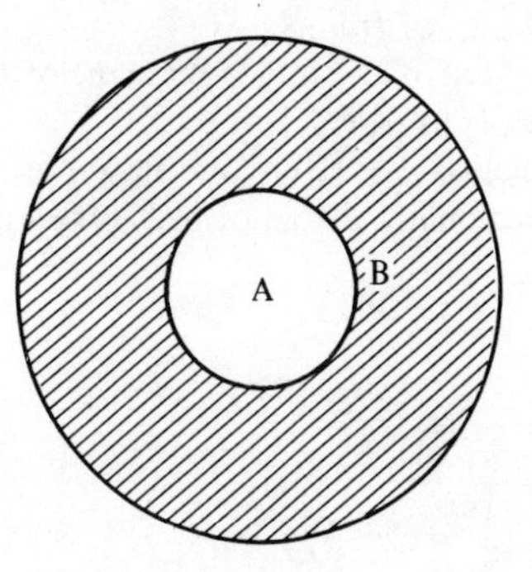

   (a) Find the area of A, in terms of π
   (b) Find the ratio $\dfrac{\text{area of B}}{\text{area of A}}$.
4. In Figure 1 a circle of radius 4 cm is inscribed in a square. In Figure 2 a square is inscribed in a circle of radius 4 cm.
   Calculate the shaded area in each diagram.

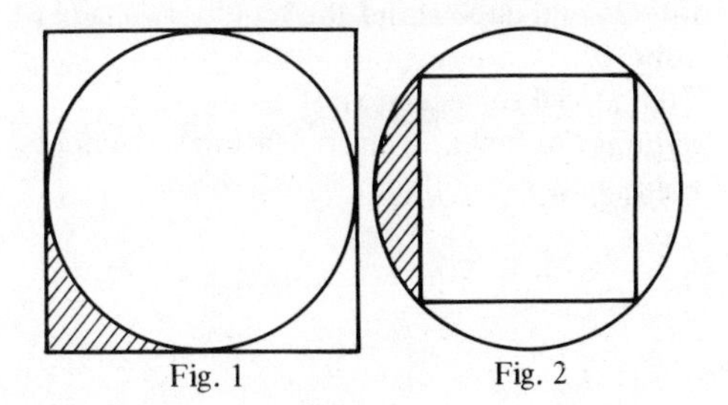

5. Given that OA = 10 cm and $A\hat{O}B = 70^\circ$ (where O is the centre of the circle), calculate
   (a) the arc length AB
   (b) the area of minor sector AOB.

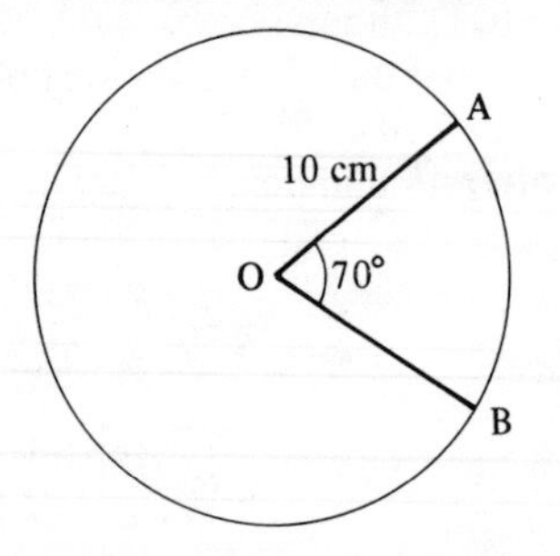

6. The points X and Y lie on the circumference of a circle, of centre O and radius 8 cm, where $X\hat{O}Y = 80^\circ$. Calculate
   (a) the length of the minor arc XY
   (b) the length of the chord XY
   (c) the area of sector XOY
   (d) the area of triangle XOY
   (e) the area of the minor segment of the circle cut off by XY.
7. Given that ON = 10 cm and minor arc MN = 18 cm, calculate the angle $M\hat{O}N$ (shown as x°).

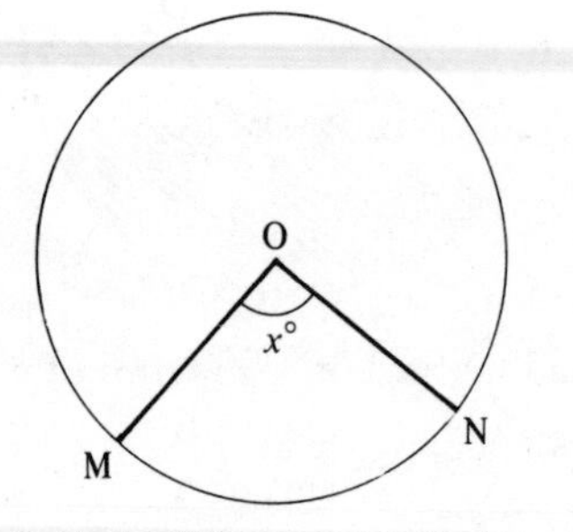

8. A cylinder of radius 8 cm has a volume of 2 litres. Calculate the height of the cylinder.
9. Calculate
   (a) the volume of a sphere of radius 6 cm
   (b) the radius of a sphere whose volume is $800\ \text{cm}^3$.
10. A sphere of radius 5 cm is melted down and made into a solid cube. Find the length of a side of the cube.
11. The curved surface area of a solid circular cylinder of height 8 cm is $100\ \text{cm}^2$. Calculate the volume of the cylinder.
12. A cone has base radius 5 cm and vertical height 10 cm, correct to the nearest cm. Calculate the maximum and minimum possible volumes of the cone, consistent with this data.
13. Calculate the radius of a hemispherical solid whose total surface area is $48\pi\ \text{cm}^2$.
14. Calculate:
   (a) the area of an equilateral triangle of side 6 cm.
   (b) the area of a regular hexagon of side 6 cm.
   (c) the volume of a regular hexagonal prism of length 10 cm, where the side of the hexagon is 12 cm.
15. Ten spheres of radius 1 cm are immersed in liquid contained in a vertical cylinder of radius 6 cm. Calculate the rise in the level of the liquid in the cylinder.
16. A cube of side 10 cm is melted down and made into ten identical spheres. Calculate the surface area of one of the spheres.
17. Calculate the distance in km:
   (a) from A (0°N, 65°W) due West to B (0°N, 100°W)
   (b) from C (40°N, 20°E) due West to D (40°N, 30°W)
   (c) from E (72°N, 10°W) due South to F (10°N, 10°W).
   [Radius of the Earth = 6400 km.]
18. Calculate the distance in nautical miles:
   (a) from G (17°S, 10°E) due North to H (20°N, 10°E)
   (b) from I (50°N, 19°E) due East to J (50°N, 70°E)
   (c) from K (65°N, 60°W) to L (80°N, 120°E) over the North pole.
19. An aircraft flies from point X (55°N, 15°E) for a distance of 1500 nautical miles. Find its new position
   (a) if it flies due North
   (b) if it flies due East.
20. The distance from A to B over the North pole is 2400 nautical miles. If A is the point (70°N, 110°E), calculate
   (a) the longitude of B
   (b) the latitude of B.

EXAMINATION EXERCISE 3B

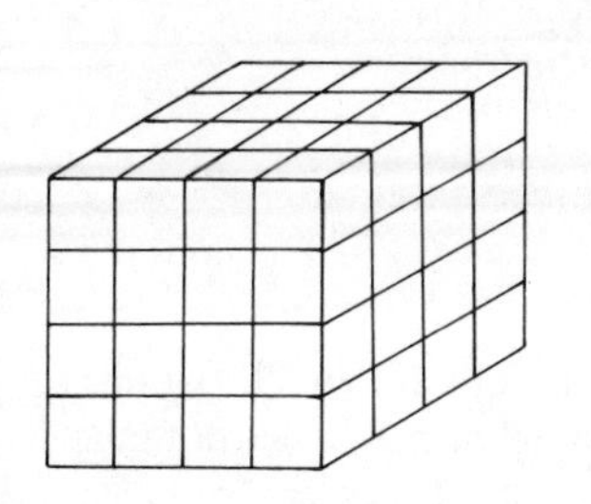

1. Sixty-four unpainted cubes, each of edge 2 cm, are arranged to form a large cube as in the figure on the left. The figure is not drawn to scale.
All the faces of the large cube are painted blue.
   (a) Calculate the length of an edge of the large cube.
   (b) Calculate the volume of the large cube.
   (c) Calculate the total surface area which is painted blue.
   (d) Find the number of 2 cm cubes which have
      (i) exactly three faces painted blue,
      (ii) no face painted blue,
      (iii) exactly one face painted blue,
      (iv) exactly two faces painted blue.
   (e) Find the total area of the unpainted faces for all the sixty-four 2 cm cubes.
[C 16+]

2. (a) A piece of thin wire of length a cm is bent to form the four sides of a square. Find the area of the square in terms of a.
   (b) A second piece of wire of length b cm is bent to form the circumference of a circle. Find the radius and the area of the circle in terms of b and π.
   (c) A third piece of wire of length c cm is bent into the shape of the perimeter of a semi-circle (including both the semi-circular arc and the diameter). Find the radius and the area of the semi-circle in terms of c and π.
[O & C]

3. A mug is in the form of a right circular cylinder closed at one end. The internal radius of the mug is 5 cm, and the internal height is 8 cm.
Taking π to be 3·142, calculate
   (i) the volume of water which the mug will hold,
   (ii) the total surface area of the inside of the mug,
   (iii) the depth of the water, correct to the nearest millimetre, if 360 cm^3 of water are poured into the mug.
[C]

4. In a photograph of an eclipse, the sun and the moon each appear as a circular disc of diameter 2·00 cm.
The centres of the circles are A and B. For each circle the circumference of one passes through the centre of the other, as shown in the diagram. The two circles cut at C and D and AB cuts CD at E.

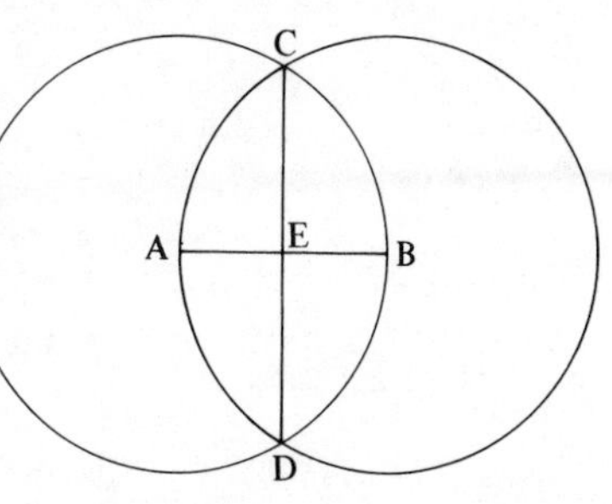

   (i) Give reasons why triangle ABC is equilateral.
   (ii) Calculate (a) the length of CE;
   (b) the area of triangle ACD.
   (iii) (a) For the circle with centre A, calculate the area of the sector ACBD (bounded by the radii AC and AD).
   (b) Hence calculate the area common to the two circles and express this area as a percentage of the area of one circle.
[SMP]

5. Oil is being discharged through a pipe at the rate of 50·4 kg per minute. Express this rate in grams per second.

   The density of the oil is 0·8 g/cm^3. Calculate in cubic centimetres the volume of oil discharged per second.

   The oil is discharged into cylindrical containers of height 80 cm and base radius 24 cm. Calculate the time required to fill one such container.

   [Take π as 3·142.] [O & C]

6. During a storm, the depth of the rainfall was 15·4 mm. The rain which fell on a horizontal roof measuring 7·5 m by 3·6 m was collected in a cylindrical tank of radius 35 cm which was empty before the storm began. Calculate
   (a) the area, in cm^2, of the roof;
   (b) the volume, in cm^3, of the rain which fell on the roof.
   Taking π as $\frac{22}{7}$, find
   (c) the area, in cm^2, of the cross section of the tank;
   (d) the height, in cm, of the rain water in the tank.
   Given that a watering can holds five litres, how many times could it be filled completely from the rain water? [L]

7. The diagram shows a circular window in a church. Semi-circles are drawn on each side of a square ABCD such that the semi-circles touch the circumference of the outer circle at P, Q, R and S. Given that AB = $2x$ cm, find, in terms of x, the area of the outer circle and the area of the shaded region. Show that the ratio of these areas is $8\pi : (\pi - 2)$.

   Given that the area of the shaded portion is 3000 cm^2, calculate, to the nearest cm, the radius OP of the outer circle.

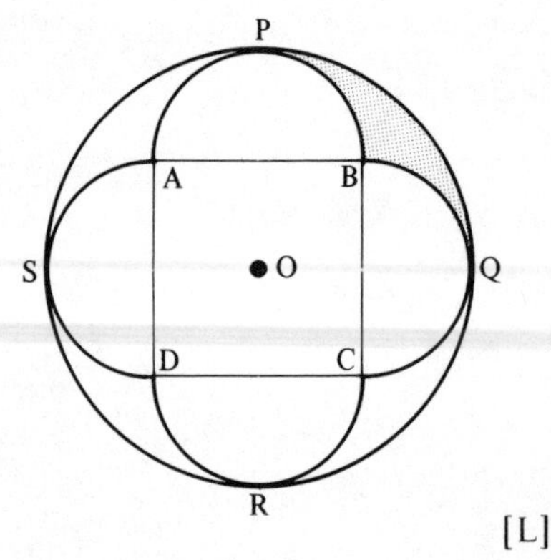

[L]

8. The wood used to make a pencil is in the form of a regular hexagonal prism with a circular hole drilled symmetrically along the axis of the prism, as shown in the diagram.

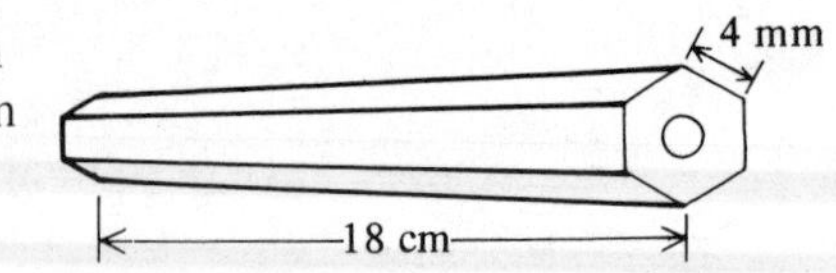

   Each side of the hexagon is of length 4 mm, the diameter of the circular hole is 2 mm and the length of the pencil is 18 cm.
   Calculate, correct to three significant figures:
   (i) the height of an equilateral triangle with sides each of length 4 mm,
   (ii) the area of a regular hexagon with sides each of length 4 mm,
   (iii) the area of a circle of diameter 2 mm (take π as 3·142),
   (iv) the volume of wood in the pencil in mm^3.
   Assuming that each cubic centimetre of the wood has a mass of 0·7 g, calculate the mass of the wood in the pencil correct to two significant figures. [JMB]

9. A certain parallel of north latitude has a circumference of 33 180 km. Calculate the radius of this parallel of latitude.

   Assuming the radius of the earth to be 6370 km, calculate also the latitude of all points on this parallel.

If the distance between two points A and B, measured along this parallel of latitude, is 4424 km, calculate the difference in longitude between the points A and B. If the longitude of A is 10° E, state two possible values for the longitude of B. [Take π as 3·142.] [O & C]

10. A, B, C, D are four points on the surface of the Earth and N is the North pole.

(i) Calculate the shorter distance, in nautical miles, between A (0° N, 40° W) and B (0° N, 50° E), measured along the equator.

(ii) An aircraft flies from A to B, then to N and finally back to A, always taking the shortest route between any two points. Find the total distance travelled in nautical miles.

(iii) Calculate the distance, to the nearest nautical mile, between C (56° N, 40° W) and D (56° N, 50° E), measured along the circle of latitude.

(iv) Another aircraft travels from A due North to C, then due East to D and finally due South to B. Find in nautical miles the total distance travelled, and the total flying time to the nearest hour, given that the aircraft flies at a steady speed of 160 knots. [C]

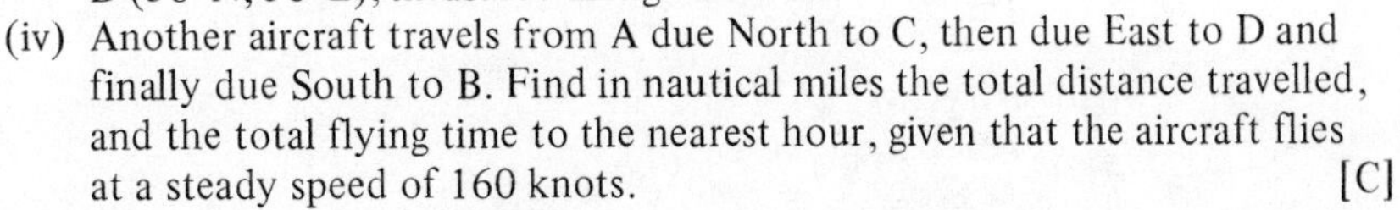

11. In the figure below, AB and CD are common tangents to the circles centred at P and Q, PR is parallel to AB and N and M lie on their respective circles.

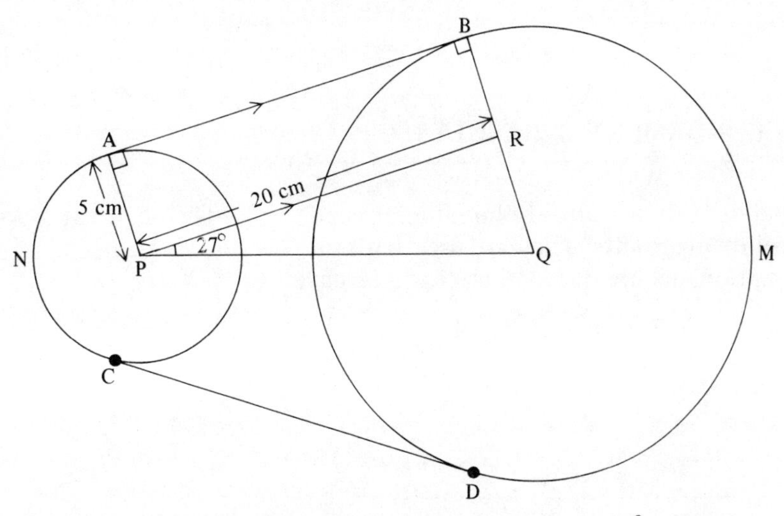

Given that AP = 5 cm, PR = 20 cm and the angle RPQ = 27°, calculate

(i) the length of PQ and the radius BQ of the larger circle,

(ii) the area of the triangle APC,

(iii) the length of the arc ANC,

(iv) the total length of the perimeter of the figure bounded by the tangents AB and CD and the arcs ANC and BMD.

(Take π to be 3·142.) [AEB]

4 Geometry

Pythagoras (569–500 B.C.) was one of the first of the great mathematical names in Greek antiquity. He settled in southern Italy and formed a mysterious brotherhood with his students who were bound by an oath not to reveal the secrets of numbers and who exercised great influence. They laid the foundations of arithmetic through geometry but failed to resolve the concept of irrational numbers. The work of these and others was brought together by Euclid at Alexandria in a book called 'The Elements' which was still studied in English schools as recently as 1900.

4.1 FUNDAMENTAL RESULTS

The student should already be familiar with the following results. They are used later in this section and are quoted here for reference.

The angles on a straight line add up to 180°:

$$\hat{x} + \hat{y} + \hat{z} = 180°$$

The angles at a point add up to 360°:

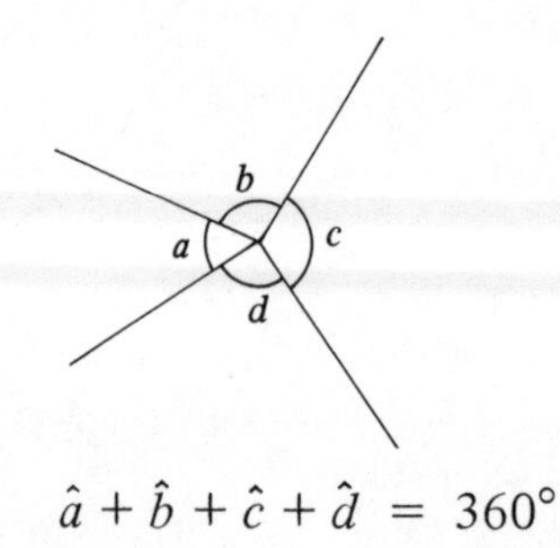

$$\hat{a} + \hat{b} + \hat{c} + \hat{d} = 360°$$

(i) The angle sum of a triangle is 180°.
(ii) The angle sum of a quadrilateral is 360°.

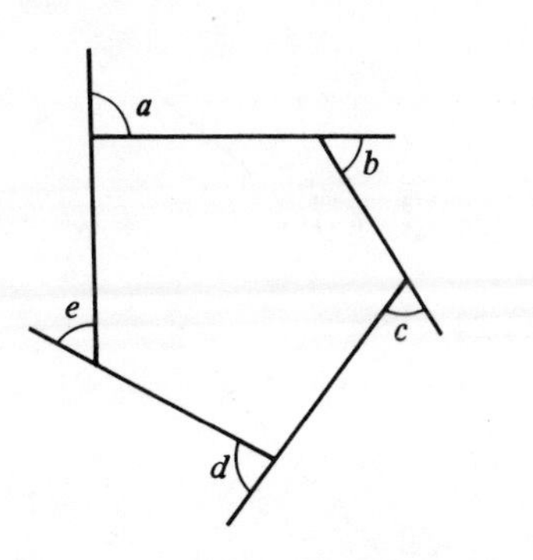

(iii) The exterior angles of a polygon add up to 360° ($\hat{a} + \hat{b} + \hat{c} + \hat{d} + \hat{e} = 360°$)

(iv) The sum of the interior angles of a polygon is $(2n - 4) \times 90°$ where n is the number of sides of the polygon.

Example 1

Find the angles marked with letters.

(a) AB is a straight line

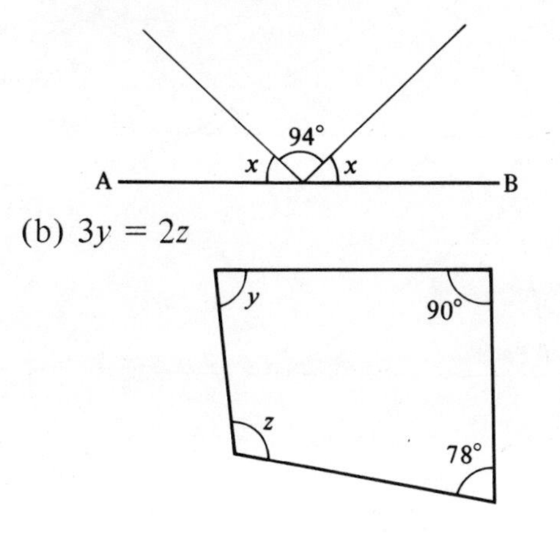

(b) $3y = 2z$

(a) $x + 94 + x = 180$ (Angles on a straight line)

$$2x = 86$$
$$x = 43°$$

(b) $y + z + 78 + 90 = 360$ (angle sum of a quadrilateral)

$$y + z = 192$$
$$2y + 2z = 384$$

But $3y = 2z$

$\therefore \quad 5y = 384 \qquad y = 76{\cdot}8°$

$z = 115{\cdot}2°$

Example 2

Find the angles marked

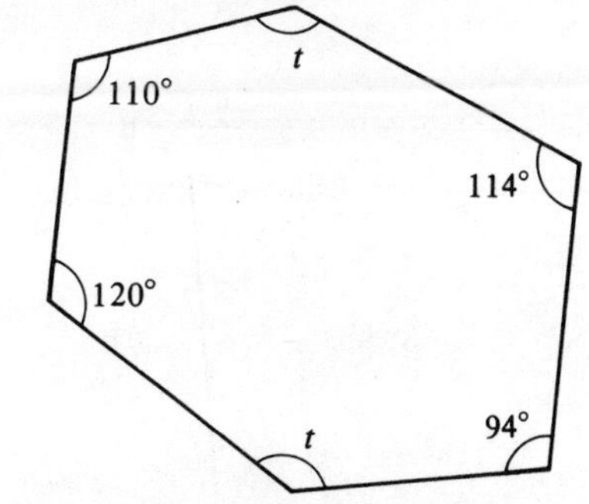

The sum of the interior angles

$= (2n - 4) \times 90°$

where n is the number of sides of the polygon. In this case $n = 6$.

$$\therefore \quad 110 + 120 + 94 + 114 + 2t = (2 \times 6 - 4) \times 90$$
$$438 + 2t = 720$$
$$2t = 282$$
$$t = 141°$$

Exercise 1

Find the angles marked with letters. (AB is always a straight line.)

1.

A 25° a 60° B

2.

41° 40° b 50° A B

3.

140° 120° c

4.

d 132° 55° 96°

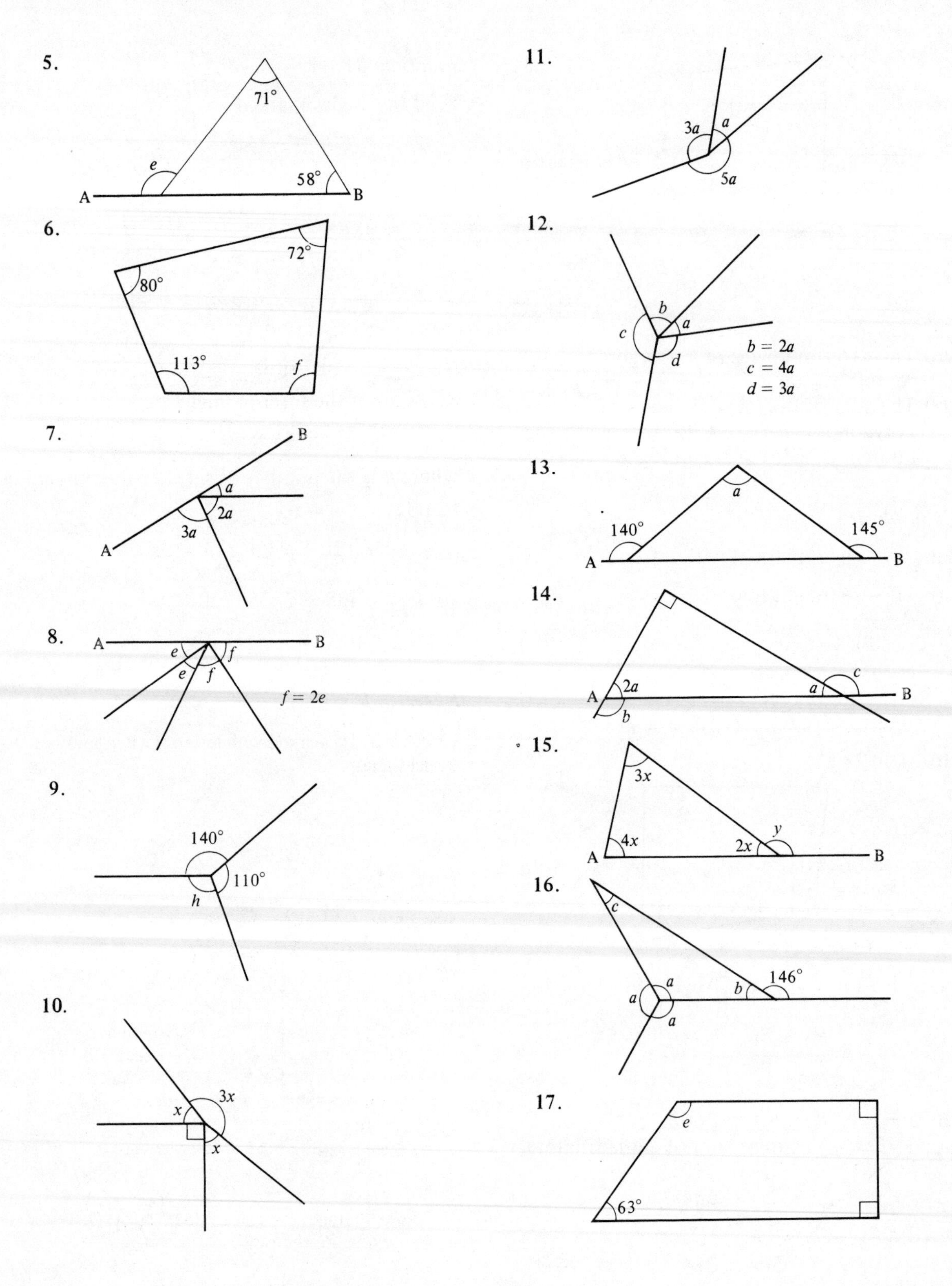
5.
71°
e
58°
A
B
6.
72°
80°
113°
f
7.
B
a
2a
3a
A
8.
A
B
e
f
e
f
f = 2e
9.
140°
110°
h
10.
x
3x
x
11.
3a
a
5a
12.
b
a
c
d
b = 2a
c = 4a
d = 3a
13.
a
140°
145°
A
B
14.
c
A
2a
a
B
b
15.
3x
4x
2x
y
A
B
16.
c
a
a
b
146°
a
17.
e
63°

18. 19. 20.

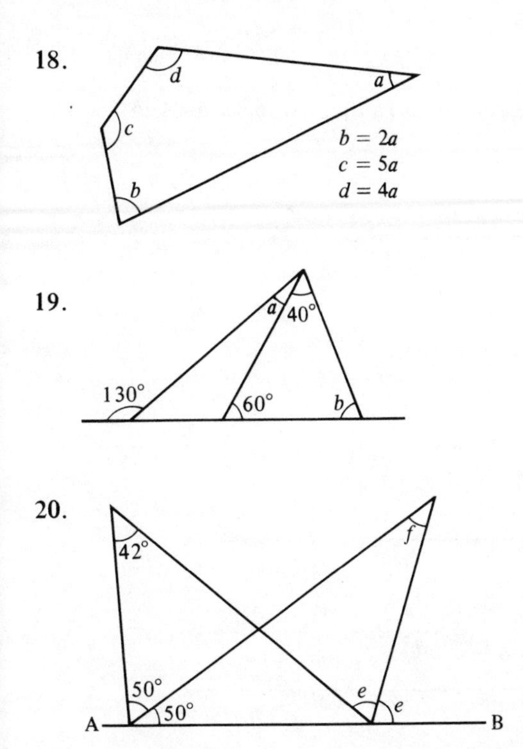

21. Calculate the largest angle of a triangle in which one angle is eight times each of the others.

22. In $\triangle$ABC, $\hat{A}$ is a right angle and D is a point on AC such that BD bisects $\hat{B}$. If $B\hat{D}C = 100°$, calculate $\hat{C}$.

23. WXYZ is a quadrilateral in which $\hat{W} = 108°$, $\hat{X} = 88°$, $\hat{Y} = 57°$ and $W\hat{X}Z = 31°$. Calculate $W\hat{Z}X$ and $X\hat{Z}Y$.

24. In quadrilateral ABCD, AB produced is perpendicular to DC produced. If $\hat{A} = 44°$ and $\hat{C} = 148°$, calculate $\hat{D}$ and $\hat{B}$.

25. Find angles a and b for the regular pentagon below.

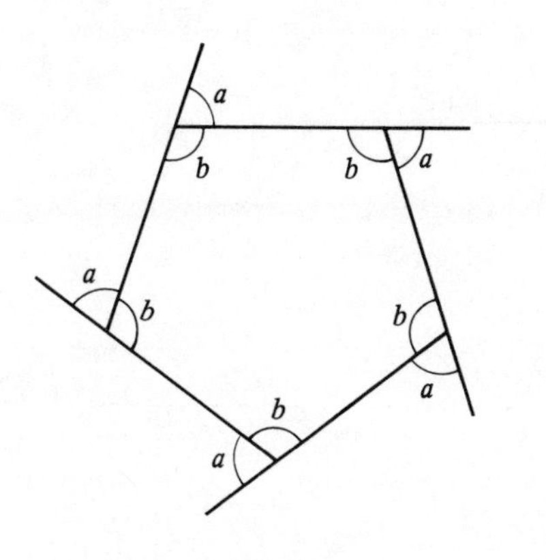

26. Find x and y.

27. Find a.

28. Find m.

29. Find a.

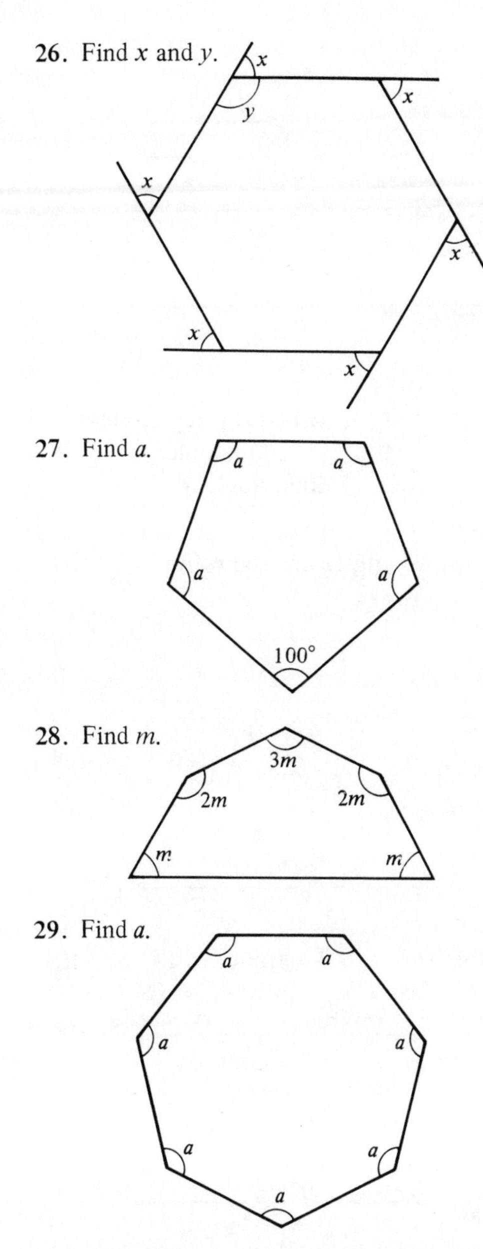

30. Calculate the number of sides of a regular polygon whose interior angles are each 156°.

31. Calculate the number of sides of a regular polygon whose interior angles are each 150°.

32. Calculate the number of sides of a regular polygon whose exterior angles are each 40°.

33. In a regular polygon each interior angle is 140° greater than each exterior angle. Calculate the number of sides of the polygon.

34. In a regular polygon each interior angle is 120° greater than each exterior angle. Calculate the number of sides of the polygon.

Parallel lines

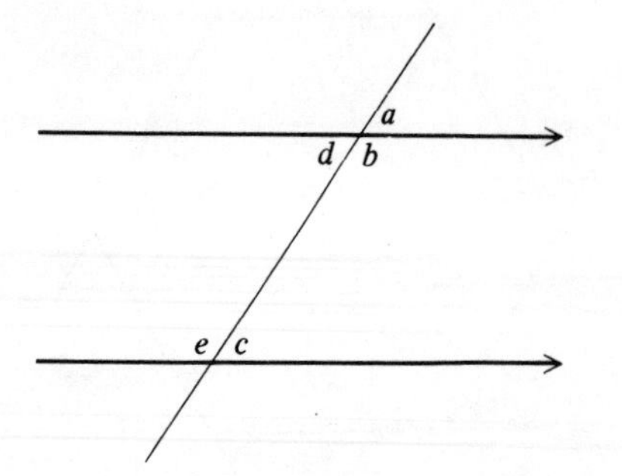

(i) $\hat{a} = \hat{c}$ (corresponding angles)
(ii) $\hat{c} = \hat{d}$ (alternate angles)
(iii) $\hat{b} + \hat{c} = 180°$ (allied angles)

Remember: 'The acute angles are the same and the obtuse angles are the same.'

Example 3

Find the angle x.

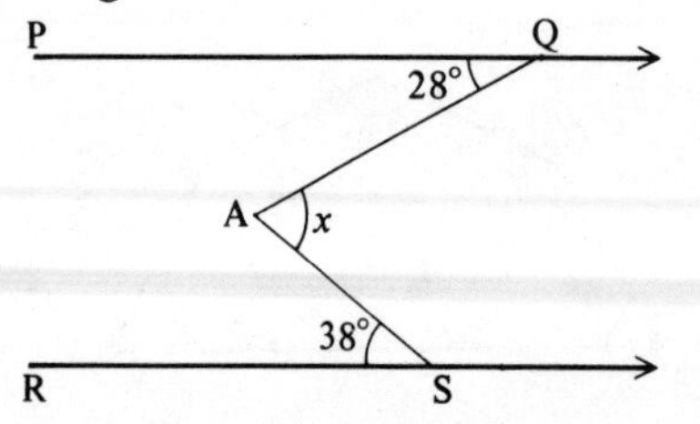

Draw a line through A parallel to PQ and RS and label it AB.

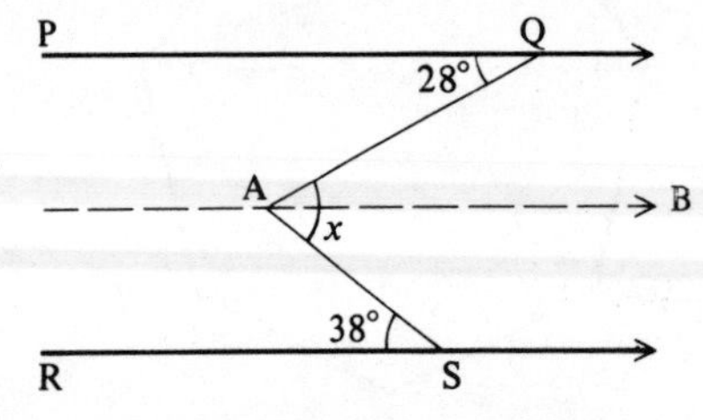

$$Q\hat{A}B = 28° \text{ (alternate angles)}$$
$$S\hat{A}B = 38° \text{ (alternate angles)}$$
$$\therefore \quad Q\hat{A}S = x = 28 + 38$$
$$x = 66°$$

Exercise 2

In questions **1** to **18** find the angles marked.

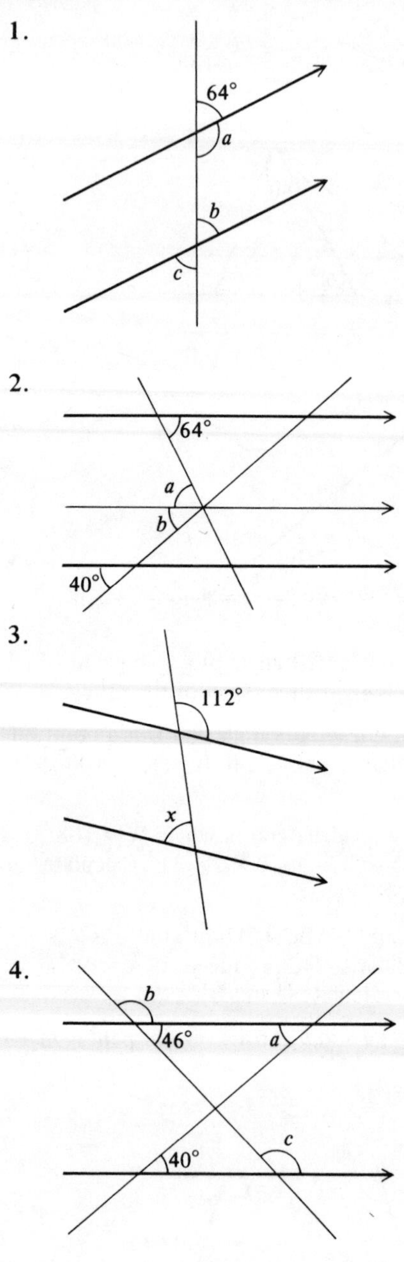

5.

6.

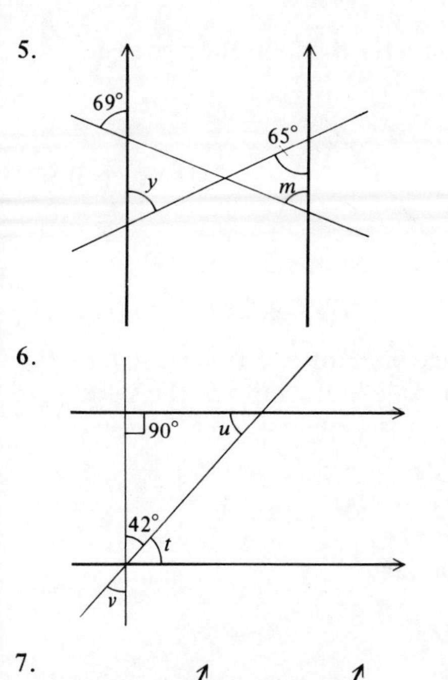

7.

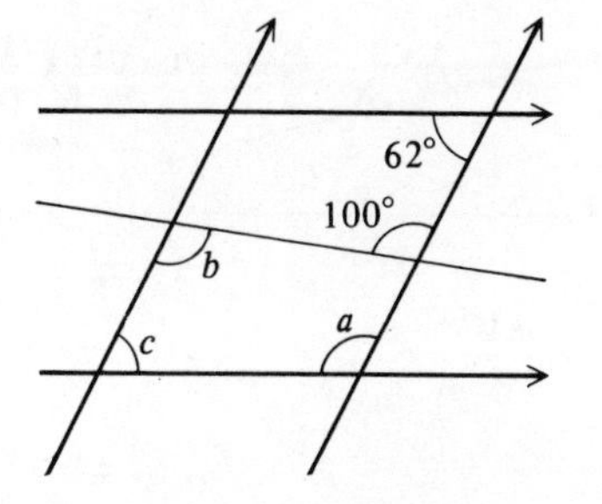

8.

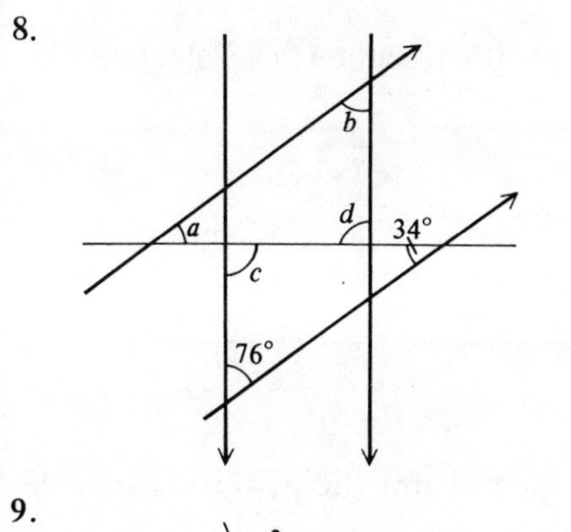

9.

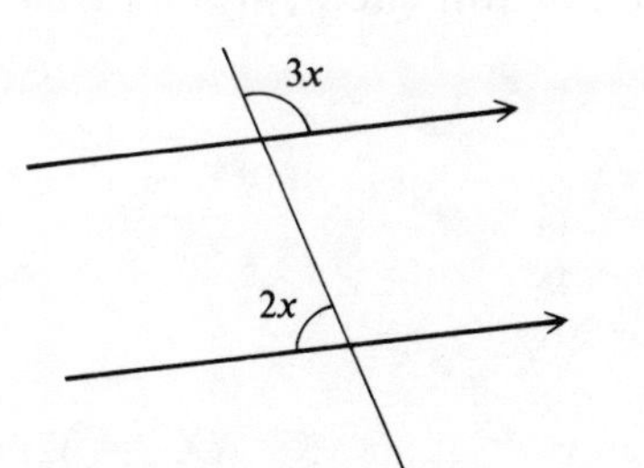

10.

11.

12.

13.

14.

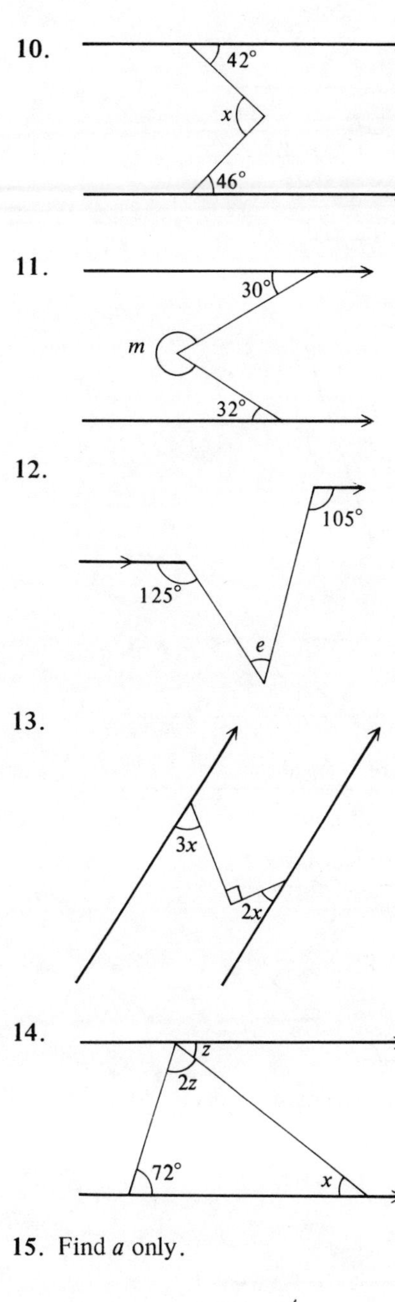

15. Find a only.

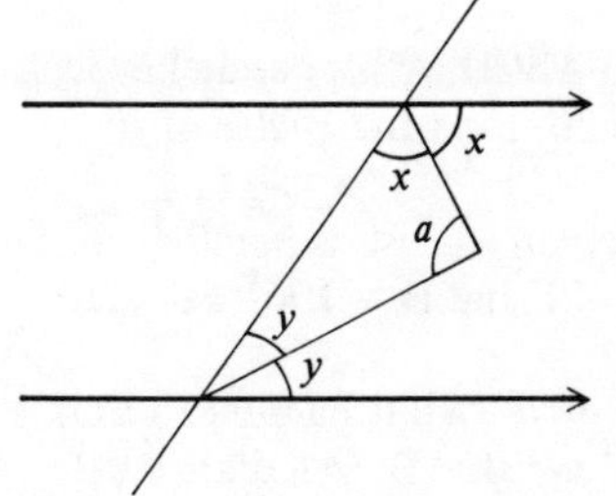

16.

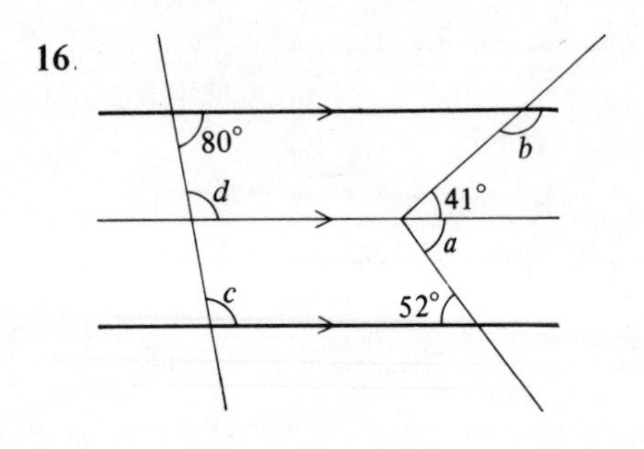

17.

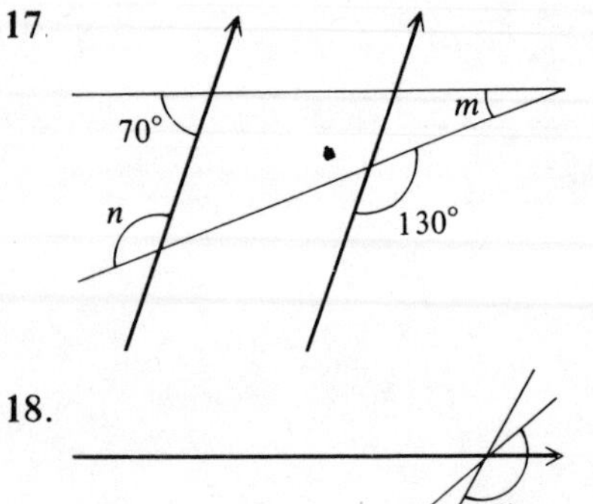

18.

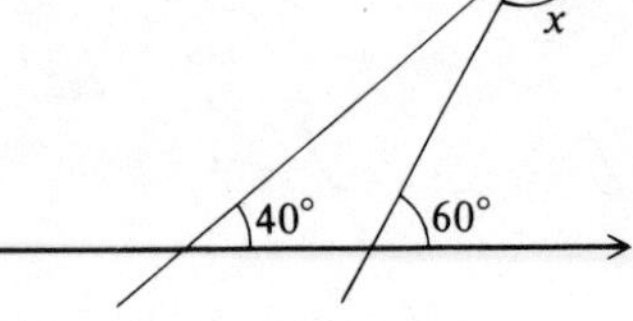

19. Find a formula involving a, b and c.

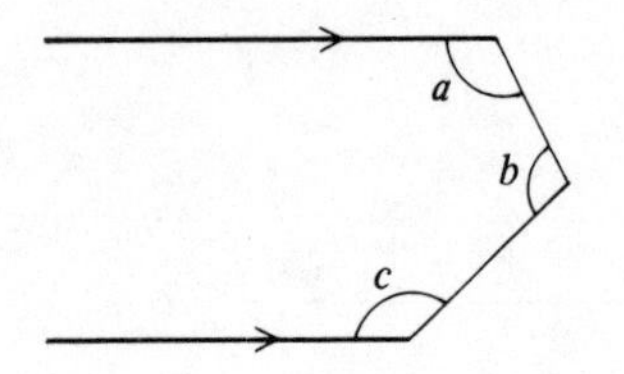

20. Find a formula involving l, m and n.

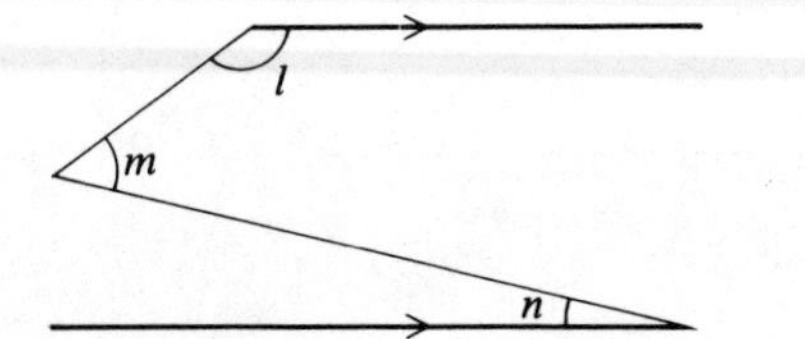

21. In a quadrilateral ABCD, AB is parallel to DC, $\hat{A} = 4\hat{D}$ and $\hat{B} = 2\hat{C}$. Find the angles of the quadrilateral.
22. ABCDE is a pentagon and BC is parallel to ED. If $\hat{A} = 54°$, $\hat{B} = 140°$ and $\hat{D} = 106°$, calculate $B\hat{C}D$ and $A\hat{E}D$.
23. In a pentagon ABCDE, AB is parallel to ED, $\hat{C} = 40°$, $\hat{E} = 72°$ and $\hat{B} = \hat{D}$. Calculate $\hat{A}$, $\hat{B}$ and $\hat{D}$.

The angle bisector theorem for a triangle

(i) Interior angle

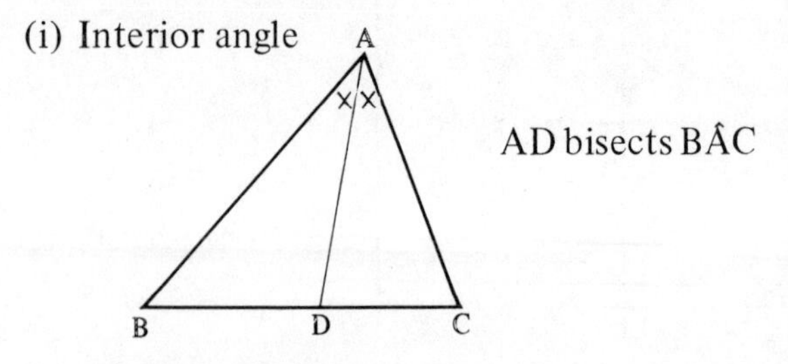

The angle bisector of a triangle divides the opposite side in the ratio of the sides containing the angle

i.e.

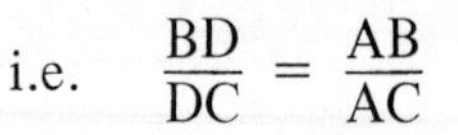

$$\frac{BD}{DC} = \frac{AB}{AC}$$

(ii) Exterior angle

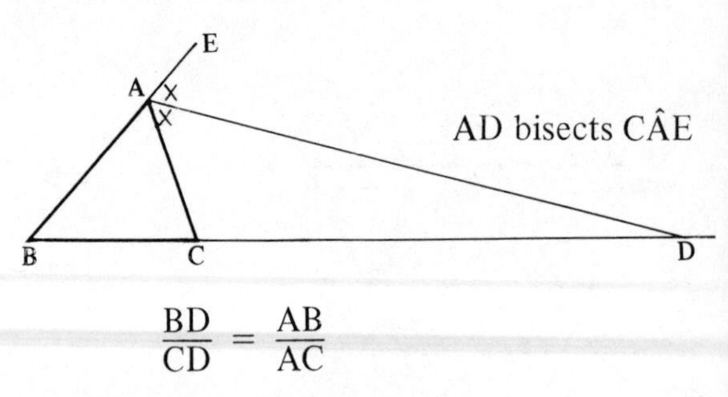

$$\frac{BD}{CD} = \frac{AB}{AC}$$

Example 4

In the diagram BX bisects $A\hat{B}C$. Calculate the length AX.

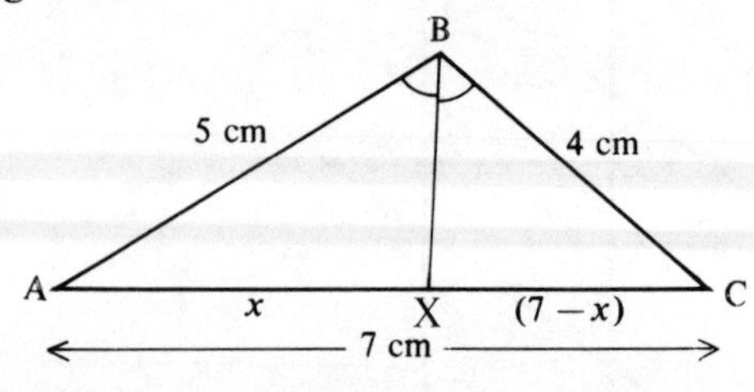

Let AX = x.
Using the angle bisector theorem for a triangle:

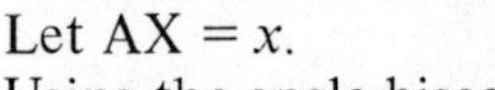

$$\frac{5}{4} = \frac{x}{7-x}$$

$$\begin{aligned} 5(7-x) &= 4x \\ 35 - 5x &= 4x \\ 35 &= 9x \\ \frac{35}{9} &= x \end{aligned}$$

$\therefore$ AX $= 3\frac{8}{9}$ cm.

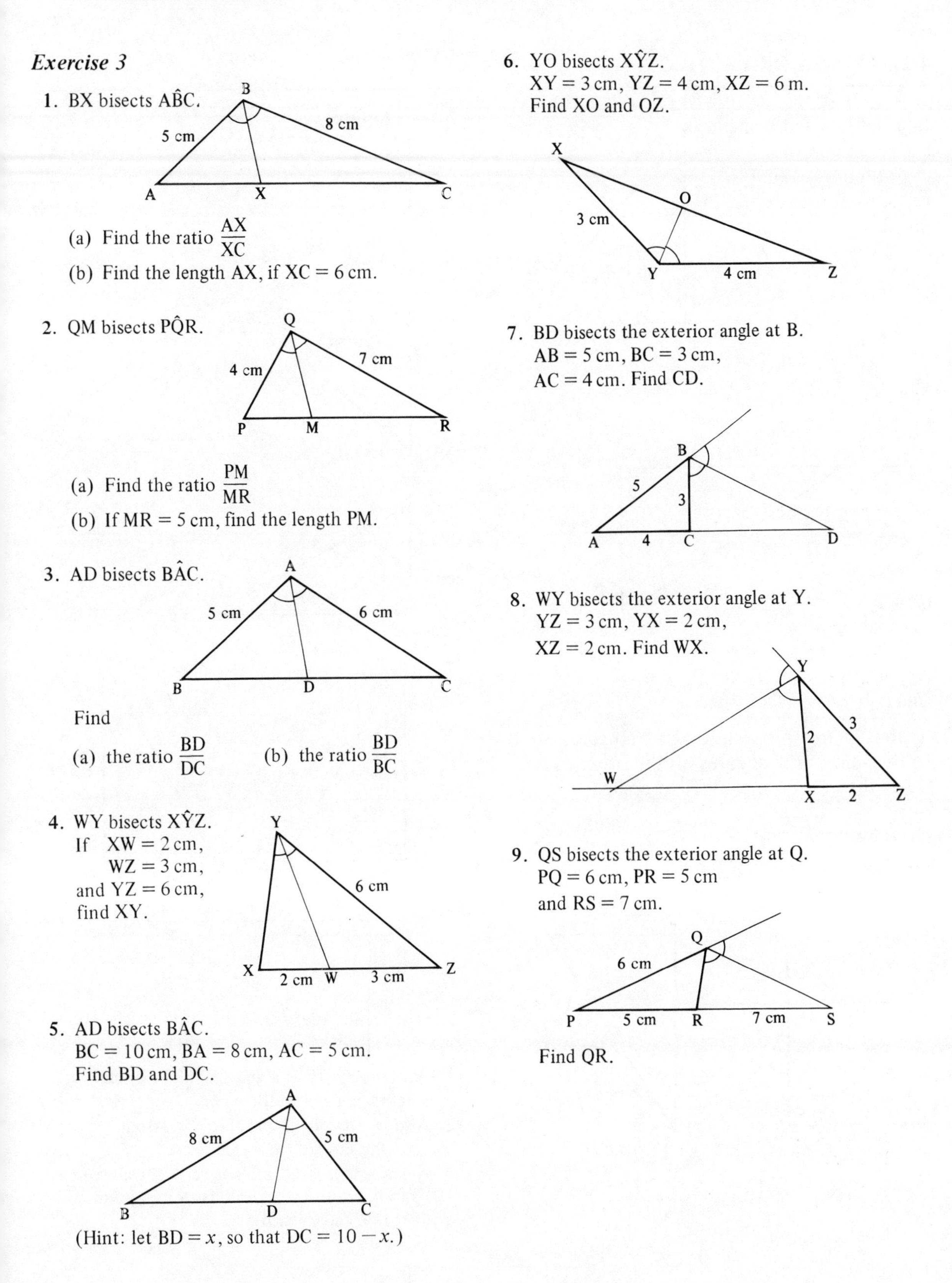

Exercise 3

1. BX bisects $A\hat{B}C$.

   (a) Find the ratio $\frac{AX}{XC}$

   (b) Find the length AX, if XC = 6 cm.

2. QM bisects $P\hat{Q}R$.

   (a) Find the ratio $\frac{PM}{MR}$

   (b) If MR = 5 cm, find the length PM.

3. AD bisects $B\hat{A}C$.

   Find

   (a) the ratio $\frac{BD}{DC}$ (b) the ratio $\frac{BD}{BC}$

4. WY bisects $X\hat{Y}Z$.
   If XW = 2 cm,
   WZ = 3 cm,
   and YZ = 6 cm,
   find XY.

5. AD bisects $B\hat{A}C$.
   BC = 10 cm, BA = 8 cm, AC = 5 cm.
   Find BD and DC.

   (Hint: let BD = x, so that DC = $10 - x$.)

6. YO bisects $X\hat{Y}Z$.
   XY = 3 cm, YZ = 4 cm, XZ = 6 m.
   Find XO and OZ.

7. BD bisects the exterior angle at B.
   AB = 5 cm, BC = 3 cm,
   AC = 4 cm. Find CD.

8. WY bisects the exterior angle at Y.
   YZ = 3 cm, YX = 2 cm,
   XZ = 2 cm. Find WX.

9. QS bisects the exterior angle at Q.
   PQ = 6 cm, PR = 5 cm
   and RS = 7 cm.

   Find QR.

4.2 PYTHAGORAS' THEOREM

In a right-angled triangle the square on the hypotenuse is equal to the sum of the squares on the other two sides.

$$a^2 + b^2 = c^2$$

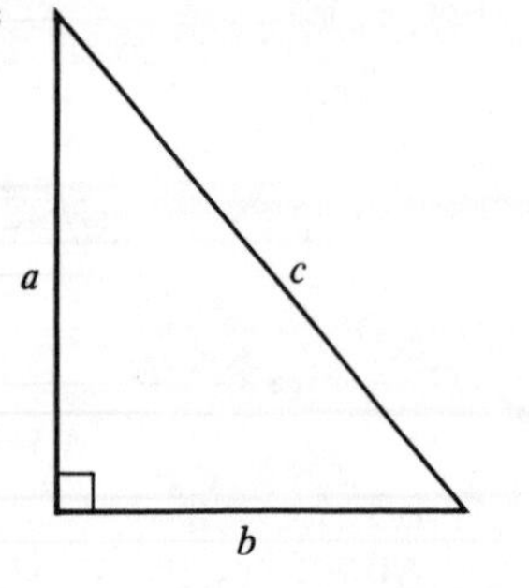

Example 1

Find the side marked d.

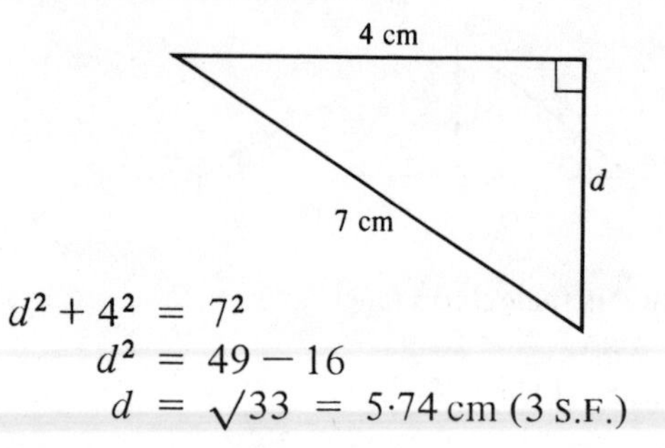

$$d^2 + 4^2 = 7^2$$
$$d^2 = 49 - 16$$
$$d = \sqrt{33} = 5{\cdot}74 \text{ cm (3 S.F.)}$$

The *converse* is also true:

'If the square on one side of a triangle is equal to the sum of the squares on the other two sides, then the triangle is right-angled.'

Exercise 4

In questions **1** to **10**, find x. All the lengths are in cm.

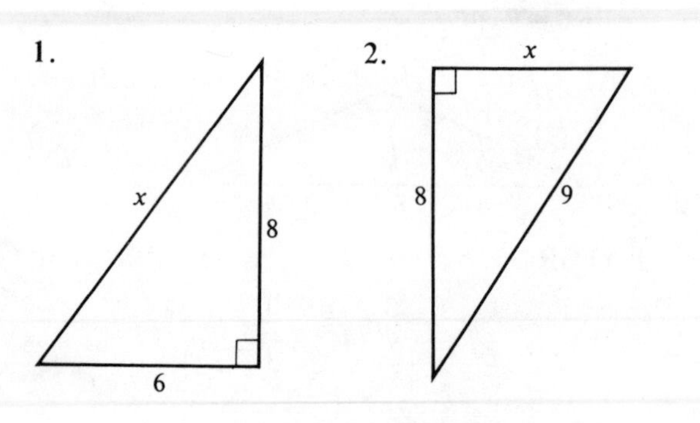

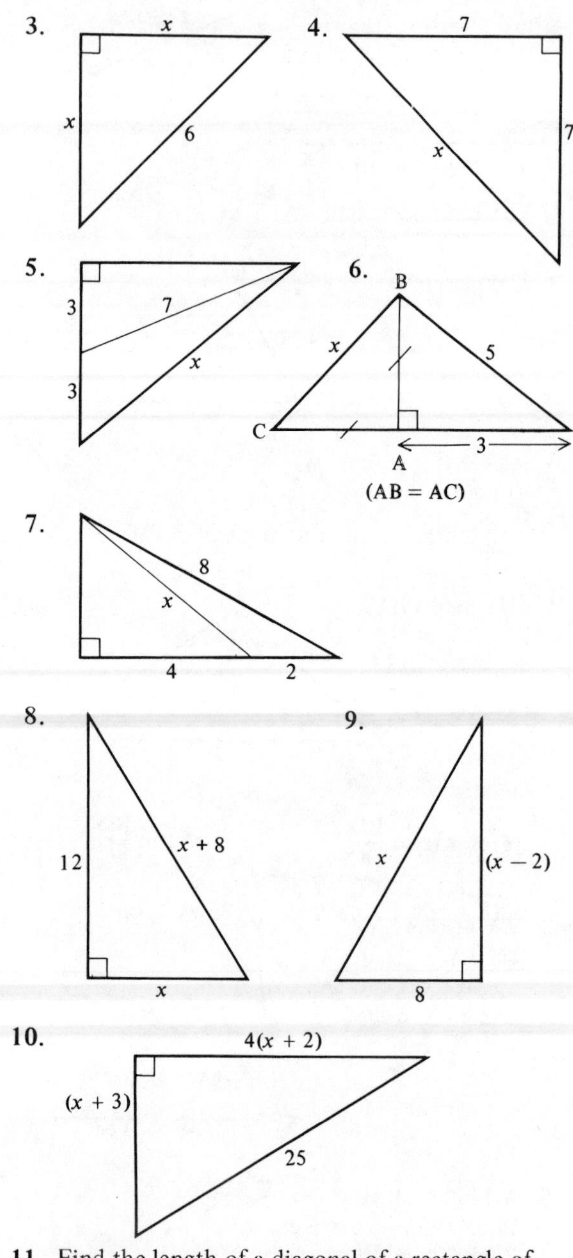

11. Find the length of a diagonal of a rectangle of length 9 cm and width 4 cm.

12. A square has diagonals of length 10 cm. Find the sides of the square.

13. A 4 m ladder rests against a vertical wall with its foot 2 m from the wall. How far up the wall does the ladder reach?

14. A ship sails 20 km due North and then 35 km due East. How far is it from its starting point?
15. Find the length of a diagonal of a rectangular box of length 12 cm, width 5 cm and height 4 cm.
16. Find the length of a diagonal of a rectangular room of length 5 m, width 3 m and height 2·5 m.
17. Find the height of a rectangular box of length 8 cm, width 6 cm where the length of a diagonal is 11 cm.
18. An aircraft flies equal distances South-East and then South-West to finish 120 km due South of its starting-point. How long is each part of its journey?
19. The diagonal of a rectangle exceeds the length by 2 cm. If the width of the rectangle is 10 cm, find the length.
20. A cone has base radius 5 cm and *slant* height 11 cm. Find its vertical height.
21. It is possible to find the sides of a right-angled triangle, with lengths which are whole numbers, by substituting different values of x into the expressions: (a) $2x^2 + 2x + 1$
    (b) $2x^2 + 2x$
    (c) $2x + 1$
    ((a) represents the hypotenuse, (b) and (c) the other two sides.)
    (i) Find the sides of the triangles when $x = 1, 2, 3, 4$ and 5.
    (ii) Confirm that $(2x + 1)^2 + (2x^2 + 2x)^2 = (2x^2 + 2x + 1)^2$
22. The diagram represents the starting position (AB) and the finishing position (CD) of a ladder as it slips. The ladder is leaning against a vertical wall.

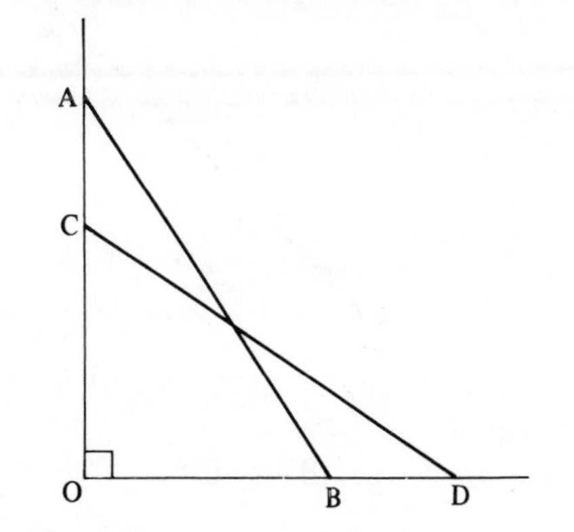

Given: $AC = x$, $OC = 4AC$, $BD = 2AC$ and $OB = 5$ m.
Form an equation in x, find x and hence find the length of the ladder.

4.3 CONGRUENCY

Two plane figures are congruent if one fits exactly on the other. The four types of congruence for triangles are as follows:

(a) Two sides and the included angle (S.A.S)

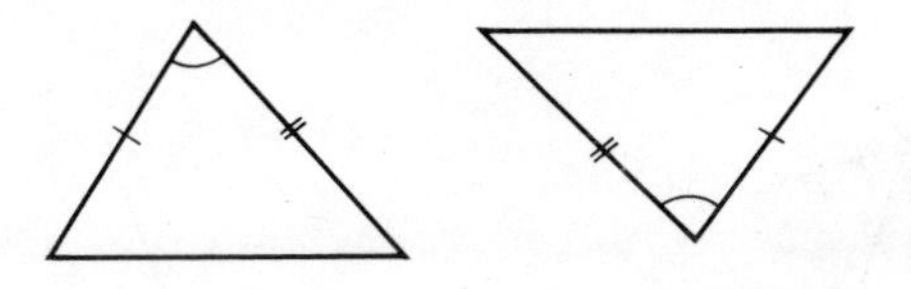

(b) Two angles and a corresponding side (A.A.S)

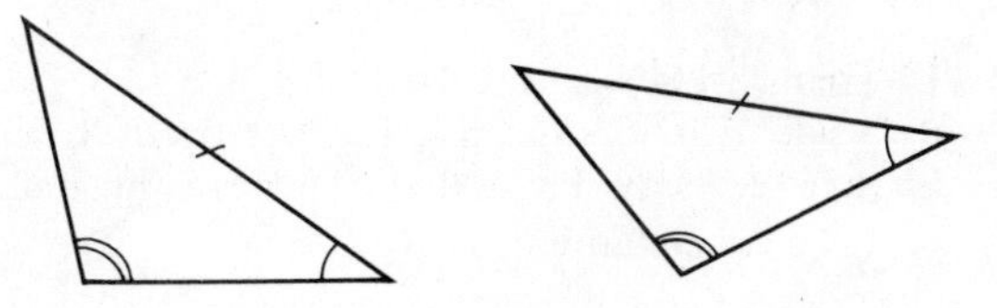

(c) Three sides (S.S.S)

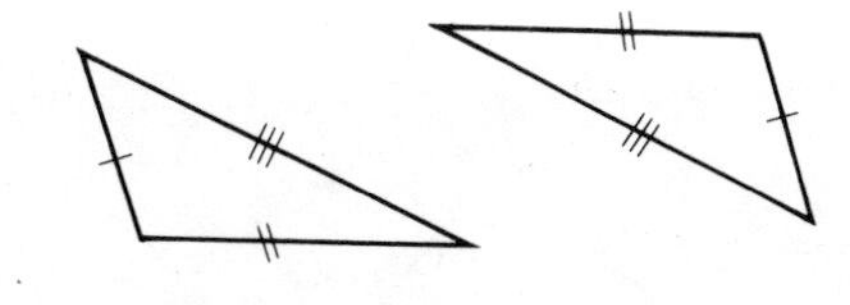

(d) Right angle, hypotenuse and one other side (R.H.S)

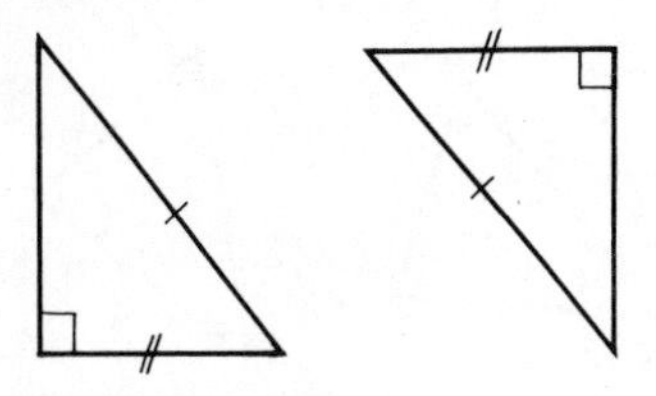

Example 1

In triangle ABC, AB = AC, BX bisects $A\hat{B}C$ and CY bisects $B\hat{C}A$.

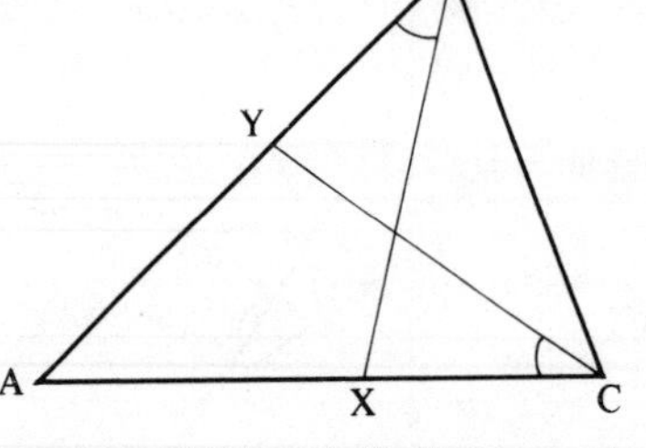

(a) Prove that triangles ABX and ACY are congruent.
(b) Hence prove that BX = CY.

(a) $A\hat{B}C = B\hat{C}A$ (isosceles triangle ABC)

$\therefore$ $A\hat{B}X = A\hat{C}Y$ (both angles bisected)

$B\hat{A}C$ is common to triangles ABX and ACY.

AB = AC (given).

$\therefore$ Triangles ABX and ACY are congruent (A.A.S.)

(b) Since triangles ABX and ACY are congruent and since BX and CY are both opposite $B\hat{A}C$, we deduce that BX = CY.

Exercise 5

For questions **1** to **10**, decide whether the pair of triangles are congruent. If they are congruent, state which conditions for congruency are satisfied.

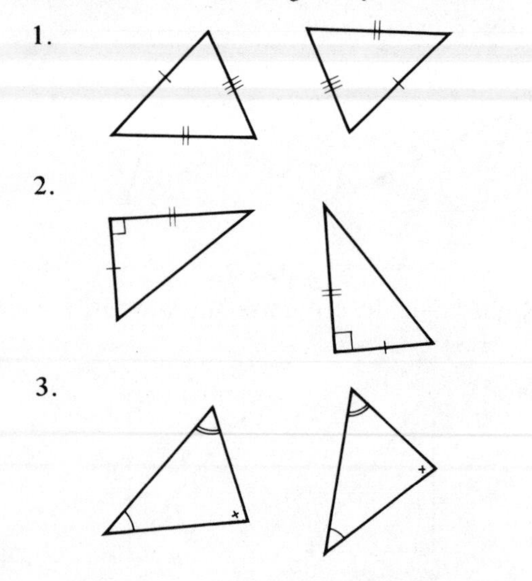

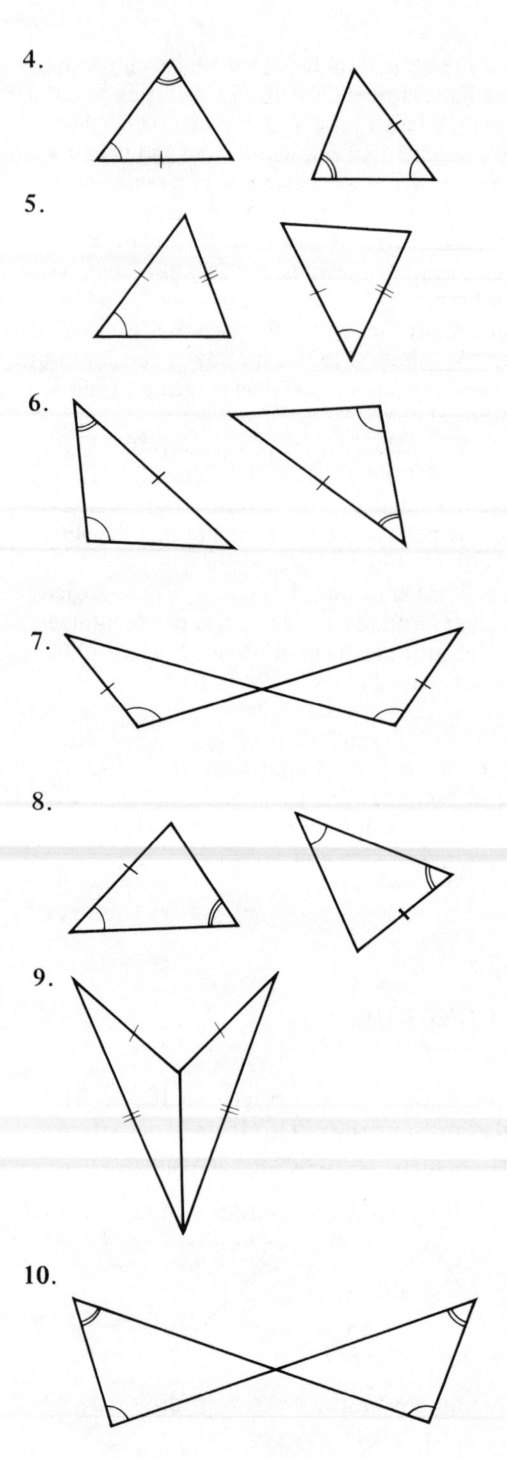

11. Triangle LMN is isosceles with LM = LN; X and Y are points on LM, LN respectively such that LX = LY. Prove that triangles LMY and LNX are congruent.

12. ABCD is a quadrilateral and a line through A parallel to BC meets DC at X. If $\hat{D} = \hat{C}$, prove that $\triangle ADX$ is isosceles.
13. XYZ is a triangle with XY = XZ. The bisectors of angles Y and Z meet the opposite sides in M and N respectively. Prove that YM = ZN.
14. In the diagram, DX = XC, DV = ZC and the lines AB and DC are parallel. Prove that
    (a) AX = BX
    (b) AC = BD
    (c) triangles DBZ and CAV are congruent.

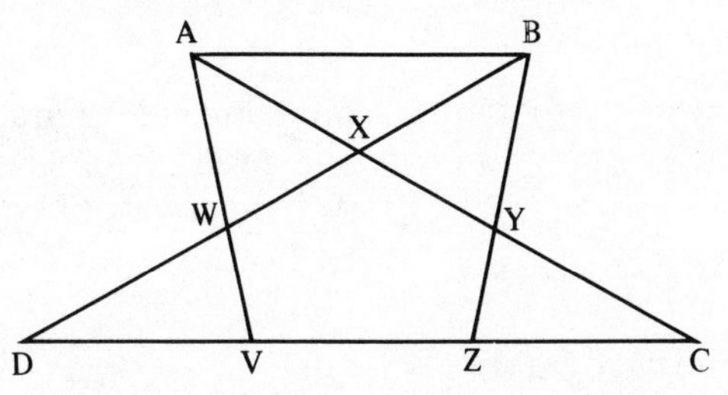

15. Points L and M on the side YZ of a triangle XYZ are drawn so that L is between Y and M. Given that XY = XZ and $Y\hat{X}L = M\hat{X}Z$, prove that YL = MZ.
16. Squares AMNB and AOPC are drawn on the sides of triangle ABC, so that they lie outside the triangle. Prove that MC = OB.
17. In the diagram, N lies on a side of the square ABCD, AM and LC are perpendicular to DN. Prove that
    (a) $A\hat{D}N = L\hat{C}D$ (b) AM = LD

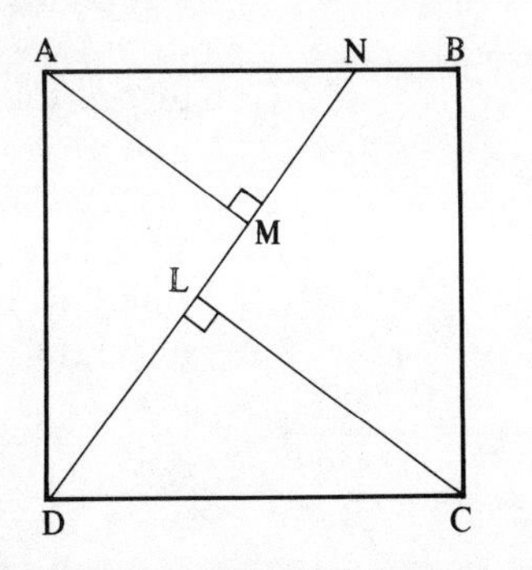

18. ABCD is a rectangle and X is a point inside the rectangle such that XA = XB. Prove that:
    (i) $X\hat{A}D = X\hat{B}C$ (ii) XD = XC.
19. In the diagram, $L\hat{M}N = O\hat{N}M = 90°$. P is the mid-point of MN, MN = 2ML and MN = NO. Prove that
    (a) the triangles MNL and NOP are congruent
    (b) $O\hat{P}N = L\hat{N}O$
    (c) $L\hat{Q}O = 90°$.

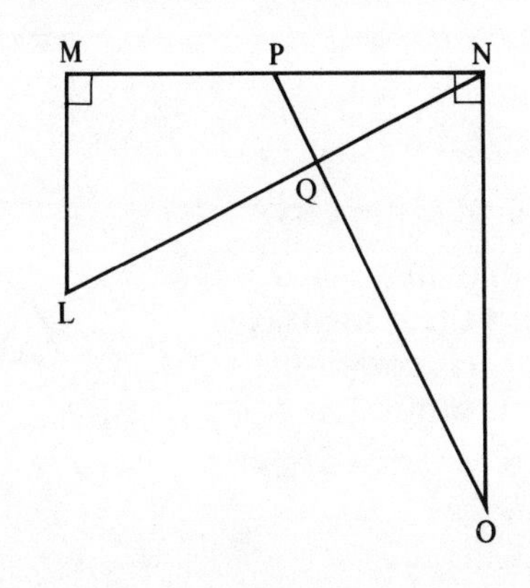

20. PQRS is a parallelogram in which the bisectors of the angles P and Q meet at X. Prove that the angle PXQ is a right angle.

4.4 SYMMETRY

(a) Line symmetry

The letter A has one line of symmetry, shown dotted.

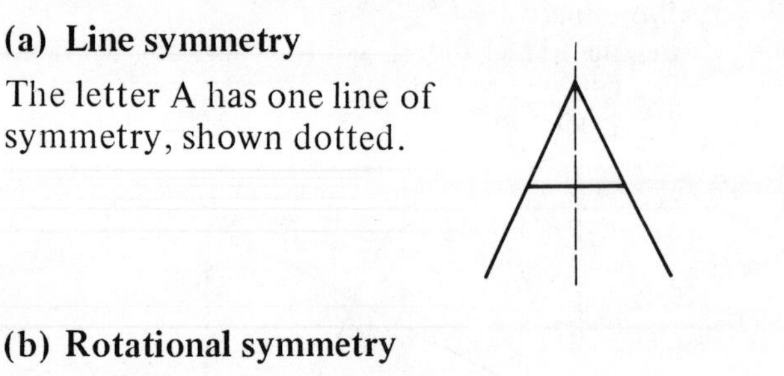

(b) Rotational symmetry

The shape may be turned about O into three identical positions. It has rotational symmetry of order 3.

Example 1

For the shapes given:
Find (a) the number of lines of symmetry
(b) the order of rotational symmetry.

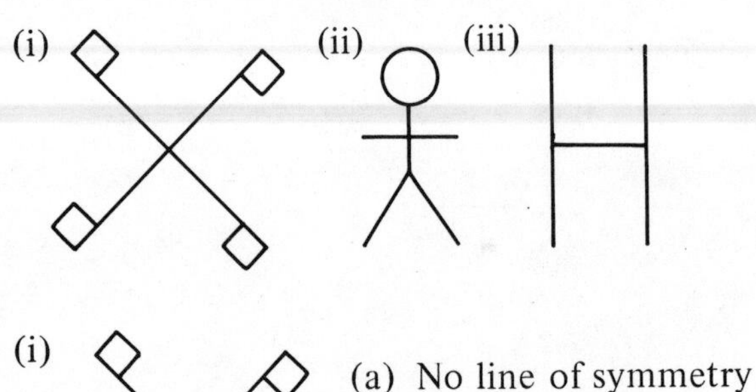

(i)

(a) No line of symmetry
(b) Order of rotational symmetry = 4

(ii)

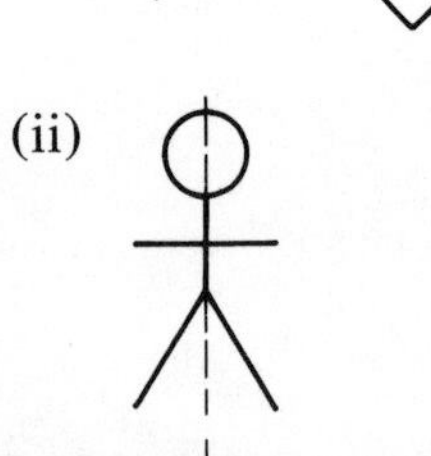

(a) 1 line of symmetry
(b) No rotational symmetry (i.e. order: 1)

(iii)

(a) 2 lines of symmetry
(b) Order of rotational symmetry = 2

(c) Quadrilaterals

1. *Square*

all sides are equal, all angles 90°, opposite sides parallel; diagonals bisect at right angles.

2. *Rectangle*

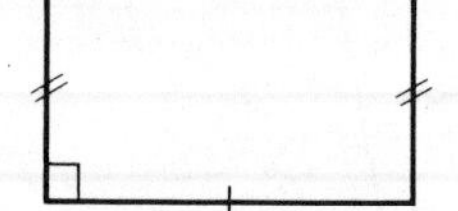

opposite sides parallel and equal, all angles 90°, diagonals bisect each other.

3. *Parallelogram*

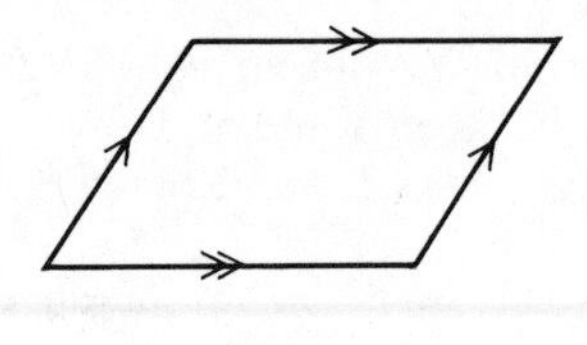

opposite sides parallel and equal, opposite angles equal, diagonals bisect each other (but not equal)

4. *Rhombus*

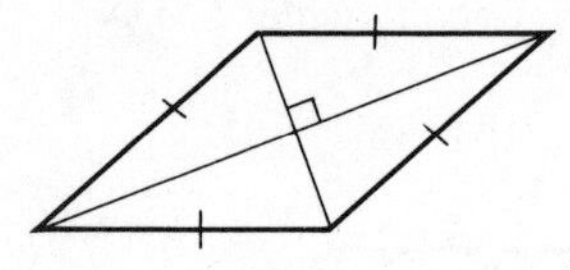

a parallelogram with all sides equal, diagonals bisect each other at right angles and bisect angles.

5. *Trapezium*

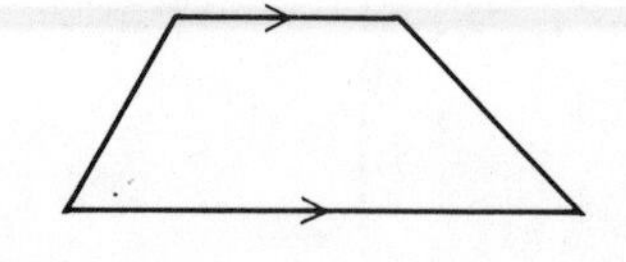

one pair of sides are parallel.

6. *Kite*

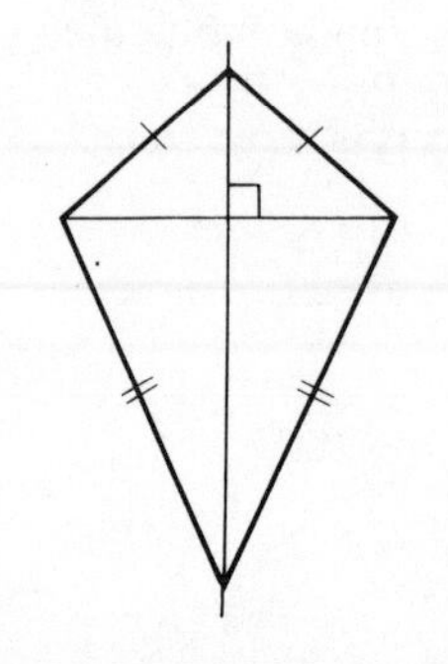

two pairs of adjacent sides equal, diagonals meet at right angles bisecting one of them.

Exercise 6

1. For each shape state:
   (a) the number of lines of symmetry
   (b) the order of rotational symmetry.

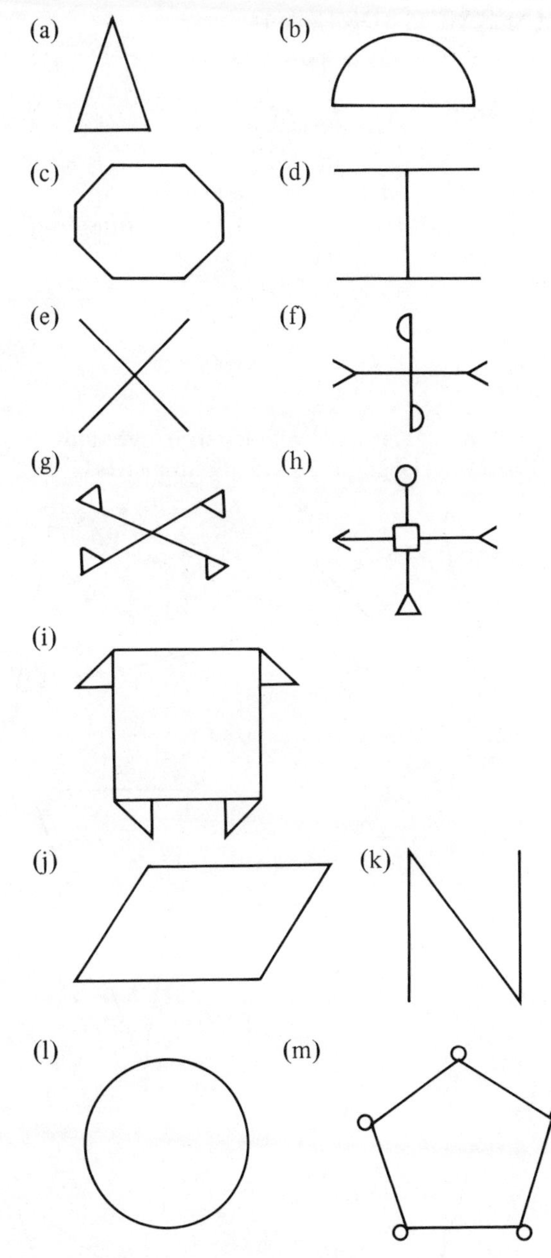

2. For each of the following shapes, find:
   (a) the number of lines of symmetry
   (b) the order of rotational symmetry.

   square; rectangle; parallelogram; rhombus; trapezium; kite; equilateral triangle; regular hexagon.

In questions **3** to **13**, begin by drawing a diagram.

3. In a rectangle KLMN, $L\hat{N}M = 34^\circ$. Calculate:
   (a) $K\hat{L}N$ (b) $K\hat{M}L$
4. In a trapezium ABCD; $A\hat{B}D = 35^\circ$, $B\hat{A}D = 110^\circ$ and AB is parallel to DC. Calculate:
   (a) $A\hat{D}B$ (b) $B\hat{D}C$.
5. In a parallelogram WXYZ, $W\hat{X}Y = 72^\circ$, $Z\hat{W}Y = 80^\circ$. Calculate:
   (a) $W\hat{Z}Y$ (b) $X\hat{W}Z$ (c) $W\hat{Y}X$
6. In a kite ABCD, AB = AD; BC = CD; $C\hat{A}D = 40^\circ$ and $C\hat{B}D = 60^\circ$. Calculate:
   (a) $B\hat{A}C$ (b) $B\hat{C}A$ (c) $A\hat{D}C$
7. In a rhombus ABCD, $A\hat{B}C = 64^\circ$. Calculate:
   (a) $B\hat{C}D$ (b) $A\hat{D}B$ (c) $B\hat{A}C$
8. In a rectangle WXYZ, M is the mid-point of WX and $Z\hat{M}Y = 70^\circ$. Calculate:
   (a) $M\hat{Z}Y$ (b) $Y\hat{M}X$
9. In a trapezium ABCD, AB is parallel to DC, AB = AD, BD = DC and $B\hat{A}D = 128^\circ$. Find:
   (a) $A\hat{B}D$ (b) $B\hat{D}C$ (c) $B\hat{C}D$
10. In a parallelogram KLMN, KL = KM and $K\hat{M}L = 64^\circ$. Find:
    (a) $M\hat{K}L$ (b) $K\hat{N}M$ (c) $L\hat{M}N$
11. In a kite PQRS with PQ = PS and RQ = RS, $Q\hat{R}S = 40^\circ$ and $Q\hat{P}S = 100^\circ$. Find:
    (a) $Q\hat{S}R$ (b) $P\hat{S}Q$ (c) $P\hat{Q}R$
12. In a rhombus PQRS, $R\hat{P}Q = 54^\circ$. Find:
    (a) $P\hat{R}Q$ (b) $P\hat{S}R$ (c) $R\hat{Q}S$
13. In a kite PQRS, $R\hat{P}S = 2\,P\hat{R}S$, PQ = QS = PS and QR = RS. Find:
    (a) $Q\hat{P}S$ (b) $P\hat{R}S$ (c) $Q\hat{S}R$ (d) $P\hat{Q}R$

4.5 SIMILARITY

Two triangles are similar if they have the same angles.
For other shapes, not only must corresponding angles be equal, but also corresponding sides must be in the same proportion.

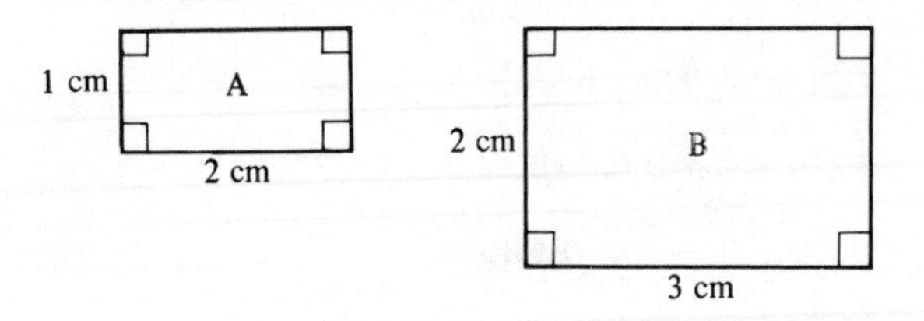

The two rectangles A and B are *not* similar even though they have the same angles.

Example 1

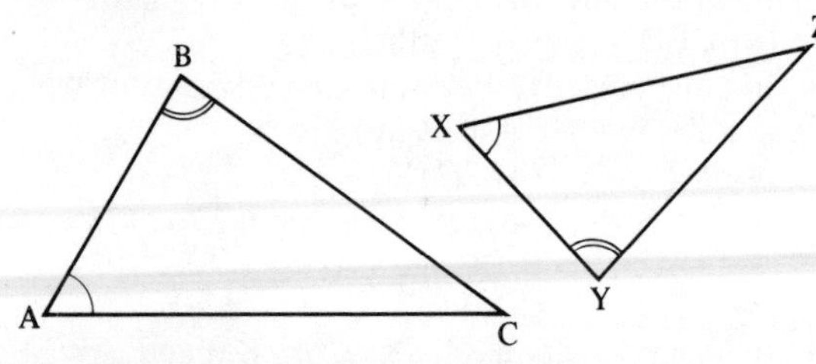

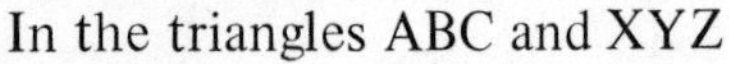

In the triangles ABC and XYZ

$\hat{A} = \hat{X}$ and $\hat{B} = \hat{Y}$

so the triangles are similar. ($\hat{C}$ must be equal to $\hat{Z}$.)

We have $\frac{BC}{YZ} = \frac{AC}{XZ} = \frac{AB}{XY}$

Note: BC and YZ are opposite $\hat{A}$ (= $\hat{X}$)
AC and XZ are opposite $\hat{B}$ (= $\hat{Y}$)
AB and XY are opposite $\hat{C}$ (= $\hat{Z}$)

Example 2

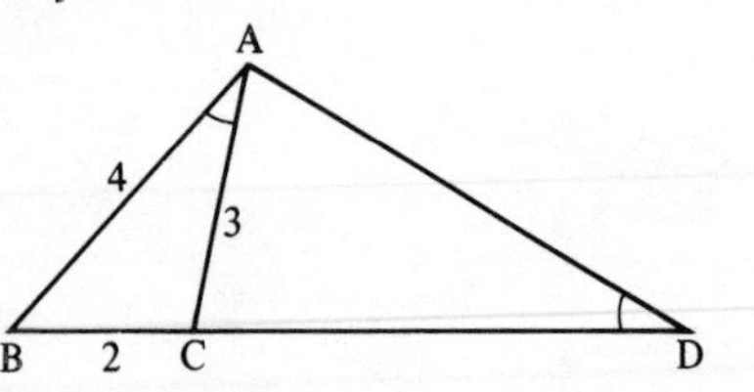

If AB = 4 cm, AC = 3 cm, BC = 2 cm and $B\hat{A}C = A\hat{D}B$, find CD.

In the triangles ABC and ABD
(a) angle B is common
(b) $B\hat{A}C = A\hat{D}B$ (given)

$\therefore$ the triangles ABC and DBA are similar.

so $\frac{BC}{AB} = \frac{AB}{BD} \left(= \frac{AC}{AD}\right)$

$\frac{2}{4} = \frac{4}{BD}$ $\quad\therefore$ BD = 8 cm
$\quad\therefore$ CD = 6 cm.

Exercise 7

Find the sides marked with letters in questions **1** to **11**; all lengths are given in centimetres.

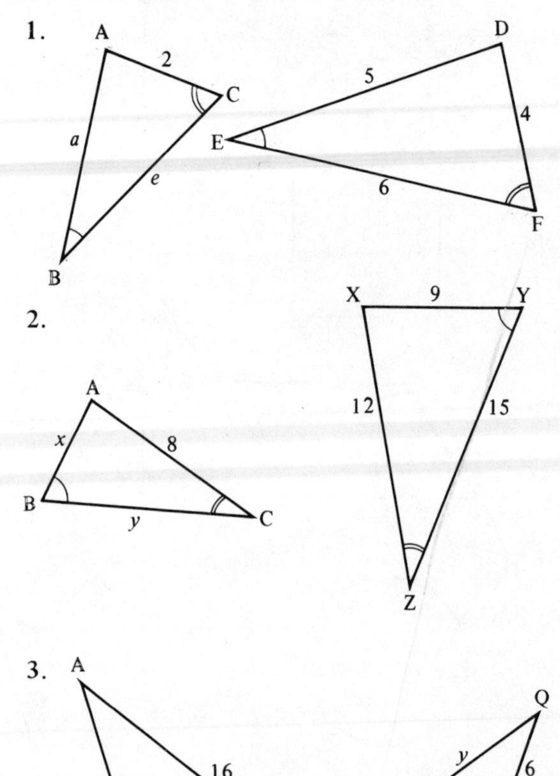

4.

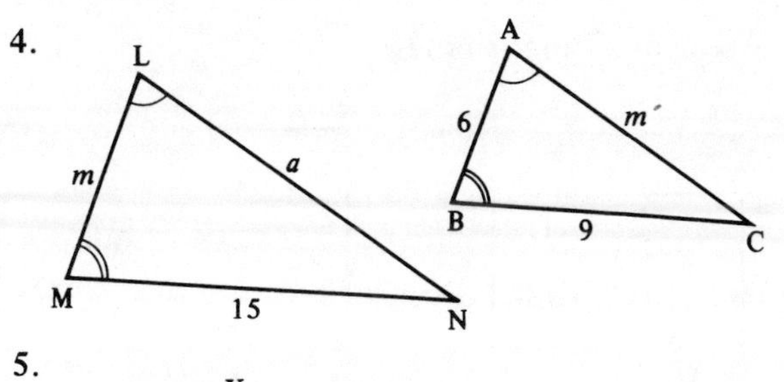

5.

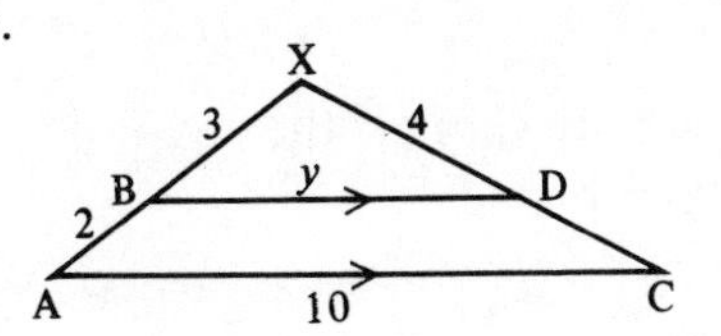

6.

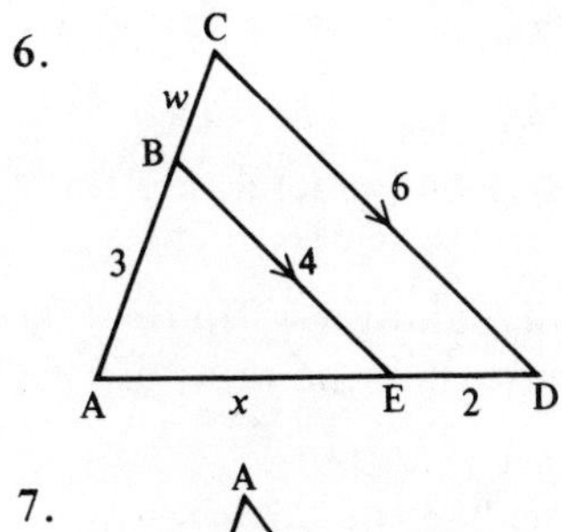

7.

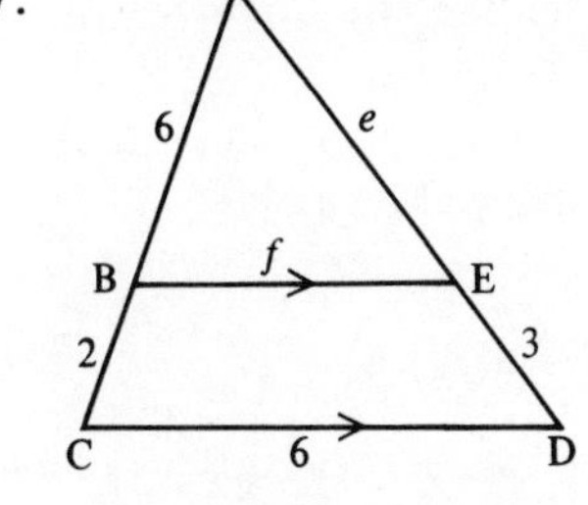

8.

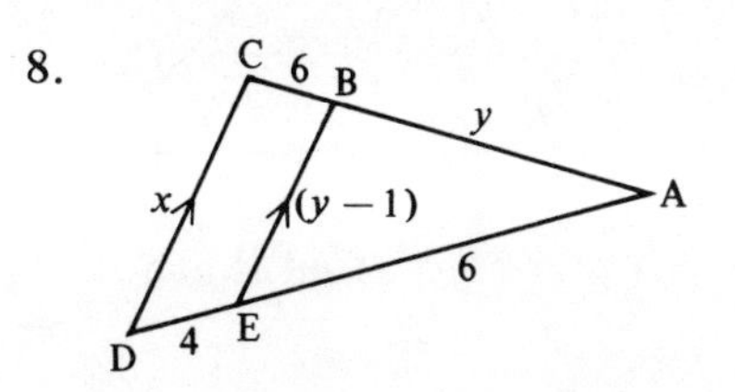

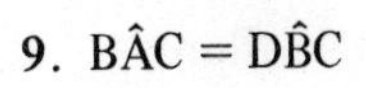

9. $B\hat{A}C = D\hat{B}C$

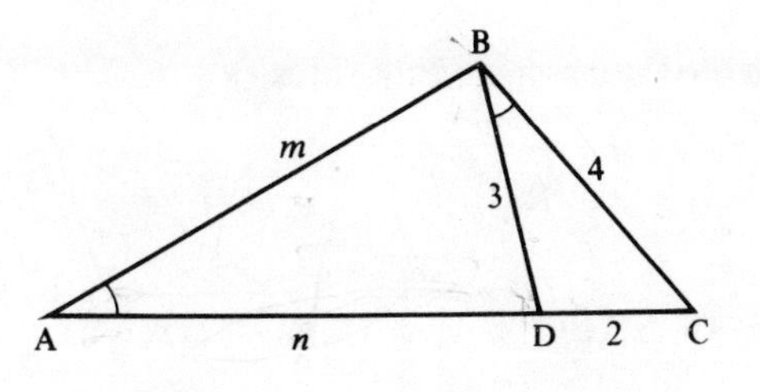

10.

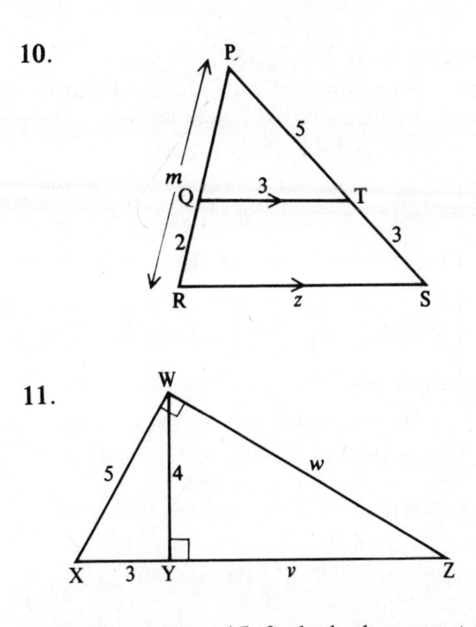

11.

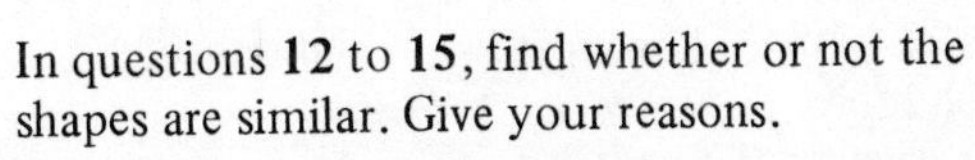

In questions **12** to **15**, find whether or not the shapes are similar. Give your reasons.

12.

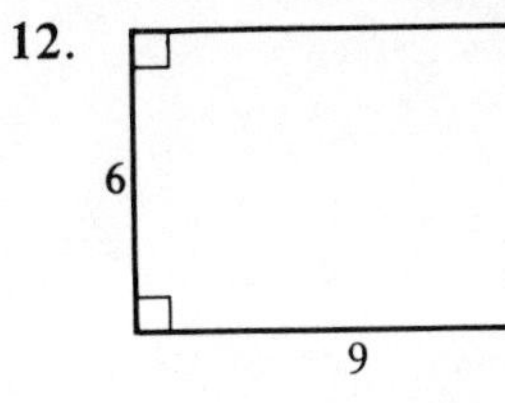

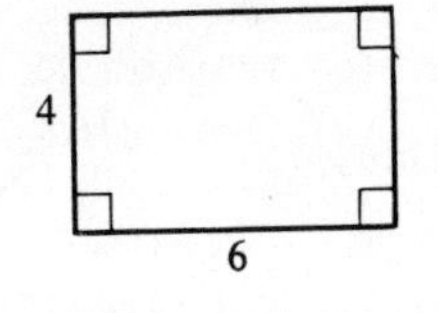

13.

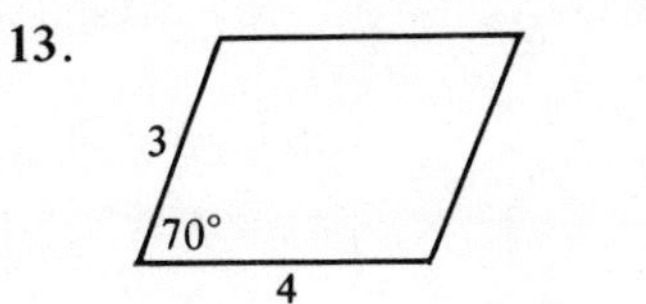

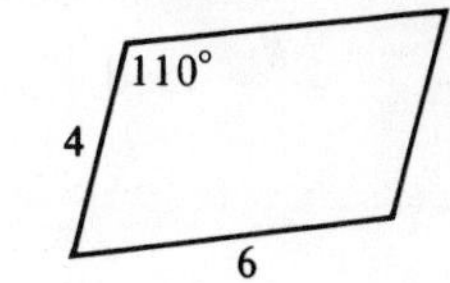

14.

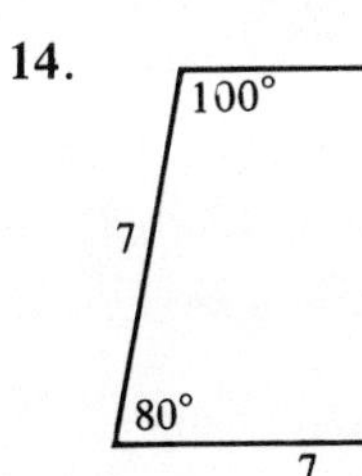

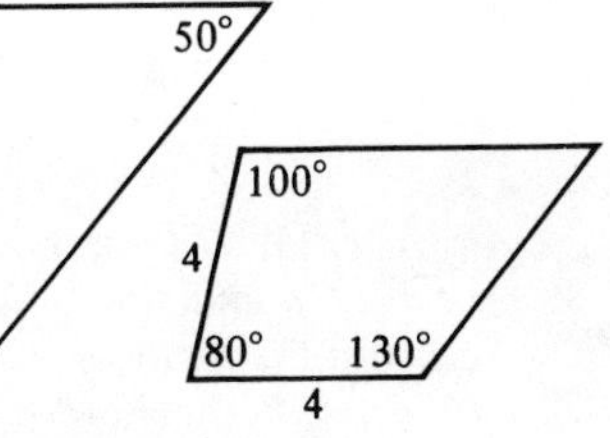

15.

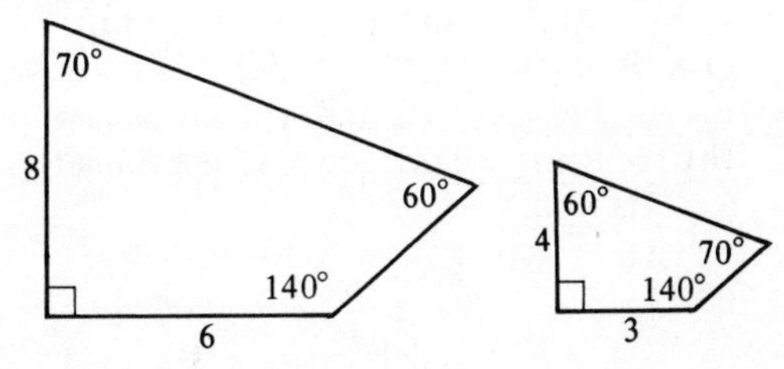

16. The drawing shows a rectangular picture 16 cm × 8 cm surrounded by a border of width 4 cm. Are the two rectangles similar?

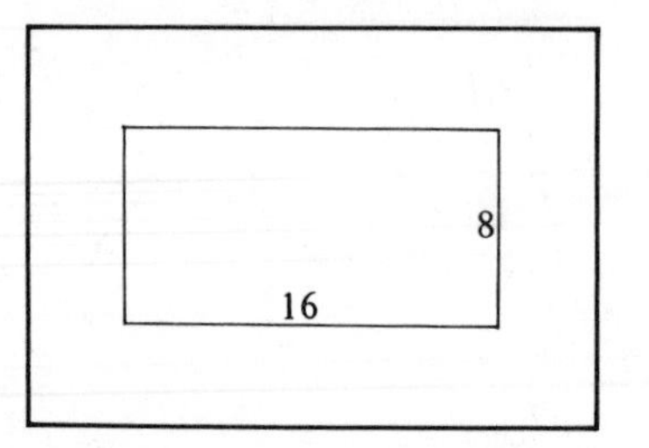

17. The diagonals of a trapezium ABCD intersect at O. AB is parallel to DC, AB = 3 cm and DC = 6 cm. If CO = 4 cm and OB = 3 cm, find AO and DO.

18. A tree of height 4 m casts a shadow of length 6·5 m. Find the height of a house casting a shadow 26 m long.

19. Which of the following *must* be similar to each other.
(a) Two equilateral triangles.
(b) Two rectangles.
(c) Two isosceles triangles.
(d) Two squares.
(e) Two regular pentagons.
(f) Two kites.
(g) Two rhombuses.
(h) Two circles.

20. In the diagram $A\hat{B}C = A\hat{D}B = 90°$, AD = p and DC = q.

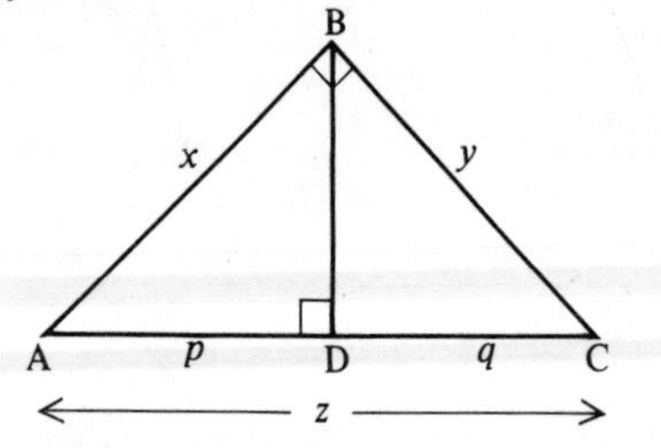

(a) Use similar triangles to show that $x^2 = pz$
(b) Find a similar expression for y^2
(c) Add the expressions for x^2 and y^2 and hence prove Pythagoras' theorem.

21. In a triangle ABC, a line is drawn parallel to BC to meet AB at D and AC at E. DC and BE meet at X. Prove that
(a) the triangles ADE and ABC are similar
(b) the triangles DXE and BXC are similar
(c) $\frac{AD}{AB} = \frac{EX}{XB}$

Areas of similar shapes

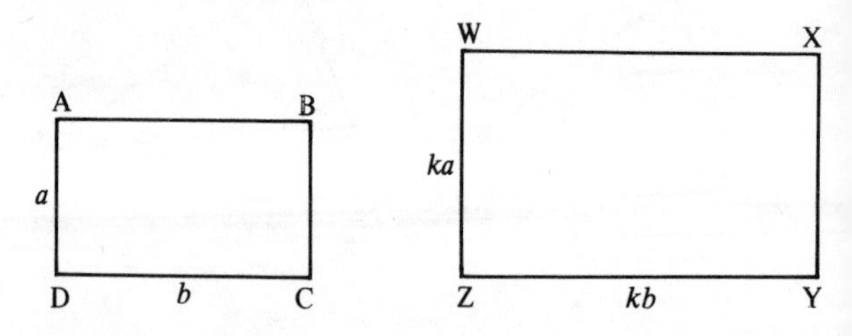

The two rectangles are similar, the ratio of corresponding sides being k.

area of ABCD $= ab$
area of WXYZ $= ka \,.\, kb = k^2ab.$

$$\therefore \quad \frac{\text{area WXYZ}}{\text{area ABCD}} = \frac{k^2ab}{ab} = k^2$$

This illustrates an important general rule for all similar shapes:

If two figures are similar and the ratio of corresponding sides is k, then the ratio of their areas is k^2.

Note: k is sometimes called the *linear scale factor.*
This result also applies for the surface areas of similar three dimensional objects.

Example 3

XY is parallel to BC

$$\frac{AB}{AX} = \frac{3}{2}$$

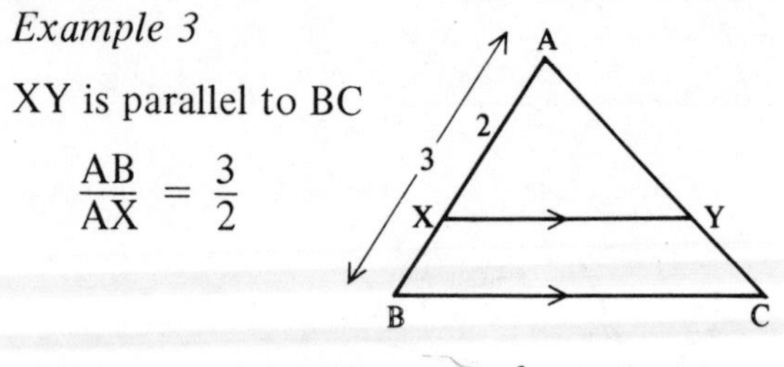

If the area of ΔAXY = 4 cm², find the area of ΔABC.

The triangles ABC and AXY are similar.

$$\text{Ratio of corresponding sides } (k) = \frac{3}{2}$$

$$\therefore \quad \text{Ratio of areas } (k^2) = \frac{9}{4}$$

$$\therefore \quad \text{Area of } \Delta ABC = \frac{9}{4} \times (\text{area of } \Delta AXY)$$

$$= \frac{9}{4} \times (4) = 9\,\text{cm}^2.$$

Example 4

Two similar triangles have areas of 18 cm^2 and 32 cm^2 respectively. If the base of the smaller triangle is 6 cm, find the base of the larger triangle.

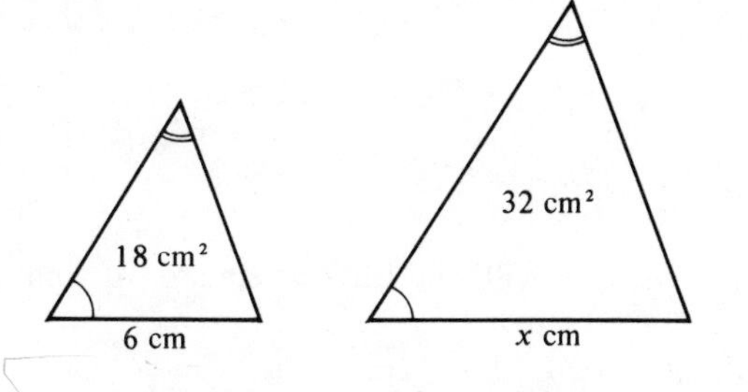

$$\text{Ratio of areas } (k^2) = \frac{32}{18} = \frac{16}{9}$$

$$\therefore \quad \text{Ratio of corresponding sides } (k) = \sqrt{\left(\frac{16}{9}\right)} = \frac{4}{3}$$

$$\therefore \quad \text{Base of larger triangle} = 6 \times \frac{4}{3} = 8 \text{ cm.}$$

Exercise 8

In this exercise a number written inside a figure represents the area of the shape in cm^2. Numbers on the outside give linear dimensions in cm.
In questions **1** to **6** find the unknown area A. In each case the shapes are similar.

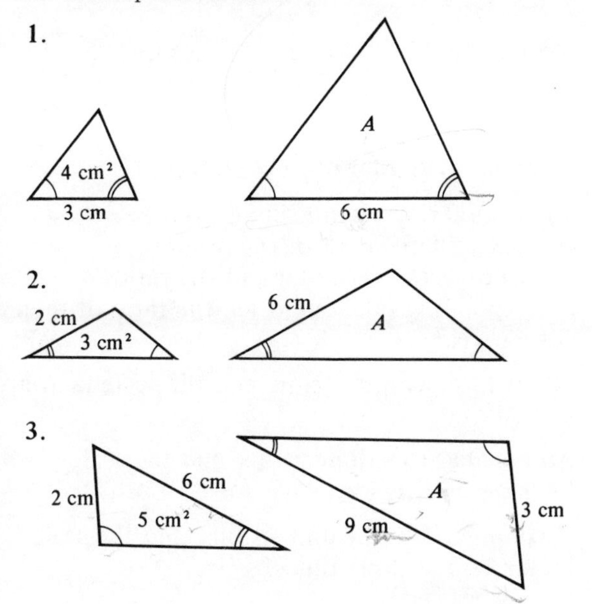

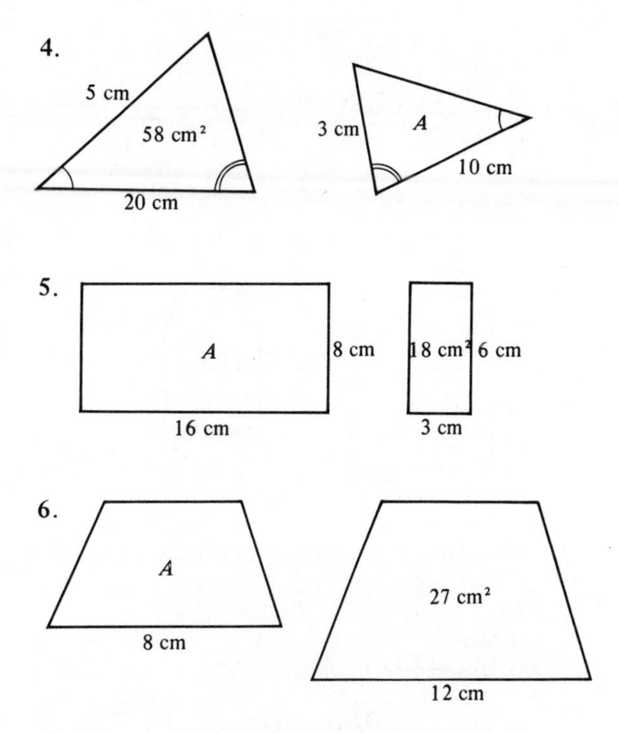

In questions **7** to **12**, find the lengths marked for each pair of similar shapes.

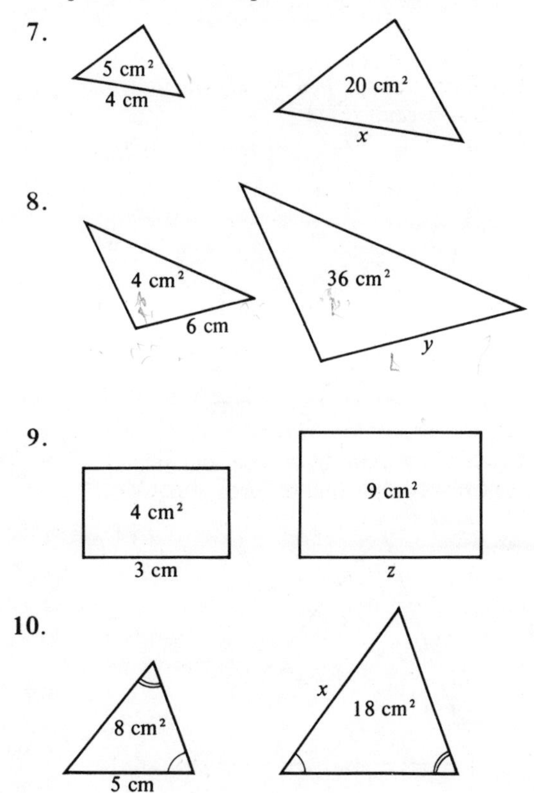

11.

12.

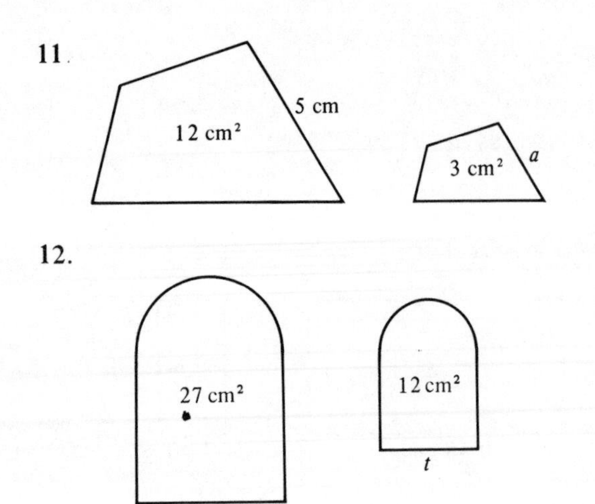

13. Given AD = 3 cm, AB = 5 cm and area of ΔADE = 6 cm^2.
Find:
(a) area of ΔABC
(b) area of DECB.

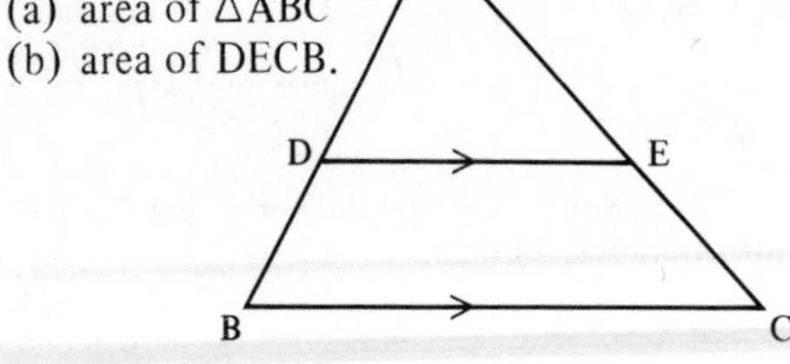

14. Given XY = 5 cm, MY = 2 cm and area of ΔMYN = 4 cm^2. Find:
(a) area of ΔXYZ (b) area of MNZX.

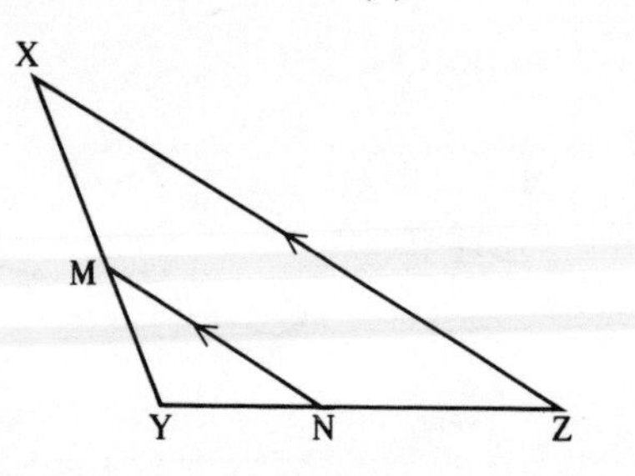

15. Given XY = 2 cm, BC = 3 cm and area of XYCB = 10 cm^2, find the area of ΔAXY.

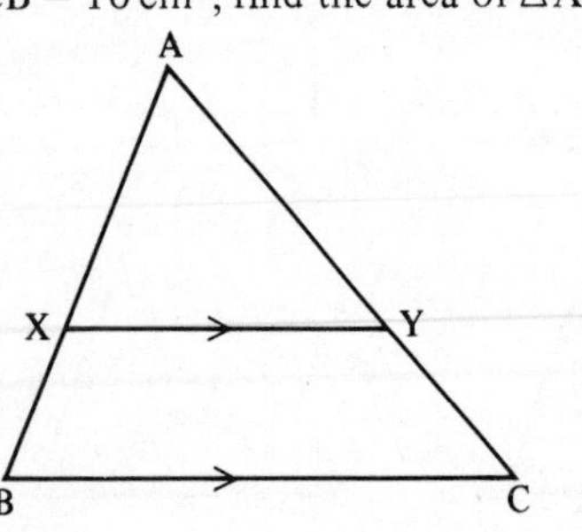

16. Given KP = 3 cm, area of ΔKOP = 2 cm^2 and area of OPML = 16 cm^2, find the length of PM.

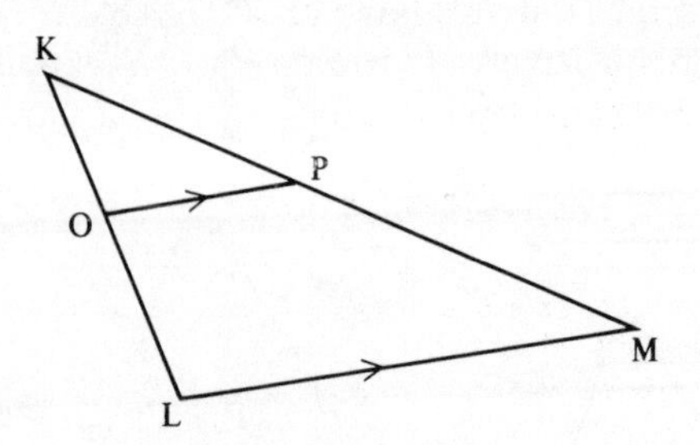

17. The triangles ABC and EBD are similar (AC and DE are *not* parallel).
If AB = 8 cm, BE = 4 cm and the area of ΔDBE = 6 cm^2, find the area of ΔABC.

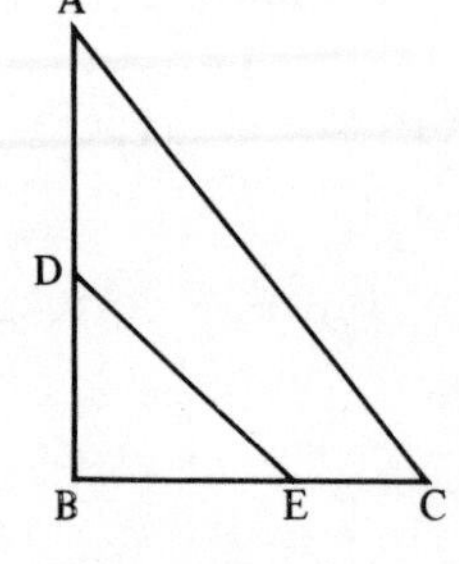

18. Given AZ = 3 cm, ZC = 2 cm, MC = 5 cm, BM = 3 cm. Find:
(a) XY
(b) YZ
(c) the ratio of areas AXY : AYZ
(d) the ratio of areas AXY : ABM

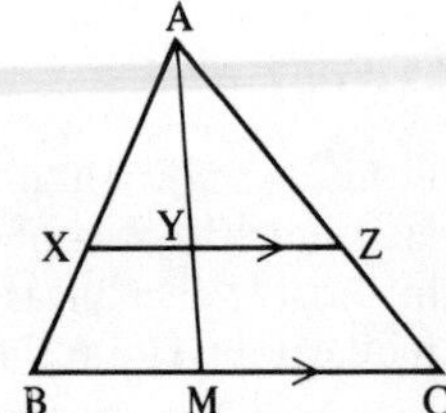

Volumes of similar objects

When solid objects are similar, one is an accurate enlargement of the other.
If two objects are similar and the ratio of corresponding sides is k, then the ratio of their volumes is k^3.

A line has one dimension, and the scale factor is used once.

An area has two dimensions, and the scale factor is used twice.

A volume has three dimensions, and the scale factor is used three times.

Example 5

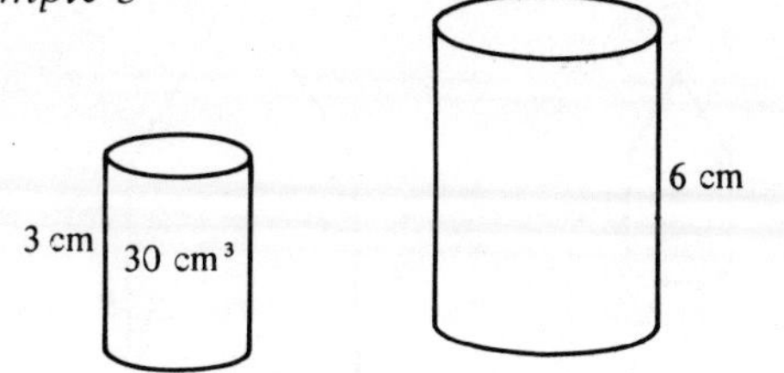

Two similar cylinders have heights of 3 cm and 6 cm respectively. If the volume of the smaller cylinder is 30 cm^3, find the volume of the larger cylinder.

$$\text{ratio of heights } (k) \text{ (linear scale factor)} = \frac{6}{3} = 2$$

$$\therefore \quad \text{ratio of volumes } (k^3) = 2^3 = 8$$

$$\text{and volume of larger cylinder} = 8 \times 30 = 240 \text{ cm}^3.$$

Example 6

Two similar spheres made of the same material have weights of 32 kg and 108 kg respectively. If the radius of the larger sphere is 9 cm, find the radius of the smaller sphere.

We may take the ratio of weights to be the same as the ratio of volumes.

$$\text{ratio of volumes } (k^3) = \frac{32}{108} = \frac{8}{27}$$

$$\text{ratio of corresponding lengths } (k) = \sqrt[3]{\left(\frac{8}{27}\right)} = \frac{2}{3}.$$

$$\therefore \quad \text{Radius of smaller sphere} = \frac{2}{3} \times 9 = 6 \text{ cm}.$$

Exercise 9

In this exercise, the objects are similar and a number written inside a figure represents the volume of the object in cm^3.
Numbers on the outside give linear dimensions in cm.
In questions **1** to **8**, find the unknown volume V.

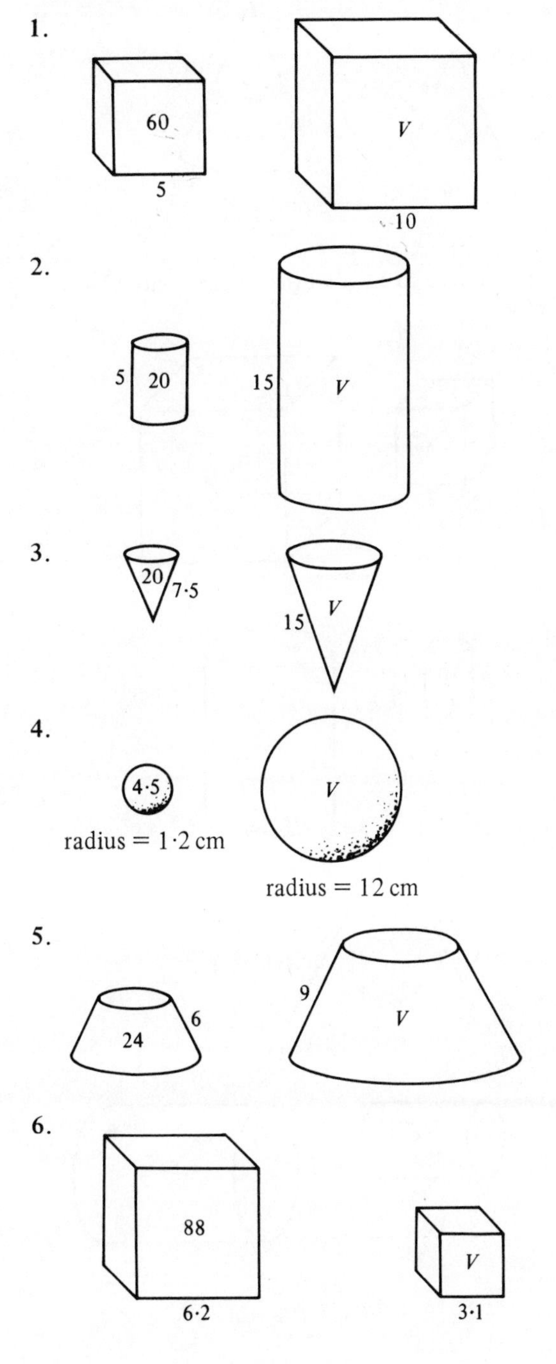

7.

8.

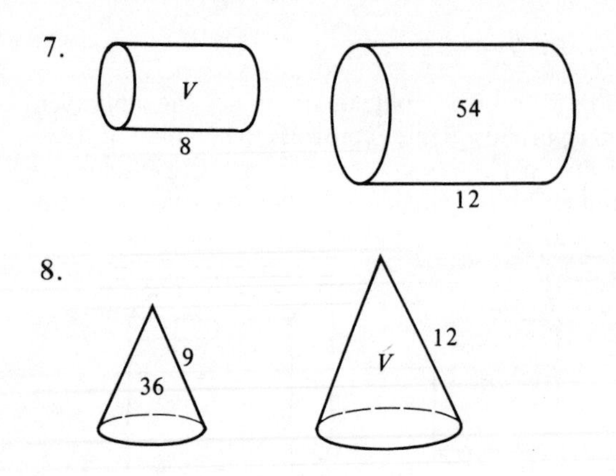

In questions **9** to **14**, find the lengths marked by a letter.

9.

10.

11.

12.

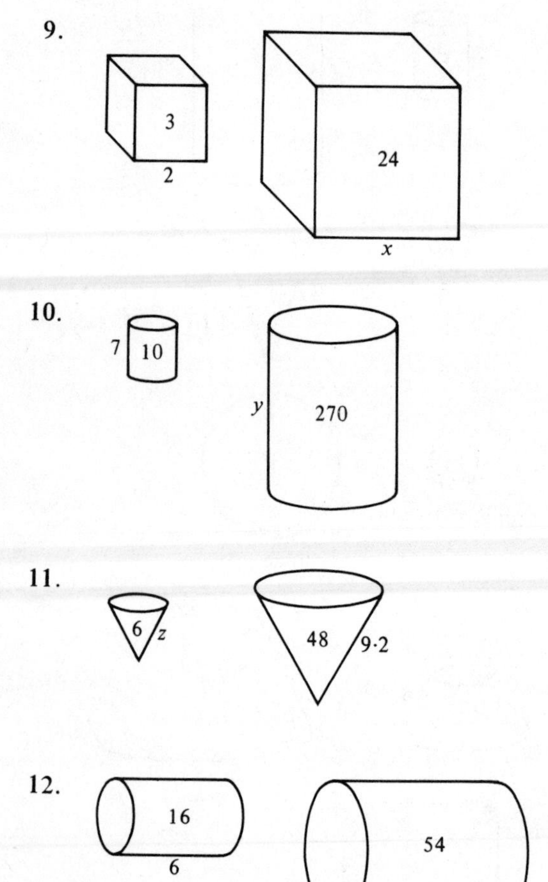

13.

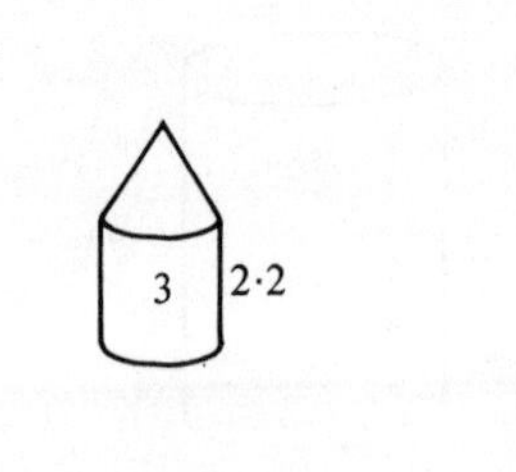

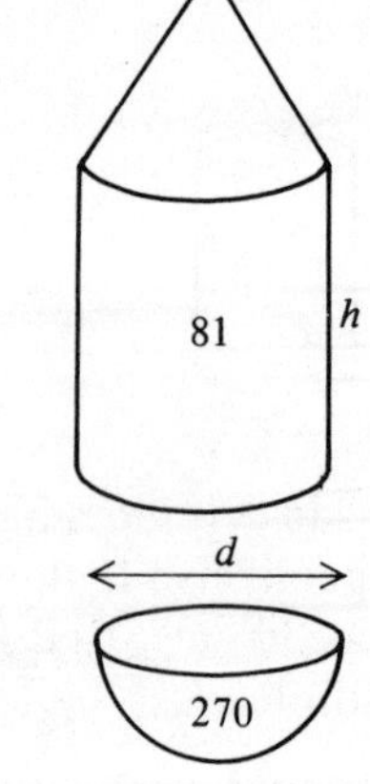

14. 3 80 d 270

15. Two similar jugs have heights of 4 cm and 6 cm respectively. If the capacity of the smaller jug is 50 cm^3, find the capacity of the larger jug.
16. Two similar cylindrical tins have base radii of 6 cm and 8 cm respectively. If the capacity of the larger tin is 252 cm^3, find the capacity of the small tin.
17. Two solid metal spheres have masses of 5 kg and 135 kg respectively. If the radius of the smaller one is 4 cm, find the radius of the larger one.
18. Two similar cones have surface areas in the ratio 4:9. Find the ratio of:
    (a) their lengths, (b) their volumes.
19. The area of the bases of two similar glasses are in the ratio 4:25. Find the ratio of their volumes.
20. Two similar solids have volumes V_1 and V_2 and corresponding sides of length x_1 and x_2. State the ratio $V_1 : V_2$ in terms of x_1 and x_2.
21. Two solid spheres have surface areas of 5 cm^2 and 45 cm^2 respectively and the mass of the smaller sphere is 2 kg. Find the mass of the larger sphere.
22. The masses of two similar objects are 24 kg and 81 kg respectively. If the surface area of the larger object is 540 cm^2, find the surface area of the smaller object.
23. A cylindrical can has a circumference of 40 cm and a capacity of 4·8 litres. Find the capacity of a similar cylinder of circumference 50 cm.
24. A container has a surface area of 5000 cm^2 and a capacity of 12·8 litres. Find the surface area of a similar container which has a capacity of 5·4 litres.
25. A full size snooker ball has a diameter of $2\frac{1}{16}$ inches and weighs 133·1 g. Calculate the weight of a snooker ball of diameter $1\frac{7}{8}$ inches, assuming that both balls are made of the same material.

4.6 CIRCLE THEOREMS

(a) The angle subtended at the centre of a circle is twice the angle subtended at the circumference.

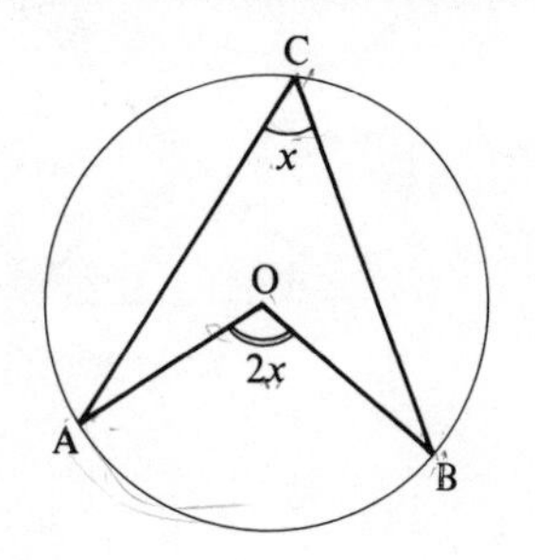

$A\hat{O}B = 2 \times A\hat{C}B$

Proof:
Draw the straight line COD.
Let $A\hat{C}O = y$
and $B\hat{C}O = z$.

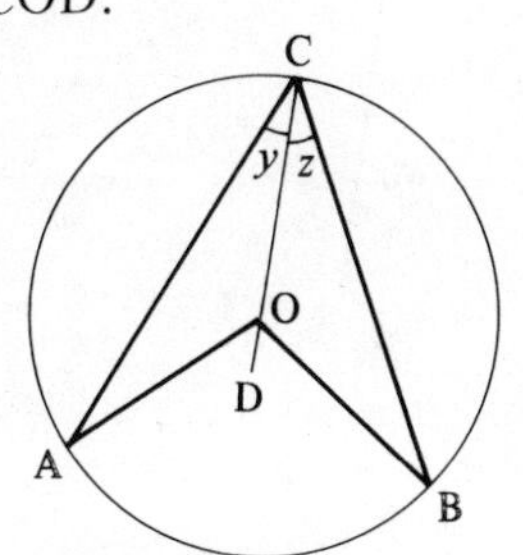

In triangle AOC,

$$AO = OC \quad \text{(radii)}$$
$$\therefore \; O\hat{C}A = O\hat{A}C \quad \text{(isosceles triangle)}$$
$$\therefore \; C\hat{O}A = 180 - 2y \quad \text{(angle sum of triangle)}$$
$$\therefore \; A\hat{O}D = 2y \quad \text{(angles on a straight line)}$$

Similarly from triangle COB, we find

$$D\hat{O}B = 2z$$
Now $A\hat{C}B = y + z$
and $A\hat{O}B = 2y + 2z$
$\therefore$ $A\hat{C}B = 2 \times A\hat{O}B$ as required.

(b) Angles subtended by an arc in the same segment of a circle are equal.

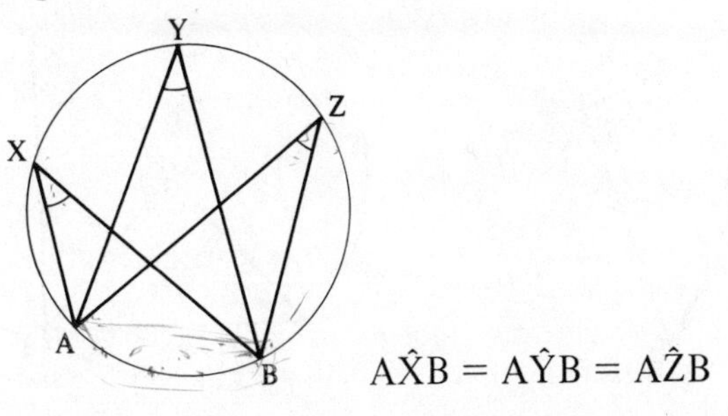

$A\hat{X}B = A\hat{Y}B = A\hat{Z}B$

Example 1

Given $A\hat{B}O = 50°$, find $B\hat{C}A$.

Triangle OBA is isosceles (OA = OB).

$\therefore$ $O\hat{A}B = 50°$
$\therefore$ $B\hat{O}A = 80°$ (angle sum of a triangle)
$\therefore$ $B\hat{C}A = 40°$ (angle at the centre)

Example 2

Given $B\hat{D}C = 62°$ and $D\hat{C}A = 44°$, find $B\hat{A}C$ and $A\hat{B}D$.

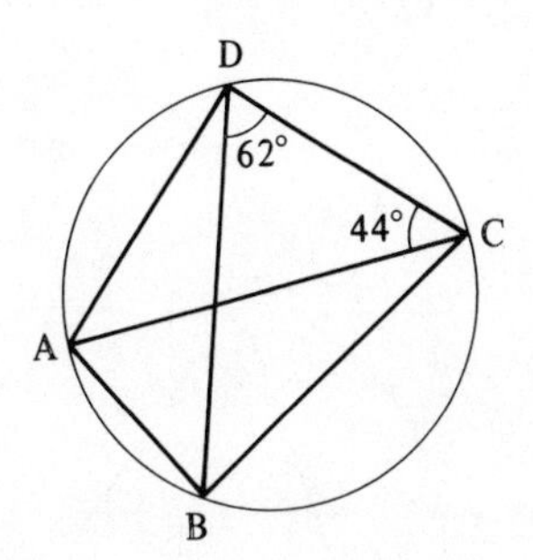

$B\hat{D}C = B\hat{A}C$ (both subtended by arc BC)
$\therefore$ $B\hat{A}C = 62°$

$D\hat{C}A = A\hat{B}D$ (both subtended by arc DA)
$\therefore$ $A\hat{B}D = 44°$.

Exercise 10

Find the angles marked with letters.
A line passes through the centre only when point O is shown.

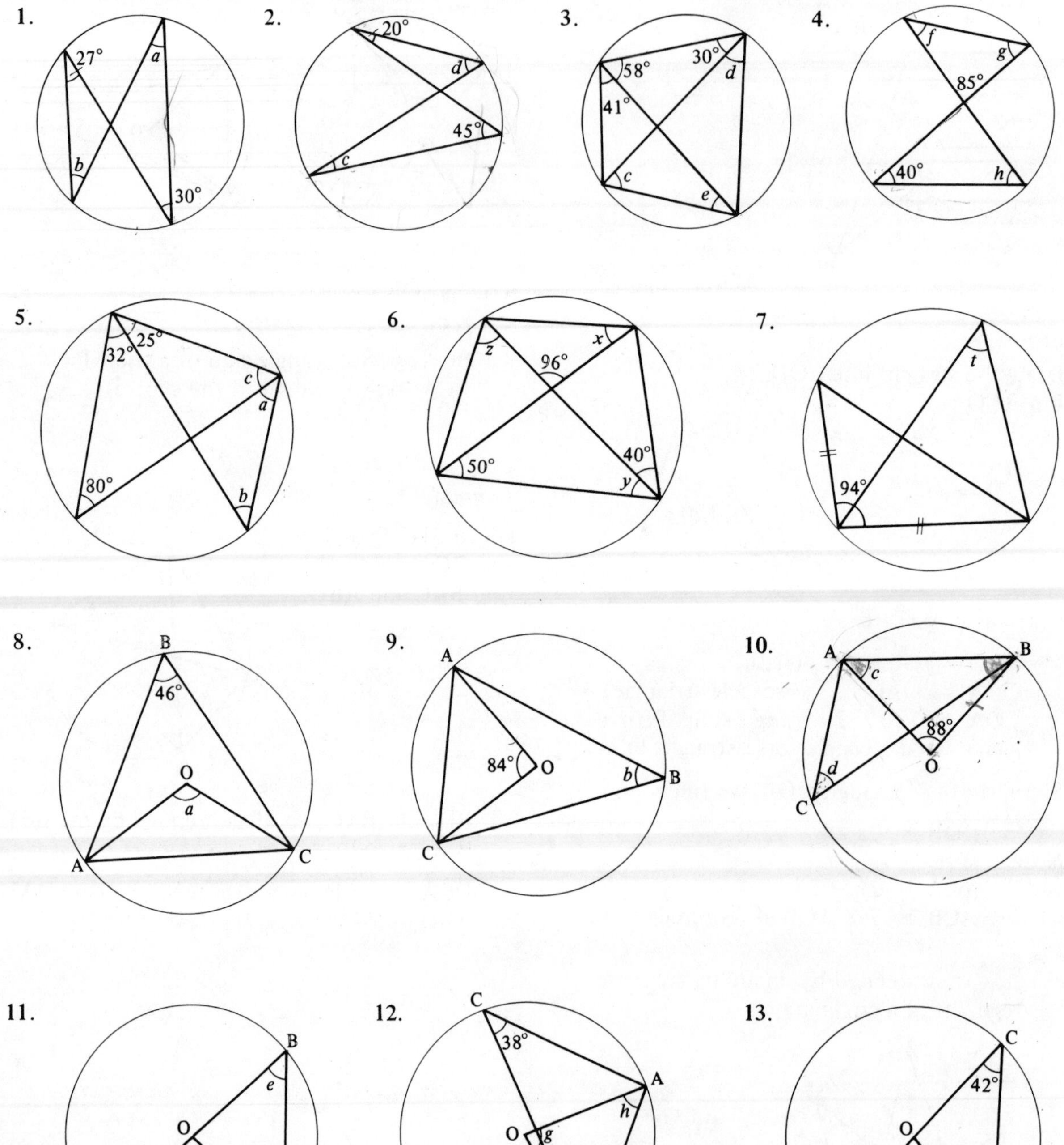

11.

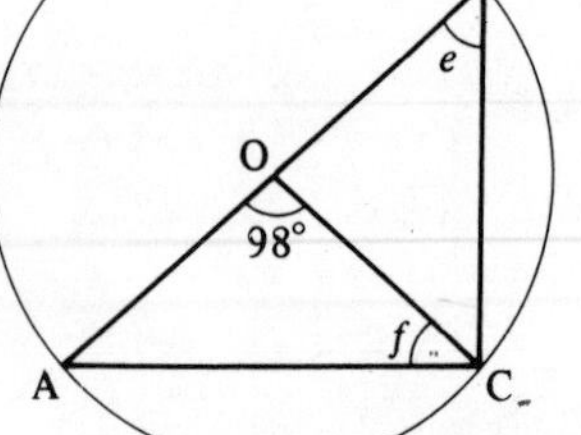

12.

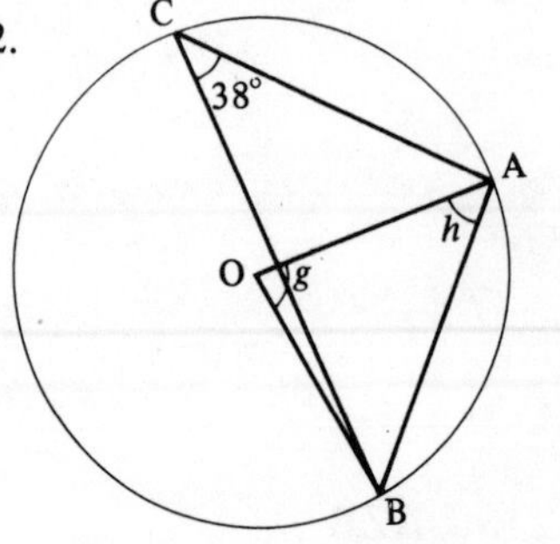

13.

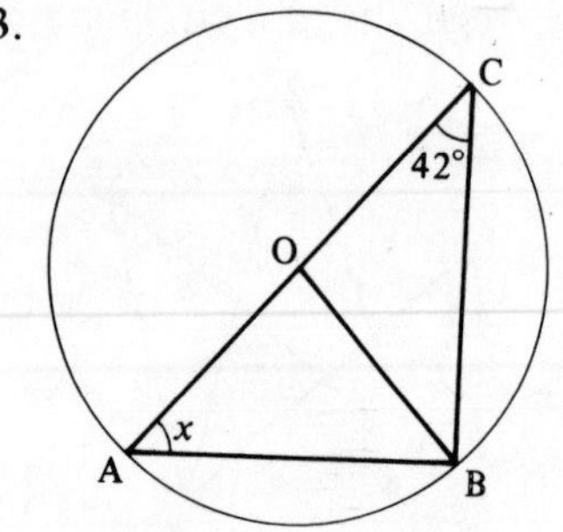

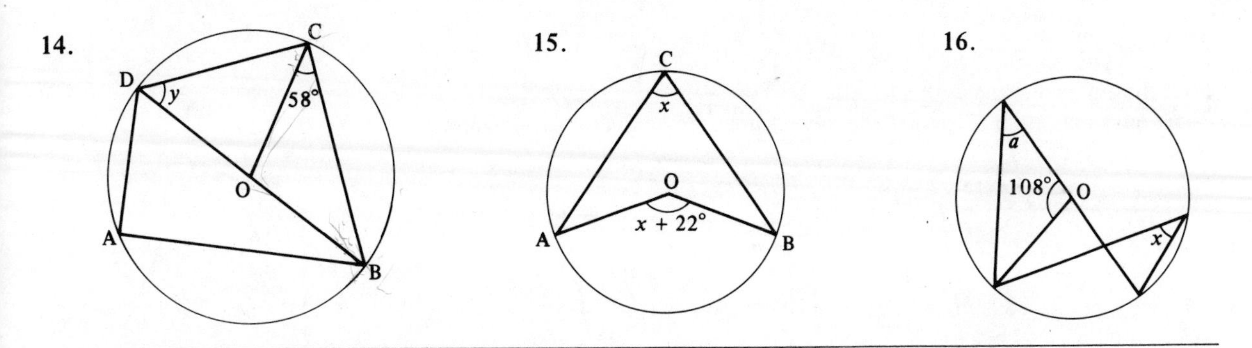

(c) The opposite angles in a cyclic quadrilateral add up to 180° (the angles are supplementary).

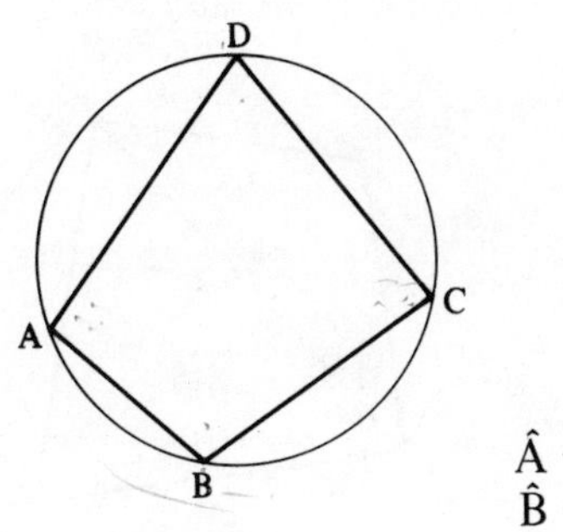

$$\hat{A} + \hat{C} = 180°$$
$$\hat{B} + \hat{D} = 180°$$

Proof:

Draw radii OA and OC.
Let $A\hat{D}C = x$
and $A\hat{B}C = y$.

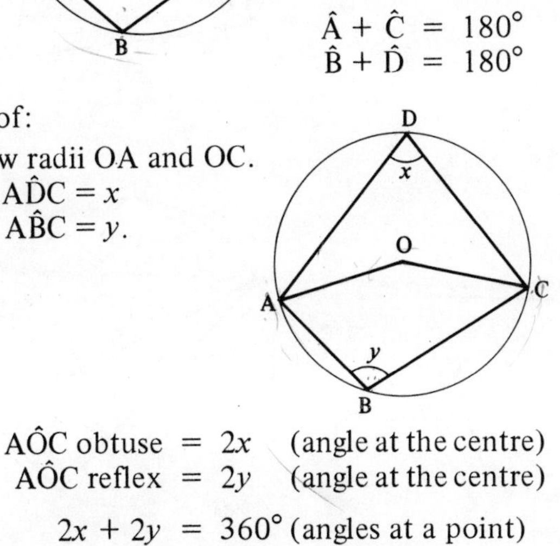

$A\hat{O}C$ obtuse $= 2x$ (angle at the centre)
$A\hat{O}C$ reflex $= 2y$ (angle at the centre)

$\therefore \quad 2x + 2y = 360°$ (angles at a point)

$\therefore \quad x + y = 180°$ as required.

(d) The angle in a semi-circle is a right angle.

In the diagram,
AB is a diameter

$A\hat{C}B = 90°$.

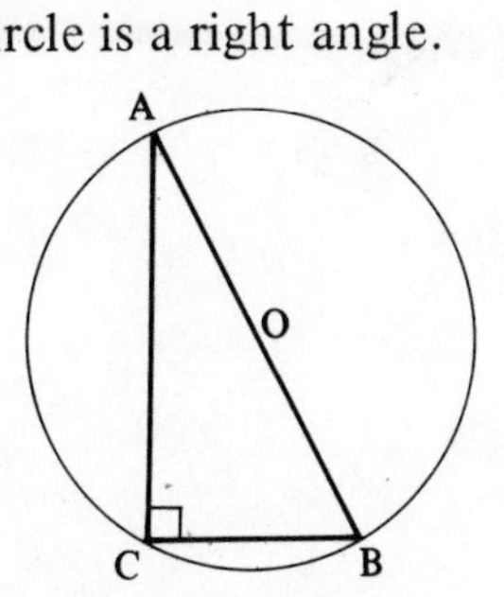

Example 3

Find a and x.

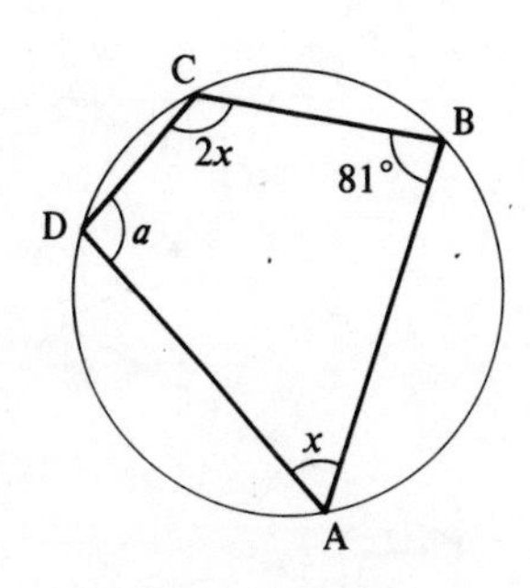

$a = 180° - 81°$ (opposite angles of a cyclic quadrilateral)

$\therefore \quad a = 99°$

$x + 2x = 180°$ (opposite angles of a cyclic quadrilateral)

$3x = 180°$
$x = 60°$

Example 4

Find b.

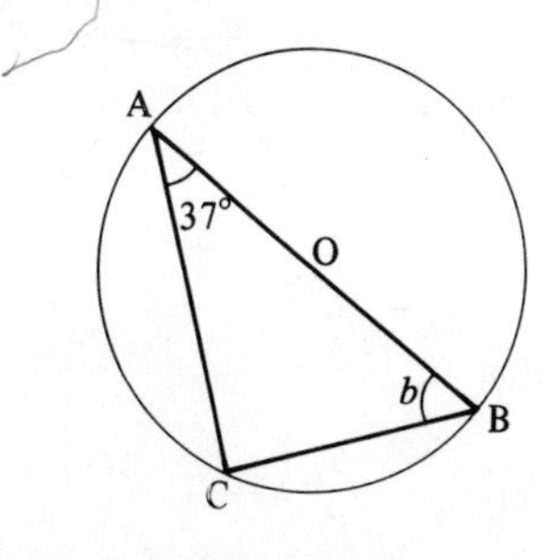

$A\hat{C}B = 90°$ (angle in a semi-circle)

$\therefore \quad b = 180° - (90 + 37)°$
$= 53°$.

Exercise 11

Find the angles marked with a letter.
Some questions are made easier when extra lines are drawn on the diagram.

1.

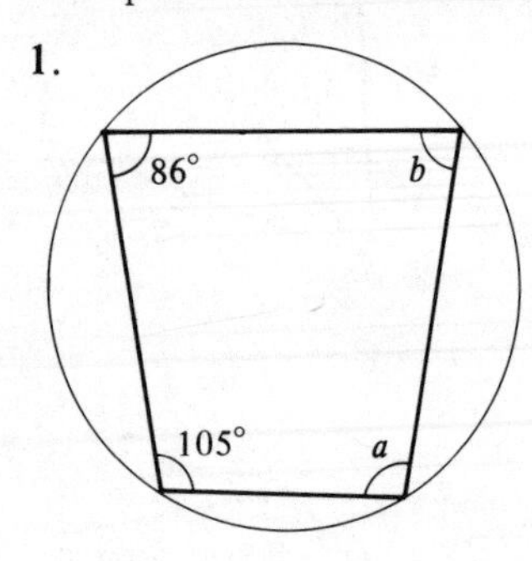

2.

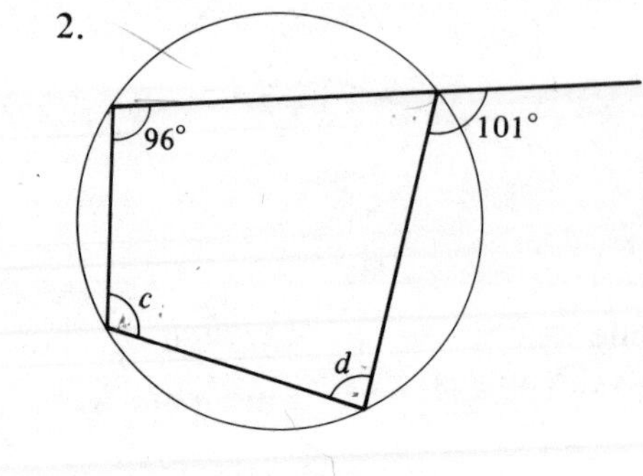

3.

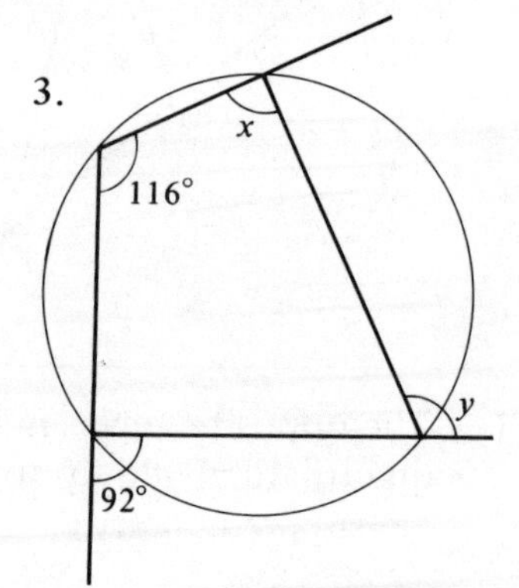

4.

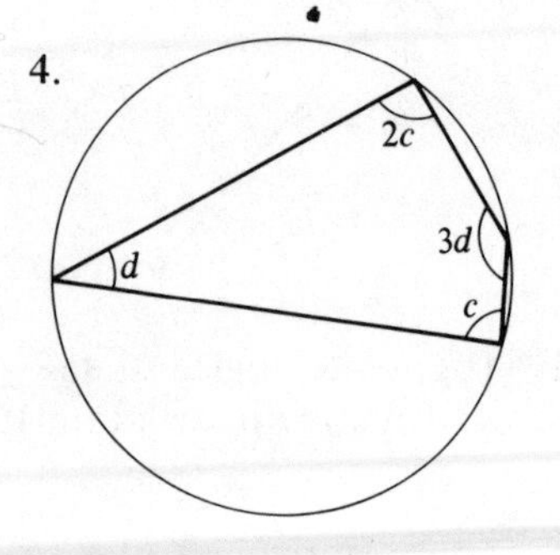

5.

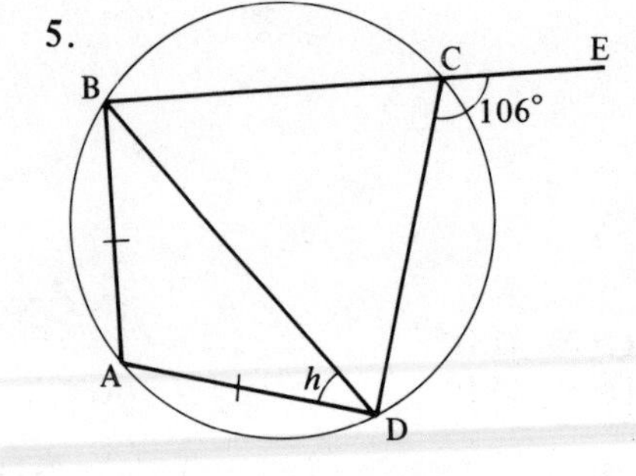

6.

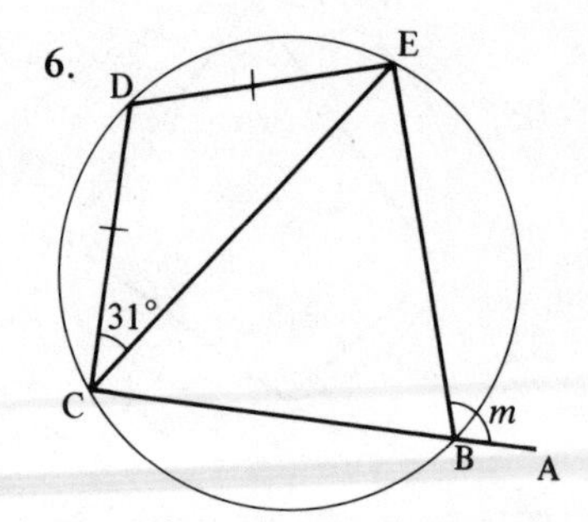

7.

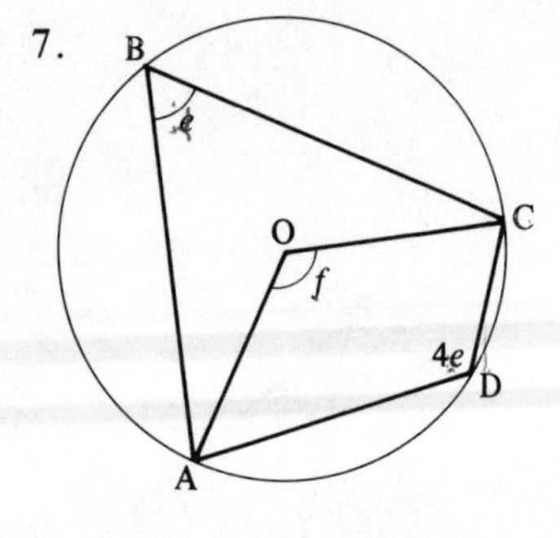

8.

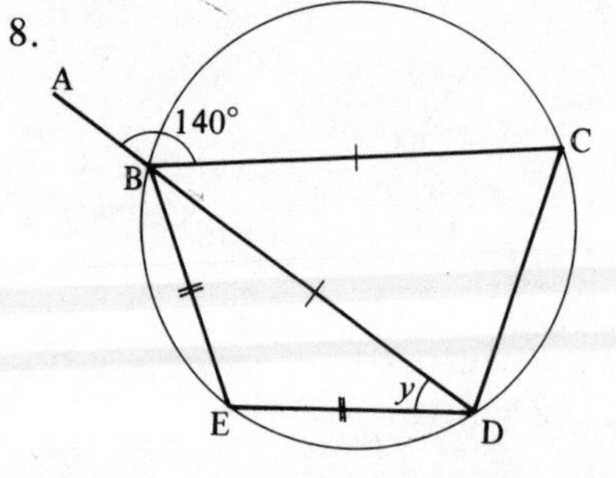

9.

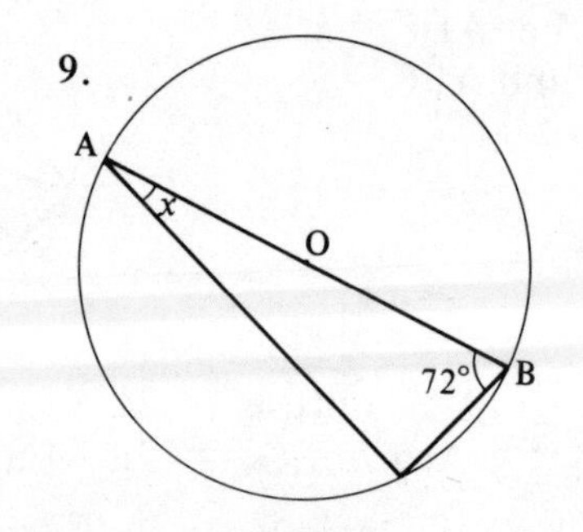

10.

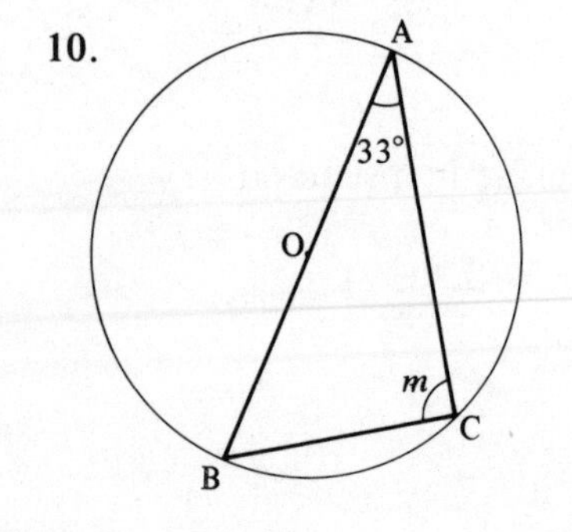

11.

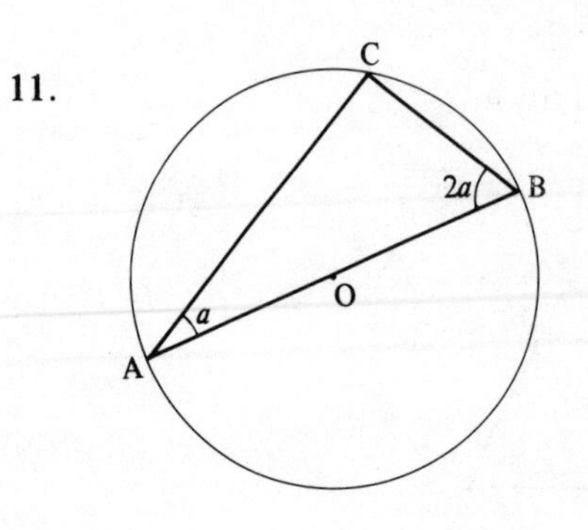

12.

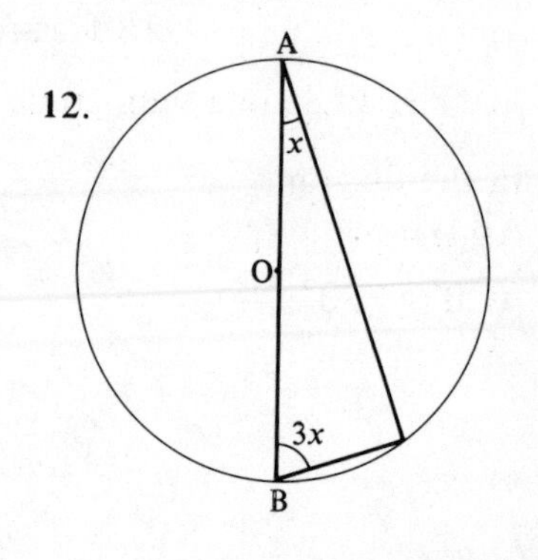

13.

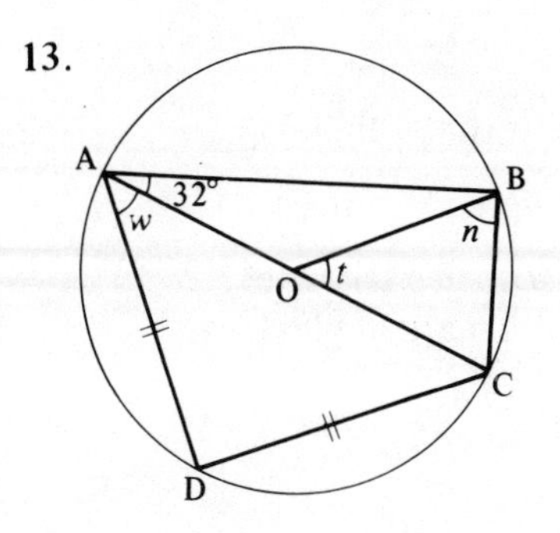

14.

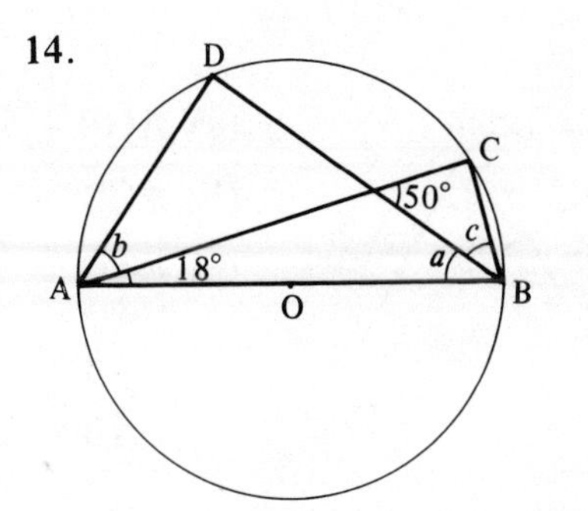

15.

16.

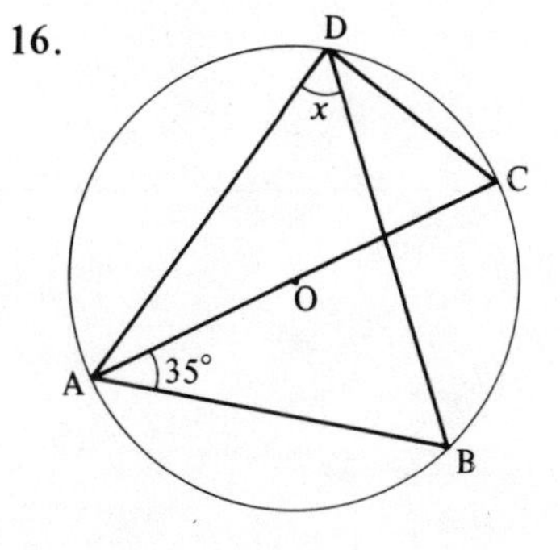

17.

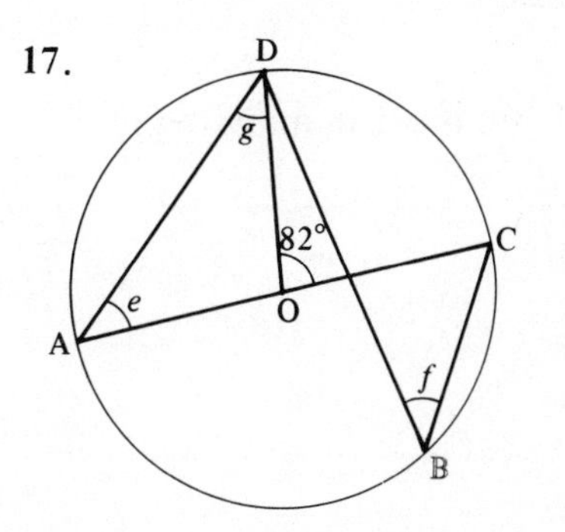

18.

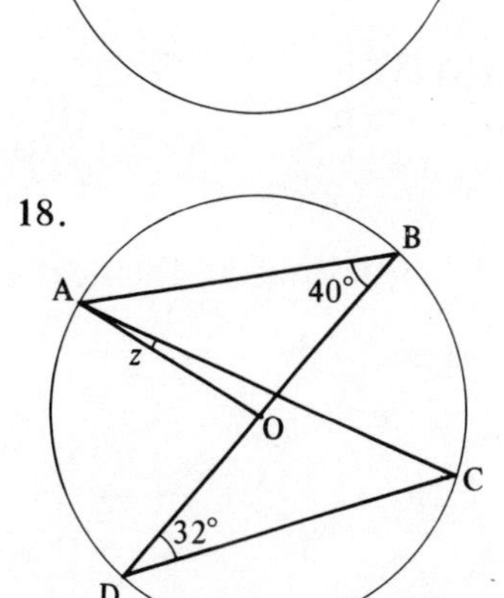

19.

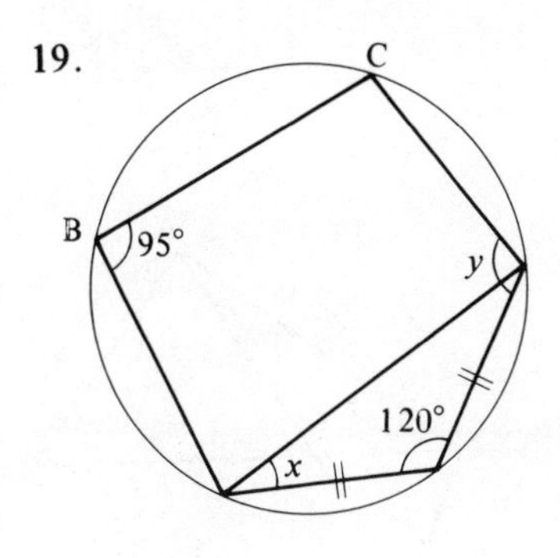

20.

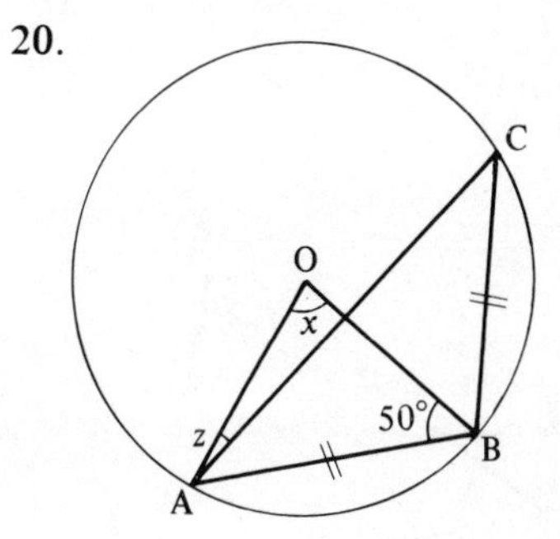

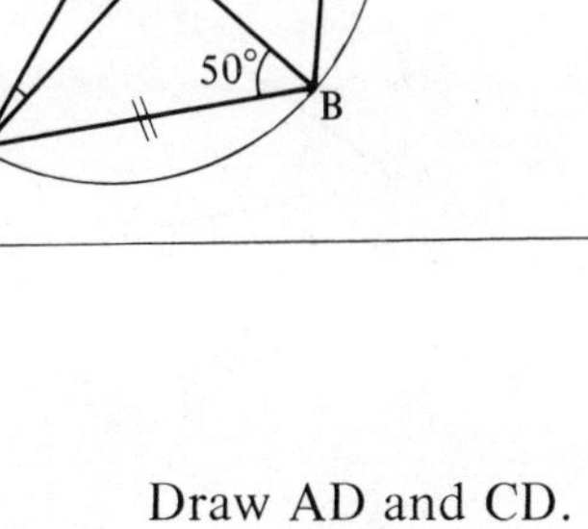

Example 5

O is the centre of the circle and $A\hat{O}C = 140°$. Find $A\hat{B}C$.

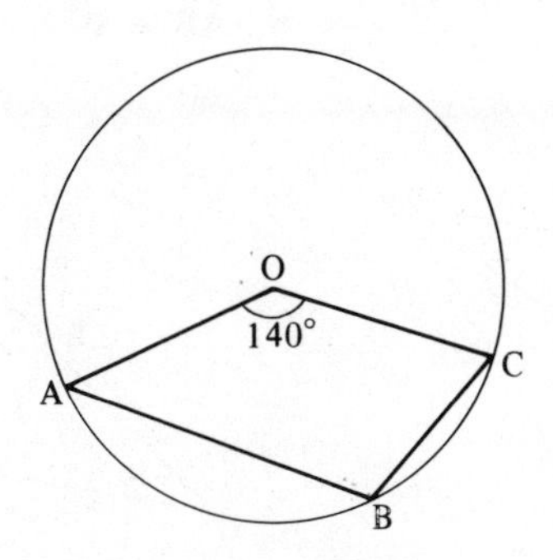

Draw AD and CD.

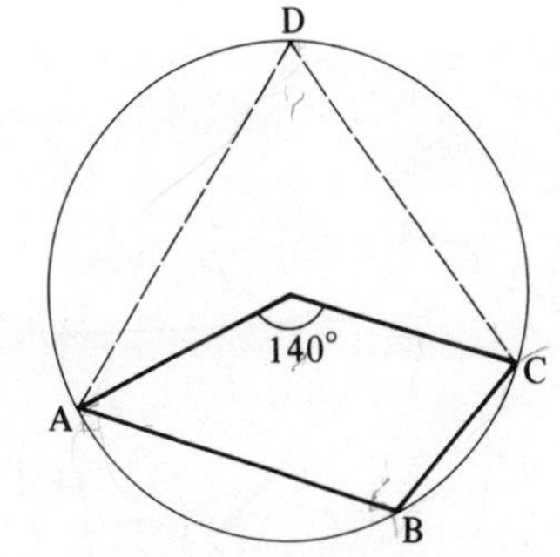

Now $A\hat{D}C = 70°$ (angle at centre)

$\therefore$ $A\hat{B}C = 110°$ (opposite angles in a cyclic quadrilateral)

Example 6

AO is parallel to BC and $O\hat{B}C = 58°$.

Find
(a) $A\hat{O}B$
(b) $A\hat{C}B$
(c) $C\hat{A}B$

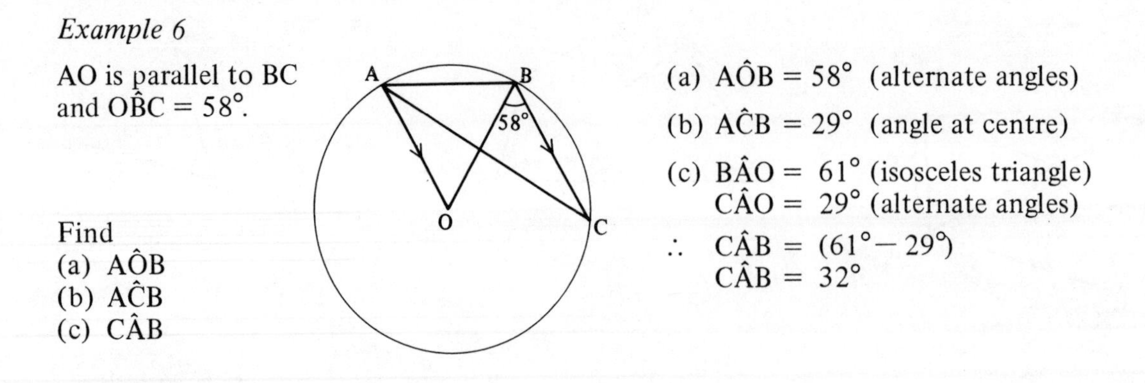

(a) $A\hat{O}B = 58°$ (alternate angles)

(b) $A\hat{C}B = 29°$ (angle at centre)

(c) $B\hat{A}O = 61°$ (isosceles triangle)
$C\hat{A}O = 29°$ (alternate angles)
$\therefore\ C\hat{A}B = (61° - 29°)$
$C\hat{A}B = 32°$

The questions in the next exercise are more demanding.

Exercise 12

For questions **1** to **18**, find the angles marked with a letter.

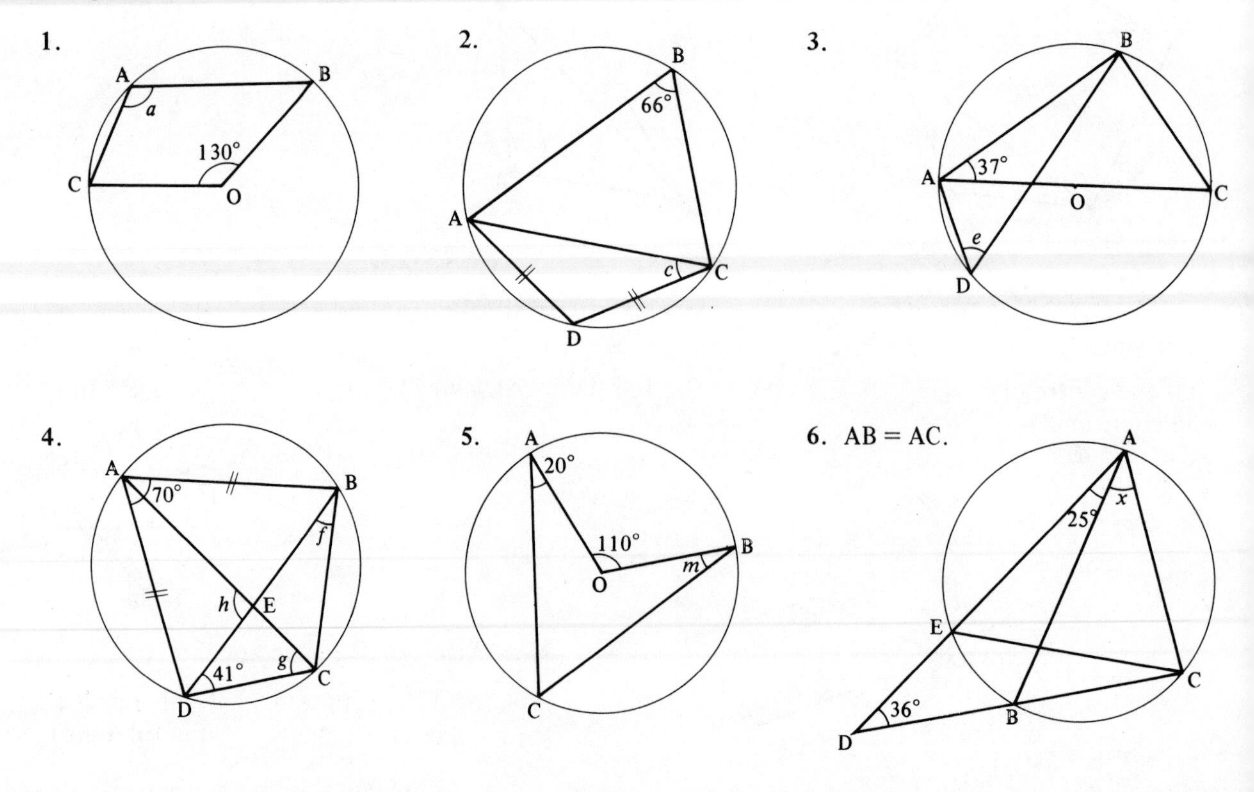

7.

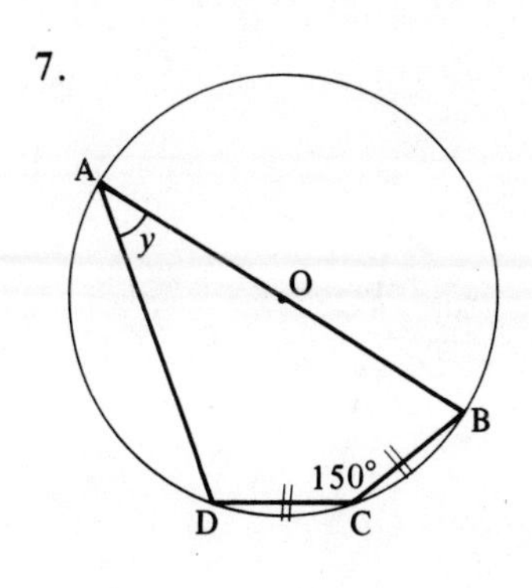

8.

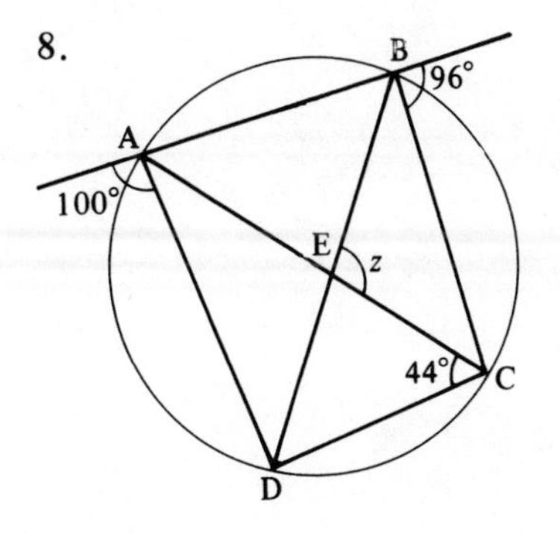

9.

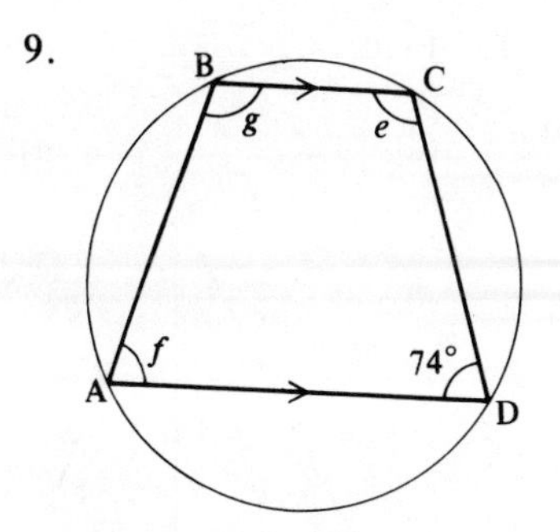

10.

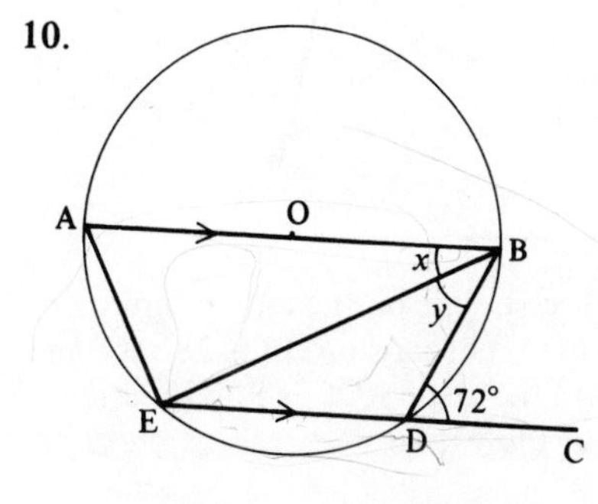

11.

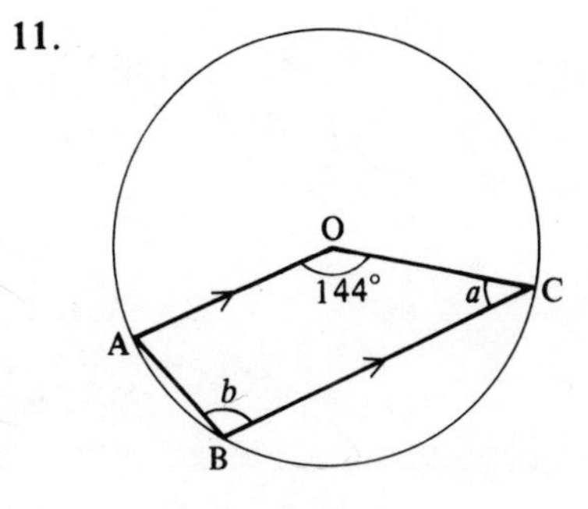

12.

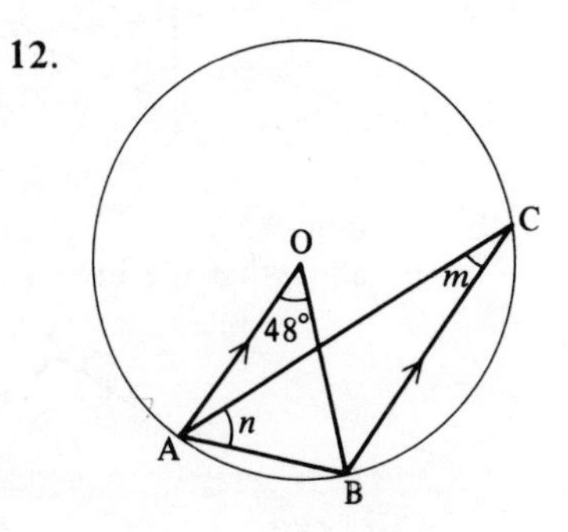

13.

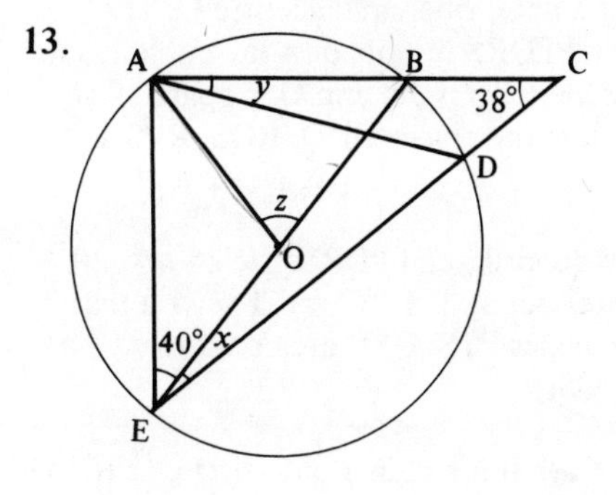

14.

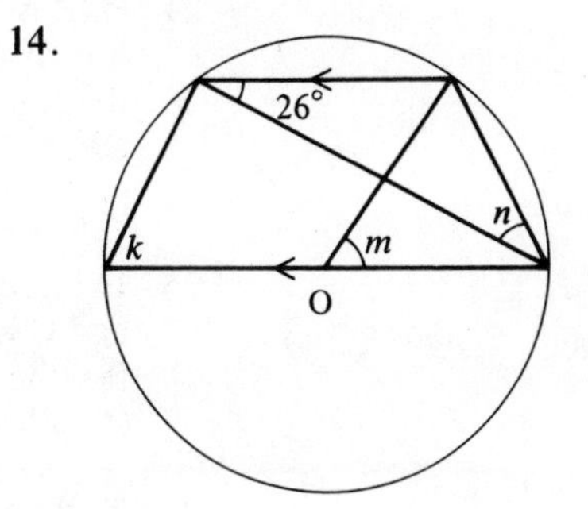

15.

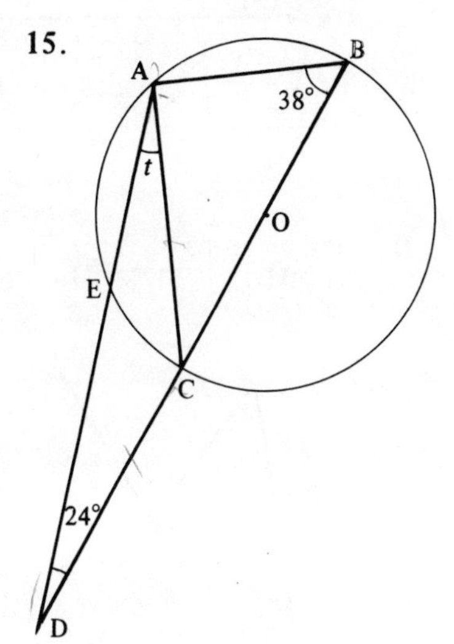

16.

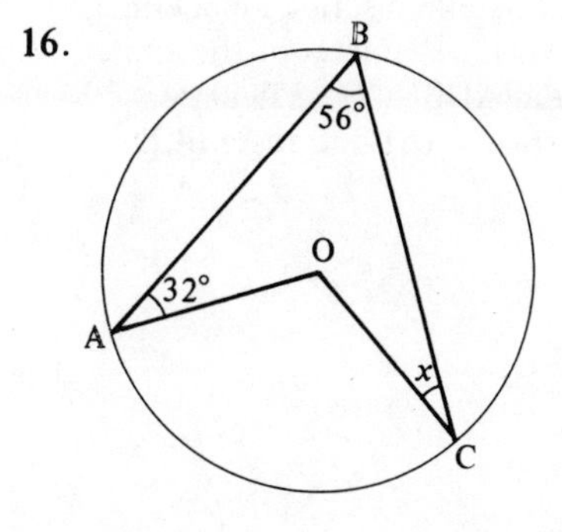

17.

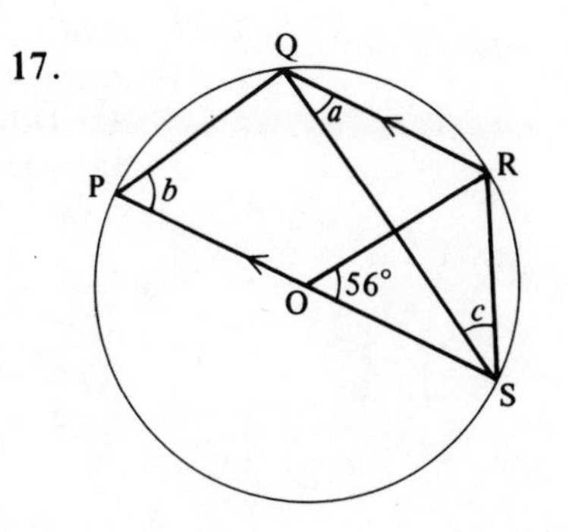

18.

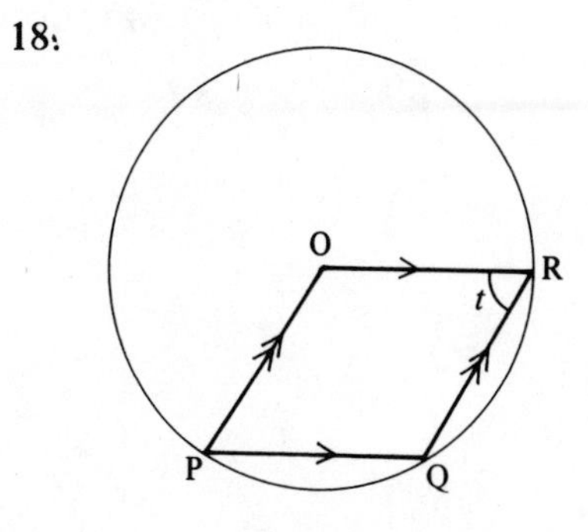

19. Find:
(a) $A\hat{B}O$; (b) $A\hat{C}D$;
(c) $D\hat{O}C$; (d) $A\hat{D}C$.

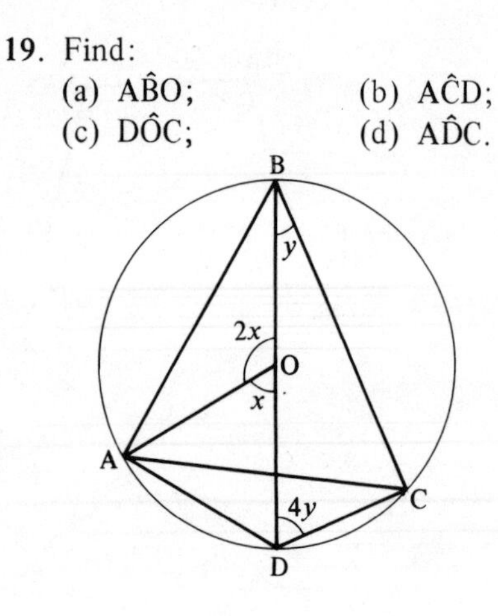

20. (a) Find $W\hat{X}Y$.
(b) Show that WY bisects $X\hat{W}Z$.

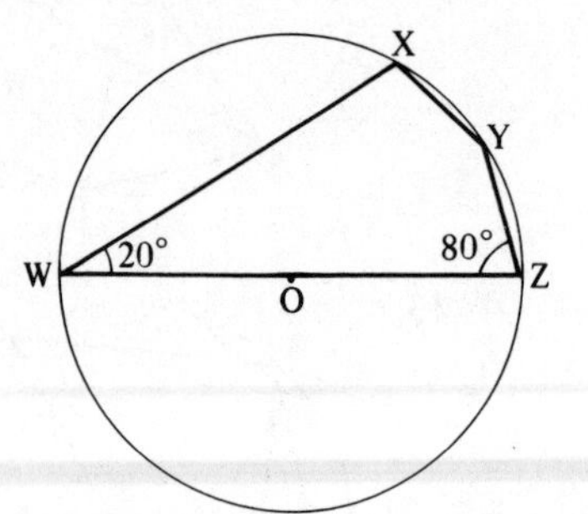

21. Find, in terms of z,
(a) $N\hat{M}O$ (b) $M\hat{O}N$ (c) $N\hat{O}P$.

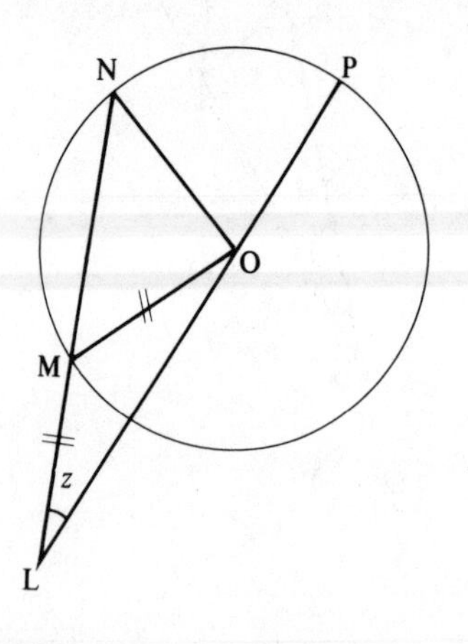

22. (a) Find $W\hat{X}Y$ in terms of b.
(b) If $a = 3b$, calculate b.

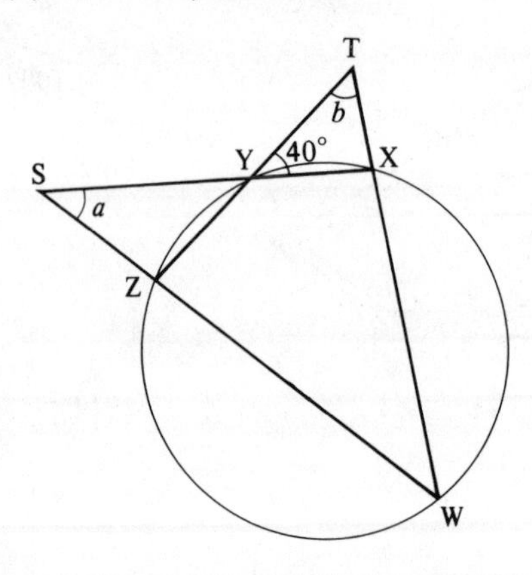

23. A, B, C are three points on the circumference of a circle, centre O. The angle AOB is 28°, the angle BOC is 52° and the angle AOC is 80°. Find the three angles of the triangle ABC.

24. POR is a diameter of a circle, centre O. The quadrilateral PQRS inscribed in the circle is such that the angle RPQ is 38° and OS is parallel to QR. Calculate the angles ORQ, ROS, RPS and QRS.

25. In a cyclic quadrilateral PQRS, the chords PR and QS intersect at T. If PT = QT, prove that
(a) the triangles PTS, QTR are congruent
(b) PR = QS.

26. X, Y and Z are three points on a circle, centre O. Angle XOZ = 144°, Y lies between X and Z on the minor arc XZ, and the arc XY is three times as long as the arc YZ. OY meets the line XZ at A. Calculate
(a) the angle ZXO (b) the angle XOY
(c) the angle OAZ.

27. ABCD is a quadrilateral inscribed in a circle, centre O, and AD is a diameter of the circle. If $C\hat{D}B = 40°$ and $A\hat{D}B = 29°$, calculate
(a) the angle ABC (b) the angle BCD.

Tangents to circles

(a) The angle between a tangent and the radius drawn to the point of contact is 90°.

$A\hat{B}O = 90°$

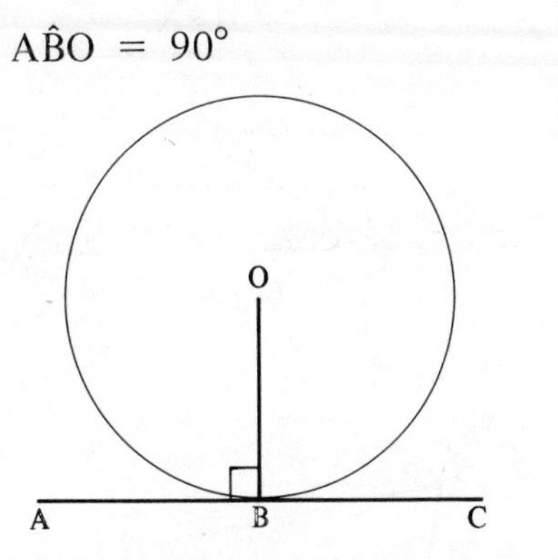

(b) From any point outside a circle just two tangents to the circle may be drawn and they are of equal length.

TA = TB

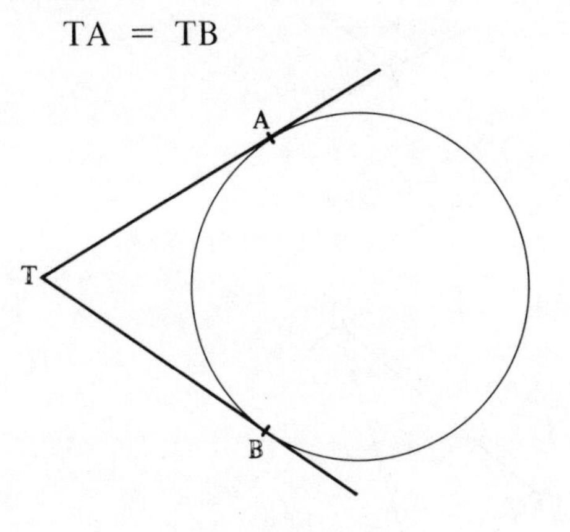

(c) Alternate segment theorem.
The angle between a tangent and a chord through the point of contact is equal to the angle subtended by the chord in the alternate segment.

$T\hat{A}B = B\hat{C}A$
and $S\hat{A}C = C\hat{B}A$

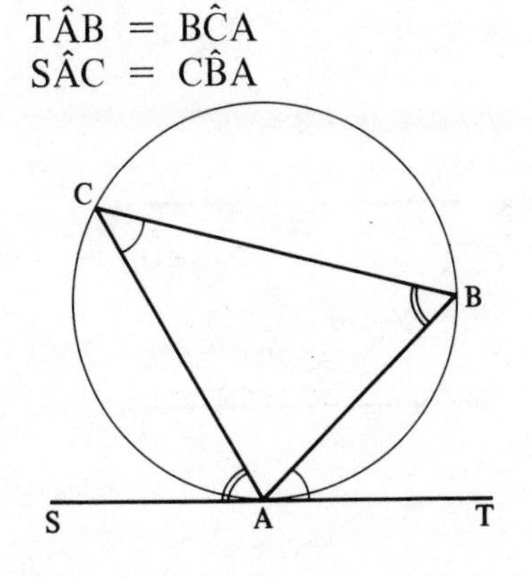

Example 1

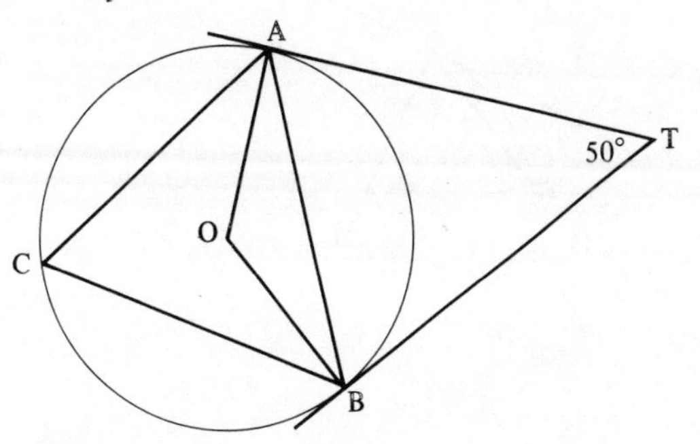

TA and TB are tangents to the circle, centre O.
Given $A\hat{T}B = 50°$, find
(a) $A\hat{B}T$
(b) $O\hat{B}A$
(c) $A\hat{C}B$

(a) $\triangle TBA$ is isosceles (TA = TB)

$\therefore \quad A\hat{B}T = \frac{1}{2}(180 - 50) = 65°$

(b) $O\hat{B}T = 90°$ (tangent and radius)

$\therefore \quad O\hat{B}A = 90 - 65$
$= 25°$

(c) $A\hat{C}B = A\hat{B}T$ (alternate segment theorem)

$A\hat{C}B = 65°$.

Exercise 13

For questions **1** to **12**, find the angles marked with a letter.

1.

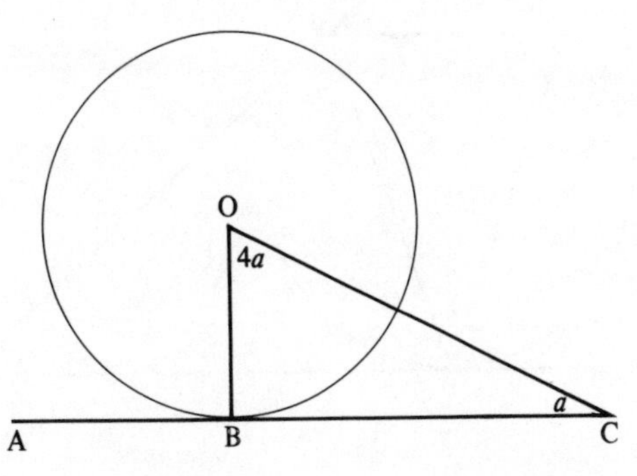

2.

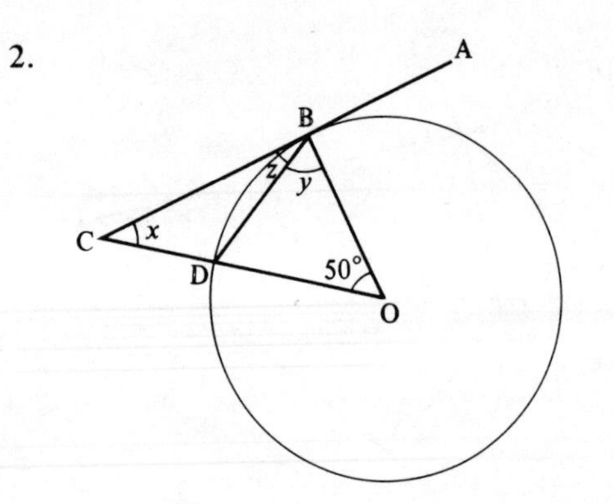
A
B
z
y
C
x
D
50°
O

3.

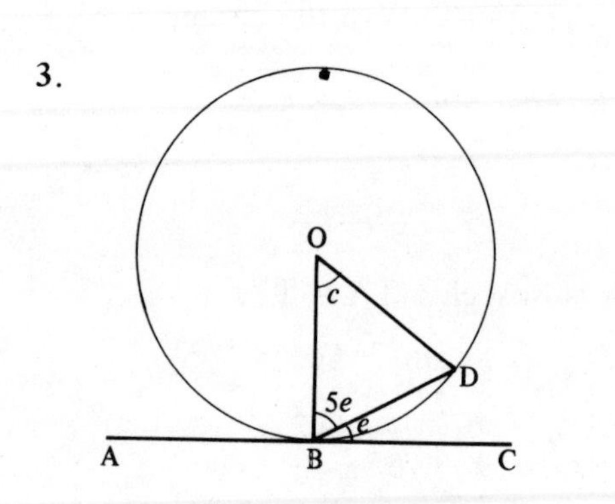
O
c
D
5e
e
A
B
C

4.

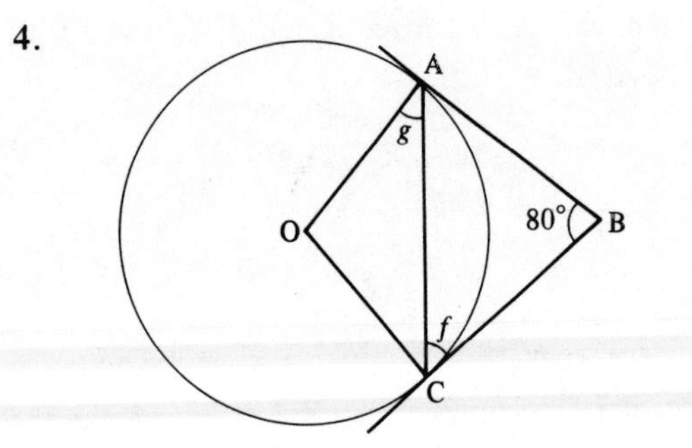
A
g
O
80°
B
f
C

5.

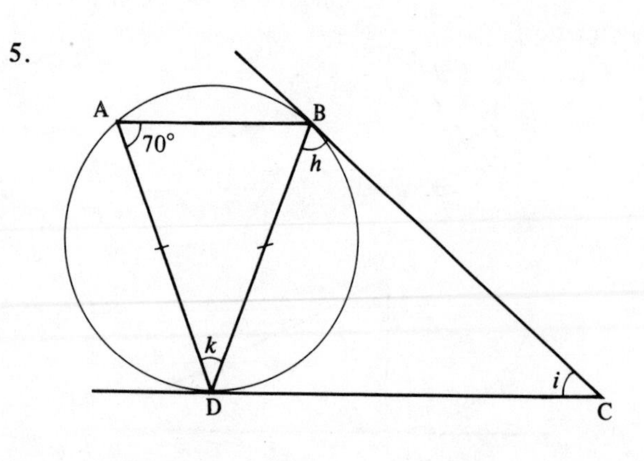
A
70°
B
h
k
i
D
C

6.

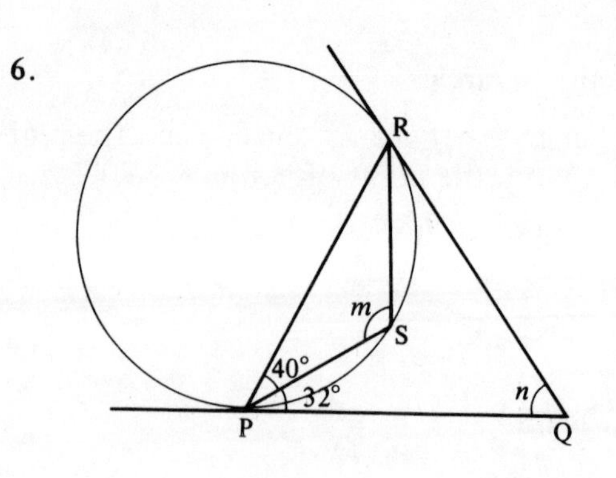
R
m
S
40°
32°
n
P
Q

7.

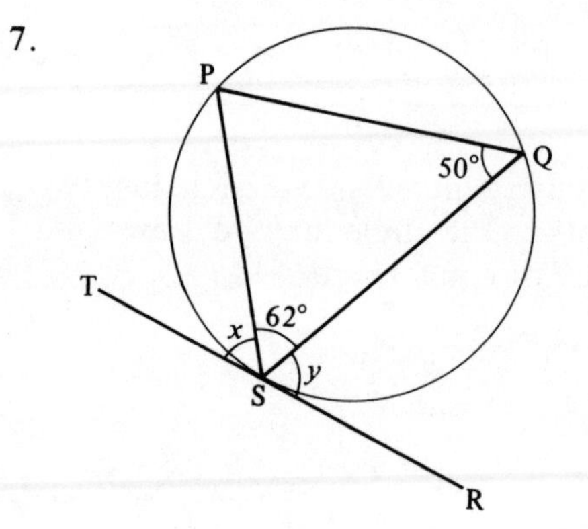
P
50°
Q
T
x
62°
y
S
R

8.

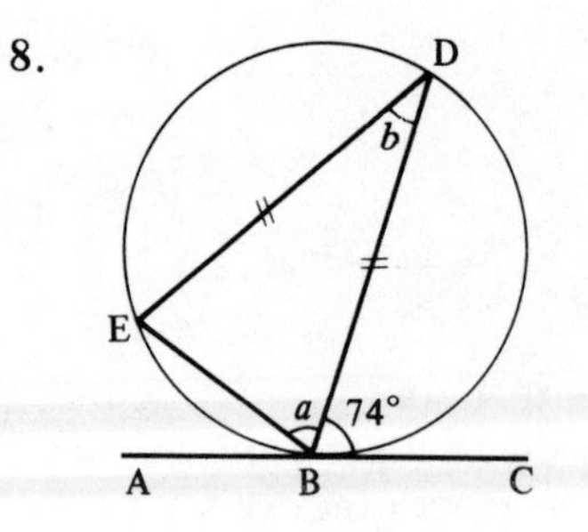
D
b
E
a
74°
A
B
C

9.

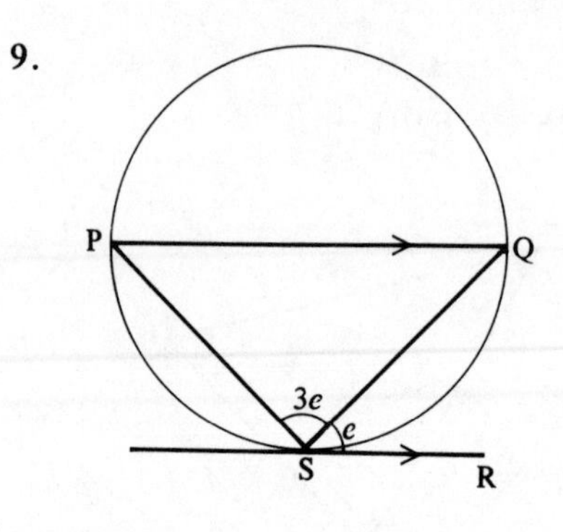
P
Q
3e
e
S
R

10.

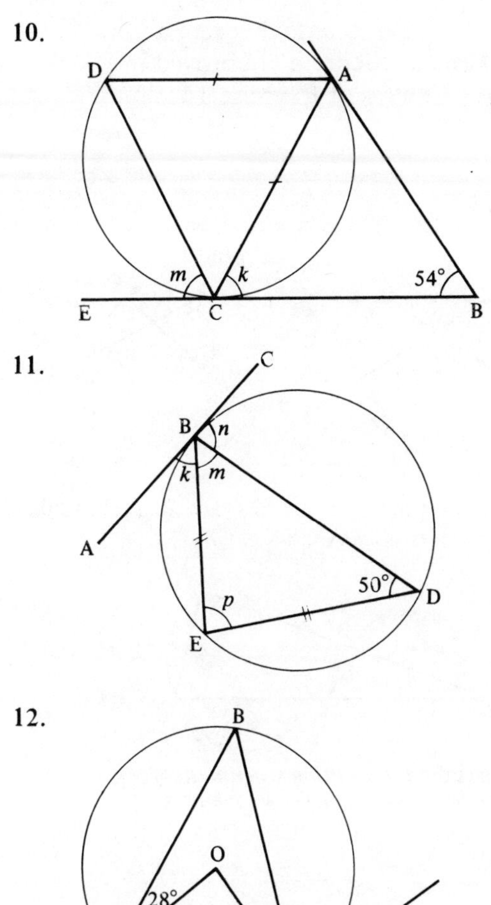

11.

12.

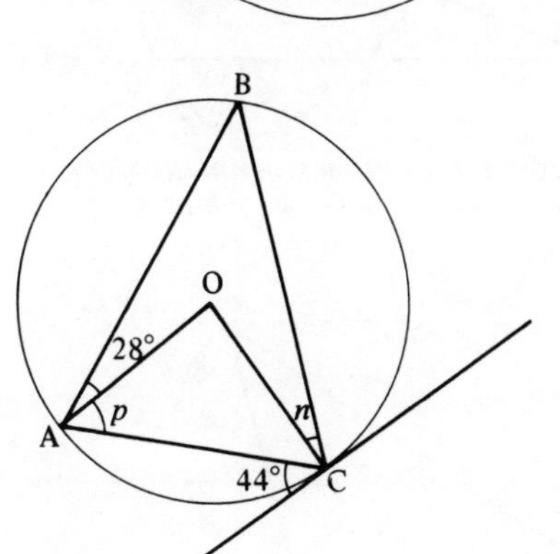

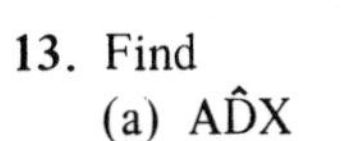

13. Find
(a) AD̂X (b) AB̂C (c) BĈD

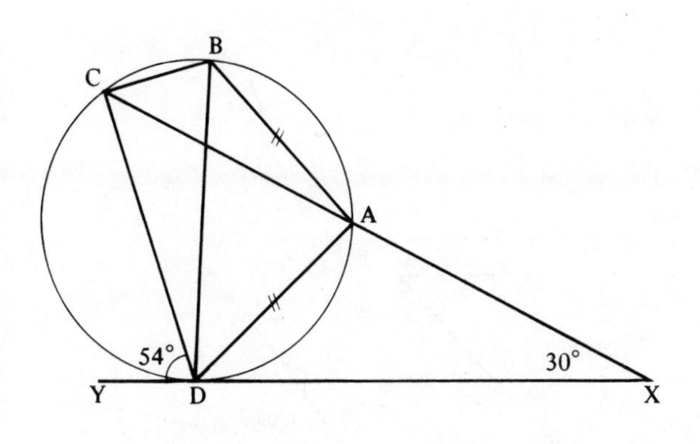

14. Find, in terms of p,
(a) BÂC (b) XĈA (c) AĈO

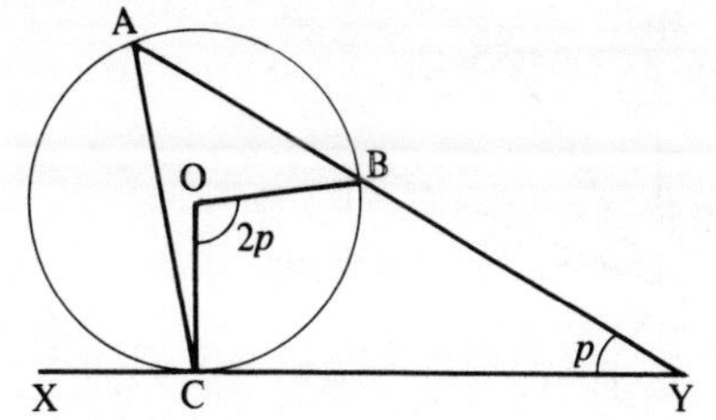

15. Find x, y and z.

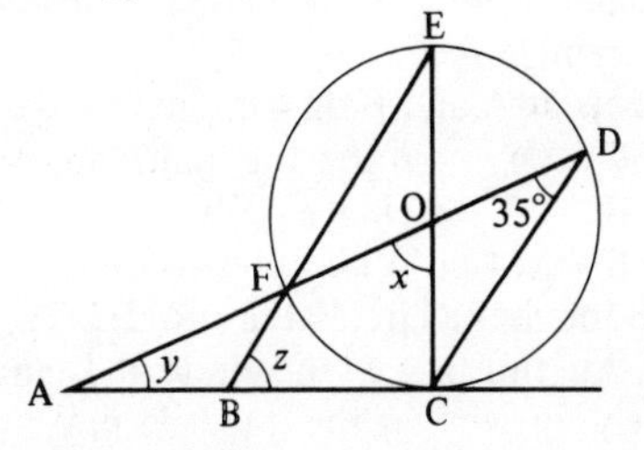

16. Given that KL = LN, LP bisects KL̂N and MK̂N = a
(a) prove that ΔKLQ is isosceles
(b) find LQ̂M and LM̂Q in terms of a.

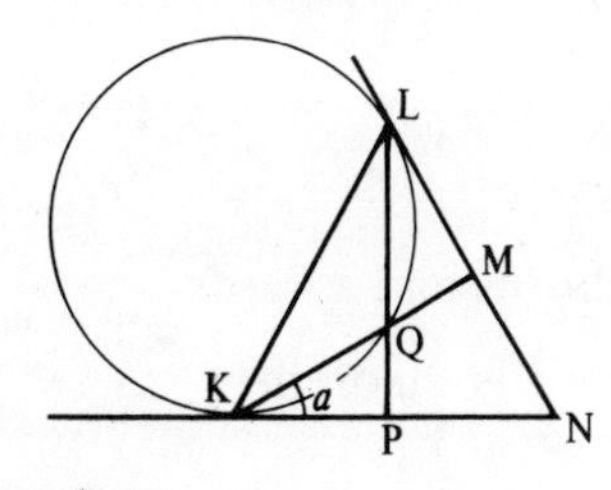

17. Show that:
(a) YŴX = VŴZ
(b) the triangles VWZ and YWX are similar
(c) VW.WX = YW.WZ.

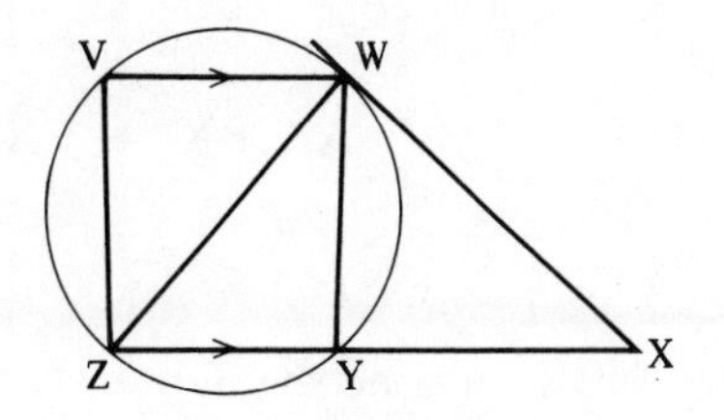

18. Given that BOC is a diameter and that $A\hat{D}C = 90°$, prove that AC bisects $B\hat{C}D$.

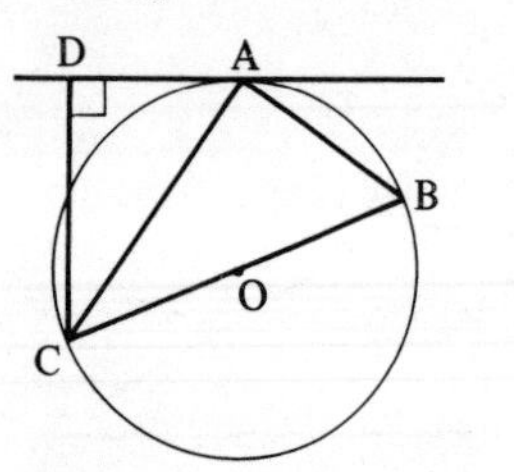

19. The angles of a triangle are 50°, 60° and 70°, and a circle touches the sides at A, B, C. Calculate the angles of triangle ABC.
20. The tangents at A and B on a circle intersect at T, and C is any point on the major arc AB.
    (a) If $A\hat{T}B = 52°$, calculate $A\hat{C}B$.
    (b) If $A\hat{C}B = x$, find $A\hat{T}B$ in terms of x.
21. Line ATB touches a circle at T and TC is a diameter. AC and BC cut the circle at D and E respectively. Prove that the quadrilateral ADEB is cyclic.
22. Two circles touch externally at T. A chord of the first circle XY is produced and touches the other at Z. The chord ZT of the second circle, when produced, cuts the first circle at W. Prove that $X\hat{T}W = Y\hat{T}Z$.

Intersecting chords and secants

Intersecting chords theorem

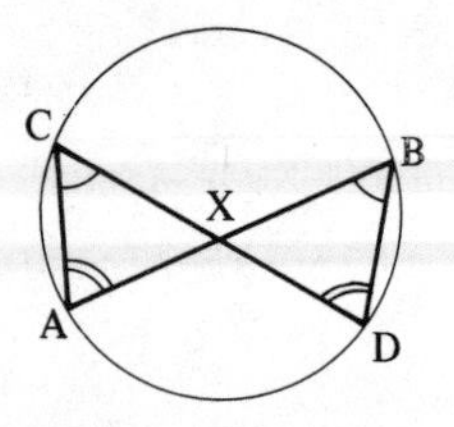

$$AX.BX = CX.DX$$

Proof:
In triangles AXC and BXD:

$A\hat{C}X = D\hat{B}X$ (same segment)
$C\hat{A}X = B\hat{D}X$ (same segment)

∴ the triangles AXC and BXD are similar.

$$\frac{AX}{DX} = \frac{CX}{BX}$$

or $AX.BX = CX.DX$

The intersecting secants theorem and the secant/tangent theorem are proved in questions **3** and **4** of Exercise 14.

Exercise 14

1. Find x.
2. Find x.

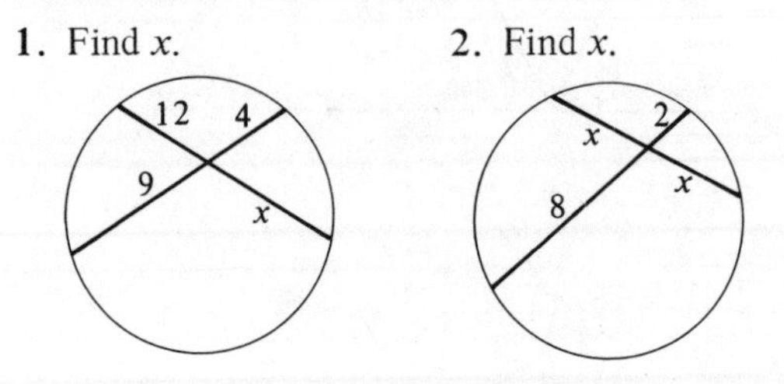

3. Show that the triangles AXC and BXD are similar. Hence show that AX.BX = CX.DX.

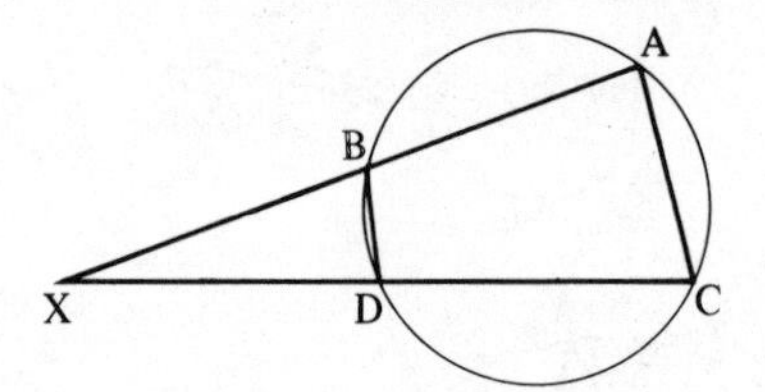

This is the *intersecting secants theorem.*

4. Show that the triangles ATC and BTC are similar. Hence show that $AT.BT = CT^2$.

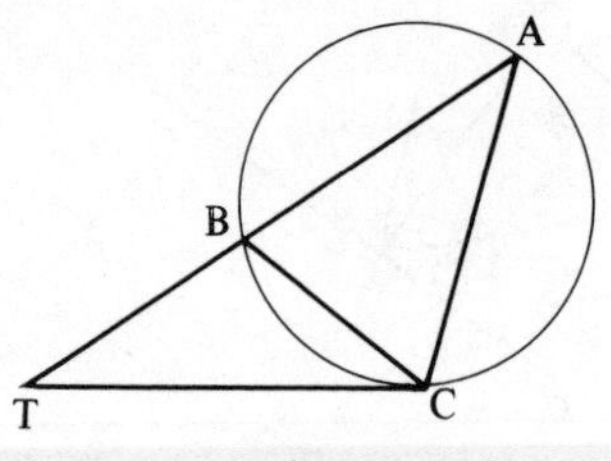

This is the *secant/tangent theorem.*

In questions **5** to **8**, find x.

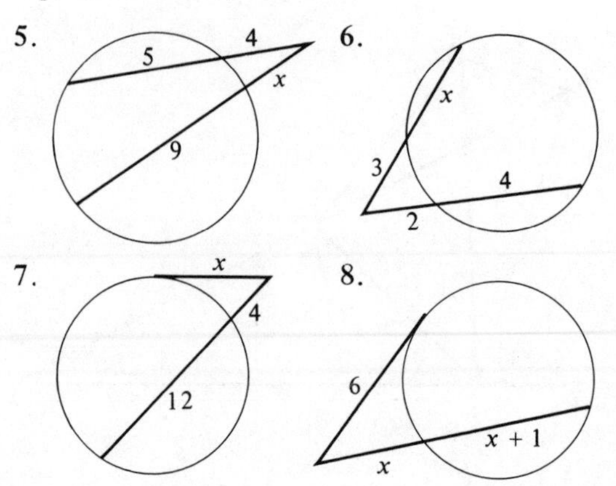

9. Two chords of a circle KL and MN intersect at X, and KL is produced to T. Given that KX = 6 cm, XL = 4 cm, MX = 8 cm and LT = 8 cm, calculate
(a) NX
(b) the length of the tangent from T to the circle
(c) the ratio of the areas of ΔKXM to ΔLXM
(d) the ratio of the areas of ΔKXM to ΔLXN.

10. Chords AB and DC of a circle are produced to meet outside the circle at T. A tangent is drawn from T to touch the circle at E. Given AB = 5 cm, BT = 4 cm, and DC = 9 cm, calculate
(a) CT (b) TE
(c) the ratio of the areas of ΔADT to ΔBCT
(d) the ratio of the areas of ΔBET to ΔAET.

4.7 CONSTRUCTIONS AND LOCI

When the word 'construct' is used, the diagram in question should be drawn using *ruler* and *compasses* only.
Four basic constructions are shown below.

(a) Perpendicular bisector of a line joining two points.

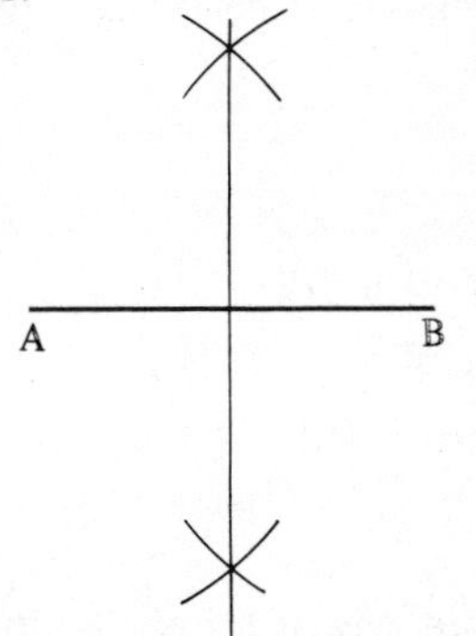

(b) Perpendicular from a point to a line.

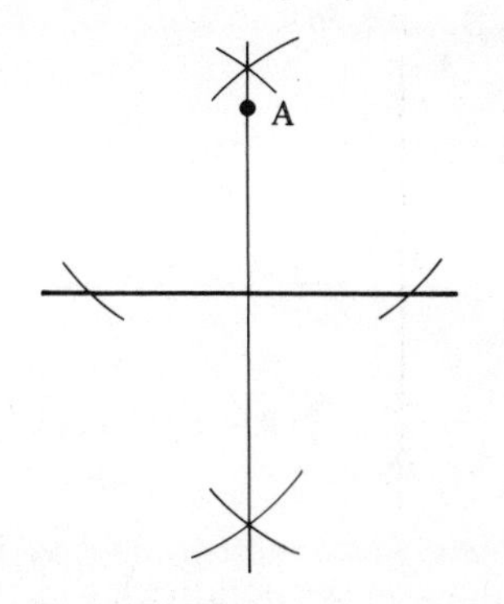

(c) Bisector of an angle.

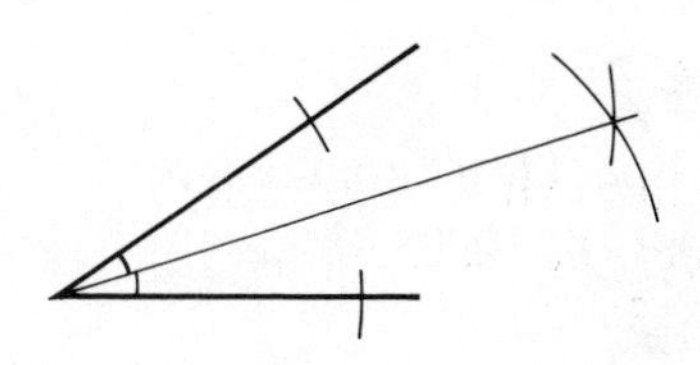

(d) 60° angle construction.

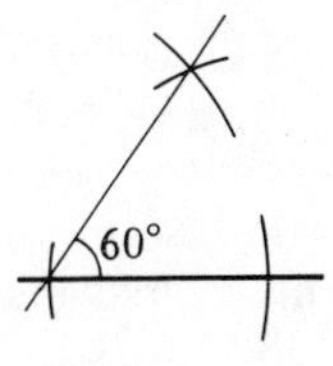

Exercise 15

1. Construct a triangle ABC in which AB = 8 cm, AC = 6 cm and BC = 5 cm. Measure the angle $A\hat{C}B$.
2. Construct a triangle PQR in which PQ = 10 cm, PR = 7 cm and RQ = 6 cm. Measure the angle $R\hat{P}Q$.
3. Construct an equilateral triangle of side 7 cm.
4. Draw a line AB of length 10 cm. Construct the perpendicular bisector of AB.
5. Draw two lines AB and AC of length 8 cm, where $B\hat{A}C$ is approximately 40°. Construct the line which bisects $B\hat{A}C$.
6. Draw a line AB of length 12 cm and draw a point X approximately 6 cm above the middle of the line. Construct the line through X which is perpendicular to AB.
7. Construct an equilateral triangle ABC of side 9 cm. Construct a line through A to meet BC at 90° at the point D. Measure the length AD.
8. Construct the triangles shown and measure the lengths x and y.

(a) (b)

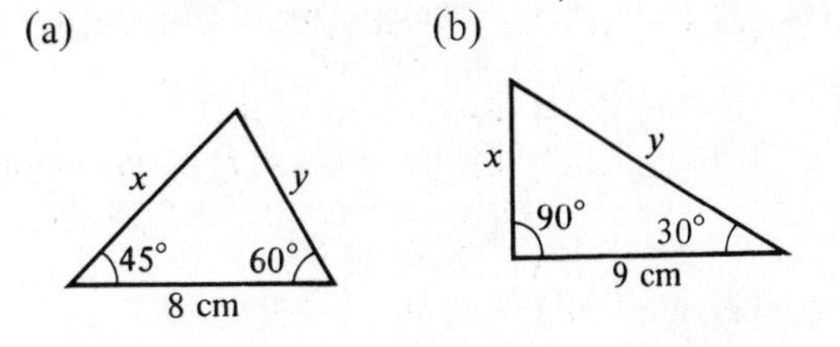

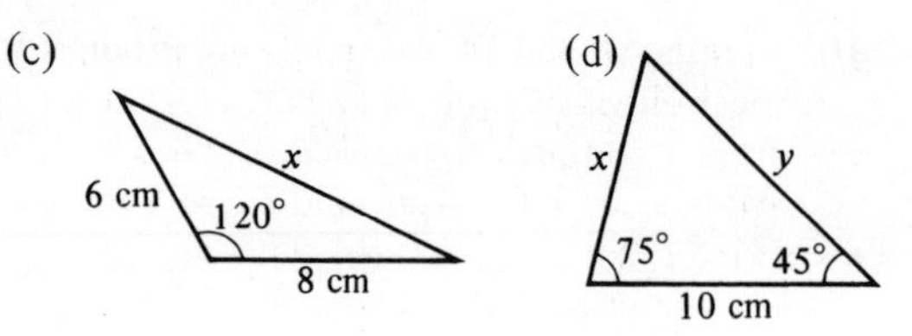

9. (a) Construct a triangle ABC in which AB
AB = 8 cm, AC = 6 cm and BC = 5 cm.
(b) Construct the perpendicular bisectors of AB and AC and hence construct the circumcircle of triangle ABC.

10. (a) Construct a triangle XYZ in which
XZ = 10 cm, XY = 8 cm and YZ = 7 cm.
(b) Construct the circumcircle of triangle XYZ and measure the radius of the circle.

11. (a) Construct a triangle PQR in which
PQ = 11 cm, PR = 9 cm and RQ = 7 cm.
(b) Construct the bisectors of angles QPR and RQP and hence draw the inscribed circle of triangle PQR.

12. (a) Construct a triangle XYZ in which
XY = 10 cm, XZ = 9 cm and YZ = 8 cm.
(b) Construct the inscribed circle of triangle XYZ.
(c) Construct the circumscribed circle of triangle XYZ.

13. Construct a parallelogram WXYZ in which
WX = 10 cm, WZ = 6 cm and XŴZ = 60°.
By construction, find the point A that lies on ZY and is equidistant from lines WZ and WX.
Measure the length WA.

14. (a) Draw a line OX = 10 cm and construct an angle XOY = 60°.
(b) Bisect the angle XOY and mark a point A on the bisector so that OA = 7 cm.
(c) Construct a circle with centre A to touch OX and OY and measure the radius of the circle.

15. (a) Construct a triangle PQR with PQ = 8 cm,
PR = 12 cm and PQ̂R = 90°.
(b) Construct the bisector of QP̂R.
(c) Construct the perpendicular bisector of PR and mark the point X where this line meets the bisector of QP̂R.
(d) Measure the length PX.

16. (a) Construct a triangle ABC in which AB = 8 cm,
AC = 6 cm and BC = 9 cm.
(b) Construct the bisector of BÂC.
(c) Construct the line through C perpendicular to CA and mark the point X where this line meets the bisector of BÂC.
(d) Measure the lengths CX and AX.

17. Draw four lines so that they form four triangles as in the diagram.

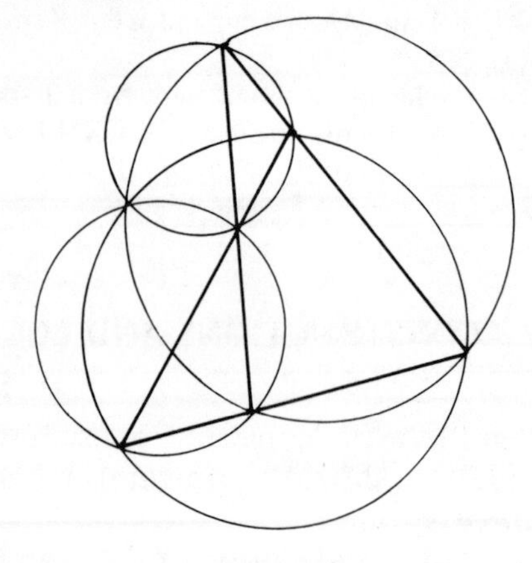

Construct the circumcircle of each of the four triangles to demonstrate Wallace's theorem that the four circles always pass through one point.

The locus of a point

The locus of a point is the path which it describes as it moves.

Example 1

Draw a line AB of length 8 cm.
Construct the locus of a point P which moves so that BÂP = 90°.

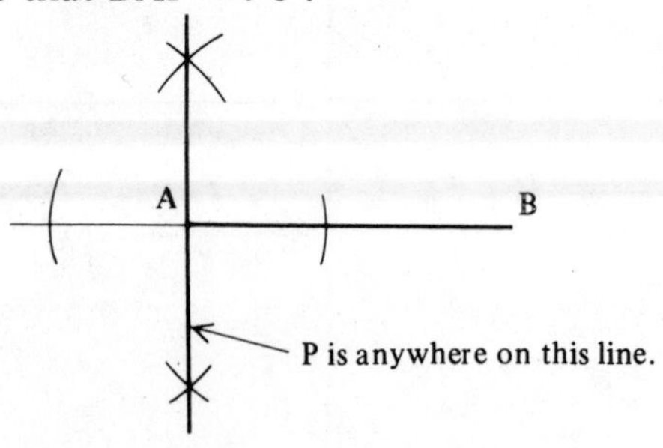

Construct the perpendicular at A.
This line is the locus of P.

Example 2

Draw a line AB 8 cm in length.
Construct the locus of a point P such that AP̂B = 45°.

(a) Draw AB = 8 cm.
(b) Construct perpendiculars AC and BD at A and B such that $B\hat{A}C = A\hat{B}D = 90°$.
(c) Construct the bisectors of $B\hat{A}C$ and $A\hat{B}D$ to meet at O.
(d) Draw the segment of a circle, centre O and radius OA.
(e) Since $A\hat{O}B = 90°$, the angle subtended by AB on the arc of the circle is always 45°. (Angle at centre is twice angle at circumference.)

Hence the locus of P is the major arc of the circle.

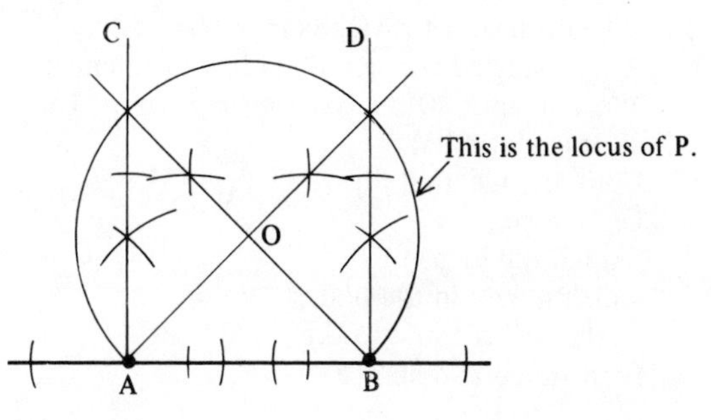

Exercise 16

1. Draw a line XY of length 10 cm. Construct the locus of a point which is equidistant from X and Y.
2. Draw two lines AB and AC of length 8 cm, where $B\hat{A}C$ is approximately 70°. Construct the locus of a point which is equidistant from the lines AB and AC.
3. Draw a circle, centre O, of radius 5 cm and draw a radius OA. Construct the locus of a point P which moves so that $O\hat{A}P = 90°$.
4. Draw a line AB of length 10 cm and construct the circle with diameter AB. Indicate the locus of a point P which moves so that $A\hat{P}B = 90°$.
5. Construct triangle ABP where AB = 8 cm, $A\hat{B}P = 45°$ and $B\hat{A}P = 60°$. Measure $A\hat{P}B$. Construct the locus of a point X which moves so that $A\hat{X}B = 75°$. (Hint: construct the circumcircle of $\triangle$ABP.)
6. (a) Construct a triangle PQR where PQ = 10 cm, $Q\hat{P}R = 30°$ and $P\hat{Q}R = 90°$.
   (b) Construct the locus of a point X which moves so that $P\hat{X}Q = 60°$.
7. Construct a triangle PQR with PQ = 10 cm, $Q\hat{P}R = 30°$ and $P\hat{Q}R = 60°$. Construct the locus of a point X such that $P\hat{X}Q = 90°$.
8. (a) Draw a line AB 8 cm in length and construct the segment of a circle to contain an angle of 45°.
   (b) Construct a line through A to cut the segment of the circle such that it makes an angle of 60° with AB.
   (c) Hence draw triangle ABC with AB = 8 cm, $B\hat{A}C = 60°$ and $A\hat{C}B = 45°$.
   (d) Measure and state the length BC.
9. (a) Construct a triangle ABC in which base BC = 11 cm, AB = 9 cm and CA = 8 cm.
   (b) Construct an equilateral triangle BCD with D on the same side of BC as A.
   (c) Construct the perpendicular bisectors of BD and DC to meet at O.
   (d) Construct the locus of a point X such that $B\hat{X}C = 60°$ and X lies on the same side of BC as A.
   (e) Construct the circle with AC as diameter.
   (f) Label the point Y such that $A\hat{Y}C = 90°$ and $B\hat{Y}C = 60°$.
10. Construct a rectangle ABCD with AB = 9 cm and BC = 5 cm. Construct a square with the same area as the rectangle. Measure the length of a side of the square.

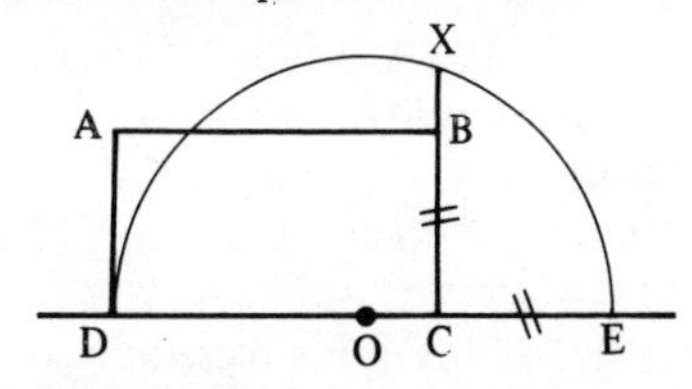

Hint: $CX^2 = DC.CE$ (intersecting chords theorem)

$\therefore\ CX^2 = DC.CB$.

11. Construct a rectangle PQRS with PQ = 10 cm and QR = 7 cm. Construct a square with the same area as the rectangle. Measure the length of a side of the square.
12. Draw a circle, centre O and radius 4 cm. Draw a line through O and mark A where OA = 10 cm. Construct the circle with diameter OA. Hence construct two tangents to the first circle, both of which pass through A.

Sometimes we need to *describe* the locus of a point which moves in a plane or in three dimensional space.

Example 3

Describe the locus of a point which moves in three dimensional space and is always 4 cm from a fixed point O.

The locus is the surface of a sphere, centre O and radius 4 cm.

Exercise 17

In questions **1** to **13**, sketch and describe the following loci:

1. A point which moves in a plane and is equidistant from two fixed points X and Y in the plane.
2. A point which moves in a plane and is equidistant from two intersecting lines PQ and PR in the plane.
3. The loci of points in the plane of triangle ABC which are
   (a) equidistant from A and B
   (b) equidistant from AB and AC
   (c) 5 cm from C.
4. A point which moves inside rectangle ABCD with AB = 10 cm and AD = 5 cm, so that it is equidistant from AD and DC.
5. A point P which moves in a plane containing two fixed points A and B so that:
   (a) $B\hat{A}P = 90°$ (b) $A\hat{P}B = 90°$.
6. A point P which moves in a plane containing two fixed points M and N (when MN = 20 cm) so that the area of triangle MNP is 100 cm^2.
7. The centre of circles all of which touch a given fixed line at a given point.
8. A point which moves in three dimensional space and is equidistant from two fixed points.
9. A point which moves in space and is always 10 cm from an infinitely long straight line.
10. A spherical ball of radius 5 cm rolls around the perimeter of a square of side 15 cm. Sketch the locus of the centre of the sphere.
11. A point which moves in space so that it is equidistant from three fixed points A, B and C which form a triangle.
12. A point P which moves in a plane containing two fixed points X and Y so that
   (a) $X\hat{P}Y = 65°$ (b) $X\hat{P}Y > 65°$.
13. A point which moves in a plane and is always 15 cm from a fixed point X, not in the plane.
14. A circle, centre O, radius 5 cm, is inscribed inside a square ABCD. Point P moves so that $OP \leqslant 5$ cm and $BP \leqslant 5$ cm. Shade the set of points indicating where P can be.
15. Construct a triangle ABC where AB = 9 cm, BC = 7 cm and AC = 5 cm. Sketch and describe:
   (a) the locus of points within the triangle which are equidistant from AB and AC.
   (b) Shade the set of points within the triangle which are less than 5 cm from B and are also nearer to AC than to AB.
16. A right-angled triangle KLM has hypotenuse KM of length 20 cm. Sketch the locus of L for all triangles KLM.
17. Copy the diagram shown. OC = 4 cm. Sketch the locus of P which moves in the plane so that PC is equal to the perpendicular distance from P to the line AB. This locus is called a *parabola*.

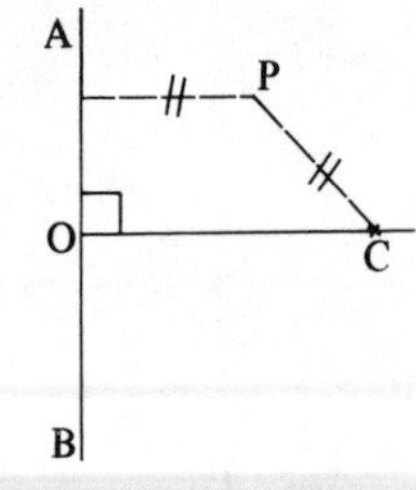

18. A rod OA of length 60 cm rotates about O at a constant rate of 1 revolution per minute. An ant, with good balance, walks along the rod at a speed of 1 cm per second. Sketch the locus of the ant for 1 minute after it leaves O.

REVISION EXERCISE 4A

1. ABCD is a parallelogram and AE bisects angle A. Prove that DE = BC.

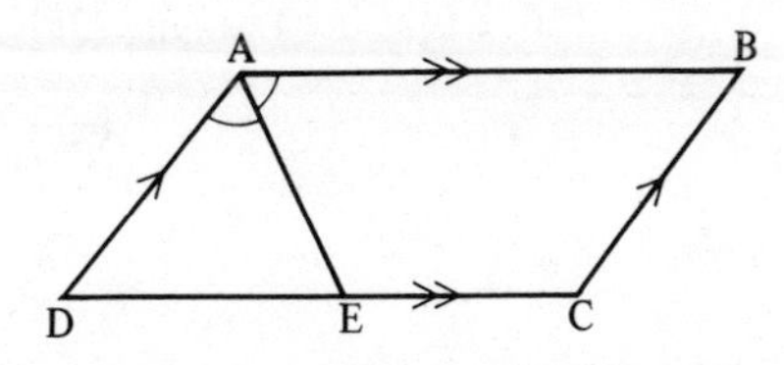

2. In a triangle PQR, $P\hat{Q}R = 50°$ and point X lies on PQ such that QX = XR. Calculate $Q\hat{X}R$.

3. (a) ABCDEF is a regular hexagon. Calculate $F\hat{D}E$.
   (b) ABCDEFGH is a regular eight-sided polygon. Calculate $A\hat{G}H$.
   (c) Each interior angle of a regular polygon measures 150°. How many sides has the polygon?

4. In triangle ABC, AD bisects $B\hat{A}C$. Calculate the length DC.

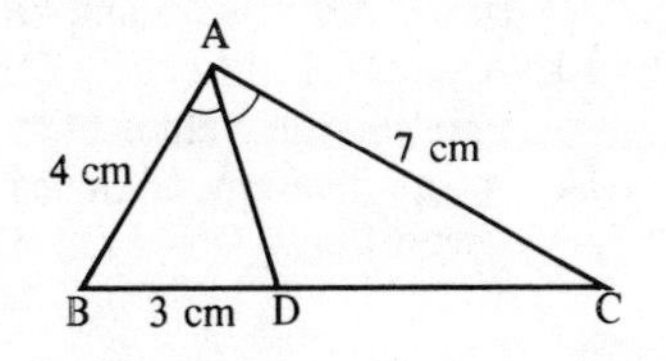

5. In the rectangle ABCD, AB = 8 cm and BC = 6 cm. The bisector of angle $A\hat{C}B$ meets AB at X. Calculate the length AX.

6. In the quadrilateral PQRS, PQ = QS = QR, PS is parallel to QR and $Q\hat{R}S = 70°$. Calculate
   (a) $R\hat{Q}S$
   (b) $P\hat{Q}S$.

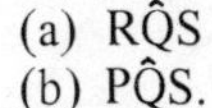

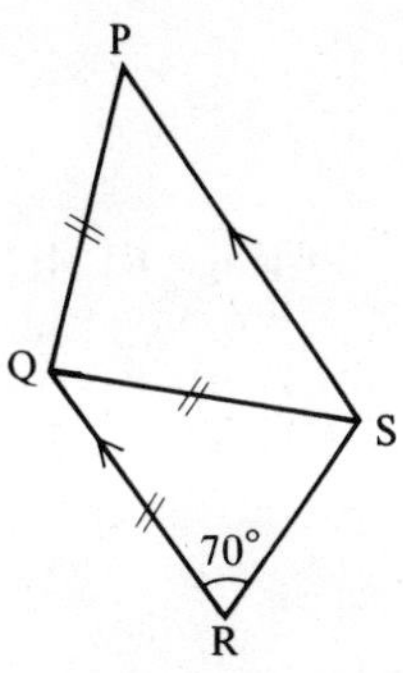

7. Find x.

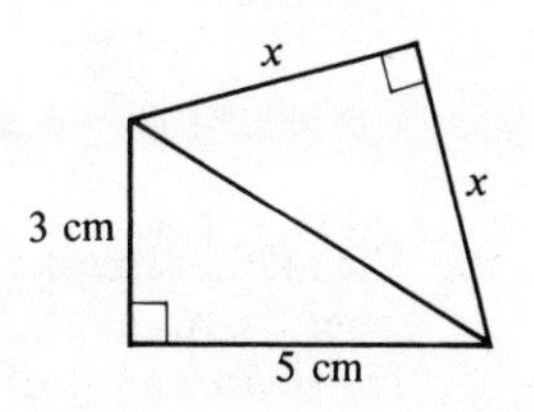

8. In the triangle ABC, AB = 7 cm, BC = 8 cm and $A\hat{B}C = 90°$. Point P lies inside the triangle such that BP = PC = 5 cm. Calculate
   (i) the perpendicular distance from P to BC
   (ii) the length AP.

9. ABC is a triangle in which AB = AC. The points D and E lie on the lines AB and AC respectively so that AD = AE. Prove that the triangles ADC and AEB are congruent.
10. In the diagram PQRS is a parallelogram, PX bisects $Q\hat{P}S$ and RY bisects $Q\hat{R}S$. Prove that the triangles QPX and RYS are congruent.

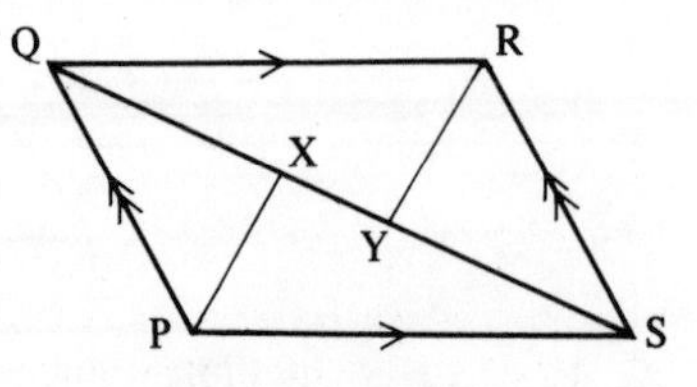

11. In triangle PQR the bisector of $P\hat{Q}R$ meets PR at S and the point T lies on PQ such that ST is parallel to RQ.
    (a) Prove that QT = TS
    (b) Prove that the triangles PTS and PQR are similar
    (c) Given that PT = 5 cm and TQ = 2 cm, calculate the length of QR.
12. In the quadrilateral ABCD, AB is parallel to DC and $D\hat{A}B = D\hat{B}C$.
    (a) Prove that the triangles ABD and DBC are similar.
    (b) If AB = 4 cm and DC = 9 cm, calculate the length of BD.
13. A rectangle 11 cm by 6 cm is similar to a rectangle 2 cm by x cm. Find the two possible values of x.
14. In the diagram, triangles ABC and EBD are similar but DE is *not* parallel to AC. Given that AD = 5 cm, DB = 3 cm and BE = 4 cm, calculate the length of BC.

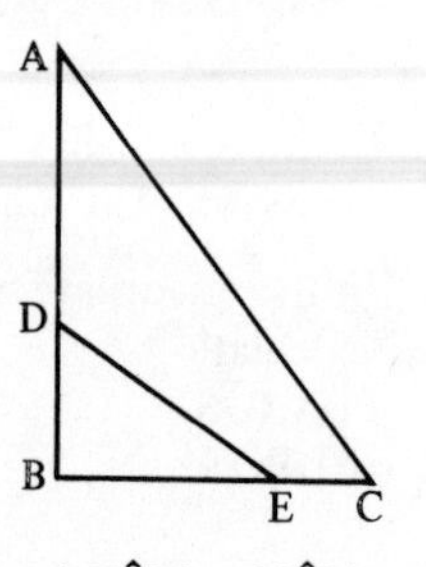

15. In triangle KLM, KL = LM and $L\hat{K}M = N\hat{L}M$.
    (a) Prove that LN = NM.
    (b) Prove that triangles KLM and LNM are similar.
    (c) Given that KL = 8 cm and KM = 12 cm, calculate the length of LN.

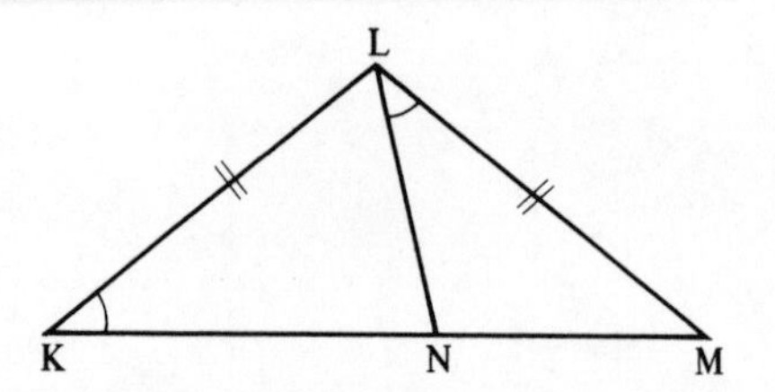

16. The radii of two spheres are in the ratio 2 : 5. The volume of the smaller sphere is 16 cm^3. Calculate the volume of the larger sphere.
17. The surface areas of two similar jugs are 50 cm^2 and 450 cm^2 respectively.
    (a) If the height of the larger jug is 10 cm, find the height of the smaller jug.
    (b) If the volume of the smaller jug is 60 cm^3, find the volume of the larger jug.

18. A car is an enlargement of a model, the scale factor being 10.
    (a) If the windscreen of the model has an area of $100\,\text{cm}^2$, find the area of the windscreen on the actual car (answer in m^2).
    (b) If the capacity of the boot of the car is $1\,\text{m}^3$, find the capacity of the boot on the model (answer in cm^3).
19. Find the angles marked with letters. (O is the centre of the circle.)

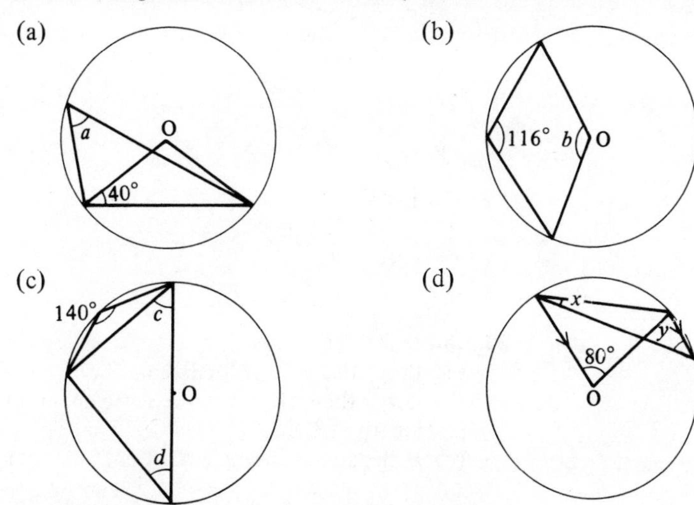

20. Find the angles stated. DE is a tangent in each case.
    (a) Find DÊA.

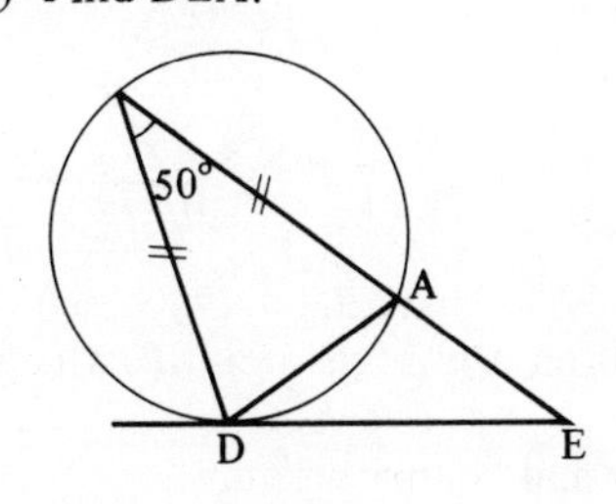

    (b) Find (i) DÂB (ii) BD̂E (iii) DB̂A (iv) ED̂A

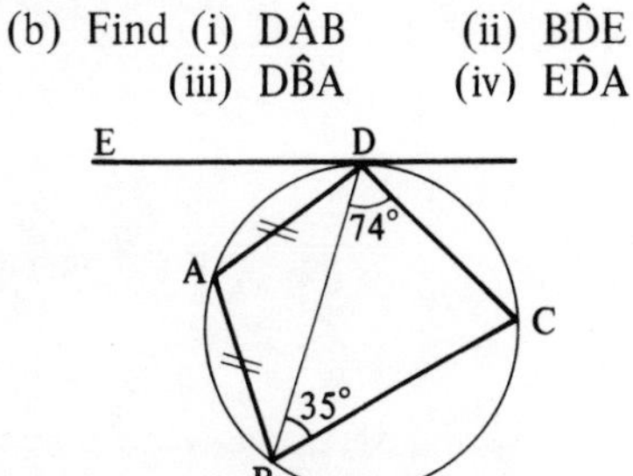

    (c) Find (i) DB̂A (ii) AD̂C (iii) ED̂A

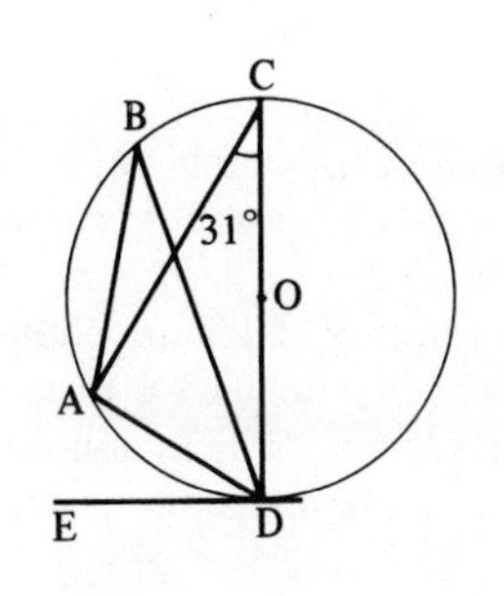

    (d) DBE is a tangent. Find (i) BÂD (ii) AB̂D (iii) OÊD

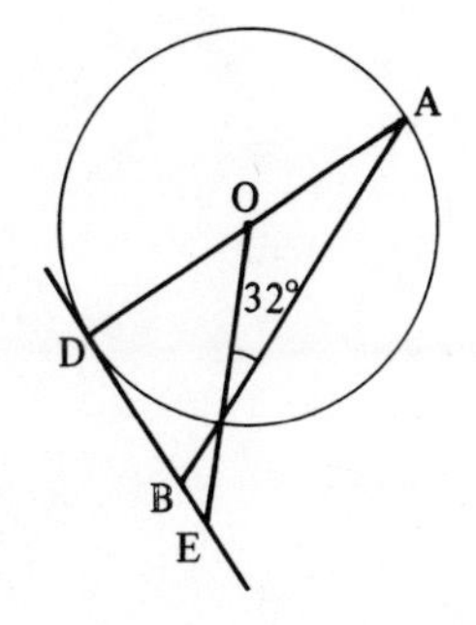

21. Find the length x.

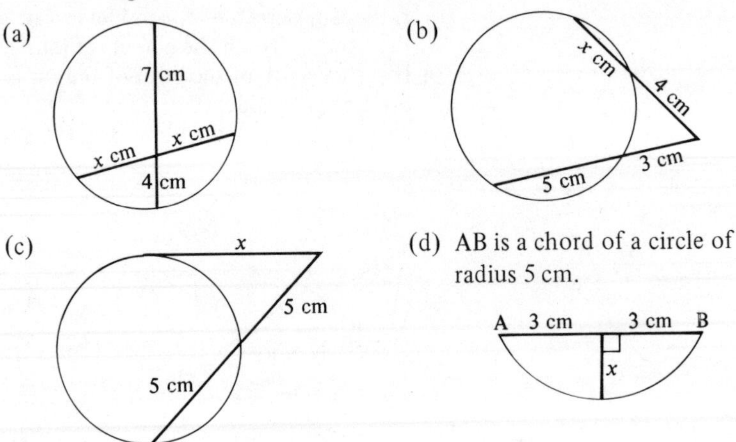

22. In the diagram below,
(a) Prove that triangles ABX and DCX are similar.
(b) Given that AB = 8 cm, AX = 6 cm and DC = 4 cm, calculate the length of DX.
(c) If the area of triangle ABX is 15 cm^2 calculate the area of triangle DCX.

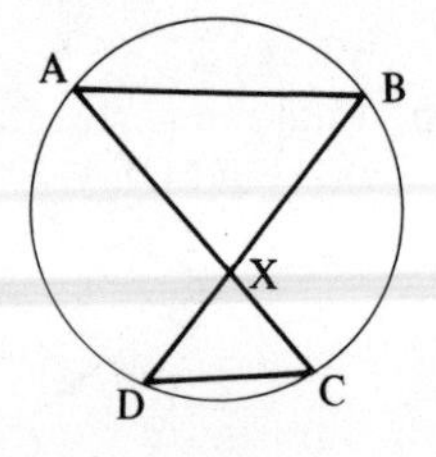

23. ABCD is a cyclic quadrilateral in which AB = BC and $A\hat{B}C = 70°$. AD produced meets BC produced at the point P, where $A\hat{P}B = 30°$. Calculate
(a) $A\hat{D}B$ (b) $A\hat{B}D$.

24. Using ruler and compasses only,
(a) Construct the triangle ABC in which AB = 7 cm, BC = 5 cm and AC = 6 cm.
(b) Construct the circle which passes through A, B and C and measure the radius of this circle.

25. Construct:
(a) the triangle XYZ in which XY = 10 cm, YZ = 11 cm and XZ = 9 cm.
(b) the locus of points, inside the triangle, which are equidistant from the lines XZ and YZ.
(c) the locus of points which are equidistant from Y and Z.
(d) the circle which touches YZ at its mid-point and also touches XZ.

26. The fixed points M and N lie in a given plane and are 10 cm apart. Describe the complete locus of a point P which moves in the given plane so that
(a) MP = NP
(b) $M\hat{P}N = 55°$
(c) the area of triangle MPN = 30 cm^2.

EXAMINATION EXERCISE 4B

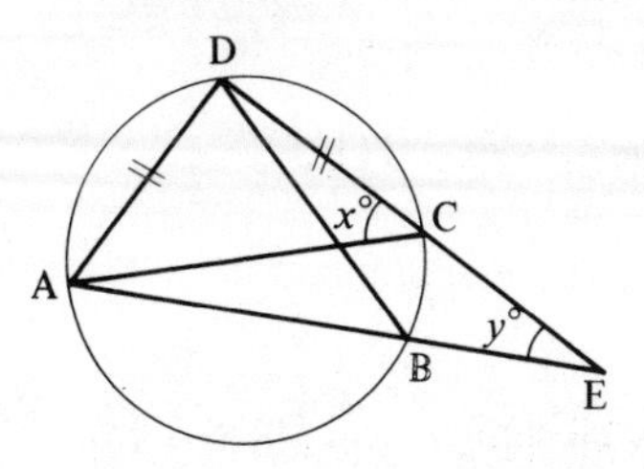

1. The diagram shows a cyclic quadrilateral ABCD with DA = DC. The sides AB and DC, when produced, meet at E. Given that the angle ACD $= x°$ and the angle BEC $= y°$, express in terms of x and y the angles CAD, ABD, DBC, BCE, ADC and CAB.

   Given that AB = 7 cm, DC = 6 cm, BE = 4 cm and CE = l cm, show that the value of l is given by

$$l^2 + 6l - 44 = 0$$

   and hence find the value of l correct to 2 significant figures. [AEB]

2. In the diagram (not drawn to scale) TA and TB are the tangents to the circle at the points A and B respectively. TBR is a straight line. The straight lines AB and QP intersect at K. Angle QAB = 27°, angle PBT = 40° and angle ATB = 50°.

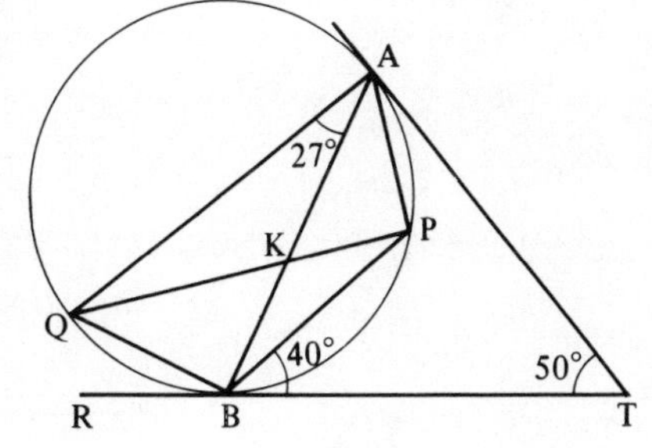

   Calculate

   (i) the size of the angle PAB,
   (ii) the size of the angle RBQ,
   (iii) the size of the angle PBA,
   (iv) the size of the angle AQP,
   (v) the size of the angle AKP.

   [NEA]

3. (a) In the diagram, AB is a diameter of the circle centre O, $A\hat{P}Q = 35°$ and $A\hat{B}R = 22°$. Calculate
   (i) $A\hat{B}Q$,
   (ii) $Q\hat{R}A$,
   (iii) $A\hat{O}R$.

   (b) Construct, in a single diagram, on a sheet of plain paper:
   (i) a triangle LMN with sides LM = 7 cm, LN = 8 cm and MN = 6 cm,
   (ii) the locus of points which are 5 cm from L,
   (iii) the locus of points which are equidistant from MN and LN.

   Given that $\mathscr{E}$ = {P: P lies inside ΔLMN},
   X = {P: LP $<$ 5 cm}
   and Y = {P: P is nearer to MN than to LN},
   indicate the region $X \cap Y$ by suitable shading on your diagram. [C]

4. ABCDE is a five-sided figure (not regular) in which BC = AE, CD = DE and $\angle$BCD = $\angle$DEA. Prove that
   (i) BD = AD,
   (ii) $\angle$ABC = $\angle$EAB,
   (iii) AC = BE. [JMB]

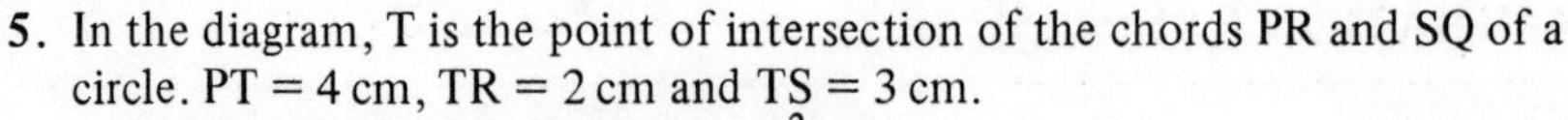

5. In the diagram, T is the point of intersection of the chords PR and SQ of a circle. PT = 4 cm, TR = 2 cm and TS = 3 cm.
   (a) Prove that the length of TQ is $2\frac{2}{3}$ cm.
   (b) Prove that ΔPTS is similar to ΔQTR.
   (c) Given that the area of ΔPTS is 3 cm^2, find the area of ΔQTR.
   (d) Find the value of the ratio

$$\frac{\text{area of } \Delta \text{PTQ}}{\text{area of } \Delta \text{RTS}}.$$

[L]

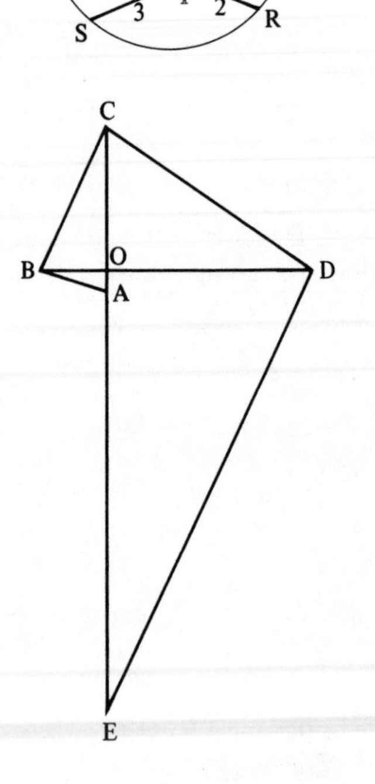

6. The four triangles AOB, BOC, COD and DOE are all similar, and OA = 1 cm, OB = 2 cm.
   (a) Find, in cm^2, the area of the whole figure.
   (b) Find the length, in cm correct to one decimal place, of BE.
   (c) Prove that BC is parallel to ED.
   (d) Find, to the nearest degree, the size of the angle BED. [L]

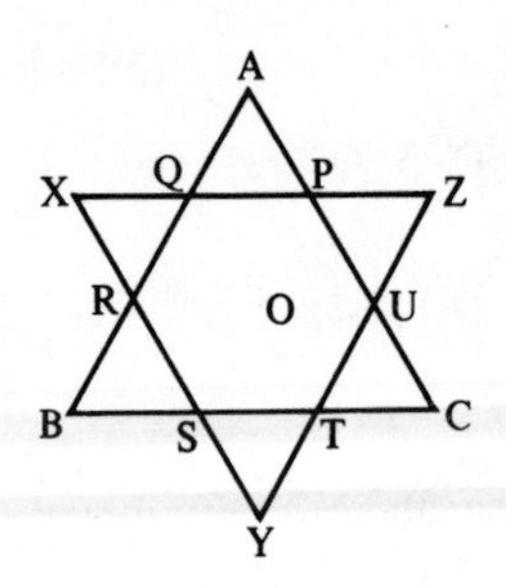

7. The diagram shows a six pointed star which is formed by drawing an equilateral triangle on each side of the regular hexagon PQRSTU. The centre of the hexagon is O.
   (a) Prove that ΔXYZ is congruent to ΔABC.
   (b) Show that the area of the six-pointed star is twice the area of the hexagon PQRSTU.
   (c) If ON is drawn from O perpendicular to QP to meet QP at N, find the value of the ratio $\frac{\text{ON}}{\text{OA}}$.
   (d) Find the value of the ratio

$$\frac{\text{area of circle AXBYCZ}}{\text{area of circle inscribed in hexagon PQRSTU}}.$$

[The inscribed circle is that which touches each side of the hexagon.] [L]

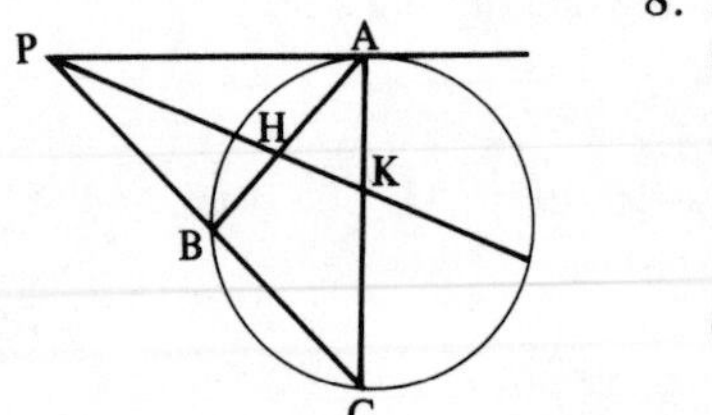

8. In the diagram, PA is a tangent to the circle through A, B and C and PBC is a straight line. The bisector of $\angle$APB meets AB at H and AC at K. Prove that
   (i) ΔAHK is isosceles,
   (ii) $\frac{\text{PA}}{\text{PB}} = \frac{\text{PC}}{\text{PA}}$,
   (iii) $\frac{\text{AH}}{\text{HB}} = \frac{\text{KC}}{\text{AK}}$,
   (iv) $\text{AH}^2 = \text{HB} \times \text{KC}$. [JMB]

9. In triangle ABC, D is the mid-point of BC and X is the mid-point of AD. The straight line through X parallel to BA meets CA at M and BC at N. Prove that:
   (i) AM $= \frac{1}{4}$AC,
   (ii) MN $= \frac{3}{4}$AB.
   Find the ratio of the area of trapezium AMNB to the area of triangle ABC.
   [O]

10. Construct the triangle XYZ in which XY = 5 cm, $\hat{X} = 60°$ and $\hat{Y} = 90°$. Measure, and write down, the length of YZ.
    On the same diagram,
    (i) construct the circumcircle of ΔXYZ,
    (ii) construct, on the same side of XY as Z, the locus of the point P such that the area of ΔXYP equals half the area of ΔXYZ,
    (iii) mark, and label clearly, a point Q such that X$\hat{Q}$Y $= 30°$ and the area of ΔXYQ is half the area of ΔXYZ.
    Given that M is a point such that X$\hat{M}$Y $= 30°$, find the largest possible area of ΔXMY.
    [C]

11. (i) Construct a triangle XYZ, with XY = YZ = ZX = 4·5 cm.
    (ii) Construct the angle XYW $= 90°$ such that W lies on XZ produced.
    (iii) Explain clearly, but briefly (and without measurements), why
        (a) the angle XWY $= 30°$,
        (b) XZ = ZW.
    (iv) Construct the angle YXT = the angle YXZ, such that T lies on WY produced.
    (v) Construct the locus of each of two points, P and Q, given that
        (a) PX = PY,
        (b) the angle XQY $= 30°$,
        indicating clearly the complete path of each locus.
    [AEB]

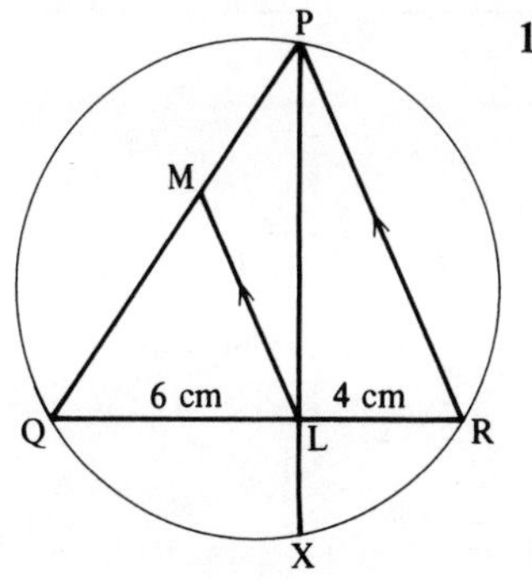

12. In the diagram, PL bisects the angle QPR and PL produced meets the circumcircle of triangle PQR at X. LM is parallel to RP.
    PR = 8 cm, QL = 6 cm and LR = 4 cm.
    (i) Calculate the length QP.
    (ii) Calculate the length LM.
    (iii) Calculate the ratio of the area of triangle PQL to the area of triangle PQR.
    (iv) Calculate the ratio of the area of triangle QLM to the area of triangle QRP.
    (v) If PL $= x$ cm and LX $= y$ cm, find the relationship between x and y.
    [AEB]

5 Algebra 2

Girolamo Cardan (1501–1576) was a colourful character who became Professor of Mathematics at Milan. As well as being a distinguished academic, he was an astrologer, a physician, a gambler and a heretic, yet he received a pension from the Pope. His mathematical genius enabled him to open up the general theory of cubic and quartic equations, although a method for solving cubic equations which he claimed as his own was pirated from Niccolo Tartaglia.

5.1 ALGEBRAIC FRACTIONS

Simplifying fractions

Example 1

Simplify: (a) $\frac{32}{56}$; (b) $\frac{3a}{5a^2}$

(a) Write $\frac{32}{56}$ as $\frac{8 \times 4}{8 \times 7}$ and cancel the two 8's.

$\therefore \quad \frac{32}{56} = \frac{\cancel{8} \times 4}{\cancel{8} \times 7} = \frac{4}{7}$

(b) Write $\frac{3a}{5a^2}$ as $\frac{3 \times a}{5 \times a \times a}$ and cancel the two a's.

$\therefore \quad \frac{3a}{5a^2} = \frac{3 \times a}{5 \times a \times a} = \frac{3}{5a}$.

Example 2

Simplify: (a) $\frac{10 + 35}{100}$; (b) $\frac{3y + y^2}{6y}$.

In both cases, factorise where possible.

(a) $\frac{10 + 35}{100} = \frac{\cancel{5}(2 + 7)}{\cancel{5} \times 20} = \frac{9}{20}$

(b) $\frac{\cancel{y}(3 + y)}{6\cancel{y}} = \frac{3 + y}{6}$

N.B. Do *not* cancel the 3 and the 6.

Exercise 1

Simplify as far as possible.

1. $\frac{25}{35}$ **2.** $\frac{84}{96}$ **3.** $\frac{75}{20}$

4. $\frac{10 + 12}{4}$ **5.** $\frac{9 + 36}{9}$ **6.** $\frac{35 - 20}{20}$

7. $\frac{21-7}{14}$ 8. $\frac{12+18}{6+24}$ 9. $\frac{8x^2}{x}$

10. $\frac{5y^2}{y}$ 11. $\frac{y}{2y}$ 12. $\frac{8x^2}{2x}$

13. $\frac{11x}{12x}$ 14. $\frac{7a}{9a}$ 15. $\frac{5b^2}{6b}$

16. $\frac{2x}{4y}$ 17. $\frac{6y}{3y}$ 18. $\frac{5ab}{10b}$

19. $\frac{8ab^2}{12ab}$ 20. $\frac{11x}{12x^2}$ 21. $\frac{7a^2b}{35ab^2}$

22. $\frac{(2a)^2}{4a}$ 23. $\frac{7yx}{8xy}$ 24. $\frac{-3x}{5x^2}$

25. $\frac{5xy^2z^2}{10xz^2}$ 26. $\frac{8ya^2}{12a^3}$ 27. $\frac{12m^2n^2}{15mn^3}$

28. $\frac{108x^2ab}{60ax}$ 29. $\frac{-7x}{-11x^2y}$ 30. $\frac{-3xy}{-4x^2}$

31. $\frac{-70am}{-20a^2}$ 32. $\frac{(4ab)^2}{12ab^2}$ 33. $\frac{(-2a)^2}{6ab}$

34. $\frac{63xy^2}{54x^2y}$ 35. $\frac{4ab^2c^3}{3a^3b^2c}$ 36. $\frac{3\frac{1}{2}(ab)^2}{7ab^2}$

37. $\frac{42(xy)^2}{35xy^2}$ 38. $\frac{-60xyz^2}{15x^2z^2}$ 39. $\frac{10(a^2b)^2}{0{\cdot}1ab^2}$

The next example shows how a common mistake is made in cancelling.

Example 3

Simplify $\frac{3x+9}{6x}$.

Suppose $x = 2$; then the expression becomes

$$\frac{(3 \times 2)+9}{6 \times 2} = \frac{15}{12}.$$

If the two 2's are incorrectly cancelled,

we obtain $\frac{3+9}{6} = \frac{12}{6}$

which is the *wrong* answer.

The expression $\frac{3x+9}{6x}$ must be written

$\frac{\cancel{3}(x+3)}{\cancel{3} \times 2x}$ and this becomes $\frac{(x+3)}{2x}$.

A number or a letter can only be cancelled if it occurs in *all* terms in both the numerator and the denominator.

Example 4

Simplify: (a) $\frac{8x+7xy}{4x}$; (b) $\frac{6y^2+3x}{3x}$.

(a) $\frac{8x+7xy}{4x} = \frac{\cancel{x}(8+7y)}{4\cancel{x}} = \frac{8+7y}{4}$

(b) $\frac{6y^2+3x}{3x} = \frac{\cancel{3}(2y^2+x)}{\cancel{3}x} = \frac{2y^2+x}{3}$.

Neither of these answers can be simplified further.

Exercise 2

Simplify where possible.

1. $\frac{5x+2x^2}{3x}$ 2. $\frac{9x+3}{3x}$ 3. $\frac{25+7}{25}$

4. $\frac{4a+5a^2}{5a}$ 5. $\frac{3x}{4x-x^2}$ 6. $\frac{5ab}{15a+10a^2}$

7. $\frac{3x-x^2}{2x}$ 8. $\frac{5x+4}{8x}$ 9. $\frac{84+60}{48-36}$

10. $\frac{3x^2+2x}{4x^2+x}$ 11. $\frac{12x+6}{6y}$ 12. $\frac{8-4x}{4x}$

13. $\frac{xy+x^2y}{x}$ 14. $\frac{5x+10y}{15xy}$ 15. $\frac{3(2x^2+5x)}{6x}$

16. $\frac{10x^2}{5x^2-3x}$ 17. $\frac{5x^3+4x^2}{x(3x^2-2x)}$ 18. $\frac{9+18x}{6x}$

19. $\frac{2x^2-x}{2x}$ 20. $\frac{18a-3ab}{6a^2}$ 21. $\frac{4ab+8a^2}{2ab}$

22. $\frac{18a^2bc-27ab^2c}{9ac}$ 23. $\frac{5-15x^2}{5x}$

24. $\frac{a+4a^2}{ab+ab^2}$ 25. $\frac{4+8x+8x^2}{4x}$

26. $\frac{54mn^2-27m^2n}{18(mn)^2}$ 27. $\frac{4abc-12a^2bc^2}{8abc^2}$

28. $\frac{6xy^2+9x^2y-15xy}{3xy}$ 29. $\frac{18a^2d+12ad}{24ad^2-30a^2d}$

30. $\frac{7xz-28zx^2}{(xz)^2+x^2z}$

More complicated expressions require the techniques of factorising developed earlier on page 39.

Example 5

Simplify:

(a) $\dfrac{x^2+x-6}{x^2+2x-3}$; (b) $\dfrac{x^2+3x-10}{x^2-4}$

(a) $\dfrac{x^2+x-6}{x^2+2x-3} = \dfrac{(x-2)\cancel{(x+3)}}{\cancel{(x+3)}(x-1)} = \dfrac{x-2}{x-1}$

(b) $\dfrac{x^2+3x-10}{x^2-4} = \dfrac{\cancel{(x-2)}(x+5)}{\cancel{(x-2)}(x+2)} = \dfrac{x+5}{x+2}$

Example 6

Write $\dfrac{7-x}{-4-x}$ in an easier form involving fewer negative signs.

Multiply 'top' and 'bottom' of the fraction by (−1).

i.e. $\dfrac{-1(7-x)}{-1(-4-x)} = \dfrac{-7+x}{4+x} = \dfrac{x-7}{x+4}$.

The value of the fraction is unchanged.

Example 7

Write $\dfrac{x+\frac{1}{2}}{x+\frac{1}{3}}$ without the fractions.

Multiply 'top' and 'bottom' by 6.

i.e. $\dfrac{6(x+\frac{1}{2})}{6(x+\frac{1}{3})} = \dfrac{6x+3}{6x+2}$

Exercise 3

Simplify as far as possible.

1. $\dfrac{x^2+2x}{x^2-3x}$
2. $\dfrac{x^2-3x}{x^2-2x-3}$
3. $\dfrac{x^2+4x}{2x^2-10x}$
4. $\dfrac{x^2+6x+5}{x^2-x-2}$
5. $\dfrac{x^2-4x-21}{x^2-5x-14}$
6. $\dfrac{x^2+7x+10}{x^2-4}$
7. $\dfrac{2x^2+x}{4x^2-1}$
8. $\dfrac{2x^2-5x-3}{2x^2-3x-9}$
9. $\dfrac{3x^2+x-4}{9x^2-16}$
10. $\dfrac{-5+x}{-3-x}$
11. $\dfrac{4-x}{-2}$
12. $\dfrac{-x}{7-x}$
13. $\dfrac{-5-x}{-1+x}$
14. $\dfrac{x+\frac{1}{x}}{x}$
15. $\dfrac{2x-\frac{1}{x}}{\frac{1}{x}}$
16. $\dfrac{x-\frac{1}{2}}{\frac{1}{2}}$
17. $\dfrac{3x+\frac{1}{4}}{\frac{1}{4}}$
18. $\dfrac{-x}{3-x}$
19. $\dfrac{3x-\frac{1}{x}}{\frac{1}{x}}$
20. $\dfrac{5x-\frac{1}{3}}{\frac{1}{6}}$
21. $\dfrac{\frac{1}{4}-x}{\frac{1}{2}}$
22. $\dfrac{3x+\frac{1}{x}}{x+\frac{2}{x}}$
23. $\dfrac{-7-2x}{2-x}$
24. $\dfrac{x+\frac{1}{3}}{2x-\frac{1}{2}}$
25. $\dfrac{2-x}{x-2}$
26. $\dfrac{4-2x}{x-2}$
27. $\dfrac{x-\frac{1}{2}}{2-4x}$
28. $\dfrac{\frac{1}{4}-x}{2x+\frac{1}{2}}$
29. $\dfrac{-\frac{1}{2}+x}{\frac{1}{4}-x}$
30. $\dfrac{x-\frac{2}{3}}{1\frac{2}{3}+x}$
31. $\dfrac{-\frac{1}{x}}{1-\frac{2}{x}}$
32. $\dfrac{3+\frac{2}{x}}{x}$
33. $\dfrac{x-\frac{4}{x}}{x-2}$
34. $\dfrac{2x-3}{4x-\frac{9}{x}}$
35. $\dfrac{1-\frac{1}{x^2}}{\frac{1}{x^2}}$
36. $\dfrac{1-\frac{4}{x^2}}{1-\frac{2}{x}}$
37. $\dfrac{1+\frac{2}{x}-\frac{3}{x^2}}{1+\frac{3}{x}-\frac{4}{x^2}}$
38. $\dfrac{2+\frac{1}{x}-\frac{1}{x^2}}{2-\frac{7}{x}+\frac{3}{x^2}}$

39. $\dfrac{1-\frac{4}{x}-\frac{5}{x^2}}{2-\frac{9}{x}-\frac{5}{x^2}}$

40. $\dfrac{x\left(1-\frac{9}{x^2}\right)}{x\left(1-\frac{3}{x}\right)}$

Addition and subtraction of algebraic fractions

Example 8

Write as a single fraction:

(a) $\frac{2}{3}+\frac{3}{4}$; (b) $\frac{2}{x}+\frac{3}{y}$

Compare these two workings line for line.

(a) $\frac{2}{3}+\frac{3}{4}$; the L.C.M. of 3 and 4 is 12.

$$\therefore\quad \frac{2}{3}+\frac{3}{4} = \frac{8}{12}+\frac{9}{12} = \frac{17}{12}$$

(b) $\frac{2}{x}+\frac{3}{y}$; the L.C.M. of x and y is xy.

$$\therefore\quad \frac{2}{x}+\frac{3}{y} = \frac{2y}{xy}+\frac{3x}{xy} = \frac{2y+3x}{xy}$$

Exercise 4

Simplify the following.

1. $\frac{2}{5}+\frac{1}{5}$ **2.** $\frac{2x}{5}+\frac{x}{5}$

3. $\frac{2}{x}+\frac{1}{x}$ **4.** $\frac{1}{7}+\frac{3}{7}$

5. $\frac{x}{7}+\frac{3x}{7}$ **6.** $\frac{1}{7x}+\frac{3}{7x}$

7. $\frac{5}{8}+\frac{1}{4}$ **8.** $\frac{5x}{8}+\frac{x}{4}$

9. $\frac{5}{8x}+\frac{1}{4x}$ **10.** $\frac{2}{3}+\frac{1}{6}$

11. $\frac{2x}{3}+\frac{x}{6}$ **12.** $\frac{2}{3x}+\frac{1}{6x}$

13. $\frac{3}{4}+\frac{2}{5}$ **14.** $\frac{3x}{4}+\frac{2x}{5}$

15. $\frac{3}{4x}+\frac{2}{5x}$ **16.** $\frac{3}{4}-\frac{2}{3}$

17. $\frac{3x}{4}-\frac{2x}{3}$ **18.** $\frac{3}{4x}-\frac{2}{3x}$

19. $\frac{1}{2}-\frac{1}{3}$ **20.** $\frac{x}{2}-\frac{x}{3}$

21. $\frac{1}{2x}-\frac{1}{3x}$ **22.** $\frac{3x}{5}+\frac{2x}{3}$

23. $\frac{4x}{7}-\frac{x}{2}$ **24.** $\frac{3}{4x}-\frac{1}{6x}$

25. $\frac{2}{3}+\frac{1}{2}+\frac{3}{4}$ **26.** $\frac{x}{3}+\frac{3x}{2}+\frac{3x}{4}$

27. $\frac{2}{3x}+\frac{1}{2x}+\frac{3}{4x}$ **28.** $\frac{4}{5}+\frac{1}{2}-\frac{3}{4}$

29. $\frac{5}{7x}-\frac{5}{6x}$ **30.** $\frac{3}{2y}+\frac{1}{5y}$

31. $\frac{3}{2x}+\frac{7}{3x}-\frac{1}{6x}$ **32.** $\frac{5}{3a}+\frac{1}{4a}-\frac{1}{6a}$

Example 9

Write each of the following as a single fraction:

(a) $\frac{x+1}{3}-\frac{x-2}{2}$; (b) $\frac{3}{x+2}+\frac{2}{x-1}$.

(a) $\frac{x+1}{3}-\frac{x-2}{2}$.

The L.C.M. of 3 and 2 is 6.

$$\therefore\quad \frac{x+1}{3}-\frac{x-2}{2} = \frac{2(x+1)}{6}-\frac{3(x-2)}{6} = \frac{2x+2-3x+6}{6} = \frac{8-x}{6}$$

(b) $\dfrac{3}{x+2} + \dfrac{2}{x-1}$.

The L.C.M. of $(x + 2)$ and $(x - 1)$ is $(x + 2)(x - 1)$.

$$\therefore \quad \frac{3}{x+2} + \frac{2}{x-1}$$

$$= \frac{3(x-1)}{(x+2)(x-1)} + \frac{2(x+2)}{(x-1)(x+2)}$$

$$= \frac{3x - 3 + 2x + 4}{(x+2)(x-1)}$$

$$= \frac{5x+1}{(x+2)(x-1)}.$$

Multiplication and division

The method is similar to that used when multiplying and dividing ordinary numerical fractions.

Example 10

Simplify:

(a) $\dfrac{2a}{3x} \times \dfrac{2x^2}{5a}$ (b) $\dfrac{(x-1)}{(x+2)} \div \dfrac{2(x-1)}{(x+3)}$

(a) $\dfrac{2a}{3x} \times \dfrac{2x^2}{5a} = \dfrac{4ax^2}{15ax} = \dfrac{4x}{15}$

(b) $\dfrac{(x-1)}{(x+2)} \div \dfrac{2(x-1)}{(x+3)} = \dfrac{(x-1)}{(x+2)} \times \dfrac{(x+3)}{2(x-1)}$

$$= \frac{(x+3)}{2(x+2)}$$

Exercise 5

Write as a single fraction:

1. $\dfrac{x}{2} + \dfrac{x+1}{3}$
2. $\dfrac{x-1}{3} + \dfrac{x+2}{4}$
3. $\dfrac{2x-1}{5} + \dfrac{x+3}{2}$
4. $\dfrac{x+1}{3} - \dfrac{(2x+1)}{4}$
5. $\dfrac{x-3}{3} - \dfrac{(x-2)}{5}$
6. $\dfrac{2x+1}{7} - \dfrac{(x+2)}{2}$
7. $\dfrac{1}{x} + \dfrac{2}{x+1}$
8. $\dfrac{3}{x-2} + \dfrac{4}{x}$
9. $\dfrac{5}{x-2} + \dfrac{3}{x+3}$
10. $\dfrac{7}{x+1} - \dfrac{3}{x+2}$
11. $\dfrac{2}{x+3} - \dfrac{5}{x-1}$
12. $\dfrac{3}{x-2} - \dfrac{4}{x+1}$
13. $\dfrac{x}{x+1} + \dfrac{3x}{x-1}$
14. $\dfrac{3x}{x+2} - \dfrac{2x}{x-1}$
15. $\dfrac{x}{2} + \dfrac{x+3}{x+2}$
16. $\dfrac{x}{3} - \dfrac{(x+1)}{x-3}$
17. $\dfrac{1}{x(x-1)} + \dfrac{3}{x(x+2)}$
18. $\dfrac{3}{(x-1)(x+2)} + \dfrac{5}{(x+2)(x-3)}$
19. $\dfrac{4}{x^2+x-2} + \dfrac{1}{x^2+2x-3}$
20. $\dfrac{1}{x^2-3x} + \dfrac{1}{x^2+2x}$

Exercise 6

Simplify the following:

1. $\dfrac{3x}{2} \times \dfrac{2a}{3x}$
2. $\dfrac{5mn}{3} \times \dfrac{2}{n}$
3. $\dfrac{3y^2}{x^2} \times \dfrac{2x}{9y}$
4. $\dfrac{5ab^2}{2} \times \dfrac{3}{2a^2b}$
5. $\dfrac{2}{a} \div \dfrac{a}{2}$
6. $\dfrac{4x}{3} \div \dfrac{x}{2}$
7. $\dfrac{x-1}{2} \div \dfrac{x+2}{2}$
8. $\dfrac{x}{5} \times \dfrac{y^2}{x^2}$
9. $\dfrac{x-1}{4} \times \dfrac{x}{2x-2}$
10. $\dfrac{x^2-4}{3} \times \dfrac{9}{x-2}$
11. $\dfrac{5}{x^2-1} \times \dfrac{x+1}{2}$
12. $\dfrac{x}{x^2-9} \div \dfrac{2x}{x-3}$
13. $\dfrac{x}{x^2+x-6} \times \dfrac{x+3}{x}$
14. $\dfrac{3(x-1)}{x} \div \dfrac{6}{xy}$
15. $\dfrac{4}{x^2+2x} \div \dfrac{8}{x^2-4}$
16. $\dfrac{2}{x+1} + \dfrac{3}{x-1}$
17. $\dfrac{4}{x-2} - \dfrac{3}{x+3}$
18. $\dfrac{7}{x} + \dfrac{3}{x+5}$
19. $\dfrac{x^2-9}{x^2-4} \times \dfrac{2x-4}{2}$
20. $\dfrac{x}{x-1} - \dfrac{3x}{x-2}$
21. $\dfrac{3xy}{5y^2} \div \dfrac{x^2y}{10y}$
22. $\dfrac{x^2-1}{x} \times \dfrac{x^2}{x-1}$
23. $\dfrac{3ab}{c} \times \dfrac{2c^2}{a^2b}$
24. $4xyz \times \dfrac{3y}{x^2}$

25. $\dfrac{4}{x+1} - \dfrac{3}{x}$

26. $\dfrac{3}{2x} - \dfrac{1}{3x}$

27. $\dfrac{3x}{y} + \dfrac{5y}{x}$

28. $\dfrac{4}{x-2} + \dfrac{3}{2(x-2)}$

29. $\dfrac{x+\dfrac{1}{x}}{x} \times \dfrac{x}{y}$

30. $\dfrac{3}{x} + \dfrac{2}{x^2+x}$

31. $\dfrac{4mn^2}{x} \div \dfrac{8(mn)^2}{x^2}$

32. $\dfrac{9xy}{z} \times \dfrac{z^2}{(xy)^2}$

33. $\dfrac{x}{y} \times \dfrac{xy}{z} \times \dfrac{z}{x^2}$

34. $\dfrac{ab}{x} \times \dfrac{xb^2}{a^2} \times \dfrac{a^2}{x}$

35. $\dfrac{x-1}{y} \times \dfrac{y^2}{x^2-1}$

36. $\dfrac{9ac^2}{8b} \div \dfrac{6ac}{4b^2}$

37. $\dfrac{x^2+7x}{x^2-1} \times \dfrac{x+1}{x+7}$

38. $\dfrac{\dfrac{x}{y}}{z} \times \dfrac{z^3}{x}$

5.2 CHANGING THE SUBJECT OF A FORMULA

The operations involved in solving ordinary linear equations are exactly the same as the operations required in changing the subject of a formula.

Example 1

(a) Solve the equation $3x + 1 = 12$.
(b) Make x the subject of the formula $Mx + B = A$.

(a)
$$3x + 1 = 12$$
$$3x = 12 - 1$$
$$x = \frac{12-1}{3} = \frac{11}{3}$$

(b)
$$Mx + B = A$$
$$Mx = A - B$$
$$x = \frac{A-B}{M}$$

Example 2

(a) Solve the equation $3(y-2) = 5$.
(b) Make y the subject of the formula $x(y-a) = e$.

(a)
$$3(y-2) = 5$$
$$3y - 6 = 5$$
$$3y = 11$$
$$y = \tfrac{11}{3}$$

(b)
$$x(y-a) = e$$
$$xy - xa = e$$
$$xy = e + xa$$
$$y = \frac{e+xa}{x}$$

Exercise 7

Make x the subject of the following:

1. $2x = 5$
2. $7x = 21$
3. $Ax = B$
4. $Nx = T$
5. $Mx = K$
6. $xy = 4$
7. $Bx = C$
8. $4x = D$
9. $9x = T + N$
10. $Ax = B - R$
11. $Cx = R + T$
12. $Lx = N - R^2$
13. $R - S^2 = Nx$
14. $x + 5 = 7$
15. $x + 10 = 3$
16. $x + A = T$
17. $x + B = S$
18. $N = x + D$
19. $M = x + B$
20. $L = x + D^2$
21. $N^2 + x = T$
22. $L + x = N + M$
23. $Z + x = R - S$
24. $x - 5 = 2$
25. $x - R = A$
26. $x - A = E$
27. $F = x - B$
28. $F^2 = x - B^2$
29. $x - D = A + B$
30. $x - E = A^2$

Make y the subject of the following:

31. $L = y - B$
32. $N = y - T$
33. $3y + 1 = 7$
34. $2y - 4 = 5$
35. $Ay + C = N$
36. $By + D = L$
37. $Dy + E = F$
38. $Ny - F = H$
39. $Yy - Z = T$
40. $Ry - L = B$
41. $Vy + m = Q$
42. $ty - m = n + a$
43. $qy + n = s - t$
44. $ny - s^2 = t$
45. $V^2y + b = c$
46. $r = ny - 6$
47. $s = my + d$
48. $t = my - b$
49. $j = my + c$
50. $2(y+1) = 6$
51. $3(y-1) = 5$
52. $A(y+B) = C$
53. $D(y+E) = F$
54. $h(y+n) = a$
55. $b(y-d) = q$
56. $n = r(y+t)$
57. $t(y-4) = b$
58. $z = S(y+t)$
59. $s = v(y-d)$
60. $g = m(y+n)$

Example 3

(a) Solve the equation $\frac{3a+1}{2} = 4$.

(b) Make a the subject of the formula $\frac{na+b}{m} = n$.

(a) $$\frac{3a+1}{2} = 4$$
$$3a + 1 = 8$$
$$3a = 7$$
$$a = \frac{7}{3}$$

(b) $$\frac{na+b}{m} = n$$
$$na + b = mn$$
$$na = mn - b$$
$$a = \frac{mn-b}{n}$$

Example 4

(a) Solve the equation $3 - 2a = 7$.

(b) Make a the subject of the formula $x - na = y$.

(a) $3 - 2a = 7$

Make the 'a' term positive.
$$3 = 7 + 2a$$
$$3 - 7 = 2a$$
$$\frac{3-7}{2} = a$$
$$-2 = a$$

(b) $x - na = y$.

Make the 'a' term positive
$$x = y + na$$
$$x - y = na$$
$$\frac{x-y}{n} = a$$

Exercise 8

Make a the subject.

1. $\frac{a}{4} = 3$
2. $\frac{a}{5} = 2$
3. $\frac{a}{D} = B$
4. $\frac{a}{B} = T$
5. $\frac{a}{N} = R$
6. $b = \frac{a}{m}$
7. $\frac{a-2}{4} = 6$
8. $\frac{a-A}{B} = T$
9. $\frac{a-D}{N} = A$
10. $\frac{a+Q}{N} = B^2$
11. $g = \frac{a-r}{e}$
12. $\frac{2a+1}{5} = 2$
13. $\frac{Aa+B}{C} = D$
14. $\frac{na+m}{p} = q$
15. $\frac{ra-t}{S} = v$
16. $\frac{za-m}{q} = t$
17. $\frac{m+Aa}{b} = c$
18. $A = \frac{Ba+D}{E}$
19. $n = \frac{ea-f}{h}$
20. $q = \frac{ga+b}{r}$
21. $6 - a = 2$
22. $7 - a = 9$
23. $5 = 7 - a$
24. $A - a = B$
25. $C - a = E$
26. $D - a = H$
27. $n - a = m$
28. $t = q - a$
29. $b = s - a$
30. $v = r - a$
31. $t = m - a$
32. $5 - 2a = 1$
33. $T - Xa = B$
34. $M - Na = Q$
35. $V - Ma = T$
36. $L = N - Ra$
37. $r = v^2 - ra$
38. $t^2 = w - na$
39. $n - qa = 2$
40. $\frac{3-4a}{2} = 1$
41. $\frac{5-7a}{3} = 2$
42. $\frac{B-Aa}{D} = E$
43. $\frac{D-Ea}{N} = B$
44. $\frac{h-fa}{b} = x$
45. $\frac{v^2-ha}{C} = d$
46. $\frac{M(a+B)}{N} = T$
47. $\frac{f(Na-e)}{m} = B$
48. $\frac{T(M-a)}{E} = F$
49. $\frac{y(x-a)}{z} = t$
50. $\frac{k^2(m-a)}{x} = x$

Example 5

(a) Solve the equation $\frac{4}{z} = 7$.

(b) Make z the subject of the formula $\frac{n}{z} = k$.

(a) $\frac{4}{z} = 7$

$4 = 7z \qquad \frac{4}{7} = z$

(b) $\frac{n}{z} = k$

$n = kz \qquad \frac{n}{k} = z$

Example 6

Make t the subject of the formula $\frac{x}{t} + m = a$.

$\frac{x}{t} = a - m$

$x = (a-m)t \qquad \frac{x}{(a-m)} = t$

Exercise 9

Make a the subject.

1. $\frac{7}{a} = 14$
2. $\frac{5}{a} = 3$
3. $\frac{B}{a} = C$
4. $\frac{T}{a} = X$
5. $\frac{M}{a} = B$
6. $m = \frac{n}{a}$
7. $t = \frac{v}{a}$
8. $\frac{n}{a} = \sin 20°$
9. $\frac{7}{a} = \cos 30°$
10. $\frac{B}{a} = x$
11. $\frac{5}{a} = \frac{3}{4}$
12. $\frac{N}{a} = \frac{B}{D}$
13. $\frac{H}{a} = \frac{N}{M}$
14. $\frac{t}{a} = \frac{b}{e}$
15. $\frac{v}{a} = \frac{m}{s}$
16. $\frac{t}{b} = \frac{m}{a}$
17. $\frac{5}{a+1} = 2$
18. $\frac{7}{a-1} = 3$
19. $\frac{B}{a+D} = C$
20. $\frac{Q}{a-C} = T$
21. $\frac{V}{a-T} = D$
22. $\frac{L}{Ma} = B$
23. $\frac{N}{Ba} = C$
24. $\frac{m}{ca} = d$
25. $t = \frac{b}{c-a}$
26. $x = \frac{z}{y-a}$

Make x the subject.

27. $\frac{2}{x} + 1 = 3$
28. $\frac{5}{x} - 2 = 4$
29. $\frac{A}{x} + B = C$
30. $\frac{V}{x} + G = H$
31. $\frac{r}{x} - t = n$
32. $q = \frac{b}{x} + d$
33. $t = \frac{m}{x} - n$
34. $h = d - \frac{b}{x}$
35. $C - \frac{d}{x} = e$
36. $r - \frac{m}{x} = e^2$
37. $t^2 = b - \frac{n}{x}$
38. $\frac{d}{x} + b = mn$
39. $\frac{M}{x+q} - N = 0$
40. $\frac{Y}{x-c} - T = 0$
41. $3M = M + \frac{N}{P+x}$
42. $A = \frac{B}{c+x} - 5A$
43. $\frac{K}{Mx} + B = C$
44. $\frac{z}{xy} - z = y$
45. $\frac{m^2}{x} - n = -p$
46. $t = w - \frac{q}{x}$

Example 7

Make x the subject of the formula
(a) $\sqrt{(x^2 + A)} = B$
(b) $(Ax - B)^2 = M$
(c) $\sqrt{(R - x)} = T$.

(a) $\sqrt{(x^2 + A)} = B$

$x^2 + A = B^2$ (square both sides)

$x^2 = B^2 - A$

$x = \pm\sqrt{(B^2 - A)}$

(b)
$$\begin{aligned} (Ax-B)^2 &= M \\ Ax-B &= \pm\sqrt{M} \text{ (square root both sides)} \\ Ax &= B\pm\sqrt{M} \\ x &= \frac{B\pm\sqrt{M}}{A} \end{aligned}$$

(c)
$$\begin{aligned} \sqrt{(R-x)} &= T \\ R-x &= T^2 \\ R &= T^2+x \\ R-T^2 &= x \end{aligned}$$

Exercise 10

Make x the subject.

1. $\sqrt{x}=2$
2. $\sqrt{(x+1)}=5$
3. $\sqrt{(x-2)}=3$
4. $\sqrt{(x+A)}=B$
5. $\sqrt{(x+C)}=D$
6. $\sqrt{(x-E)}=H$
7. $\sqrt{(ax+b)}=c$
8. $\sqrt{(x-m)}=a$
9. $b=\sqrt{(gx-t)}$
10. $r=\sqrt{(b-x)}$
11. $\sqrt{(d-x)}=t$
12. $b=\sqrt{(x-d)}$
13. $c=\sqrt{(n-x)}$
14. $f=\sqrt{(b-x)}$
15. $g=\sqrt{(c-x)}$
16. $\sqrt{(M-Nx)}=P$
17. $\sqrt{(Ax+B)}=\sqrt{D}$
18. $\sqrt{(x-D)}=A^2$
19. $x^2=g$
20. $x^2+1=17$
21. $x^2=B$
22. $x^2+A=B$
23. $x^2-A=M$
24. $b=a+x^2$
25. $C-x^2=m$
26. $n=d-x^2$
27. $mx^2=n$
28. $b=ax^2$

Make k the subject.

29. $\frac{kz}{a}=t$
30. $ak^2-t=m$
31. $n=a-k^2$
32. $\sqrt{(k^2-4)}=6$
33. $\sqrt{(k^2-A)}=B$
34. $\sqrt{(k^2+y)}=x$
35. $t=\sqrt{(m+k^2)}$
36. $2\sqrt{(k+1)}=6$
37. $A\sqrt{(k+B)}=M$
38. $\sqrt{\left(\frac{M}{k}\right)}=N$
39. $\sqrt{\left(\frac{N}{k}\right)}=B$
40. $\sqrt{(a-k)}=b$
41. $\sqrt{(a^2-k^2)}=t$
42. $\sqrt{(m-k^2)}=x$
43. $2\pi\sqrt{(k+t)}=4$
44. $A\sqrt{(k+1)}=B$
45. $\sqrt{(ak^2-b)}=C$
46. $a\sqrt{(k^2-x)}=b$
47. $k^2+b=x^2$
48. $\frac{k^2}{a}+b=c$
49. $\sqrt{(c^2-ak)}=b$
50. $\frac{m}{k^2}=a+b$

Example 8

Make x the subject of the formula
(a) $Ax-B=Cx+D$
(b) $x+a=\frac{x+b}{c}$

(a)
$$\begin{aligned} Ax-B &= Cx+D \\ Ax-Cx &= D+B \\ x(A-C) &= D+B \text{ (factorise)} \\ x &= \frac{D+B}{A-C} \end{aligned}$$

(b)
$$\begin{aligned} x+a &= \frac{x+b}{c} \\ c(x+a) &= x+b \\ cx+ca &= x+b \\ cx-x &= b-ca \\ x(c-1) &= b-ca \text{ (factorise)} \\ x &= \frac{b-ca}{c-1} \end{aligned}$$

Exercise 11

Make y the subject.

1. $5(y-1)=2(y+3)$
2. $7(y-3)=4(3-y)$
3. $Ny+B=D-Ny$
4. $My-D=E-2My$
5. $ay+b=3b+by$
6. $my-c=e-ny$
7. $xy+4=7-ky$
8. $Ry+D=Ty+C$
9. $ay-x=z+by$
10. $m(y+a)=n(y+b)$
11. $x(y-b)=y+d$
12. $\frac{a-y}{a+y}=b$
13. $\frac{1-y}{1+y}=\frac{c}{d}$
14. $\frac{M-y}{M+y}=\frac{a}{b}$
15. $m(y+n)=n(n-y)$
16. $y+m=\frac{2y-5}{m}$
17. $y-n=\frac{y+2}{n}$
18. $y+b=\frac{ay+e}{b}$
19. $\frac{ay+x}{x}=4-y$
20. $c-dy=e-ay$
21. $y(a-c)=by+d$
22. $y(m+n)=a(y+b)$
23. $t-ay=s-by$
24. $\frac{y+x}{y-x}=3$
25. $\frac{v-y}{v+y}=\frac{1}{2}$
26. $y(b-a)=a(y+b+c)$
27. $\sqrt{\left(\frac{y+x}{y-x}\right)}=2$
28. $\sqrt{\left(\frac{z+y}{z-y}\right)}=\frac{1}{3}$
29. $\sqrt{\left[\frac{m(y+n)}{y}\right]}=p$
30. $n-y=\frac{4y-n}{m}$

Example 9

Make w the subject of the formula

$$\sqrt{\left(\frac{w}{w+a}\right)} = c.$$

Squaring both sides, $\dfrac{w}{w+a} = c^2$

Multiplying by $(w+a)$,

$$\begin{aligned} w &= c^2(w+a) \\ w &= c^2w + c^2a \\ w - c^2w &= c^2a \\ w(1-c^2) &= c^2a \\ w &= \frac{c^2a}{1-c^2} \end{aligned}$$

Exercise 12

Make the letter in square brackets the subject.

1. $ax + by + c = 0$ [x]
2. $\sqrt{\{a(y^2 - b)\}} = e$ [y]
3. $\dfrac{\sqrt{(k-m)}}{n} = \dfrac{1}{m}$ [k]
4. $a - bz = z + b$ [z]
5. $\dfrac{x+y}{x-y} = 2$ [x]
6. $\sqrt{\left(\dfrac{a}{z} - c\right)} = e$ [z]
7. $lm + mn + a = 0$ [n]
8. $t = 2\pi\sqrt{\left(\dfrac{d}{g}\right)}$ [d]
9. $t = 2\pi\sqrt{\left(\dfrac{d}{g}\right)}$ [g]
10. $\sqrt{(x^2 + a)} = 2x$ [x]
11. $\sqrt{\left\{\dfrac{b(m^2+a)}{e}\right\}} = t$ [m]
12. $\sqrt{\left(\dfrac{x+1}{x}\right)} = a$ [x]
13. $a + b - mx = 0$ [m]
14. $\sqrt{(a^2 + b^2)} = x^2$ [a]
15. $\dfrac{a}{k} + b = \dfrac{c}{k}$ [k]
16. $a - y = \dfrac{b+y}{a}$ [y]
17. $G = 4\pi\sqrt{(x^2 + T^2)}$ [x]
18. $M(ax + by + c) = 0$ [y]
19. $x = \sqrt{\left(\dfrac{y-1}{y+1}\right)}$ [y]
20. $a\sqrt{\left(\dfrac{x^2-n}{m}\right)} = \dfrac{a^2}{b}$ [x]
21. $\dfrac{M}{N} + E = \dfrac{P}{N}$ [N]
22. $\dfrac{Q}{P-x} = R$ [x]
23. $\sqrt{(z - ax)} = t$ [a]
24. $e + \sqrt{(x+f)} = g$ [x]
25. $\dfrac{m(ny - e^2)}{p} + n = 5n$ [y]

5.3 VARIATION

Direct variation

There are several ways of expressing a relationship between two quantities x and y. Here are some examples.

x varies as y

x varies directly as y

x is proportional to y

These three all mean the same and they are written in symbols as follows.

$$x \propto y$$

The '$\propto$' sign can always be replaced by '$= k$' where k is a constant:

$$x = ky$$

Suppose $x = 3$ when $y = 12$;
then $3 = k \times 12$
and $k = \frac{1}{4}$

We can then write $x = \frac{1}{4}y$, and this allows us to find the value of x for any value of y, and *vice versa*.

Example 1

y varies as z, and $y = 2$ when $z = 5$; find
(a) the value of y when $z = 6$
(b) the value of z when $y = 5$

Because $y \propto z$, then $y = kz$ where k is a constant.

$y = 2$ when $z = 5$

$\therefore \quad 2 = k \times 5$

$k = \frac{2}{5}$

So $y = \frac{2}{5}z$.

(a) When $z = 6$, $y = \frac{2}{5} \times 6 = 2\frac{2}{5}$.

(b) When $y = 5$, $5 = \frac{2}{5}z$

$z = \frac{25}{2} = 12\frac{1}{2}$.

Example 2

The value V of a diamond is proportional to the square of its weight W. If a diamond weighing 10 grams is worth £200, find
(a) the value of a diamond weighing 30 grams
(b) the weight of a diamond worth £5000.

$V \propto W^2$

or $V = kW^2$ where k is a constant.

$V = 200$ when $W = 10$

$\therefore \quad 200 = k \times 10^2$

$k = 2$

So $V = 2W^2$

(a) When $W = 30$,

$V = 2 \times 30^2 = 2 \times 900$

$V = £1800$

So a diamond of weight 30 grams is worth £1800.

(b) When $V = 5000$,

$5000 = 2 \times W^2$

$W^2 = \dfrac{5000}{2} = 2500$

$W = \sqrt{2500} = 50$

So a diamond of value £5000 weighs 50 grams.

Exercise 13

1. Rewrite the statement connecting each pair of variables using a constant k instead of '$\propto$'.

(a) $S \propto e$ (b) $v \propto t$ (c) $x \propto z^2$
(d) $y \propto \sqrt{x}$ (e) $T \propto \sqrt{L}$ (f) $C \propto r$
(g) $A \propto r^2$ (h) $V \propto r^3$

2. y varies as t. If $y = 6$ when $t = 4$, calculate
(a) the value of y, when $t = 6$
(b) the value of t, when $y = 4$.

3. z is proportional to m. If $z = 20$ when $m = 4$, calculate
(a) the value of z, when $m = 7$
(b) the value of m, when $z = 55$.

4. A varies directly as r^2. If $A = 12$, when $r = 2$, calculate
(a) the value of A, when $r = 5$
(b) the value of r, when $A = 48$.

5. Given that $z \propto x$, copy and complete the table.

x	1	3		$5\frac{1}{2}$
z	4		16	

6. Given that $V \propto r^3$, copy and complete the table.

r	1	2		$1\frac{1}{2}$
V	4		256	

7. Given that $w \propto \sqrt{h}$, copy and complete the table.

h	4	9		$2\frac{1}{4}$
w	6		15	

8. s is proportional to $(v - 1)^2$. If $s = 8$, when $v = 3$, calculate
(a) the value of s, when $v = 4$
(b) the value of v, when $s = 2$.

9. m varies as $(d + 3)$. If $m = 28$ when $d = 1$, calculate
(a) the value of m, when $d = 3$
(b) the value of d, when $m = 49$.

10. The pressure of the water P at any point below the surface of the sea varies as the depth of the point below the surface d. If the pressure is 200 newtons/cm^2 at a depth of 3 m, calculate the pressure at a depth of 5 m.

11. The distance d through which a stone falls from rest is proportional to the square of the time taken t. If the stone falls 45 m in 3 seconds, how far will it fall in 6 seconds?
How long will it take to fall 20 m?

12. The energy E stored in an elastic band varies as the square of the extension x. When the elastic is extended by 3 cm, the energy stored is 243 joules. What is the energy stored when the extension is 5 cm?
What is the extension when the stored energy is 36 joules?

13. In the first few days of its life, the length of an earthworm l is thought to be proportional to the square root of the number of hours n which have elapsed since its birth. If a worm is 2 cm long after 1 hour, how long will it be after 4 hours?
How long will it take to grow to a length of 14 cm?

14. It is well known that the number of golden eggs which a goose lays in a week varies as the cube root of the average number of hours of sleep she has. When she has 8 hours sleep, she lays 4 golden eggs. How long does she sleep when she lays 5 golden eggs?

15. The resistance to motion of a car is proportional to the square of the speed of the car. If the resistance is 4000 newtons at a speed of 20 m/s, what is the resistance at a speed of 30 m/s?
At what speed is the resistance 6250 newtons?

16. A road research organisation recently claimed that the damage to road surfaces was proportional to the fourth power of the axle load. The axle load of a 44-ton HGV is about 15 times that of a car.
Calculate the ratio of the damage to road surfaces made by a 44-ton HGV and a car.

Inverse variation

There are several ways of expressing an inverse relationship between two variables,

x varies inversely as y

x is inversely proportional to y.

We write $x \propto \frac{1}{y}$ for both statements and proceed using the method outlined in the previous section.

Example 3

z is inversely proportional to t^2 and $z = 4$ when $t = 1$. Calculate
(a) z when $t = 2$
(b) t when $z = 16$.

We have $z \propto \frac{1}{t^2}$

or $z = k \times \frac{1}{t^2}$ (k is a constant)

$z = 4$ when $t = 1$, $\therefore 4 = k\left(\frac{1}{1^2}\right)$

so $k = 4$

$\therefore \quad z = 4 \times \frac{1}{t^2}$

(a) when $t = 2$, $z = 4 \times \frac{1}{2^2} = 1$.

(b) when $z = 16$, $16 = 4 \times \frac{1}{t^2}$

$$16t^2 = 4$$
$$t^2 = \tfrac{1}{4}$$
$$t = \pm\tfrac{1}{2}.$$

Exercise 14

1. Rewrite the statements connecting the variables using a constant of variation, k.
(a) $x \propto \frac{1}{y}$ (b) $s \propto \frac{1}{t^2}$ (c) $t \propto \frac{1}{\sqrt{q}}$
(d) m varies inversely as w
(e) z is inversely proportional to t^2.

2. b varies inversely as e. If $b = 6$ when $e = 2$, calculate
(a) the value of b when $e = 12$
(b) the value of e when $b = 3$.

3. q varies inversely as r. If $q = 5$ when $r = 2$, calculate
(a) the value of q when $r = 4$
(b) the value of r when $q = 20$.

4. x is inversely proportional to y^2. If $x = 4$ when $y = 3$, calculate
(a) the value of x when $y = 1$
(b) the value of y when $x = 2\frac{1}{4}$.

5. R varies inversely as v^2. If $R = 120$ when $v = 1$, calculate
(a) the value of R when $v = 10$
(b) the value of v when $R = 30$.

6. T is inversely proportional to x^2. If $T = 36$ when $x = 2$, calculate
   (a) the value of T when $x = 3$
   (b) the value of x when $T = 1{\cdot}44$.
7. p is inversely proportional to $\sqrt{y}$. If $p = 1{\cdot}2$ when $y = 100$, calculate
   (a) the value of p when $y = 4$
   (b) the value of y when $p = 3$.
8. y varies inversely as z. If $y = \frac{1}{8}$ when $z = 4$, calculate
   (a) the value of y when $z = 1$
   (b) the value of z when $y = 10$.

9. Given that $z \propto \frac{1}{y}$, copy and complete the table:

y	2	4		$\frac{1}{4}$
z	8		16	

10. Given that $v \propto \frac{1}{t^2}$, copy and complete the table:

t	2	5		10
v	25		$\frac{1}{4}$	

11. Given that $r \propto \frac{1}{\sqrt{x}}$, copy and complete the table:

x	1	4		
r	12		$\frac{3}{4}$	2

12. e varies inversely as $(y - 2)$. If $e = 12$ when $y = 4$, find
   (a) e when $y = 6$ (b) y when $e = \frac{1}{2}$.
13. M is inversely proportional to the square of l. If $M = 9$ when $l = 2$, find
   (a) M when $l = 10$ (b) l when $M = 1$.
14. Given $z = \frac{k}{x^n}$, find k and n, then copy and complete the table.

x	1	2	4	
z	100	$12\frac{1}{2}$		$\frac{1}{10}$

15. Given $y = \frac{k}{\sqrt[n]{v}}$, find k and n, then copy and complete the table.

v	1	4	36	
y	12	6		$\frac{3}{25}$

16. The volume V of a given mass of gas varies inversely as the pressure P. When $V = 2\,\text{m}^3$, $P = 500\,\text{N/m}^2$. Find the volume when the pressure is $400\,\text{N/m}^2$. Find the pressure when the volume is $5\,\text{m}^3$.
17. The number of hours N required to dig a certain hole is inversely proportional to the number of men available x. When 6 men are digging, the hole takes 4 hours. Find the time taken when 8 men are available. If it takes $\frac{1}{2}$ hour to dig the hole, how many men are there?
18. The life expectancy L of a rat varies inversely as the square of the density d of poison distributed around his home. When the density of poison is $1\,\text{g/m}^2$ the life expectancy is 50 days. How long will he survive if the density of poison is
   (a) $5\,\text{g/m}^2$? (b) $\frac{1}{2}\,\text{g/m}^2$?
19. The force of attraction F between two magnets varies inversely as the square of the distance d between them. When the magnets are 2 cm apart, the force of attraction is 18 newtons. How far apart are they if the attractive force is 2 newtons?

Joint variation

Example 4

z varies jointly as p and q. If $z = 10$ when $p = 1$ and $q = 2$, calculate
(a) the value of z when $p = 2$ and $q = 3$
(b) the value of p when $z = 20$ and $q = 1$.

$z \propto pq$

$\therefore\ z = k\,pq$, where k is a constant.

$z = 10$, when $p = 1$ and $q = 2$

$\therefore\ 10 = k \times 1 \times 2$

$\therefore\ k = 5$

and $z = 5\,pq$

(a) When $p = 2, q = 3$

$z = 5 \times 2 \times 3$

$z = 30$

(b) When $z = 20, q = 1$

$20 = 5 \times p \times 1$

$p = 4$

Example 5

y varies as x and inversely as z^2. Find an equation connecting y, x and z.

We have $y \propto x$ and $y \propto \frac{1}{z^2}$

so $y \propto \frac{x}{z^2}$ or $y = k\frac{x}{z^2}$

Exercise 15

1. v varies jointly as t and x. If $v = 12$ when $t = 1$ and $x = 3$, calculate
   (a) the value of v when $t = 3$ and $x = 2$
   (b) the value of t when $v = 20$ and $x = 1$.
2. s varies as x and inversely as y. If $s = 24$ when $x = 2$ and $y = 1$, calculate
   (a) the value of s when $x = 6$ and $y = 2$
   (b) the value of y when $s = 36$ and $x = 2$.
3. w is proportional to x^2 and inversely proportional to t. If $w = 12$ when $x = 2$ and $t = 2$, find
   (a) w when $x = 3$ and $t = 3$
   (b) x when $w = 10$ and $t = 15$.
4. m is proportional to d and inversely proportional to the square root of s. If $m = 2$ when $d = 6$ and $s = 9$, find
   (a) m when $d = 5$ and $s = 4$
   (b) s when $m = 2$ and $d = 12$.
5. y varies jointly as x and z and inversely as t. If $y = 12$ when $x = 3$, $z = 10$ and $t = 5$, find y when $x = 5$, $z = 2$ and $t = 1$.
6. x varies jointly as y and t, and inversely as n^2. If $x = 15$ when $y = 2$, $t = 12$ and $n = 4$, find x when $y = 3$, $t = 8$ and $n = 3$. Find also y when $x = 8$, $t = 2$ and $n = 1$.
7. If $y \propto x$ and $x \propto z^2$, how does y vary with z?
8. If $x \propto y^2$ and $y \propto z^2$, how does x vary with z?
9. If $z \propto x^2$ and $x \propto \frac{1}{t}$, how does z vary with t?
10. If $x \propto yz$ and $y \propto z^2$, how does x vary with z?
11. If $y \propto \frac{z}{x}$ and $x \propto z^3$, how does y vary with z?
12. The mass m of a solid sphere varies jointly as its density d and the cube of its radius r. When the radius of the sphere is 1 cm and its density is 8 g/cm³ the mass of the sphere is 32 g. Find the mass of a sphere of radius 2 cm and density 6 g/cm³.

5.4 INDICES

Rules of indices

1. $a^n \times a^m = a^{n+m}$ e.g. $7^2 \times 7^4 = 7^6$
2. $a^n \div a^m = a^{n-m}$ e.g. $6^6 \div 6^2 = 6^4$
3. $(a^n)^m = a^{nm}$ e.g. $(3^2)^5 = 3^{10}$

Also, $a^{-n} = \frac{1}{a^n}$ e.g. $5^{-2} = \frac{1}{5^2}$

$a^{\frac{1}{n}}$ means 'the nth root of a' e.g. $9^{\frac{1}{2}} = \sqrt[2]{9}$

$a^{\frac{m}{n}}$ means 'the nth root of a raised to the power m'

e.g. $4^{\frac{3}{2}} = (\sqrt{4})^3 = 8$

Example 1

Simplify (a) $x^7 . x^{13}$ (b) $x^3 \div x^7$ (c) $(x^4)^3$ (d) $(3x^2)^3$ (e) $(2x^{-1})^2 \div x^{-5}$ (f) $3y^2 \times 4y^3$

(a) $x^7 . x^{13} = x^{7+13} = x^{20}$

(b) $x^3 \div x^7 = x^{3-7} = x^{-4} = \frac{1}{x^4}$

(c) $(x^4)^3 = x^{12}$

(d) $(3x^2)^3 = 3^3 . (x^2)^3 = 27x^6$

(e) $(2x^{-1})^2 \div x^{-5} = 4x^{-2} \div x^{-5}$
$= 4x^{(-2--5)}$
$= 4x^3$.

(f) $3y^2 \times 4y^3 = 12y^5$.

Example 2

Evaluate (a) $9^{\frac{1}{2}}$ (b) 5^{-1}
(c) $4^{-\frac{1}{2}}$ (d) $25^{\frac{3}{2}}$
(e) $(5^{\frac{1}{2}})^3 . 5^{\frac{1}{2}}$ (f) 7^0

(a) $9^{\frac{1}{2}} = \sqrt{9} = 3$

(b) $5^{-1} = \frac{1}{5}$

(c) $4^{-\frac{1}{2}} = \dfrac{1}{4^{\frac{1}{2}}} = \dfrac{1}{\sqrt{4}} = \dfrac{1}{2}$

(d) $25^{\frac{3}{2}} = (\sqrt{25})^3 = 5^3 = 125$

(e) $(5^{\frac{1}{2}})^3 . 5^{\frac{1}{2}} = 5^{\frac{3}{2}} . 5^{\frac{1}{2}} = 5^2$
$= 25$

(f) $7^0 = 1 \left[\text{consider } \dfrac{7^3}{7^3} = 7^{3-3} = 7^0 = 1\right]$

Remember. $a^0 = 1$ for any non-zero value of a.

Exercise 16

Express in index form

1. $3 \times 3 \times 3 \times 3$
2. $4 \times 4 \times 5 \times 5 \times 5$
3. $3 \times 7 \times 7 \times 7$
4. $2 \times 2 \times 2 \times 7$
5. $\dfrac{1}{10 \times 10 \times 10}$
6. $\dfrac{1}{2 \times 2 \times 3 \times 3 \times 3}$
7. $\sqrt{15}$
8. $\sqrt[3]{3}$
9. $\sqrt[5]{10}$
10. $(\sqrt{5})^3$

Simplify

11. $x^3 \times x^4$
12. $y^6 \times y^7$
13. $z^7 \div z^3$
14. $z^{50} \times z^{50}$
15. $m^3 \div m^2$
16. $e^{-3} \times e^{-2}$
17. $y^{-2} \times y^4$
18. $w^4 \div w^{-2}$
19. $y^{\frac{1}{2}} \times y^{\frac{1}{2}}$
20. $(x^2)^5$
21. 17^0
22. $w^{-3} \times w^{-2}$
23. $w^{-7} \times w^2$
24. $x^3 \div x^{-4}$
25. $(a^2)^4$
26. $(k^{\frac{1}{2}})^6$
27. $e^{-4} \times e^4$
28. $x^{-1} \times x^{30}$
29. $(y^4)^{\frac{1}{2}}$
30. $(x^{-3})^{-2}$
31. $z^2 \div z^{-2}$
32. $t^{-3} \div t$
33. $(2x^3)^2$
34. $(4y^5)^2$
35. $2x^2 \times 3x^2$
36. $5y^3 \times 2y^2$
37. $5a^3 \times 3a$
38. $(2a)^3$
39. $3x^3 \div x^3$
40. $8y^3 \div 2y$
41. $10y^2 \div 4y$
42. $8a \times 4a^3$
43. $(2x)^2 \times (3x)^3$
44. $4z^4 \times z^{-7}$
45. $6x^{-2} \div 3x^2$
46. $5y^3 \div 2y^{-2}$
47. $(x^2)^{\frac{1}{2}} \div (x^{\frac{1}{3}})^3$
48. $7w^{-2} \times 3w^{-1}$
49. $(2n)^4 \div 8n^0$
50. $4x^{\frac{3}{2}} \div 2x^{\frac{1}{2}}$

Exercise 17

Evaluate the following

1. $3^2 \times 3$
2. 100^0
3. 3^{-2}
4. $(5^{-1})^{-2}$
5. $4^{\frac{1}{2}}$
6. $16^{\frac{1}{2}}$
7. $81^{\frac{1}{2}}$
8. $8^{\frac{1}{3}}$
9. $9^{\frac{3}{2}}$
10. $27^{\frac{1}{3}}$
11. $9^{-\frac{1}{2}}$
12. $8^{-\frac{1}{3}}$
13. $1^{\frac{5}{2}}$
14. $25^{-\frac{1}{2}}$
15. $1000^{\frac{1}{3}}$
16. $2^{-2} \times 2^5$
17. $2^4 \div 2^{-1}$
18. $8^{\frac{2}{3}}$
19. $27^{-\frac{2}{3}}$
20. $4^{-\frac{3}{2}}$
21. $36^{\frac{1}{2}} \times 27^{\frac{1}{3}}$
22. $10\,000^{\frac{1}{4}}$
23. $100^{\frac{3}{2}}$
24. $(100^{\frac{1}{2}})^{-3}$
25. $(9^{\frac{1}{2}})^{-2}$
26. $(-16{\cdot}371)^0$
27. $81^{\frac{1}{4}} \div 16^{\frac{1}{4}}$
28. $(5^{-4})^{\frac{1}{2}}$
29. $1000^{-\frac{1}{3}}$
30. $(4^{-\frac{1}{2}})^2$
31. $8^{-\frac{2}{3}}$
32. $100^{\frac{5}{2}}$
33. $1^{\frac{4}{3}}$
34. 2^{-5}
35. $(0{\cdot}01)^{\frac{1}{2}}$
36. $(0{\cdot}04)^{\frac{1}{2}}$
37. $(2{\cdot}25)^{\frac{1}{2}}$
38. $(7{\cdot}63)^0$
39. $3^5 \times 3^{-3}$
40. $(3\frac{3}{8})^{\frac{1}{3}}$
41. $(1\frac{1}{9})^{-\frac{1}{2}}$
42. $(\frac{1}{8})^{-2}$
43. $(\frac{1}{1000})^{\frac{2}{3}}$
44. $(\frac{9}{25})^{-\frac{1}{2}}$
45. $(10^{-6})^{\frac{1}{3}}$
46. $7^2 \div (7^{\frac{1}{2}})^4$
47. $(0{\cdot}0001)^{-\frac{1}{2}}$
48. $\dfrac{9^{\frac{1}{2}}}{4^{-\frac{1}{2}}}$
49. $\dfrac{25^{\frac{3}{2}} \times 4^{\frac{1}{2}}}{9^{-\frac{1}{2}}}$
50. $(-\frac{1}{7})^2 \div (-\frac{1}{7})^3$

Example 3

Simplify (a) $(2a)^3 \div (9a^2)^{\frac{1}{2}}$
(b) $(3ac^2)^3 \times 2a^{-2}$
(c) $(2x)^2 \div 2x^2$

(a) $(2a)^3 \div (9a^2)^{\frac{1}{2}} = 8a^3 \div 3a$
$= \frac{8}{3}a^2$

(b) $(3ac^2)^3 \times 2a^{-2} = 27a^3c^6 \times 2a^{-2}$
$= 54ac^6$

(c) $(2x)^2 \div 2x^2 = 4x^2 \div 2x^2$
$= 2$

Exercise 18

Rewrite without brackets

1. $(5x^2)^2$
2. $(7y^3)^2$
3. $(10ab)^2$
4. $(2xy^2)^2$
5. $(4x^2)^{\frac{1}{2}}$
6. $(9y)^{-1}$
7. $(x^{-2})^{-1}$
8. $(2x^{-2})^{-1}$
9. $(5x^2y)^0$
10. $(\frac{1}{2}x)^{-1}$
11. $(3x)^2 \times (2x)^2$
12. $(5y)^2 \div y$
13. $(2x^{\frac{1}{2}})^4$
14. $(3y^{\frac{1}{3}})^3$
15. $(5x^0)^2$
16. $[(5x)^0]^2$
17. $(7y^0)^2$
18. $[(7y)^0]^2$
19. $(2x^2y)^3$
20. $(10xy^3)^2$

Simplify the following:

21. $(3x^{-1})^2 \div 6x^{-3}$
22. $(4x)^{\frac{1}{2}} \div x^{\frac{3}{2}}$
23. $x^2y^2 \times xy^3$
24. $4xy \times 3x^2y$
25. $10x^{-1}y^3 \times xy$
26. $(3x)^2 \times (\frac{1}{9}x^2)^{\frac{1}{2}}$
27. $z^3yx \times x^2yz$
28. $(2x)^{-2} \times 4x^3$
29. $(3y)^{-1} \div (9y^2)^{-1}$
30. $(xy)^0 \times (9x)^{\frac{3}{2}}$
31. $(x^2y)(2xy)(5y^3)$
32. $(4x^{\frac{1}{2}}) \times (8x^{\frac{3}{2}})$
33. $5x^{-3} \div 2x^{-5}$
34. $[(3x^{-1})^{-2}]^{-1}$
35. $(2a)^{-2} \times 8a^4$
36. $(abc^2)^3$
37. Write in the form 2^p (e.g. $4 = 2^2$)
    (a) 32 (b) 128 (c) 64 (d) 1
38. Write in the form 3^q
    (a) $\frac{1}{27}$ (b) $\frac{1}{81}$ (c) $\frac{1}{3}$ (d) $9 \times \frac{1}{81}$

Evaluate, with $x = 16$ and $y = 8$.

39. $2x^{\frac{1}{2}} \times y^{\frac{1}{3}}$
40. $x^{\frac{1}{4}} \times y^{-1}$
41. $(y^2)^{\frac{1}{6}} \div (9x)^{\frac{1}{2}}$
42. $(x^2y^3)^0$
43. $x + y^{-1}$
44. $x^{-\frac{1}{2}} + y^{-1}$
45. $y^{\frac{1}{3}} \div x^{\frac{3}{4}}$
46. $(1000y)^{\frac{1}{3}} \times x^{-\frac{5}{2}}$
47. $(x^{\frac{1}{4}} + y^{-1}) \div x^{\frac{1}{4}}$
48. $x^{\frac{1}{2}} - y^{\frac{2}{3}}$
49. $(x^{\frac{3}{4}}y)^{-\frac{1}{3}}$
50. $\left(\frac{x}{y}\right)^{-2}$

Solve the equations for x

51. $2^x = 8$
52. $3^x = 81$
53. $5^x = \frac{1}{5}$
54. $10^x = \frac{1}{100}$
55. $3^{-x} = \frac{1}{27}$
56. $4^x = 64$
57. $6^{-x} = \frac{1}{6}$
58. $100\,000^x = 10$
59. $12^x = 1$
60. $10^x = 0{\cdot}0001$
61. $2^x + 3^x = 13$
62. $(\frac{1}{2})^x = 32$
63. $5^{2x} = 25$
64. $1\,000\,000^{3x} = 10$

5.5 FACTOR THEOREM

If $(x - a)$ is a *factor* of a function of x, $f(x)$, then $f(a) = 0$.

This provides a method for solving quadratic and cubic equations where one solution is a simple integer or fraction.

Example 1

Find a factor of $f(x)$ where:
(a) $f(x) = x^2 + 9x - 22$
(b) $f(x) = x^3 + 5x^2 + 7x + 2$

(a)
$$\begin{aligned} f(1) &= 1^2 + 9 - 22 \neq 0 \\ f(-1) &= 1 - 9 - 22 \neq 0 \\ f(2) &= 4 + 18 - 22 = 0 \end{aligned}$$

$\therefore$ $(x - 2)$ is a factor of $x^2 + 9x - 22$.

(b)
$$\begin{aligned} f(1) &= 1 + 5 + 7 + 2 \neq 0 \\ f(-1) &= -1 + 5 - 7 + 2 \neq 0 \\ f(2) &= 8 + 20 + 14 + 2 \neq 0 \\ f(-2) &= -8 + 20 - 14 + 2 = 0 \end{aligned}$$

$\therefore$ $(x + 2)$ is a factor of $x^3 + 5x^2 + 7x + 2$.

Exercise 19

In questions **1** to **10**, find one factor of each expression.

1. $x^2 + 10x - 24$
2. $x^2 - 11x - 42$
3. $2x^2 - x - 10$
4. $3x^2 + 2x - 21$
5. $x^3 + x^2 - 7x - 7$
6. $x^3 - 3x^2 - 5x + 15$
7. $x^3 + 2x^2 + x + 2$
8. $2x^3 + 2x^2 - 7x - 7$
9. $2x^3 + 6x^2 + x + 3$
10. $x^4 - 2x^3 - 7x + 14$

In questions **11** to **20**, find the value of k using the information given.

11. $(x + 1)$ is a factor of $(8x^2 - x + k)$
12. $(x - 3)$ is a factor of $(7x^2 - 20x + k)$
13. $(x + 6)$ is a factor of $(3x^2 + kx - 30)$
14. $(x + 1)$ is a factor of $(x^3 + 7x + k)$
15. $(2x - 1)$ is a factor of $(2x^3 - x^2 + 4x + k)$
16. $(2x + 1)$ is a factor of $(2x^3 + x^2 + kx - 3)$
17. $(x - 3)$ is a factor of $(x^3 + kx^2 - 5x + 15)$
18. $(x - 2)$ is a factor of $(x^3 - 4x^2 + kx + k + 5)$
19. $(2x - 3)$ is a factor of $(2x^4 + kx^3 + 4x - 6)$
20. $(3x + 1)$ is a factor of $(3x^4 + kx^3 - 6x - 2)$

Example 2

Divide $(x^3 + 5x^2 + 4x - 4)$ by $(x + 2)$.

$$
\begin{array}{r}
x + 2\,)\,x^3 + 5x^2 + 4x - 4\,(\,x^2 + 3x - 2 \\
\text{(a)}\quad x^3 + 2x^2 \qquad\qquad\qquad \\
3x^2 + 4x \qquad\qquad \\
\text{(b)}\quad 3x^2 + 6x \qquad\qquad \\
-2x - 4 \qquad \\
\text{(c)}\quad -2x - 4 \qquad \\
0 \qquad
\end{array}
$$

(a) x 'into' x^3 'goes' x^2 [on the right]
(b) x 'into' $3x^2$ 'goes' $3x$ [on the right]
(c) x 'into' $-2x$ 'goes' -2 [on the right]

so $x^3 + 5x^2 + 4x - 4 = (x + 2)(x^2 + 3x - 2)$

Example 3

Solve the equation $f(x) = 0$ where
$f(x) = x^3 + 13x^2 + 6x - 72$.

(a) Find one factor using the factor theorem.
(b) Find the other factors by division.

(a) $f(1) = 1 + 13 + 6 - 72 \neq 0$
$f(-1) = -1 + 13 - 6 - 72 \neq 0$
$f(2) = 8 + 52 + 12 - 72 = 0$

$\therefore$ $(x - 2)$ is a factor.

(b)
$$
\begin{array}{r}
x - 2\,)\,x^3 + 13x^2 + 6x - 72\,(\,x^2 + 15x + 36 \\
x^3 - 2x^2 \qquad\qquad\qquad\qquad \\
15x^2 + 6x \qquad\qquad \\
15x^2 - 30x \qquad\qquad \\
36x - 72 \qquad \\
36x - 72 \qquad \\
0 \qquad
\end{array}
$$

so $f(x) = (x - 2)(x^2 + 15x + 36)$
$= (x - 2)(x + 3)(x + 12)$

for $f(x) = 0$
$(x - 2)(x + 3)(x + 12) = 0$

$\therefore$ $x = 2$ or $x = -3$ or $x = -12$

Exercise 20

1. Divide $(x^3 + 4x^2 + 5x + 2)$ by $(x + 2)$
2. Divide $(x^3 + x^2 - 9x - 9)$ by $(x - 3)$
3. Divide $(2x^3 + 3x^2 - 1)$ by $(x + 1)$
4. Divide $(x^3 + 5x^2 + 4x - 4)$ by $(x + 2)$
5. Divide $(2x^3 - x^2 + 2x - 1)$ by $(2x - 1)$
6. Divide $(x^4 - 2x^3 - 7x + 14)$ by $(x - 2)$

Solve the following equations:

7. $2x^3 + x^2 - 13x + 6 = 0$
8. $x^3 - 2x^2 - 5x + 6 = 0$
9. $2x^3 + x^2 - 8x - 4 = 0$
10. $2x^3 + 3x^2 - 32x + 15 = 0$
11. $x^3 - 4x^2 + x + 6 = 0$
12. $2x^3 + 11x^2 + 17x + 6 = 0$
13. $2x^3 = x^2 - 2x + 1$
14. $3x^3 + 2x^2 = 61x - 20$
15. $x^3 + 26x = 9x^2 + 24$
16. $x^4 + 10x^3 + 13x^2 - 96x - 180 = 0$
17. $x^4 - 2x^3 - 7x + 14 = 0$
18. $x^4 + 3x^3 - 3x - 9 = 0$
19. $2x^3 - x^2 = 4x - 2$
20. $x^5 + 2x^4 - x - 2 = 0$
21. One of the solutions published by Cardan in 1545 for the solution of cubic equations is given below.
For an equation in the form $x^3 + px = q$

$$x = \sqrt[3]{\left[\sqrt{\left(\frac{p}{3}\right)^3 + \left(\frac{q}{2}\right)^2} + \frac{q}{2}\right]} - \sqrt[3]{\left[\sqrt{\left(\frac{p}{3}\right)^3 + \left(\frac{q}{2}\right)^2} - \frac{q}{2}\right]}$$

Use the formula to solve the following equations, giving answers to 4 S.F. where necessary.
(a) $x^3 + 7x = -8$
(b) $x^3 + 6x = 4$
(c) $x^3 + 3x = 2$
(d) $x^3 + 9x - 2 = 0$

5.6 ITERATIVE METHODS

These methods are particularly well suited when a computer or calculator is available.

Example 1

Suppose we want to find $\sqrt{11}$.
Let our first approximation be a_1.

(a) Try $a_1 = 3$.
Since $3^2 = 9$, we deduce that 3 is smaller than $\sqrt{11}$.
We also deduce that $\frac{11}{3}$ is larger than $\sqrt{11}$.

(b) For our second approximation a_2, try the *average* of 3 and $\frac{11}{3}$.

i.e. $a_2 = \frac{1}{2}(3 + \frac{11}{3}) \approx 3{\cdot}333$.

(c) Now, $3{\cdot}333^2 > 11$

$\therefore \quad \dfrac{11}{3{\cdot}333} < \sqrt{11}$ by the argument above.

(d) Find the average of $3{\cdot}333$ and $\dfrac{11}{3{\cdot}333}$

$$\text{i.e. } a_3 = \frac{1}{2}\left(3{\cdot}333 + \frac{11}{3{\cdot}333}\right)$$

$$= 3{\cdot}317$$

This is already $\sqrt{11}$ correct to four significant figures. This is an example of an iterative method.

The iterative formula for finding the square root of a number N is

$$a_{n+1} = \frac{1}{2}\left(a_n + \frac{N}{a_n}\right)$$

where a_n is the approximation after n repetitions of the formula and a_{n+1} is the next approximation.

Exercise 21

1. Start by making a suitable 'guess' and then use the formula above to find the square root of the following numbers, correct to four significant figures.
(a) 8 (b) 18 (c) 39
(d) 7·63 (e) 111 (f) 413

2. The iterative formula for finding the cube root of a number N is

$$a_{n+1} = \frac{1}{3}\left(2a_n + \frac{N}{a_n^2}\right)$$

Use the formula to find the cube root of the following numbers, correct to four significant figures.
(a) 9 (b) 28 (c) 70
(d) 0·962 (e) 1050

3. The iterative formula for finding the fifth root of a number N is

$$a_{n+1} = \frac{1}{5}\left(4a_n + \frac{N}{a_n^4}\right)$$

Find the fifth root of the following numbers correct to four significant figures.
(a) 35 (b) 260 (c) 115 000
(d) 27

4. It is possible to find a solution to the equation $x^3 - 4x^2 - 20 = 0$ by using the iterative formula

$$x_{n+1} = \frac{20}{x_n^2} + 4$$

Find a solution, correct to three significant figures. (Try $x_1 = 1$.)

5. The equation $x^4 - 3x^3 - 10 = 0$ may be solved approximately by using the formula

$$x_{n+1} = \frac{10}{x_n^3} + 3$$

Find a solution, correct to three significant figures. (Try $x_1 = 1$.)

6. Find the value of $\frac{1}{13}$ correct to six decimal places by using the iterative formula
$x_{n+1} = x_n(2 - 13x_n)$.

7. One solution of the quadratic equation $x^2 - 10x + 5 = 0$ may be found by using the iterative formula

$$x_{n+1} = 10 - \frac{5}{x_n}$$

Find this solution, correct to four significant figures by using as an initial approximation:
(a) $x_1 = 8$ (b) $x_1 = -7$ (c) $x_1 = 100$

You will notice that each time the formula gives the same solution but fails to find the other solution. This illustrates a weakness of this type of simple 'first order' formula.

8. Use the formula

$$x_{n+1} = \frac{1}{2}\left(3 + \frac{1}{x_n}\right)$$

to find a solution of the equation

$$2x^2 - 3x - 1 = 0.$$

Again try a wide variety of initial approximations.

9. The following formula may be used to calculate the value of π to a high degree of accuracy.

$$\pi = 16\left(\frac{1}{5} - \frac{1}{3}\times\frac{1}{5^3} + \frac{1}{5}\times\frac{1}{5^5} - \frac{1}{7}\times\frac{1}{5^7} + \ldots\right)$$
$$-4\left(\frac{1}{239} - \frac{1}{3}\times\frac{1}{239^3} + \frac{1}{5}\times\frac{1}{239^5} - \frac{1}{7}\times\frac{1}{239^7} + \ldots\right)$$

Use the series to obtain a value of π correct to 6 decimal places.

10. Compare the values for π obtained by the following approximations:

(a) Between $\frac{22}{7}$ and $\frac{223}{71}$ (Archimedes ~ 250 B.C.)

(b) $\frac{355}{113}$ (Tsu Ch'ung-Chih ~ 460)

(c) $\sqrt{\left(\sqrt{\left(\frac{2143}{22}\right)}\right)}$ (Ramanujan ~ 1914).

5.7 INEQUALITIES

$x < 4$ means 'x is *less than* 4'
$y > 7$ means 'y is *greater than* 7'
$z \leqslant 10$ means 'z is *less than or equal to* 10'
$t \geqslant -3$ means 't is *greater than or equal to* -3'

Solving inequalities

We follow the same procedure used for solving equations except that when we multiply or divide by a *negative* number the inequality is *reversed*.

e.g. $4 > -2$
but multiplying by -2,
$-8 < 4$

Example 1

Solve the inequalities
(a) $2x - 1 > 5$
(b) $5 - 3x \leqslant 1$

(a) $2x - 1 > 5$
$2x > 5 + 1$
$x > \frac{6}{2}$
$x > 3$

(b) $5 - 3x \leqslant 1$
$5 \leqslant 1 + 3x$
$5 - 1 \leqslant 3x$
$\frac{4}{3} \leqslant x$

Exercise 22

Introduce one of the symbols $<, >$ or $=$ between each pair of numbers.

1. $-2, 1$
2. $(-2)^2, 1$
3. $\frac{1}{4}, \frac{1}{5}$
4. $0{\cdot}2, \frac{1}{5}$
5. $10^2, 2^{10}$
6. $\frac{1}{4}, 0{\cdot}4$
7. 40%, $0{\cdot}4$
8. $(-1)^2, (-\frac{1}{2})^2$
9. $5^2, 2^5$
10. $3\frac{1}{3}, \sqrt{10}$
11. $\pi^2, 10$
12. $-\frac{1}{3}, -\frac{1}{2}$
13. $2^{-1}, 3^{-1}$
14. 50%, $\frac{1}{5}$
15. 1%, 100^{-1}

State whether the following are true or false:

16. $0{\cdot}7^2 > \frac{1}{2}$
17. $10^3 = 30$
18. $\frac{1}{8} > 12\%$
19. $(0{\cdot}1)^3 = 0{\cdot}0001$
20. $(-\frac{1}{5})^0 = -1$
21. $\frac{1}{5^2} > \frac{1}{2^5}$
22. $(0{\cdot}2)^2 < (0{\cdot}3)^2$
23. $\frac{6}{7} > \frac{7}{8}$
24. $0{\cdot}1^2 > 0{\cdot}1$

Solve the following inequalities:

25. $x-3>10$ **26.** $x+1<0$
27. $5>x-7$ **28.** $2x+1\leqslant 6$
29. $3x-4>5$ **30.** $10\leqslant 2x-6$
31. $5x<x+1$ **32.** $2x\geqslant x-3$
33. $4+x<-4$ **34.** $3x+1<2x+5$
35. $2(x+1)>x-7$ **36.** $7<15-x$
37. $9>12-x$ **38.** $4-2x\leqslant 2$
39. $3(x-1)<2(1-x)$ **40.** $7-3x<0$

The number line

The inequality $x<4$ is represented on the number line as

−3 −2 −1 0 1 2 3 4 5

$x\geqslant -2$ is shown as

−3 −2 −1 0 1 2 3 4 5 6

In the first case, 4 is *not* included so we have ○.
In the second case, −2 *is* included so we have ●.

$-1\leqslant x<3$ is shown as

−3 −2 −1 0 1 2 3 4 5 6

Example 2

Find the solution set for $2x-1<20$, given that x is a prime number.

$$2x-1<20$$
$$2x<21$$
$$x<10\tfrac{1}{2}$$

But x is a prime number, so the solution set is $\{2, 3, 5, 7\}$.

Exercise 23

For questions **1** to **25**, solve each inequality and show the result on a number line.

1. $2x+1>11$ **2.** $3x-4\leqslant 5$
3. $2<x-4$ **4.** $6\geqslant 10-x$
5. $8<9-x$ **6.** $8x-1<5x-10$
7. $2x>0$ **8.** $1<3x-11$
9. $4-x>6-2x$ **10.** $\frac{x}{3}<-1$
11. $1<x<4$ **12.** $-2\leqslant x\leqslant 5$
13. $1\leqslant x<6$ **14.** $0\leqslant 2x<10$
15. $-3\leqslant 3x\leqslant 21$ **16.** $x^2<4$
17. $x^2\leqslant 16$ **18.** $x^2>1$
19. $x^2\geqslant 9$ **20.** $1<2x+1<9$
21. $10\leqslant 2x\leqslant x+9$ **22.** $x<3x+2<2x+6$
23. $x\leqslant 2x-1\leqslant x+5$ **24.** $x<3x-1<2x+7$
25. $x-10<2(x-1)<x$

(Hint: in questions **20** to **25**, solve the two inequalities separately.)

For questions **26** to **35**, find the solution set, subject to the given condition.

26. $3a+1<20$; a is a positive integer
27. $b-1\geqslant 6$; b is a prime number less than 20
28. $2e-3<21$; e is a positive even number
29. $1<z<50$; z is a square number
30. $0<3x<40$; x is divisible by 5
31. $2x>-10$; x is a negative integer
32. $x+1<2x<x+13$; x is an integer
33. $x^2<100$; x is a positive square number
34. $0\leqslant 2z-3\leqslant z+8$; z is a prime number
35. $\frac{a}{2}+10>a$; a is a positive even number.

36. State the smallest integer n for which $4n>19$.
37. Find an integer value of x such that $2x-7<8<3x-11$.
38. Find an integer value of y such that $3y-4<12<4y-5$.
39. Find any value of z such that $9<z+5<10$.
40. Find any value of p such that $9<2p+1<11$.
41. Find a simple fraction q such that $\frac{4}{9}<q<\frac{5}{9}$.
42. Find an integer value of a such that $a-3\leqslant 11\leqslant 2a+10$.
43. State the largest prime number z for which $3z<66$.
44. Find a simple fraction r such that $\frac{1}{3}<r<\frac{2}{3}$.
45. Find the largest prime number p such that $p^2<400$.
46. Illustrate on a number line the solution set of each pair of simultaneous inequalities:
(a) $x<6$; $-3\leqslant x\leqslant 8$
(b) $x>-2$; $-4<x<2$
(c) $2x+1\leqslant 5$; $-12\leqslant 3x-3$
(d) $3x-2<19$; $2x\geqslant -6$

Graphical display

It is useful to represent inequalities on a graph, particularly where two variables are involved.

Example 3

Draw a sketch graph and shade the area which represents the set of points that satisfy each of these inequalities.
(a) $x > 2$ (b) $1 \leqslant y \leqslant 5$ (c) $x + y \leqslant 8$

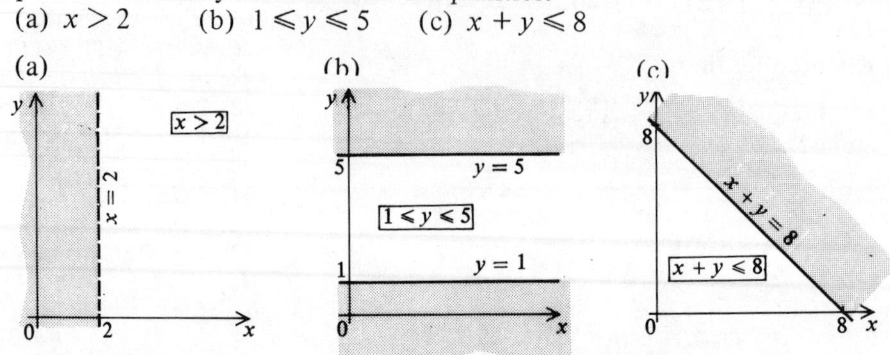

In each graph, the unwanted region is shaded so that the region representing the set of points is left clearly visible.

In (a), the line $x = 2$ is shown as a broken line to indicate that the points on the line are *not* included.

In (b) and (c), the lines $y = 1$, $y = 5$ and $x + y = 8$ are shown as solid lines because points on the line *are* included in the solution set.

An inequality can thus be regarded as a set of points, for example, the unshaded region in (c) may be described as

$$\{(x, y): x + y \leqslant 8\}$$

i.e. the set of points (x, y) such that $x + y \leqslant 8$.

Exercise 24

In questions **1** to **9** describe the region left unshaded.

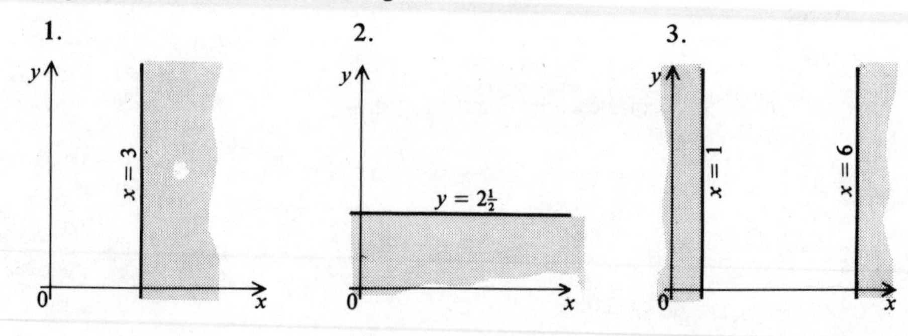

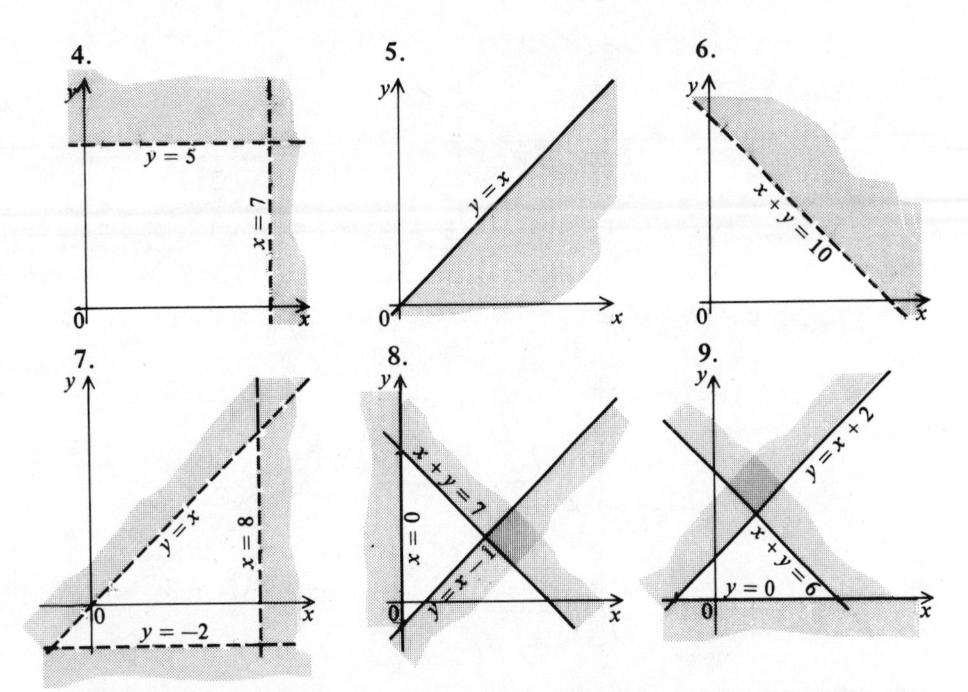

For questions **10** to **31**, draw a sketch graph similar to those above and indicate the set of points which satisfy the inequalities by shading the unwanted regions.

10. $2 \leqslant x \leqslant 7$

11. $0 \leqslant y \leqslant 3\frac{1}{2}$

12. $-2 < x < 2$

13. $x < 6$ and $y \leqslant 4$

14. $0 < x < 5$ and $y < 3$

15. $1 \leqslant x \leqslant 6$ and $2 \leqslant y \leqslant 8$

16. $-3 < x < 0$ and $-4 < y < 2$

17. $y \leqslant x$

18. $x + y < 5$

19. $y > x + 2$ and $y < 7$

20. $x \geqslant 0$ and $y \geqslant 0$ and $x + y \leqslant 7$

21. $x \geqslant 0$ and $x + y < 10$ and $y > x$

22. $8 \geqslant y \geqslant 0$ and $x + y > 3$

23. $x + 2y < 10$ and $x \geqslant 0$ and $y \geqslant 0$

24. $3x + 2y \leqslant 18$ and $x \geqslant 0$ and $y \geqslant 0$

25. $x \geqslant 0, y \geqslant x - 2, x + y \leqslant 10$

26. $3x + 5y \leqslant 30$ and $y > \dfrac{x}{2}$

27. $y \geqslant \dfrac{x}{2}, y \leqslant 2x$ and $x + y \leqslant 8$

28. $x \geqslant 0, y \geqslant x - 2$ and $x + y \leqslant 10$

29. $2x + y \geqslant 10, x + y \leqslant 10$ and $y \geqslant 0$

30. $2x + 3y < 16, y > x$ and $x > -2$

31. $3y > x, y < 3x$ and $x < 7$

In questions **32** to **40**, describe the region left unshaded.

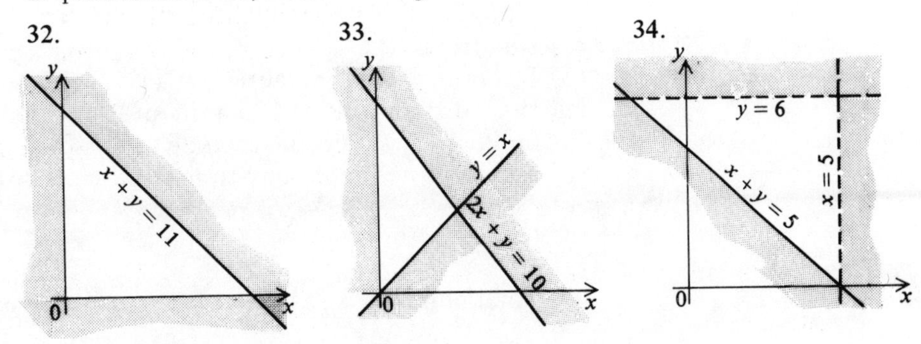

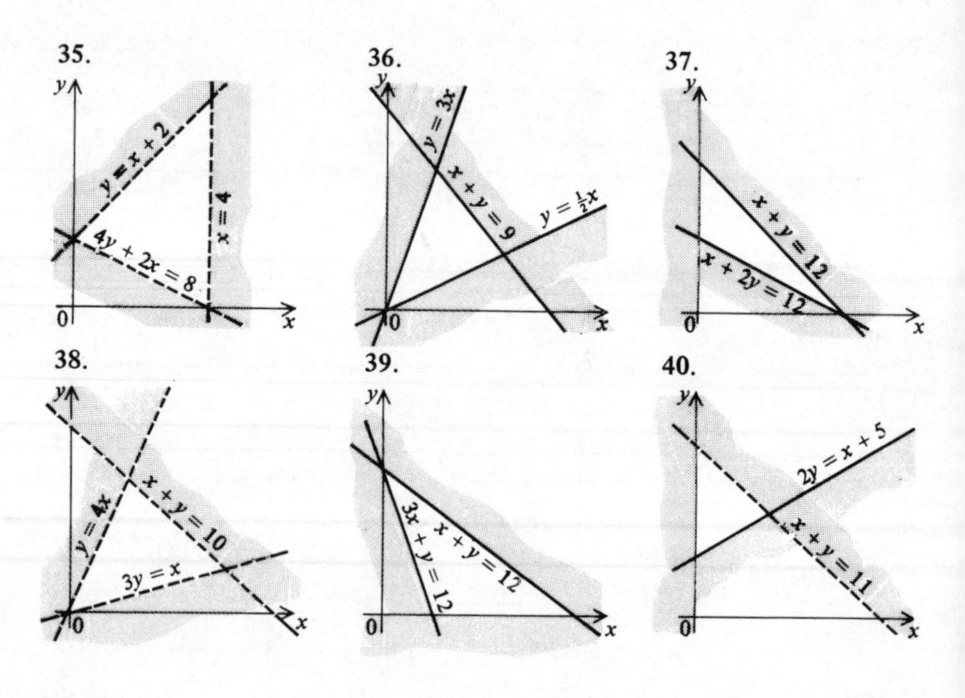

5.8 LINEAR PROGRAMMING

In most linear programming problems, there are two stages:

1. to interpret the information given as a series of simultaneous inequalities and display them graphically.
2. to investigate some characteristic of the points in the unshaded solution set.

Example 1

A shopkeeper buys two types of dog food for his shop: Bruno at 40p a tin and Blaze at 60p a tin. He has £15 available and decides to buy at least 30 tins altogether. He also decides that at least one third of the tins should be Blaze. He buys x tins of Bruno and y tins of Blaze.

(a) Write down three inequalities which correspond to the above conditions.

(b) Illustrate these inequalities on a graph, as shown on the next page.

(c) He makes a profit of 10p a tin on Bruno and a profit of 20p a tin on Blaze. Assuming he can sell all his stock, find how many tins of each type he should buy to maximise his profit and find that profit.

(a) Cost. $40x + 60y \leqslant 1500$ $\quad 2x + 3y \leqslant 75$ . . . [line A on graph]

Total number. $\quad x + y \geqslant 30$ . . . [line B on graph]

At least one third Blaze. $\dfrac{y}{x} \geqslant \dfrac{1}{2}$ $\quad 2y \geqslant x$ . . . [line C on graph]

(b) The graph below shows these three equations.

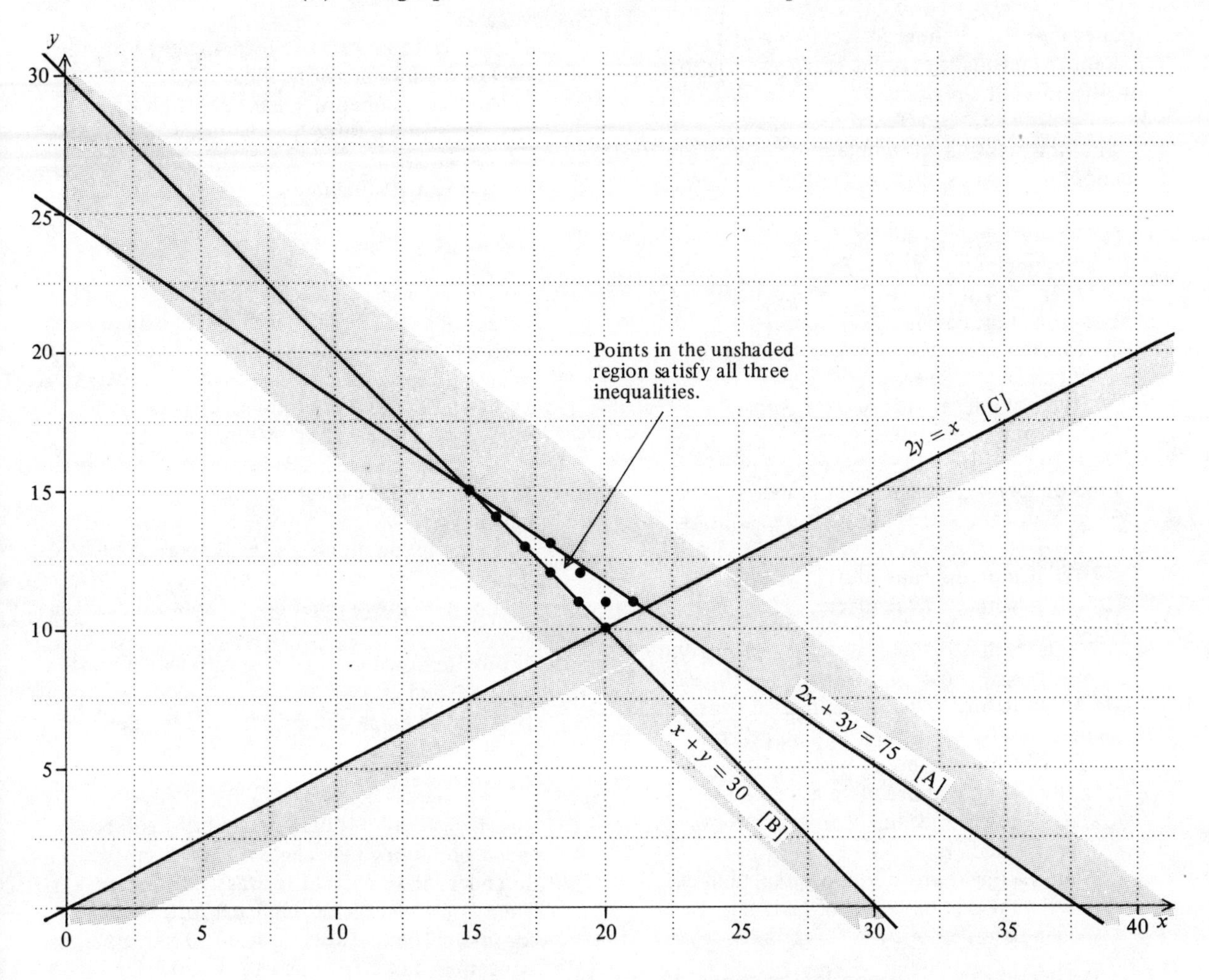

(c) The table below shows the points on the graph in the unshaded region together with the corresponding figure for the profit.

The points marked * will clearly not provide a maximum profit.

x	15	16	17*	18	19	19*	20*	20*	21
y	15	14	13	13	12	11	11	10	11
profit	150 + 300 = 450p	160 + 280 = 440p		180 + 260 = 440p	190 + 240 = 430p				210 + 220 = 430p

Conclusion: he should buy 15 tins of Bruno and 15 tins of Blaze. His maximised profit is then 450p.

Exercise 25

For questions **1** to **3**, draw an accurate graph to represent the inequalities listed, using shading to show the unwanted regions.

1. $x+y \leqslant 11; y \geqslant 3; y \leqslant x$.
   Find the point having whole number coordinates and satisfying these inequalities which gives
   (a) the maximum value of $x+4y$
   (b) the minimum value of $3x+y$
2. $3x+2y > 24; x+y < 12; y < \frac{1}{2}x; y > 1$.
   Find the point having whole number coordinates and satisfying these inequalities which gives
   (a) the maximum value of $2x+3y$
   (b) the minimum value of $x+y$
3. $3x+2y \leqslant 60; x+2y \leqslant 30; x \geqslant 10; y \geqslant 0$.
   Find the point having whole number coordinates and satisfying these inequalities which gives
   (a) the maximum value of $2x+y$
   (b) the maximum value of xy
4. A boy is given £1·20 to buy some peaches and apples. Peaches cost 20p each, apples 10p each. He is told to buy at least 6 individual fruits, but he must not buy more apples than peaches.
   Let x be the number of peaches he buys.
   Let y be the number of apples he buys.
   (a) Write down three inequalities which must be satisfied.
   (b) Draw a linear programming graph and use it to list the combinations of fruit that are open to him.
5. A girl is told to buy some melons and oranges. Melons are 50p each and oranges 25p each, and she has £2 to spend. She must not buy more than 2 melons and she must buy at least 4 oranges. She is also told to buy at least 6 fruits altogether.
   Let x be the number of melons.
   Let y be the number of oranges.
   (a) Write down four inequalities which must be satisfied.
   (b) Draw a graph and use it to list the combinations of fruit that are open to her.
6. A chef is going to make some fruit cakes and sponge cakes. He has plenty of all ingredients except for flour and sugar. He has only 2000 g of flour and 1200 g of sugar.
   A fruit cake uses 500 g of flour and 100 g of sugar.
   A sponge cake uses 200 g of flour and 200g of sugar.
   He wishes to make *more than* 4 cakes altogether.
   Let the number of fruit cakes be x.
   Let the number of sponge cakes be y.
   (a) Write down three inequalities which must be satisfied.
   (b) Draw a graph and use it to list the possible combinations of fruit cakes and sponge cakes which he can make.
7. A man has a spare time job spraying cars and vans. Vans take 2 hours each and cars take 1 hour each. He has 14 hours available per week. He has an agreement with one firm to do 2 of their vans every week. Apart from that he has no fixed work.
   His permission to use his back garden contains the clause that he must do at least twice as many cars as vans.
   Let x be the number of vans sprayed each week.
   Let y be the number of cars sprayed each week.
   (a) Write down three inequalities which must be satisfied.
   (b) Draw a graph and use it to list the possible combinations of vehicles which he can spray each week.
8. The manager of a football team has £100 to spend on buying new players. He can buy defenders at £6 each or forwards at £8 each. There must be at least 6 of each sort. To cover for injuries he must buy at least 13 players altogether. Let x represent the number of defenders he buys and y the number of forwards.
   (a) In what ways can he buy players?
   (b) If the wages are £10 per week for each defender and £20 per week for each forward, what is the combination of players which has the lowest wage bill?
9. A tennis-playing golfer has £15 to spend on golf balls (x) costing £1 each and tennis balls (y) costing 60p each. He must buy at least 16 altogether and he must buy *more* golf balls than tennis balls.
   (a) What is the greatest number of balls he can buy?
   (b) After using them, he can sell golf balls for 10p each and tennis balls for 20p each. What is his maximum possible income from sales?

10. A travel agent has to fly 1000 people and 35 000 kg of baggage from London to Paris. Two types of aircraft are available: A which takes 100 people and 2000 kg of baggage, or B which takes 60 people and 3000 kg of baggage. He can use no more than 16 aircraft altogether. Write down three inequalities which must be satisfied if he uses x of A and y of B.
   (a) What is the smallest number of aircraft he could use?
   (b) If the hire charge for each aircraft A is £10 000 and for each aircraft B is £12 000, find the cheapest option available to him.
   (c) If the hire charges are altered so that each A costs £10 000 and each B costs £20 000, find the cheapest option now available to him.

11. A farmer has to transport 20 people and 32 sheep to a market. He can use either Fiats (x) which take 2 people and 1 sheep, or Rolls Royces (y) which take 2 people and 4 sheep. He must not use more than 15 cars altogether.
   (a) What is the lowest total numbers of cars he could use?
   (b) If it costs £10 to hire each Fiat and £30 for each Rolls Royce, what is the *cheapest* solution?

12. A shop owner wishes to buy up to 20 televisions for stock. He can buy either type A for £150 each or type B for £300 each. He has a total of £4500 he can spend. He must have at least 6 of each type in stock. If he buys x of type A and y of type B, write down 4 inequalities which must be satisfied and represent the information on a graph.
   (a) If he makes a profit of £40 on each of type A and £100 on each of type B, how many of each should he buy for maximum profit?
   (b) If the profit is £80 on each of type A and £100 on each of type B, how many of each should he buy now?

13. A farmer needs to buy up to 25 cows for a new herd. He can buy either brown cows (x) at £50 each or black cows (y) at £80 each and he can spend a total of no more than £1600. He must have at least 9 of each type. On selling the cows he makes a profit of £50 on each brown cow and £60 on each black cow. How many of each sort should he buy for maximum profit?

14. The manager of a car park allows $10\,m^2$ of parking space for each car and $30\,m^2$ for each lorry. The total space available is $300\,m^2$. He decides that the maximum number of vehicles at any time must not exceed 20 and he also insists that there must be at least as many cars as lorries. If the number of cars is x and the number of lorries is y, write down three inequalities which must be satisfied.
   (a) If the parking charge is £1 for a car and £5 for a lorry, find how many vehicles of each kind he should admit to maximise his income.
   (b) If the charges are changed to £2 for a car and £3 for a lorry, find how many of each kind he would be advised to admit.

REVISION EXERCISE 5A

1. Express the following as single fractions:
   (a) $\frac{x}{4} + \frac{x}{5}$
   (b) $\frac{1}{2x} + \frac{2}{3x}$
   (c) $\frac{x+2}{2} + \frac{x-4}{3}$
   (d) $\frac{7}{x-1} - \frac{2}{x+3}$

2. (a) Factorise $x^2 - 4$
   (b) Simplify $\frac{3x-6}{x^2-4}$

3. Given that $s - 3t = rt$, express
   (a) s in terms of r and t
   (b) r in terms of s and t
   (c) t in terms of s and r.

4. (a) Given that $x - z = 5y$, express z in terms of x and y.
   (b) Given that $mk + 3m = 11$, express m in terms of k.
   (c) For the formula $T = C\sqrt{z}$, express z in terms of T and C.

5. It is given that $y = \frac{k}{x}$ and that $1 \leqslant x \leqslant 10$.
   (a) If the smallest possible value of y is 5, find the value of the constant k.
   (b) Find the largest possible value of y.
6. Given that y varies as x^2 and that $y = 36$ when $x = 3$, find
   (a) the value of y when $x = 2$
   (b) the value of x when $y = 64$.
7. (a) Evaluate (i) $9^{\frac{1}{2}}$ (ii) $8^{\frac{2}{3}}$ (iii) $16^{-\frac{1}{2}}$
   (b) Find x, given that
   (i) $3^x = 81$ (ii) $7^x = 1$.
8. List the integer values of x which satisfy.
   (a) $2x - 1 < 20 < 3x - 5$
   (b) $5 < 3x + 1 < 17$.
9. Given that $t = k\sqrt{(x+5)}$, express x in terms of t and k.
10. Given that $z = \frac{3y+2}{y-1}$, express y in terms of z.
11. Given that $y = \frac{k}{k+w}$
   (a) Find the value of y when $k = \frac{1}{2}$ and $w = \frac{1}{3}$
   (b) Express w in terms of y and k.
12. On a suitable sketch graph, identify clearly the region A defined by $x \geqslant 0, x + y \leqslant 8$ and $y \geqslant x$.
13. Without using tables, calculate the value of
   (a) $9^{-\frac{1}{2}} + (\frac{1}{8})^{\frac{1}{3}} + (-3)^0$
   (b) $(1000)^{-\frac{1}{3}} - (0{\cdot}1)^2$
14. Solve the cubic equations
   (a) $x^3 + x^2 - 9x - 9 = 0$
   (b) $4x^3 + 8x^2 - x - 2 = 0$
   (c) $x^3 - 2x^2 - 2x - 3 = 0$
15. Make x the subject of the following formulae
   (a) $x + a = \frac{2x-5}{a}$
   (b) $cz + ax + b = 0$
   (c) $a = \sqrt{\left(\frac{x+1}{x-1}\right)}$
16. The shaded region A is formed by the lines $y = 2, y = 3x$ and $x + y = 6$. Write down the three inequalities which define A.

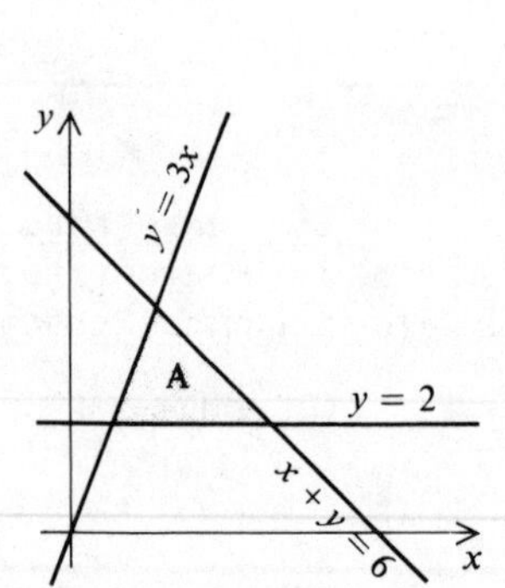

17. The shaded region B is formed by the lines $x = 0, y = x - 2$ and $x + y = 7$.

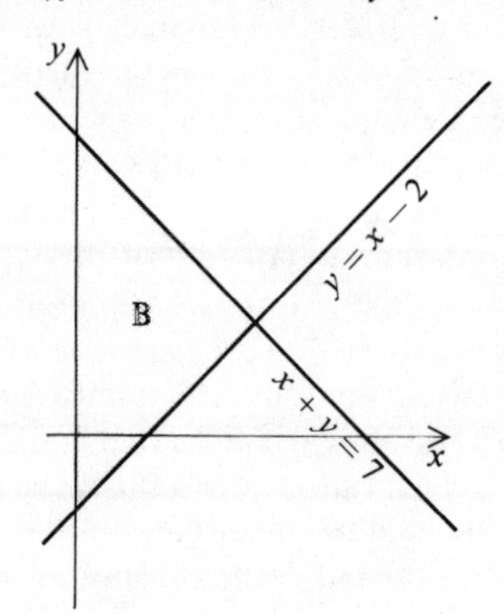

Write down the three inequalities which define B.

18. In the diagram below, the solution set $-1 \leqslant x < 2$ is shown on a number line.

   −1 0 1 2 3 4

   Illustrate, on similar diagrams, the solution set of the following pairs of simultaneous inequalities.
   (a) $x > 2, x \leqslant 7$
   (b) $4 + x \geqslant 2, x + 4 < 10$
   (c) $2x + 1 \geqslant 3, x - 3 \leqslant 3$.
19. Write the following as single fractions
   (a) $\frac{3}{x} + \frac{1}{2x}$
   (b) $\frac{3}{a-2} + \frac{1}{a^2-4}$
   (c) $\frac{3}{x(x+1)} - \frac{2}{x(x-2)}$
20. p varies jointly as the square of t and inversely as s. Given that $p = 5$ when $t = 1$ and $s = 2$, find a formula for p in terms of t and s.

EXAMINATION EXERCISE 5B

1. It is given that $x = \dfrac{t^2}{t-2}$ and $y = \dfrac{t^2}{t+2}$.

(a) Express each of the following as a single fraction in terms of t:
(i) $x + y$ (ii) $x - y$ (iii) xy.

(b) Given that $x = 9$, find the two possible values of y. [O & C]

2. Given the formula $y = ax + \frac{1}{2}bx^2$

(i) calculate y, given that $a = 60, x = 2, b = 3$
(ii) calculate b, given that $y = 72, a = 10, x = 4$
(iii) calculate the two values of x, given that $y = 24, a = 22, b = 4$
(iv) make b the subject of the formula. [JMB]

3. It is given that $x + \dfrac{1}{x} = 1$, where x is not zero. Show that $x^2 = x - 1$ and $x^3 = x^2 - x$. Use these two expressions to show that $x^3 = -1$.
Using this value for x^3, find the value of x^6 and show that $x^7 = x$. Hence show that $x^6 + \dfrac{1}{x^6} = 2$ and $x^7 + \dfrac{1}{x^7} = 1$. Deduce the value of $x^{60} + \dfrac{1}{x^{60}}$ and of $x^{61} + \dfrac{1}{x^{61}}$. [C 16+]

4. Given that $p = \dfrac{2x-1}{x-2}$ and $q = \dfrac{x+2}{x-1}$,

show that $p - q = \dfrac{x^2 - 3x + 5}{x^2 - 3x + 2}$.

Calculate the values of x, if any, which satisfy the following equations:
(i) $q = 0$ (ii) $p = 1$
(iii) $p - q = 1$ (iv) $p - q = 2$. [JMB]

5. Given that $x = 3t - 1$ and $y = t + 7$,

(i) find x and y when $t = 1\frac{1}{2}$,
(ii) find t and y when $x = -7$,
(iii) express in its simplest form $\dfrac{7x + y}{3y - x}$ in terms of t.

Given that $x^2 + y^2 = 164$, form an equation in t and hence find the possible values of t, x and y. [AEB]

6. (a) Two variables x and y are such that $y = 2x^2 - 9x + 4$.
Calculate the values of x for which
(i) $y = 0$ (ii) $y = 4$
(iii) $y = 2$, giving your answers correct to 2 decimal places.

(b) When a space satellite orbits the earth the force F attracting it towards the earth is inversely proportional to the square of its distance R from the centre of the earth. Express F in terms of R and a constant of variation k.
Hence calculate
(i) the value of k if $F = 56$ when $R = 35$,
(ii) the value of R if $F = 350$. [AEB]

7. (a) Add together the two fractions

$$\frac{2}{x-5} \text{ and } \frac{4}{3-x},$$

and simplify your answer.

(b) Solve the equation

$$\frac{2x-14}{8x-15-x^2} = 1,$$

giving your answers correct to one decimal place.

(c) Sketch the graph of

$$y = x^2 - 6x + 1.$$

Show clearly on your graph the coordinates of the points where the graph cuts the x-axis. [L]

8. (a) Given that

$$y = xp + x^2q,$$

and that $y = -1$ when $x = 1$, and that $y = 2$ when $x = 2$, calculate the values of p and q.

(b) Given that

$$y = \frac{3x}{5a-2x},$$

express x in terms of a and y.

(c) Given that y varies inversely as x and that $y = 8$ when $x = 5$, calculate the value of y when $x = 4$. [AEB]

9. Show on one sheet of graph paper the region within which all the following inequalities are satisfied:

$$2x - y \geqslant 0, \quad y - x \geqslant 0, \quad x + y \leqslant 13, \quad 3x + y \leqslant 24.$$

(A suitable scale is 2 cm for 1 unit on each axis.)

(i) Find the greatest values of $2x + y$ and of $x + 2y$ subject to the above restrictions, and state the values of x and y at which these occur.

(ii) Find the greatest values of $2x + y$ and of $x + 2y$ subject to the above restrictions if x and y are confined to integral values, and again state the values of x and y at which these occur. [O]

10. A landscape designer has £500 to spend on planting trees and shrubs to landscape an area of 2000 m^2. For a tree he plans to allow 50 m^2 and for a shrub 5 m^2. Planting a tree will cost £10 and a shrub £2.

(i) If he uses x trees and y shrubs, show that two inequalities (other than $x \geqslant 0$ and $y \geqslant 0$) satisfied by x and y are $5x + y \leqslant 250$ and $10x + y \leqslant 400$.

(ii) Show these inequalities on a graph, using scales of 1 cm to 5 trees and 1 cm to 50 shrubs, shading the unwanted region.

(iii) If he plants 80 shrubs, what is the maximum number of trees he can plant?

(iv) If he plants 10 shrubs for every tree, what is the maximum number of trees he can plant? [SMP]

6 Trigonometry

Leonard Euler (1707–1783) was born near Basel in Switzerland but moved to St Petersburg in Russia and later to Berlin. He had an amazing facility for figures but delighted in speculating in the realms of pure intellect. In trigonometry he introduced the use of small letters for the sides and capitals for the angles of a triangle. He also wrote r, R and s for the radius of the inscribed and of the circumscribed circles and the semi-perimeter, giving the beautiful formula $4rRs = abc$.

6.1 USE OF TABLES

NATURAL SINES $\quad \sin x°$

$x°$	0′ 0·0°	6′ 0·1°	12′ 0·2°	18′ 0·3°	24′ 0·4°	30′ 0·5°	36′ 0·6°	42′ 0·7°	48′ 0·8°	54′ 0·9°	1′ ADD*	2′	3′	4′	5′
0°	0·0000	0017	0035	0052	0070	0087	0105	0122	0140	0157	3	6	9	12	15
1	0·0175	0192	0209	0227	0244	0262	0279	0297	0314	0332	3	6	9	11	14
2	0·0349	0366	0384	0401	0419	0436	0454	0471	0488	0506	3	6	9	11	14
3	0·0523	0541	0558	0576	0593	0610	0628	0645	0663	0680	3	6	9	11	14
4	0·0698	0715	0732	0750	0767	0785	0802	0819	0837	0854	3	6	9	11	14
5	0·0872	0889	0906	0924	0941	0958	0976	0993	1011	1028	3	6	9	111	14
6	0·1045	1063	1080	1097	1115	1132	1149	1167	1184	1201	3	6	9	11	14

Example 1

Find the sine of 14·4° (written sin 14·4°)

(a) Locate 14° in the left-hand column of the table of natural sines.

(b) Move across the 14° row until in the 0·4° column you read 0·2487

$$\sin 14{\cdot}4° = 0{\cdot}2487$$

Example 2

Find the tangent of $57° 20'$.
(a) Locate $57°$ in the left-hand column of the table of natural tangents, and move across to the $18'$ column. The number here is 1·5577.
(b) The number for the extra 2 minutes comes from the $2'$ column on the far right. Here the number is 20.
(c) Add the two numbers together to give $\tan 57° 20' = 1{\cdot}5597$.

N.B. When using the cosine table, the figure from the far right is subtracted.

e.g. $\cos 48° 20' = 0{\cdot}6652 - 0{\cdot}0004$
$= 0{\cdot}6648$

Example 3

Find the cosine of $17{\cdot}6°$ using a pocket calculator.
(a) Make sure the calculator is operating with degrees (not radians)
(b) Insert 17·6
(c) Press the COS button.

$\cos 17{\cdot}6° = 0{\cdot}9532$ (to four figures)

Exercise 1

For the following, find
(a) the sine of the angle
(b) the cosine of the angle
(c) the tangent of the angle.

1. 17°	**2.** 59°	**3.** 81·2°
4. 72·8°	**5.** 17·3°	**6.** 60°
7. 7°	**8.** 8·9°	**9.** 45·1°
10. 3·2°	**11.** 65·9°	**12.** 17·9°
13. 63·7°	**14.** 35·9°	**15.** 81·4°
16. 18° 12′	**17.** 27° 30′	**18.** 49° 8′
19. 20° 6′	**20.** 27° 19′	**21.** 47° 23′
22. 16° 45′	**23.** 78° 19′	**24.** 8° 11′
25. 19° 19′	**26.** 23° 10′	**27.** 81° 20′
28. 57° 15′	**29.** 81° 10′	**30.** 15° 8′

Example 4

Find the angle which has a sine of 0·8039, i.e. if $\sin x° = 0{\cdot}8039$, find x.
(a) In the sine table, locate 0·8039.
(b) Read $53{\cdot}5°$ or $53° 30'$.

Example 5

If $\tan y = 0{\cdot}4738$, find y (in degrees and minutes).
(a) In the tangent table, locate the nearest number *below* 0·4738, in this case 0·4727.
(b) We need another 11 to make 0·4738 and this corresponds to $3'$ from the far right column.
(c) Read $y = 25° 18' + 3'$ $\quad y = 25° 21'$

Example 6

If $\cos x = 0{\cdot}4441$, find x.
(a) In the cosine table, locate the nearest number *above* 0·4441, in this case 0·4446.
(b) We need to subtract 5 to make 0·4441 and this corresponds to $2'$ from the far right column.
(c) Read $x = 63° 36' + 2'$ $\quad x = 63° 38'$

Example 7

If $\cos z = 0{\cdot}6519$, find z using a pocket calculator.
(a) Make sure the calculator is operating with degrees (not radians).
(b) Insert 0·6519.
(c) Press INV COS

$z = 49{\cdot}3°$ (to one decimal place)

Exercise 2

Give the answers
(a) as a decimal, to one decimal place
(b) in degrees and minutes.

Find the angle which has a sine of:

1. 0·7242	**2.** 0·9489	**3.** 0·3551
4. 0·9781	**5.** 0·9438	**6.** 0·0140
7. 0·2893	**8.** 0·3990	**9.** 0·9412
10. 0·9931	**11.** 0·0265	**12.** 0·1352

Find the angle which has a cosine of:

13. 0·5850	**14.** 0·3502	**15.** 0·9478
16. 0·8721	**17.** 0·0244	**18.** 0·2773
19. 0·5400	**20.** 0·9170	**21.** 0·8190
22. 0·9683	**23.** 0·5529	**24.** 0·4091

Find the angle which has a tangent of:

25. 0·3346	**26.** 0·5914	**27.** 1·3564
28. 2·0057	**29.** 0·0507	**30.** 1·2938
31. 0·3006	**32.** 0·5898	**33.** 1·1757
34. 1·3286	**35.** 2·5275	**36.** 0·5919

6.2 RIGHT-ANGLED TRIANGLES

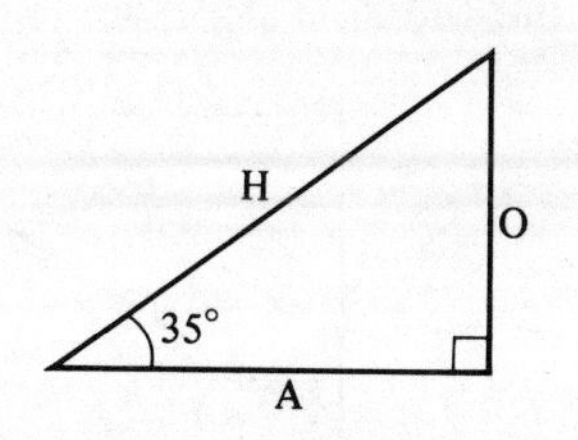

The side opposite the right angle is called the hypotenuse (we will use H). It is the longest side.
The side opposite the marked angle of 35° is called the opposite (we will use O).
The other side is called the adjacent (we will use A).

Example 1

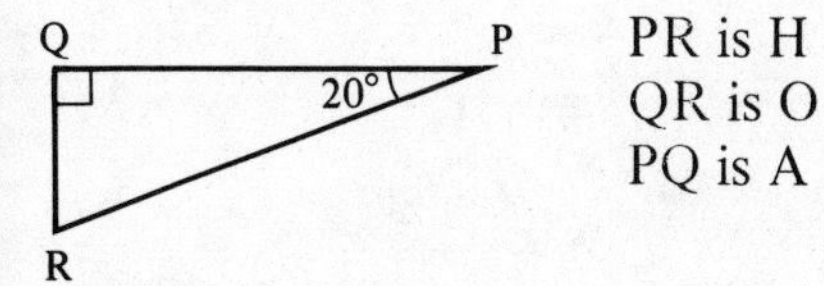

PR is H
QR is O
PQ is A

Notice that the side marked O, for opposite, depends on where the angle is marked.

Exercise 3

Use the letters to describe H, O and A as in the example above.

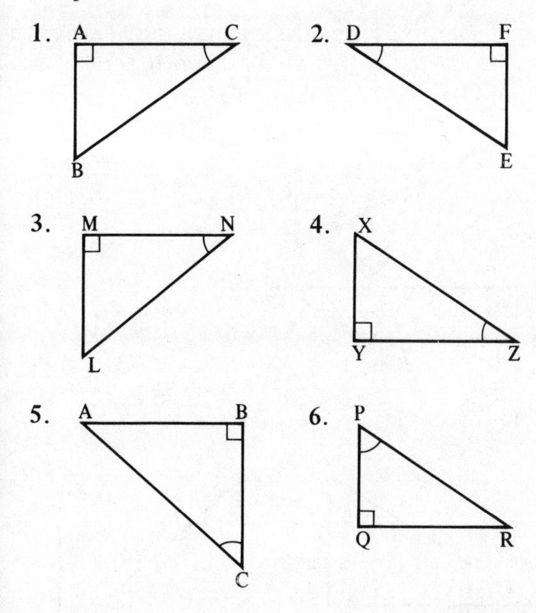

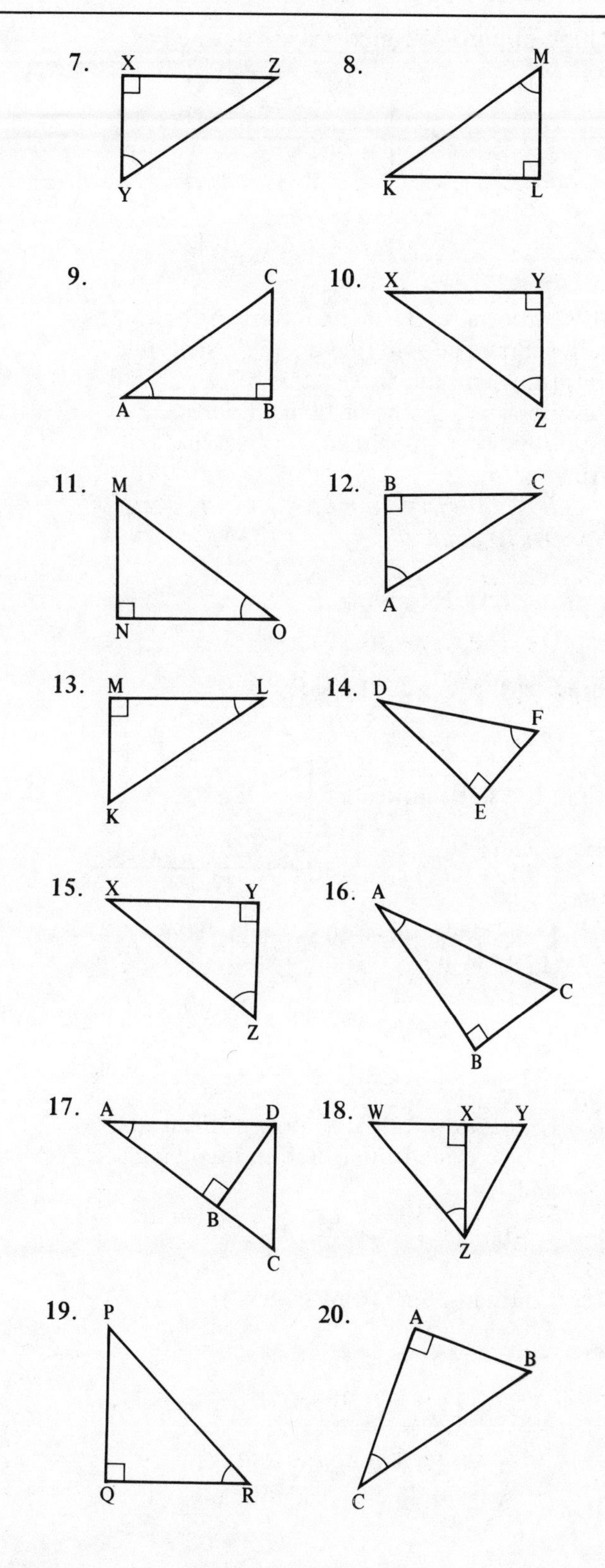

Sine, cosine and tangent

Three important functions are defined as follows:

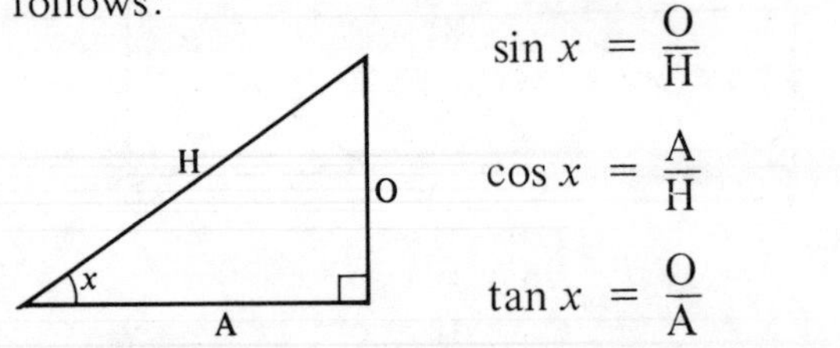

$$\sin x = \frac{O}{H}$$

$$\cos x = \frac{A}{H}$$

$$\tan x = \frac{O}{A}$$

It is important to get the letters in the right order. Some people find a simple sentence helpful where the first letters of each word describe sine, cosine or tangent and Hypotenuse, Opposite and Adjacent.
An example is:

Silly Old Harry Caught A Herring Trawling Off Afghanistan.

e.g. S O H : $\sin = \frac{O}{H}$

Finding the length of a side

Example 2

Find the side marked x

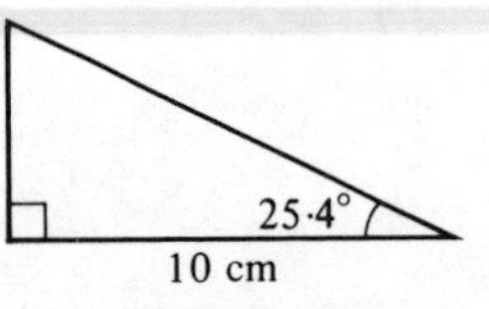

(a) Label the sides of the triangle H, O, A (in brackets)

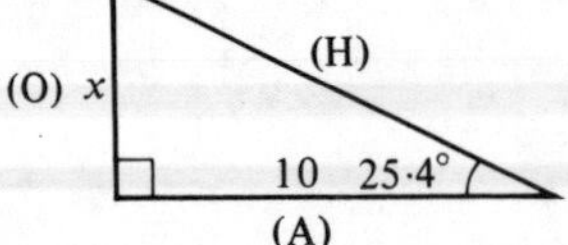

(b) In this example, we know nothing about H so we need the function involving O and A.

$$\tan 25{\cdot}4° = \frac{O}{A} = \frac{x}{10}$$

(c) Find tan 25·4° from tables

$$0{\cdot}4748 = \frac{x}{10}$$

(d) Solve for x

$$x = 10 \times 0{\cdot}4748 = 4{\cdot}748$$
$$x = 4{\cdot}75 \text{ cm (3 significant figures)}$$

Example 3

Find the side marked z.

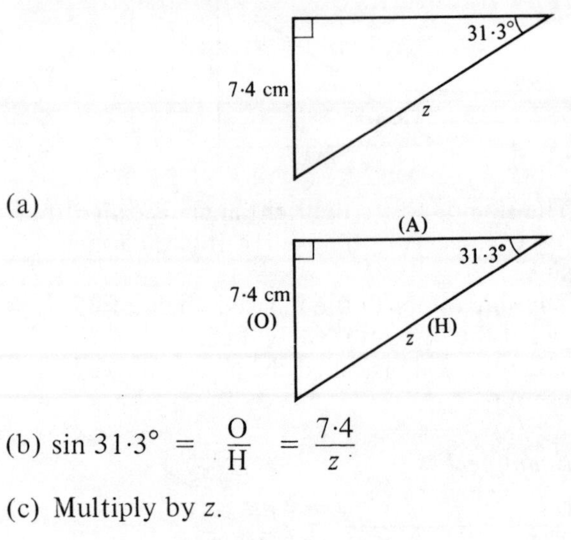

(b) $\sin 31{\cdot}3° = \frac{O}{H} = \frac{7{\cdot}4}{z}$

(c) Multiply by z.

$$z \times (\sin 31{\cdot}3°) = 7{\cdot}4$$

$$z = \frac{7{\cdot}4}{\sin 31{\cdot}3}$$

(d) On a calculator, press the keys as follows:

[7·4] [÷] [31·3] [SIN] [=]

$$z = 14{\cdot}2 \text{ cm (to 3 S.F.)}$$

Exercise 4

In questions **1** to **22**, find the length of the side marked marked with a letter. Give your answers to three significant figures.

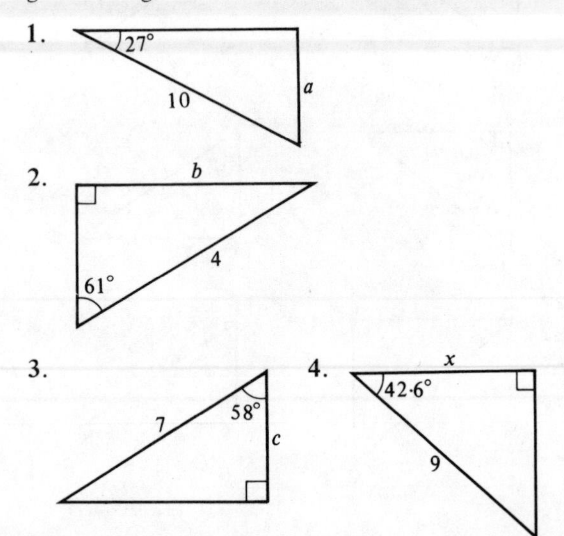

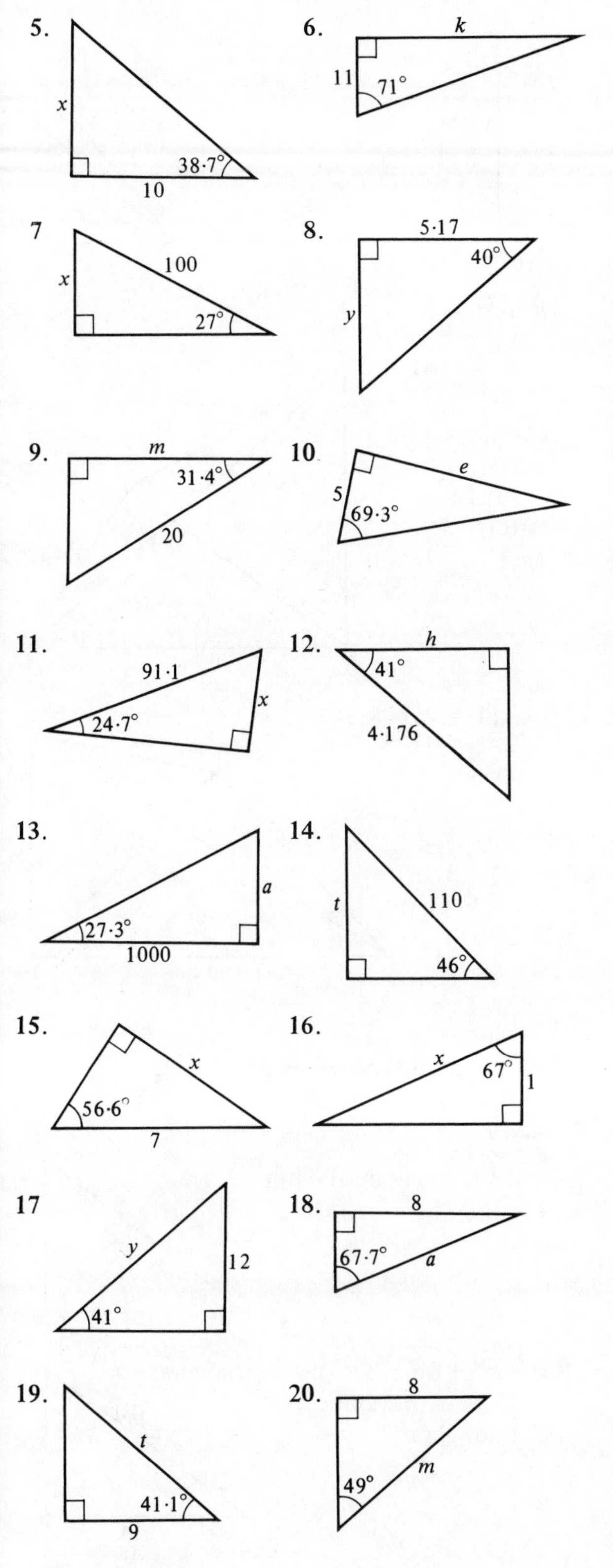

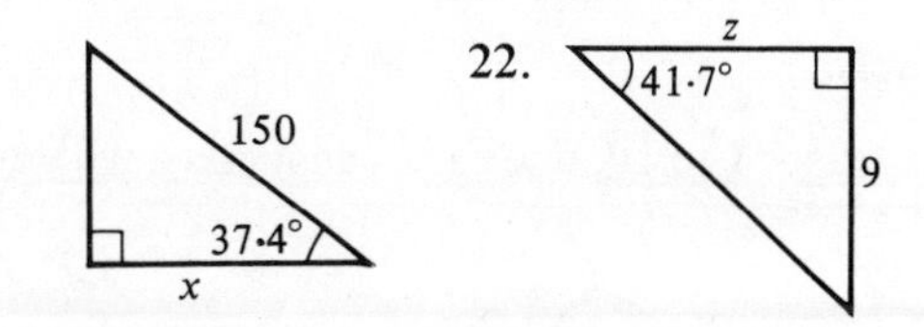

In questions **23** to **34**, the triangle has a right angle at the middle letter.

23. In ΔABC, $\hat{C} = 40°$, BC = 4 cm. Find AB
24. In ΔDEF, $\hat{F} = 35{\cdot}3°$, DF = 7 cm. Find ED
25. In ΔGHI, $\hat{I} = 70°$, GI = 12 m. Find HI
26. In ΔJKL, $\hat{L} = 55°$, KL = 8·21 m. Find JK
27. In ΔMNO, $\hat{M} = 42{\cdot}6°$, MO = 14 cm. Find ON
28. In ΔPQR, $\hat{P} = 28°$, PQ = 5·071 m. Find PR
29. In ΔSTU, $\hat{S} = 39°$, TU = 6 cm. Find SU
30. In ΔVWX, $\hat{X} = 17°$, WV = 30·7 m. Find WX
31. In ΔABC, $\hat{A} = 14°\,17'$, BC = 14 m. Find AC
32. In ΔKLM, $\hat{K} = 72°\,50'$, KL = 5·04 cm. Find LM
33. In ΔPQR, $\hat{R} = 31°\,43'$, QR = 0·81 cm. Find PR
34. In ΔXYZ, $\hat{X} = 81°\,4'$, YZ = 52·6 m. Find XY

Example 4

Find the length marked x.

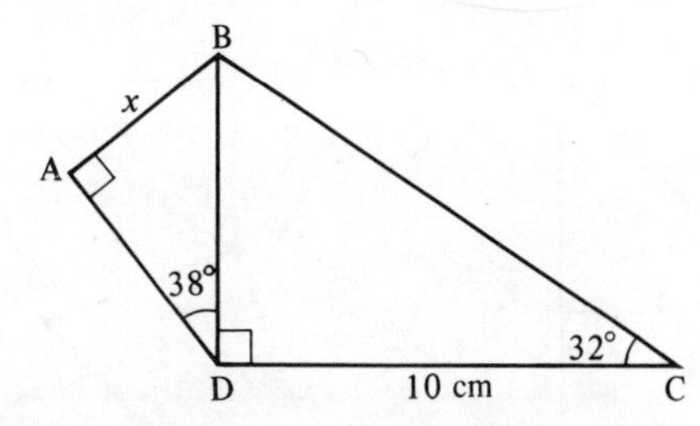

(a) Find BD from triangle BDC

$$\tan 32° = \frac{BD}{10}$$

$$\therefore \quad BD = 10 \times \tan 32° \quad \ldots [1]$$

(b) Now find x from triangle ABD

$$\sin 38° = \frac{x}{BD}$$

$$\therefore \quad x = BD \times \sin 38°$$

$$x = 10 \times \tan 32° \times \sin 38 \quad \text{(from [1])}$$

$$x = 3{\cdot}85 \text{ cm (to 3 S.F.)}$$

Notice that BD was *not* calculated from [1].

It is better to do all the multiplications at one time.

Exercise 5

In questions **1** to **10**, find each side marked with a letter.

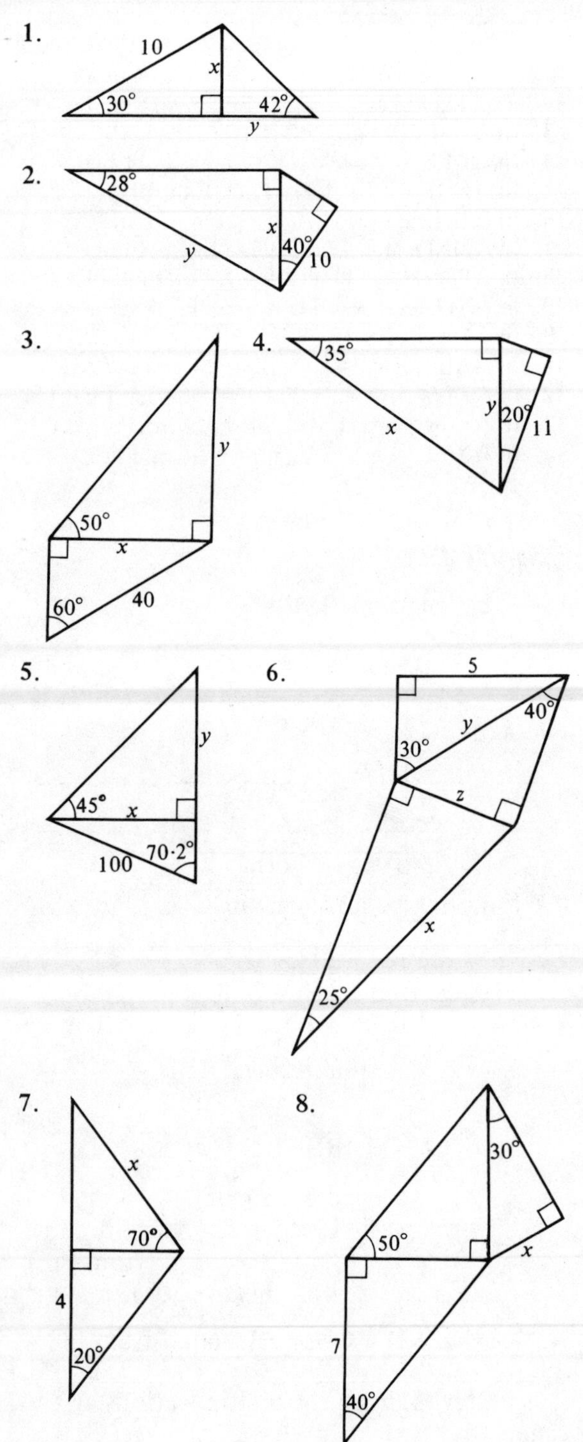

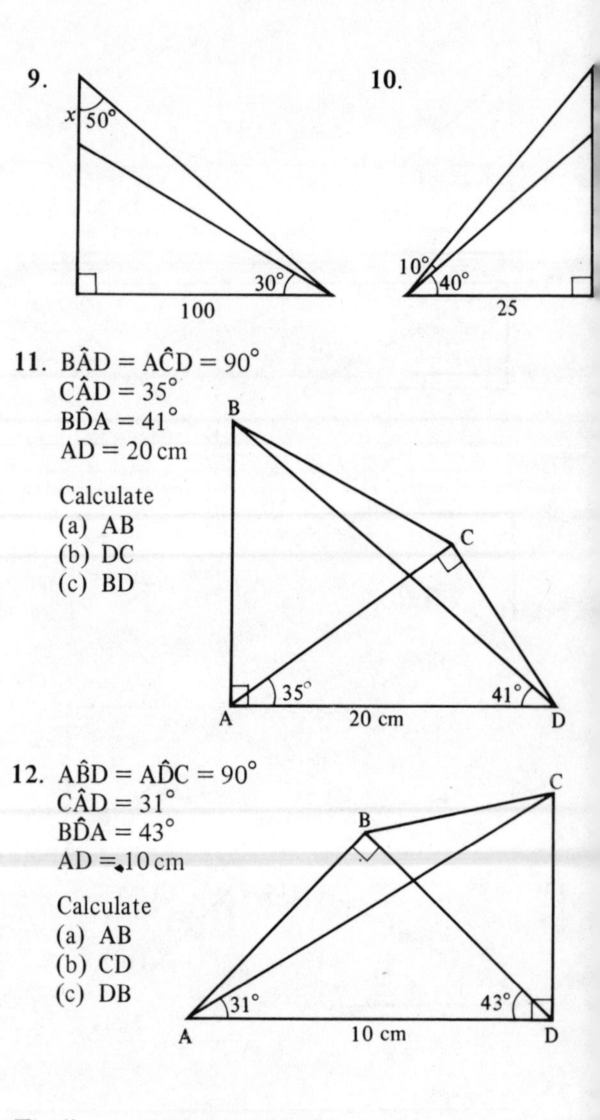

11. $B\hat{A}D = A\hat{C}D = 90^\circ$
$C\hat{A}D = 35^\circ$
$B\hat{D}A = 41^\circ$
$AD = 20\text{ cm}$

Calculate
(a) AB
(b) DC
(c) BD

12. $A\hat{B}D = A\hat{D}C = 90^\circ$
$C\hat{A}D = 31^\circ$
$B\hat{D}A = 43^\circ$
$AD = 10\text{ cm}$

Calculate
(a) AB
(b) CD
(c) DB

Finding an unknown angle

Example 5

Find the angle marked m.

5
m
4

(a) Label the sides of the triangle H, O, A in relation to angle m.

5 (H)
(O)
m
4 (A)

(b) In this example, we do not know 'O' so we need the cosine.

$$\cos m = \left(\frac{A}{H}\right) = \frac{4}{5}$$

(c) Change $\frac{4}{5}$ to a decimal: $\frac{4}{5} = 0{\cdot}8$

(d) $\cos m = 0{\cdot}8$
Find angle m from the cosine table

$$m = 36{\cdot}9^\circ \quad (36^\circ\, 52')$$

Note: On a calculator, angles can be found as follows:

If $\cos m = \frac{4}{5}$

(a) Press [4] [÷] [5] [=]

(b) Press [INV] and then [COS]
This will give the angle as $36{\cdot}86989765^\circ$. We require the angle to 1 place of decimals so $m = 36{\cdot}9^\circ$.

Exercise 6

For questions **1** to **15**, find the angle marked with a letter. All lengths are in cm.

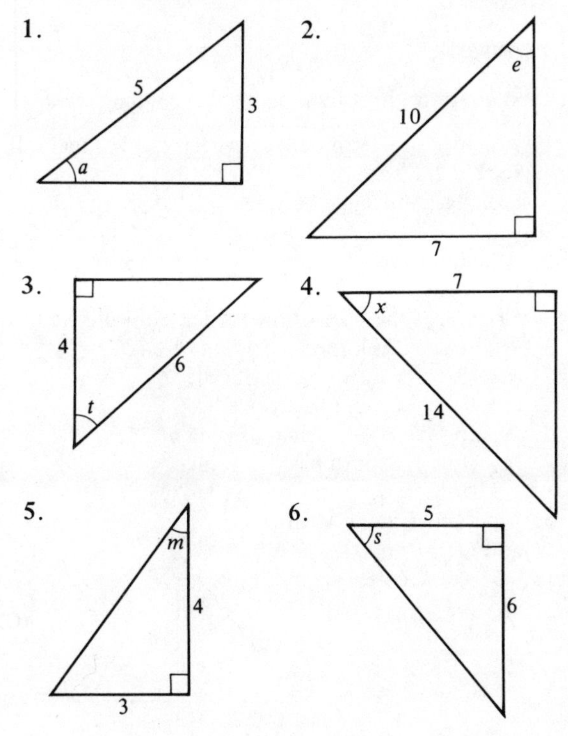

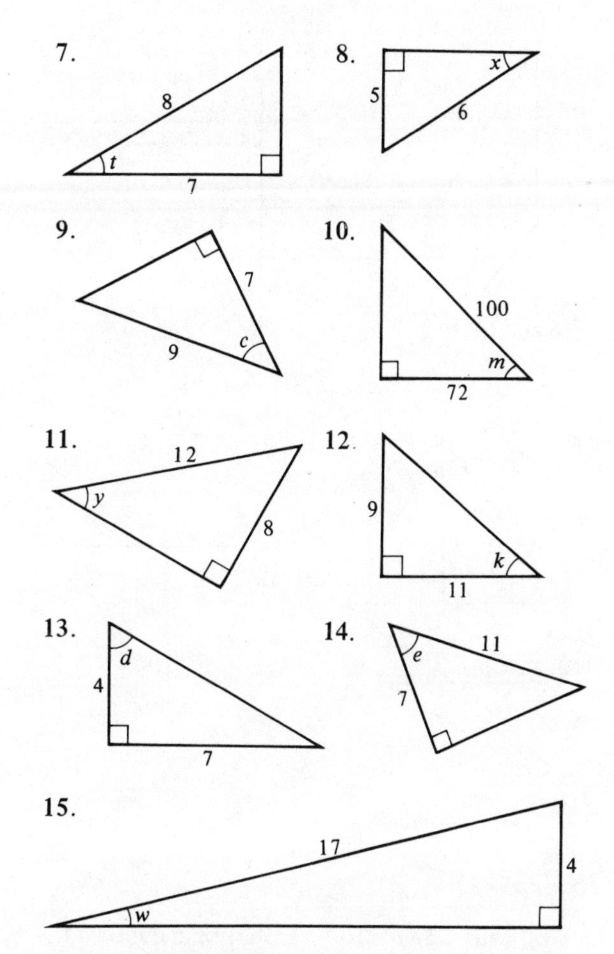

In questions **16** to **20**, the triangle has a right angle at the middle letter.

16. In $\triangle$ABC, BC = 4, AC = 7. Find $\hat{A}$.
17. In $\triangle$DEF, EF = 5, DF = 10. Find $\hat{F}$.
18. In $\triangle$GHI, GH = 9, HI = 10. Find $\hat{I}$.
19. In $\triangle$JKL, JL = 5, KL = 3. Find $\hat{J}$.
20. In $\triangle$MNO, MN = 4, NO = 5. Find $\hat{M}$.

In questions **21** to **26**, find the angle x.

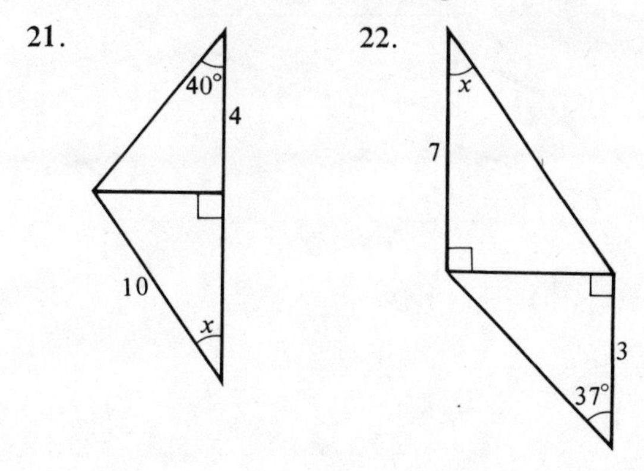

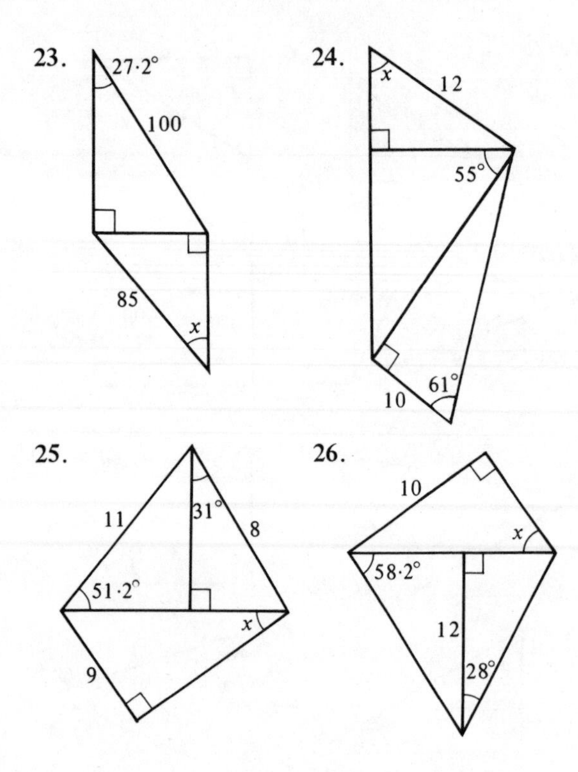

Example 6

A ship sails 22 km from A on a bearing of 042°, and a further 30 km on a bearing of 090° to arrive at B. What is the distance and bearing of B from A?

(a) Draw a clear diagram and label extra points as shown.

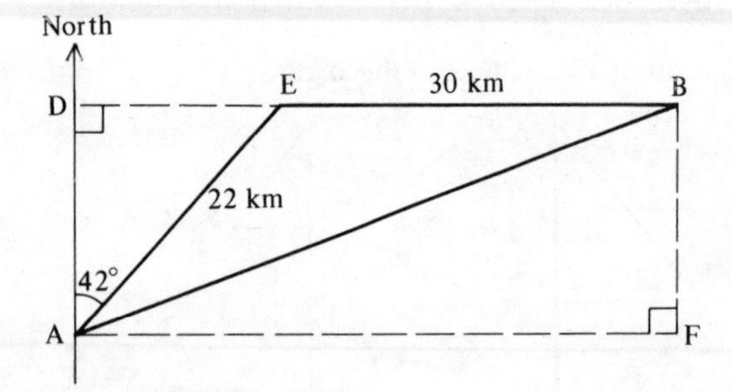

(b) Find DE and AD.

(i) $\sin 42^\circ = \dfrac{DE}{22}$

$\therefore \quad DE = 22 \times \sin 42^\circ = 14{\cdot}72$ km

(ii) $\cos 42^\circ = \dfrac{AD}{22}$

$\therefore \quad AD = 22 \times \cos 42^\circ = 16{\cdot}35$ km

(c) Using triangle ABF,

$AB^2 = AF^2 + BF^2$ (Pythagoras' Theorem)

and $AF = DE + EB$

$AF = 14{\cdot}72 + 30 = 44{\cdot}72$ km

and $BF = AD = 16{\cdot}35$ km.

$\therefore \quad AB^2 = 44{\cdot}72^2 + 16{\cdot}35^2$
$= 2267{\cdot}2$

$AB = 47{\cdot}6$ km (to 3 S.F.)

(d) The bearing of B from A is given by the angle DAB.
But $D\hat{A}B = A\hat{B}F$.

$\tan A\hat{B}F = \dfrac{AF}{BF} = \dfrac{44{\cdot}72}{16{\cdot}35}$

$= 2{\cdot}7352$

$\therefore \quad A\hat{B}F = 69{\cdot}9^\circ$.

B is 47·6 km from A on a bearing of 069·9°.

Exercise 7

In this exercise, start by drawing a clear diagram.

1. A ladder of length 6 m leans against a vertical wall so that the base of the ladder is 2 m from the wall. Calculate the angle between the ladder and the wall.
2. A ladder of length 8 m rests against a wall so that the angle between the ladder and the wall is 31°. How far is the base of the ladder from the wall?
3. A ship sails 35 km on a bearing of 042°.
   (a) How far north has it travelled?
   (b) How far east has it travelled?
4. A ship sails 200 km on a bearing of 243·7°.
   (a) How far south has it travelled?
   (b) How far west has it travelled?
5. Find TR if PR = 10 m and QT = 7 m.

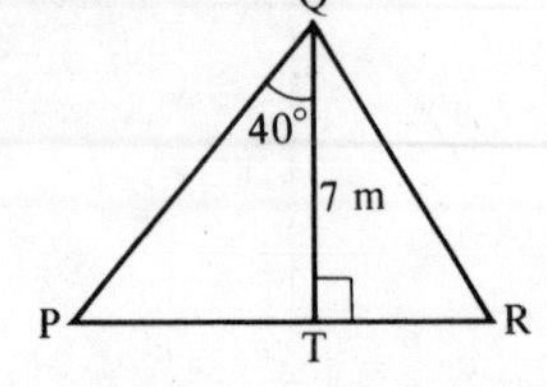

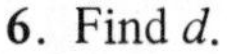

6. Find d.

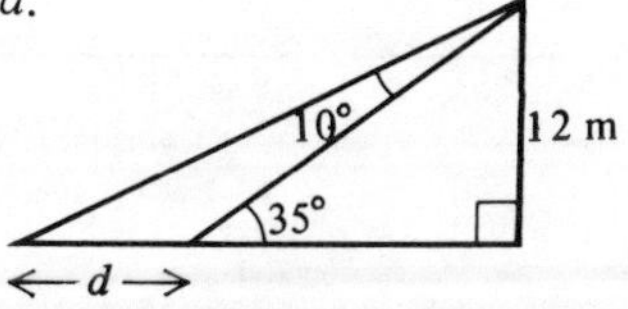

7. An aircraft flies 400 km from a point O on a bearing of 025° and then 700 km on a bearing of 080° to arrive at B.
(a) How far north of O is B?
(b) How far east of O is B?
(c) Find the distance and bearing of B from O.

8. An aircraft flies 500 km on a bearing of 100° and then 600 km on a bearing of 160°. Find the distance and bearing of the finishing point from the starting point.

For questions **9** to **12**, plot the points for each question on a sketch graph with x- and y-axes drawn to the same scale.

9. For the points A (5, 0) and B (7, 3), calculate the angle between AB and the x-axis.
10. For the points C (0, 2) and D (5, 9), calculate the angle between CD and the y-axis.
11. For the points A (3, 0), B (5, 2) and C (7, −2), calculate the angle BAC.
12. For the points P (2, 5), Q (5, 1) and R (0, −3), calculate the angle PQR.

13. From the top of a tower of height 75 m, a guard sees two prisoners, both due West of him. If the angles of depression of the two prisoners are 10° and 17°, calculate the distance between them.
14. An isosceles triangle has sides of length 8 cm, 8 cm and 5 cm. Find the angle between the two equal sides.
15. The angles of an isosceles triangle are 66°, 66° and 48°. If the shortest side of the triangle is 8·4 cm, find the length of one of the two equal sides.
16. A chord of length 12 cm subtends an angle of 78·2° at the centre of a circle. Find the radius of the circle.
17. Find the acute angle between the diagonals of a rectangle whose sides are 5 cm and 7 cm.
18. A kite flying at a height of 55 m is attached to a string which makes an angle of 55° with the horizontal. What is the length of the string?
19. A boy is flying a kite from a string of length 150 m. If the string is taut and makes an angle of 67° with the horizontal, what is the height of the kite?

20. A rocket flies 10 km vertically, then 20 km at an angle of 15° to the vertical and finally 60 km at an angle of 26° to the vertical. Calculate the vertical height of the rocket at the end of the third stage.
21. Find x, given AD = BC = 6 m.

22. Find x.

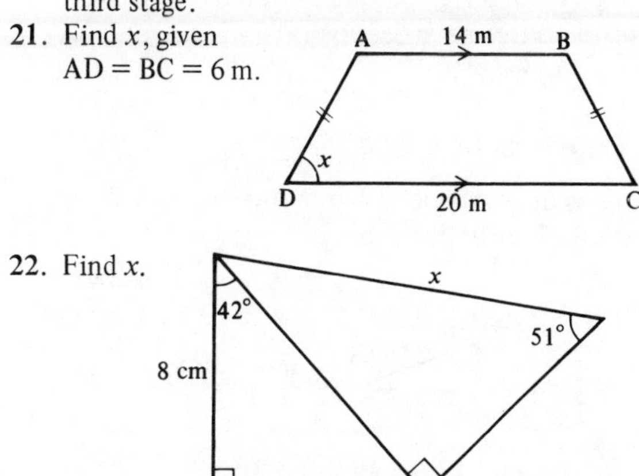

23. Ants can hear each other up to a range of 2 m. An ant A, 1 m from a wall sees his friend B about to be eaten by a spider. If the angle of elevation of B from A is 62°, will the spider have a meal or not? (Assume B escapes if he hears A calling.)
24. A hedgehog wishes to cross a road without being run over. He observes the angle of elevation of a lamp post on the other side of the road to be 27° from the edge of the road and 15° from a point 10 m back from the road. How wide is the road? If he can run at 1 m/s, how long will he take to cross?
If cars are travelling at 20 m/s, how far apart must they be if he is to survive?
25. From a point 10 m from a vertical wall, the angles of elevation of the bottom and the top of a statue of Sir Isaac Newton, set in the wall, are 40° and 52°. Calculate the length of the statue.
26. A rectangular paving stone 3 m by 1 m rests against a vertical wall as shown.

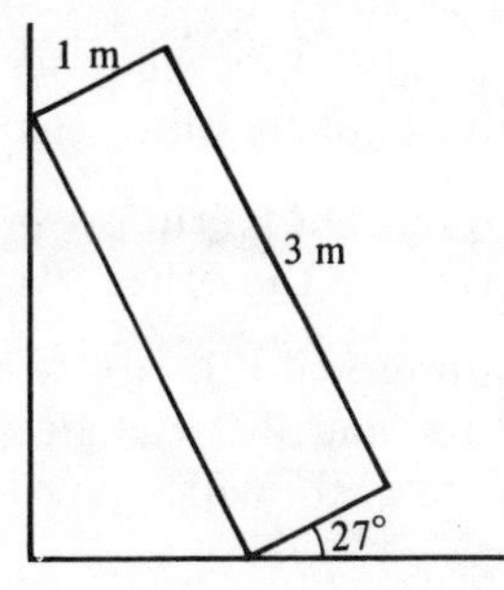

What is the height of the highest point of the stone above the ground?

6.3 ANGLES GREATER THAN 90°

Although we cannot have an angle of more than 90° in a right-angled triangle, it is still useful to define sine, cosine and tangent for these angles.

The quadrant method

Angles are measured from the line OX in an anticlockwise direction.

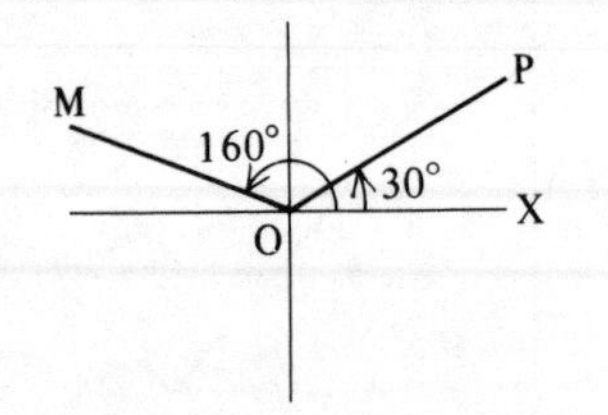

e.g. OP makes an angle of 30° with OX,
OM makes an angle of 160° with OX.

The line from O can be in one of four quadrants.
The quadrants are marked:

A for All
S for Sine
T for Tangent
C for Cosine

S A
T C

For a line in quadrant A, *all* the ratios sin, cos and tan are positive.

For a line in quadrant S, the *sine* of the angle is positive and the other two negative.

For a line in quadrant T, the *tangent* of the angle is positive and the other two negative.

For a line in quadrant C, the *cosine* of the angle is positive and the other two negative.

Finally, the numerical value of the angle's ratio is equal to that of the angle between the line and 180° or 360°, whichever is the nearer.

Example 1

Find cos 130°.

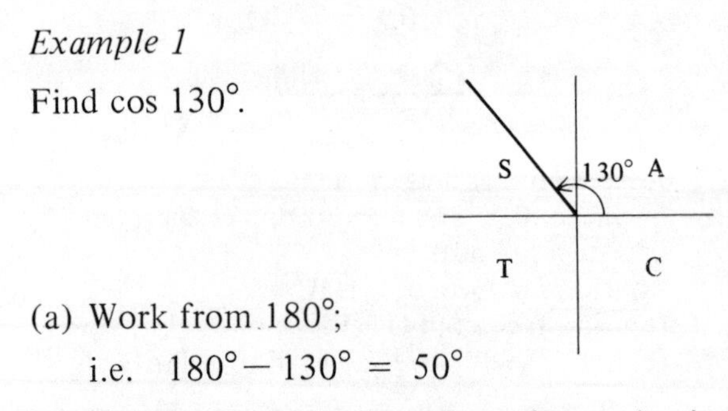

(a) Work from 180°;
i.e. $180° - 130° = 50°$

(b) The line is in quadrant S, so that cosine is negative;

$$\cos 130° = -\cos 50°$$
$$= -0{\cdot}6428$$

You may be able to check this on your calculator.

Example 2

Find tan 200°.

S A
200°
T C

(a) Work from 180°;
i.e. $200° - 180° = 20°$

(b) The line is in quadrant T, so that tangent is positive;

$$\tan 200° = +\tan 20°$$
$$= 0{\cdot}3640$$

Example 3

Find sin 315°.

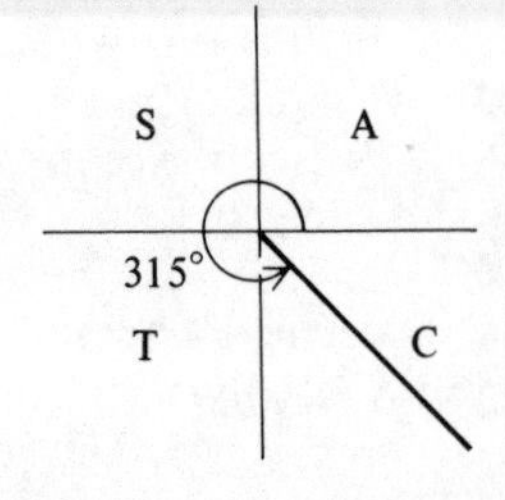

(a) Work from 360°;
i.e. $360° - 315° = 45°$

(b) The line is in quadrant C, so that sine is negative;

$$\sin 315° = -\sin 45°$$
$$= -0{\cdot}7071$$

Exercise 8

Find the value of the following.

1. cos 160° 2. tan 300° 3. sin 130°
4. sin 290° 5. cos 147° 6. cos 311°
7. sin 216° 8. tan 279° 9. tan 111°
10. cos 107° 11. sin 190° 12. cos 93°
13. tan 210° 14. sin 340° 15. sin 11·3°
16. cos 201° 17. cos 305·4° 18. tan 95°
19. sin 165·6° 20. cos 306·2° 21. tan 75°
22. sin 181° 23. sin 242° 24. cos 97°
25. tan 163·2° 26. tan 301° 27. tan 214·5°
28. sin 200° 29. cos 358° 30. sin 113·3°

In questions **31** to **45**, find two angles in the range 0° to 360° which satisfy the equation; e.g. if $\sin x = 0{\cdot}5$, $x = 30°$ or $x = 150°$.

31. $\sin x = 0{\cdot}8$ 32. $\sin x = 0{\cdot}4210$
33. $\cos x = 0{\cdot}497$ 34. $\cos x = 0{\cdot}230$
35. $\tan x = 1{\cdot}175$ 36. $\sin x = 0{\cdot}9690$
37. $\tan x = 1$ 38. $\cos x = 0{\cdot}859$
39. $\tan x = 2{\cdot}106$ 40. $\cos x = 0{\cdot}749$
41. $\sin x = -0{\cdot}75$ 42. $\tan x = -1{\cdot}24$
43. $\cos x = -0{\cdot}866$ 44. $\tan x = -2$
45. $\sin x = -0{\cdot}9521$

46. (a) Find the sine of all the angles 0°, 10°, 20°, 30°, ... 350°, 360°.
    (b) Draw a graph of $y = \sin x$ for $0° \leqslant x \leqslant 360°$. Use a scale of 1 cm to 20° on the x-axis and 5 cm to 1 unit on the y-axis.
47. (a) Find the cosine of all the angles 0°, 10°, 20°, 30°, ... 350°, 360°.
    (b) Draw a graph of $y = \cos x$ for $0° \leqslant x \leqslant 360°$. Use a scale of 1 cm to 20° on the x-axis and 5 cm to 1 unit on the y-axis.
48. Draw a graph of $y = \tan x$ for $0° \leqslant x \leqslant 360°$. Use a scale of 1 cm to 20° on the x-axis and 1 cm to 1 unit on the y-axis.

6.4 THE SINE RULE

The sine rule enables us to calculate sides and angles in some triangles where there is not a right angle.

In △ABC, we use the convention that
  a is the side opposite $\hat{A}$
  b is the side opposite $\hat{B}$, etc.

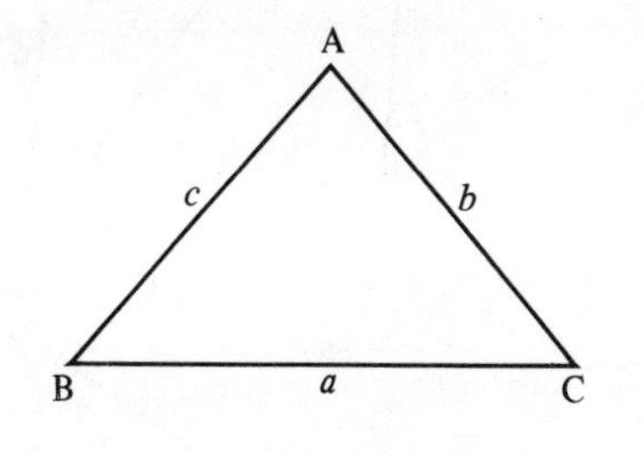

Sine rule

either $$\frac{a}{\sin A} = \frac{b}{\sin B} = \frac{c}{\sin C} \quad \ldots [1]$$

or $$\frac{\sin A}{a} = \frac{\sin B}{b} = \frac{\sin C}{c} \quad \ldots [2]$$

Use [1] when finding a *side*,

and [2] when finding an *angle*.

Example 1

Find c.

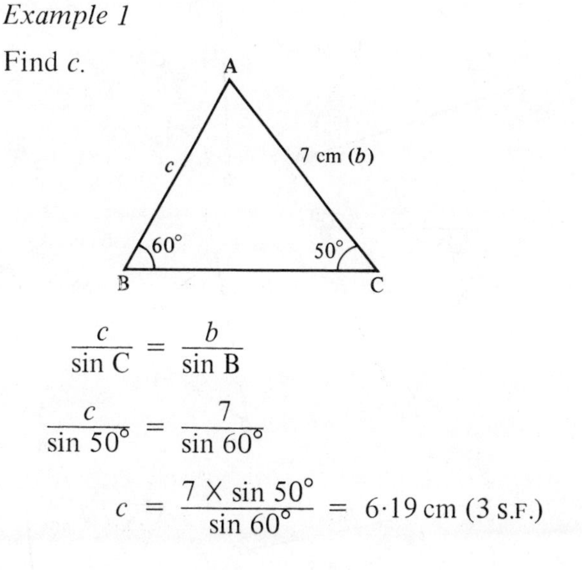

$$\frac{c}{\sin C} = \frac{b}{\sin B}$$

$$\frac{c}{\sin 50°} = \frac{7}{\sin 60°}$$

$$c = \frac{7 \times \sin 50°}{\sin 60°} = 6{\cdot}19 \text{ cm (3 S.F.)}$$

Example 2

Find $\hat{B}$.

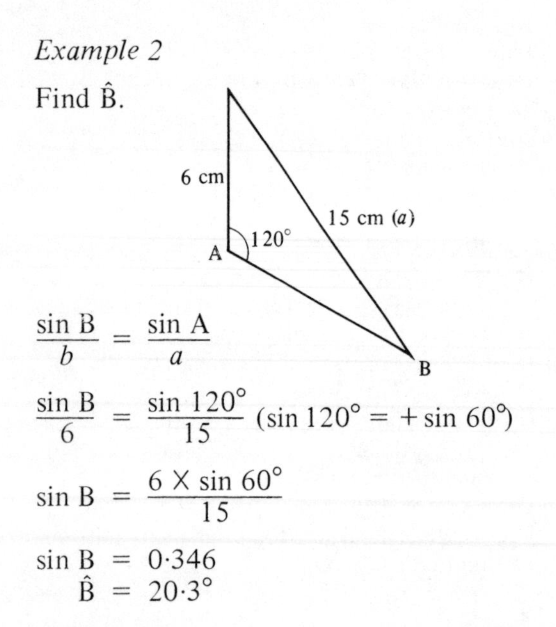

$$\frac{\sin B}{b} = \frac{\sin A}{a}$$

$$\frac{\sin B}{6} = \frac{\sin 120°}{15} \quad (\sin 120° = +\sin 60°)$$

$$\sin B = \frac{6 \times \sin 60°}{15}$$

$$\sin B = 0{\cdot}346$$

$$\hat{B} = 20{\cdot}3°$$

Exercise 9

For questions **1** to **6**, find each side marked with a letter.

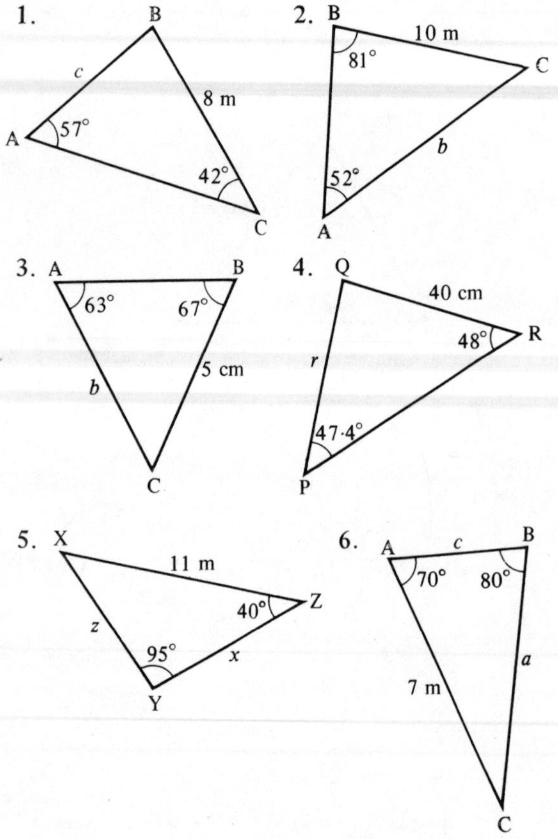

7. In $\triangle$ABC, $\hat{A} = 61°$, $\hat{B} = 47°$, AC = 7·2 cm. Find BC.

8. In $\triangle$XYZ, $\hat{Z} = 32°$, $\hat{Y} = 78°$, XY = 5·4 cm. Find XZ.

9. In $\triangle$PQR, $\hat{Q} = 100°$, $\hat{R} = 21°$, PQ = 3·1 cm. Find PR.

10. In $\triangle$LMN, $\hat{L} = 21°$, $\hat{N} = 30°$, MN = 7 cm. Find LN.

In questions **11** to **18**, find each angle marked *. All lengths are in centimetres.

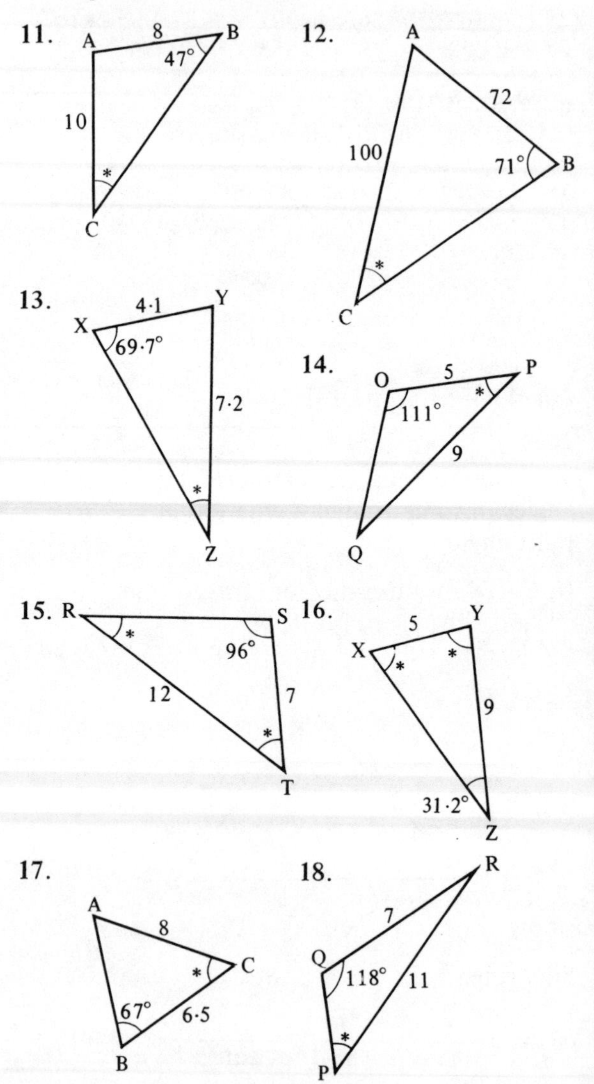

19. In $\triangle$ABC, $\hat{A} = 62°$, BC = 8, AB = 7. Find $\hat{C}$.

20. In $\triangle$XYZ, $\hat{Y} = 97{\cdot}3°$, XZ = 22, XY = 14. Find $\hat{Z}$.

21. In $\triangle$DEF, $\hat{D} = 58°$, EF = 7·2, DE = 5·4. Find $\hat{F}$.

22. In $\triangle$LMN, $\hat{M} = 127{\cdot}1°$, LN = 11·2, LM = 7·3. Find $\hat{L}$.

6.5 THE COSINE RULE

We use the cosine rule when we have either
(a) two sides and the included angle or
(b) all three sides.

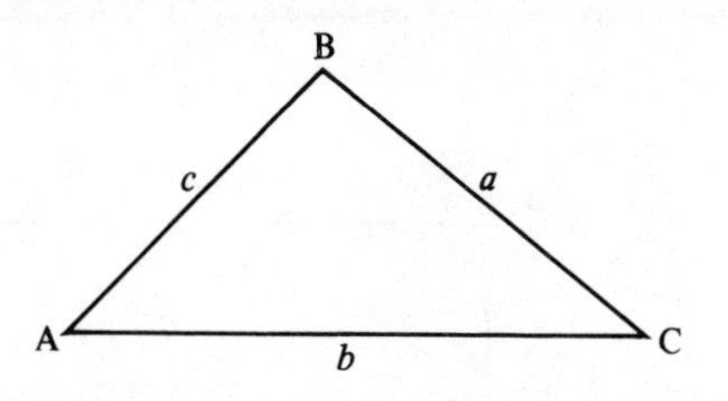

There are two forms.

1. To find the length of a side.

$$a^2 = b^2 + c^2 - (2bc \cos A)$$
$$\text{or } b^2 = c^2 + a^2 - (2ac \cos B)$$
$$\text{or } c^2 = a^2 + b^2 - (2ab \cos C)$$

2. To find an angle when given all three sides.

$$\cos A = \frac{b^2 + c^2 - a^2}{2\,bc}$$

$$\text{or } \cos B = \frac{a^2 + c^2 - b^2}{2\,ac}$$

$$\text{or } \cos C = \frac{a^2 + b^2 - c^2}{2\,ab}$$

Remember: for obtuse angles the cosine is negative; e.g.

$$\cos 120° = -\cos 60°$$
$$= -0{\cdot}5$$

Example 1

Find b.

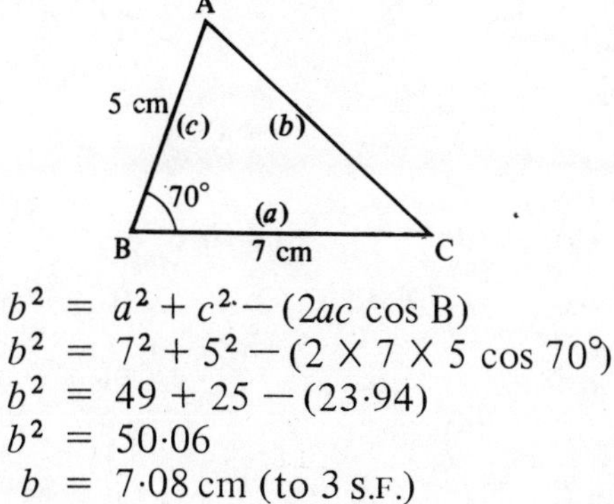

$$b^2 = a^2 + c^2 - (2ac \cos B)$$
$$b^2 = 7^2 + 5^2 - (2 \times 7 \times 5 \cos 70°)$$
$$b^2 = 49 + 25 - (23{\cdot}94)$$
$$b^2 = 50{\cdot}06$$
$$b = 7{\cdot}08 \text{ cm (to 3 S.F.)}$$

Example 2

Find AC.

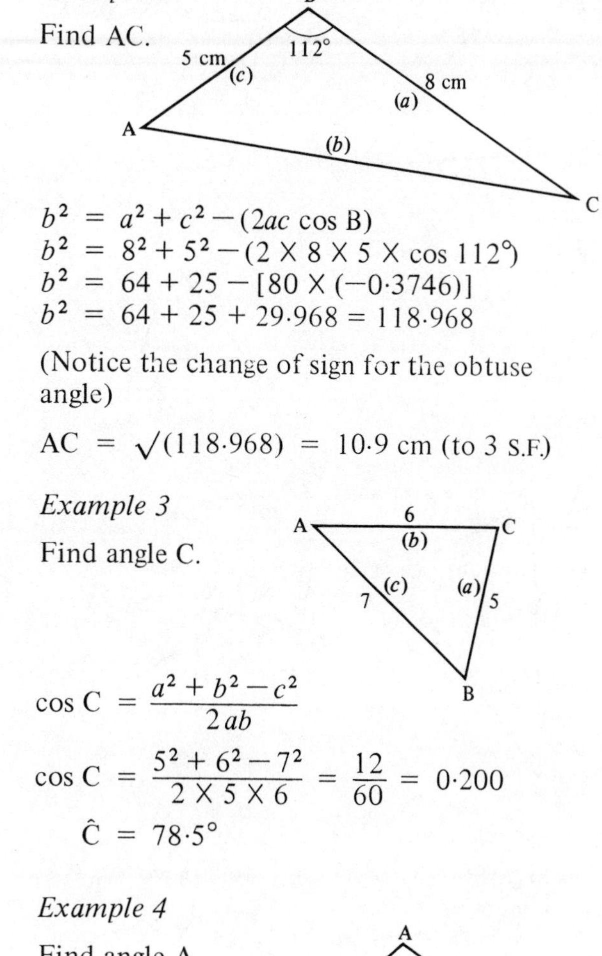

$$b^2 = a^2 + c^2 - (2ac \cos B)$$
$$b^2 = 8^2 + 5^2 - (2 \times 8 \times 5 \times \cos 112°)$$
$$b^2 = 64 + 25 - [80 \times (-0{\cdot}3746)]$$
$$b^2 = 64 + 25 + 29{\cdot}968 = 118{\cdot}968$$

(Notice the change of sign for the obtuse angle)

$$AC = \sqrt{(118{\cdot}968)} = 10{\cdot}9 \text{ cm (to 3 S.F.)}$$

Example 3

Find angle C.

$$\cos C = \frac{a^2 + b^2 - c^2}{2\,ab}$$

$$\cos C = \frac{5^2 + 6^2 - 7^2}{2 \times 5 \times 6} = \frac{12}{60} = 0{\cdot}200$$

$$\hat{C} = 78{\cdot}5°$$

Example 4

Find angle A.

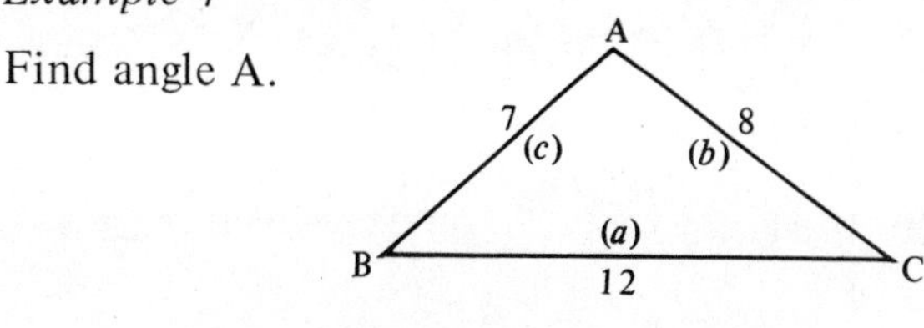

$$\cos A = \frac{b^2 + c^2 - a^2}{2\,bc}$$

$$\cos A = \frac{8^2 + 7^2 - 12^2}{2 \times 8 \times 7} = \frac{-31}{112} = -0{\cdot}277$$

$$\hat{A} = 106{\cdot}1°$$

The minus sign tells us the angle is obtuse

$$\cos 73{\cdot}9 = 0{\cdot}277$$
$$\therefore \quad \text{angle A} = (180 - 73{\cdot}9)° = 106{\cdot}1°$$

Exercise 10

Find the sides marked *.

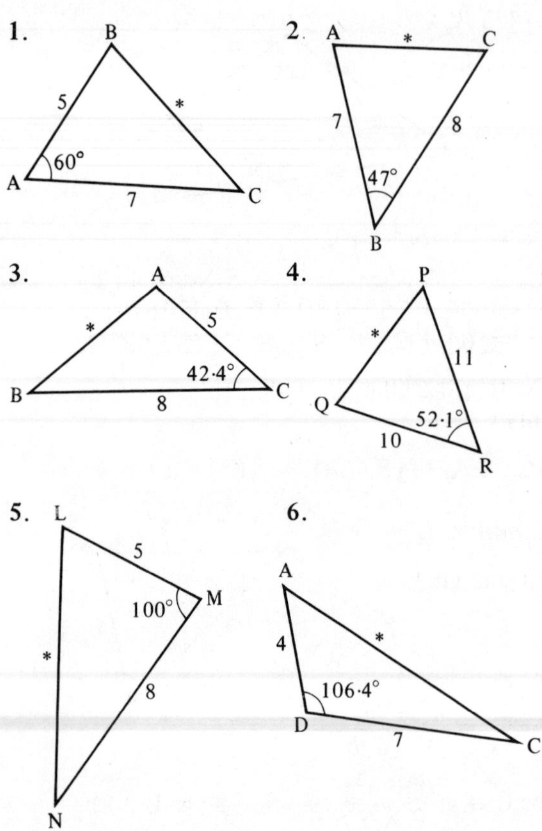

7. In ΔABC, AB = 4 cm, AC = 7 cm, $\hat{A} = 57°$. Find BC.

8. In ΔXYZ, XY = 3 cm, YZ = 3 cm, $\hat{Y} = 90°$. Find XZ.

9. In ΔLMN, LM = 5·3 cm, MN = 7·9 cm, $\hat{M} = 127°$. Find LN.

10. In ΔPQR, $\hat{Q} = 117°$, PQ = 80 cm, QR = 100 cm. Find PR.

In questions **11** to **16**, find the angles marked *.

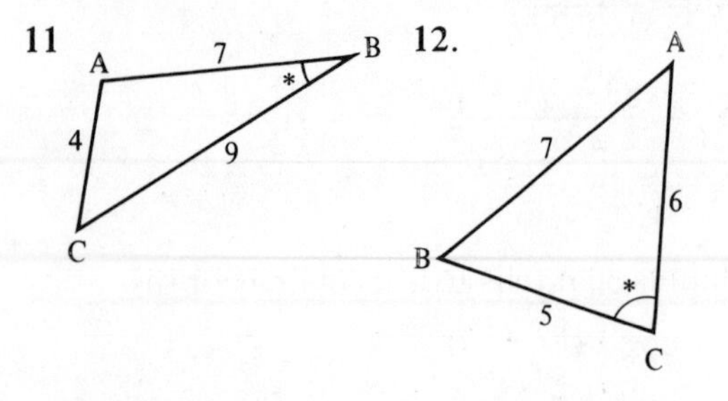

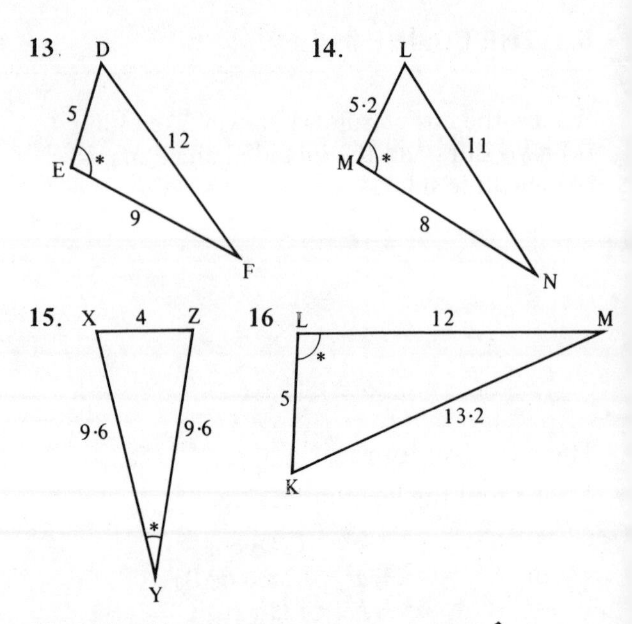

17. In ΔABC, $a = 4·3$, $b = 7·2$, $c = 9$. Find $\hat{C}$.

18. In ΔDEF, $d = 30$, $e = 50$, $f = 70$. Find $\hat{E}$.

19. In ΔPQR, $p = 8$, $q = 14$, $r = 7$. Find $\hat{Q}$.

20. In ΔLMN, $l = 7$, $m = 5$, $n = 4$. Find $\hat{N}$.

21. In ΔXYZ, $x = 5·3$, $y = 6·7$, $z = 6·14$. Find $\hat{Z}$.

22. In ΔABC, $a = 4·1$, $c = 6·3$, $\hat{B} = 112°10'$. Find b.

23. In ΔPQR, $r = 0·72$, $p = 1·14$, $\hat{Q} = 94°33'$. Find q.

24. In ΔLMN, $n = 7·206$, $l = 6·3$, $\hat{L} = 51°10'$, $\hat{N} = 63°$. Find m.

Some triangles are solved using a combination of the sine and cosine rules.

Example 5

Find all the sides and angles of the triangle given.

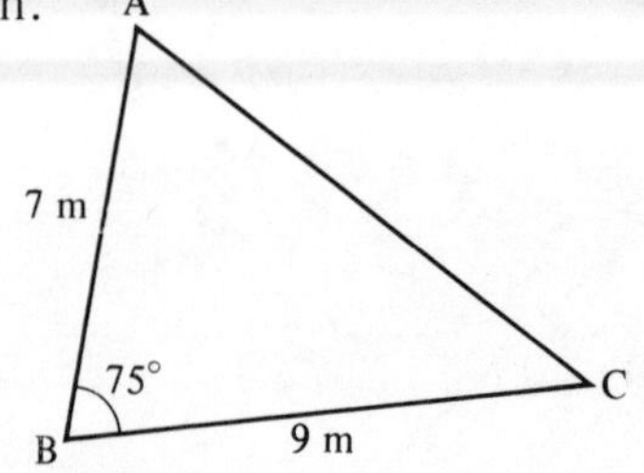

(a) Find AC using the cosine rule

$$AC^2 = 7^2 + 9^2 - (2 \times 7 \times 9 \cos 75°)$$
$$AC^2 = 49 + 81 - (32·61)$$
$$AC^2 = 97·39$$
$$AC = 9·869$$
$$AC = 9·87 \text{ m (to 3 S.F.)}$$

(b) Find $\hat{A}$ using the sine rule.

$$\frac{\sin A}{9} = \frac{\sin 75^\circ}{9{\cdot}869}$$

Notice that we work with 9·869 to avoid accumulating errors.

$$\sin A = \frac{9 \times \sin 75^\circ}{9{\cdot}869}$$

$$\sin A = 0{\cdot}8809$$

$$\therefore \quad \hat{A} = 61{\cdot}7^\circ$$

(c) Find $\hat{C}$ using the angle sum of a triangle

$$\therefore \quad \hat{C} = 180^\circ - (75^\circ + 61{\cdot}7^\circ)$$

$$\hat{C} = 43{\cdot}3^\circ$$

AC = 9·87 m, $\hat{A} = 61{\cdot}7^\circ$ and $\hat{C} = 43{\cdot}3^\circ$.

Exercise 11

Find the sides and angles marked *.

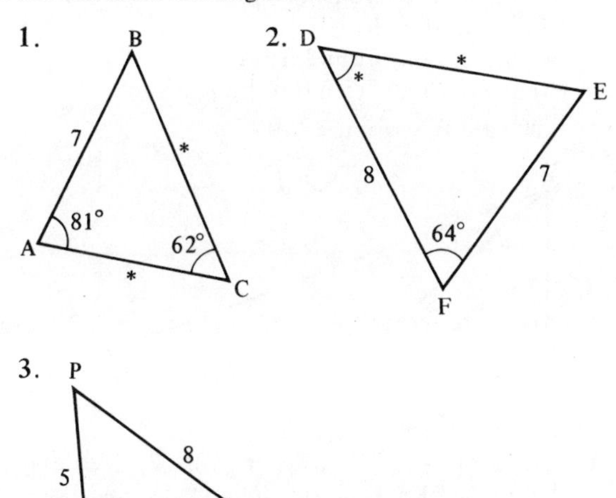

4. In ΔXYZ, XY = 5 cm, YZ = 6 cm, $\hat{X} = 31^\circ$. Find $\hat{Z}$ and XZ.
5. In ΔLMN, LM = 8 m, MN = 9 m, $\hat{M} = 72^\circ$. Find LN and $\hat{L}$.
6. In ΔABC, AC = 11 cm, $\hat{B} = 80^\circ$, $\hat{C} = 41^\circ$. Find AB and BC.
7. In ΔXYZ, XZ = 11 cm, YX = 6 m, YZ = 8 cm. Find $\hat{X}$, $\hat{Y}$ and $\hat{Z}$.

In questions **8** to **11**, find each angle and side marked *.

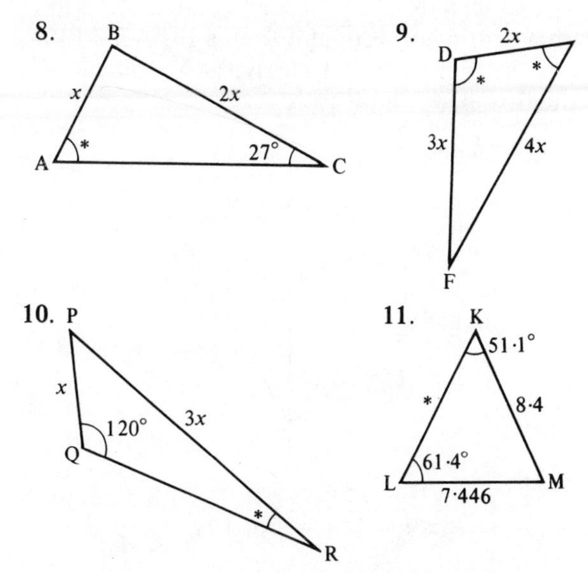

12. In ΔABC, AB = 112 m, AC = 175 m, $B\hat{A}C = 75^\circ 16'$. Find BC.

13. In ΔDEF, EF = 19·2 cm, $\dfrac{DF}{\sin E} = 28{\cdot}4$. Find $\hat{D}$.

14. In ΔGHI, GH = 15·7 cm, $\dfrac{HI}{\sin G} = 23{\cdot}1$. Find $\hat{I}$.

In questions **15** to **18** find the sides and angles marked *.

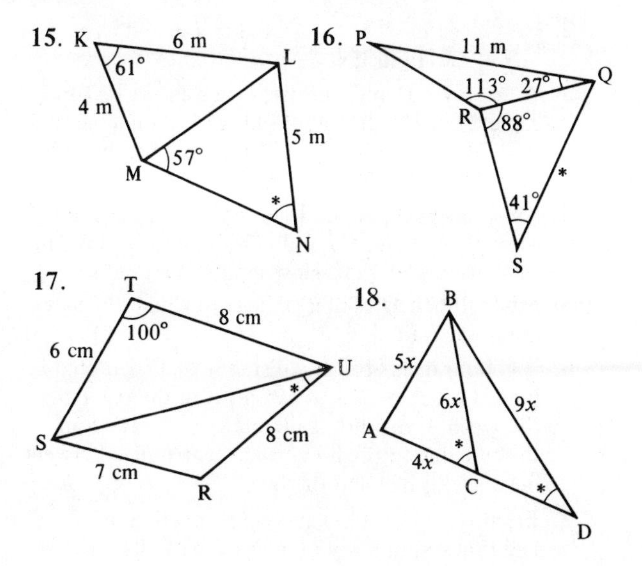

(ACD is a straight line)

Example 6

A ship sails from a port P a distance of 7 km on a bearing of 306° and then a further 11 km on a bearing of 070° to arrive at X. Calculate the distance from P to X.

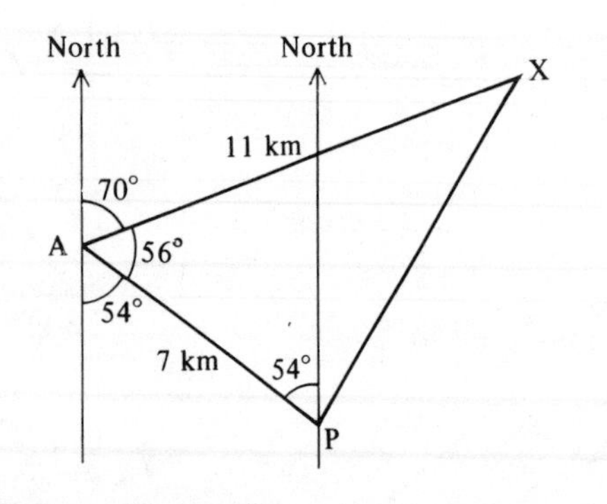

$$PX^2 = 7^2 + 11^2 - (2 \times 7 \times 11 \times \cos 56°)$$
$$= 49 + 121 - (86{\cdot}12)$$
$$PX^2 = 83{\cdot}88$$
$$PX = 9{\cdot}16 \text{ km (to 3 s.f.)}$$

The distance from P to X is 9·16 km.

Exercise 12

Start each question by drawing a large, clear diagram.

1. In triangle PQR $\hat{Q} = 72°$, $\hat{R} = 32°$ and PR = 12 cm. Find PQ.
2. In triangle LMN $\hat{M} = 84°$, LM = 7 m and MN = 9 m. Find LN.
3. A destroyer D and a cruiser C leave port P at the same time. The destroyer sails 25 km on a bearing 040° and the cruiser sails 30 km on a bearing of 320°. How far apart are the ships?
4. Two honeybees A and B leave the hive H at the same time; A flies 27 m due south and B flies 9 m on a bearing of 111°. How far apart are they?
5. Find all the angles of a triangle in which the sides are in the ratio 5:6:8.
6. A golfer hits his ball B a distance of 170 m on a hole H which measures 195 m from the tee T to the green. If his shot is directed 10° away from the true line to the hole, find the distance between between his ball and the hole.
7. From A, B lies 11 km away on a bearing of 041° and C lies 8 km away on a bearing of 341°. Find
   (a) the distance between B and C
   (b) the bearing of B from C.
8. From a lighthouse L an aircraft carrier A is 15 km away on a bearing of 112° and a submarine S is 26 km away on a bearing of 200°. Find
   (a) the distance between A and S
   (b) the bearing of A from S.
9. Calculate all the sides and angles in the diagram.

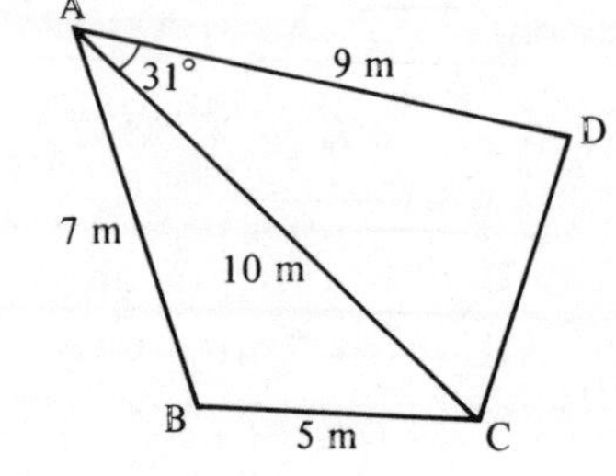

10. An aircraft flies from its base 200 km on a bearing 162°, then 350 km on a bearing 260°, and then returns directly to base. Calculate the length and bearing of the return journey.
11. Two lightships L and M are 30 km apart and M is due west of L. A sinking ship S is on a bearing 140° from M and 246° from L. Calculate the distances of S from M and from L.
12. B is 12 km from A on a bearing 060°; C is on a bearing of 130° from A and is on a bearing of 220° from B. Calculate
    (a) the distance from C to A
    (b) the distance from C to B.
13. Calculate WX, given YZ = 15 m.

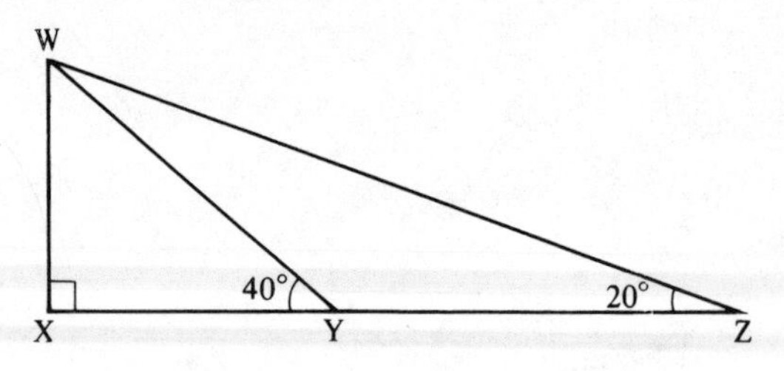

14. A golfer hits his ball a distance of 127 m so that it finishes 31 m from the hole. If the length of the hole is 150 m, calculate the angle between the line of his shot and the direct line to the hole.
15. A tasty-looking mouse can see two hungry kestrels, W to the West of him and E to the East, W is 25 m away at an elevation of 50° and E is 32 m away at an elevation of 30°. The mouse knows that if the kestrels are less than 40 m apart they will see each other and have a game with each other rather than eat him. Find the distance between the kestrels and hence work out out what will happen.

16. Find (a) AE
(b) $E\hat{A}C$
If the line BCD is horizontal, find the angle of elevation of E from A.

17. Two points A and B on opposite sides of a tower are 160 m apart. The angles of elevation are 30° and 38°. How high is the tower?

18. Town M is 15 km due north of town L. Town N is 7 km from M, 10 km from L and somewhere to the east of the line LM. Find
(a) the bearing of N from M
(b) the bearing of N from L.

19. Town Y is 9 km due north of town Z. Town X is 8 km from Y, 5 km from Z and somewhere to the west of the line YZ.
(a) Draw triangle XYZ and find angle YZX
(b) During an earthquake, town X moves due south until it is due west of Z. Find how far it has moved.

20.

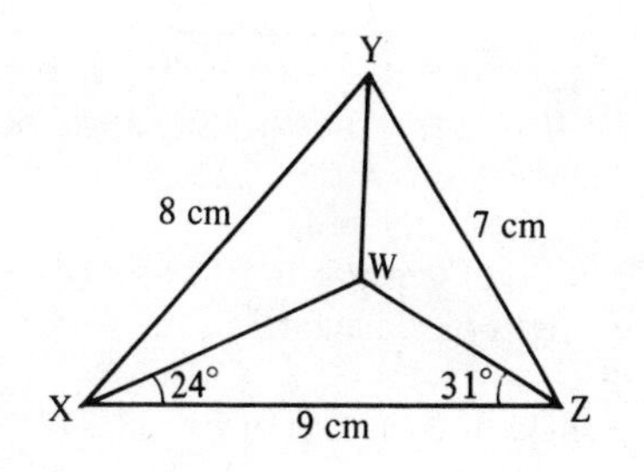

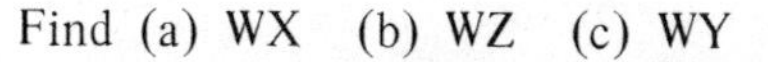
Find (a) WX (b) WZ (c) WY

6.6 THREE-DIMENSIONAL PROBLEMS

Always draw a large, clear diagram. It is often helpful to redraw the triangle which contains the length or angle to be found.

Example 1

A rectangular box with top WXYZ and base ABCD has AB = 6 cm, BC = 8 cm and WA = 3 cm.

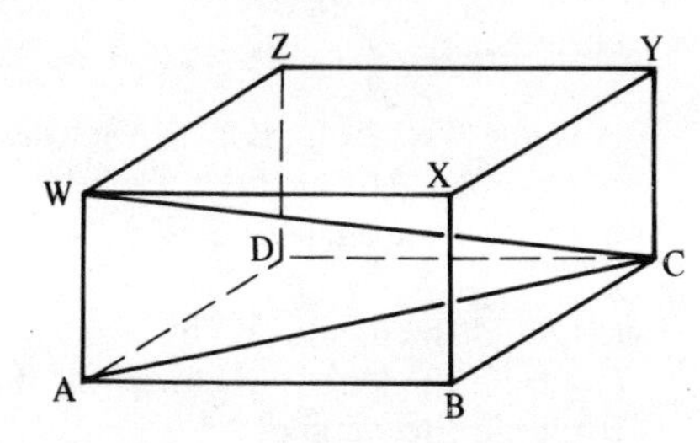

Calculate
(a) the length of AC
(b) the angle between WC and AC.

(a) Redraw triangle ABC

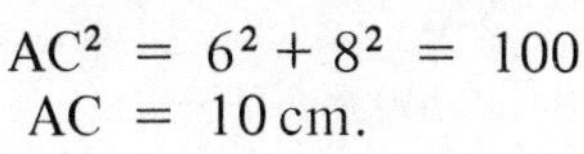
$$AC^2 = 6^2 + 8^2 = 100$$
$$AC = 10\text{ cm.}$$

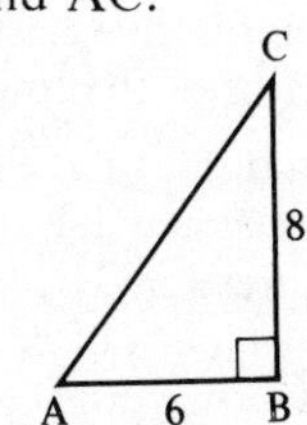

(b) Redraw triangle WAC

Let $W\hat{C}A = \theta$
$$\tan\theta = \frac{3}{10}$$
$$\theta = 16{\cdot}7°$$

The angle between WC and AC is 16·7°.

Exercise 13

1. In the rectangular box shown, find
(a) AC
(b) AR
(c) the angle between AC and AR.

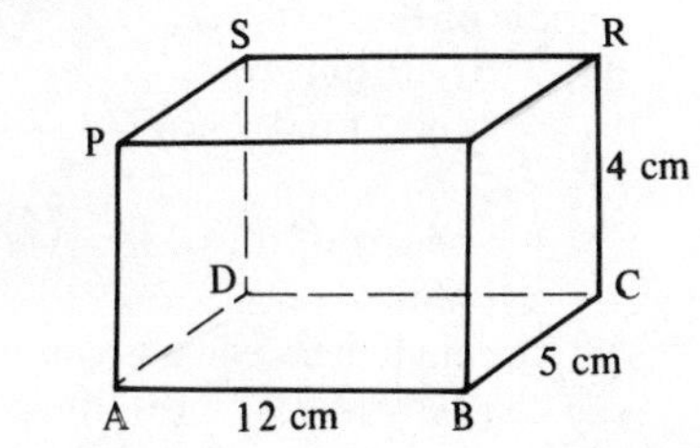

2. A vertical pole BP stands at one corner of a horizontal rectangular field as shown.

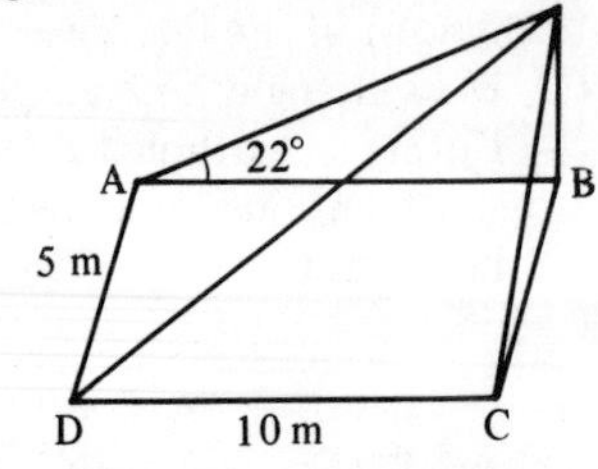

If AB = 10 m, AD = 5 m and the angle of elevation of P from A is 22°, calculate
(a) the height of the pole
(b) the angle of elevation of P from C
(c) the length of a diagonal of the rectangle ABCD
(d) the angle of elevation of P from D.

3. In the cube shown, find
(a) BD
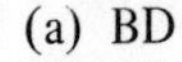
(b) AS
(c) BS
(d) the angle SBD
(e) the angle ASB

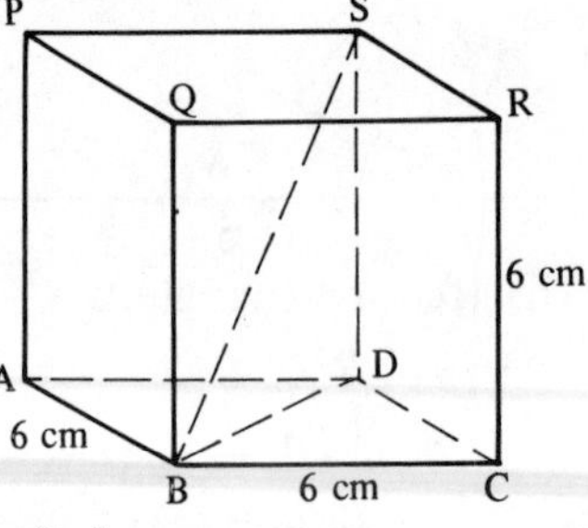

4. In the cuboid shown, find
(a) WY
(b) DY
(c) WD
(d) the angle WDY

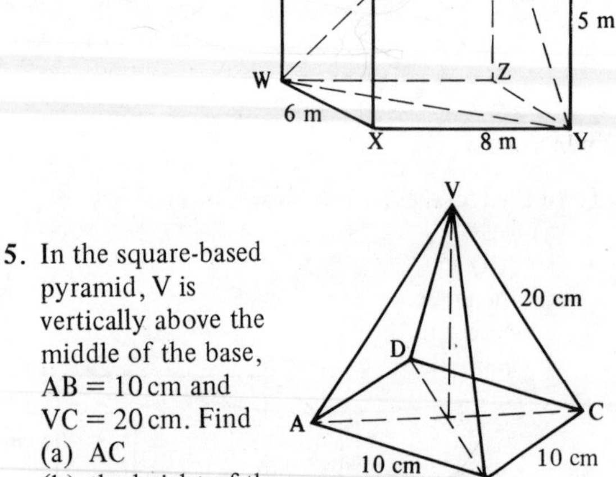

5. In the square-based pyramid, V is vertically above the middle of the base, AB = 10 cm and VC = 20 cm. Find
(a) AC
(b) the height of the pyramid
(c) the angle between VC and the base ABCD
(d) the angle AVB (e) the angle AVC.

6. In the wedge shown, PQRS is perpendicular to ABRQ; PQRS and ABRQ are rectangles with AB = QR = 6 m, BR = 4 m, RS = 2 m. Find
(a) BS (b) AS
(c) angle BSR (d) angle ASR
(e) angle PAS

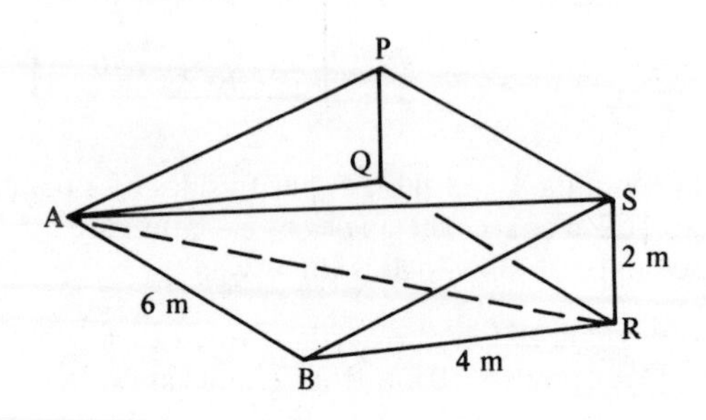

7. The edges of a box are 4 cm, 6 cm and 8 cm. Find the length of a diagonal and the angle it makes with the diagonal on the largest face.

8. In the diagram A, B and O are points in a horizontal plane and P is vertically above O, where OP = h m.

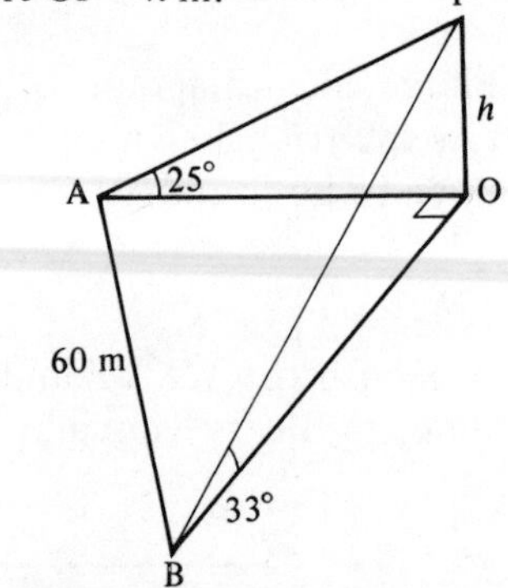

A is due West of O, B is due South of O and AB = 60 m. The angle of elevation of P from A is 25° and the angle of elevation of P from B is 33°.
(a) Find the length AO in terms of h
(b) Find the length BO in terms of h
(c) Find the value of h.

9. The angle of elevation of the top of a tower is 38° from a point A due South of it. The angle of elevation of the top of the tower from another point B, due East of the tower is 29°. Find the height of the tower if the distance AB is 50 m.

10. An observer at the top of a tower of height 15 m sees a man due West of him at an angle of depression 31°. He sees another man due South at an angle of depression 17°. Find the distance between the men.

11. The angle of elevation of the top of a tower is 27° from a point A due East of it. The angle of elevation of the top of the tower is 11° from another point B due South of the tower. Find the height of the tower if the distance AB is 40 m.

The angle between a line and a plane is the angle between the line and its *projection* in the plane.

The angle between two planes is the angle between two lines, one in each plane, which meet the line of intersection of the planes at right angles.

Example 2

VABCD is a pyramid with a rectangular base ABCD in which AB = 12 cm and BC = 5 cm.

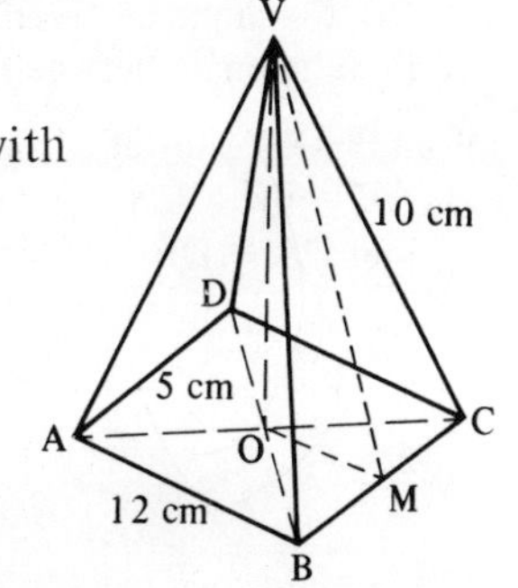

V is vertically above the centre of the rectangle and VA = VB = VC = VD = 10 cm.

Find
(a) the angle between VA and the plane ABCD
(b) the angle between the planes VBC and ABCD.

(a) Find AC from $\triangle$ABC

$$AC^2 = 12^2 + 5^2$$
$$AC = 13 \text{ cm}$$
$$\therefore \quad AO = 6{\cdot}5 \text{ cm}$$

Redraw $\triangle$OVA

The projection of VA in the plane ABCD is AO. The angle required is $V\hat{A}O$.

$$\cos V\hat{A}O = \frac{6{\cdot}5}{10} = 0{\cdot}65$$
$$\therefore \quad V\hat{A}O = 49{\cdot}5°$$

The angle between VA and the plane ABCD is 49·5°.

(b) Introduce M at the mid-point of BC. The line of intersection of the planes in question is BC. OM and VM meet this line at right angles so the angle between planes ABCD and VBC is $V\hat{M}O$.

Redraw $\triangle$VMO

$OM = \frac{1}{2}AB = 6$ cm.

From $\triangle$OVA,
$$VO^2 = 10^2 - 6{\cdot}5^2$$
$$VO = 7{\cdot}599 \text{ cm.}$$

$$\tan V\hat{M}O = \frac{7{\cdot}599}{6}$$
$$\therefore \quad V\hat{M}O = 51{\cdot}7°$$

The angle between planes VBC and ABCD is 51·7°.

Exercise 14

1. The figure shows a cuboid in which AB = 8 cm, BC = 8 cm and YC = 5 cm.

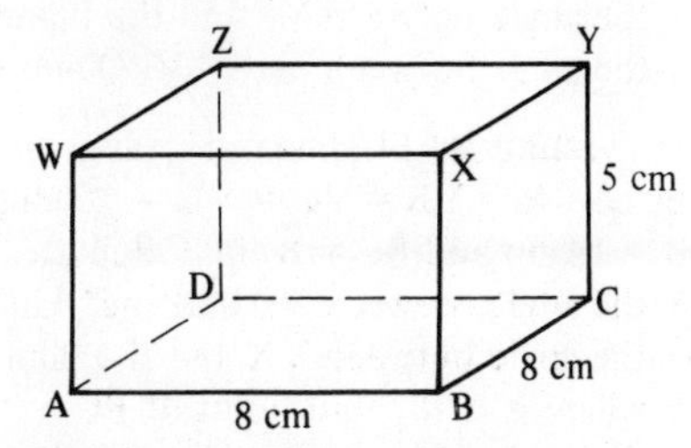

Calculate
(a) the lengths of AC and AY
(b) the angle between AY and the plane ABCD
(c) the angle between the planes WBCZ and ABCD

2. The figure shows a cuboid in which WX = 8 cm, XY = 6 cm and MY = 5 cm.

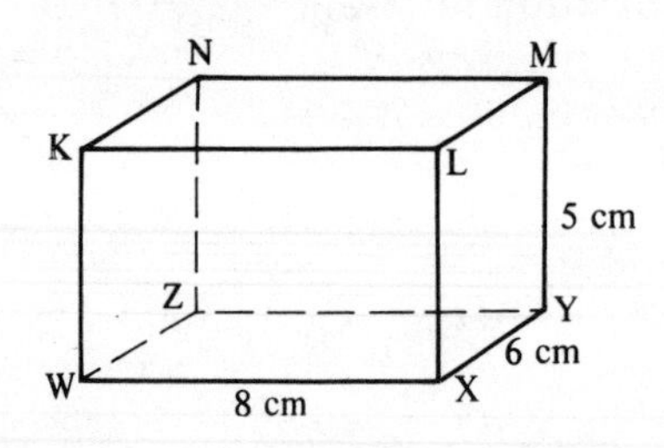

Calculate
(a) the lengths ZX and KX
(b) the angle between NX and the plane WXYZ
(c) the angle between KY and the plane KLWX
(d) the angle between the planes KXYN and WXYZ.

3. The figure shows a cuboid in which AB = 10 cm, BC = 10 cm and GC = 8 cm.

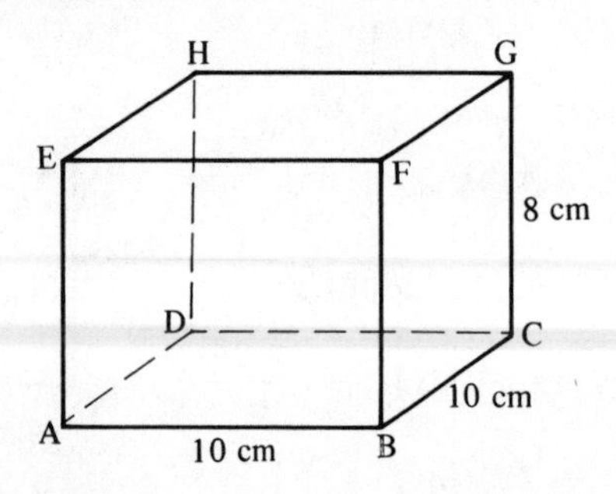

Calculate
(a) the angle between AG and the plane ABCD
(b) the angle between the planes EBCH and ABCD
(c) the angle between DB and DF.

4. The pyramid VPQRS has a square base PQRS. VP = VQ = VR = VS = 12 cm and PQ = 9 cm. Calculate
(a) the angle between VP and the plane PQRS
(b) the angle between planes VPQ and PQRS.

5. The pyramid VABCD has a rectangular base ABCD. VA = VB = VC = VD = 15 cm, AB = 14 cm and BC = 8 cm. Calculate
(a) the angle between VB and the plane ABCD
(b) the angle between VX and the plane ABCD where X is the mid-point of BC
(c) the angle between the planes VBC and ABCD.

6. The pyramid VABCD has a horizontal rectangular base ABCD in which AB = 20 m and BC = 14 m. V is vertically above B and VB = 7 m. Calculate
(a) the angle between DV and the plane ABCD
(b) the angle between AV and the plane ABCD
(c) the angle between the planes VAD and ABCD.

7. The figure shows a triangular pyramid on a horizontal base ABC, V is vertically above B where VB = 10 cm, $A\hat{B}C = 90°$ and AB = BC = 15 cm.

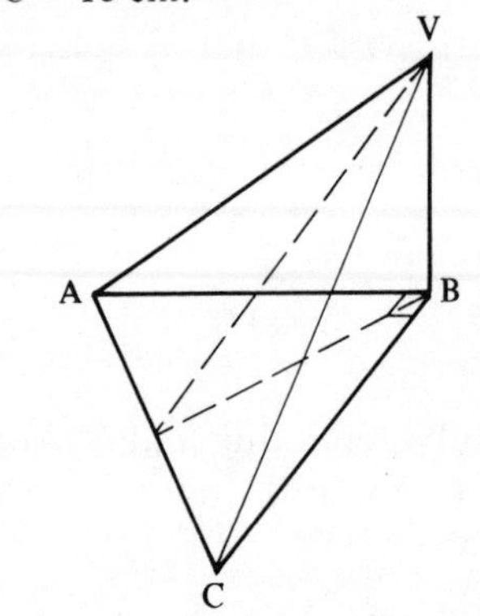

Calculate the angle between the planes AVC and ABC.

8. ABCD is a regular tetrahedron of side 10 cm. Calculate
(a) the height DO of the tetrahedron
(b) the angle between DA and the plane ABC
(c) the angle between the planes DBC and ABC.

9. ABCD is a tetrahedron with AB = BC = CA = 8 cm and DA = DB = DC = 10 cm.

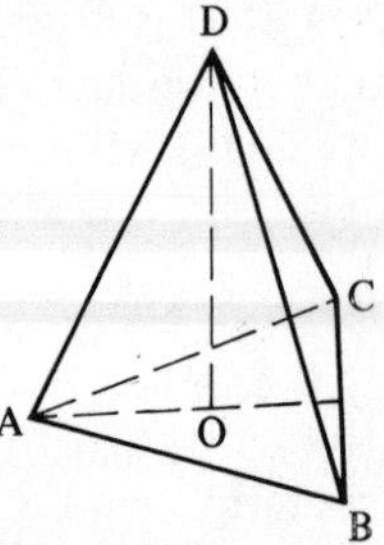

Calculate
(a) the height DO of the tetrahedron
(b) the angle between DC and the plane ABC
(c) the angle between the planes DBC and ABC.

10. ABCD is a tetrahedron with AB = BC = CA = 6 cm and DA = DB = DC = 9 cm. Calculate
(a) the height DO of the tetrahedron
(b) the angle between DA and the plane ABC
(c) the angle between the planes DAB and ABC.

REVISION EXERCISE 6A

1. Calculate the side or angle marked with a letter.

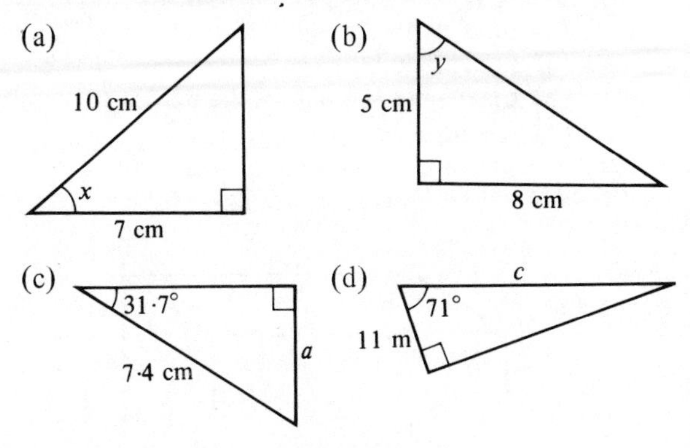

2. Given that x is an acute angle and that

$$3 \tan x - 2 = 4 \cos 35^\circ 18'$$

calculate
(a) $\tan x$
(b) the value of x in degrees and minutes.

3. In the triangle XYZ, XY = 14 cm, XZ = 17 cm and angle YXZ = 25°. A is the foot of the perpendicular from Y to XZ. Calculate
(a) the length XA (b) the length YA
(c) the angle ZYA.

4. Calculate the length of AB.

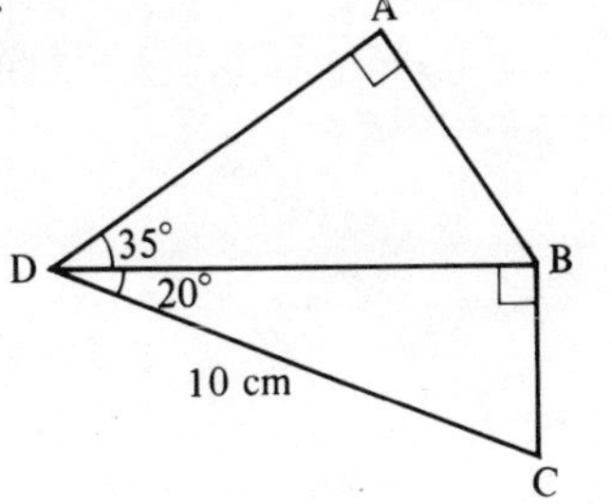

5. (a) A lies on a bearing of 040° from B. Calculate the bearing of B from A.
(b) The bearing of X from Y is 115°. Calculate the bearing of Y from X.

6. Given BD = 1 m, calculate the length AC.

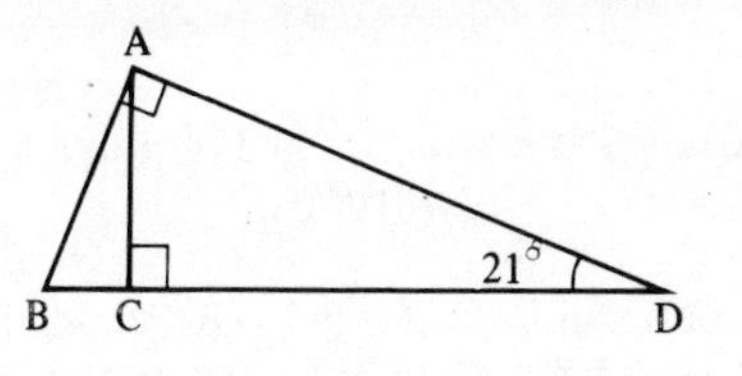

7. In the triangle PQR, angle PQR = 90° and angle RPQ = 31°. The length of PQ is 11 cm. Calculate
(a) the length of QR
(b) the length of PR
(c) the length of the perpendicular from Q to PR.

8. Find the values of
(a) $\sin 215^\circ$ (b) $\tan 306^\circ 6'$ (c) $\cos 207^\circ$
(d) $\sin 116{\cdot}6^\circ$ (e) $\tan 300^\circ$ (f) $\cos 284^\circ 12'$

9. (a) Given $\tan 35^\circ = 0{\cdot}7$, evaluate
(i) $\tan 145^\circ$
(ii) $2 \tan 215^\circ + 10 \tan 325^\circ$
(b) Given that $\cos x = -\cos 63^\circ$, find two values of x between 0° and 360°.

10. In triangle PQR, PQ = 7 cm, PR = 8 cm and QR = 9 cm. Calculate angle QPR.

11. In triangle XYZ, XY = 8 m, $\hat{X} = 57^\circ$ and $\hat{Z} = 50^\circ$. Calculate the lengths YZ and XZ.

12. In triangle ABC, $\hat{A} = 22^\circ$ and $\hat{C} = 44^\circ$.
Calculate the ratio $\dfrac{BC}{AB}$.

13. $B\hat{A}D = D\hat{C}A = 90^\circ$, $C\hat{A}D = 32{\cdot}4^\circ$, $B\hat{D}A = 41^\circ$ and AD = 100 cm.

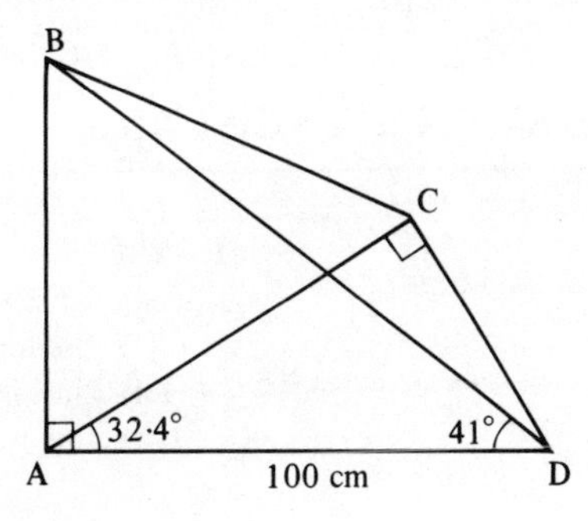

Calculate
(a) the length of AB
(b) the length of DC
(c) the length of BD.

14. Calculate the smallest angle in a triangle whose sides are of length $3x$, $4x$ and $6x$.

15. An observer at the top of a tower of height 20 m sees a man due East of him at an angle of depression of 27°. He sees another man due South of him at an angle of depression of 30°. Find the distance between the men on the ground.

16. Two lighthouses A and B are 25 km apart and A is due West of B. A submarine S is on a bearing of 137° from A and on a bearing of 170° from B. Calculate the distance of S from A and the distance of S from B.

17. Given $\cos A\hat{C}B = 0{\cdot}6$, AC = 4 cm, BC = 5 cm and CD = 7 cm, calculate the length of AB and AD.

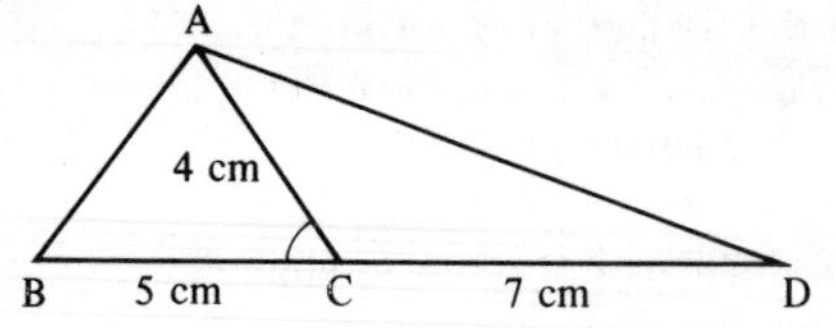

18. The figure shows a cube of side 10 cm.

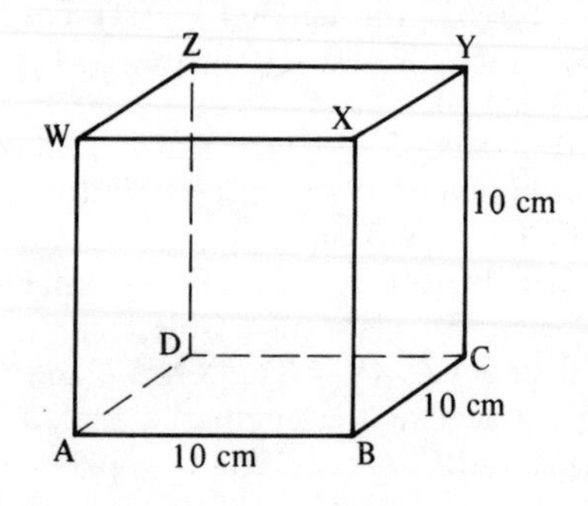

Calculate
(a) the length of AC
(b) the angle YAC
(c) the angle ZBD.

19. VABCD is a pyramid in which the base ABCD is a square of side 8 cm; V is vertically above the centre of the square and
VA = VB = VC = VD = 10 cm.
Calculate
(a) the length AC
(b) the height of V above the base
(c) the angle $V\hat{C}A$.

20. The diagram shows a rectangular block.
AY = 12 cm, AB = 8 cm, BC = 6 cm.

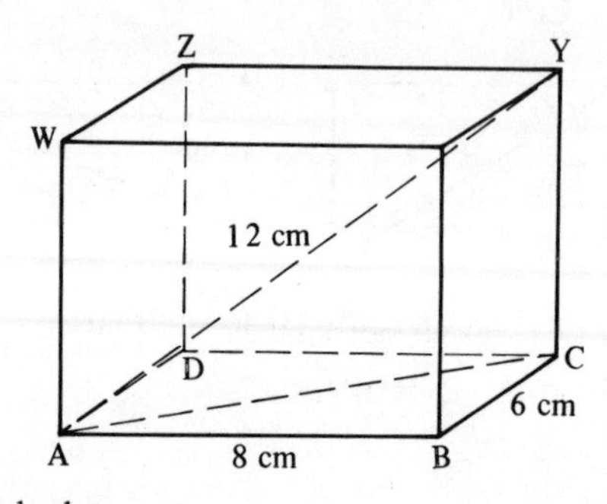

Calculate
(a) the length YC
(b) the angle between AY and the face ADZW.

EXAMINATION EXERCISE 6B

1. ABCD is a quadrilateral in which AB = 7 cm, BC = 6 cm, DA = 4 cm, the angle BAD = 60° and angle BCD = 90°. Calculate
(i) the lengths of BD and CD
(ii) the size of the angle ADC
(iii) the area of the quadrilateral ABCD.

[AEB]

2. A, B, C and D are four points on an airfield. A is due north of D, B is due east of D, and $B\hat{D}C = 39°\,16'$. The distance DA = 1·7 km, DB = 4·3 km and DC = 2·8 km. Calculate, giving angles in degrees and minutes correct to the nearest minute,
(i) $B\hat{A}D$,
(ii) the bearing of B from A,
(iii) the area of the triangular piece of land BDC,
(iv) the distance BC.

[C]

3. A power boat race is run round a triangular course marked by three buoys A, B and C. AB = 1200 m, AC = 800 m and BC = 1000 m.
(i) Calculate the size of angle ABC.
During the race a boat breaks down after it has travelled a distance 200 m from B along the leg BC, and has to be towed back directly to the start at A.
(ii) Calculate the distance the boat is towed.

[NEA]

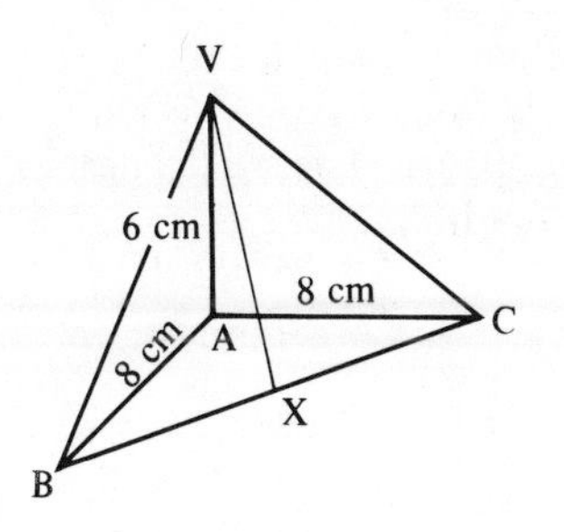

4. The diagram shows a pyramid on a triangular base. The angles VAB, VAC, BAC are all 90°, the length VA is 6 cm and the lengths AB, AC are each 8 cm. X is the mid-point of BC. Calculate
   (i) the size of the angle between the edge VC and the base ABC to the nearest 0·1°,
   (ii) the length of AX correct to three significant figures,
   (iii) the size of the angle VXA to the nearest 0·1°,
   (iv) the volume of the pyramid VABC.
   (The volume of a pyramid = $\frac{1}{3}$ base area × height.) [JMB]

5. ABCD is a quadrilateral in which the lengths of the sides AB, BC and CD are 4 cm, 5 cm and 10 cm respectively. The angle ABC is 80° and the angle ACD is 30°. Calculate
   (i) the length of AC,
   (ii) the angle ACB,
   (iii) the length of BD,
   (iv) the area of the quadrilateral ABCD.
   [O]

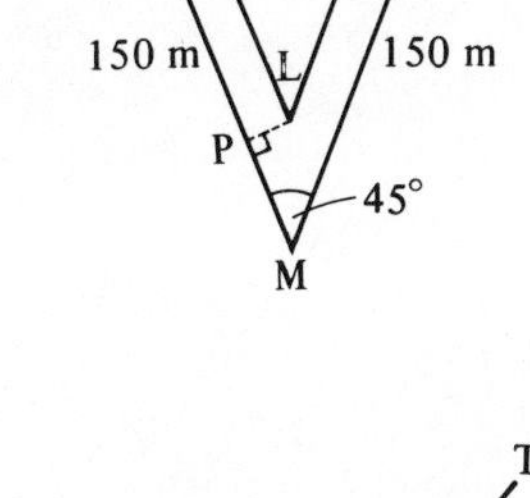

6. To commemorate the Golden Jubilee of Queen Victoria in 1887 a wood was planted on the South Downs in the shape of a V as shown on the diagram. The points A, B, C, D lie on a straight line perpendicular to the line LM. The outer edges AM and DM are 150 m long, the width of each arm is 20 m, and the angle of the V is 45°. Calculate
   (i) the lengths AD and PM,
   (ii) the length BL,
   (iii) the area of the wood.
   [SMP]

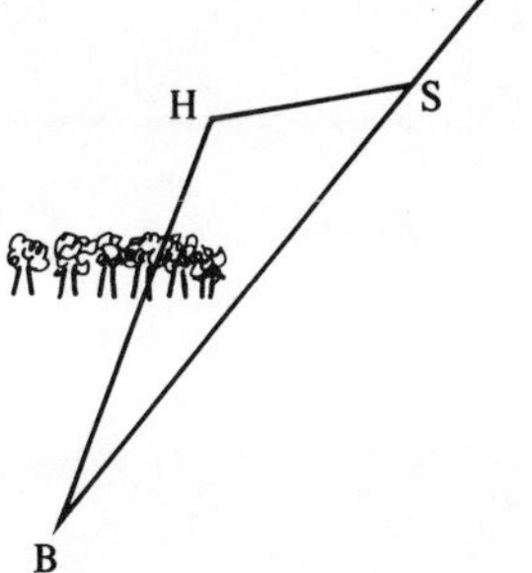

7. A golfer whose ball is at B on a level golf course has the choice of two shots. He can either hit his ball over the trees in the direction of the hole H or hit it past the end of the trees to some point S along the line BT.
   He estimates that angle HBT is 20° and know that, when he hits the ball as far as he can, BS is 120 m. In this case he estimates that angle BSH is 50°. Using these values, calculate BH.
   His estimate of angle BSH is not quite accurate and in fact BH is exactly 100 m. Angle HBT is exactly 20°. Calculate the distance he should hit his ball along BT so that it finishes as close to the hole as possible, and find how far from the hole the ball would then be. How much nearer to the hole than this would the ball be if he hits it to a point X on BH where BX is 70 m?
   [AEB]

8. Two points A and B which are 10 km apart are at sea level on a coastline. The bearing of B from A is Nx°E, where $\cos x = \frac{3}{5}$. A ship is 12 km due South of B and an aeroplane vertically above the ship is seen at an angle of elevation of 11° from A. Calculate
   (i) the distance, in kilometres, of the ship from A,
   (ii) the bearing of the ship from A,
   (iii) the height, in metres, of the aeroplane above sea level,
   (iv) the angle of elevation of the aeroplane from B.
   [JMB]

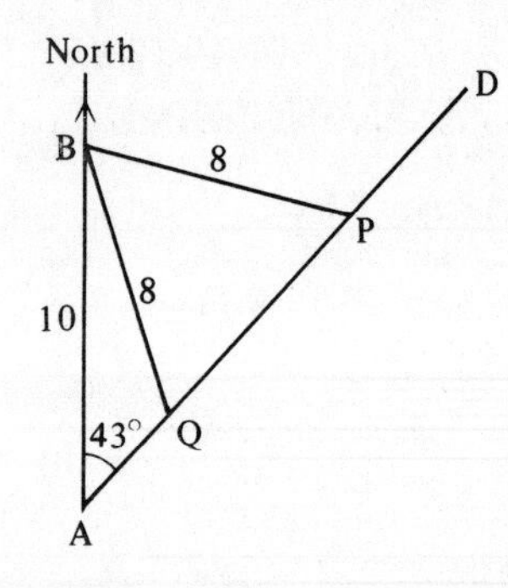

9. The diagram shows a point A which lies 10 km due south of a point B. A straight level road AD is such that the bearing of D from A is 043°. P and Q are the two points on this road which are 8 km from B. calculate, to the nearest $\frac{1}{2}^\circ$, the bearing of P from B, and the bearing of Q from B.

Calculate also the distance between P and Q. [O & C]

10. A pair of compasses has two arms OX and OY, each of which is 10 cm long. OX is straight but OY has a hinge at P, which is 2 cm from Y, so that OY may be bent if required.

(i) Calculate $\angle OXY$ when the compasses are used with OPY straight to draw a circle of radius 8 cm.

(ii) Calculate the radius of a circle which can be drawn with the compasses when OPY is straight and $\angle XOY = 28^\circ$.

(iii) A circle of radius 12 cm is to be drawn and OPY is bent so that PY is perpendicular to the paper. Calculate PX^2 and hence calculate $\angle XOP$.

[JMB]

7 Graphs

Rene Descartes (1596–1650) was one of the greatest philosophers of his time. Strangely his restless mind only found peace and quiet as a soldier and he apparently discovered the idea of 'cartesian' geometry in a dream before the battle of Prague. The word 'cartesian' is derived from his name and his work formed the link between geometry and algebra which inevitably led to the discovery of calculus. He finally settled in Holland for ten years, but later moved to Sweden where he soon died of pneumonia.

7.1 EQUATIONS AND GRAPHS

When a graph is drawn to illustrate the relation between variables, say x and y, the coordinates of the points on the curve *satisfy* the equation of the curve.

The point $(1, 4)$ lies on the curve $y = x^2 + 3$ because the equation is satisfied when we substitute $x = 1$ and $y = 4$.

An alternative notation represents graphs by *mappings* rather than equations. For example, we could draw the graph of the function defined by '$f : x \rightarrow x^2 + 3$' ['The function f such that x is mapped onto $x^2 + 3$'].
Using another notation, we could define the same function as '$f(x) = x^2 + 3$'.

It should be noted that these mathematical statements represent the same information.

e.g. $y = x + \frac{1}{x}$,

$f : x \rightarrow x + \frac{1}{x}$

and $f(x) = x + \frac{1}{x}$ are equivalent statements.

Example 1

Which of the following functions are satisfied by the point $(2, 5)$?

(a) $y = x^2 + 1$
(b) $y = 2x^2 - 2$
(c) $f : x \rightarrow \frac{x + 8}{2}$
(d) $f(x) = 12 - x^2$.

(a) When $x = 2, y = 2^2 + 1 = 5$

$\therefore$ (2, 5) does satisfy the equation.

(b) When $x = 2, y = 2(2^2) - 2 = 6$

$\therefore$ (2, 5) does *not* satisfy the equation.

(c) When $x = 2, 2 \rightarrow \dfrac{2+8}{2}$

$2 \rightarrow 5$

$\therefore$ (2, 5) does satisfy the function f.

(d) When $x = 2, f(2) = 12 - 2^2 = 8$

$\therefore$ (2, 5) does *not* satisfy the function.

Example 2

Find the equation of the straight line which is satisfied by the following points:

x	0	3	5
y	4	7	9

By inspection, the y-coordinate is four more than the x-coordinate.

$\therefore$ The equation of the line is $y = x + 4$

Exercise 1

In questions **1** to **15**, an equation is followed by several pairs of coordinates. Find which points do not lie on the curve given.

1. $y = x^2 - 4$;

x	0	2	-2	1
y	-4	0	0	3

2. $y = 3x - 5$;

x	1	0	3	-1
y	-2	-2	4	8

3. $y = 6 - x$;
$(1, 5), (7, -1), (-3, 8), (-6, 12)$.

4. $y = 4(3 - x)$;
$(0, 12), (3, 0), (-1, 12), (-2, 20)$.

5. $y = x^2 + 3x$;
$(1, 4), (-1, -2), (0, 3), (4, 30)$.

6. $v = 2u + 1$;

u	1	-2	$\frac{1}{2}$
v	3	-3	3

7. $s = 3t^2 - 1$;

t	0	-1	2
s	2	2	11

8. $v = 9 - 2t$;
$(t, v), (0, 7), (-1, 11), (3, 3)$.

9. $a = 5t + 1$;
$(t, a), (\frac{1}{2}, 3\frac{1}{2}), (-2, 9), (0, 1)$.

10. $c = 10n + 200$;
$(n, c), (10, 300), (-20, 0), (-1, 180)$.

11. $f(x) = x^2 - 2x$;
$(x, f(x)), (1, -1), (2, 0), (3, 3), (4, 6)$.

12. $g(x) = 7 - 3x$;

x	2	0	-1	4
$g(x)$	1	7	4	-5

13. $h(x) = \dfrac{12}{x}$;

x	1	3	-4	$\frac{1}{2}$
$h(x)$	12	4	-2	6

14. $f : x \rightarrow x^2 + 2$;
$(1, 3), (3, 10), (-2, 6), (-3, 10)$.

15. $g : x \rightarrow 3x - 2x^2$;
$(0, 0), (2, -2), (-1, -5), (-2, 2)$.

In questions **16** to **27**, find the relation connecting the variables given that the points lie on a straight line.

16.

x	4	7	10	12
y	1	4	7	9

17.

x	2	3	5
y	4	6	10

18.

x	3	6	12
y	1	2	4

19.

x	1	2	4	7
y	9	8	6	3

20.

x	2	4	-2
y	8	10	4

21.

x	2	3	5	10
y	5	7	11	21

22.

x	0	1	2	3
y	4	7	10	13

23.

x	2	4	6
y	10	8	6

24.

x	1	3	5	7
y	-3	1	5	9

25.

x	0	2	3	4
y	10	6	4	2

26.

t	1	2	3	4
v	3	0	-3	-6

27.

t	1	2	3	4
v	$3\frac{1}{2}$	2	$\frac{1}{2}$	-1

Find the equation of the lines in questions **28** to **30**.

28.

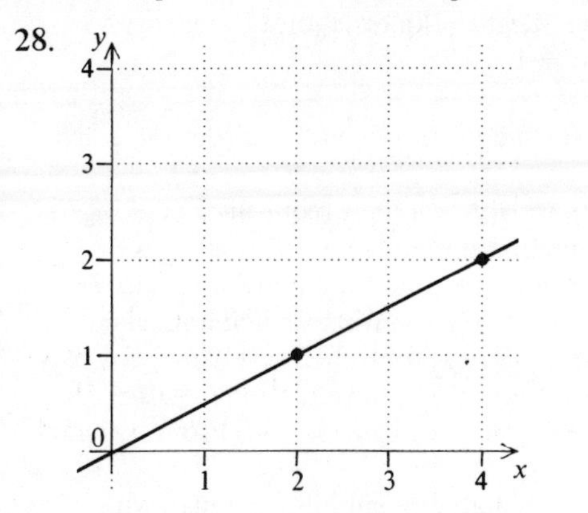

29.

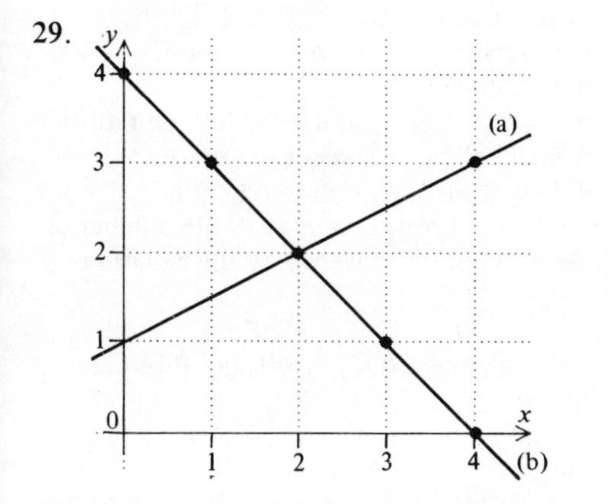

30.

31.

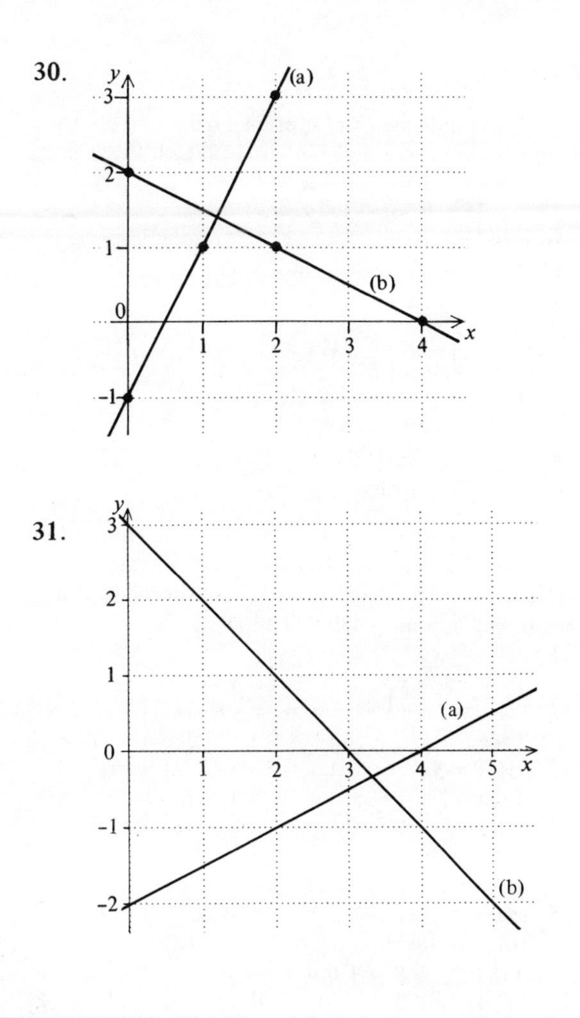

7.2 DRAWING ACCURATE GRAPHS

Example 1 .

Draw the graph of $y = 2x - 3$ for values of x from -2 to $+4$.

(a) The coordinates of points on the line are calculated in a table.

x	-2	-1	0	1	2	3	4
$2x$	-4	-2	0	2	4	6	8
-3	-3	-3	-3	-3	-3	-3	-3
y	-7	-5	-3	-1	1	3	5

(b) Draw and label axes using suitable scales.

(c) Plot the points and draw a pencil line through them. Label the line with its equation.

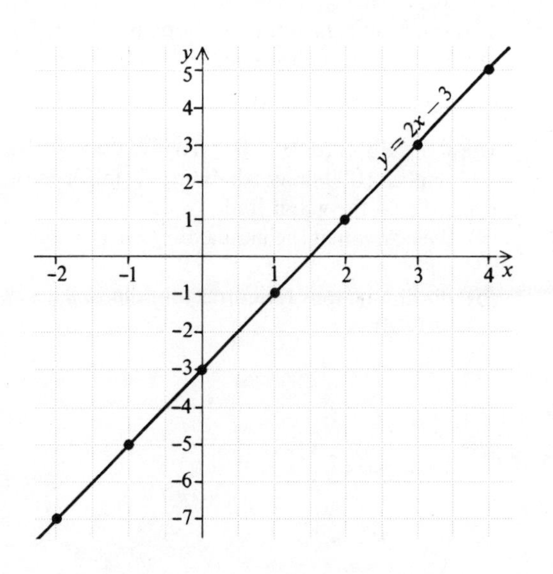

Exercise 2

Draw the following graphs, using a scale of 2 cm to 1 unit on the x-axis and 1 cm to 1 unit on the y-axis.

1. $y = 2x + 1$ for $-3 \leqslant x \leqslant 3$
2. $y = 3x - 4$ for $-3 \leqslant x \leqslant 3$
3. $y = 2x - 1$ for $-3 \leqslant x \leqslant 3$
4. $y = 8 - x$ for $-2 \leqslant x \leqslant 4$
5. $y = 10 - 2x$ for $-2 \leqslant x \leqslant 4$
6. $y = \dfrac{x+5}{2}$ for $-3 \leqslant x \leqslant 3$
7. $y = 3(x - 2)$ for $-3 \leqslant x \leqslant 3$
8. $y = \frac{1}{2}x + 4$ for $-3 \leqslant x \leqslant 3$
9. $v = 2t - 3$ for $-2 \leqslant t \leqslant 4$
10. $z = 12 - 3t$ for $-2 \leqslant t \leqslant 4$

In each question from **11** to **16**, draw the graphs on the same page and hence find the coordinates of the vertices of the polygon formed. Give the answers as accurately as your graph will allow.

11. (a) $y = x$ (b) $y = 8 - 4x$ (c) $y = 4x$
Take $-1 \leqslant x \leqslant 3$ and $-4 \leqslant y \leqslant 14$.

12. (a) $y = 2x + 1$ (b) $y = 4x - 8$ (c) $y = 1$
Take $0 \leqslant x \leqslant 5$ and $-8 \leqslant y \leqslant 12$.

13. (a) $y = 3x$ (b) $y = 5 - x$ (c) $y = x - 4$
Take $-2 \leqslant x \leqslant 5$ and $-9 \leqslant y \leqslant 8$.

14. (a) $y = -x$ (b) $y = 3x + 6$ (c) $y = 8$
(d) $x = 3\frac{1}{2}$
Take $-2 \leqslant x \leqslant 5$ and $-6 \leqslant y \leqslant 10$.

15. (a) $y = \frac{1}{2}(x - 8)$ (b) $2x + y = 6$
(c) $y = 4(x + 1)$
Take $-3 \leqslant x \leqslant 4$ and $-7 \leqslant y \leqslant 7$

16. (a) $y = 2x + 7$ (b) $3x + y = 10$
(c) $y = x$ (d) $2y + x = 4$
Take $-2 \leqslant x \leqslant 4$ and $0 \leqslant y \leqslant 13$.

17. The equation connecting the annual mileage, M miles, of a certain car and the annual running cost, £C is $C = \dfrac{M}{20} + 200$.
Draw the graph for $0 \leqslant M \leqslant 10\,000$ using scales of 1 cm for 1000 miles for M and 2 cm for £100 for C. From the graph find
(a) the cost when the annual mileage is 7200 miles,
(b) the annual mileage corresponding to a cost of £320.

18. The equation relating the cooking time t hours and the weight w kg for a joint of meat is $t = \dfrac{3w + 1}{2}$.
Draw the graph for $0 \leqslant w \leqslant 5$. From the graph find
(a) the weight of a joint requiring a cooking time of 2·8 hours,
(b) the cooking time for a joint of weight 4·4 kg.

19. Some drivers try to estimate their annual cost of repairs £c in relation to their average speed of driving s km/h using the equation $c = 6s + 50$.
Draw the graph for $0 \leqslant s \leqslant 160$. From the graph find
(a) the estimated repair bill for a man who drives at an average speed of 23 km/h,
(b) the average speed at which a motorist drives if his annual repair bill is £1000,
(c) the annual saving for a man who, on returning from a holiday, reduces his average speed of driving from 100 km/h to 65 km/h.

20. The value of a car £v is related to the number of miles n which it has travelled by the equation $v = 4500 - \dfrac{n}{20}$.
Draw the graph for $0 \leqslant n \leqslant 90\,000$. From the graph find
(a) the value of a car which has travelled 3700 miles,
(b) the number of miles travelled by a car valued at £3200,
(c) the decrease in value of a car when the mileometer reading changes from 28100 to 35700.

7.3 GRADIENTS

The gradient of a straight line is a measure of how steep it is.

Example 1

Find the gradient of the line joining the points A (1, 2) and B (6, 5).

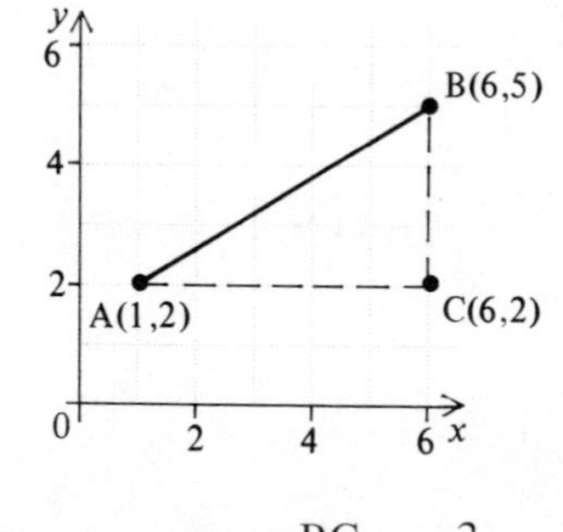

$$\text{gradient of AB} = \frac{BC}{AC} = \frac{3}{5}$$

It is possible to use the formula

$$\text{gradient} = \frac{\text{difference in } y\text{-coordinates}}{\text{difference in } x\text{-coordinates}}.$$

Example 2

Find the gradient of the line joining the points D (1, 5) and E (5, 2).

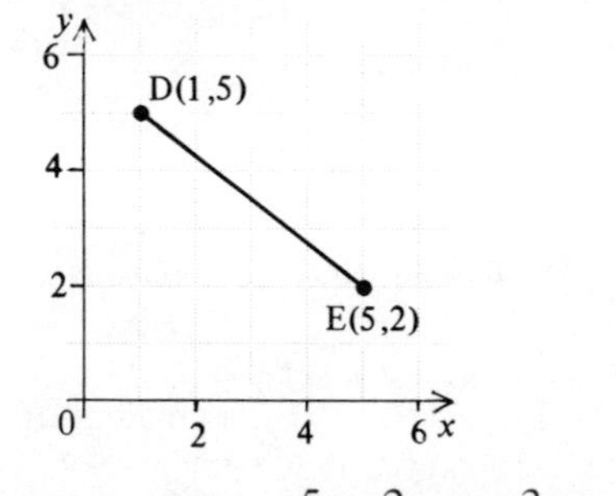

$$\text{gradient of DE} = \frac{5-2}{1-5} = \frac{3}{-4} = -\frac{3}{4}$$

Note

(a) Lines which slope upward to the right have a *positive* gradient.

(b) Lines which slope downward to the right have a *negative* gradient.

Exercise 3

Calculate the gradient of the line joining the following pairs of points.

1. $(3, 1)\,(5, 4)$
2. $(1, 1)\,(3, 5)$
3. $(3, 0)\,(4, 3)$
4. $(-1, 3), (1, 6)$
5. $(-2, -1)\,(0, 0)$
6. $(7, 5)\,(1, 6)$
7. $(2, -3)\,(1, 4)$
8. $(0, -2)\,(-2, 0)$
9. $(\frac{1}{2}, 1)\,(\frac{3}{4}, 2)$
10. $(-\frac{1}{2}, 1)\,(0, -1)$
11. $(3{\cdot}1, 2)\,(3{\cdot}2, 2{\cdot}5)$
12. $(-7, 10)\,(0, 0)$
13. $(\frac{1}{3}, 1)\,(\frac{1}{2}, 2)$
14. $(3, 4)\,(-2, 4)$
15. $(2, 5)\,(1{\cdot}3, 5)$
16. $(2, 3)\,(2, 7)$
17. $(-1, 4)\,(-1, 7{\cdot}2)$
18. $(2{\cdot}3, -2{\cdot}2)\,(1{\cdot}8, 1{\cdot}8)$
19. $(0{\cdot}75, 0)\,(0{\cdot}375, -2)$
20. $(17{\cdot}6, 1)\,(1{\cdot}4, 1)$
21. $(a, b)\,(c, d)$
22. $(m, n)\,(a, -b)$
23. $(2a, f)\,(a, -f)$
24. $(2k, -k)\,(k, 3k)$
25. $(m, 3n)\,(-3m, 3n)$
26. $\left(\frac{c}{2}, -d\right)\left(\frac{c}{4}, \frac{d}{2}\right)$

In questions **27** and **28**, find the gradient of each straight line.

27.

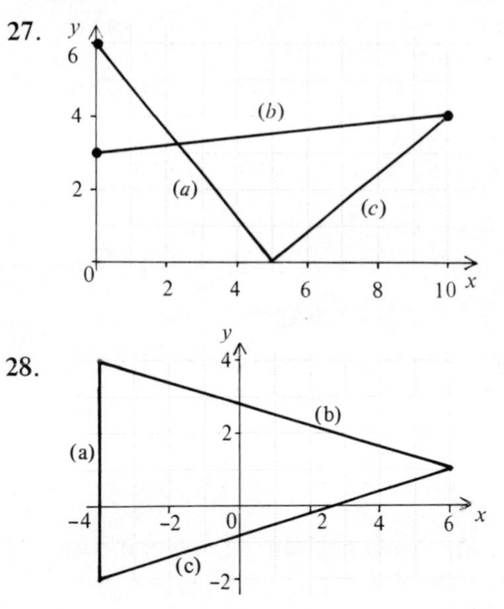

28.

29. Find the value of a if the line joining the points $(3a, 4)$ and $(a, -3)$ has a gradient of 1.

30. (a) Write down the gradient of the line joining the points $(2m, n)$ and $(3, -4)$,

(b) Find the value of n if the line is parallel to the x-axis,

(c) Find the value of m if the line is parallel to the y-axis.

7.4 THE FORM $y = mx + c$

When the equation of a straight line is written in the form $y = mx + c$, the gradient of the line is m and the intercept on the y-axis is c.

Example 1

Draw the line $y = 2x + 3$ on a *sketch* graph.

The word 'sketch' implies that we do not plot a series of points but simply show the position and slope of the line.

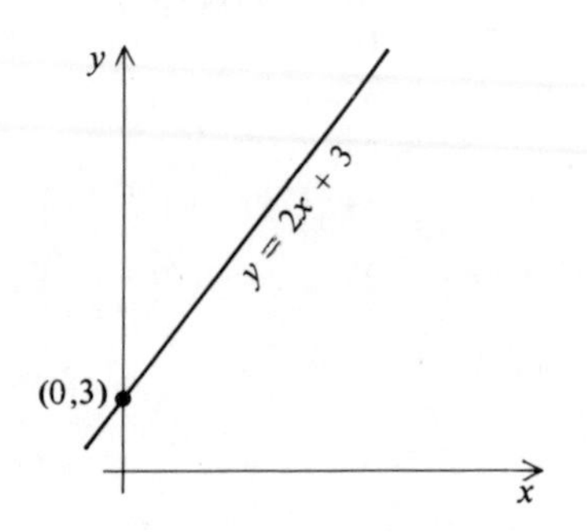

The line $y = 2x + 3$ has a gradient of 2 and cuts the y-axis at (0, 3).

Example 2

Draw the line $x + 2y - 6 = 0$ on a sketch graph.

(a) Rearrange the equation to make y the subject.

$$\begin{aligned} x + 2y - 6 &= 0 \\ 2y &= -x + 6 \\ y &= -\tfrac{1}{2}x + 3. \end{aligned}$$

(b) The line has a gradient of $-\frac{1}{2}$ and cuts the y-axis at (0, 3).

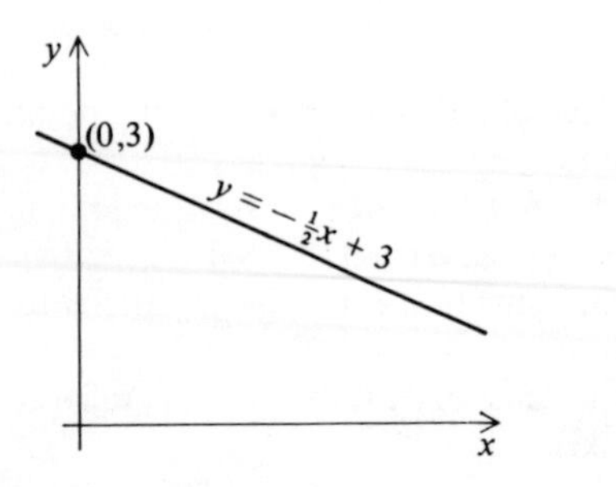

Finding the equation of a line

Example 3

Find the equation of the straight line which passes through (1, 3) and (3, 7).

(a) Let the equation of the line take the form $y = mx + c$.

The gradient, $m = \dfrac{7-3}{3-1} = 2$

so we may write the equation as

$$y = 2x + c \qquad \ldots [1]$$

(b) Since the line passes through (1, 3), substitute 3 for y and 1 for x in [1].

$$\begin{aligned} \therefore \quad 3 &= 2 \times 1 + c \\ 1 &= c \end{aligned}$$

The equation of the line is $y = 2x + 1$.

Exercise 4

In questions **1** to **20**, find the gradient of the line and the intercept on the y-axis. Hence draw a small sketch graph of each line.

1. $y = x + 3$
2. $y = x - 2$
3. $y = 2x + 1$
4. $y = 2x - 5$
5. $y = 3x + 4$
6. $y = \frac{1}{2}x + 6$
7. $y = 3x - 2$
8. $y = 2x$
9. $y = \frac{1}{4}x - 4$
10. $y = -x + 3$
11. $y = 6 - 2x$
12. $y = 2 - x$
13. $y + 2x = 3$
14. $3x + y + 4 = 0$
15. $2y - x = 6$
16. $3y + x - 9 = 0$
17. $4x - y = 5$
18. $3x - 2y = 8$
19. $10x - y = 0$
20. $y - 4 = 0$

In questions **21** to **31**, find the equation of the straight line which

21. Passes through (0, 7) at a gradient of 3
22. Passes through (0, −9) at a gradient of 2
23. Passes through (0, 5) at a gradient of −1
24. Passes through (2, 3) at a gradient of 2
25. Passes through (2, 11) at a gradient of 3
26. Passes through (4, 3) at a gradient of −1
27. Passes through (6, 0) at a gradient of $\frac{1}{2}$
28. Passes through (2, 1) and (4, 5)
29. Passes through (5, 4) and (6, 7)
30. Passes through (0, 5) and (3, 2)
31. Passes through (3, −3) and (9, −1)

In questions **32** to **34**, find the equation of each line.

32.

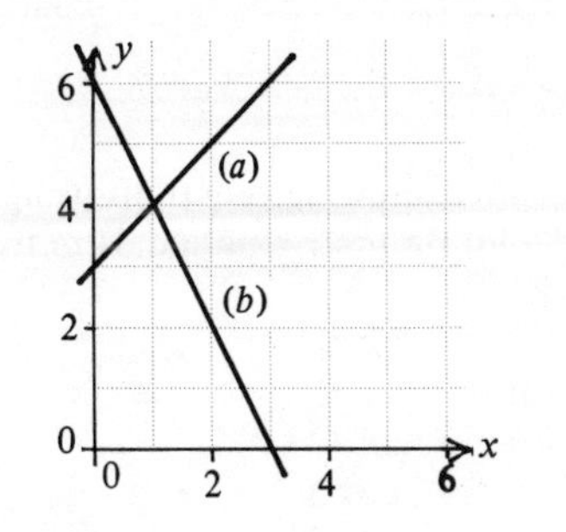

33.

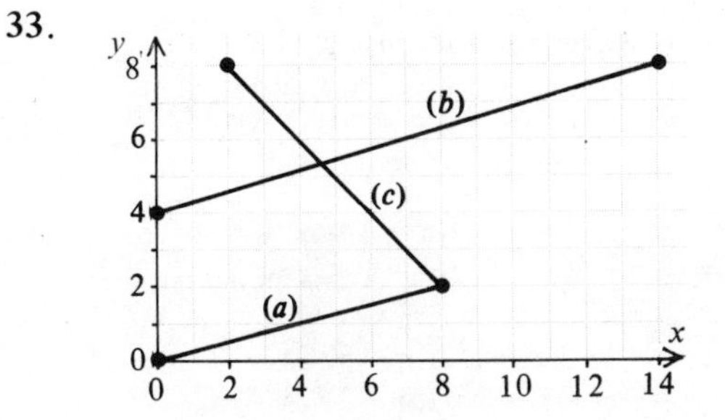

34.

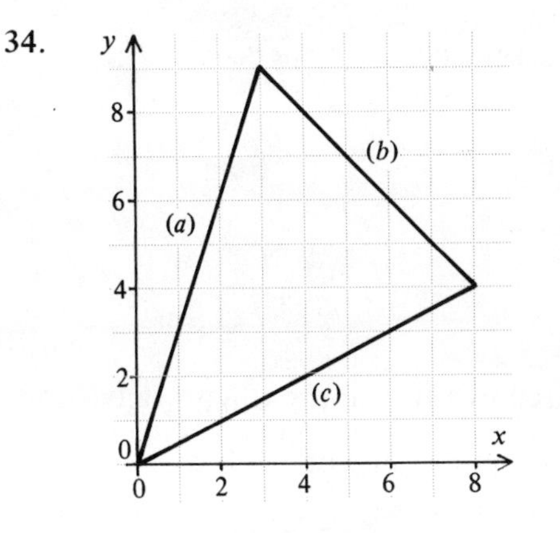

35. Variables v and t are connected by an equation of the form $v = mt + c$, where m and c are constants. In an experiment, the values of v and t were recorded as follows

t	1	3	5	7
v	8	14	20	26

Draw a graph with t on the horizontal axis and v on the vertical axis and hence find the values of m and c.

36. In an experiment, the following measurements of the variables q and t were taken.

q	0·5	1·0	1·5	2·0	2·5	3·0
t	3·85	5·0	6·1	7·0	7·75	9·1

A scientist suspects that q and t are related by an equation of the form $t = mq + c$, (m and c constants). Plot the values obtained from the experiment and draw the 'best' straight line through the points. Plot q on the horizontal axis with a scale of 4 cm to 1 unit, and t on the vertical axis with a scale of 2 cm to 1 unit. Find the gradient and intercept on the t-axis and hence estimate the values of m and c.

37. In an experiment, the following measurements of p and z were taken:

z	1·2	2·0	2·4	3·2	3·8	4·6
p	11·5	10·2	8·8	7·0	6	3·5

Plot the points on a graph with z on the horizontal axis and draw the 'best' straight line through the points. Hence estimate the values of n and k if the equation relating p and z is of the form $p = nz + k$.

38. In an experiment the following measurements of t and z were taken:

t	1·41	2·12	2·55	3·0	3·39	3·74
z	3·4	3·85	4·35	4·8	5·3	5·75

Draw a graph, plotting t^2 on the horizontal axis and z on the vertical axis, and hence confirm that the equation connecting t and z is of the form $z = mt^2 + c$. Find approximate values for m and c.

7.5 PLOTTING CURVES

Example 1

Draw the graph of the function
$y = 2x^2 + x - 6$, for $-3 \leqslant x \leqslant 3$.

(a)

x	-3	-2	-1	0	1	2	3
$2x^2$	18	8	2	0	2	8	18
x	-3	-2	-1	0	1	2	3
-6	-6	-6	-6	-6	-6	-6	-6
y	9	0	-5	-6	-3	4	15

(b) Draw and label axes using suitable scales.
(c) Plot the points and draw a smooth curve through them with a pencil.
(d) Check any points which interrupt the smoothness of the curve.
(e) Label the curve with its equation.

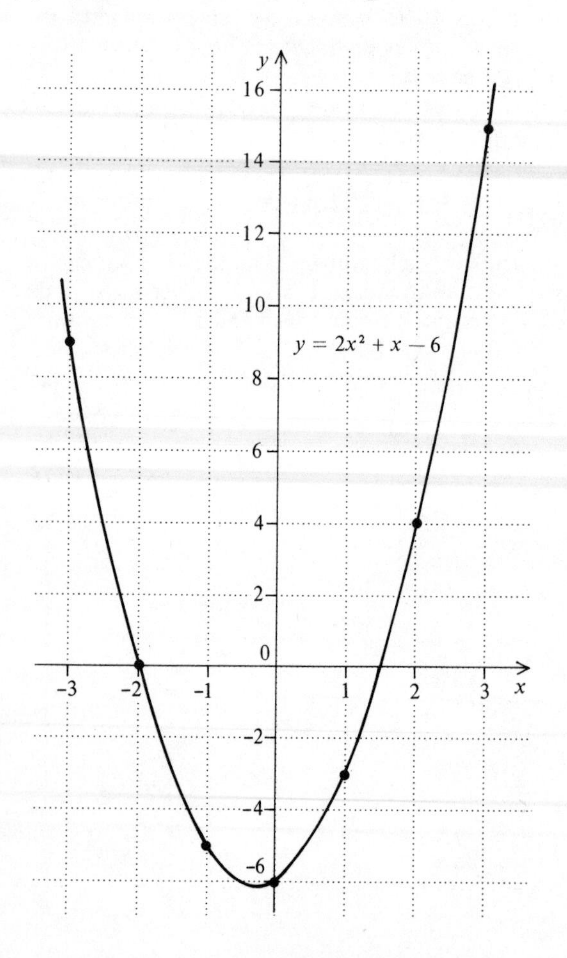

Exercise 5

Draw the graphs of the following functions using a scale of 2 cm for 1 unit on the x-axis and 1 cm for 1 unit on the y-axis.

1. $y = x^2 + 2x$, for $-3 \leqslant x \leqslant 3$.
2. $y = x^2 + 4x$, for $-3 \leqslant x \leqslant 3$
3. $y = x^2 - 3x$, for $-3 \leqslant x \leqslant 3$
4. $y = x^2 + 2$, for $-3 \leqslant x \leqslant 3$
5. $y = x^2 - 7$, for $-3 \leqslant x \leqslant 3$
6. $y = x^2 + x - 2$, for $-3 \leqslant x \leqslant 3$
7. $y = x^2 + 3x - 9$, for $-4 \leqslant x \leqslant 3$
8. $y = x^2 - 3x - 4$, for $-2 \leqslant x \leqslant 4$
9. $y = x^2 - 5x + 7$, for $0 \leqslant x \leqslant 6$
10. $y = 2x^2 - 6x$, for $-1 \leqslant x \leqslant 5$
11. $y = 2x^2 + 3x - 6$, for $-4 \leqslant x \leqslant 2$
12. $y = 3x^2 - 6x + 5$, for $-1 \leqslant x \leqslant 3$
13. $y = 2 + x - x^2$, for $-3 \leqslant x \leqslant 3$
14. $f(x) = 1 - 3x - x^2$, for $-5 \leqslant x \leqslant 2$
15. $f(x) = 3 + 3x - x^2$, for $-2 \leqslant x \leqslant 5$
16. $f(x) = 7 - 3x - 2x^2$, for $-3 \leqslant x \leqslant 3$
17. $f(x) = 6 + x - 2x^2$, for $-3 \leqslant x \leqslant 3$
18. $f: x \rightarrow 8 + 2x - 3x^2$, for $-2 \leqslant x \leqslant 3$
19. $f: x \rightarrow x(x - 4)$, for $-1 \leqslant x \leqslant 6$
20. $f: x \rightarrow (x + 1)(2x - 5)$, for $-3 \leqslant x \leqslant 3$.

Example 2

Draw the graph of $y = \dfrac{12}{x} + x - 6$, for $1 \leqslant x \leqslant 8$.
Use the graph to find approximate values for
(a) the minimum value of $\dfrac{12}{x} + x - 6$
(b) the value of $\dfrac{12}{x} + x - 6$, when $x = 2{\cdot}25$.
(c) the gradient of the tangent to the curve drawn at the point where $x = 5$.

Here is the table of values.

x	1	2	3	4	5	6	7	8	1·5
$\dfrac{12}{x}$	12	6	4	3	2·4	2	1·71	1·5	8
x	1	2	3	4	5	6	7	8	1·5
-6	-6	-6	-6	-6	-6	-6	-6	-6	-6
y	7	2	1	1	1·4	2	2·71	3·5	3·5

Notice that an 'extra' value of y has been calculated at $x = 1{\cdot}5$ because of the large difference between the y-values at $x = 1$ and $x = 2$.

(a) From the graph, the minimum value of $\frac{12}{x} + x - 6$ (i.e. y) is approximately 0·9.

(b) At $x = 2{\cdot}25$, y is approximately 1·6.

(c) The tangent AB is drawn to touch the curve at $x = 5$. The gradient of $AB = \frac{BC}{AC}$.

$$\text{gradient} = \frac{3}{8 - 2{\cdot}4} = \frac{3}{5{\cdot}6} \approx 0{\cdot}54$$

It is difficult to obtain an accurate value for the gradient of a tangent so the above result is more realistically 'approximately 0·5'.

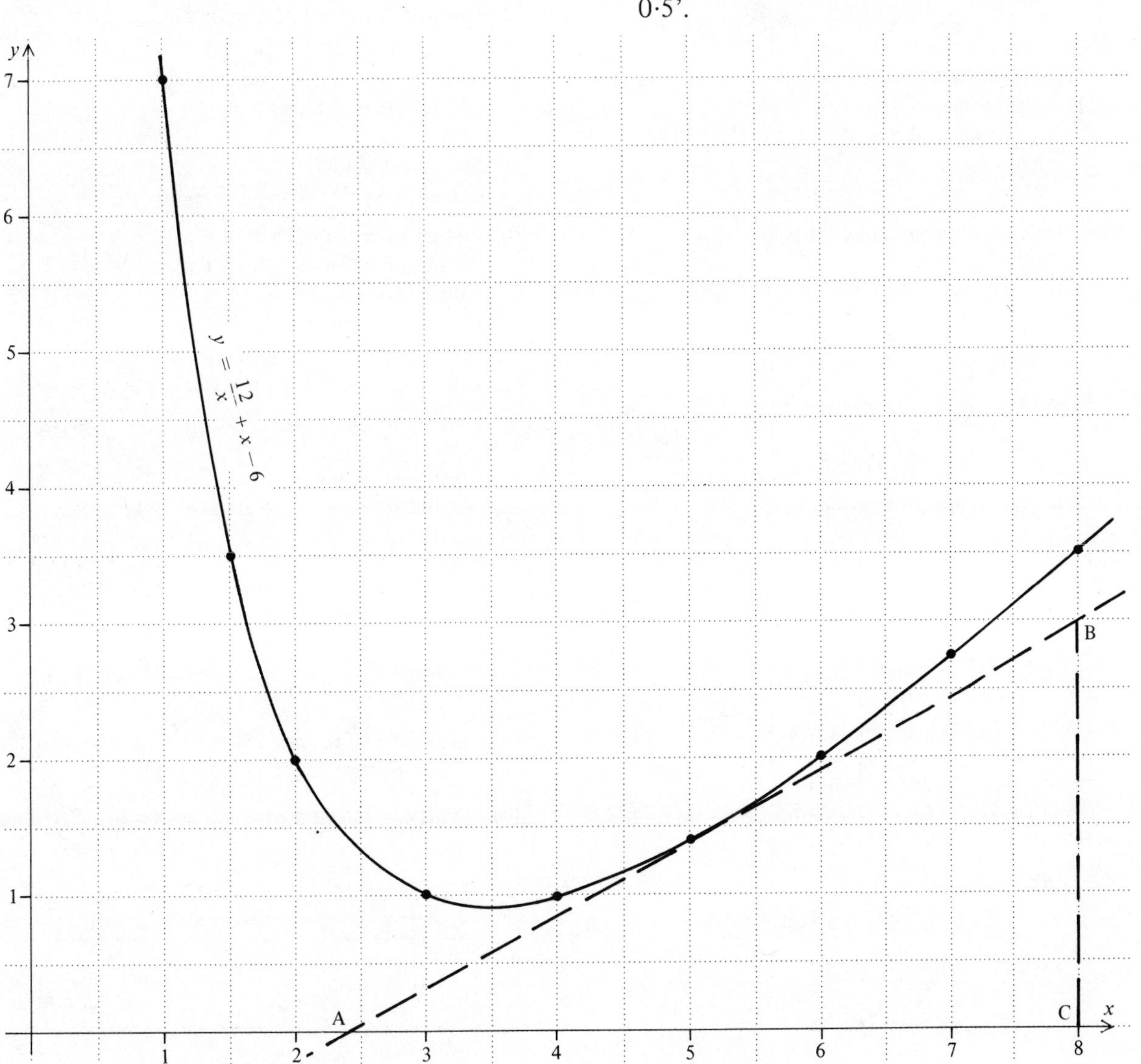

Exercise 6

Draw the following curves. The scales given are for one unit of x and y.

1. $y = x^2$, for $0 \leqslant x \leqslant 6$.
   (Scales: 2 cm for x, $\frac{1}{2}$ cm for y)
   Find
   (a) the gradient of the tangent to the curve at $x = 2$,
   (b) the gradient of the tangent to the curve at $x = 4$,
   (c) the y-value at $x = 3{\cdot}25$.

2. $y = x^2 - 3x$, for $-2 \leqslant x \leqslant 5$.
   (Scales: 2 cm for x, 1 cm for y)
   Find
   (a) the gradient of the tangent to the curve at $x = 3$,
   (b) the gradient of the tangent to the curve at $x = -1$,
   (c) the value of x where the gradient of the curve is zero.

3. $y = 5 + 3x - x^2$, for $-2 \leqslant x \leqslant 5$.
   (Scales: 2 cm for x, 1 cm for y)
   Find
   (a) the maximum value of the function $5 + 3x - x^2$,
   (b) the gradient of the tangent to the curve at $x = 2{\cdot}5$
   (c) the two values of x for which $y = 2$.

4. $y = \dfrac{12}{x}$, for $1 \leqslant x \leqslant 10$.
   (Scales: 1 cm for x and y)

5. $y = \dfrac{9}{x}$, for $1 \leqslant x \leqslant 10$.
   (Scales: 1 cm for x and y)

6. $y = \dfrac{12}{x+1}$, for $0 \leqslant x \leqslant 8$.
   (Scales: 2 cm for x, 1 cm for y)

7. $y = \dfrac{8}{x-4}$, for $-4 \leqslant x \leqslant 3{\cdot}5$.
   (Scales: 2 cm for x, 1 cm for y)

8. $y = \dfrac{15}{3-x}$, for $-4 \leqslant x \leqslant 2$.
   (Scales: 2 cm for x, 1 cm for y)

9. $y = \dfrac{x}{x+4}$, for $-3{\cdot}5 \leqslant x \leqslant 4$.
   (Scales: 2 cm for x and y)

10. $y = \dfrac{3x}{5-x}$, for $-3 \leqslant x \leqslant 4$.
    (Scales: 2 cm for x, 1 cm for y)

11. $y = \dfrac{x+8}{x+1}$, for $0 \leqslant x \leqslant 8$.
    (Scales: 2 cm for x and y)

12. $y = \dfrac{x-3}{x+2}$, for $-1 \leqslant x \leqslant 6$.
    (Scales: 2 cm for x and y)

13. $y = \dfrac{10}{x} + x$, for $1 \leqslant x \leqslant 7$.
    (Scales: 2 cm for x, 1 cm for y)

14. $y = \dfrac{12}{x} - x$, for $1 \leqslant x \leqslant 7$.
    (Scales: 2 cm for x, 1 cm for y)

15. $y = \dfrac{15}{x} + x - 7$, for $1 \leqslant x \leqslant 7$.
    (Scales: 2 cm for x and y)
    Find
    (a) the minimum value of y,
    (b) the y value when $x = 5{\cdot}5$.

16. $y = x^3 - 2x^2$, for $0 \leqslant x \leqslant 4$.
    (Scales: 2 cm for x, $\frac{1}{2}$ cm for y)
    Find
    (a) the y value at $x = 2{\cdot}5$
    (b) the x value at $y = 15$

17. $y = \frac{1}{10}(x^3 + 2x + 20)$, for $-3 \leqslant x \leqslant 3$.
    (Scales: 2 cm for x and y)
    Find
    (a) the x-value where $x^3 + 2x + 20 = 0$
    (b) the gradient of the tangent to the curve at $x = 2$.

18. Copy and complete the table for the function $y = 7 - 5x - 2x^2$, giving values of y correct to one decimal place.

x	-4	$-3{\cdot}5$	-3	$-2{\cdot}5$	-2	$-1{\cdot}5$
7	7	7		7		7
$-5x$	20	17·5		12·5		7·5
$-2x^2$	-32	$-24{\cdot}5$		$-12{\cdot}5$		$-4{\cdot}5$
y	5	0		7		10

x	-1	$-0{\cdot}5$	0	0·5	1	1·5	2
7		7		7		7	
$-5x$		2·5		$-2{\cdot}5$		$-7{\cdot}5$	
$-2x^2$		$-0{\cdot}5$		$-0{\cdot}5$		$-4{\cdot}5$	
y		9		4		-5	

Draw the graph, using a scale of 2 cm for x and

1 cm for y. Find
(a) the gradient of the tangent to the curve at $x = -2{\cdot}5$,
(b) the maximum value of y,
(c) the value of x at which this maximum value occurs.

19. Draw the graph of $y = \frac{x}{x^2+1}$, for $-6 \leqslant x \leqslant 6$.
(Scales: 1 cm for x, 10 cm for y)

20. Draw the graph of $E = \frac{5000}{x} + 3x$ for $10 \leqslant x \leqslant 80$. (Scales: 1 cm to 5 units for x and 1 cm to 25 units for E)
From the graph find,
(a) the minimum value of E,
(b) the value of x corresponding to this minimum value.
(c) the range of values of x for which E is less than 275.

21. A rectangle has a perimeter of 14 cm and length x cm. Show that the width of the rectangle is $(7-x)$ cm and hence that the area A of the rectangle is given by the formula $A = x(7-x)$. Draw the graph, plotting x on the horizontal axis with a scale of 2 cm to 1 unit, and A on the vertical axis with a scale of 1 cm to 1 unit. Take Take x from 0 to 7. From the graph find,
(a) the area of the rectangle when $x = 2{\cdot}25$ cm,
(b) the dimensions of the rectangle when its area is 9 cm^2,
(c) the maximum area of the rectangle,
(d) the length and width of the rectangle corresponding to the maximum area.
(e) what shape of rectangle has the largest area.

22. A ball is thrown in the air so that t seconds after it is thrown, its height h metres above its starting point is given by the function $h = 25t - 5t^2$. Draw the graph of the function for $0 \leqslant t \leqslant 6$, plotting t on the horizontal axis with a scale of 2 cm to 1 second, and h on the vertical axis with a scale of 2 cm for 10 metres.
Use the graph to find,
(a) the time when the ball is at its greatest height,
(b) the greatest height reached by the ball,
(c) the interval of time during which the ball is at a height of more than 30 m.

23. The velocity v m/s of a missile t seconds after launching is given by the equation $v = 54t - 2t^3$. Draw a graph, plotting t on the horizontal axis with a scale of 2 cm to 1 second, and v on the vertical axis with a scale of 1 cm for 10 m/s. Take values of t from 0 to 5.
Use the graph to find
(a) the maximum velocity reached,
(b) the time taken to accelerate to a velocity of 70 m/s,
(c) the interval of time during which the missile is travelling at more than 100 m/s.

In questions **24** to **32**, a calculator will make the working much easier.

24. Draw the graph of $y = 2^x$, for $-4 \leqslant x \leqslant 4$.
(Scales: 2 cm for x, 1 cm for y)

25. Draw the graph of $y = 3^x$, for $-3 \leqslant x \leqslant 3$.
(Scales: 2 cm for x, $\frac{1}{2}$ cm for y)
Find the gradient of the tangent to the curve at $x = 1$.

26. Consider the equation $y = \frac{1}{x}$.

When $x = \frac{1}{2}$, $y = \frac{1}{\frac{1}{2}} = 2$.

When $x = \frac{1}{100}$, $y = \frac{1}{\frac{1}{100}} = 100$.

As the denominator of $\frac{1}{x}$ gets smaller, the answer gets larger. An 'infinitely small' denominator gives an 'infinitely large' answer.

We write $\frac{1}{0} \rightarrow \infty$. '$\frac{1}{0}$ tends to an infinitely large number.'

Draw the graph of $y = \frac{1}{x}$ for $x = \pm 4, \pm 3, \pm 2, \pm 1, \pm 0{\cdot}5, \pm 0{\cdot}25$. (Scales: 2 cm for x and y)

27. Draw the graph of $y = x + \frac{1}{x}$ for $x = \pm 4, \pm 3, \pm 2, \pm 1, \pm 0{\cdot}5, \pm 0{\cdot}25$. (Scales: 2 cm for x and y)

28. Draw the graph of $y = x + \frac{1}{x^2}$ for $x = \pm 4, \pm 3, \pm 2, \pm 1, \pm 0{\cdot}5, \pm 0{\cdot}25$.
(Scales: 2 cm for x, 1 cm for y)

29. Draw the graph of $y = \frac{2^x}{x}$, for $x = \pm 4, \pm 3, \pm 2, \pm 1, \pm \frac{1}{2}, 0, 5, 6, 7$.
(Scales: 1 cm to 1 unit for x and y)

30. Draw the graph of $y = \frac{x^3}{3^x}$, for $x = -1, -\frac{1}{2}, 0, \frac{1}{2}, 1, 2, 3, 4, 5, 6, 7$.
(Scales: 2 cm to 1 unit for x, 5 cm to 1 unit for y)
(a) What is the maximum value of y?
(b) At what value of x does y have its maximum value?

31. Draw the graph of $y = \frac{x^4}{4^x}$, for $x = \pm 1, \pm\frac{3}{4}, \pm\frac{1}{2}, \pm\frac{1}{4}, 0, 1{\cdot}5, 2, 2{\cdot}5, 3, 4, 5, 6, 7$.
(Scales: 2 cm to 1 unit for x, 5 cm to 1 unit for y)
(a) For what values of x is the gradient of the function zero?
(b) For what values of x is $y = 0{\cdot}5$?

32. Draw the graph of $y = \frac{x^5}{5^x}$.
Choose the values of x which best illustrate the interesting behaviour of the function in the region of the origin.

7.6 GRAPHICAL SOLUTION OF EQUATIONS

Accurately drawn graphs enable us to find approximate solutions to a wide range of equations, many of which are impossible to solve exactly by 'conventional' methods.

Example 1

Draw the graph of the function

$$y = 2x^2 - x - 3$$

for $-2 \leqslant x \leqslant 3$. Use the graph to find approximate solutions to the following equations.

(a) $2x^2 - x - 3 = 6$
(b) $2x^2 - x = x + 5$

The table of values for $y = 2x^2 - x - 3$ is found. Note the 'extra' value at $x = \frac{1}{2}$.

x	-2	-1	0	1	2	3	$\frac{1}{2}$
$2x^2$	8	2	0	2	8	18	$\frac{1}{2}$
$-x$	2	1	0	-1	-2	-3	$-\frac{1}{2}$
-3	-3	-3	-3	-3	-3	-3	-3
y	7	0	-3	-2	3	12	-3

The graph drawn from this table is shown below.

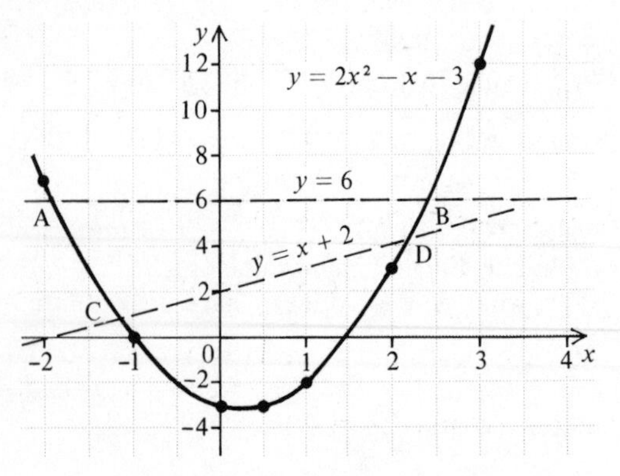

(a) To solve the equation $2x^2 - x - 3 = 6$, the line $y = 6$ is drawn. At the points of intersection (A and B), y simultaneously equals both 6 and $(2x^2 - x - 3)$.
So we may write

$$2x^2 - x - 3 = 6$$

The solutions are the x-values of the points A and B.
i.e. $x = -1{\cdot}9$ and $x = 2{\cdot}4$ approx.

(b) To solve the equation $2x^2 - x = x + 5$, we rearrange the equation to obtain the function $(2x^2 - x - 3)$ on the left-hand side.
In this case, subtract 3 from both sides.

$$2x^2 - x - 3 = x + 5 - 3$$
$$2x^2 - x - 3 = x + 2$$

If we now draw the line $y = x + 2$, the solutions of the equation are given by the x-values of C and D, the points of intersection.
i.e. $x = -1{\cdot}2$ and $x = 2{\cdot}2$ approx.

It is important to rearrange the equation to be solved so that the function already plotted is on one side.

Example 2

Assuming that the graph of $y = x^2 - 3x + 1$ has been drawn, find the equation of the line which should be drawn to solve the following equations.

(a) $x^2 - 4x + 3 = 0$
(b) $2x^2 - 6x = 5$
(c) $x + \frac{1}{x} = 4$

(a) Rearrange $x^2 - 4x + 3 = 0$ in order to obtain $(x^2 - 3x + 1)$ on the left-hand side.

$$x^2 - 4x + 3 = 0$$

add x $\quad x^2 - 3x + 3 = x$

subtract 2 $\quad x^2 - 3x + 1 = x - 2$

Therefore draw the line $y = x - 2$ to solve the equation.

(b) Rearrange $2x^2 - 6x = 5$ in order to obtain $(x^2 - 3x + 1)$ on the left-hand side.

$$2x^2 - 6x = 5$$

divide by 2 $\quad x^2 - 3x = 2\frac{1}{2}$

add 1 $\quad x^2 - 3x + 1 = 3\frac{1}{2}$

Therefore draw the line $y = 3\frac{1}{2}$ to solve the equation.

(c) Rearrange $x + \frac{1}{x} = 4$ in order to obtain $(x^2 - 3x + 1)$ on the left-hand side.

$$x + \frac{1}{x} = 4$$

multiply by x $\quad x^2 + 1 = 4x$

subtract $3x$ $\quad x^2 - 3x + 1 = x$

Therefore draw the line $y = x$ to solve the equation.

Exercise 7

1. In the diagram below, the graphs of $y = x^2 - 2x - 3$, $y = -2$ and $y = x$ have been drawn.

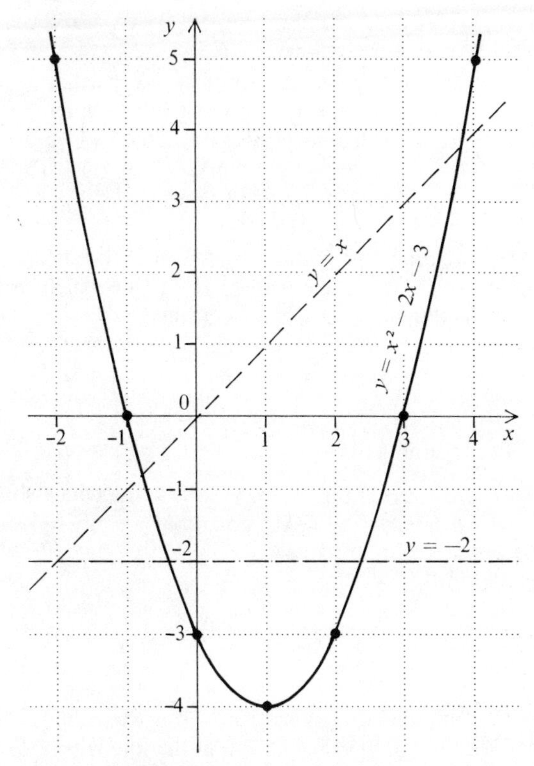

Use the graphs to find approximate solutions to the following equations.

(a) $x^2 - 2x - 3 = -2$
(b) $x^2 - 2x - 3 = x$
(c) $x^2 - 2x - 3 = 0$
(d) $x^2 - 2x - 1 = 0$

In questions **2** to **4**, use a scale of 2 cm to 1 unit for x and 1 cm to 1 unit for y.

2. Draw the graphs of the functions $y = x^2 - 3x + 10$ and $y = x + 7$ for $-1 \leqslant x \leqslant 5$. Hence find approximate solutions of the equation $x^2 - 3x + 10 = x + 7$.

3. Draw the graphs of the functions $y = x^2 + 3x$ and $y = 20 - x$ for $0 \leqslant x \leqslant 4$. Hence find an approximate solution of the equation $x^2 + 3x = 20 - x$.

4. Draw the graphs of the functions $y = 6x - x^2$ and $y = 2x + 1$ for $0 \leqslant x \leqslant 5$. Hence find approximate solutions of the equation $6x - x^2 = 2x + 1$.

In questions **5** to **9**, do *not* draw any graphs.

5. Assuming the graph of $y = x^2 - 5x$ has been drawn, find the equation of the line which should be drawn to solve the equations
(a) $x^2 - 5x = 3$ (b) $x^2 - 5x = -2$
(c) $x^2 - 5x = x + 4$ (d) $x^2 - 6x = 0$
(e) $x^2 - 5x - 6 = 0$

6. Assuming the graph of $y = x^2 + x + 1$ has been drawn, find the equation of the line which should be drawn to solve the equations
(a) $x^2 + x + 1 = 6$ (b) $x^2 + x + 1 = 0$
(c) $x^2 + x - 3 = 0$ (d) $x^2 - x + 1 = 0$
(e) $x^2 - x - 3 = 0$

7. Assuming the graph of $y = 6x - x^2$ has been drawn, find the equation of the line which should be drawn to solve the equations
(a) $4 + 6x - x^2 = 0$ (b) $4x - x^2 = 0$
(c) $2 + 5x - x^2 = 0$ (d) $x^2 - 6x = 3$
(e) $x^2 - 6x = -2$

8. Assuming the graph of $y = x + \frac{4}{x}$ has been drawn, find the equation of the line which should be drawn to solve the equations
(a) $x + \frac{4}{x} - 5 = 0$ (b) $\frac{4}{x} - x = 0$
(c) $x + \frac{4}{x} = 0{\cdot}2$ (d) $2x + \frac{4}{x} - 3 = 0$
(e) $x^2 + 4 = 3x$

9. Assuming the graph of $y = x^2 - 8x - 7$ has been drawn, find the equation of the line which should be drawn to solve the equations
(a) $x = 8 + \frac{7}{x}$ (b) $2x^2 = 16x + 9$
(c) $x^2 = 7$ (d) $x = \frac{4}{x - 8}$
(e) $2x - 5 = \frac{14}{x}$.

For questions **10** to **14**, use scales of 2 cm to 1 unit for for x and 1 cm to 1 unit for y.

10. Draw the graph of $y = x^2 - 2x + 2$ for $-2 \leqslant x \leqslant 4$. By drawing other graphs, solve the equations
(a) $x^2 - 2x + 2 = 8$
(b) $x^2 - 2x + 2 = 5 - x$
(c) $x^2 - 2x - 5 = 0$

11. Draw the graph of $y = x^2 - 7x$ for $0 \leqslant x \leqslant 7$. Draw suitable straight lines to solve the equations
(a) $x^2 - 7x + 9 = 0$
(b) $x^2 - 5x + 1 = 0$

12. Draw the graph of $y = x^2 + 4x + 5$ for $-6 \leqslant x \leqslant 1$. Draw suitable straight lines to find approximate solutions of the equations
(a) $x^2 + 3x - 1 = 0$ (b) $x^2 + 5x \div 2 = 0$

13. Draw the graph of $y = 2x^2 + 3x - 9$ for $-3 \leqslant x \leqslant 2$. Draw suitable straight lines to find approximate solutions of the equations
(a) $2x^2 + 3x - 4 = 0$
(b) $2x^2 + 2x - 9 = 1$

14. Draw the graph of $y = 2 + 3x - 2x^2$ for $-2 \leqslant x \leqslant 4$.
(a) Draw suitable straight lines to find approximate solutions of the equations,
(i) $2 + 4x - 2x^2 = 0$
(ii) $2x^2 - 3x - 2 = 0$
(b) Find the range of values of x for which $2 + 3x - 2x^2 \geqslant -5$.

15. Draw the graph of $y = \frac{18}{x}$ for $1 \leqslant x \leqslant 10$, using scales of 1 cm to one unit on both axes. Use the graph to solve approximately
(a) $\frac{18}{x} = x + 2$ (b) $\frac{18}{x} + x = 10$
(c) $x^2 = 18$

16. Draw the graph of $y = \frac{1}{2}x^2 - 6$ for $-4 \leqslant x \leqslant 4$, taking 2 cm to 1 unit on each axis.
(a) Use your graph to solve approximately the equation $\frac{1}{2}x^2 - 6 = 1$.
(b) Using tables or a calculator confirm that your solutions are approximately $\pm\sqrt{14}$ and explain why this is so.
(c) Use your graph to find the square roots of 8.

17. Draw the graph of $y = 6 - 2x - \frac{1}{2}x^3$ for $x = \pm 2, \pm 1\frac{1}{2}, \pm 1, \pm\frac{1}{2}, 0$. Take 4 cm to 1 unit for x and 1 cm to 1 unit for y.
Use your graph to find approximate solutions of the equations
(a) $\frac{1}{2}x^3 + 2x - 6 = 0$
(b) $x - \frac{1}{2}x^3 = 0$
Using tables confirm that two of the solutions to the equation in part (b) are $\pm\sqrt{2}$ and explain why this is so.

18. Draw the graph of $y = x + \frac{12}{x} - 5$ for $x = 1, 1\frac{1}{2}, 2, 3, 4, 5, 6, 7, 8$, taking 2 cm to 1 unit on each axis.
(a) From your graph find the range of values of x for which $x + \frac{12}{x} \leqslant 9$
(b) Find an approximate solution of the equation $2x - \frac{12}{x} - 12 = 0$.

19. Draw the graph of $y = 2\sin x^\circ + 1$ for $0 \leqslant x \leqslant 180^\circ$, taking 1 cm to 10° for x and 5 cm to 1 unit for y.
Find approximate solutions to the equations
(a) $2\sin x + 1 = 2{\cdot}3$
(b) $\frac{1}{(2\sin x + 1)} = 0{\cdot}5$

20. Draw the graph of $y = 2\sin x^\circ + \cos x^\circ$ for $0 \leqslant x \leqslant 180^\circ$, taking 1 cm to 10° for x and 5 cm to 1 unit for y.
(a) Solve approximately the equations
(i) $2\sin x + \cos x = 1{\cdot}5$
(ii) $2\sin x + \cos x = 0$
(b) Estimate the maximum value of y
(c) Find the value of x at which the maximum occurs.

21. Draw the graph of $y = 3\cos x^\circ - 4\sin x^\circ$ for $0^\circ \leqslant x \leqslant 220^\circ$, taking 1 cm to 10° for x and 2 cm to 1 unit for y.
(a) Solve approximately the equations
(i) $3\cos x - 4\sin x + 1 = 0$
(ii) $3\cos x = 4\sin x$
(b) Find the range of values of x for which $3\cos x - 4\sin x < -4$

22. Draw the graph of $y = 2^x$ for $-4 \leqslant x \leqslant 4$, taking 2 cm to one unit for x and 1 cm to one unit for y.
Find approximate solutions to the equations
(a) $2^x = 6$ (b) $2^x = 3x$ (c) $x2^x = 1$
Find also the approximate value of $2^{2{\cdot}5}$.

23. Draw the graph of $y = \frac{1}{x}$ for $-4 \leqslant x \leqslant 4$, taking 2 cm to one unit on each axis. Find approximate solutions to the equations
(a) $\frac{1}{x} = x + 1$
(b) $2x^2 - x - 1 = 0$

24. Draw the graph of $y = \frac{2^x}{x}$ for $-2 \leqslant x \leqslant 6$ (including $x = \pm\frac{1}{2}$), taking 2 cm to one unit for x and 1 cm to one unit for y.
(a) Find approximate solutions to the equation $\frac{2^x}{x} = x$
(b) Find the range of values of x for which $\frac{2^x}{x} > 2\frac{1}{2}$.

25. Draw the graph of $y = \frac{x^4}{4^x}$ for $-1 \leqslant x \leqslant 7$ (including $x = \pm\frac{3}{4}, \pm\frac{1}{2}, \pm\frac{1}{4}, 1{\cdot}5, 2{\cdot}5$), taking 2 cm to one unit for x and 5 cm to one unit for y.
(a) Find approximate solutions to the equations
(i) $2x^4 = 4^x$
(ii) $\frac{x^4}{4^x} = 1 - \frac{1}{4}x$.

REVISION EXERCISE 7A

1. Find which points do *not* lie on the curve given.
(a) $y = x^2 - 3x - 4$; $(1, -6), (0, 4), (2, -6), (-1, 0)$
(b) $y = x(x - 7)$; $(7, 7), (-1, 8), (0, 0), (6, -6)$
(c) $y = \frac{10}{x} + 2$; $(1, 12), (4, 5{\cdot}5), (5, 4), (-1, -12)$
(d) $y = 2^x$; $(0, 1), (1, 2), (2, 8), (4, 16)$
(e) $y = x^3 - 3x^2$; $(1, -2), (-1, 4), (2, 4), (3, 0)$

2. Find the equation of the straight line satisfied by the following points.

(a)

x	2	7	10
y	-5	0	3

(b)

x	1	2	3
y	7	9	11

(c)

x	1	2	3
y	8	6	4

(d)

x	3	4	5
y	2	$2\frac{1}{2}$	3

3. Find the gradient of the line joining each pair of points
(a) $(3, 3)\ (5, 7)$
(b) $(3, -1)\ (7, 3)$
(c) $(-1, 4)\ (1, -3)$
(d) $(2, 4)\ (-3, 4)$
(e) $(0{\cdot}5, -3)\ (0{\cdot}4, -4)$

4. Find the gradient and the intercept on the y-axis for the following lines. Draw a *sketch* graph of each line.
(a) $y = 2x - 7$
(b) $y = 5 - 4x$
(c) $2y = x + 8$
(d) $2y = 10 - x$
(e) $y + 2x = 12$
(f) $2x + 3y = 24$

5. In the diagram, the equations of the lines are $y = 3x, y = 6, y = 10 - x$ and $y = \frac{1}{2}x - 3$.

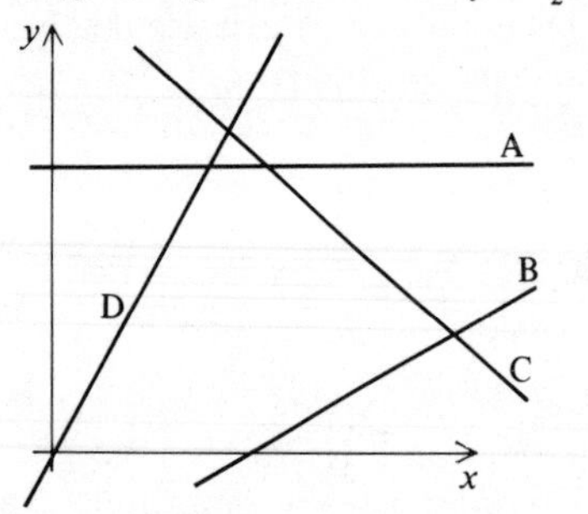

Find the equation corresponding to each line.

6. In the diagram, the equations of the lines are $2y = x - 8$, $2y + x = 8$, $4y = 3x - 16$ and $4y + 3x = 16$.

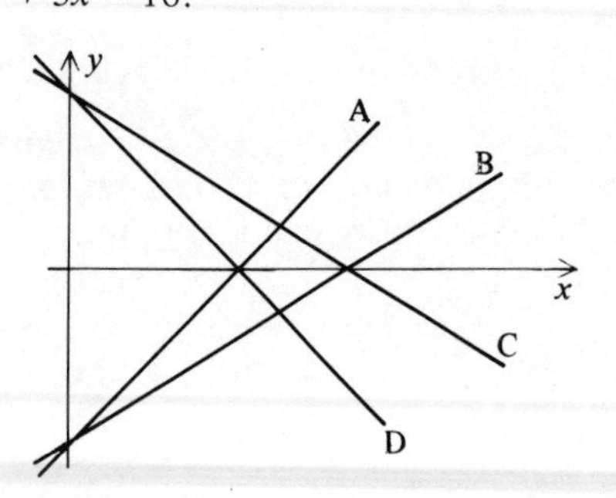

Find the equation corresponding to each line.

7. Find the equation of the line which passes through the following pairs of points,
   (a) (2, 1) (4, 5) (b) (0, 4), (−1, 1)
   (c) (2, 8) (−2, 12) (d) (0, 7) (−3, 7)

8. The sketch below represents a section of the curve $y = 8 + 7x - x^2$.

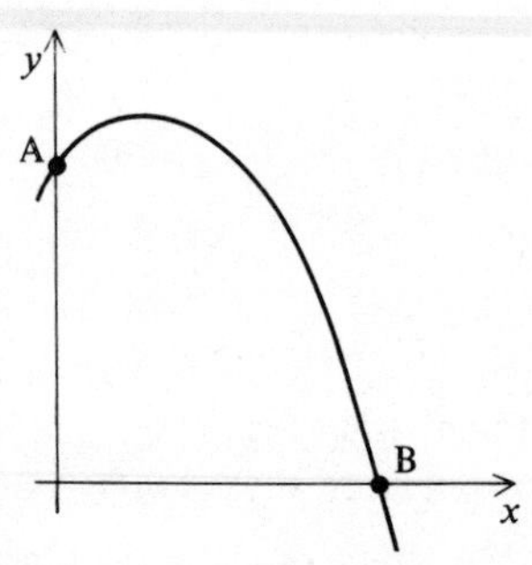

Calculate
(a) the coordinates of A and of B,
(b) the gradient of the line AB,
(c) the equation of the straight line AB.

9. The sketch represents a section of the curve $y = x^2 - 2x - 8$.

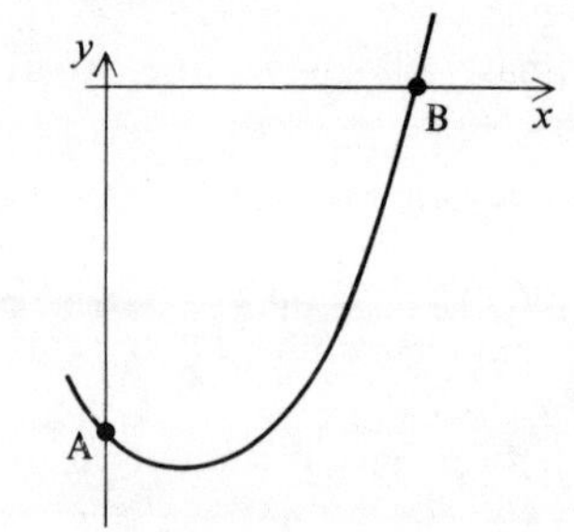

Calculate
(a) the coordinates of A and of B,
(b) the gradient of the line AB,
(c) the equation of the straight line AB.

10. Find from the diagram below
    (a) the equations of OM, ON and MN,
    (b) the area of triangle OMN.

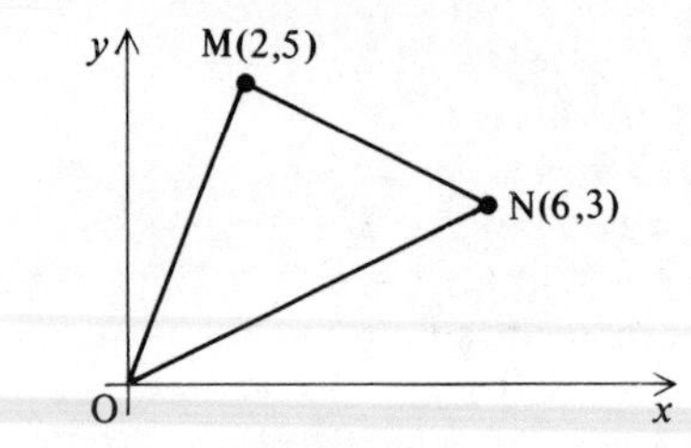

11. Find the equation of the straight line which passes through the point (5, 3) and is parallel to the line $y = 2x$.

12. Find the area of the triangle formed by the intersection of the lines $y = x$, $x + y = 10$ and $x = 0$.

13. Draw the graph of $y = 7 - 3x - 2x^2$ for $-4 \leqslant x \leqslant 2$.
    Find the gradient of the tangent to the curve at the point where the curve cuts the y-axis.

14. Draw the graph of $y = \dfrac{4000}{x} + 3x$ for $10 \leqslant x \leqslant 80$.
    Find the minimum value of y.

15. Draw the graph of $y = \dfrac{1}{x} + 2^x$ for $x = \frac{1}{4}, \frac{1}{2}, \frac{3}{4}, 1, 1\frac{1}{2}, 2$ and 3.

16. Assuming that the graph of $y = 4 - x^2$ has been drawn, find the equation of the straight line which should be drawn in order to solve the following equations
    (a) $4 - 3x - x^2 = 0$ (b) $\frac{1}{2}(4 - x^2) = 0$
    (c) $x^2 - x + 7 = 0$ (d) $\dfrac{4}{x} - x = 5$.

17. Assuming that the graph of $y = 2x - \frac{1}{x}$ has been drawn, find the equation of the straight line which should be drawn in order to solve the following equations
(a) $x - 3 - \frac{1}{x} = 0$ (b) $x = \frac{1}{2x}$
(c) $2x^2 - 3x - 1 = 0$ (d) $6x = \frac{3}{x} + 15$

18. Draw the graph of $y = 5 - x^2$ for $-3 \leqslant x \leqslant 3$, taking 2 cm to one unit for x and 1 cm to one unit for y.
Use the graph to find
(a) approximate solutions to the equation $4 - x - x^2 = 0$,
(b) the square roots of 5,
(c) the square roots of 7.

19. Draw the graph of $y = \frac{5}{x} + 2x - 3$, for $\frac{1}{2} \leqslant x \leqslant 7$, taking 2 cm to one unit for x and 1 cm to one unit for y.
Use the graph to find
(a) approximate solutions to the equation $2x^2 - 10x + 7 = 0$,
(b) the range of values of x for which $\frac{5}{x} + 2x - 3 < 6$.
(c) the minimum value of y.

20. Draw the graph of $y = 4^x$ for $-2 \leqslant x \leqslant 2$.
Use the graph to find
(a) the approximate value of $4^{1\cdot 6}$,
(b) the approximate value of $4^{-\frac{1}{3}}$,
(c) the gradient of the curve at $x = 0$,
(d) an approximate solution to the equation $4^x = 10$.

EXAMINATION EXERCISE 7B

1. (i) Copy and complete the following table.

x	1	1·5	2	3	4	5	6
$\frac{12}{x}$			6				2
$y = \frac{12}{x} - x$			4				-4

(ii) Taking as scales 2 cm to represent 1 unit on the x-axis and 2 cm to represent 2 units on the y-axis, use the calculated values to draw the graph of $y = \frac{12}{x} - x$ from $x = 1$ to $x = 6$.
(iii) Using the same scales and axes, draw the graph of the line $y = 2x + 3$, plotting points for the values $x = 0, x = 2$ and $x = 4$.
(iv) Simplify the equation $\frac{12}{x} - x = 2x + 3$ into the form $x^2 + bx + c = 0$.
Use the graph to find one root of this equation. [NEA]

2. Given that $y = 30 + 4x - 3x^2$, copy and complete the following table:

x	-3	-2	-1	0	1	2	3	4	5
y	-9	10		30		26		-2	-25

(i) Taking 2 cm to represent 1 unit on the x-axis and 2 cm to represent 10 units on the y-axis, draw the graph of $y = 30 + 4x - 3x^2$ from $x = -3$ to $x = 5$.
(ii) Using the same axes and same scales draw the graph of $y + 2x = 10$.
(iii) Write down and simplify the equation in x which can be solved by using the points of intersection of the two graphs and use your graphs to state the values of x which satisfy this equation.
(iv) By drawing a tangent, or otherwise, find the gradient of the curve $y = 30 + 4x - 3x^2$ at the point where $x = 1{\cdot}5$. [AEB]

3. A piece of wire 80 metres long is cut into two pieces. One piece is $12x$ metres long and is bent to form a rectangle of length $4x$ metres and width $2x$ metres. The other piece is bent to form a square. Find, in terms of x,

(i) the area of the rectangle,
(ii) the length of a side of the square,
(iii) the area of the square.

Show that the combined area, y square metres, of the rectangle and the square, is given by

$$y = 17x^2 - 120x + 400.$$

Corresponding values of x and y are given below.

x	1	2	3	4	5	6
y		228	193	192	225	

Calculate, and write down, the value of y when $x = 1$, and when $x = 6$.

Using a scale of 2 cm to 1 unit on the x-axis and 1 cm to 20 units on the y-axis draw the graph of $y = 17x^2 - 120x + 400$, for values of x in the range $1 \leqslant x \leqslant 6$.

Use your graph to find the value of x for which the combined area is smallest. [C]

4. A pottery makes a profit of y thousand pounds from x thousand plates where

$$y = 4x - x^2 - 1.$$

Corresponding values of y and x are given in the table below.

x	0	0·5	1·0	1·5	2·0	2·5	3·0
y	−1·0	0·75	2·0	2·75	3·0	2·75	2·0

Using a scale of 4 cm to represent one unit on each axis, draw the graph of $y = 4x - x^2 - 1$ for $0 \leqslant x \leqslant 3$.

Use your graph to find

(i) the number of plates the company should produce in order to obtain the maximum profit.
(ii) the minimum number of plates that should be produced in order to cover the cost of production.
(iii) the range of values of x for which the profit is more than £2850.

The profit is £$\frac{y}{x}$ per plate. By drawing the appropriate straight line on your graph, determine the smaller number of plates that would have to be produced in order to make a profit of £1 per plate. [C]

5. Using a scale of 2 cm to represent 1 unit on the x-axis and 4 cm to represent 1 unit on the y-axis, draw the graph of the function defined by

$$f : x \to \frac{4}{x} + \frac{x}{4} \text{ for } 1 \leqslant x \leqslant 7.$$

(a) Draw the tangent to the graph at the point where $x = 2·5$ and find its gradient.
(b) Using the same scale and axes, draw the graph of the function defined by

$$g : x \to -\tfrac{1}{3}x + 4.$$

(c) From your graph find the values of x for which $f(x) = g(x)$, and form, but do not simplify, an equation of which these values are the roots. [L]

6. Draw the graph of

$$y = 6x - x^2$$

for values of x from 0 to 6 inclusive, using a scale of 2 cm to 1 unit on each axis.
On the same axes and with the same scales draw the graph of

$$y = \frac{6}{x}$$

for values of x from 1 to 6 inclusive.
From your graphs estimate the range of positive values of x for which
(i) $6x - x^2$ is greater than 6,
(ii) $\frac{6}{x}$ is less than $6x - x^2$. [JMB]

7. Copy and complete the given table of values of the function

$$y = \frac{x^2}{2} + \frac{18}{x} - 10,$$

for the values indicated. In the table values have been corrected to one place of decimals.

x	1	1·5	2	3	4	4·5	5
$\frac{x^2}{2}$	0·5	1·1		4·5		10·1	
$\frac{18}{x}$	18	12		6		4	
-10	-10	-10		-10		-10	
y	8·5	3·1		0·5		4·1	

Draw the graph of the function for values of x from 1 to 5 using scales of 2 cm to 1 unit on both axes.
(a) From your graph, find the value of x giving the least value of y and this least value of y.
(b) By drawing a suitable straight line graph, find the positive values of x which satisfy $\frac{x^2}{2} + \frac{18}{x} - 10 = x$. [L]

8. Using a scale of 2 cm to 1 unit on the x-axis and 1 cm to 1 unit on the y-axis, draw the graph of

$$y = (x-2)(x+1)$$

for values of x from -3 to $+4$ inclusive.
Using the same axes and scales, draw the graph of the straight line

$$y = x + 3.$$

From your graphs, estimate the solutions of the following equations:
(i) $(x-2)(x+1) = -1$,
(ii) $x^2 - x - 7 = 0$,
(iii) $x^2 - 2x - 5 = 0$. [JMB]

9. Taking 2 cm to represent 1 unit on each axis, draw the graph of

$$y = \frac{6x}{x+1},$$

plotting the points for which x is $0, \frac{1}{2}, 1, 2, 3, 4$ and 5.

Using the same axes and scales, draw the graph of $y = x + 1$.

From your graphs, find

(i) the range of positive values of x for which $\frac{6x}{x+1} < \frac{7}{2}$,

(ii) the range of positive values of x for which $\frac{6x}{x+1} > x + 1$,

(iii) the solutions of the equation $6x = (x + 1)^2$. [O & C]

10. Measurements taken from a metronome are tabulated below:

Distance of top of weight from pivot x (mm)	35	51	67	75	83	91	99
Indicated frequency y (beats per minute)	200	160	120	100	80	60	40

(i) Taking scales of 1 cm to 10 mm on the x-axis and 1 cm to 10 beats per minute on the y-axis, plot these points and draw a line through them.

(ii) From your graph, estimate

(a) the frequency for a distance of 60 mm,

(b) the distance to produce a frequency of 90 beats per minute.

(iii) By considering two points on the line, or otherwise, obtain its equation. [SMP]

8 Matrices and transformations

Albert Einstein (1879–1955) working as a patent office clerk in Berne, was responsible for the greatest advance in mathematical physics of this century. His theories of relativity, put forward in 1905 and 1915 were based on the postulate that the velocity of light is absolute: mass, length and even time can only be measured relative to the observer and undergo transformation when studied by another observer. His formula $E = mc^2$ laid the foundations of nuclear physics, a fact that he came to deplore in its application to warfare. In 1933 he moved from Nazi Germany and settled in America.

8.1 MATRIX OPERATIONS

Order

The order of a matrix describes the matrix in terms of the number of rows and columns which it has.

Example 1

$$\begin{pmatrix} 1 & 2 & 4 \\ -1 & 0 & 3 \end{pmatrix}$$

This matrix has 2 rows and 3 columns, and has order 2×3.

Addition and subtraction

Matrices of the same order are added (or subtracted) by adding (or subtracting) the corresponding elements in each matrix.

Example 2

$$\begin{pmatrix} 2 & -4 \\ 3 & 0 \end{pmatrix} + \begin{pmatrix} 3 & 5 \\ -1 & 7 \end{pmatrix} = \begin{pmatrix} 5 & 1 \\ 2 & 7 \end{pmatrix}$$

Multiplication by a number

Each element of the matrix is multiplied by the multiplying number.

Example 3

$$3 \times \begin{pmatrix} 2 & -1 \\ 1 & 4 \end{pmatrix} = \begin{pmatrix} 6 & -3 \\ 3 & 12 \end{pmatrix}$$

Multiplication by another matrix

For 2 X 2 matrices,

$$\begin{pmatrix} a & b \\ c & d \end{pmatrix}\begin{pmatrix} w & x \\ y & z \end{pmatrix} = \begin{pmatrix} aw+by & ax+bz \\ cw+dy & cx+dz \end{pmatrix}$$

The same process is used for matrices of other orders.

Example 4

Perform the following multiplications.

(a) $\begin{pmatrix} 3 & 2 \\ 4 & 1 \end{pmatrix}\begin{pmatrix} 2 & 1 \\ 1 & 5 \end{pmatrix}$ (b) $\begin{pmatrix} 2 & 1 & -2 \\ 0 & 1 & 3 \end{pmatrix}\begin{pmatrix} 1 & 0 \\ 1 & -2 \\ 4 & 3 \end{pmatrix}$

(a) $$\begin{pmatrix} 3 & 2 \\ 4 & 1 \end{pmatrix}\begin{pmatrix} 2 & 1 \\ 1 & 5 \end{pmatrix} = \begin{pmatrix} 6+2 & 3+10 \\ 8+1 & 4+5 \end{pmatrix} = \begin{pmatrix} 8 & 13 \\ 9 & 9 \end{pmatrix}$$

(b) $$\begin{pmatrix} 2 & 1 & -2 \\ 0 & 1 & 3 \end{pmatrix}\begin{pmatrix} 1 & 0 \\ 1 & -2 \\ 4 & 3 \end{pmatrix} = \begin{pmatrix} 2+1-8 & 0-2-6 \\ 0+1+12 & 0-2+9 \end{pmatrix} = \begin{pmatrix} -5 & -8 \\ 13 & 7 \end{pmatrix}$$

Matrices may be multiplied only if they are *compatible*. The number of *columns* in the left-hand matrix must equal the number of *rows* in the right-hand matrix.

Matrix multiplication is not commutative, i.e. for square matrices **A** and **B**, the product **AB** does not necessarily equal the product **BA**.

Exercise 1

In questions **1** to **36**, the matrices have the following values.

$\mathbf{A} = \begin{pmatrix} 2 & -1 \\ 3 & 4 \end{pmatrix}$; $\mathbf{B} = \begin{pmatrix} 0 & 5 \\ 1 & -2 \end{pmatrix}$; $\mathbf{C} = \begin{pmatrix} 4 & 3 \\ 1 & -2 \end{pmatrix}$; $\mathbf{D} = \begin{pmatrix} 1 & 5 & 1 \\ 4 & -6 & 1 \end{pmatrix}$;

$\mathbf{E} = \begin{pmatrix} 1 & 0 \\ -1 & 1 \\ 2 & 5 \end{pmatrix}$; $\mathbf{F} = (4 \quad 5)$; $\mathbf{G} = \begin{pmatrix} 4 \\ 1 \\ 3 \end{pmatrix}$; $\mathbf{H} = \begin{pmatrix} 0 & 1 & -2 \\ 3 & -4 & 5 \end{pmatrix}$;

$\mathbf{J} = \begin{pmatrix} 3 \\ 1 \end{pmatrix}$; $\mathbf{K} = \begin{pmatrix} 1 & -3 \\ 0 & 1 \\ -7 & 0 \end{pmatrix}$

Calculate the resultant value for each question where possible.

1. A + B **2.** D + H **3.** J + F **4.** B − C
5. 2F **6.** 3B **7.** K − E **8.** 2A + B
9. G − J **10.** C + B + A **11.** 2E − 3K **12.** $\frac{1}{2}$A − B
13. AB **14.** BA **15.** BC **16.** CB
17. DG **18.** AJ **19.** HK **20.** (AB)C
21. A(BC) **22.** AF **23.** CK **24.** GF
25. B(2A) **26.** (D + H)G **27.** JF **28.** FJ
29. (A − C)D **30.** A^2 **31.** A^4 **32.** E^2
33. KH **34.** (CA)J **35.** ED **36.** B^4

In questions **37** to **46**, find the value of the letters.

37. $\begin{pmatrix} 2 & x \\ y & 7 \end{pmatrix} + \begin{pmatrix} 4 & y \\ -3 & 2 \end{pmatrix} = \begin{pmatrix} x & 9 \\ z & 9 \end{pmatrix}$ **38.** $\begin{pmatrix} x & 2 \\ -1 & -2 \\ w & 3 \end{pmatrix} + \begin{pmatrix} x & y \\ y & -3 \\ v & 5 \end{pmatrix} = \begin{pmatrix} 8 & z \\ x & w \\ w & 8 \end{pmatrix}$

39. $\begin{pmatrix} a & b \\ c & 0 \end{pmatrix} - \begin{pmatrix} 2 & 5 \\ -3 & d \end{pmatrix} = 2\begin{pmatrix} 1 & a \\ b & -1 \end{pmatrix}$ 40. $\begin{pmatrix} x & 3 \\ -2 & y \end{pmatrix}\begin{pmatrix} 2 \\ 1 \end{pmatrix} = \begin{pmatrix} 5 \\ 0 \end{pmatrix}$

41. $\begin{pmatrix} 2 & 0 \\ 0 & -3 \end{pmatrix}\begin{pmatrix} m \\ n \end{pmatrix} = \begin{pmatrix} 10 \\ 1 \end{pmatrix}$ 42. $\begin{pmatrix} p & 2 & -1 \\ q & -2 & 2q \end{pmatrix}\begin{pmatrix} 2 \\ 1 \\ 3 \end{pmatrix} = \begin{pmatrix} 5 \\ -10 \end{pmatrix}$

43. $\begin{pmatrix} 3 & 0 \\ 2 & x \end{pmatrix}\begin{pmatrix} y & z \\ 4 & 0 \end{pmatrix} = \begin{pmatrix} 6 & -3 \\ 8 & w \end{pmatrix}$ 44. $\begin{pmatrix} 3y & 3z \\ 2y+4x & 2z \end{pmatrix} = \begin{pmatrix} 6 & -3 \\ 8 & w \end{pmatrix}$

45. $\begin{pmatrix} 2 & e \\ a & 3 \end{pmatrix} + k\begin{pmatrix} 3 & 1 \\ 0 & -2 \end{pmatrix} = \begin{pmatrix} 8 & 6 \\ -3 & -1 \end{pmatrix}$ 46. $\begin{pmatrix} 4 & 0 \\ 1 & m \end{pmatrix}\begin{pmatrix} n & p \\ -2 & 0 \end{pmatrix} = \begin{pmatrix} 20 & 12 \\ -1 & q \end{pmatrix}$

47. If $\mathbf{A} = \begin{pmatrix} 1 & 0 \\ 3 & 2 \end{pmatrix}$, $\mathbf{B} = \begin{pmatrix} x & 0 \\ 1 & 3 \end{pmatrix}$, and $\mathbf{AB} = \mathbf{BA}$, find x.

48. If $\mathbf{X} = \begin{pmatrix} k & 2 \\ 2 & -k \end{pmatrix}$ and $\mathbf{X}^2 = 5\begin{pmatrix} 1 & 0 \\ 0 & 1 \end{pmatrix}$, find k.

49. $\mathbf{B} = \begin{pmatrix} 3 & 3 \\ -1 & -1 \end{pmatrix}$
(a) Find k if $\mathbf{B}^2 = k\mathbf{B}$
(b) Find m if $\mathbf{B}^4 = m\mathbf{B}$

50. $\mathbf{A} = \begin{pmatrix} 5 & 5 \\ -2 & -2 \end{pmatrix}$
(a) Find n if $\mathbf{A}^2 = n\mathbf{A}$
(b) Find q if $\mathbf{A}^3 = q\mathbf{A}$

8.2 THE INVERSE OF A MATRIX

The inverse of a matrix $\mathbf{A}$ is written $\mathbf{A}^{-1}$, and the inverse exists if

$$\mathbf{AA}^{-1} = \mathbf{A}^{-1}\mathbf{A} = \mathbf{I}$$

where $\mathbf{I}$ is called the identity matrix.

Only square matrices possess an inverse.

For 2 × 2 matrices, $\mathbf{I} = \begin{pmatrix} 1 & 0 \\ 0 & 1 \end{pmatrix}$

For 3 × 3 matrices, $\mathbf{I} = \begin{pmatrix} 1 & 0 & 0 \\ 0 & 1 & 0 \\ 0 & 0 & 1 \end{pmatrix}$, etc.

If $\mathbf{A} = \begin{pmatrix} a & b \\ c & d \end{pmatrix}$, the inverse $\mathbf{A}^{-1}$ is given by

$$\mathbf{A}^{-1} = \frac{1}{(ad - cb)}\begin{pmatrix} d & -b \\ -c & a \end{pmatrix}$$

Here, the number $(ad - cb)$ is called the *determinant* of the matrix and is written $|\mathbf{A}|$.

If $|\mathbf{A}| = 0$, then the matrix has no inverse.

Example 1

Find the inverse of $\mathbf{A} = \begin{pmatrix} 3 & -4 \\ 1 & -2 \end{pmatrix}$.

$$\mathbf{A}^{-1} = \frac{1}{[3(-2) - 1(-4)]}\begin{pmatrix} -2 & 4 \\ -1 & 3 \end{pmatrix}$$
$$= \frac{1}{-2}\begin{pmatrix} -2 & 4 \\ -1 & 3 \end{pmatrix}$$

Check: $\mathbf{A}^{-1}\mathbf{A} = \frac{1}{-2}\begin{pmatrix} -2 & 4 \\ -1 & 3 \end{pmatrix}\begin{pmatrix} 3 & -4 \\ 1 & -2 \end{pmatrix}$
$$= \frac{-1}{2}\begin{pmatrix} -2 & 0 \\ 0 & -2 \end{pmatrix}$$
$$= \begin{pmatrix} 1 & 0 \\ 0 & 1 \end{pmatrix}$$

Multiplying by the inverse of a matrix gives the same result as dividing by the matrix: the effect is similar to ordinary algebraic operations.

e.g. if $\mathbf{AB} = \mathbf{C}$
$\mathbf{A}^{-1}\mathbf{AB} = \mathbf{A}^{-1}\mathbf{C}$
$\mathbf{B} = \mathbf{A}^{-1}\mathbf{C}$

Exercise 2

In questions **1** to **15**, find the inverse of the matrix.

1. $\begin{pmatrix} 4 & 1 \\ 3 & 1 \end{pmatrix}$ **2.** $\begin{pmatrix} 1 & 2 \\ 2 & 5 \end{pmatrix}$ **3.** $\begin{pmatrix} 3 & 4 \\ 1 & 2 \end{pmatrix}$ **4.** $\begin{pmatrix} 5 & 2 \\ 1 & 1 \end{pmatrix}$ **5.** $\begin{pmatrix} 2 & -2 \\ -1 & 2 \end{pmatrix}$

6. $\begin{pmatrix} 4 & -3 \\ -1 & 2 \end{pmatrix}$ **7.** $\begin{pmatrix} 2 & 1 \\ -2 & 3 \end{pmatrix}$ **8.** $\begin{pmatrix} 0 & -3 \\ 2 & 4 \end{pmatrix}$ **9.** $\begin{pmatrix} -1 & -2 \\ 1 & -3 \end{pmatrix}$ **10.** $\begin{pmatrix} 2 & 4 \\ 1 & 2 \end{pmatrix}$

11. $\begin{pmatrix} 3 & -2 \\ 1 & 4 \end{pmatrix}$ **12.** $\begin{pmatrix} -3 & 1 \\ 2 & 1 \end{pmatrix}$ **13.** $\begin{pmatrix} 2 & -3 \\ 1 & -4 \end{pmatrix}$ **14.** $\begin{pmatrix} 7 & 0 \\ -5 & 1 \end{pmatrix}$ **15.** $\begin{pmatrix} 2 & 1 \\ -2 & -4 \end{pmatrix}$

16. If $\mathbf{B} = \begin{pmatrix} 2 & 4 \\ 1 & 3 \end{pmatrix}$ and $\mathbf{AB} = \mathbf{I}$, find $\mathbf{A}$. **17.** Find $\mathbf{Y}$ if $\mathbf{Y}\begin{pmatrix} -2 & 0 \\ 3 & 1 \end{pmatrix} = \begin{pmatrix} 1 & 0 \\ 0 & 1 \end{pmatrix}$.

18. If $\begin{pmatrix} 2 & -3 \\ 0 & 4 \end{pmatrix} + \mathbf{X} = \begin{pmatrix} 1 & 0 \\ 0 & 1 \end{pmatrix}$, find $\mathbf{X}$.

19. Find $\mathbf{B}$ if $\mathbf{A} = \begin{pmatrix} 2 & -2 \\ -1 & 3 \end{pmatrix}$ and $\mathbf{AB} = \begin{pmatrix} 4 & -2 \\ 0 & 7 \end{pmatrix}$.

20. If $\begin{pmatrix} 3 & -3 \\ 2 & 5 \end{pmatrix} - \mathbf{X} = \begin{pmatrix} 1 & 0 \\ 0 & 1 \end{pmatrix}$, find $\mathbf{X}$.

21. Find $\mathbf{M}$ if $\begin{pmatrix} 1 & 1 \\ -2 & 1 \end{pmatrix}\mathbf{M} = 2\begin{pmatrix} 1 & 0 \\ 0 & 1 \end{pmatrix}$.

22. $\mathbf{A} = \begin{pmatrix} 2 & -3 \\ 0 & 1 \end{pmatrix}$ and $\mathbf{B} = \begin{pmatrix} 1 & -1 \\ -1 & 3 \end{pmatrix}$. Find (a) $\mathbf{AB}$, (b) $\mathbf{A}^{-1}$, (c) $\mathbf{B}^{-1}$. Show that $(\mathbf{AB})^{-1} = \mathbf{B}^{-1}\mathbf{A}^{-1}$.

23. If $\mathbf{M} = \begin{pmatrix} 3 & 1 \\ 2 & -1 \end{pmatrix}$ and $\mathbf{MN} = \begin{pmatrix} 7 & -9 \\ -2 & -6 \end{pmatrix}$, find $\mathbf{N}$.

24. $\mathbf{A} = \begin{pmatrix} 2 & 1 \\ 1 & 1 \end{pmatrix}$; $\mathbf{C} = \begin{pmatrix} 11 \\ 7 \end{pmatrix}$. If $\mathbf{B}$ is a (2×1) matrix such that $\mathbf{AB} = \mathbf{C}$, find $\mathbf{B}$.

25. Find x if the determinant of $\begin{pmatrix} x & 3 \\ 1 & 2 \end{pmatrix}$ is (a) 5, (b) -1, (c) 0.

26. If the matrix $\begin{pmatrix} 1 & -2 \\ x & 4 \end{pmatrix}$ has no inverse, what is the value of x?

27. The elements of a (2×2) matrix consist of four different numbers. Find the largest possible value of the determinant of this matrix if the numbers are
(a) 1, 3, 5, 9 (b) -1, 2, 3, 4.

8.3 SOLUTION OF SIMULTANEOUS EQUATIONS

Example 1 .

Solve the simultaneous equations

$$2x - y = 7$$
$$x + 2y = 1$$

(a) Write the equations in matrix form.

$$\mathbf{A}\begin{pmatrix} x \\ y \end{pmatrix} = \begin{pmatrix} 7 \\ 1 \end{pmatrix}$$

i.e. $\begin{pmatrix} 2 & -1 \\ 1 & 2 \end{pmatrix}\begin{pmatrix} x \\ y \end{pmatrix} = \begin{pmatrix} 7 \\ 1 \end{pmatrix}$

(b) Find the inverse of matrix **A**.

$$\mathbf{A}^{-1} = \frac{1}{4-(-1)}\begin{pmatrix} 2 & 1 \\ -1 & 2 \end{pmatrix} = \frac{1}{5}\begin{pmatrix} 2 & 1 \\ -1 & 2 \end{pmatrix}$$

(c) Pre-multiply both sides of the matrix equation by $\mathbf{A}^{-1}$ so that

$$\mathbf{A}^{-1}\mathbf{A}\begin{pmatrix} x \\ y \end{pmatrix} = \mathbf{A}^{-1}\begin{pmatrix} 7 \\ 1 \end{pmatrix}$$

i.e. $\begin{pmatrix} x \\ y \end{pmatrix} = \mathbf{A}^{-1}\begin{pmatrix} 7 \\ 1 \end{pmatrix}$

$$\therefore \begin{pmatrix} x \\ y \end{pmatrix} = \frac{1}{5}\begin{pmatrix} 2 & 1 \\ -1 & 2 \end{pmatrix}\begin{pmatrix} 7 \\ 1 \end{pmatrix} = \frac{1}{5}\begin{pmatrix} 15 \\ -5 \end{pmatrix}$$

$$\begin{pmatrix} x \\ y \end{pmatrix} = \begin{pmatrix} 3 \\ -1 \end{pmatrix}$$

So $x = 3$ and $y = -1$.

Exercise 3

Solve questions **1** to **12** using matrices.

1. $2x + y = 7$, $x + 2y = 8$
2. $3x - y = 13$, $x + 2y = 2$
3. $3x + 4y = 0$, $x + 2y = 2$
4. $2x - y = 6$, $4x + 3y = -13$
5. $5x - 2y = -15$, $2x + 3y = -6$
6. $3x + 4y = 3$, $5x - y = 3\frac{1}{12}$
7. $3a + 2b = 2$, $7a - 4b = 48$
8. $a - 3b = -24$, $5a - 2b = -29$
9. $3m - 2n = -10$, $m + 8n = 1$
10. $7m - n - 5 = 0$, $2m + 3n + 15 = 0$
11. $p = 2q + 16$, $3p - q = 13$
12. $2p + q = 1{\cdot}1$, $3p - 4q = -2{\cdot}2$
13. If $\mathbf{M} = \begin{pmatrix} 3 & 5 \\ 2 & 4 \end{pmatrix}$, $\mathbf{X} = \begin{pmatrix} s \\ t \end{pmatrix}$ and $\mathbf{Y} = \begin{pmatrix} 23 \\ 18 \end{pmatrix}$, such that $\mathbf{MX} = \mathbf{Y}$, find s and t.
14. $\mathbf{BZ} = \mathbf{A}$ where $\mathbf{B} = \begin{pmatrix} 3 & 0 \\ 2 & -1 \end{pmatrix}$, $\mathbf{A} = \begin{pmatrix} -6 \\ -5 \end{pmatrix}$; find matrix **Z**.
15. Does the matrix $\begin{pmatrix} 3 & 2 \\ 6 & 4 \end{pmatrix}$ have an inverse? On a sketch graph, draw the lines $3x + 2y = 6$ and $6x + 4y = 18$; hence explain why there are no solutions to the simultaneous equations
    $3x + 2y = 6$
    $6x + 4y = 18$.
16. Find which of the following pairs of simultaneous equations have no solutions.
    (a) $3x - y = 5$, $9x - 3y = 10$
    (b) $4x + y = 3$, $8x - 2y = 5$
    (c) $x + 2y = 0$, $4x + 8y = 1$
    (d) $y = 2x - 3$, $4x - 2y + 5 = 0$
17. Confirm that the second matrix below is the inverse of the first.
    $$\begin{pmatrix} 1 & 1 & 1 \\ 1 & 2 & 3 \\ 1 & 4 & 9 \end{pmatrix}; \begin{pmatrix} 3 & -\frac{5}{2} & \frac{1}{2} \\ -3 & 4 & -1 \\ 1 & -\frac{3}{2} & \frac{1}{2} \end{pmatrix}$$
    Hence solve these simultaneous equations.
    $$x + y + z = 6$$
    $$x + 2y + 3z = 14$$
    $$x + 4y + 9z = 36.$$
18. $$\begin{pmatrix} 3 & 0 & 1 \\ 1 & 2 & 1 \\ 0 & 1 & 1 \end{pmatrix}; \frac{1}{4}\begin{pmatrix} 1 & 1 & -2 \\ -1 & 3 & -2 \\ 1 & -3 & 6 \end{pmatrix}$$
    Given that the second matrix above is the inverse of the first, solve these simultaneous equations.
    (a) $3x + z = 4$, $x + 2y + z = 6$, $y + z = 1$
    (b) $x + y - 2z = 12$, $-x + 3y - 2z = 20$, $x - 3y + 6z = -4$.

8.4 SIMPLE TRANSFORMATIONS

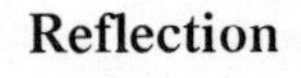

Reflection

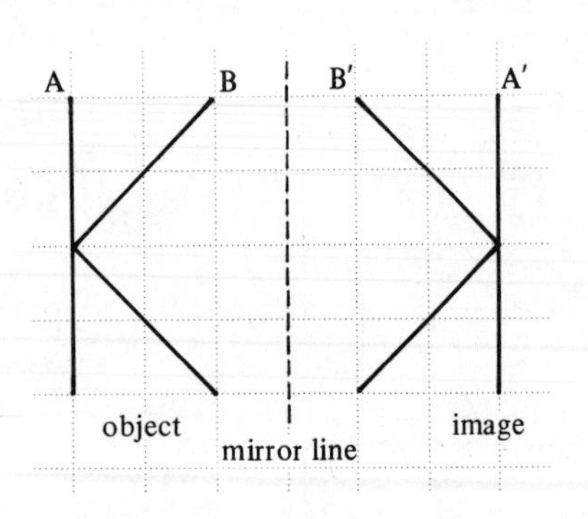

Every point on the object is the same perpendicular distance from the mirror line as the corresponding point on the image.

The perpendicular distance must be measured from the mirror line as shown below: X′, Y′ and Z′ are the images of X, Y and Z.

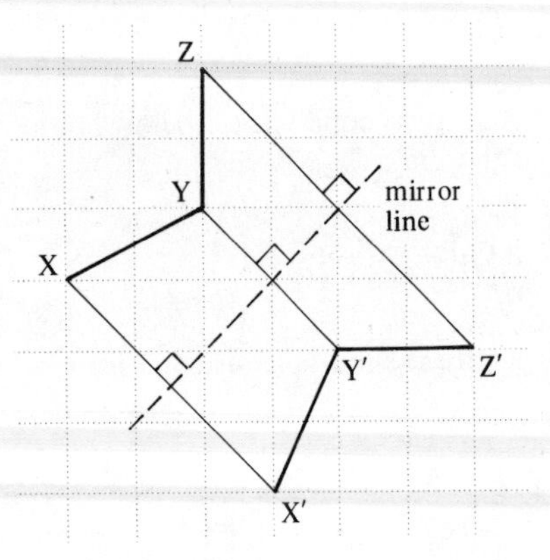

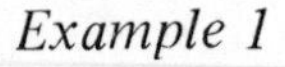

Example 1

Draw the triangle ABC and its image after reflection in mirror line M_1. Then draw the image of this new triangle after reflection in the mirror line M_2.

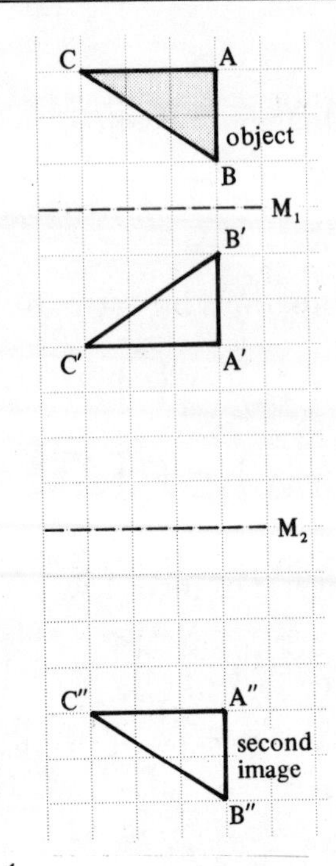

Exercise 4

In questions **1** to **7**, draw the object and the mirror line on squared paper; then draw the image of the shape.

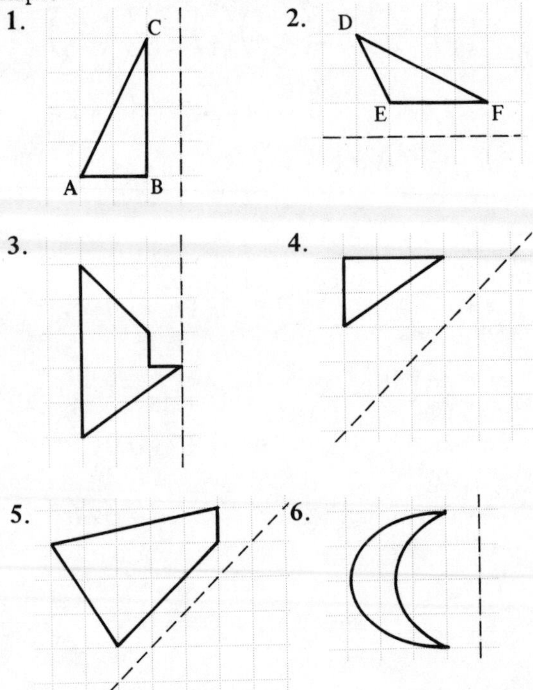

7.

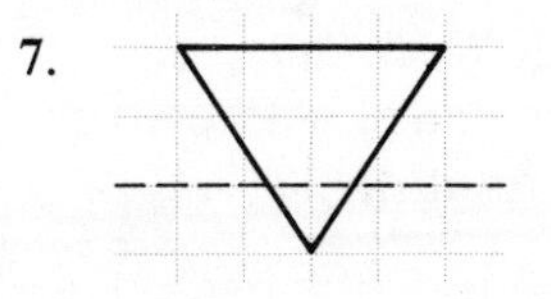

In questions **8** to **13**, draw the object and the two mirror lines M_1 and M_2 on squared paper.
(a) Draw the image I_1 formed by reflection in M_1,
(b) Draw the image I_2 formed by reflecting I_1 in M_2.

8.

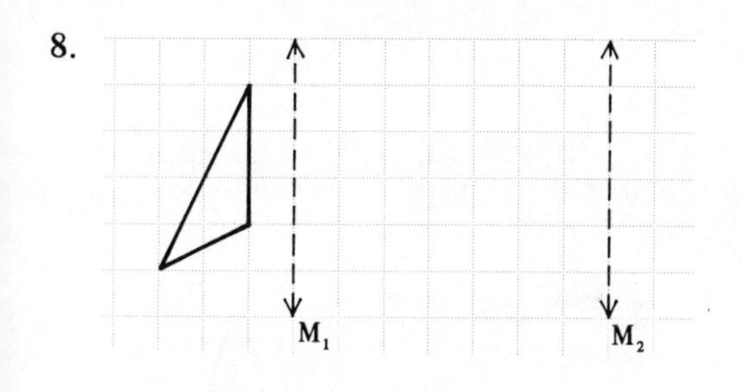

9.

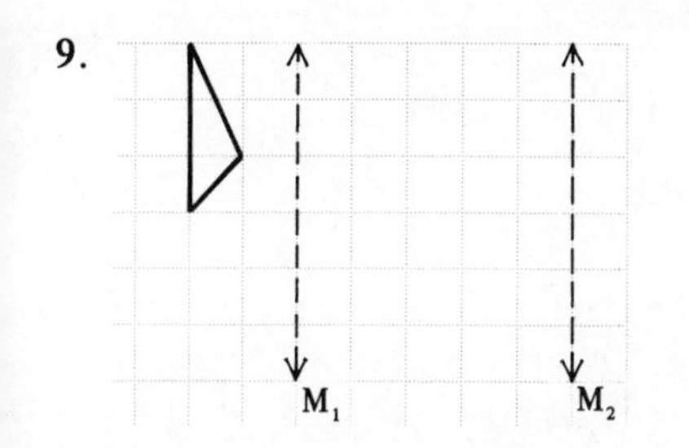

10.

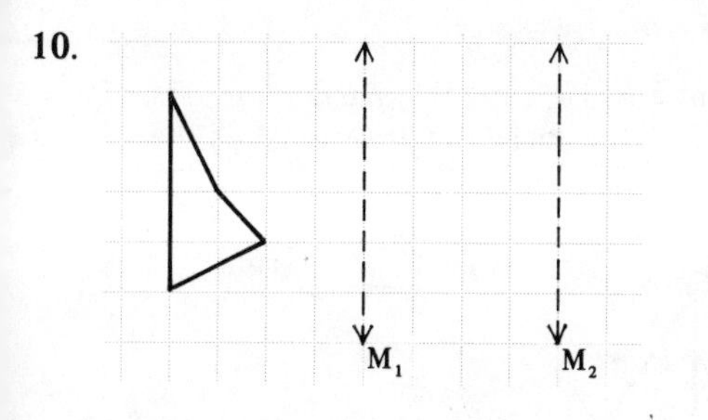

11.

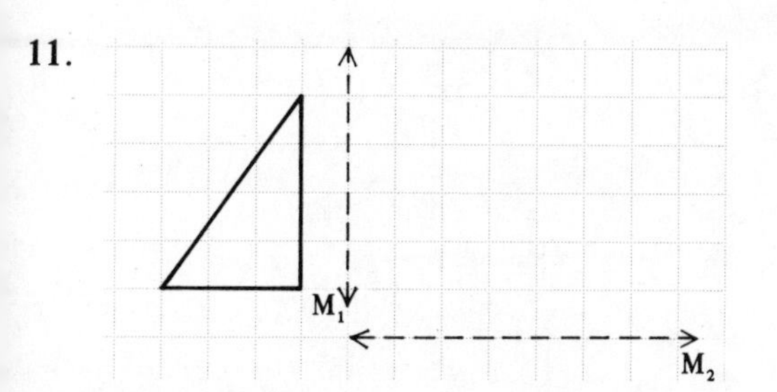

12.

13.

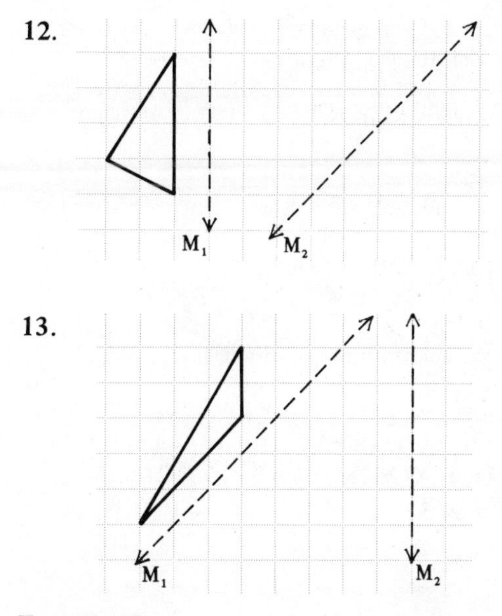

Exercise 5

1. On graph paper draw x- and y-axes for values from 0 to 10.
   (a) Mark on your graph the points A = (2, 4), B = (5, 6), C = (6, 9), D = (1, 8), and E = (7, 0).
   (b) If a mirror is standing on the line $x = 5$, where would the reflection of A, B, C, D and E appear to be?
   (c) If a mirror is standing on the line $y = 5$, where would the reflection of A, B, C, D and E appear to be?
   (d) If a mirror is standing on the line $y = x$, where would the reflection of A, B, C, D and E appear to be?
2. On graph paper, draw x- and y-axes in the middle of a page for values from -12 to $+12$. Draw the triangle ABC at A(2, 4), B(6, 8) and C(9, 0). Find the image of ABC under reflection in the following lines:
   (a) $x = 0$, (b) $y = 0$, (c) $x = 2$,
   (d) $y = -2$, (e) $y = x$.
3. On graph paper, draw x- and y-axes in the middle of a page for values from -16 to $+16$. Draw the object triangle ABC at A(0, 3), B(−3, 9), and C(−10, 7). Find the equation of the mirror line if the image of ABC is at the following:
   (a) A′B′C′ where A′ = (0, 3), B′ = (3, 9), C′ = (10, 7),
   (b) A″B″C″ where A″ = (6, 3), B″ = (9, 9), C″ = (16, 7),
   (c) A*B*C* where A* = (3, 0), B* = (9, −3), C* = (7, −10).

4. Draw x- and y-axes for values from -16 to $+16$. Draw the triangle ABC at A(2, 4), B(6, 8), C(9, 0). Find the image of ABC under reflection in the following lines:
   (a) $y = x + 2$, (b) $y = -x$,
   (c) $x + y = 4$.
5. Draw x- and y-axes for values from -16 to $+16$. Draw the triangle PQR at P(1, 4), Q(-2, 10) and R(8, 11). Find the image of PQR under reflection in the following lines:
   (a) $y = x - 1$, (b) $y + x = 6$.
6. Find the equation of the mirror line in the following:
   (a) Object A(2, 2), B(10, 2), C(7, 8).
   Image A′(0, 0), B′(0, -8), C′(-6, -5),
   (b) Object X(2, 2), Y(7, 8), Z(5, -3).
   Image X′(-1, 5), Y′(5, 10), Z′(-6, 8).

Rotation

Example 2

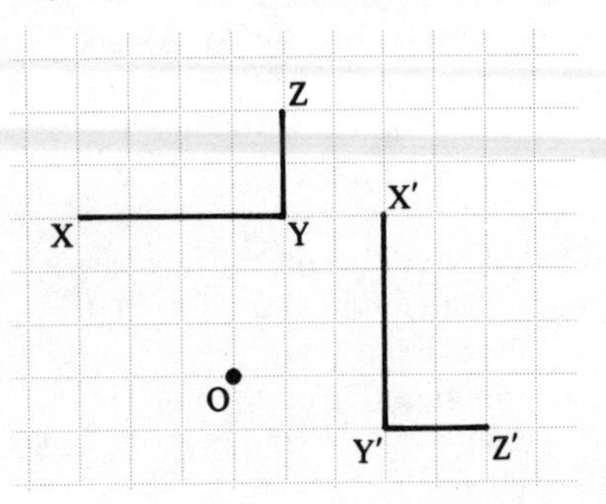

The letter L has been rotated through 90° clockwise about the centre O. The angle, direction, and centre are needed to fully describe a rotation.

We say that the object *maps* onto the image. Here,

X maps onto X′
Y maps onto Y′
Z maps onto Z′

In this work, a clockwise rotation is negative and an anticlockwise rotation is positive: in Example 2, the letter L has been rotated through $-90°$. The angle, the direction, and the centre of rotation can be found using tracing paper and a sharp pencil placed where you think the centre of rotation is.

For more accurate work, draw the perpendicular perpendicular bisector of the line joining two corresponding points, e.g. Y and Y′. Repeat for another pair of corresponding points. The centre of rotation is at the intersection of the two perpendicular bisectors.

Exercise 6

In questions **1** to **4**, copy the diagram on squared paper and draw the image of the shape after a rotation of 90° clockwise about O.

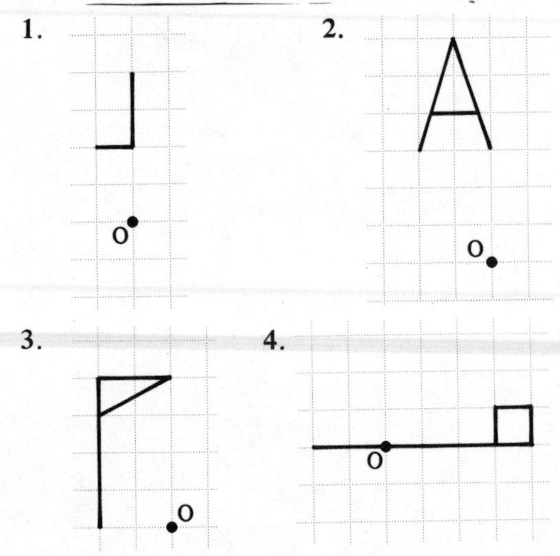

In questions **5** to **10**, copy the diagram on squared paper and find the angle, the direction, and the centre of the rotation.

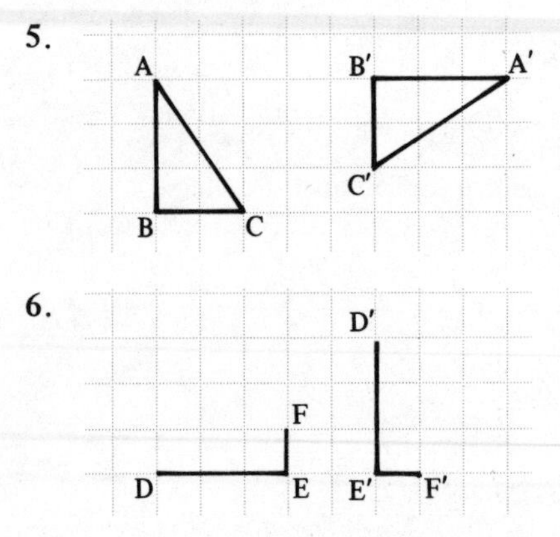

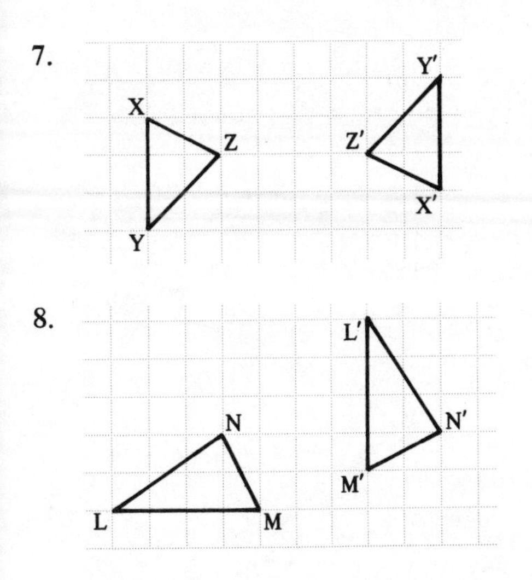

Exercise 7

1. Draw the x- and y-axes in the middle of a page so that x- and y- can take values from -8 to $+8$.
   (a) Draw the object triangle ABC at A(1, 3), B(1, 6), C(3, 6), rotate ABC through 90° clockwise about (0, 0), mark A′B′C′.
   (b) Draw the object triangle DEF at D(3, 3), E(6, 3), F(6, 1), rotate DEF through 90° clockwise about (0, 0), mark D′E′F′.
   (c) Draw the object triangle PQR at P(−4, 7), Q(−4, 5), R(−1, 5), rotate PQR through 90° anticlockwise about (0, 0), mark P′Q′R′.

In questions **2** to **5**, draw x- and y-axes for values from -14 to $+14$.

2. Draw triangle ABC at A(2, 4), B(6, 10), C(12, 4). Find the image of ABC under the following rotations:
   (a) +90°, centre (0, 0); label the image A′B′C′,
   (b) −90°, centre (−4, 4); label the image A″B″C″,
   (c) 180°, centre (2, 0); label the image A*B*C*.
3. Draw the triangle PQR at P(−3, 1), Q(−8, 5), R(−10, −2). Find the image of PQR under the following rotations:
   (a) +90°, centre (0, 0); label the image P′Q′R′,
   (b) 180°, centre (1, −4); label the image P″Q″R″,
   (c) −90°, centre (1, 1); label the image P*Q*R*.
4. Draw triangle ABC at A(2, −1), B(2, 10), C(7, −2).
   (a) Rotate ABC through +90° about (0, 0). Label this triangle A′B′C′,
   (b) Rotate A′B′C′ through 180° about (−4, −1). Label this triangle A″B″C″,
   (c) Rotate A″B″C″ through 180° about B″. Label this triangle A*B*C*.
5. Draw triangle ABC at A(2, 4), B(6, 10), C(12, 4). Find the image of ABC under the following rotations:
   (a) +60°, centre (0, 0); label it A′B′C′,
   (b) +120°, centre (−1, 1); label it A″B″C″,
   (c) −45°, centre (0, 6); label it A*B*C*.
6. Draw the x- and y-axes for values from -8 to $+8$. Draw the object triangle A(3, 1), B(6, 1), C(6, 3) and shade it. Describe fully the rotation which maps ABC onto each of the following:
   (a) A′(−1, 3), B′(−1, 6), C′(−3, 6),
   (b) A′(1, 1), B′(−2, 1), C′(−2, −1),
   (c) A′(3, −1), B′(3, −4), C′(5, −4),
   (d) A′(4, 4), B′(1, 4), C′(1, 2),
   (e) A′(−1, −5), B′(−1, −8), C′(1, −8),
   (f) A′(−5, −1), B′(−5, 2), C′(−7, 2).

Translation

The triangle ABC below has been transformed onto the triangle A′B′C′ by a *translation*.

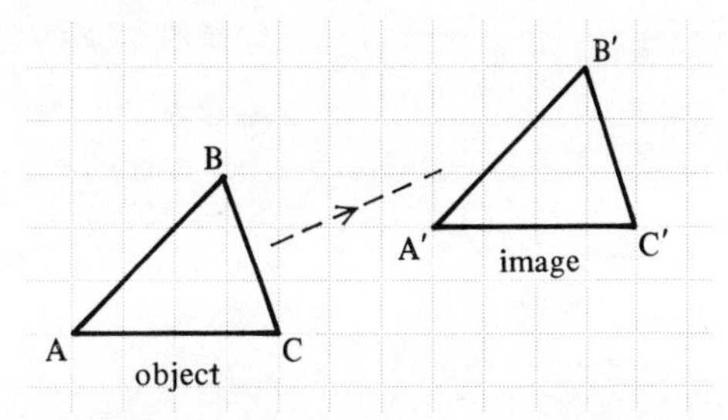

Here the translation is 7 squares to the right and 2 squares up the page. The translation can be described by a column vector.

In this case the translation is $\begin{pmatrix} 7 \\ 2 \end{pmatrix}$.

Example 3

Describe the translation XYZ → X′Y′Z′ below by means of a column vector.

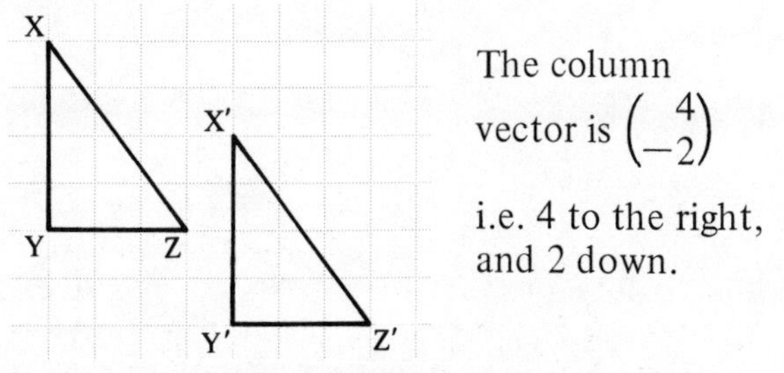

The column vector is $\begin{pmatrix} 4 \\ -2 \end{pmatrix}$

i.e. 4 to the right, and 2 down.

In a translation, every point moves the same distance in the same direction. There is no turning.

Exercise 8

For questions **1** to **5**, write down the column vector for each translation.

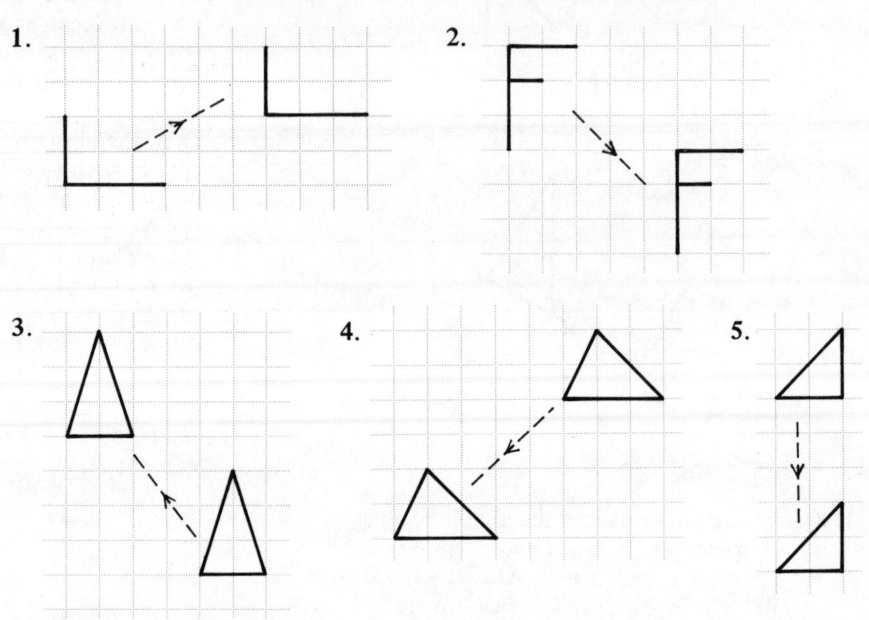

6. Make a copy of the diagram below and write down the column vector for each of the following translations.

(a) D onto A (b) B onto F (c) E onto A (d) A onto C
(e) E onto C (f) C onto B (g) F onto E (h) B onto C.

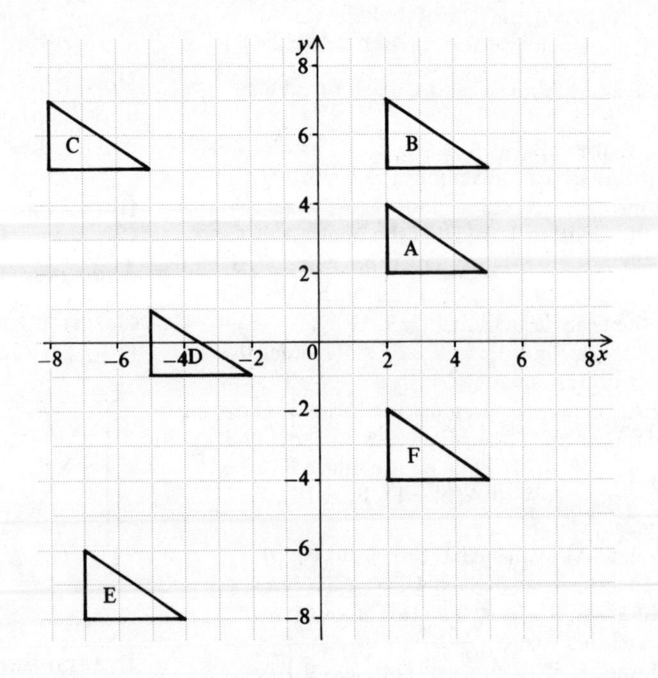

For questions **7** to **23**, draw x- and y-axes with values from -8 to 8. Draw object triangle ABC at A(-4, -1), B(-4, 1), C(-1, -1) and shade it.
Draw the image of ABC under the translations described by the vectors below.
For each question, write down the new coordinates of point C.

7. $\begin{pmatrix}6\\3\end{pmatrix}$ **8.** $\begin{pmatrix}6\\7\end{pmatrix}$ **9.** $\begin{pmatrix}9\\-4\end{pmatrix}$ **10.** $\begin{pmatrix}1\\7\end{pmatrix}$ **11.** $\begin{pmatrix}5\\-6\end{pmatrix}$

12. $\begin{pmatrix}-2\\5\end{pmatrix}$ **13.** $\begin{pmatrix}-2\\-4\end{pmatrix}$ **14.** $\begin{pmatrix}0\\-7\end{pmatrix}$ **15.** $\begin{pmatrix}7\\0\end{pmatrix}$

16. $\begin{pmatrix}3\\1\end{pmatrix}$ followed by $\begin{pmatrix}3\\2\end{pmatrix}$ **17.** $\begin{pmatrix}3\\2\end{pmatrix}$ followed by $\begin{pmatrix}3\\1\end{pmatrix}$

18. $\begin{pmatrix}3\\4\end{pmatrix}$ followed by $\begin{pmatrix}-2\\3\end{pmatrix}$ **19.** $\begin{pmatrix}3\\3\end{pmatrix}$ followed by $\begin{pmatrix}6\\4\end{pmatrix}$

20. $\begin{pmatrix}-2\\3\end{pmatrix}$ followed by $\begin{pmatrix}3\\4\end{pmatrix}$ **21.** $\begin{pmatrix}-2\\0\end{pmatrix}$ followed by $\begin{pmatrix}0\\3\end{pmatrix}$

22. $\begin{pmatrix}3\\1\end{pmatrix}$ followed by $\begin{pmatrix}4\\-2\end{pmatrix}$ followed by $\begin{pmatrix}2\\-3\end{pmatrix}$

23. $\begin{pmatrix}5\\0\end{pmatrix}$ followed by $\begin{pmatrix}0\\-3\end{pmatrix}$ followed by $\begin{pmatrix}-7\\-1\end{pmatrix}$

Enlargement

In the diagram below, the letter T has been enlarged by a scale factor of 2 using the point O as the centre of the enlargement.

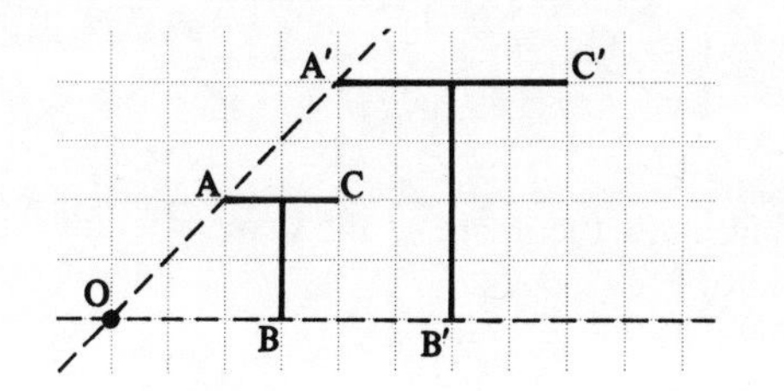

Notice that $OA' = 2 \times OA$
$OB' = 2 \times OB$

The scale factor and the centre of enlargement are both required to describe an enlargement.

Example 4

Draw the image of triangle ABC under an enlargement scale factor of $\frac{1}{2}$ using O as centre of enlargement.

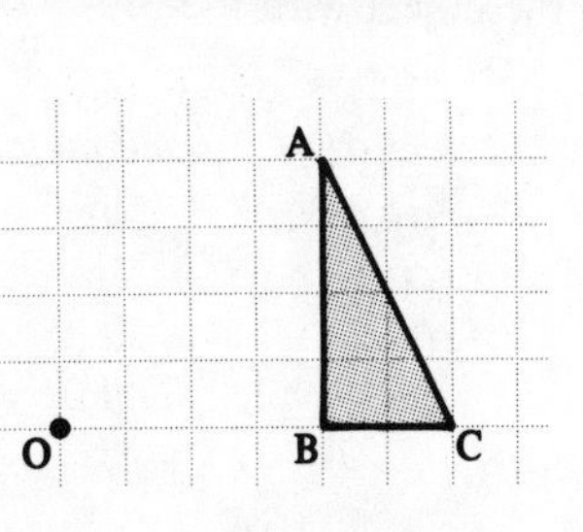

(a) Draw lines through OA, OB and OC.

(b) Mark A′ so that $OA' = \frac{1}{2}OA$
Mark B′ so that $OB' = \frac{1}{2}OB$
Mark C′ so that $OC' = \frac{1}{2}OC$.

(c) Join A′B′C′.

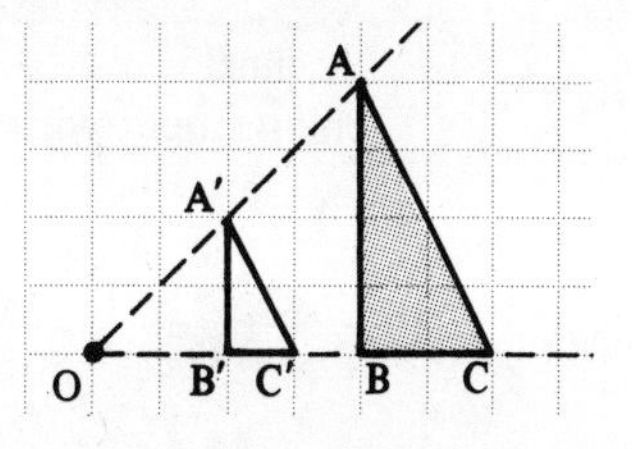

Remember always to measure the lengths from O, not from A, B, or C.

Example 5

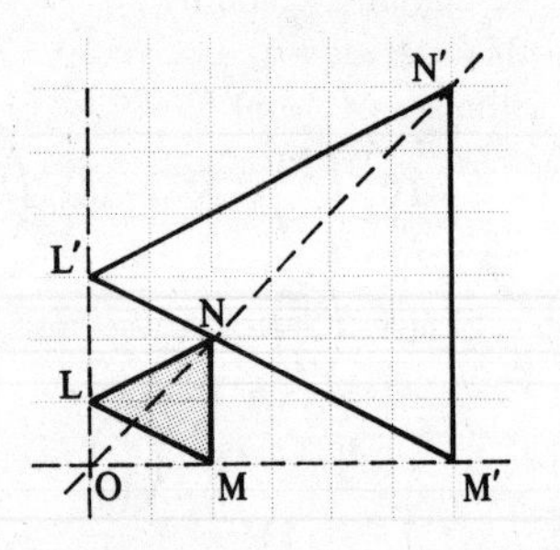

Find the centre of enlargement and the scale factor which maps triangle LMN onto triangle L′M′N′.

(a) Draw lines through LL′, MM′ and NN′. The centre of enlargement O is where these lines intersect.

(b) To find the scale factor, measure the distance from O to any point on the image and the distance from O to the corresponding point on the object.

e.g. OM′ = 6
OM = 2

The scale factor is $\frac{OM'}{OM} = 3$.

Note that any length on the image is three times the corresponding length on the object, e.g. L′N′ = 3LN.

Exercise 9

In questions **1** to **3**, copy the diagram on squared paper and draw the image of the triangle after an enlargement with centre O and scale factor 2.

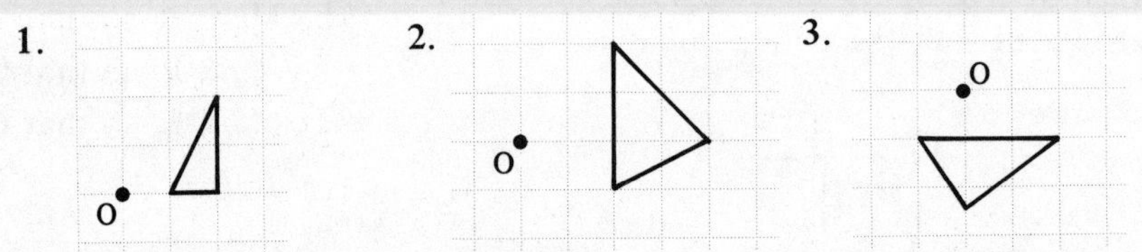

In questions **4** to **6**, copy the diagram and draw the image of the shape after an enlargement with centre O and scale factor −2.

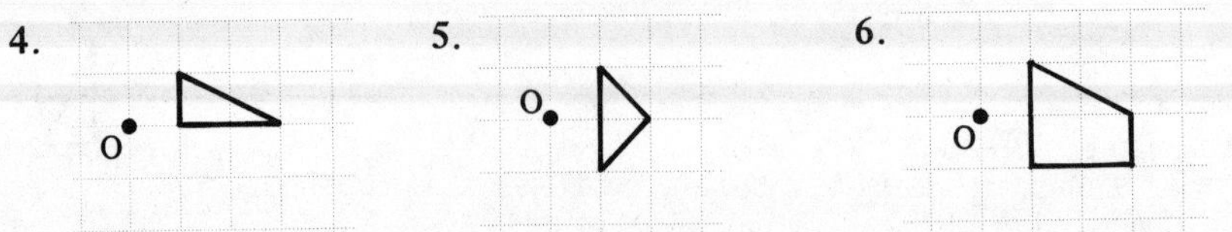

Answer questions **7** to **30** on graph paper taking x and y from 0 to 15. The vertices of the object are given in coordinate form.

In questions **7** to **11**, enlarge the object with the centre of enlargement and scale factor indicated.

	object	*centre*	*scale factor*
7.	(2, 4) (4, 2) (5, 5)	(0, 0)	+2
8.	(2, 4) (4, 2) (5, 5)	(1, 2)	+2
9.	(1, 1) (4, 2) (2, 3)	(1, 1)	+3
10.	(4, 4) (7, 6) (9, 3)	(7, 4)	+2
11.	(1, 1) (4, 1) (4, 3) (1, 3)	(2, 0)	+2

In questions **12** to **17**, plot the object and image and find the centre of enlargement and the scale factor for each enlargement.

	object	*image*
12.	A(2, 1), B(5, 1), C(3, 3)	A′(4, 2), B′(10, 2), C′(6, 6)
13.	A(2, 1), B(5, 1), C(3, 3)	A′(2, 1), B′(11, 1), C′(5, 7)
14.	A(4, 4), B(7, 4), C(7, 6), D(4, 6)	A′(1, 2), B′(7, 2), C′(7, 6), D′(1, 6)
15.	A(4, 4), B(7, 4), C(7, 6), D(4, 6)	A′(1, 0), B′(10, 0), C′(10, 6), D′(1, 6)
16.	A(2, 5), B(9, 3), C(5, 9)	A′($6\frac{1}{2}$, 7), B′(10, 6), C′(8, 9)
17.	A(6, 4), B(10, 4), C(10, 8), D(6, 8)	A′(7, 3), B′(13, 3), C′(13, 9), D′(7, 9)

In questions **18** to **26**, enlarge the object using the centre of enlargement and scale factor indicated.

	object	*centre*	*scale factor*
18.	(1, 2) (13, 2) (1, 10)	(0, 0)	$+\frac{1}{2}$
19.	(1, 2) (13, 2) (1, 10)	(11, 10)	$+\frac{1}{2}$
20.	(3, 3) (6, 3) (6, 9) (3, 9)	(3, 3)	$+\frac{1}{3}$
21.	(3, 3) (6, 3) (6, 9) (3, 9)	(0, 0)	$+\frac{1}{3}$
22.	(10, 3) (10, 5) (10, 7) (12, 7) (12, 5) (10, 5)	(8, 5)	-2
23.	(7, 3) (9, 3) (7, 8)	(5, 5)	-1
24.	(1, 1) (3, 1) (3, 2) (1, 2)	(4, 3)	-2
25.	(12, 1) (15, 1) (15, 3) (12, 3)	(12, 3)	-3
26.	(9, 2) (14, 2) (14, 6) (9, 6)	(7, 4)	$-\frac{1}{2}$

In questions **27** to **30**, draw the object and its image and describe fully the enlargement in each case.

	object	*image*
27.	A(2, 2), B(4, 4), C(2, 6)	A′(11, 8), B′(7, 4), C′(11, 0)
28.	A(1, 0), B(5, 1), C(1, 3)	A′(10, 9), B′(2, 7), C′(10, 3)
29.	A(7, 0), B(7, 3), C(13, 3), D(13, 0)	A′(3, 8), B′(3, 7), C′(1, 7), D′(1, 8)
30.	A(0, 6), B(4, 6), C(3, 0)	A′(12, 6), B′(8, 6), C′(9, 12)

8.5 COMBINED TRANSFORMATIONS

It is convenient to denote transformations by a symbol.

Let **A** denote 'reflection in line $x = 3$'

and **B** denote 'translation $\begin{pmatrix}2\\1\end{pmatrix}$'.

Perform **A** on XYZ

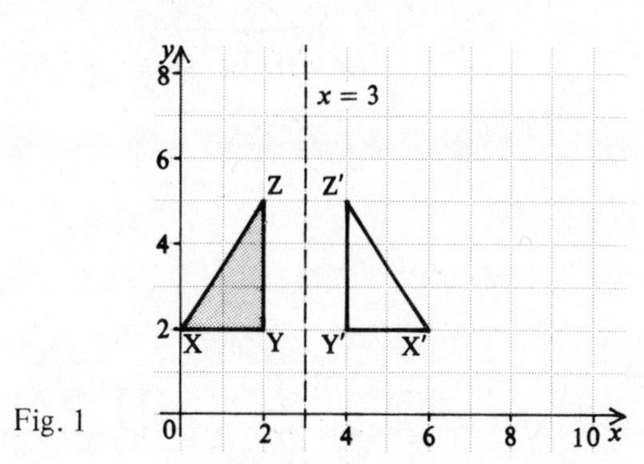

Fig. 1

X′Y′Z′ is the image of XYZ under the reflection in $x = 3$

i.e. **A**(XYZ) = X′Y′Z′

A(XYZ) means 'perform the transformation **A** on triangle XYZ'.

Perform **B** on X′Y′Z′

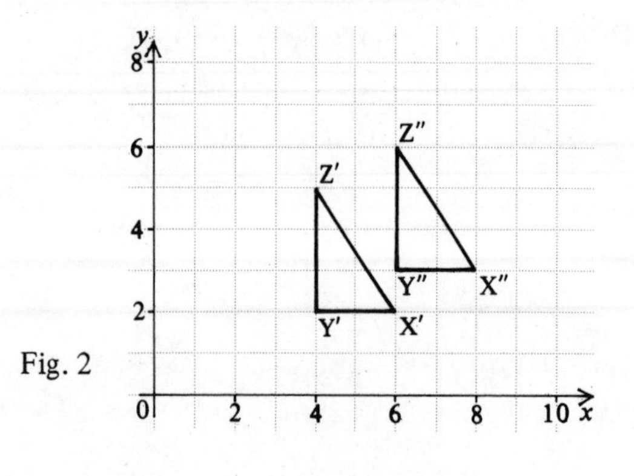

Fig. 2

From Fig. 2 we can see that

B(X′Y′Z′) = X″Y″Z″

The effect of going from XYZ to X″Y″Z″ may be written

BA(XYZ) = X″Y″Z″

It is very important to notice that **BA**(XYZ) means do **A** first and then **B**.

Now let **C** denote 'rotation +90° on centre (6, 6)'.

Perform **C** on X″Y″Z″

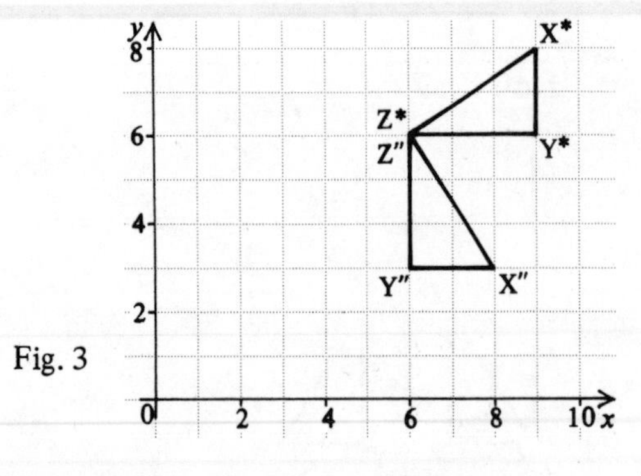

Fig. 3

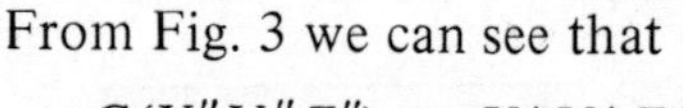

From Fig. 3 we can see that

C(X″Y″Z″) = X*Y*Z*

Starting from XYZ and performing first **A**, then **B**, then **C**, we have arrived at X*Y*Z*.

Thus **CBA**(XYZ) = X*Y*Z*

We can imagine the transformations 'queuing up' to operate on XYZ with **A** first, **B** second in line, and **C** last.

Repeated transformations

XX(P) means 'perform transformation **X** on P and then perform **X** on the image'.

It may be written $\mathbf{X}^2$(P)

Similarly **TTT**(P) = $\mathbf{T}^3$(P).

Inverse transformations

If translation **T** has vector $\begin{pmatrix} 3 \\ -2 \end{pmatrix}$, the translation which has the opposite effect has vector $\begin{pmatrix} -3 \\ 2 \end{pmatrix}$. This is written $\mathbf{T}^{-1}$.

If rotation **R** denotes 90° clockwise rotation about (0, 0), then $\mathbf{R}^{-1}$ denotes 90° *anti*clockwise rotation about (0, 0).

The *inverse* of a transformation is the transformation which takes the *image* back to the object.

Note
For all reflections, the inverse is the same reflection.
e.g. if **X** is reflection in $x = 0$, then $\mathbf{X}^{-1}$ is also reflection in $x = 0$.

The symbol $\mathbf{T}^{-3}$ means $(\mathbf{T}^{-1})^3$ i.e. perform $\mathbf{T}^{-1}$ three times.

Exercise 10

Draw x- and y-axes with values from -8 to $+8$ and plot the point P(3, 2).
R denotes 90° clockwise rotation about (0, 0);
X denotes reflection in $x = 0$.
H denotes 180° rotation about (0, 0);
T denotes translation $\begin{pmatrix}3\\2\end{pmatrix}$.

For each question, write down the coordinates of the final image of P.

1. **R**(P)	2. **TR**(P)	3. **T**(P)	4. **RT**(P)	5. **TH**(P)
6. **XT**(P)	7. **HX**(P)	8. **XX**(P)	9. $\mathbf{R}^{-1}$(P)	10. $\mathbf{T}^{-1}$(P)
11. $\mathbf{X}^3$(P)	12. $\mathbf{T}^{-2}$(P)	13. $\mathbf{R}^2$(P)	14. $\mathbf{T}^{-1}\mathbf{R}^2$(P)	15. **THX**(P)
16. $\mathbf{R}^3$(P)	17. $\mathbf{TX}^{-1}$(P)	18. $\mathbf{T}^3\mathbf{X}$(P)	19. $\mathbf{T}^2\mathbf{H}^{-1}$(P)	20. **XTH**(P)

Exercise 11

For questions **1** to **24**, make a copy of the diagram shown; then copy and complete the data and write down the equivalent single transformation if

$\mathbf{M}_1$ denotes reflection in $x = 0$
$\mathbf{M}_2$ denotes reflection in $y = 0$
$\mathbf{M}_3$ denotes reflection in $y = x$
$\mathbf{M}_4$ denotes reflection in $y = -x$
$\mathbf{R}_a$ denotes rotation +90° about (0, 0)
$\mathbf{R}_b$ denotes rotation 180° about (0, 0)
$\mathbf{R}_c$ denotes rotation −90° about (0, 0)
I is the identity: image is same as object.

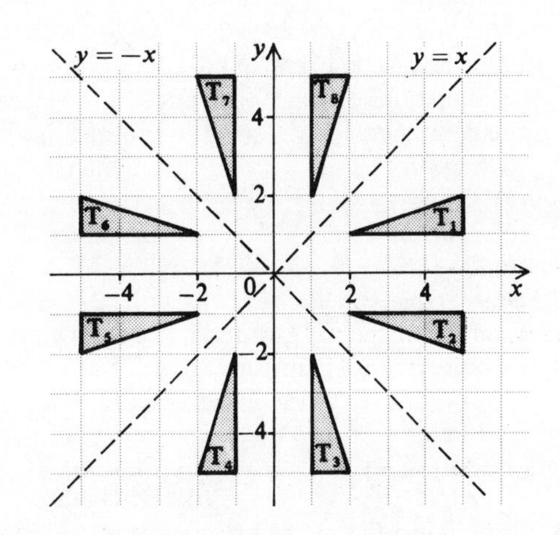

For example,

$\mathrm{T}_1 \xrightarrow{\mathbf{M}_1} \xrightarrow{\mathbf{M}_2}$ becomes $\mathrm{T}_1 \xrightarrow{\mathbf{M}_1} \mathrm{T}_6 \xrightarrow{\mathbf{M}_2} \mathrm{T}_5$

The equivalent single transformation $\mathrm{T}_1 \to \mathrm{T}_5$ is $\mathbf{R}_b$

1. $\mathrm{T}_1 \xrightarrow{\mathbf{M}_1} \xrightarrow{\mathbf{M}_2}$
2. $\mathrm{T}_2 \xrightarrow{\mathbf{M}_4} \xrightarrow{\mathbf{R}_a}$
3. $\mathrm{T}_3 \xrightarrow{\mathbf{M}_1} \xrightarrow{\mathbf{R}_a}$
4. $\mathrm{T}_5 \xrightarrow{\mathbf{M}_1} \xrightarrow{\mathbf{R}_a}$
5. $\mathrm{T}_8 \xrightarrow{\mathbf{M}_1} \xrightarrow{\mathbf{R}_a}$
6. $\mathrm{T}_3 \xrightarrow{\mathbf{R}_a} \xrightarrow{\mathbf{M}_1}$
7. $\mathrm{T}_6 \xrightarrow{\mathbf{R}_a} \xrightarrow{\mathbf{M}_1}$
8. $\mathrm{T}_5 \xrightarrow{\mathbf{R}_a} \xrightarrow{\mathbf{M}_3}$
9. $\mathrm{T}_7 \xrightarrow{\mathbf{M}_1} \xrightarrow{\mathbf{M}_3}$
10. $\mathrm{T}_4 \xrightarrow{\mathbf{R}_b} \xrightarrow{\mathbf{R}_c}$
11. $\mathrm{T}_6 \xrightarrow{\mathbf{R}_c} \xrightarrow{\mathbf{M}_4}$
12. $\mathrm{T}_5 \xrightarrow{\mathbf{M}_4} \xrightarrow{\mathbf{M}_3}$
13. $\mathrm{T}_5 \xrightarrow{\mathbf{M}_3} \xrightarrow{\mathbf{M}_4}$
14. $\mathrm{T}_8 \xrightarrow{\mathbf{M}_2} \xrightarrow{\mathbf{M}_2}$
15. $\mathrm{T}_2 \xrightarrow{\mathbf{R}_c} \xrightarrow{\mathbf{M}_1}$
16. $\mathrm{T}_5 \xrightarrow{\mathbf{M}_2} \xrightarrow{\mathbf{M}_4} \xrightarrow{\mathbf{R}_c}$
17. $\mathrm{T}_1 \xrightarrow{\mathbf{M}_1} \xrightarrow{\mathbf{R}_b} \xrightarrow{\mathbf{M}_3}$
18. $\mathrm{T}_1 \xrightarrow{\mathbf{R}_a} \xrightarrow{\mathbf{M}_3} \xrightarrow{\mathbf{R}_b}$
19. $\mathrm{T}_3 \xrightarrow{\mathbf{R}_c} \xrightarrow{\mathbf{M}_3} \xrightarrow{\mathbf{R}_b}$
20. $\mathrm{T}_4 \xrightarrow{\mathbf{M}_2} \xrightarrow{\mathbf{M}_3} \xrightarrow{\mathbf{M}_4}$
21. $\mathrm{T}_7 \xrightarrow{\mathbf{M}_3} \xrightarrow{\mathbf{R}_a} \xrightarrow{\mathbf{M}_1}$
22. $\mathrm{T}_6 \xrightarrow{\mathbf{M}_2} \xrightarrow{\mathbf{M}_4} \xrightarrow{\mathbf{R}_a}$
23. $\mathrm{T}_8 \xrightarrow{\mathbf{R}_b} \xrightarrow{\mathbf{R}_c} \xrightarrow{\mathbf{M}_2}$
24. $\mathrm{T}_2 \xrightarrow{\mathbf{M}_1} \xrightarrow{\mathbf{R}_a} \xrightarrow{\mathbf{M}_2}$

25. Is the equivalent single transformation the same if you change the order of the transformations?

Exercise 12

In this exercise, transformations **A**, **B**, . . . **H**, are as follows:

A denotes reflection in $x = 2$
B denotes 180° rotation, centre (1, 1)
C denotes translation $\begin{pmatrix} -6 \\ 2 \end{pmatrix}$
D denotes reflection in $y = x$
E denotes reflection in $y = 0$
F denotes translation $\begin{pmatrix} 4 \\ 3 \end{pmatrix}$
G denotes 90° rotation clockwise, centre (0, 0)
H denotes enlargement, scale factor $+\frac{1}{2}$, centre (0, 0)

Draw x- and y-axes with values from -8 to $+8$.

1. Draw triangle LMN at L(2, 2), M(6, 2), N(6, 4). Find the image of LMN under the following combinations of transformations. Write down the coordinates of the image of point L in each case:
   (a) **CA**(LMN) (b) **ED**(LMN)
   (c) **DB**(LMN) (d) **BE**(LMN)
   (e) **EB**(LMN).
2. Draw triangle PQR at P(2, 2), Q(6, 2), R(6, 4). Find the image of PQR under the following combinations of transformations. Write down the coordinates of the image of point P in each case:
   (a) **AF**(PQR) (b) **CG**(PQR)
   (c) **AG**(PQR) (d) **HE**(PQR).
3. Write down the inverses of the transformations **A**, **B**, . . . **H**.
4. Draw triangle JKL at J(-2, 2), K(-2, 5), L(-4, 5). Find the image of JKL under the following transformations. Write down the coordinates of the image of point J in each case.
   (a) $\mathbf{C}^{-1}$ (b) $\mathbf{F}^{-1}$ (c) $\mathbf{G}^{-1}$
   (d) $\mathbf{D}^{-1}$ (e) $\mathbf{A}^{-1}$.
5. Draw triangle PQR at P(-2, 4), Q(-2, 1), R(-4, 1). Find the image of PQR under the following combinations of transformations. Write down the coordinates of the image of point P in each case.
   (a) $\mathbf{DF}^{-1}$(PQR) (b) $\mathbf{EC}^{-1}$(PQR)
   (c) $\mathbf{D}^2\mathbf{F}$(PQR) (d) **GA**(PQR)
   (e) $\mathbf{C}^{-1}\mathbf{G}^{-1}$(PQR).
6. Draw triangle XYZ at X(-2, 4), Y(-2, 1), Z(-4, 1). Find the image of XYZ under the following combinations of transformations and state the equivalent single transformation in each case.
   (a) $\mathbf{G}^2\mathbf{E}$(XYZ) (b) **CB**(XYZ)
   (c) **DA**(XYZ).
7. Draw triangle OPQ at O(0, 0), P(0, 2), Q(3, 2). Find the image of OPQ under the following combinations of transformations and state the equivalent single transformation in each case.
   (a) **DE**(OPQ) (b) **FC**(OPQ)
   (c) **DEC**(OPQ) (d) **DFE**(OPQ).
8. Draw triangle LMN at L(-2, 4), M(-4, 1), N(-2, 1). Find the image of LMN under the following combinations of transformations. Write down the coordinates of the image of point L in each case.
   (a) **HE**(LMN) (b) $\mathbf{EAG}^{-1}$(LMN)
   (c) **EDA**(LMN) (d) $\mathbf{BG}^2\mathbf{E}$(LMN).
9. Draw triangle RST at R(-4, -1), S($-2\frac{1}{2}$, -2), T(-4, -4). Find the image of RST under the following combinations of transformations and state the equivalent single transformation in each case.
   (a) **EAG**(RST) (b) **FH**(RST)
   (c) **GF**(RST).

8.6 TRANSFORMATIONS USING MATRICES

Example 1

Find the image of triangle ABC, with A(1, 1), B(3, 1), C(3, 2), under the transformation represented by the matrix $\mathbf{M} = \begin{pmatrix} 1 & 0 \\ 0 & -1 \end{pmatrix}$.

(a) Write the coordinates of A as a column vector and multiply this vector by **M**.

$$\overset{\mathbf{M}}{\begin{pmatrix} 1 & 0 \\ 0 & -1 \end{pmatrix}} \overset{\mathrm{A}}{\begin{pmatrix} 1 \\ 1 \end{pmatrix}} = \overset{\mathrm{A}'}{\begin{pmatrix} 1 \\ -1 \end{pmatrix}}$$

A′, the image of A, has coordinates (1, −1).

(b) Repeat for B and C.

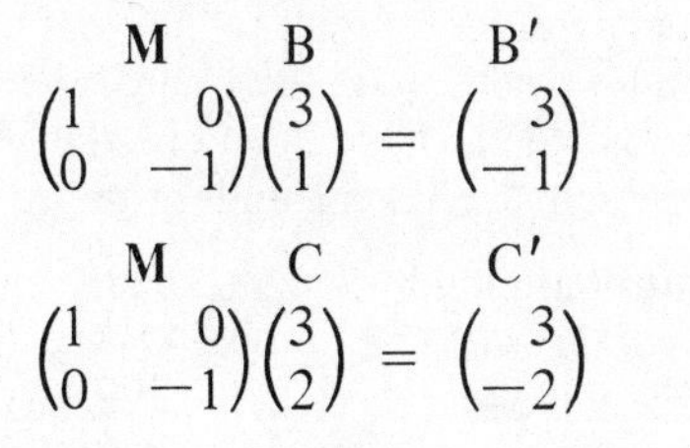

$$\overset{\mathbf{M}}{\begin{pmatrix} 1 & 0 \\ 0 & -1 \end{pmatrix}} \overset{\mathrm{B}}{\begin{pmatrix} 3 \\ 1 \end{pmatrix}} = \overset{\mathrm{B}'}{\begin{pmatrix} 3 \\ -1 \end{pmatrix}}$$

$$\overset{\mathbf{M}}{\begin{pmatrix} 1 & 0 \\ 0 & -1 \end{pmatrix}} \overset{\mathrm{C}}{\begin{pmatrix} 3 \\ 2 \end{pmatrix}} = \overset{\mathrm{C}'}{\begin{pmatrix} 3 \\ -2 \end{pmatrix}}$$

(c) Plot A′(1, −1), B′(3, −1) and C′(3, −2).

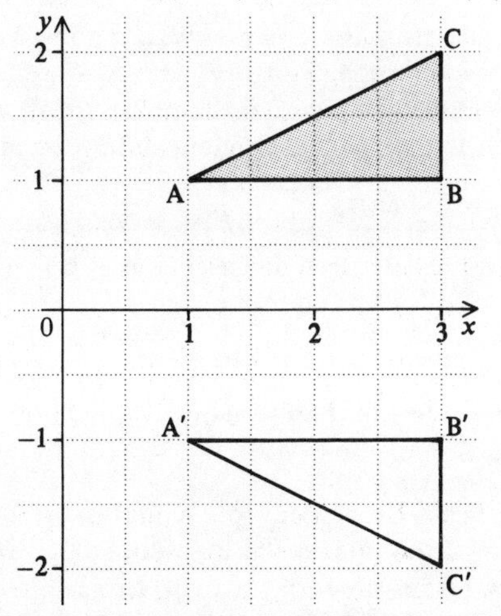

The transformation is a reflection in the x-axis.

Example 2

Find the image of L(1, 1), M(1, 3), N(2, 3) under the transformation represented by the matrix $\begin{pmatrix} 0 & -1 \\ 1 & 0 \end{pmatrix}$.

A quicker method is to write the three vectors for L, M and N in a single 2 × 3 matrix, and then perform the multiplication

$$\begin{pmatrix} 0 & -1 \\ 1 & 0 \end{pmatrix} \overset{\mathrm{L}\ \ \mathrm{M}\ \ \mathrm{N}}{\begin{pmatrix} 1 & 1 & 2 \\ 1 & 3 & 3 \end{pmatrix}} = \overset{\mathrm{L}'\ \ \mathrm{M}'\ \ \mathrm{N}'}{\begin{pmatrix} -1 & -3 & -3 \\ 1 & 1 & 2 \end{pmatrix}}$$

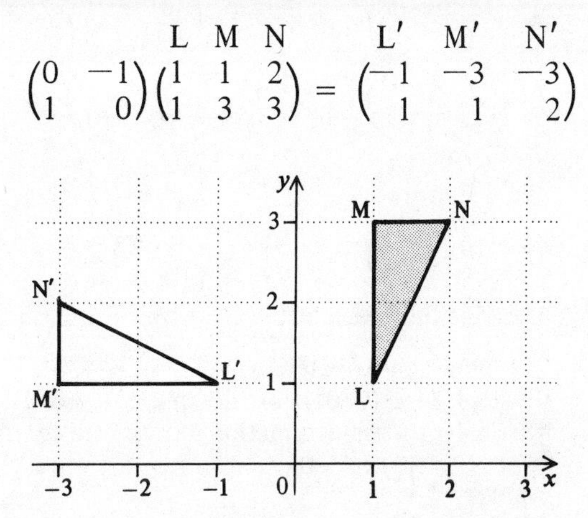

The transformation is a rotation, +90°, centre (0, 0).

Exercise 13

For questions **1** to **5**, draw x- and y-axes so that x and y can take values from −8 to +8.

1. Draw the triangle A(2, 2), B(6, 2), C(6, 4). Find its image under the transformations represented by the following matrices.

(a) $\begin{pmatrix} 0 & -1 \\ 1 & 0 \end{pmatrix}$ (b) $\begin{pmatrix} -1 & 0 \\ 0 & 1 \end{pmatrix}$ (c) $\begin{pmatrix} 1 & 0 \\ 0 & -1 \end{pmatrix}$

(d) $\begin{pmatrix} 0 & 1 \\ 1 & 0 \end{pmatrix}$ (e) $\begin{pmatrix} \frac{1}{2} & 0 \\ 0 & \frac{1}{2} \end{pmatrix}$.

2. Plot the object and image for the following:

	Object	*Matrix*
(a)	P(4, 2), Q(4, 4), R(0, 4)	$\begin{pmatrix} 2 & 0 \\ 0 & 2 \end{pmatrix}$
(b)	P(4, 2), Q(4, 4), R(0, 4)	$\begin{pmatrix} -\frac{1}{2} & 0 \\ 0 & -\frac{1}{2} \end{pmatrix}$
(c)	A(−6, 8), B(−2, 8), C(−2, 6)	$\begin{pmatrix} 0 & -1 \\ -1 & 0 \end{pmatrix}$
(d)	P(4, 2), Q(4, 4), R(0, 4)	$\begin{pmatrix} -2 & 0 \\ 0 & -2 \end{pmatrix}$

Describe each as a *single* transformation.

3. Answer all parts of this question on one graph. Draw a trapezium at K(2, 2), L(2, 5), M(5, 8), N(8, 8). Find the images of KLMN under the transformations described by the following matrices:

$A = \begin{pmatrix} 1 & 0 \\ 0 & -1 \end{pmatrix}$ $E = \begin{pmatrix} 0 & -1 \\ -1 & 0 \end{pmatrix}$

$B = \begin{pmatrix} -1 & 0 \\ 0 & 1 \end{pmatrix}$ $F = \begin{pmatrix} -1 & 0 \\ 0 & -1 \end{pmatrix}$

$C = \begin{pmatrix} 0 & 1 \\ 1 & 0 \end{pmatrix}$ $G = \begin{pmatrix} 0 & -1 \\ 1 & 0 \end{pmatrix}$

$D = \begin{pmatrix} 0 & 1 \\ -1 & 0 \end{pmatrix}$ $H = \begin{pmatrix} 1 & 0 \\ 0 & 1 \end{pmatrix}$

Describe fully each of the eight transformations.

4. (a) Draw a quadrilateral at A(3, 4), B(4, 0), C(3, 1), D(0, 0). Find the image of ABCD under the transformation represented by the matrix $\begin{pmatrix} -2 & 0 \\ 0 & -2 \end{pmatrix}$.

Find the ratio $\left(\frac{\text{area of image}}{\text{area of object}}\right)$.

(b) Repeat (a) using the quadrilateral K(−4, 1), L(−2, 4), M(0, 1), N(−1, 0).

5. The matrix $\mathbf{R} = \begin{pmatrix} \cos\theta & -\sin\theta \\ \sin\theta & \cos\theta \end{pmatrix}$ represents a positive rotation of $\theta°$ about the origin. Find the matrix which represents a rotation of:
(a) 90° (b) 180° (c) 30° (d) −90°
(e) 60° (f) 150° (g) 45° (h) 53·1°
Confirm your results for parts (a), (e), (h) by applying the matrix to the quadrilateral O(0, 0), A(0, 2), B(4, 2), C(4, 0).

6. Using the matrix **R** given in question 5, find the angle of rotation for the following,

(a) $\begin{pmatrix} 0 & -1 \\ 1 & 0 \end{pmatrix}$ (b) $\begin{pmatrix} 0{\cdot}8 & -0{\cdot}6 \\ 0{\cdot}6 & 0{\cdot}8 \end{pmatrix}$

(c) $\begin{pmatrix} 0{\cdot}5 & 0{\cdot}866 \\ -0{\cdot}866 & 0{\cdot}5 \end{pmatrix}$ (d) $\begin{pmatrix} 0{\cdot}6 & 0{\cdot}8 \\ -0{\cdot}8 & 0{\cdot}6 \end{pmatrix}$

(e) $\begin{pmatrix} 0{\cdot}707 & 0{\cdot}707 \\ -0{\cdot}707 & 0{\cdot}707 \end{pmatrix}$

Confirm your results by applying each matrix to the quadrilateral O(0, 0), A(0, 2), B(4, 2), C(4, 0).

7. (a) Draw axes so that both x and y can take values from −5 to +15.
(b) Draw triangle ABC at A(2, 1), B(7, 1), C(2, 4).
(c) Find the image of ABC under the transformation represented by the matrix $\begin{pmatrix} 1 & -1 \\ 1 & 1 \end{pmatrix}$ and plot the image on the graph.
(d) The transformation is a rotation followed by an enlargement. Calculate the angle of the rotation and the scale factor of the enlargement.

8. (a) Draw axes so that both x and y can take values from −5 to +15.
(b) Draw triangle PQR at P(2, 1), Q(7, 1), R(2, 4).
(c) Find the image of PQR under the transformation represented by the matrix $\begin{pmatrix} 2 & 1 \\ -1 & 2 \end{pmatrix}$ and plot the image on the graph.
(d) The transformation is a rotation followed by an enlargement. Calculate the angle of the rotation and the scale factor of the enlargement.

9. (a) On graph paper, draw the triangle T whose vertices are (2, 2), (6, 2) and (6, 4).
(b) Draw the image U of T under the transformation whose matrix is $\begin{pmatrix} 0 & 1 \\ 1 & 0 \end{pmatrix}$.
(c) Draw the image V of T under the transformation whose matrix is $\begin{pmatrix} 1 & 0 \\ 0 & -1 \end{pmatrix}$.
(d) Describe the single transformation which would map U onto V.

10. (a) Find the images of the points (1, 0), (2, 1), (3, −1), (−2, 3) under the transformation with matrix $\begin{pmatrix} 1 & 3 \\ 2 & 6 \end{pmatrix}$.
(b) Show that the images lie on a straight line, and find its equation.

11. The transformation with matrix $\begin{pmatrix} 2 & 3 \\ 6 & 9 \end{pmatrix}$ maps every point in the plane onto a line. Find the equation of the line.

12. Using a scale of 1 cm to one unit in each case draw x- and y-axes, taking values of x from −4 to +6 and values of y from 0 to 12.
(a) Draw and label the quadrilateral OABC with O(0, 0), A(2, 0), B(4, 2), C(0, 2).
(b) Find and draw the image of OABC under the transformation whose matrix is **R**, where $\mathbf{R} = \begin{pmatrix} 2{\cdot}4 & -1{\cdot}8 \\ 1{\cdot}8 & 2{\cdot}4 \end{pmatrix}$.
(c) Calculate, in surd form, the lengths OB and O′B′.
(d) Calculate the angle AOA′.

(e) Given that the transformation **R** consists of a rotation about O followed by an enlargement, state the angle of the rotation and the scale factor of the enlargement.

Example 3

Apply the shear represented by the matrix $\begin{pmatrix} 1 & 2 \\ 0 & 1 \end{pmatrix}$ to the rectangle A(0, 0), B(0, 2), C(3, 2), D(3, 0).

$$\begin{pmatrix} 1 & 2 \\ 0 & 1 \end{pmatrix} \overset{\begin{matrix} A & B & C & D \end{matrix}}{\begin{pmatrix} 0 & 0 & 3 & 3 \\ 0 & 2 & 2 & 0 \end{pmatrix}} = \overset{\begin{matrix} A' & B' & C' & D' \end{matrix}}{\begin{pmatrix} 0 & 4 & 7 & 3 \\ 0 & 2 & 2 & 0 \end{pmatrix}}$$

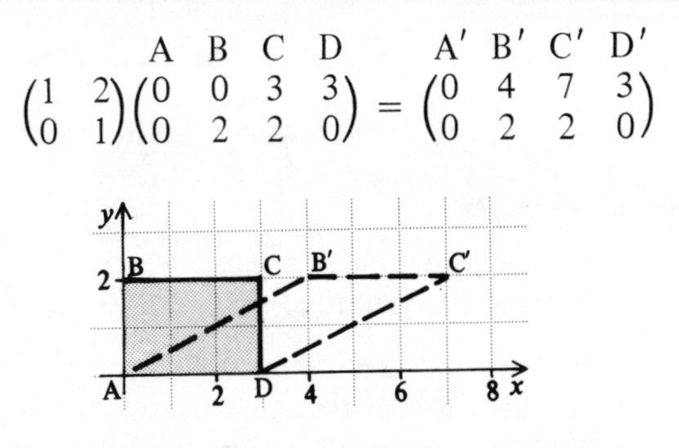

The line AD, on the x-axis is the invariant line.

Shear

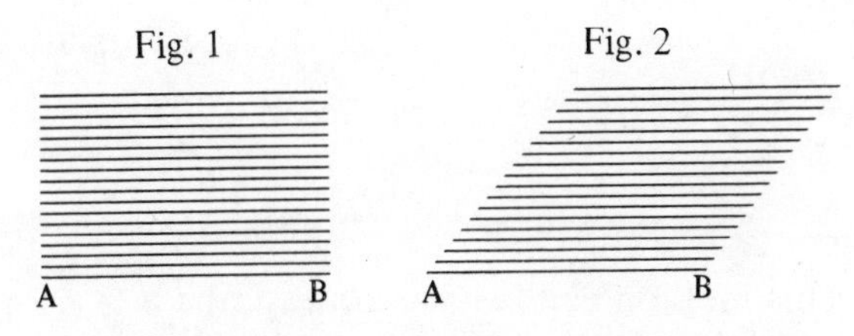

Fig. 1 shows a pack of cards stacked neatly into a pile.
Fig. 2 shows the same pack after a *shear* has been performed.

Note
(a) the card AB, at the bottom has not moved (we say the line AB is invariant).
(b) the distance moved by any card depends on its distance from the base card. The card at the top moves twice as far as the card in the middle.
(c) the height of the pile of cards remains constant. (This means that the area of the cross-section stays the same.)

A shear is fully described if we know
(a) the invariant line
(b) the image of a point not on the invariant line.

It is convenient to describe a shear using a matrix.

$\begin{pmatrix} 1 & k \\ 0 & 1 \end{pmatrix}$ is a shear with the x-axis invariant

$\begin{pmatrix} 1 & 0 \\ k & 1 \end{pmatrix}$ is a shear with the y-axis invariant.

Example 4

Apply the shear $\begin{pmatrix} 1 & 0 \\ 1 & 1 \end{pmatrix}$ to the triangle L(−2, 0), M(0, 2), N(5, 0).

$$\begin{pmatrix} 1 & 0 \\ 1 & 1 \end{pmatrix} \overset{\begin{matrix} L & M & N \end{matrix}}{\begin{pmatrix} -2 & 0 & 5 \\ 0 & 2 & 0 \end{pmatrix}} = \overset{\begin{matrix} L' & M' & N' \end{matrix}}{\begin{pmatrix} -2 & 0 & 5 \\ -2 & 2 & 5 \end{pmatrix}}$$

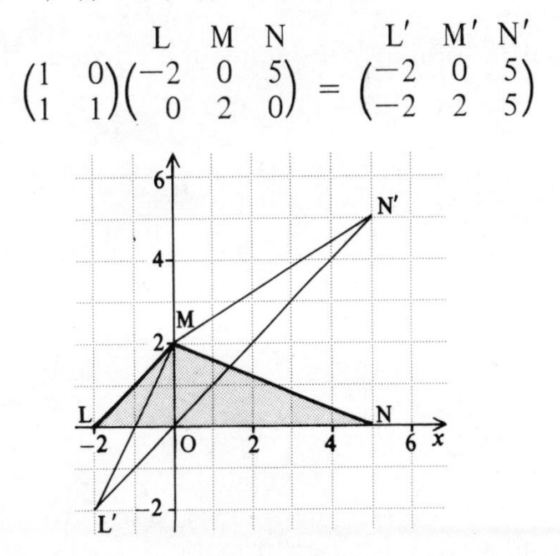

The line OM, on the y-axis, is the invariant line. Notice that points L and N, on opposite sides of the invariant line, move in opposite directions. Again the displacement is proportional to the distance from the invariant line.

Stretch

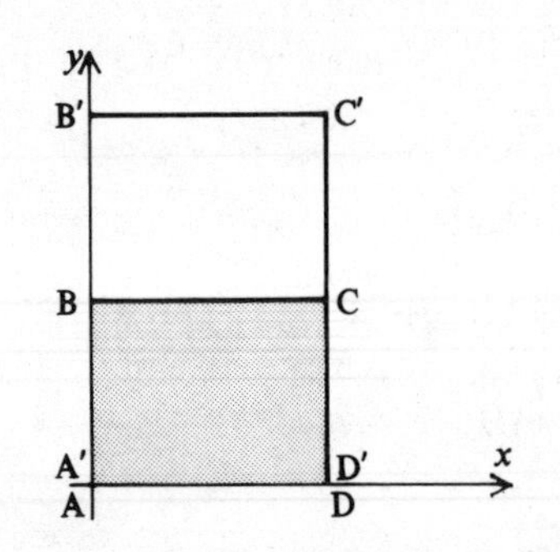

The rectangle ABCD has been *stretched* in the direction of the y-axis so that A′B′ is twice AB.

A stretch is fully described if we know
(a) the direction of the stretch,
(b) the ratio of corresponding lengths.

The matrix $\begin{pmatrix} k & 0 \\ 0 & 1 \end{pmatrix}$ represents a stretch parallel to the x-axis where the ratio of corresponding lengths is k.

Describing a matrix using base vectors

It is possible to describe a transformation in matrix form by considering the effect on the *base vectors* $\begin{pmatrix} 1 \\ 0 \end{pmatrix}$ and $\begin{pmatrix} 0 \\ 1 \end{pmatrix}$.

We will let $\begin{pmatrix} 1 \\ 0 \end{pmatrix}$ be I and $\begin{pmatrix} 0 \\ 1 \end{pmatrix}$ be J.

The *columns* of a matrix give us the images of I and J after the transformation.

Example 5

Describe the transformation with matrix $\begin{pmatrix} 0 & 1 \\ -1 & 0 \end{pmatrix}$.

Column $\begin{pmatrix} 0 \\ -1 \end{pmatrix}$ represents I′ (the image of I).

Column $\begin{pmatrix} 1 \\ 0 \end{pmatrix}$ represents J′ (the image of J)

$$\begin{matrix} \text{I}' & \text{J}' \end{matrix}$$
$$\begin{pmatrix} 0 & 1 \\ -1 & 0 \end{pmatrix}$$

Draw I, J, I′ and J′ on a diagram.

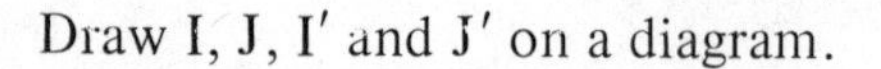

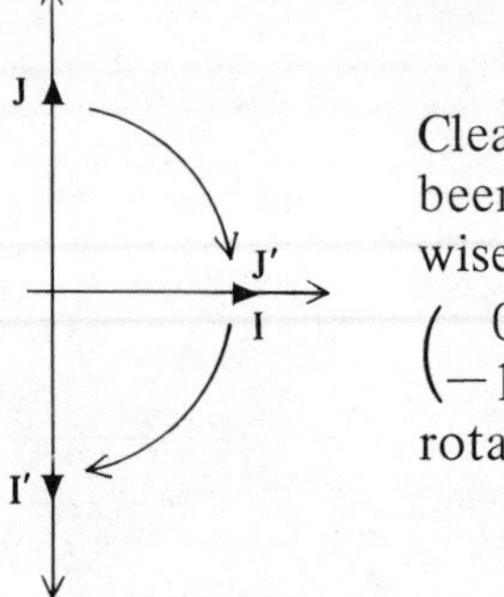

Clearly both I and J have been rotated 90° clockwise about the origin.
$\begin{pmatrix} 0 & 1 \\ -1 & 0 \end{pmatrix}$ represents a rotation of $-90°$.

This method can be used to describe a reflection, rotation, enlargement, shear or stretch in which the origin remains fixed.

Exercise 14

1. In the diagram, OABC has been mapped onto OA′B′C by a shear.

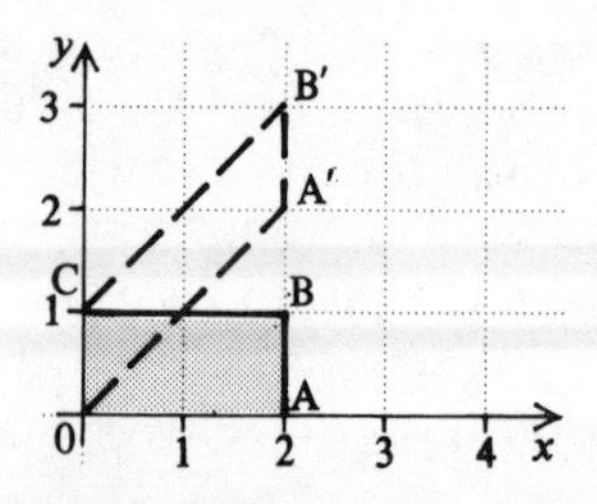

What is the invariant line of the shear?

2. What is the equation of the invariant line for the shear which maps OAB onto OAB′?

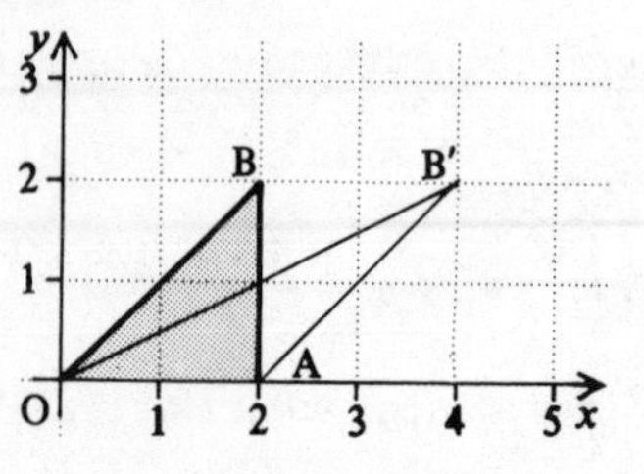

In questions **3** and **4**, draw axes for values of x from -6 to $+9$ and for values of y from -2 to $+5$.

3. Find the coordinates of the image of each of the following shapes under the shear represented by the matrix $\begin{pmatrix}1 & 2\\0 & 1\end{pmatrix}$.
Draw each object and image together on a diagram.
(a) (0, 0) (0, 3) (2, 3) (2, 0)
(b) (0, 0) (−2, 0) (−2, −2) (0, −2)
(c) (0, 0) (2, 2) (3, 0)
(d) (1, 1) (1, 3) (3, 3) (3, 1).
What is the invariant line for this shear?

4. Find the coordinates of the image of each of the following shapes under the shear represented by the matrix $\begin{pmatrix}1 & 0\\1 & 1\end{pmatrix}$.
Draw each object and image together on a diagram.
(a) (0, 0) (3, 0) (3, 2) (0, 2)
(b) (0, 0) (2, 3) (0, 4)
(c) (0, 2) (−2, 0) (−2, 3)
(d) (2, 0) (2, −1) (−1, −1) (−1, 0).
What is the invariant line for this shear?

In questions **5** to **12**, plot the rectangle ABCD at A(0, 0), B(0, 2), C(3, 2), D(3, 0). Find and draw the image of ABCD under the transformation given and describe the transformation fully.

5. $\begin{pmatrix}2 & 0\\0 & 1\end{pmatrix}$ **6.** $\begin{pmatrix}3 & 0\\0 & 1\end{pmatrix}$ **7.** $\begin{pmatrix}1 & 0\\0 & 2\end{pmatrix}$

8. $\begin{pmatrix}1\frac{1}{2} & 0\\0 & 1\end{pmatrix}$ **9.** $\begin{pmatrix}1 & 1\\0 & 1\end{pmatrix}$ **10.** $\begin{pmatrix}-2 & 0\\0 & 1\end{pmatrix}$

11. $\begin{pmatrix}1 & 0\\0 & 3\end{pmatrix}$ **12.** $\begin{pmatrix}\frac{1}{2} & 0\\0 & 1\end{pmatrix}$

In questions **13** to **24**, use base vectors to describe the transformation represented by each matrix.

13. $\begin{pmatrix}0 & -1\\1 & 0\end{pmatrix}$ **14.** $\begin{pmatrix}-1 & 0\\0 & 1\end{pmatrix}$ **15.** $\begin{pmatrix}0 & -1\\-1 & 0\end{pmatrix}$

16. $\begin{pmatrix}0 & 1\\1 & 0\end{pmatrix}$ **17.** $\begin{pmatrix}2 & 0\\0 & 2\end{pmatrix}$ **18.** $\begin{pmatrix}\frac{1}{2} & 0\\0 & \frac{1}{2}\end{pmatrix}$

19. $\begin{pmatrix}3 & 0\\0 & 1\end{pmatrix}$ **20.** $\begin{pmatrix}1 & 1\\0 & 1\end{pmatrix}$ **21.** $\begin{pmatrix}1 & 0\\2 & 1\end{pmatrix}$

22. $\begin{pmatrix}1 & 0\\0 & 2\end{pmatrix}$ **23.** $\begin{pmatrix}-2 & 0\\0 & -2\end{pmatrix}$ **24.** $\begin{pmatrix}-\frac{1}{2} & 0\\0 & -\frac{1}{2}\end{pmatrix}$

25. (a) Find and draw the image of the square (0, 0), (1, 1), (0, 2), (−1, 1) under the transformation represented by the matrix $\begin{pmatrix}4 & 3\\-3 & -2\end{pmatrix}$.
(b) Show that the transformation is a shear and find the equation of the invariant line.

26. (a) Find and draw the image of the square (0, 0), (1, 1), (0, 2), (−1, 1) under the shear represented by the matrix $\begin{pmatrix}0{\cdot}5 & -0{\cdot}5\\0{\cdot}5 & 1{\cdot}5\end{pmatrix}$.
(b) Find the equation of the invariant line.

27. Find and draw the image of the square (0, 0), (1, 0), (1, 1), (0, 1) under the transformation represented by the matrix $\begin{pmatrix}3 & 0\\0 & 2\end{pmatrix}$.
This transformation is called a two-way stretch.

28. Find and draw the image of the square (0, 0), (1, 0), (1, 1), (0, 1) under the transformation represented by the matrix $\begin{pmatrix}2 & 0\\0 & 4\end{pmatrix}$.
Describe the transformation.

29. Under a shear, the image of (1, 1) is (2, 1) and the image of (2, 2) is (4, 2). Find the invariant line and the matrix which represents the shear.

30. Under a shear, the image of (1, 1) is (1, 3) and the image of (3, 1) is (3, 7). Find the invariant line and the matrix which represents the shear.

In questions **31** to **42**, use base vectors to write down the matrix which represents each of the transformations.

31. Rotation +90° about (0, 0).
32. Reflection in $y = x$.
33. Reflection in $x = 0$.
34. Rotation 180° about (0, 0).
35. Enlargement, centre (0, 0), scale factor 3.

36. Reflection in $y = -x$.
37. Enlargement, centre (0, 0), scale factor −2.
38. Stretch, parallel to y-axis, scale factor 4.
39. Reflection in $y = 0$.
40. Rotation −90° about (0, 0).
41. Enlargement, centre (0, 0), scale factor $\frac{1}{2}$.
42. Stretch, parallel to x-axis, scale factor 3.

8.7 GENERAL TRANSFORMATIONS

Sometimes we need a vector as well as a matrix to describe a transformation.

Example 1

Let $\mathbf{R} = \begin{pmatrix} 0 & 1 \\ -1 & 0 \end{pmatrix}$ and $\mathbf{T} = \begin{pmatrix} 3 \\ -2 \end{pmatrix}$.

Let transformation **X** be '**R** followed by **T**'

i.e. **X** is given by $\begin{pmatrix} x \\ y \end{pmatrix} \rightarrow \begin{pmatrix} 0 & 1 \\ -1 & 0 \end{pmatrix}\begin{pmatrix} x \\ y \end{pmatrix} + \begin{pmatrix} 3 \\ -2 \end{pmatrix}$

Apply **X** to the rectangle A(0, 0), B(0, 1), C(2, 1), D(2, 0).

(a) perform **R**: $\begin{pmatrix} 0 & 1 \\ -1 & 0 \end{pmatrix}\overset{\begin{matrix} A & B & C & D \end{matrix}}{\begin{pmatrix} 0 & 0 & 2 & 2 \\ 0 & 1 & 1 & 0 \end{pmatrix}}$

$= \overset{\begin{matrix} A' & B' & C' & D' \end{matrix}}{\begin{pmatrix} 0 & 1 & 1 & 0 \\ 0 & 0 & -2 & -2 \end{pmatrix}}$

perform **R** on ABCD

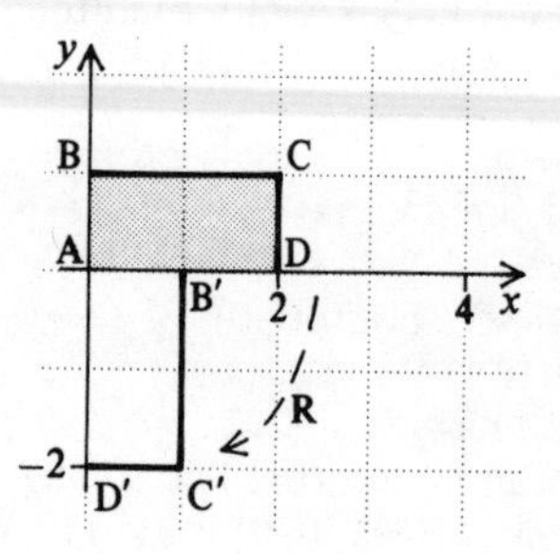

perform **T** on A′B′C′D′

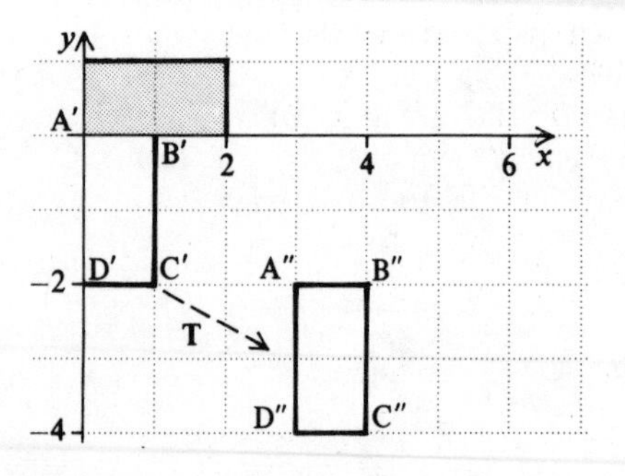

The transformation **X** has taken the rectangle ABCD to rectangle A″B″C″D″.

Area scale factor

The *determinant* of a matrix gives the *area scale factor* of the transformation.

Example 2

(a) Draw triangle ABC at A(1, 1), B(4, 1), C(1, 3) and find the image of ABC under the transformation with matrix $\mathbf{M} = \begin{pmatrix} 2 & 0 \\ 0 & 2 \end{pmatrix}$.

(b) Find the ratio, $\dfrac{\text{area of image}}{\text{area of object}}$.

(c) Write down the determinant of **M** and compare your answer with the ratio in part (b).

(a)

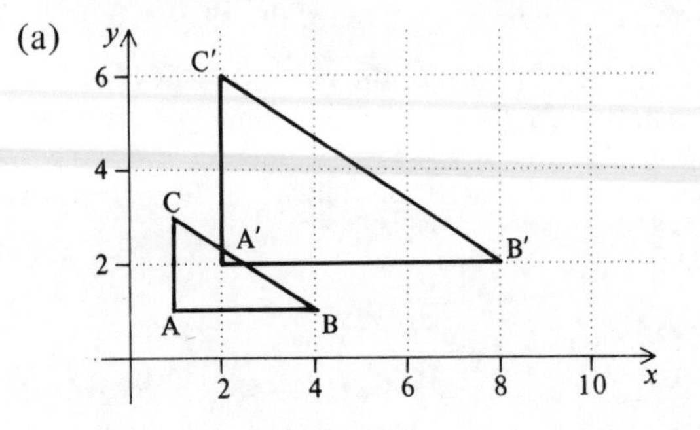

$$\begin{pmatrix} 2 & 0 \\ 0 & 2 \end{pmatrix}\overset{\begin{matrix} A & B & C \end{matrix}}{\begin{pmatrix} 1 & 4 & 1 \\ 1 & 1 & 3 \end{pmatrix}} = \overset{\begin{matrix} A' & B' & C' \end{matrix}}{\begin{pmatrix} 2 & 8 & 2 \\ 2 & 2 & 6 \end{pmatrix}}$$

(b) Area of image = 12 square units
Area of object = 3 square units.

$$\therefore \quad \frac{\text{area of image}}{\text{area of object}} = \frac{12}{3} = 4$$

(c) The determinant of $\mathbf{M} = (2 \times 2) - (0 \times 0)$
$= 4$

The area scale factor is the same as the determinant of the matrix.

Exercise 15

1. Draw the rectangle (0, 0), (0, 1), (2, 1), (2, 0) and its image under the following transformations and describe the *single* transformation which each represents:
   (a) $\begin{pmatrix}0 & 1\\1 & 0\end{pmatrix}\begin{pmatrix}x\\y\end{pmatrix}+\begin{pmatrix}1\\-1\end{pmatrix}$
   (b) $\begin{pmatrix}1 & 0\\0 & -1\end{pmatrix}\begin{pmatrix}x\\y\end{pmatrix}+\begin{pmatrix}0\\2\end{pmatrix}$
   (c) $\begin{pmatrix}0 & 1\\-1 & 0\end{pmatrix}\begin{pmatrix}x\\y\end{pmatrix}+\begin{pmatrix}4\\0\end{pmatrix}$
   (d) $\begin{pmatrix}3 & 0\\0 & 3\end{pmatrix}\begin{pmatrix}x\\y\end{pmatrix}+\begin{pmatrix}-4\\2\end{pmatrix}$

2. (a) Draw L(1, 1), M(3, 3), N(4, 1) and its image L′M′N′ under the matrix $\mathbf{A}=\begin{pmatrix}1 & 0\\0 & -1\end{pmatrix}$.
   (b) Find and draw the image of L′M′N′ under matrix $\mathbf{B}=\begin{pmatrix}0 & 1\\-1 & 0\end{pmatrix}$ and label it L″M″N″.
   (c) Calculate the matrix product **BA**,
   (d) Find the image of LMN under the matrix **BA**, and compare with the result of performing **A** and then **B**.

3. (a) (i) Draw P(0, 0), Q(2, 2), R(4, 0) and its image P′Q′R′ under matrix $\mathbf{A}=\begin{pmatrix}2 & 0\\0 & 2\end{pmatrix}$.
   (ii) Find and draw the image of P′Q′R′ under matrix $\mathbf{B}=\begin{pmatrix}1 & 1\\0 & 1\end{pmatrix}$ and label it P″Q″R″.
   (iii) Calculate the matrix product **BA**.
   (iv) Find the image of PQR under the matrix **BA**, and compare with the result of performing **A** and then **B**.
   (b) Repeat part (a) using $\mathbf{A}=\begin{pmatrix}0 & -1\\-1 & 0\end{pmatrix}$ and $\mathbf{B}=\begin{pmatrix}1 & 0\\1 & 1\end{pmatrix}$.

4. (a) Draw D(1, 2), E(1, 5), F(3, 4) and its image D′E′F′ under matrix $\mathbf{X}=\begin{pmatrix}1 & 0\\0 & -1\end{pmatrix}$.
   (b) Find and draw D″E″F″, the image of D′E′F′ under matrix $\mathbf{Y}=\begin{pmatrix}0 & -1\\-1 & 0\end{pmatrix}$.
   (c) Find and draw D*E*F*, the image of D″E″F″ under matrix $\mathbf{Z}=\begin{pmatrix}0 & 1\\1 & 0\end{pmatrix}$.
   (d) Calculate the matrix products **YX** and **Z(YX)**.
   (e) Find the image of DEF under matrix **ZYX** and compare it with D*E*F*.

5. (a) Draw A(1, 1), B(4, 1), C(4, 3) and its image A′B′C′ under matrix $\mathbf{P}=\begin{pmatrix}-1 & 0\\0 & 1\end{pmatrix}$.
   (b) Find and draw A″B″C″, the image of A′B′C′ under matrix $\mathbf{Q}=\begin{pmatrix}1 & 0\\0 & 1\end{pmatrix}$.
   (c) Find and draw A*B*C*, the image of A″B″C″ under matrix $\mathbf{R}=\begin{pmatrix}0 & -1\\1 & 0\end{pmatrix}$.
   (d) Calculate the matrix products **QP** and **R(QP)**,
   (e) Find the image of ABC under matrix **RQP** and compare it with A*B*C*.

6. (a) Draw L(1, 1), M(3, 3), N(4, 1) and its image L′M′N′ under matrix $\mathbf{K}=\begin{pmatrix}2 & 0\\0 & 2\end{pmatrix}$.
   Find $\mathbf{K}^{-1}$, the inverse of **K**, and now find the image of L′M′N′ under $\mathbf{K}^{-1}$.
   (b) Repeat part (a) with $\mathbf{K}=\begin{pmatrix}1 & 2\\0 & 1\end{pmatrix}$.
   (c) Repeat part (a) with $\mathbf{K}=\begin{pmatrix}3 & 0\\0 & 1\end{pmatrix}$.

7. The image (x', y') of a point (x, y) under a transformation is given by
   $\begin{pmatrix}x'\\y'\end{pmatrix}=\begin{pmatrix}3 & 0\\1 & -2\end{pmatrix}\begin{pmatrix}x\\y\end{pmatrix}+\begin{pmatrix}2\\5\end{pmatrix}$
   (a) Find the coordinates of the image of the point (4, 3),
   (b) The image of the point (m, n) is the point (11, 7). Write down two equations involving m and n and hence find the values of m and n.
   (c) The image of the point (h, k) is the point (5, 10). Find the values of h and k.

8. Write down the area scale factor for each of the following matrices.
   (a) $\begin{pmatrix}2 & 0\\0 & 2\end{pmatrix}$ (b) $\begin{pmatrix}3 & 1\\1 & 1\end{pmatrix}$ (c) $\begin{pmatrix}-1 & 0\\0 & -1\end{pmatrix}$
   (d) $\begin{pmatrix}0{\cdot}5 & 1\\-1 & 2\end{pmatrix}$ (e) $\begin{pmatrix}1 & 3\\0 & 1\end{pmatrix}$ (f) $\begin{pmatrix}1 & 0\\\frac{1}{2} & 1\end{pmatrix}$

9. What is the area scale factor for any shear?

10. (a) Draw the triangle (0, 0), (2, 2), (4, 0) and its image under the transformation with matrix $\begin{pmatrix}2 & 2\\1 & 2\end{pmatrix}$.
    (b) Calculate the area of the object and image.

11. (a) Draw the object triangle (0, 0), (2, 2), (4, 0) and its image under each of the following matrices.

(i) $\begin{pmatrix} 3 & 1 \\ -1 & 1 \end{pmatrix}$ (ii) $\begin{pmatrix} \frac{1}{2} & -2 \\ 1 & 4 \end{pmatrix}$

(iii) $\begin{pmatrix} 2{\cdot}4 & -1{\cdot}8 \\ 1{\cdot}8 & 2{\cdot}4 \end{pmatrix}$

(b) Use the area scale factor of the matrix to calculate the area of the image in each case.

12. (a) Find and draw the image of the square (0, 0), (1, 1), (0, 2), (−1, 1) under the shear with matrix $\begin{pmatrix} 4 & 3 \\ -3 & -2 \end{pmatrix}$.

(b) Calculate the area of the image.

13. Draw A(0, 2), B(2, 2), C(0, 4) and its image under an enlargement, A′(2, 2), B′(6, 2), C′(2, 6).

(a) What is the centre of enlargement?

(b) Find the image of ABC under an enlargement, scale factor 2, centre (0, 0).

(c) Find the translation which maps this image onto A′B′C′.

(d) What is the matrix **X** and vector **v** which represents an enlargement scale factor 2, centre (−2, 2)?

14. Draw A(0, 1), B(1, 1), C(1, 3) and its image under a reflection A′(4, 1), B′(3, 1), C′(3, 3).

(a) What is the equation of the mirror line?

(b) Find the image of ABC under a reflection in the line $x = 0$.

(c) Find the translation which maps this image onto A′B′C′.

(d) What is the matrix **X** and vector **v** which represents a reflection in the line $x = 2$?

15. Use the same approach as in questions **13** and **14** to find the matrix **X** and vector **v** which represents the following transformations. (Start by drawing an object and its image under the transformation.)

(a) Enlargement scale factor 2, centre (1, 3),

(b) Enlargement scale factor 2, centre ($\frac{1}{2}$, 1),

(c) Reflection in $y = x + 3$,

(d) Rotation 180°, centre ($1\frac{1}{2}$, $2\frac{1}{2}$),

(e) Reflection in $y = 1$,

(f) Rotation −90°, centre (2, −2).

REVISION EXERCISE 8A

1. $\mathbf{A} = \begin{pmatrix} 3 & 2 \\ 1 & 4 \end{pmatrix}$, $\mathbf{B} = \begin{pmatrix} -1 & 3 \\ 0 & 2 \end{pmatrix}$.

Express as a single matrix

(a) 2**A** (b) **A** − **B** (c) $\frac{1}{2}$**A**

(d) **AB** (e) $\mathbf{B}^2$

2. Evaluate

(a) $\begin{pmatrix} -3 & 0 \\ 1 & 2 \end{pmatrix}\begin{pmatrix} 3 & \frac{1}{3} \\ 1 & \frac{1}{2} \end{pmatrix}$ (b) $\begin{pmatrix} 3 \\ 1 \end{pmatrix}\begin{pmatrix} 4 & 2 \end{pmatrix}$

(c) $\begin{pmatrix} 3 & -2 \\ 4 & 1 \end{pmatrix} + 2\begin{pmatrix} 3 & 0 \\ -1 & -4 \end{pmatrix}$

3. $\mathbf{A} = \begin{pmatrix} 4 & 2 \\ 1 & 1 \end{pmatrix}$, $\mathbf{B} = \begin{pmatrix} 1 & 5 \end{pmatrix}$, $\mathbf{C} = \begin{pmatrix} -1 \\ 3 \end{pmatrix}$.

(a) Determine **BC** and **CB**,

(b) If $\mathbf{AX} = \begin{pmatrix} 8 & 20 \\ 3 & 7 \end{pmatrix}$, where **X** is a (2 × 2) matrix, determine **X**.

4. (a) Find the inverse of the matrix $\begin{pmatrix} 2 & -1 \\ 3 & 5 \end{pmatrix}$.

(b) Solve the simultaneous equations

$$2x - y = 8$$
$$3x + 5y = -1$$

5. The determinant of the matrix $\begin{pmatrix} 3 & 2 \\ x & -1 \end{pmatrix}$ is −9.

Find the value of x and write down the inverse of the matrix.

6. $\mathbf{A} = \begin{pmatrix} 2 & 0 \\ 1 & 2 \end{pmatrix}$; h and k are numbers so that $\mathbf{A}^2 = h\mathbf{A} + k\mathbf{I}$, where $\mathbf{I} = \begin{pmatrix} 1 & 0 \\ 0 & 1 \end{pmatrix}$.

Find the values of h and k.

7. $\mathbf{M} = \begin{pmatrix} a & 1 \\ 1 & -a \end{pmatrix}$.

(a) Find the values of a if $\mathbf{M}^2 = 17\begin{pmatrix} 1 & 0 \\ 0 & 1 \end{pmatrix}$.

(b) Find the values of a if $|\mathbf{M}| = -10$.

8. Using the diagram below, describe the transformations for the following,
(a) $T_1 \to T_6$ (b) $T_4 \to T_5$ (c) $T_8 \to T_2$
(d) $T_4 \to T_1$ (e) $T_8 \to T_4$ (f) $T_6 \to T_8$

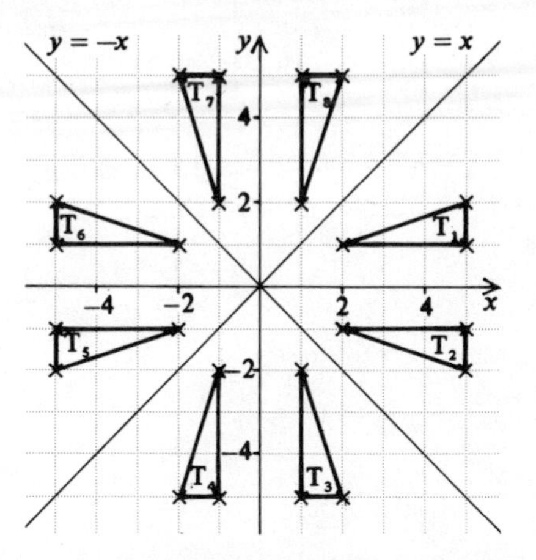

9. Find the coordinates of the image of (1, 4) under
(a) a clockwise rotation of 90° about (0, 0),
(b) a reflection in the line $y = x$,
(c) a translation which maps (5, 3) onto (1, 1).

10. **M** is a reflection in the line $x + y = 0$.
R is an anticlockwise rotation of 90° about (0, 0).
T is a translation which maps (−1, −1) onto (2, 0).
Find the image of the point (3, 1) under
(a) **M** (b) **R** (c) **T**
(d) **MR** (e) **RT** (f) **TMR**

11. **A** is a rotation of 180° about (0, 0).
B is a reflection in the line $x = 3$.
C is a translation which maps (3, −1) onto (−2, −1).
Find the image of the point (1, −2) under
(a) **A** (b) $\mathbf{A}^2$ (c) **BC**
(d) $\mathbf{C}^{-1}$ (e) **ABC** (f) $\mathbf{C}^{-1}\mathbf{B}^{-1}\mathbf{A}^{-1}$

12. Draw x- and y-axes with values from −8 to +8.
Draw triangle A(1, −1), B(3, −1), C(1, −4).
Find the image of ABC under the following enlargements
(a) scale factor 2, centre (5, −1)
(b) scale factor 2, centre (0, 0)
(c) scale factor $\frac{1}{2}$, centre (1, 3)
(d) scale factor $-\frac{1}{2}$, centre (3, 1)
(e) scale factor −2, centre (0, 0).

13. On a graph draw x- and y-axes with values from −8 to +8 and draw triangle A(−1, 1), B(−4, 1), C(−1, 3).
Find the image of ABC under the following enlargements.
(a) scale factor 2, centre (−1, 4)
(b) scale factor −2, centre (0, 0)
(c) scale factor $\frac{1}{2}$, centre (5, 1)
(d) scale factor −1, centre (0, 4).

14. The matrix $\begin{pmatrix} 0 & -1 \\ 1 & 0 \end{pmatrix}$ represents the transformation **X**.
(a) Find the image of (5, 2) under **X**
(b) Find the image of (−3, 4) under **X**
(c) Describe the transformation **X**.

15. Draw x- and y-axes with values from −8 to +8.
Draw triangle A(2, 2), B(6, 2), C(6, 4).
Find the image of ABC under the transformations represented by the matrices,
(a) $\begin{pmatrix} 0 & -1 \\ 1 & 0 \end{pmatrix}$ (b) $\begin{pmatrix} 1 & 0 \\ 0 & -1 \end{pmatrix}$ (c) $\begin{pmatrix} -1 & 0 \\ 0 & -1 \end{pmatrix}$
(d) $\begin{pmatrix} 0 & 1 \\ -1 & 0 \end{pmatrix}$ (e) $\begin{pmatrix} 0 & -1 \\ -1 & 0 \end{pmatrix}$.
Describe each transformation.

16. Describe the transformations represented by the following matrices.
(a) $\begin{pmatrix} 0 & 1 \\ 1 & 0 \end{pmatrix}$ (b) $\begin{pmatrix} -1 & 0 \\ 0 & 1 \end{pmatrix}$ (c) $\begin{pmatrix} 3 & 0 \\ 0 & 3 \end{pmatrix}$
(d) $\begin{pmatrix} 1 & 2 \\ 0 & 1 \end{pmatrix}$ (e) $\begin{pmatrix} 1 & 0 \\ 1 & 1 \end{pmatrix}$ (f) $\begin{pmatrix} 3 & 0 \\ 0 & 1 \end{pmatrix}$

17. Write down the matrices which describe the following transformations.
(a) Rotation 180°, centre (0, 0),
(b) Reflection in the line $y = 0$,
(c) Enlargement scale factor 4, centre (0, 0),
(d) Reflection in the line $x = -y$,
(e) Clockwise rotation 90°, centre (0, 0).

18. Transformation **N**, which is given by

$$\begin{pmatrix} x' \\ y' \end{pmatrix} = \begin{pmatrix} 2 & 0 \\ 0 & 2 \end{pmatrix}\begin{pmatrix} x \\ y \end{pmatrix} + \begin{pmatrix} 5 \\ -2 \end{pmatrix},$$

is composed of two single transformations.
(a) Describe each of the transformations,
(b) Find the image of the point (3, −1) under **N**,
(c) Find the image of the point $(-1, \frac{1}{2})$ under **N**,
(d) Find the point which is mapped by **N** onto the point (7, 4).

19. **A** is the reflection in the line $y = x$.
B is the reflection in the y-axis.
Find the matrix which represents
(a) **A** (b) **B** (c) **AB** (d) **BA**.
Describe the single transformations **AB** and **BA**.

20. Describe the single transformation which maps

(a) $\triangle ABC$ onto $\triangle DEF$ (b) $\triangle ABC$ onto $\triangle PQR$
(c) $\triangle ABC$ onto $\triangle XYZ$.

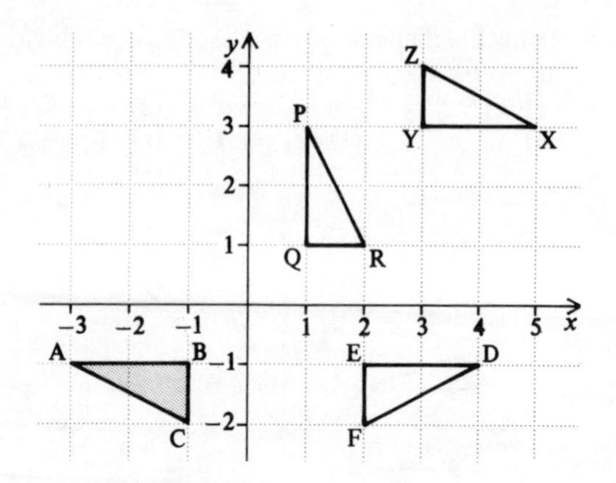

EXAMINATION EXERCISE 8B

1. On graph paper, using axes such that $-8 \leqslant x \leqslant 10$ and $-10 \leqslant y \leqslant 8$, and using a scale of 1 cm to represent 1 unit on each axis, construct $\triangle ABC$ in which A is (4, 0), B is (4, 2) and C is (0, 2). **R** is a positive (anti-clockwise) quarter-turn about the point (5, 3) and **S** is a positive quarter-turn about the point (−1, −2).

(a) Construct the images of $\triangle ABC$ under each of the transformations **R**, **S**, **RS** and **SR**. Label your images clearly.

(b) Each of the transformations **RS** and **SR** is equivalent to a single transformation. Describe fully each of these single transformations.

(c) Describe fully the transformation **M** such that **MSR** = **RS**. [L]

2. The square ABCD has vertices A(1, 1), B(2, 1), C(2, 2), D(1, 2) and the square PQRS has vertices P(−1, −1), Q(−2, −1), R(−2, −2), S(−1, −2). Describe in geometrical terms

(i) the transformation **T** which maps A, B, C, D on to P, Q, R, S respectively,
(ii) the transformation **U** which maps A, B, C, D on to P, S, R, Q respectively,
(iii) the transformation **V** which maps A, B, C, D on to R, S, P, Q respectively,
(iv) the transformation **W** which maps A, B, C, D on to Q, R, S, P respectively.

Write down the 2 × 2 matrices corresponding to the transformations **T** and **U**. [JMB]

3. (i) Using a scale of 1 cm to 1 unit and marking values on the x and y axes from $^-8$ to $^+5$, plot and label the following points: A($^-6$, 2), B($^-2$, 4), C(3, $^-1$), D(2, $^-4$).
Draw triangles OBA and OCD, where O is the origin.

(ii) A transformation **Y** consists of a reflection followed by an enlargement, centre O. **Y** maps triangle OCD onto triangle OBA.
By measurement of two appropriate lengths, determine the linear scale factor of the enlargement correct to two significant figures.

(iii) (a) Calculate the areas of triangle OCD and triangle OBA.
(b) Find the area scale factor of the enlargement.
(c) Hence find a more accurate value of the linear scale factor.

(iv) Describe geometrically the transformation $\mathbf{Y}^2$. [SMP]

4. (i) Using a scale of 1 cm to represent 1 unit on each axis, draw x and y axes, taking values of x from -8 to 12 and values of y from -6 to 14.
Draw and label the triangle X, with vertices (2, 4), (4, 4) and (4, 1).
(ii) The single transformation **U** maps the triangle X onto the triangle **U**(X) which has vertices (6, 12), (12, 12) and (12, 3).
Draw and label the triangle **U**(X) and describe fully the transformation **U**.
(iii) The transformation **R** is a clockwise rotation of 90° about the origin.
Draw and label the triangle **R**(X).
(iv) The transformation **T** is the translation $\begin{pmatrix}-8\\4\end{pmatrix}$.
Draw and label the triangle **T**(X) and the triangle **RT**(X).
(v) The single transformation **V** is represented by the matrix $\begin{pmatrix}0 & -1\\-1 & 0\end{pmatrix}$.
Draw and label the triangle **V**(X) and describe fully the transformation **V**.
[C]

5. A transformation is defined by $\mathbf{T}: \begin{pmatrix}x\\y\end{pmatrix} \to \begin{pmatrix}2{\cdot}4 & -1{\cdot}8\\1{\cdot}8 & 2{\cdot}4\end{pmatrix}\begin{pmatrix}x\\y\end{pmatrix}$.
(i) Given that O = (0, 0), A = (4, 0), B = (4, 4), C = (0, 4), find the image O′A′B′C′ of the square OABC under this transformation and represent the square and its image on a diagram, using a scale of 1 cm: 1 unit on each axis.
(ii) Calculate the length of OA′ and the size of angle AOA′.
(iii) Given that **T** is equivalent to a rotation about O followed by an enlargement, state the angle of the rotation and the scale factor of the enlargement.
(iv) Find the area of O′A′B′C′ and the length of A′C′. [JMB]

6. (a) Write down the inverse of $\begin{pmatrix}4 & 3\\3 & 2\end{pmatrix}$, and, hence or otherwise, determine p, q, r, and s such that

$$\begin{pmatrix}4 & 3\\3 & 2\end{pmatrix}\begin{pmatrix}p & q\\r & s\end{pmatrix} = \begin{pmatrix}1 & 2\\3 & 4\end{pmatrix}.$$

(b) Given that $\mathbf{M} = \begin{pmatrix}1 & 0\\0 & -1\end{pmatrix}$ and $\mathbf{R} = \begin{pmatrix}0 & -1\\1 & 0\end{pmatrix}$, identify the geometrical transformation each represents.
Evaluate as 2 × 2 matrices (i) **RM**, (ii) **MR**, and identify fully the geometrical transformation each represents. [O]

7. (i) (a) Draw x and y axes for values of x from $^-8$ to $^+16$ and values of y from $^-16$ to $^+16$, using a scale of 1 cm to 2 units. On your diagram draw and label the triangle whose vertices are P(6, $^-8$), Q(2, 14) and R(9, 13).
(b) The reflection of PQR in the line $y = 0$ is the triangle $P_1Q_1R_1$. On your diagram draw and label this triangle.
(ii) Triangle $P_2Q_2R_2$ is the image of PQR under the reflection whose matrix is $\begin{pmatrix}0{\cdot}28 & 0{\cdot}96\\0{\cdot}96 & {}^-0{\cdot}28\end{pmatrix}$.
(a) Calculate the coordinates of P_2, Q_2 and R_2.
(b) On your diagram draw and label $P_2Q_2R_2$.
Draw also the mirror line of this reflection, and write down its equation.
(iii) $P_1Q_1R_1$ may be mapped directly onto $P_2Q_2R_2$ by a rotation about the origin. By measurement from your diagram, or otherwise, find the angle of this rotation. [SMP]

8. The triangle ABC has vertices A(2, 0), B(4, 4) and C(0, 1).
The triangle PQR has vertices P(8, −2), Q(4, 0) and R(7, −4).
The triangle LMN has vertices L(−2, −7), M(−6, −9) and N(−3, −5).
Draw these triangles on graph paper, using a scale of 1 cm to 1 unit on each axis, and label the vertices.
ΔABC can be mapped onto ΔPQR by an anticlockwise rotation about the origin followed by a translation.
(i) State the angle of rotation.
(ii) Find the matrix which represents this rotation.
(iii) Find the column vector of the translation.
(iv) Given that ΔABC can be mapped onto ΔPQR by a single rotation, find the coordinates of the centre of this rotation.
(v) Given that ΔABC can be mapped onto ΔLMN by a translation of $\begin{pmatrix} 0 \\ -3 \end{pmatrix}$ followed by a reflection in the mirror line m, draw the line m on your graph and label it clearly.
(vi) Find the equation of m. [C]

9. The triangle T_1 has vertices O(0, 0), A(0, 2), B(⁻4, 0).
Under the transformation defined by the matrix **M**, where $\mathbf{M} = \begin{pmatrix} 0 & -2 \\ 2 & 0 \end{pmatrix}$, the image of T_1 is T_2. Similarly the image of T_2 under **M** is T_3, and so on.
(i) Calculate the co-ordinates of the vertices of T_2 and T_3, and draw and label T_1, T_2 and T_3 on a single diagram, taking a scale of 1 cm to 2 units on each axis.
(ii) The transformation represented by **M** is a combination of a quarter turn about the origin and another simple transformation. Describe precisely the other transformation.
(iii) Draw T_0, the image of T_1 under the transformation defined by $\mathbf{M}^{-1}$, the inverse of **M**.
(iv) The matrix $\mathbf{M}^n$ will transform T_0 to T_n. What is the smallest positive value of n for which $\mathbf{M}^n$ represents an enlargement? [SMP]

10. (a) A transformation T, which is given by

$$\begin{pmatrix} x \\ y \end{pmatrix} \to \begin{pmatrix} 0 & 1 \\ 1 & 0 \end{pmatrix}\begin{pmatrix} x \\ y \end{pmatrix} + \begin{pmatrix} -5 \\ 5 \end{pmatrix},$$

is composed of two single transformations.
(i) Give a full geometrical description of each of these two transformations.
(ii) Find the point (x, y) which is mapped by T onto the point (−2, 1).
(iii) Find two points which are unchanged by the transformation T.
(iv) Give a full geometrical description of T as a single transformation.

(b) For the transformation P, given by

$$\begin{pmatrix} x \\ y \end{pmatrix} \to \begin{pmatrix} 1 & 2 \\ 3 & 6 \end{pmatrix}\begin{pmatrix} x \\ y \end{pmatrix},$$

show that *every* point in the plane is mapped onto a line and state the equation of the line. [C 16+]

9 Sets, vectors and functions

Bertrand Russell (1872–1970) tried to reduce all mathematics to formal logic. He showed that the idea of a set of all sets which are not members of themselves leads to contradictions. He wrote to Gottlieb Frege just as he was putting the finishing touches to a book that represented his life's work, pointing out that Frege's work was invalidated. Russell's elder brother, the second Earl Russell, showed great foresight in 1903 by queueing overnight outside the vehicle licensing office in London to have his car registered as A1.

9.1 SETS

1. ∩ 'intersection'

   $A \cap B$ is shaded.

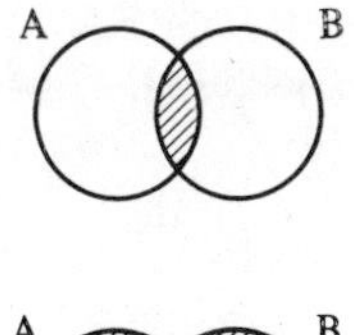

2. ∪ 'union'

   $A \cup B$ is shaded.

3. ⊂ 'is a subset of'

   $A \subset B$
   [$B \not\subset A$ means 'B is *not* a a subset of A']

4. ⊃ 'contains'

   $E \supset D$
   or 'D is a subset of E'

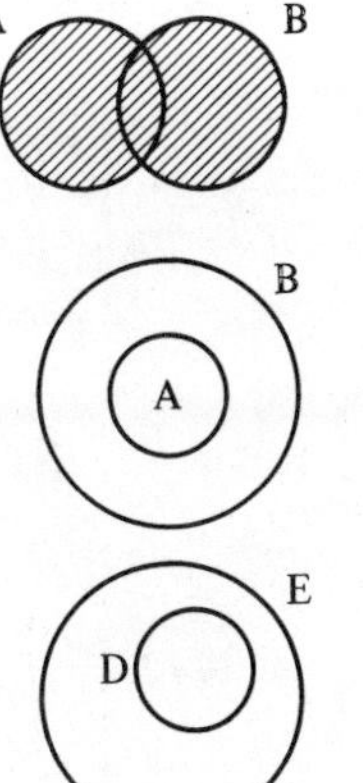

5. ∈ 'is a member of'
   'belongs to'

   $b \in X$
   [$e \notin X$ means 'e is not a member of set X']

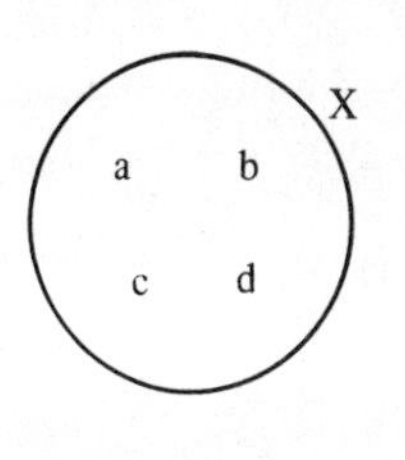

6. ℰ 'universal set'

   ℰ = {a, b, c, d, e}

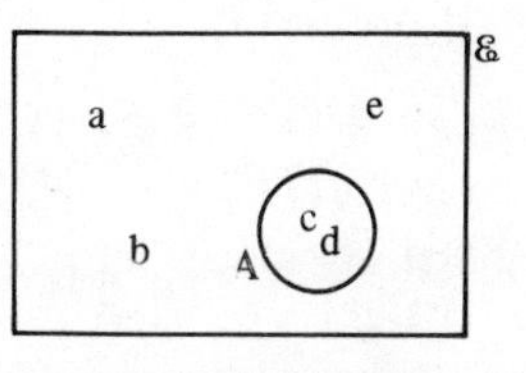

7. A′ 'complement of'
   'not in A'
   A′ is shaded
   (A ∪ A′ = ℰ)

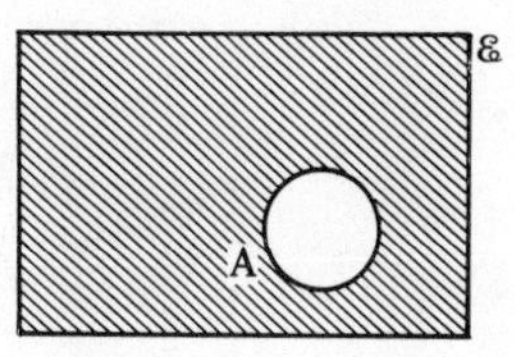

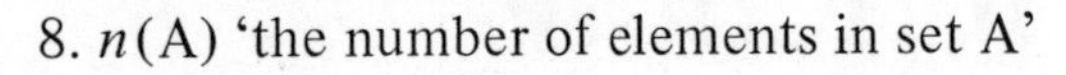

8. $n(A)$ 'the number of elements in set A'

$n(A) = 3$

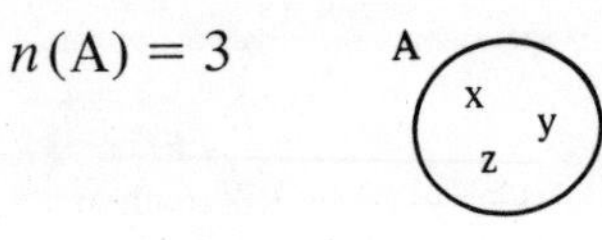

9. $A = \{x : x \text{ is an integer}, 2 \leqslant x \leqslant 9\}$

The [set of] elements x [such that] x is an integer and $2 \leqslant x \leqslant 9$.
The set A is $\{2, 3, 4, 5, 6, 7, 8, 9\}$.

10. $\emptyset$ or $\{\ \}$ 'empty set'

(Note $\emptyset \subset A$ for any set A)

Example 1

If $\mathcal{E} = \{1, 2, 3 \ldots 12\}$, $A = \{2, 3, 4, 5, 6\}$ and $B = \{2, 4, 6, 8, 10\}$, find
(a) $A \cup B$ (b) $A \cap B$
(c) A' (d) $n(A \cup B)$.

(a) $A \cup B = \{2, 3, 4, 5, 6, 8, 10\}$.

(b) $A \cap B = \{2, 4, 6\}$.

(c) $A' = \{1, 7, 8, 9, 10, 11, 12\}$.

(d) $n(A \cup B) = 7$.

Example 2

If $\mathcal{E} = \{1, 2, 3, 4, 5, 6, 7, 8, 9, 10\}$, $A = \{1, 2, 3, 7, 8\}$, $B = \{3, 4, 5, 8\}$, $C = \{5, 6, 7, 8\}$.
Find
(a) $B \cap C$ (b) $A' \cap B$
(c) $B \cap C'$ (d) $(B \cup C)'$
(e) $n(A \cup C)$

Illustrate the sets on a Venn diagram.

(a) $B \cap C = \{5, 8\}$ (b) $A' \cap B = \{4, 5\}$
(c) $B \cap C' = \{3, 4\}$
(d) $(B \cup C)' = \{1, 2, 9, 10\}$
(e) $n(A \cup C) = 7$.

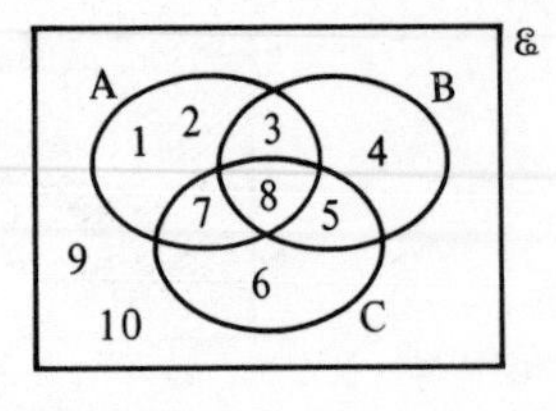

Exercise 1

In this exercise, be careful to use set notation only when the answer *is* a set.

1. If $M = \{1, 2, 3, 4, 5, 6, 7, 8\}$, $N = \{5, 7, 9, 11, 13\}$.
Find
(a) $M \cap N$ (b) $M \cup N$
(c) $n(N)$ (d) $n(M \cup N)$.
State whether true or false:
(e) $5 \in M$ (f) $7 \in (M \cup N)$
(g) $N \subset M$ (h) $\{5, 6, 7\} \subset M$.

2. If $A = \{2, 3, 5, 7\}$, $B = \{1, 2, 3, \ldots, 9\}$.
Find
(a) $A \cap B$ (b) $A \cup B$
(c) $n(A \cap B)$ (d) $\{1, 4\} \cap A$
State whether true or false:
(e) $A \in B$ (f) $A \subset B$
(g) $9 \subset B$ (h) $3 \in (A \cap B)$

3. If $X = \{1, 2, 3, \ldots 10\}$, $Y = \{2, 4, 6, \ldots 20\}$ and $Z = \{x : x \text{ is an integer}, 15 \leqslant x \leqslant 25\}$.
Find
(a) $X \cap Y$ (b) $Y \cap Z$
(c) $X \cap Z$ (d) $n(X \cup Y)$
(e) $n(Z)$ (f) $n(X \cup Z)$
State whether true or false:
(g) $5 \in Y$ (h) $20 \in X$
(i) $n(X \cap Y) = 5$ (j) $\{15, 20, 25\} \subset Z$.

4. If $D = \{1, 3, 5\}$, $E = \{3, 4, 5\}$, $F = \{1, 5, 10\}$.
Find
(a) $D \cup E$ (b) $D \cap F$
(c) $n(E \cap F)$ (d) $(D \cup E) \cap F$
(e) $(D \cap E) \cup F$ (f) $n(D \cup F)$
State whether true or false:
(g) $D \subset (E \cup F)$ (h) $3 \in (E \cap F)$
(i) $4 \notin (D \cap E)$.

5. Find
(a) $n(E)$ (b) $n(F)$ (c) $E \cap F$
(d) $E \cup F$ (e) $n(E \cup F)$ (f) $n(E \cap F)$.

&
E F
a b c
e d

6. Find
(a) $n(M \cap N)$ (b) $n(N)$ (c) $M \cup N$
(d) $M' \cap N$ (e) $N' \cap M$ (f) $(M \cap N)'$
(g) $M \cup N'$ (h) $N \cup M'$ (i) $M' \cup N'$

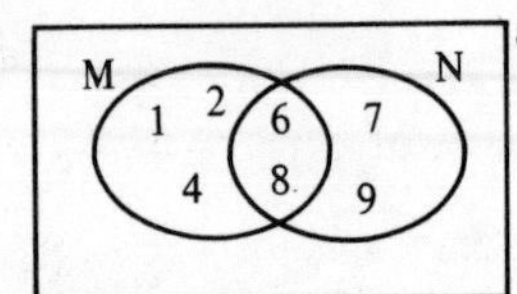

7. Find
(a) $n(B)$ (b) $n(A \cup B)$ (c) $n(\mathcal{E})$
(d) $A \cap B'$ (e) B' (f) A'
(g) $A' \cap B$ (h) $n(B')$ (i) $(A \cap B)'$

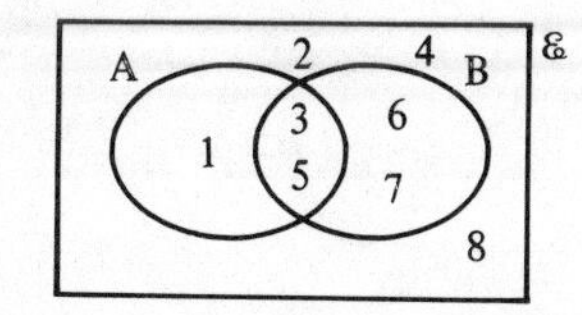

8. Find
(a) Y' (b) $n(X \cup Y)$ (c) $n(X \cup Y)'$
(d) $X' \cap Y'$ (e) $n(\mathcal{E})$ (f) $X \cup Y'$
State whether true or false:
(g) $a \in Y'$ (h) $c \in (X \cup Y)$ (i) $\{d, e\} \subset X$.

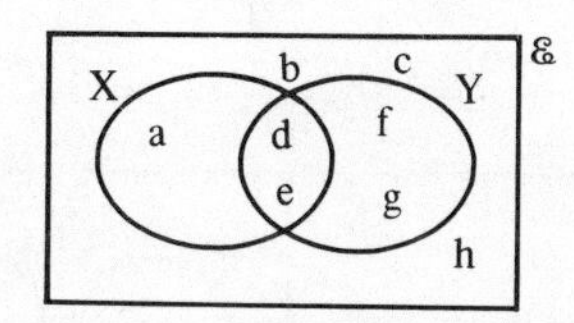

9. Illustrate the sets A and B on a Venn diagram where $\mathcal{E} = \{1, 2, \ldots 7\}$, $A = \{1, 2, 3, 7\}$, $B = \{3, 4, 5, 6, 7\}$.

10. Illustrate the sets A and B on a Venn diagram where $\mathcal{E} = \{a, b, \ldots, h\}$, $A = \{b, c, g\}$, $B = \{c, e, f, g\}$.
Find
(a) B' (b) $(A \cup B)$
(c) $(B \cup A)'$ (d) $n(A \cap B)'$.

11. If $\mathcal{E} = \{1, 2, \ldots, 10\}$,
$A = \{1, 2, 5, 9\}$,
$B = \{2, 3, 4, 5\}$,
$C = \{4, 5, 6, 7\}$.
Find
(a) $B \cap C$ (b) $A \cap (B \cap C)$
(c) $A \cup C$ (d) $A' \cap B$
(e) $B \cap C'$ (f) $(B \cup C)'$
(g) $n(A \cap B \cap C)$ (h) $n(A \cup B \cup C)$.

12. If $\mathcal{E} = \{1, 2, \ldots 12\}$,
$L = \{2, 3, 4, 5, 9, 12\}$,
$M = \{4, 5, 6, 8\}$,
$N = \{1, 5, 9, 10\}$.
Find
(a) $L \cap N$ (b) $(L \cup M) \cap N$
(c) $L \cup M \cup N$ (d) $(L \cup M \cup N)'$
(e) $L' \cap M$ (f) $N \cap M'$
(g) $n(L \cap M \cap N)$ (h) $n(L' \cap M')$.

Example 3

On a Venn diagram, shade the regions
(a) $A \cap C$ (b) $(B \cap C) \cap A'$
where A, B, C are intersecting sets.

(a) $A \cap C$

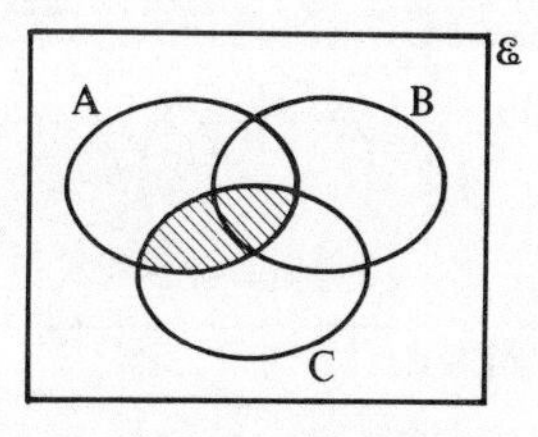

(b) $(B \cap C) \cap A'$
[find $(B \cap C)$ first]

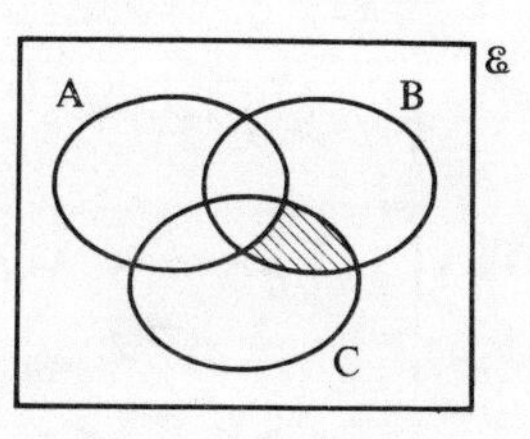

Example 4

On a Venn diagram, shade the regions
(a) $(A \cap B) \cup C$ (b) $(A \cup C) \cap B$
where A, B, C are intersecting sets.

(a) $(A \cap B) \cup C$

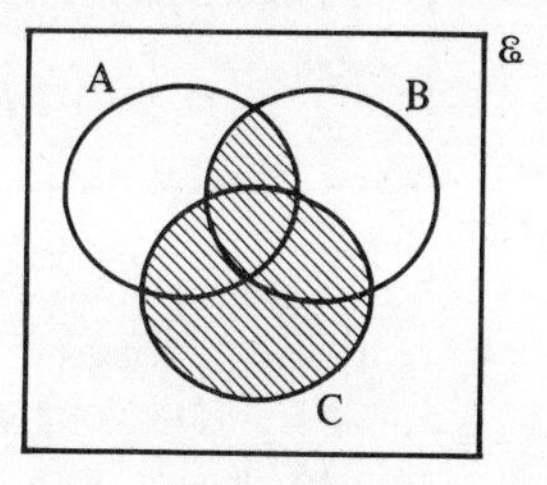

(b) $(A \cup C) \cap B$

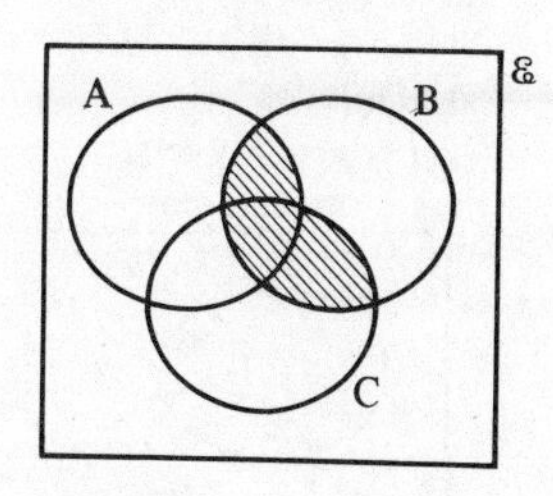

In both cases identify the term in brackets first.

Exercise 2

1. Draw six diagrams similar to Figure 1 and shade the following sets:

(a) $A \cap B$ (b) $A \cup B$ (c) A'
(d) $A' \cap B$ (e) $B' \cap A$ (f) $(B \cup A)'$.

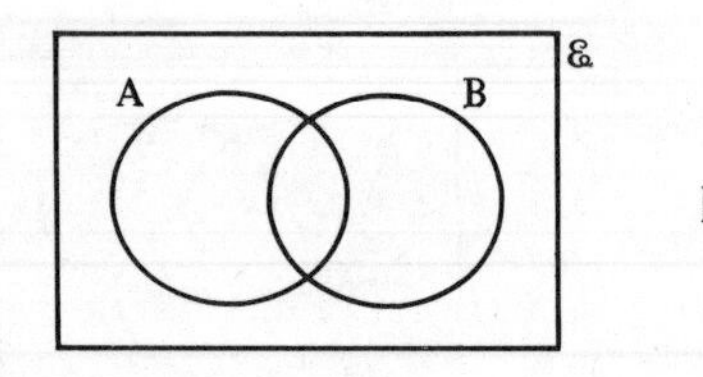

Figure 1

2. Draw four diagrams similar to Figure 2 and shade the following sets:
(a) $A \cap B$ (b) $A \cup B$ (c) $B' \cap A$
(d) $(B \cup A)'$

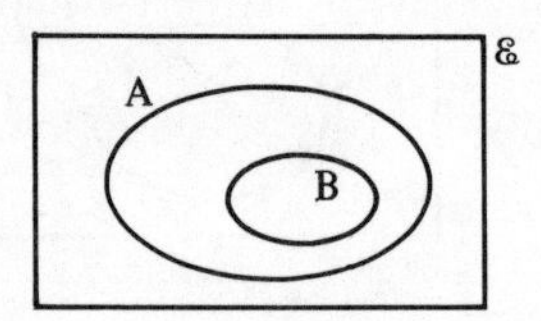

Figure 2

3. Draw four diagrams similar to Figure 3 and shade the following sets:
(a) $A \cup B$ (b) $A \cap B$ (c) $A \cap B'$
(d) $(B \cup A)'$

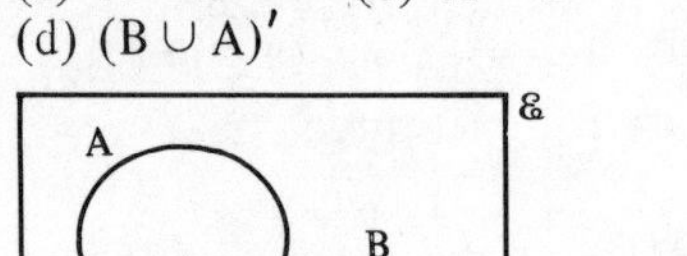

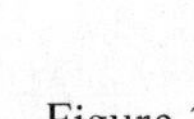

Figure 3

4. Draw eleven diagrams similar to Figure 4 and shade the following sets:
(a) $A \cap B$ (b) $A \cup C$ (c) $A \cap (B \cap C)$
(d) $(A \cup B) \cap C$ (e) $B \cap (A \cup C)$
(f) $A \cap B'$ (g) $A \cap (B \cup C)'$
(h) $(B \cup C) \cap A$ (i) $C' \cap (A \cap B)$
(j) $(A \cup C) \cup B'$ (k) $(A \cup C) \cap (B \cap C)$

Figure 4

5. Draw nine diagrams similar to Figure 5 and shade the following sets:
(a) $(A \cup B) \cap C$ (b) $(A \cap B) \cup C$
(c) $(A \cup B) \cup C$ (d) $A \cap (B \cup C)$
(e) $A' \cap C$ (f) $C' \cap (A \cup B)$
(g) $(A \cap B) \cap C$ (h) $(A \cap C) \cup (B \cap C)$
(i) $(A \cup B \cup C)'$

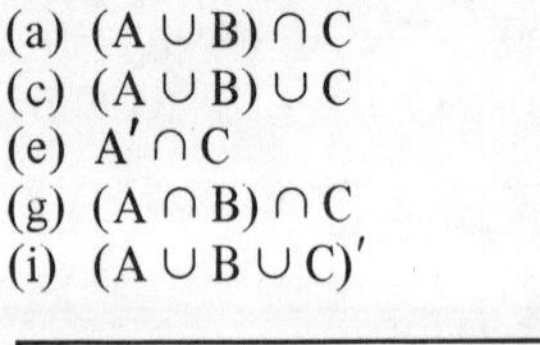

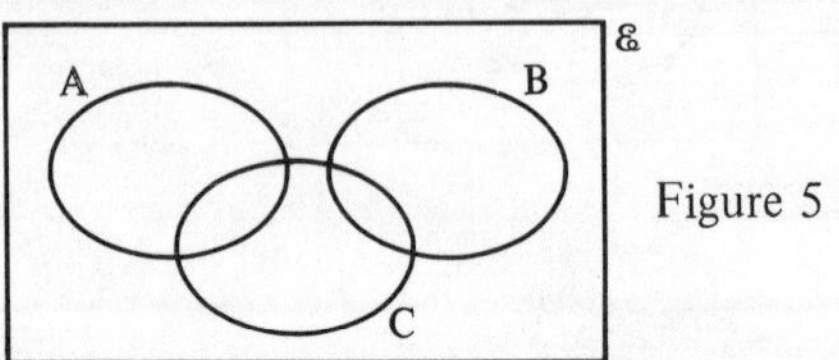

Figure 5

6. Copy each diagram and shade the region indicated.

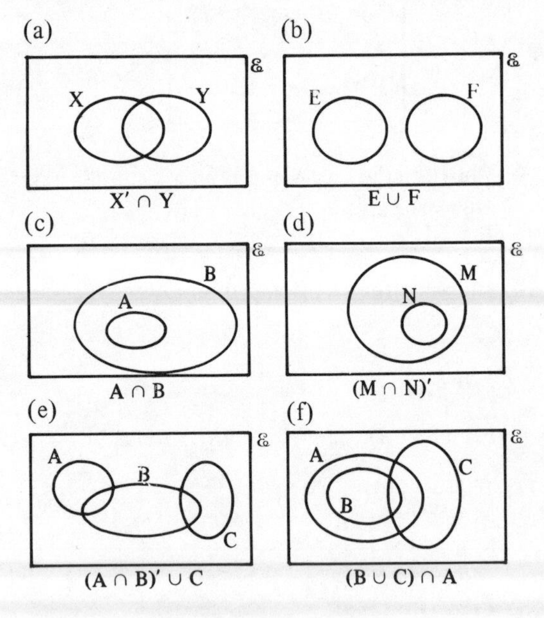

7. Describe the region shaded.

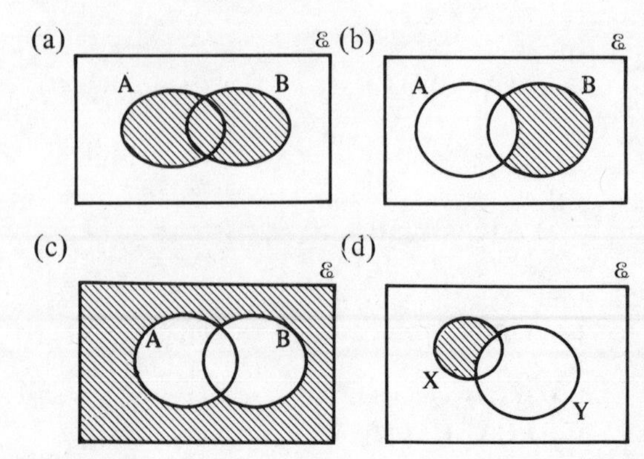

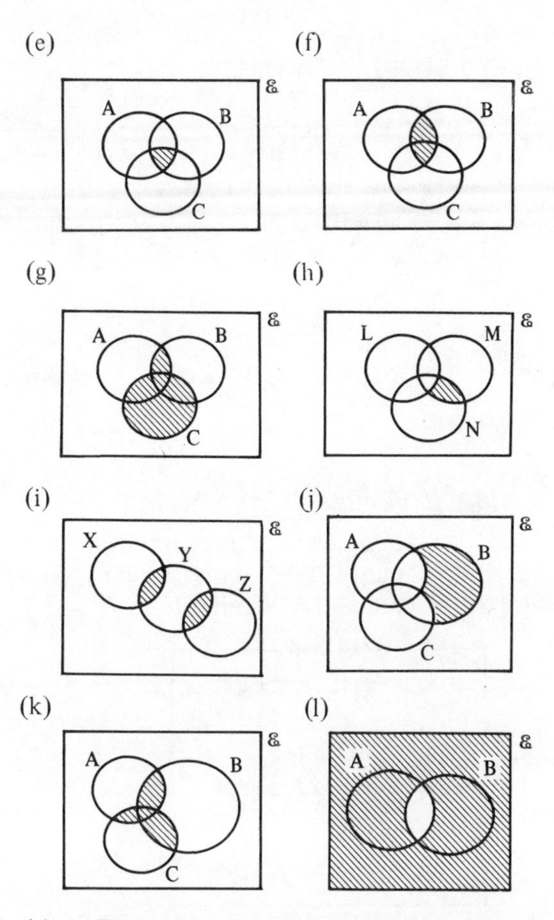

8. (a) In Figure 6, each cross represents one element.

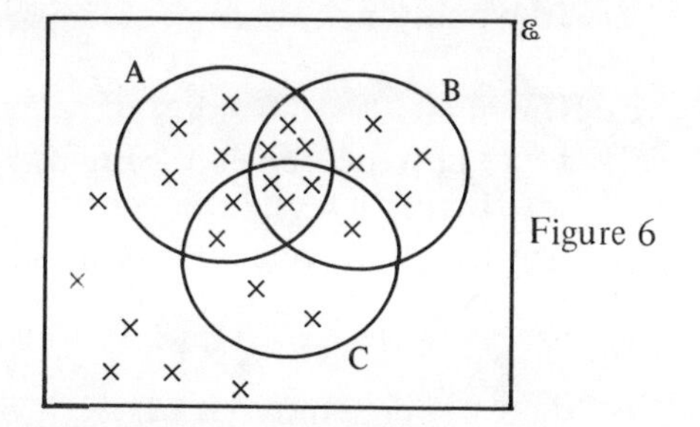

Figure 6

Write down the number of elements in each of the following:

(i) A	(ii) $B \cap C$
(iii) $A \cup C$	(iv) C'
(v) $(A \cup C)'$	(vi) $(A \cap B) \cap C$
(vii) $(A \cup B)'$	(viii) $(A \cup C) \cap B$
(ix) $(A \cap B) \cup C$	(x) $(A \cap B)'$
(xi) $A \cup B \cup C$	(xii) $(B \cup C) \cap A$

(xiii) $(A \cap B) \cup (B \cap C)$
(xiv) $(A \cup B \cup C)'$
(xv) $(A \cup B) \cap C'$

(b) Repeat part (a) using Figure 7.

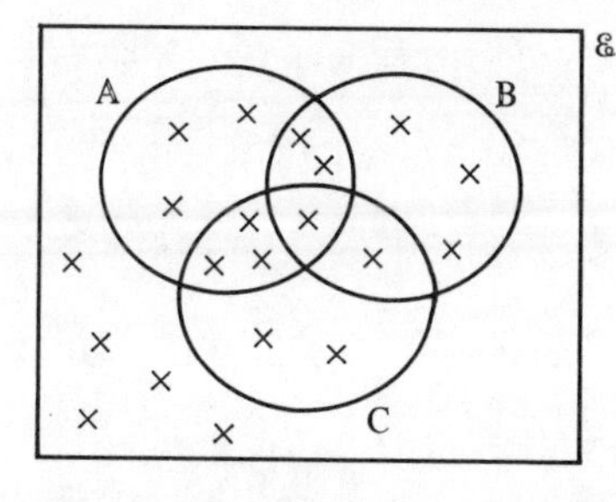

Figure 7

9. Use a Venn diagram to simplify the following:

(a) $A \cap (A \cup B)$	(b) $A \cup (A \cap B)$
(c) $(P \cup Q) \cup Q$	(d) $(X \cap Y) \cap X$
(e) $A \cap (B \cup B')$	(f) $(A' \cup B) \cap A$
(g) $A' \cap (A \cup B)$	(h) $(A \cap B') \cup B$

(i) $(A \cap B) \cup (A \cap B')$

Example 5

If A and B are two intersecting sets such that $n(A) = 13$, $n(B) = 12$ and $n(A \cap B) = 5$, find $n(A \cup B)$.

Draw a Venn diagram and start with $A \cap B$.

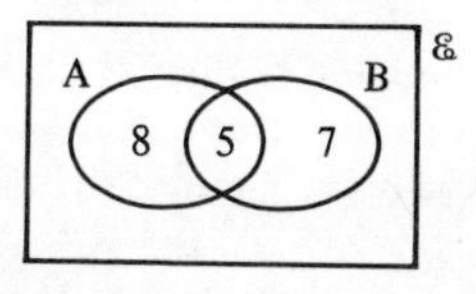

$n(A \cup B) = 20.$

Example 6

If $n(A) = 15$ and $n(B) = 7$, find the greatest and the least possible values of

(a) $n(A \cap B)$ (b) $n(A \cup B)$

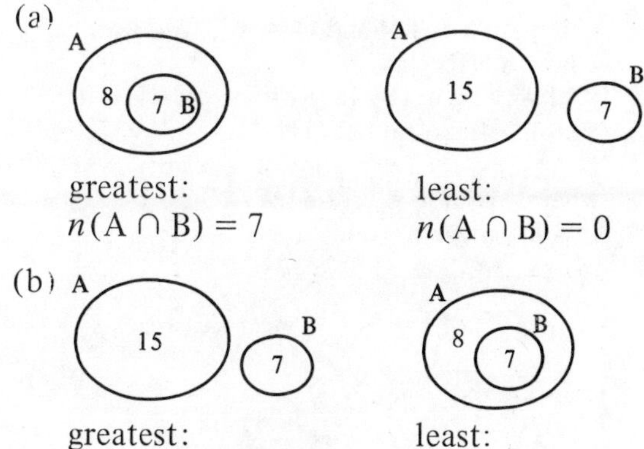

Exercise 3

1. If $n(\mathrm{A}) = 15$, $n(\mathrm{B}) = 13$ and $n(\mathrm{A} \cap \mathrm{B}) = 6$, find $n(\mathrm{A} \cup \mathrm{B})$.
2. If $n(\mathrm{X}) = 9$, $n(\mathrm{Y}) = 6$ and $n(\mathrm{X} \cap \mathrm{Y}) = 4$ find $n(\mathrm{X} \cup \mathrm{Y})$.
3. If $n(\mathrm{M}) = 14$, $n(\mathrm{S}) = 8$ and $n(\mathrm{M} \cap \mathrm{S}) = 3$, find $n(\mathrm{M} \cup \mathrm{S})$.
4. If $n(\mathrm{A}) = 10$ and $n(\mathrm{B}) = 4$, find the greatest *and* least possible values of
   (a) $n(\mathrm{A} \cap \mathrm{B})$ (b) $n(\mathrm{A} \cup \mathrm{B})$.
5. If $n(\&) = 20$, $n(\mathrm{A}) = 6$ and $n(\mathrm{B}) = 10$, find the greatest *and* least possible values of
   (a) $n(\mathrm{A} \cap \mathrm{B})$ (b) $n(\mathrm{A} \cup \mathrm{B})$.
6. If $n(\&) = 16$, $n(\mathrm{A}) = 10$ and $n(\mathrm{B}) = 8$, find the greatest *and* least possible values of
   (a) $n(\mathrm{A} \cap \mathrm{B})$ (b) $n(\mathrm{A} \cup \mathrm{B})$.
7. If $n(\&) = 21$, $n(\mathrm{X}) = 15$ and $n(\mathrm{Y}) = 9$, find the greatest *and* least possible values of
   (a) $n(\mathrm{X} \cap \mathrm{Y})$ (b) $n(\mathrm{X} \cup \mathrm{Y})$.
8. If $n(\mathrm{P}) = 5$, $n(\mathrm{Q}) = 7$ and $n(\mathrm{R}) = 4$, find the greatest possible value of
   (a) $n(\mathrm{P} \cup (\mathrm{Q} \cap \mathrm{R}))$ (b) $n(\mathrm{P} \cap (\mathrm{Q} \cup \mathrm{R}))$.
9. If & = {parallelograms}, C = {rhombuses}, D = {rectangles}, describe $\mathrm{C} \cap \mathrm{D}$.
10. If I = {isosceles triangles} and E = {equilateral triangles}, describe $\mathrm{I} \cap \mathrm{E}$ and $\mathrm{I} \cup \mathrm{E}$.
11. Draw three sets A, B and C on a Venn diagram such that $\mathrm{A} \cap \mathrm{B} = \emptyset$ and $\mathrm{A} \cup \mathrm{C} = \mathrm{A}$.
12. Draw three sets X, Y and Z on a Venn diagram such that $(\mathrm{X} \cup \mathrm{Z}) \subset \mathrm{Y}$ and $\mathrm{X} \cap \mathrm{Z} \neq \emptyset$.
13. Sets P and Q are such that $n(\mathrm{P} \cup \mathrm{Q}) = 40$, $n(\mathrm{P} \cap \mathrm{Q}) = 12$ and $n(\mathrm{P}) = 27$. Find $n(\mathrm{Q})$.
14. Copy the Venn diagram and show a set X such that $\mathrm{X} \cup \mathrm{A} = \mathrm{A}$ and $\mathrm{X} \cap \mathrm{B} = \emptyset$.

    A B

15. If $\mathrm{A} = \{x : x \geqslant 10\}$ and $\mathrm{B} = \{x : x \leqslant 16\}$ where x is an integer, find $\mathrm{A} \cap \mathrm{B}$.
16. If $\mathrm{P} = \{x : x \leqslant 12\}$ and $\mathrm{Q} = \{x : x \geqslant 3\}$ where x is an integer, find $\mathrm{P} \cap \mathrm{Q}$.
17. If $\mathrm{M} = \{y : y < 10\}$ and $\mathrm{N} = \{y : y > 4\}$ where y is an integer, find $\mathrm{M} \cap \mathrm{N}$.
18. If $\mathrm{A} = \{x : 0 < x < 8\}$ and $\mathrm{B} = \{x : 3 < x < 11\}$ where x is an integer, find $\mathrm{A} \cap \mathrm{B}$.
19. If & = {integers}, find
    (a) $\{x : 1 \leqslant x \leqslant 7\} \cup \{x : 3 \leqslant x \leqslant 12\}$
    (b) $\{x : 2 \leqslant x \leqslant 9\} \cap \{x : 6 \leqslant x \leqslant 11\}$.
20. If & = {integers}, find
    (a) $\{x : 4 \leqslant x \leqslant 10\} \cup \{x : 7 \leqslant x \leqslant 15\}$
    (b) $\{x : 2 < x < 11\} \cap \{x : 5 < x < 10\}$.
21. If & = {integers}, find
    (a) $\{x : 0 < x < 12\} \cap \{x : 7 < x < 10\}$
    (b) $\{x : 5 \leqslant x \leqslant 20\} \cup \{x : 21 \leqslant x \leqslant 30\}$.
22. If & = {integers}, find
    (a) $\{x : 1 \leqslant x \leqslant 7\} \cup \{x : 10 < x < 16\}$
    (b) $\{x : x > 50\} \cap \{x : x < 70\}$.

Example 7

In the Venn diagram,

E = {people who enjoy English}
F = {people who enjoy French}
H = {people who enjoy History}

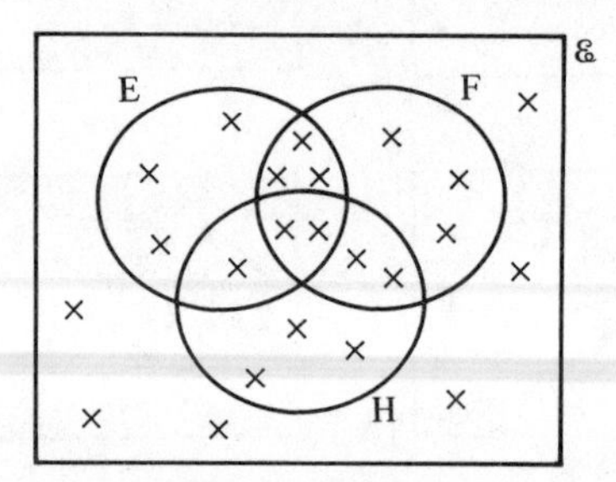

(a) How many people enjoy History?
(b) How many people enjoy English and French?
(c) How many people enjoy only French?
(d) How many people enjoy English, French and History?
(e) How many people enjoy English and History but not French?

(a) Eight people enjoy History.
(b) Five people enjoy English and French.
(c) Three people enjoy only French.
(d) Two people enjoy English, French and History.
(e) One person enjoys English and History but not French.

Exercise 4

1. In the Venn diagram,
   & = {people in an hotel}
   B = {people who like bacon}
   E = {people who like eggs}

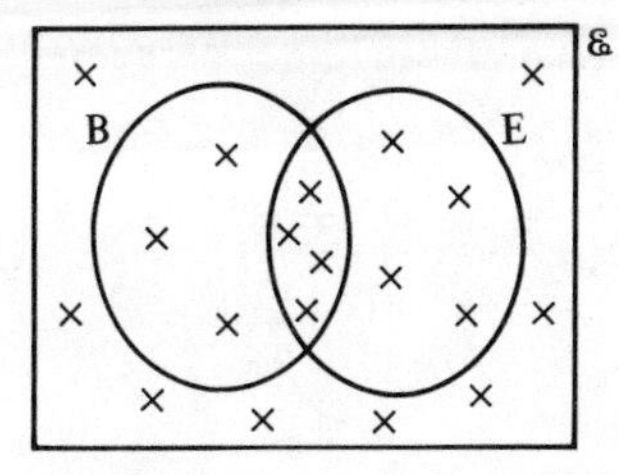

   (a) How many people like bacon?
   (b) How many people like eggs but not bacon?
   (c) How many people like bacon and eggs?
   (d) How many people are in the hotel?
   (e) How many people like neither bacon nor eggs?

2. In the Venn diagram,
   & = {boys in the fourth form}
   R = {members of the rugby team}
   C = {members of the cricket team}

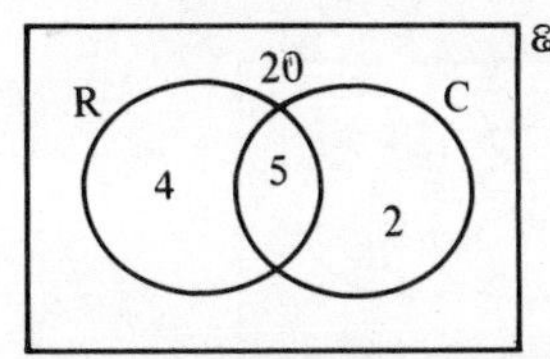

   (a) How many are in the rugby team?
   (b) How many are in both teams?
   (c) How many are in the rugby team but not in the cricket team?
   (d) How many are in neither team?
   (e) How many are there in the fourth form?

3. In the Venn diagram,
   & = {cars in a street}
   B = {blue cars}
   L = {cars with left hand drive}
   F = {cars with four doors}

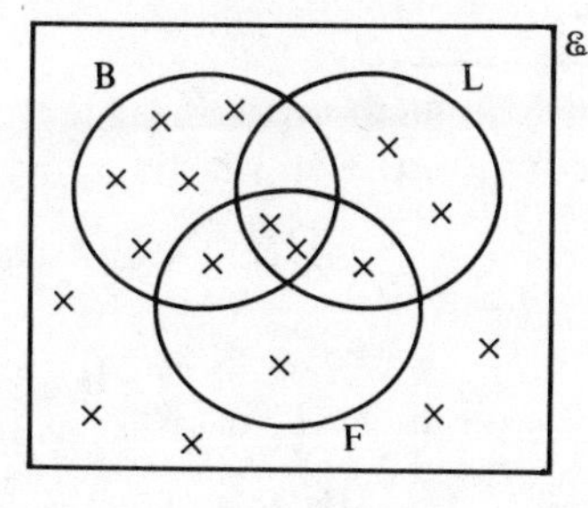

   (a) How many cars are blue?
   (b) How many blue cars have four doors?
   (c) How many cars with left hand drive have four doors?
   (d) How many blue cars have left hand drive?
   (e) How many cars are in the street?
   (f) How many blue cars with left hand drive do not have four doors?

4. In the Venn diagram,
   & = {houses in the street}
   C = {houses with central heating}
   T = {houses with a colour T.V.}
   G = {houses with a garden}

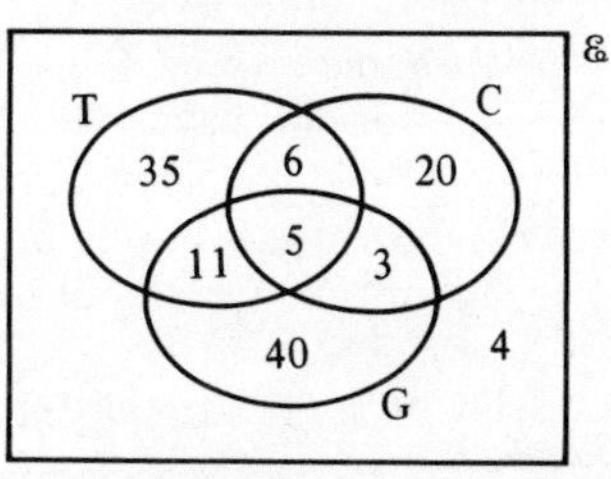

   (a) How many houses have gardens?
   (b) How many houses have a colour T.V. and central heating?
   (c) How many houses have a colour T.V. and central heating and a garden?
   (d) How many houses have a garden but not a T.V. or central heating?
   (e) How many houses have a T.V. and a garden but not central heating?
   (f) How many houses are there in the street?

5. In the Venn diagram
   & = {children in a mixed school}
   G = {girls in the school}
   S = {children who can swim}
   L = {children who are left-handed}

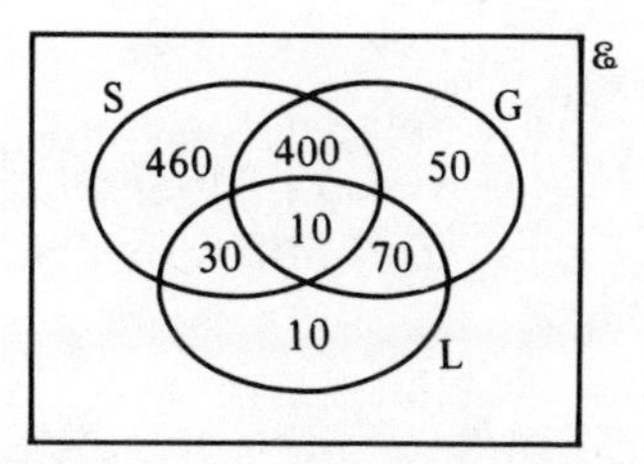

   (a) How many left-handed children are there?
   (b) How many girls cannot swim?
   (c) How many boys can swim?
   (d) How many girls are left-handed?
   (e) How many boys are left-handed?
   (f) How many left-handed girls can swim?
   (g) How many boys are there in the school?

6. In the Venn diagram,
   $\mathcal{E}$ = {pupils at a mixed school}
   S = {pupils in the second form}
   B = {boys}
   D = {pupils who have school dinners}

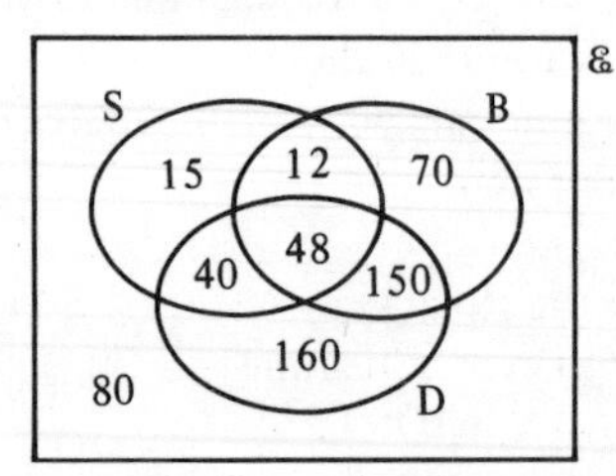

   (a) How many boys are there?
   (b) How many girls are there?
   (c) How many second formers have school dinners?
   (d) How many second form girls have dinners?
   (e) How many second form boys do not have dinners?
   (f) How many boys not in the second year do not have dinners?
   (g) How many second year boys do have dinners?
   (h) How many girls in the school are not in the second year?

7. $\mathcal{E} = \{x : x$ is an integer and $4 \leqslant x \leqslant 25\}$
   $A = \{x : x$ is divisible by $5\}$
   $B = \{x : x$ is a perfect square$\}$
   $C = \{x : x$ is a prime number$\}$
   Find
   (a) A (b) B (c) C
   (d) $A \cap B$ (e) $A \cap C$ (f) $n(C')$

8. $P = \{x : x$ is an integer and $40 \leqslant x \leqslant 150\}$
   $Q = \{x : \sqrt{x}$ is a positive integer$\}$
   $R = \{x : x$ is an integer and $x^3 \in P\}$.
   (a) List the members of the set R,
   (b) Find $P \cap Q$,
   (c) Find the value of $n(P)$.

9. In the Venn diagram, the number of elements are as shown.

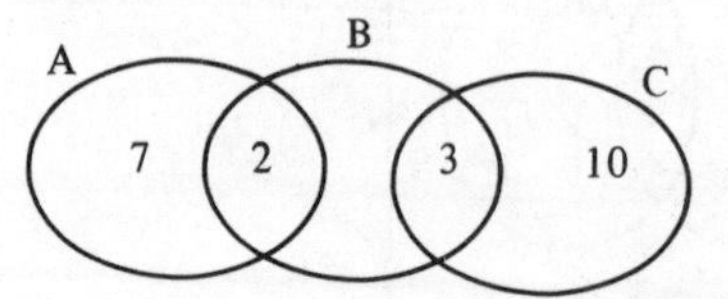

   Copy the Venn diagram.
   If $\mathcal{E} = A \cup B \cup C$ and $n(\mathcal{E}) = 26$, fill in the missing number and find
   (a) $n(A \cap C)$ (b) $n(A \cup C)$ (c) $n(A' \cap C')$.

10. $A * B$ means $A' \cap B$. Draw four Venn diagrams like the one shown and shade the areas which represent
   (a) $A * B$
   (b) $A * B'$
   (c) $A' * B$
   (d) $(A * B) \cup (A \cap B)$.

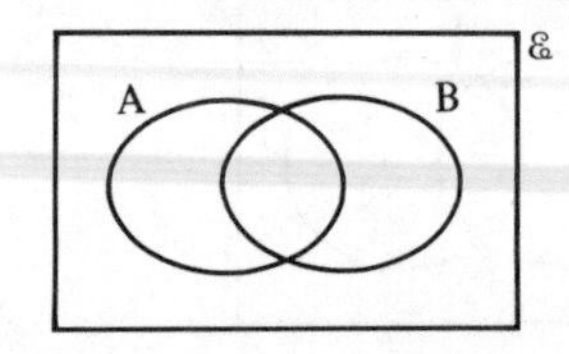

9.2 LOGICAL PROBLEMS

Example 1

In a form of 30 girls, 18 play netball and 14 play hockey, whilst 5 play neither. Find the number who play both netball and hockey.

Let $\mathcal{E}$ = {girls in the form}
N = {girls who play netball}
H = {girls who play hockey}

and x = the number of girls who play both netball and hockey

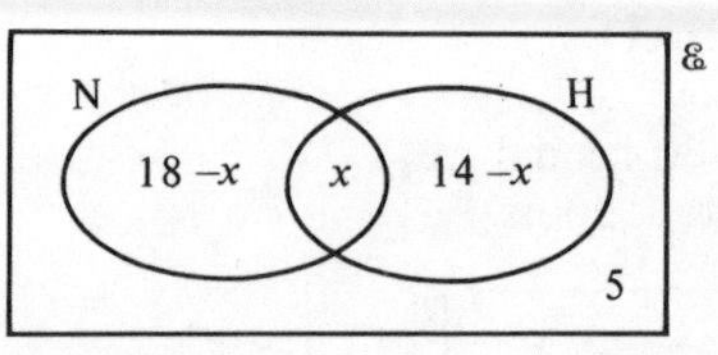

The number of girls in each portion of the universal set is shown in the Venn diagram.

Since
$$n(\mathcal{E}) = 30$$
$$18 - x + x + 14 - x + 5 = 30$$
$$37 - x = 30$$
$$x = 7$$

$\therefore$ seven girls play both netball and hockey.

Example 2

If A = {sheep}
B = {sheep dogs}
C = {'intelligent' animals}
D = {animals which make good pets}

(a) Express the following sentences in set language
(i) No sheep are 'intelligent' animals.
(ii) All sheep dogs make good pets.
(iii) Some sheep make good pets.

(b) Interpret the following statements
(i) $B \subset C$.
(ii) $B \cup C = D$.

(a) (i) $A \cap C = \emptyset$.
(ii) $B \subset D$.
(iii) $A \cap D \neq \emptyset$.

(b) (i) All sheep dogs are intelligent animals.
(ii) Animals which make good pets are either sheep dogs or 'intelligent' animals (or both).

Exercise 5

1. In the Venn diagram, $n(A) = 10$, $n(B) = 13$, $n(A \cap B) = x$ and $n(A \cup B) = 18$.

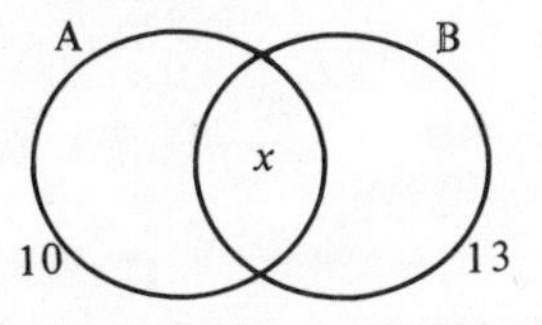

(a) Write in terms of x the number of elements in A but not in B,
(b) Write in terms of x the number of elements in B but not in A,
(c) Add together the number of elements in the three parts of the diagram to obtain the equation $10 - x + x + 13 - x = 18$,
(d) Hence find the number of elements in both A and B.

2. In the Venn diagram $n(A) = 21$, $n(B) = 17$, $n(A \cap B) = x$ and $n(A \cup B) = 29$.

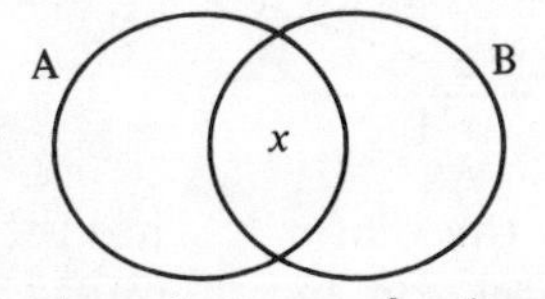

(a) Write down in terms of x the number of elements in each part of the diagram.
(b) Form an equation and hence find x.

3. The sets M and N intersect such that $n(M) = 31$, $n(N) = 18$ and $n(M \cup N) = 35$. How many elements are in both M and N?
4. The sets P and Q intersect such that $n(P) = 11$, $n(Q) = 29$ and $n(P \cup Q) = 37$. How many elements are in both P and Q?
5. The sets A and B intersect such that $n(A \cap B) = 7$, $n(A) = 20$ and $n(B) = 23$. Find $n(A \cup B)$.
6. Twenty boys in a form all play either football or basketball (or both). If thirteen play football and ten play basketball, how many play both sports?
7. Of the 53 staff at a school, 36 drink tea, 18 drink coffee and 10 drink neither tea nor coffee. How many drink both tea and coffee?
8. Of the 32 pupils in a class, 18 play golf, 16 play the piano and 7 play both. How many play neither?
9. Of the pupils in a class, 15 can spell 'parallel', 14 can spell 'Pythagoras', 5 can spell both words and 4 can spell neither. How many pupils are there in the class?
10. In a school, students must take at least one of these subjects: Maths, Physics or Chemistry. In a group of 50 students, 7 take all three subjects, 9 take Physics and Chemistry only, 8 take Maths and Physics only and 5 take Maths and Chemistry only. Of these 50 students, x take Maths only, x take Physics only and $x + 3$ take Chemistry only.
Draw a Venn diagram, find x, and hence find the number taking Maths.
11. All of 60 different vitamin pills contain at least one of the vitamins A, B and C. Twelve have A only, 7 have B only, and 11 have C only.
If 6 have all three vitamins and there are x having A and B only, B and C only and A and C only, how many pills contain vitamin A?
12. The 'O' level results of the 30 members of the Welsh Rugby squad were as follows:
All 30 players passed at least two subjects,
18 players passed at least three subjects, and
3 players passed four subjects or more. Calculate
(a) how many passed exactly two subjects,
(b) what fraction of the squad passed exactly three subjects.
13. In a group of 59 people, some are wearing hats, gloves or scarves (or a combination of these), 4 are wearing all three, 7 are wearing just a hat and gloves, 3 are wearing just gloves and a scarf and 9 are wearing just a hat and scarf. The number wearing only a hat or only gloves is x, and the number wearing only a scarf or none of the three items is $(x - 2)$. Find x and hence the number of people wearing a hat.

14. In a street of 150 houses, three different newspapers are delivered: T, G and M. Of these, 40 receive T, 35 receive G, and 60 receive M; 7 receive T and G, 10 receive G and M and 4 receive T and M; 34 receive no paper at all. How many receive all three?
Note: If '7 receive T and G', this information does not mean 7 receive T and G *only*.

15. If S = {Scottish men}, G = {good footballers}, express the following statements in words
(a) $G \subset S$
(b) $G \cap S = \emptyset$
(c) $G \cap S \neq \emptyset$.
(Ignore the truth or otherwise of the statements.)

16. Given that & = {pupils in a school}, B = {boys}, H = {hockey players}, F = {football players}, express the following in words:
(a) $F \subset B$ (b) $H \subset B'$
(c) $F \cap H \neq \emptyset$ (d) $B \cap H = \emptyset$.
Express in set notation
(e) No boys play football,
(f) All pupils play either football or hockey.

17. If & = {living creatures}, S = {spiders}, F = {animals that fly}, T = {animals which taste nice}, express in set notation:
(a) No spiders taste nice,
(b) All animals that fly taste nice,
(c) Some spiders can fly.
Express in words:
(d) $S \cup F \cup T = \mathscr{E}$ (e) $T \subset S$.

18. & = {tigers}, T = {tigers who believe in fairies}, X = {tigers who believe in Eskimos}, H = {tigers in hospital}. Express in words:
(a) $T \subset X$, (b) $T \cup X = H$,
(c) $H \cap X = \emptyset$.
Express in set notation:
(d) All tigers in hospital believe in fairies,
(e) Some tigers believe in both fairies and Eskimos.

19. & = {school teachers}, P = {teachers called Peter}, B = {good bridge players}, W = {women teachers}. Express in words:
(a) $P \cap B = \emptyset$,
(b) $P \cup B \cup W = \mathscr{E}$,
(c) $P \cap W \neq \emptyset$.
Express in set notation:
(d) Women teachers cannot play bridge well,
(e) All good bridge players are women called Peter.

9.3 VECTORS

A vector quantity has both magnitude and direction. Problems involving forces, velocities and displacements are often made easier when vectors are used.

Addition of vectors

Vectors **a** and **b** represented by the line segments below can be added using the parallelogram rule or the 'nose-to-tail' method.

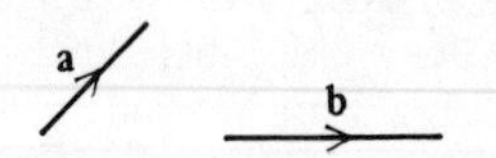

The 'tail' of vector **b** is joined to the 'nose' of vector **a**.

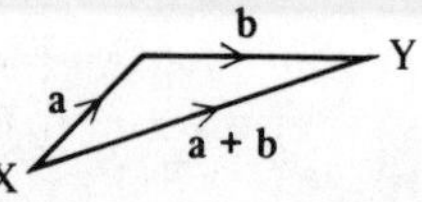

Alternatively the tail of **a** can be joined to the 'nose' of vector **b**.

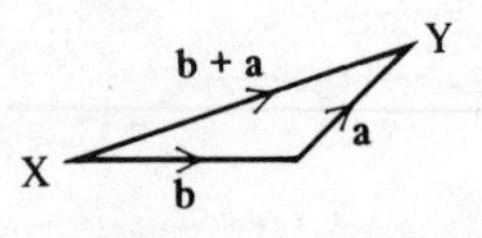

In both cases the vector $\overrightarrow{XY}$ has the same length and direction and therefore

$$\mathbf{a} + \mathbf{b} = \mathbf{b} + \mathbf{a}.$$

Multiplication by a scalar

A scalar quantity has magnitude but no direction (e.g. mass, volume, temperature). Ordinary numbers are scalars.

When vector **x** is multiplied by 2, the result is 2**x**.

x 2x

When **x** is multiplied by −3 the result is −3**x**.

x −3x

Note

(1) The negative sign reverses the direction of the vector.

(2) The result **a** − **b** is **a** + −**b**.
i.e. Subtracting **b** is equivalent to adding the negative of **b**.

Example 1

The diagram shows vectors **a** and **b**.

a b

Find $\overrightarrow{OP}$ and $\overrightarrow{OQ}$ such that

$$\overrightarrow{OP} = 3\mathbf{a} + \mathbf{b}$$

$$\overrightarrow{OQ} = -2\mathbf{a} - 3\mathbf{b}$$

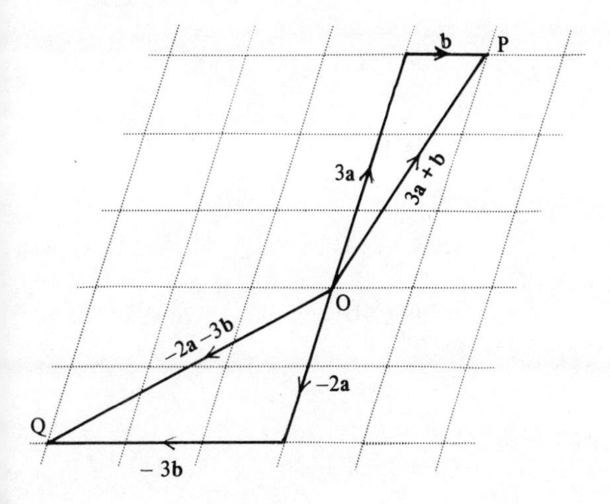

Exercise 6

In questions **1** to **26**, use the diagram below to describe the vectors given in terms of **c** and **d** where **c** = $\overrightarrow{QN}$ and **d** = $\overrightarrow{QR}$.

e.g. $\overrightarrow{QS}$ = 2**d**, $\overrightarrow{TD}$ = **c** + **d**.

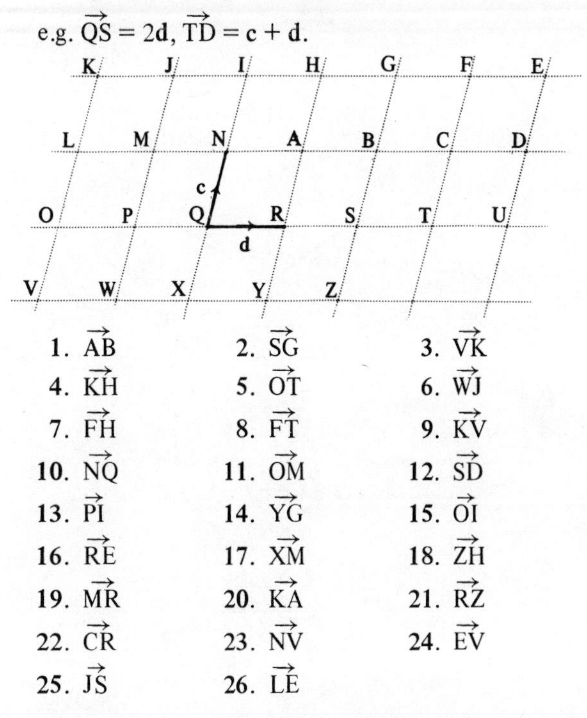

1. $\overrightarrow{AB}$	**2.** $\overrightarrow{SG}$	**3.** $\overrightarrow{VK}$
4. $\overrightarrow{KH}$	**5.** $\overrightarrow{OT}$	**6.** $\overrightarrow{WJ}$
7. $\overrightarrow{FH}$	**8.** $\overrightarrow{FT}$	**9.** $\overrightarrow{KV}$
10. $\overrightarrow{NQ}$	**11.** $\overrightarrow{OM}$	**12.** $\overrightarrow{SD}$
13. $\overrightarrow{PI}$	**14.** $\overrightarrow{YG}$	**15.** $\overrightarrow{OI}$
16. $\overrightarrow{RE}$	**17.** $\overrightarrow{XM}$	**18.** $\overrightarrow{ZH}$
19. $\overrightarrow{MR}$	**20.** $\overrightarrow{KA}$	**21.** $\overrightarrow{RZ}$
22. $\overrightarrow{CR}$	**23.** $\overrightarrow{NV}$	**24.** $\overrightarrow{EV}$
25. $\overrightarrow{JS}$	**26.** $\overrightarrow{LE}$	

In questions **27** to **38**, use the same diagram above to find vectors for the following in terms of the capital letters, starting from Q each time.

e.g. 3**d** = $\overrightarrow{QT}$, **c** + **d** = $\overrightarrow{QA}$.

27. 2**c**	**28.** 4**d**	**29.** 2**c** + **d**
30. 2**d** + **c**	**31.** 3**d** + 2**c**	**32.** 2**c** − **d**
33. −**c** + 2**d**	**34.** **c** − 2**d**	**35.** 2**c** + 4**d**
36. −**c**	**37.** −**c** − **d**	**38.** 2**c** − 2**d**

In questions **39** to **43**, write each vector in terms of **a** and/or **b**.

39. (a) $\overrightarrow{BA}$
(b) $\overrightarrow{AC}$
(c) $\overrightarrow{DB}$
(d) $\overrightarrow{AD}$

40. (a) $\overrightarrow{ZX}$
(b) $\overrightarrow{YW}$
(c) $\overrightarrow{XY}$
(d) $\overrightarrow{XZ}$

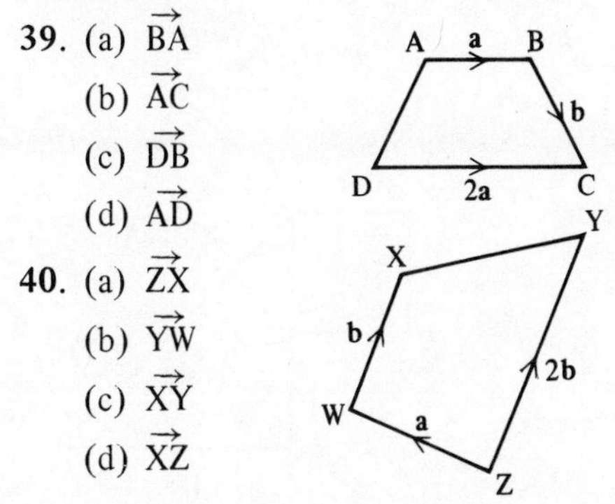

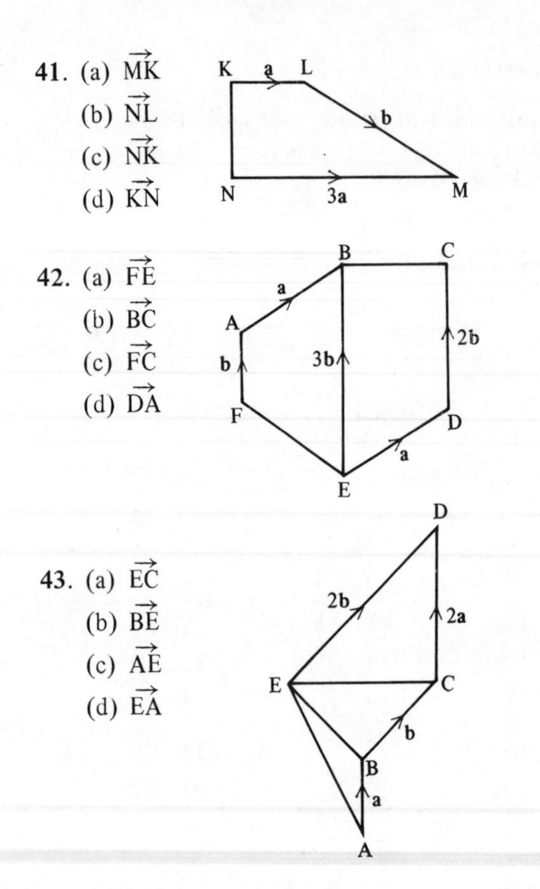

41. (a) $\overrightarrow{MK}$
(b) $\overrightarrow{NL}$
(c) $\overrightarrow{NK}$
(d) $\overrightarrow{KN}$

42. (a) $\overrightarrow{FE}$
(b) $\overrightarrow{BC}$
(c) $\overrightarrow{FC}$
(d) $\overrightarrow{DA}$

43. (a) $\overrightarrow{EC}$
(b) $\overrightarrow{BE}$
(c) $\overrightarrow{AE}$
(d) $\overrightarrow{EA}$

In questions **44** to **46**, write each vector in terms of **a**, **b** and **c**.

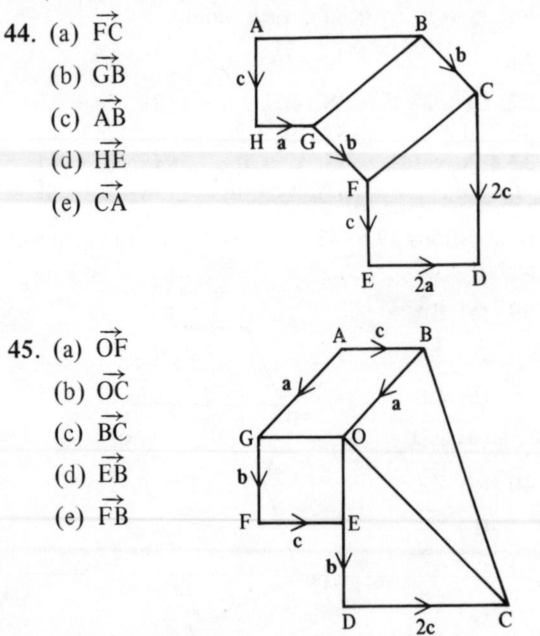

44. (a) $\overrightarrow{FC}$
(b) $\overrightarrow{GB}$
(c) $\overrightarrow{AB}$
(d) $\overrightarrow{HE}$
(e) $\overrightarrow{CA}$

45. (a) $\overrightarrow{OF}$
(b) $\overrightarrow{OC}$
(c) $\overrightarrow{BC}$
(d) $\overrightarrow{EB}$
(e) $\overrightarrow{FB}$

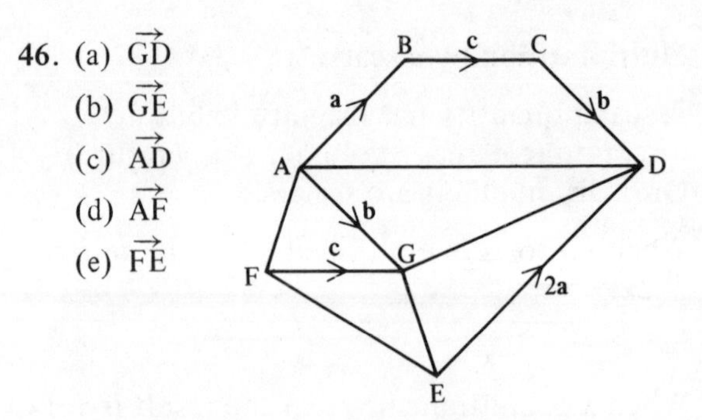

46. (a) $\overrightarrow{GD}$
(b) $\overrightarrow{GE}$
(c) $\overrightarrow{AD}$
(d) $\overrightarrow{AF}$
(e) $\overrightarrow{FE}$

Example 2

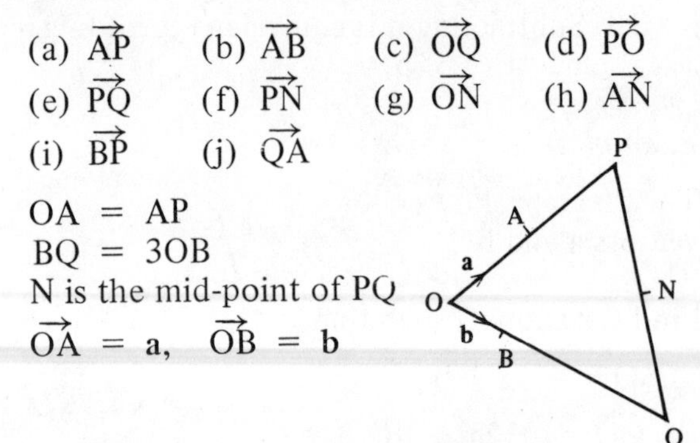

Using Figure 1, express each of the following vectors in terms of **a** and/or **b**.

(a) $\overrightarrow{AP}$ (b) $\overrightarrow{AB}$ (c) $\overrightarrow{OQ}$ (d) $\overrightarrow{PO}$
(e) $\overrightarrow{PQ}$ (f) $\overrightarrow{PN}$ (g) $\overrightarrow{ON}$ (h) $\overrightarrow{AN}$
(i) $\overrightarrow{BP}$ (j) $\overrightarrow{QA}$

OA = AP
BQ = 3OB
N is the mid-point of PQ
$\overrightarrow{OA} = \mathbf{a}$, $\overrightarrow{OB} = \mathbf{b}$

Figure 1

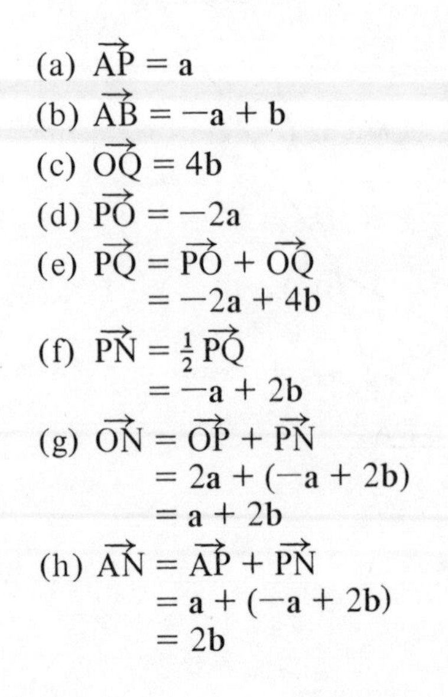

(a) $\overrightarrow{AP} = \mathbf{a}$
(b) $\overrightarrow{AB} = -\mathbf{a} + \mathbf{b}$
(c) $\overrightarrow{OQ} = 4\mathbf{b}$
(d) $\overrightarrow{PO} = -2\mathbf{a}$
(e) $\overrightarrow{PQ} = \overrightarrow{PO} + \overrightarrow{OQ}$
$= -2\mathbf{a} + 4\mathbf{b}$
(f) $\overrightarrow{PN} = \frac{1}{2}\overrightarrow{PQ}$
$= -\mathbf{a} + 2\mathbf{b}$
(g) $\overrightarrow{ON} = \overrightarrow{OP} + \overrightarrow{PN}$
$= 2\mathbf{a} + (-\mathbf{a} + 2\mathbf{b})$
$= \mathbf{a} + 2\mathbf{b}$
(h) $\overrightarrow{AN} = \overrightarrow{AP} + \overrightarrow{PN}$
$= \mathbf{a} + (-\mathbf{a} + 2\mathbf{b})$
$= 2\mathbf{b}$

(i) $\overrightarrow{BP} = \overrightarrow{BO} + \overrightarrow{OP}$
$= -\mathbf{b} + 2\mathbf{a}$

(j) $\overrightarrow{QA} = \overrightarrow{QO} + \overrightarrow{OA}$
$= -4\mathbf{b} + \mathbf{a}$

Exercise 7

In questions **1** to **6**, $\overrightarrow{OA} = \mathbf{a}$ and $\overrightarrow{OB} = \mathbf{b}$. Copy each diagram and use the information given to express the following vectors in terms of **a** and/or **b**.

(a) $\overrightarrow{AP}$ (b) $\overrightarrow{AB}$ (c) $\overrightarrow{OQ}$ (d) $\overrightarrow{PO}$

(e) $\overrightarrow{PQ}$ (f) $\overrightarrow{PN}$ (g) $\overrightarrow{ON}$ (h) $\overrightarrow{AN}$

(i) $\overrightarrow{BP}$ (j) $\overrightarrow{QA}$

1. A, B and N are mid-points of OP, OB and PQ respectively.

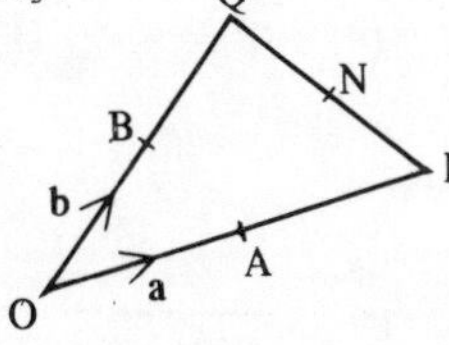

2. A and N are mid-points of OP and PQ; BQ = 2OB.

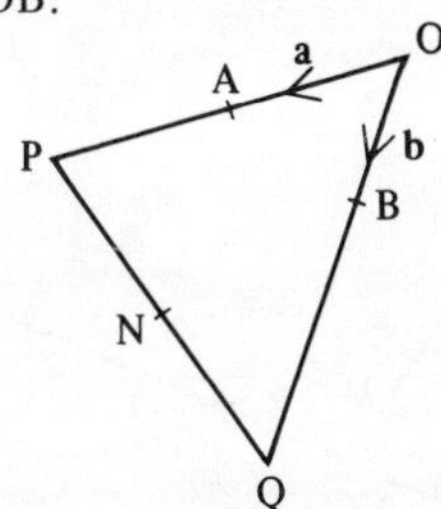

3. AP = 2OA, BQ = OB, PN = NQ.

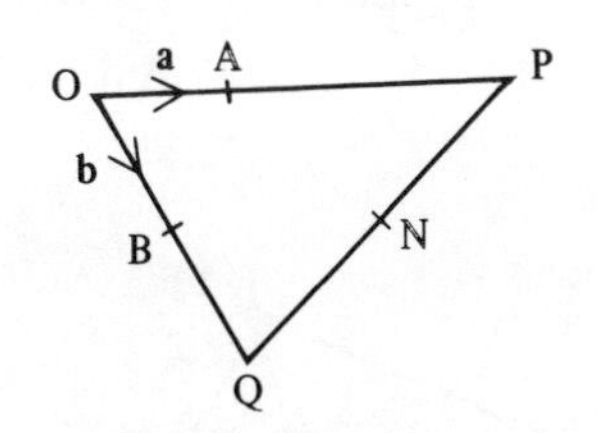

4. OA = 2AP, BQ = 3OB, PN = 2NQ.

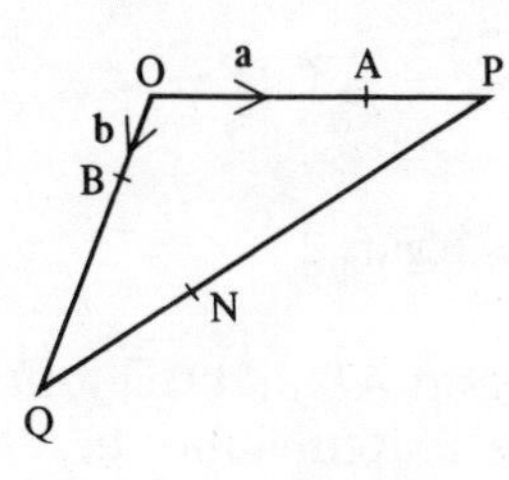

5. AP = 5OA, OB = 2BQ, NP = 2QN.

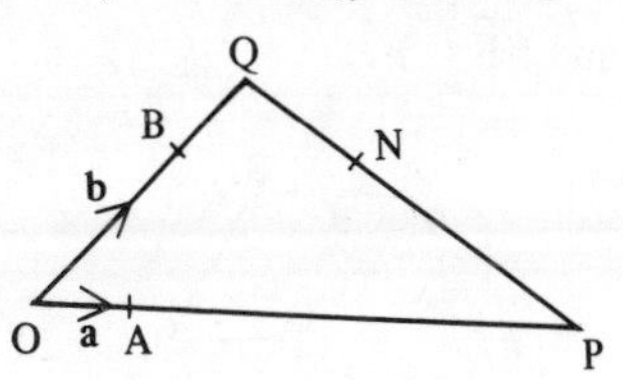

6. OA = $\frac{1}{5}$OP, OQ = 3OB, N is $\frac{1}{4}$ of the way along PQ.

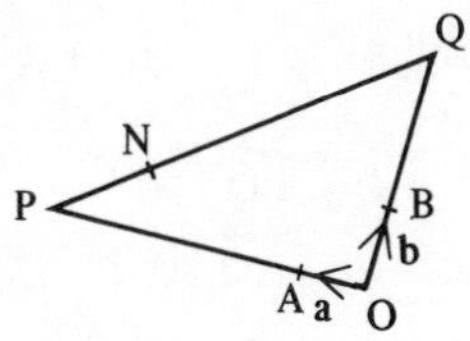

7. In ΔXYZ, the mid-point of YZ is M.
If $\overrightarrow{XY} = \mathbf{s}$ and $\overrightarrow{ZX} = \mathbf{t}$, find $\overrightarrow{XM}$ in terms of **s** and **t**.

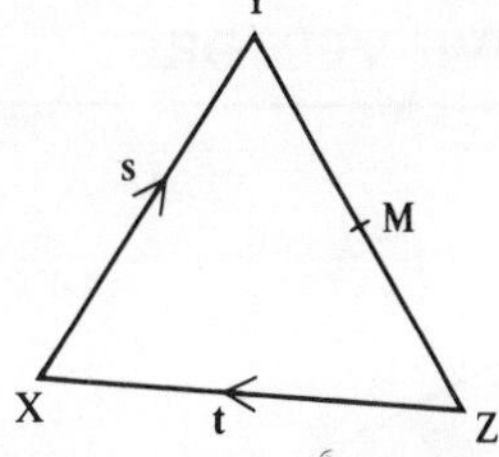

8. In ΔAOB, AM : MB = 2 : 1. If $\overrightarrow{OA} = \mathbf{a}$ and $\overrightarrow{OB} = \mathbf{b}$, find $\overrightarrow{OM}$ in terms of **a** and **b**.

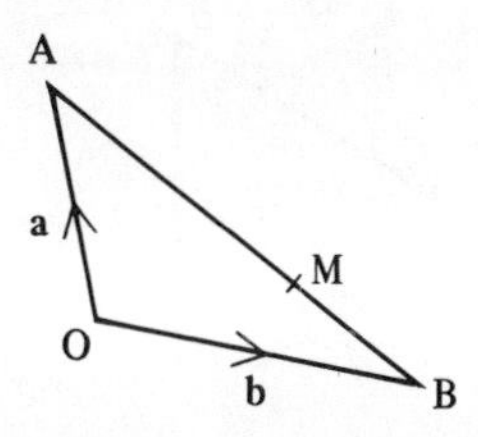

9. O is any point in the plane of the square ABCD. The vectors $\overrightarrow{OA}$, $\overrightarrow{OB}$ and $\overrightarrow{OC}$ are **a**, **b** and **c** respectively. Find the vector $\overrightarrow{OD}$ in terms of **a**, **b** and **c**.

10. ABCDEF is a regular hexagon with $\overrightarrow{AB}$ representing the vector **m** and $\overrightarrow{AF}$ representing the vector **n**. Find the vector representing $\overrightarrow{AD}$.

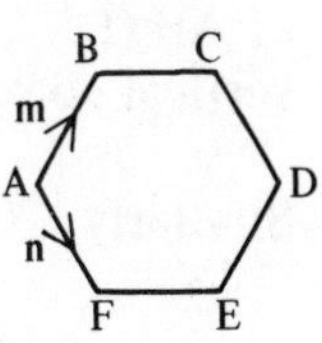

11. ABCDEF is a regular hexagon with centre O. $\vec{FA} = \mathbf{a}$ and $\vec{FB} = \mathbf{b}$.

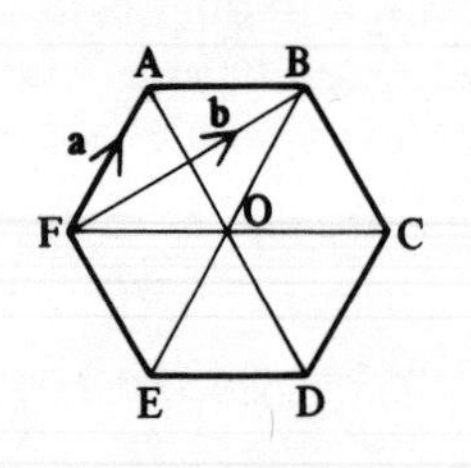

Express the following vectors in terms of **a** and/or **b**.

(a) $\vec{AB}$ (b) $\vec{FO}$ (c) $\vec{FC}$
(d) $\vec{BC}$ (e) $\vec{AO}$ (f) $\vec{FD}$

12. In the diagram, M is the mid-point of CD, BP : PM = 2 : 1, $\vec{AB} = \mathbf{x}$, $\vec{AC} = \mathbf{y}$ and AD = **z**.

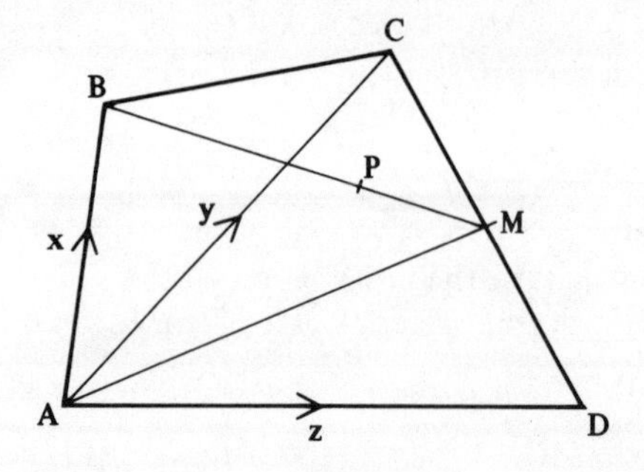

Express the following vectors in terms of **x**, **y** and **z**.

(a) $\vec{DC}$ (b) $\vec{DM}$ (c) $\vec{AM}$
(d) $\vec{BM}$ (e) $\vec{BP}$ (f) $\vec{AP}$.

9.4 COLUMN VECTORS

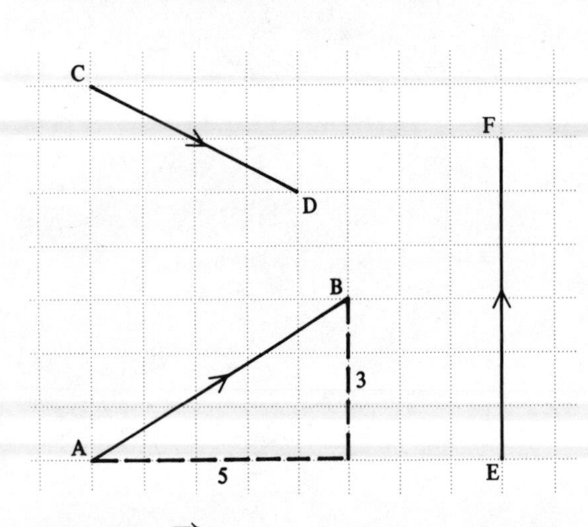

The vector $\vec{AB}$ may be written as a *column vector*. AB = $\begin{pmatrix}5\\3\end{pmatrix}$.

The top number is the horizontal component of $\vec{AB}$ (i.e. 5) and the bottom number is the vertical component (i.e. 3).

Similarly $\vec{CD} = \begin{pmatrix}4\\-2\end{pmatrix}$

$\vec{EF} = \begin{pmatrix}0\\6\end{pmatrix}$

Addition of vectors

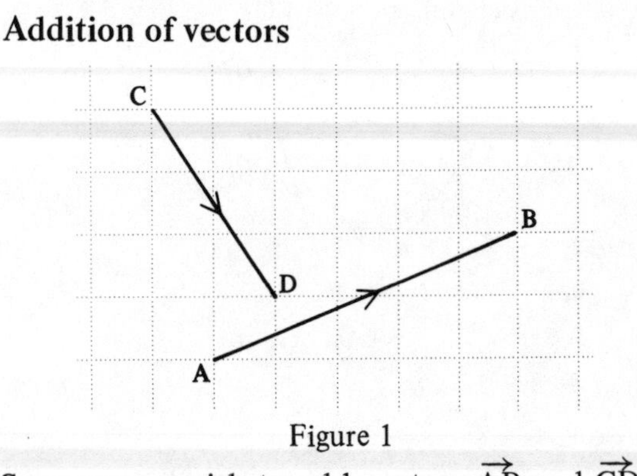

Figure 1

Suppose we wish to add vectors $\vec{AB}$ and $\vec{CD}$ in Figure 1.

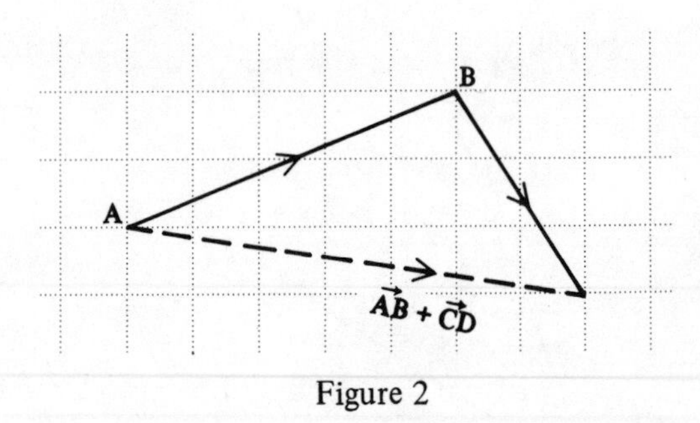

Figure 2

First move $\vec{CD}$ so that $\vec{AB}$ and $\vec{CD}$ join 'nose to tail' as in Figure 2. Remember that changing

the *position* of a vector does not change the vector. A vector is described by its length and direction.
The broken line shows the result of adding $\overrightarrow{AB}$ and $\overrightarrow{CD}$.

In column vectors,

$$\overrightarrow{AB} + \overrightarrow{CD} = \begin{pmatrix}5\\2\end{pmatrix} + \begin{pmatrix}2\\-3\end{pmatrix}$$

We see that the column vector for the broken line is $\begin{pmatrix}7\\-1\end{pmatrix}$. So we perform addition with vectors by adding together the corresponding components of the vectors.

Subtraction of vectors

Figure 3 shows $\overrightarrow{AB} - \overrightarrow{CD}$.

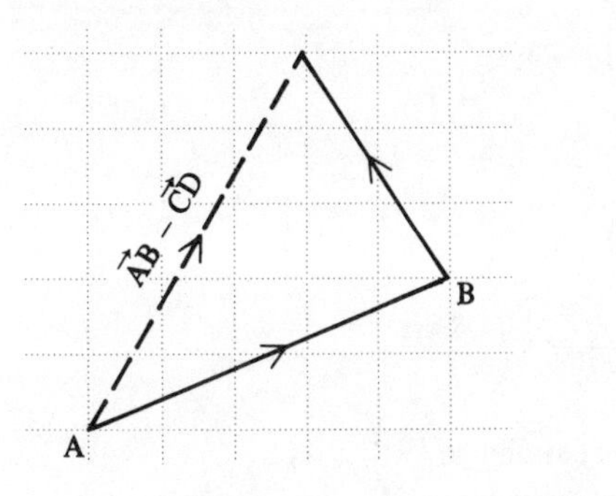

Figure 3

To subtract vector $\overrightarrow{CD}$ from $\overrightarrow{AB}$ we *add* the *negative* of $\overrightarrow{CD}$ to $\overrightarrow{AB}$.

So $\quad \overrightarrow{AB} - \overrightarrow{CD} = \overrightarrow{AB} + (-\overrightarrow{CD})$

In column vectors,

$$\overrightarrow{AB} + (-\overrightarrow{CD}) = \begin{pmatrix}5\\2\end{pmatrix} + \begin{pmatrix}-2\\3\end{pmatrix}$$
$$= \begin{pmatrix}3\\5\end{pmatrix}$$

Multiplication by a scalar

If $\mathbf{a} = \begin{pmatrix}3\\-4\end{pmatrix}$ then $2\mathbf{a} = 2\begin{pmatrix}3\\-4\end{pmatrix} = \begin{pmatrix}6\\-8\end{pmatrix}$.

Each component is multiplied by the number 2.

Parallel vectors

Vectors are parallel if they have the same direction. Both components of one vector must be in the same ratio to the corresponding components of the parallel vector.

e.g. $\begin{pmatrix}3\\-5\end{pmatrix}$ is parallel to $\begin{pmatrix}6\\-10\end{pmatrix}$.

because $\begin{pmatrix}6\\-10\end{pmatrix}$ may be written $2\begin{pmatrix}3\\-5\end{pmatrix}$.

In general the vector $k\begin{pmatrix}a\\b\end{pmatrix}$ is parallel to $\begin{pmatrix}a\\b\end{pmatrix}$.

Example 1

If $\mathbf{a} = \begin{pmatrix}3\\-2\end{pmatrix}$ and $\mathbf{b} = \begin{pmatrix}-1\\3\end{pmatrix}$, find column vectors for

(a) $\mathbf{a} + \mathbf{b}$ (b) $\mathbf{a} - \mathbf{b}$
(c) $2\mathbf{a}$ (d) $2\mathbf{a} + 3\mathbf{b}$

(a) $\mathbf{a} + \mathbf{b} = \begin{pmatrix}3\\-2\end{pmatrix} + \begin{pmatrix}-1\\3\end{pmatrix}$
$= \begin{pmatrix}2\\1\end{pmatrix}$

(b) $\mathbf{a} - \mathbf{b} = \begin{pmatrix}3\\-2\end{pmatrix} + \begin{pmatrix}1\\-3\end{pmatrix}$
$= \begin{pmatrix}4\\-5\end{pmatrix}$

(c) $2\mathbf{a} = 2\begin{pmatrix}3\\-2\end{pmatrix}$
$= \begin{pmatrix}6\\-4\end{pmatrix}$

(d) $2\mathbf{a} + 3\mathbf{b} = 2\begin{pmatrix}3\\-2\end{pmatrix} + 3\begin{pmatrix}-1\\3\end{pmatrix}$
$= \begin{pmatrix}6\\-4\end{pmatrix} + \begin{pmatrix}-3\\9\end{pmatrix}$
$= \begin{pmatrix}3\\5\end{pmatrix}$

Example 2

If $\mathbf{c} = \begin{pmatrix} 4 \\ -2 \end{pmatrix}$ and $\mathbf{d} = \begin{pmatrix} -3 \\ 1 \end{pmatrix}$, solve the following vector equations for $\mathbf{x}$.

(a) $\mathbf{x} + \mathbf{c} = \mathbf{d}$ (b) $2\mathbf{x} - \mathbf{d} = \mathbf{c}$

(c) $\mathbf{x} + \mathbf{d} = \mathbf{0}$

(a) $\mathbf{x} = \mathbf{d} - \mathbf{c}$

$= \begin{pmatrix} -3 \\ 1 \end{pmatrix} + \begin{pmatrix} -4 \\ 2 \end{pmatrix}$

$= \begin{pmatrix} -7 \\ 3 \end{pmatrix}$

(b) $2\mathbf{x} = \mathbf{c} + \mathbf{d}$

$= \begin{pmatrix} 4 \\ -2 \end{pmatrix} + \begin{pmatrix} -3 \\ 1 \end{pmatrix}$

$= \begin{pmatrix} 1 \\ -1 \end{pmatrix}$

$\therefore \quad \mathbf{x} = \begin{pmatrix} \frac{1}{2} \\ -\frac{1}{2} \end{pmatrix}$

(c) $\mathbf{x} = \mathbf{0} - \mathbf{d}$

$= \begin{pmatrix} 0 \\ 0 \end{pmatrix} + \begin{pmatrix} 3 \\ -1 \end{pmatrix}$

$= \begin{pmatrix} 3 \\ -1 \end{pmatrix}$

$\mathbf{0}$ is called the zero vector.

Exercise 8

Questions **1** to **36** refer to the following vectors.

$\mathbf{a} = \begin{pmatrix} 3 \\ 4 \end{pmatrix}$ $\mathbf{b} = \begin{pmatrix} 1 \\ 4 \end{pmatrix}$ $\mathbf{c} = \begin{pmatrix} 4 \\ -3 \end{pmatrix}$ $\mathbf{d} = \begin{pmatrix} -1 \\ 1 \end{pmatrix}$

$\mathbf{e} = \begin{pmatrix} 5 \\ 12 \end{pmatrix}$ $\mathbf{f} = \begin{pmatrix} 3 \\ -2 \end{pmatrix}$ $\mathbf{g} = \begin{pmatrix} -4 \\ -2 \end{pmatrix}$ $\mathbf{h} = \begin{pmatrix} -12 \\ 5 \end{pmatrix}$

Draw and label the following vectors on graph paper (take 1 cm to 1 unit).

1. $\mathbf{c}$ **2.** $\mathbf{f}$ **3.** $2\mathbf{b}$ **4.** $-\mathbf{a}$
5. $-\mathbf{g}$ **6.** $3\mathbf{a}$ **7.** $\frac{1}{2}\mathbf{e}$ **8.** $5\mathbf{d}$
9. $-\frac{1}{2}\mathbf{h}$ **10.** $\frac{3}{2}\mathbf{g}$ **11.** $\frac{1}{5}\mathbf{h}$ **12.** $-3\mathbf{b}$

Find the following vectors in component form.

13. $\mathbf{b} + \mathbf{h}$ **14.** $\mathbf{f} + \mathbf{g}$
15. $\mathbf{e} - \mathbf{b}$ **16.** $\mathbf{a} - \mathbf{d}$
17. $\mathbf{g} - \mathbf{h}$ **18.** $2\mathbf{a} + 3\mathbf{c}$
19. $3\mathbf{f} + 2\mathbf{d}$ **20.** $4\mathbf{g} - 2\mathbf{b}$
21. $5\mathbf{a} + \frac{1}{2}\mathbf{g}$ **22.** $\mathbf{a} + \mathbf{b} + \mathbf{c}$
23. $3\mathbf{f} - \mathbf{a} + \mathbf{c}$ **24.** $\mathbf{c} + 2\mathbf{d} + 3\mathbf{e}$

In each of the following, find $\mathbf{x}$ in component form.

25. $\mathbf{x} + \mathbf{b} = \mathbf{e}$ **26.** $\mathbf{x} + \mathbf{d} = \mathbf{a}$
27. $\mathbf{c} + \mathbf{x} = \mathbf{f}$ **28.** $\mathbf{x} - \mathbf{g} = \mathbf{h}$
29. $2\mathbf{x} + \mathbf{b} = \mathbf{g}$ **30.** $2\mathbf{x} - 3\mathbf{d} = \mathbf{g}$
31. $2\mathbf{b} = \mathbf{d} - \mathbf{x}$ **32.** $\mathbf{f} - \mathbf{g} = \mathbf{e} - \mathbf{x}$
33. $2\mathbf{x} + \mathbf{b} = \mathbf{x} + \mathbf{e}$ **34.** $3\mathbf{x} - \mathbf{b} = \mathbf{x} + \mathbf{h}$
35. $\mathbf{a} + \mathbf{b} + \mathbf{x} = \mathbf{b} + \mathbf{a}$ **36.** $2\mathbf{x} + \mathbf{e} = \mathbf{0}$ (zero vector)

37. (a) Draw and label each of the following vectors on graph paper.

$\mathbf{l} = \begin{pmatrix} -3 \\ -3 \end{pmatrix}$; $\mathbf{m} = \begin{pmatrix} 2 \\ 0 \end{pmatrix}$; $\mathbf{n} = \begin{pmatrix} 3 \\ 2 \end{pmatrix}$; $\mathbf{p} = \begin{pmatrix} 1 \\ -2 \end{pmatrix}$;

$\mathbf{q} = \begin{pmatrix} 3 \\ 0 \end{pmatrix}$; $\mathbf{r} = \begin{pmatrix} 6 \\ 4 \end{pmatrix}$; $\mathbf{s} = \begin{pmatrix} 2 \\ 2 \end{pmatrix}$; $\mathbf{t} = \begin{pmatrix} 2 \\ -4 \end{pmatrix}$;

$\mathbf{u} = \begin{pmatrix} -1 \\ -3 \end{pmatrix}$; $\mathbf{v} = \begin{pmatrix} 0 \\ 3 \end{pmatrix}$.

(b) Find four pairs of parallel vectors amongst the ten vectors.

38. State whether 'true' or 'false'.

(a) $\begin{pmatrix} 3 \\ -1 \end{pmatrix}$ is parallel to $\begin{pmatrix} 9 \\ -3 \end{pmatrix}$.

(b) $\begin{pmatrix} -2 \\ 0 \end{pmatrix}$ is parallel to $\begin{pmatrix} 4 \\ 0 \end{pmatrix}$.

(c) $\begin{pmatrix} -1 \\ 1 \end{pmatrix}$ is parallel to $\begin{pmatrix} 1 \\ -1 \end{pmatrix}$.

(d) $\begin{pmatrix} 5 \\ -15 \end{pmatrix} = 5\begin{pmatrix} 1 \\ -3 \end{pmatrix}$.

(e) $\begin{pmatrix} 4 \\ 0 \end{pmatrix}$ is parallel to $\begin{pmatrix} 0 \\ 6 \end{pmatrix}$.

(f) $\begin{pmatrix} 3 \\ -1 \end{pmatrix} + \begin{pmatrix} -4 \\ -2 \end{pmatrix} = \begin{pmatrix} -1 \\ 1 \end{pmatrix}$.

39. (a) Draw a diagram to illustrate the vector addition $\overrightarrow{AB} + \overrightarrow{CD}$.
(b) Draw a diagram to illustrate $\overrightarrow{AB} - \overrightarrow{CD}$.

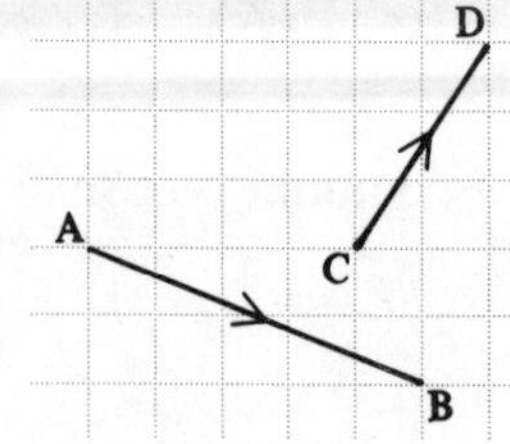

40. Draw separate diagrams to illustrate the following.
(a) $\overrightarrow{FE} + \overrightarrow{JI}$
(b) $\overrightarrow{HG} + \overrightarrow{FE}$
(c) $\overrightarrow{JI} - \overrightarrow{FE}$
(d) $\overrightarrow{HG} + \overrightarrow{JI}$.

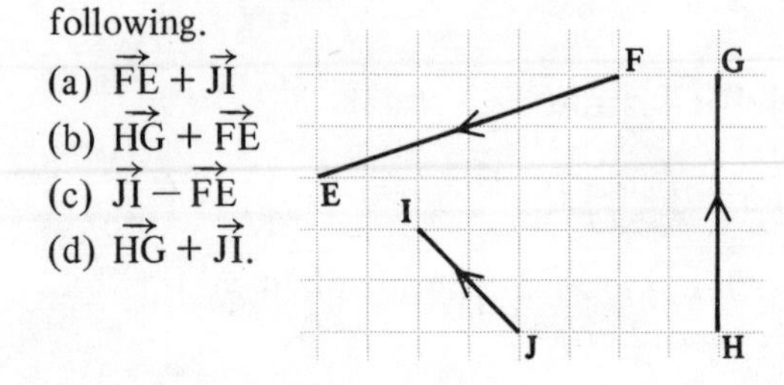

Modulus of a vector

The modulus of a vector **a** is written $|\mathbf{a}|$ and represents the length (or magnitude) of the vector.

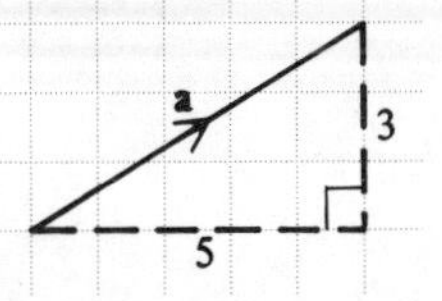

In the diagram above, $\mathbf{a} = \begin{pmatrix} 5 \\ 3 \end{pmatrix}$.

By Pythagoras' Theorem, $|\mathbf{a}| = \sqrt{(5^2 + 3^2)}$

$|\mathbf{a}| = \sqrt{34}$ units.

In general if $\mathbf{x} = \begin{pmatrix} m \\ n \end{pmatrix}$, $|\mathbf{x}| = \sqrt{(m^2 + n^2)}$.

Example 3

Find the modulus of each of the following vectors.

$\mathbf{r} = \begin{pmatrix} 3 \\ 2 \end{pmatrix}$, $\mathbf{s} = \begin{pmatrix} 0 \\ 5 \end{pmatrix}$, $\mathbf{t} = \begin{pmatrix} -2 \\ -5 \end{pmatrix}$.

$|\mathbf{r}| = \sqrt{(3^2 + 2^2)} = \sqrt{13}$ units

$|\mathbf{s}| = 5$ units

$|\mathbf{t}| = \sqrt{((-2)^2 + (-5)^2)}$
$= \sqrt{(4 + 25)}$
$= \sqrt{29}$ units.

Exercise 9

Questions **1** to **12** refer to the following vectors.

$\mathbf{a} = \begin{pmatrix} 3 \\ 4 \end{pmatrix}$ $\mathbf{b} = \begin{pmatrix} 4 \\ 1 \end{pmatrix}$ $\mathbf{c} = \begin{pmatrix} 5 \\ 12 \end{pmatrix}$ $\mathbf{d} = \begin{pmatrix} -3 \\ 0 \end{pmatrix}$

$\mathbf{e} = \begin{pmatrix} -4 \\ -3 \end{pmatrix}$ $\mathbf{f} = \begin{pmatrix} -3 \\ 6 \end{pmatrix}$

From the following, leaving the answer in square root form where necessary.

1. $|\mathbf{a}|$ **2.** $|\mathbf{b}|$ **3.** $|\mathbf{c}|$ **4.** $|\mathbf{d}|$
5. $|\mathbf{e}|$ **6.** $|\mathbf{f}|$ **7.** $|\mathbf{a}+\mathbf{b}|$ **8.** $|\mathbf{c}-\mathbf{d}|$
9. $|2\mathbf{e}|$ **10.** $|\mathbf{f}+2\mathbf{b}|$

11. (a) Find $|\mathbf{a}+\mathbf{c}|$.
(b) Is $|\mathbf{a}+\mathbf{c}|$ equal to $|\mathbf{a}| + |\mathbf{c}|$?

12. (a) Find $|\mathbf{c}+\mathbf{d}|$.
(b) Is $|\mathbf{c}+\mathbf{d}|$ equal to $|\mathbf{c}| + |\mathbf{d}|$?

13. If $\overrightarrow{AB} = \begin{pmatrix} 3 \\ -1 \end{pmatrix}$ and $\overrightarrow{BC} = \begin{pmatrix} 2 \\ 3 \end{pmatrix}$, find $|\overrightarrow{AC}|$.

14. If $\overrightarrow{PQ} = \begin{pmatrix} 5 \\ -2 \end{pmatrix}$ and $\overrightarrow{QR} = \begin{pmatrix} 0 \\ 1 \end{pmatrix}$, find $|\overrightarrow{PR}|$.

15. If $\overrightarrow{WX} = \begin{pmatrix} 1 \\ 3 \end{pmatrix}$, $\overrightarrow{XY} = \begin{pmatrix} -2 \\ 0 \end{pmatrix}$ and $\overrightarrow{YZ} = \begin{pmatrix} 2 \\ -1 \end{pmatrix}$, find $|\overrightarrow{WZ}|$.

16. Given that $\overrightarrow{OP} = \begin{pmatrix} 0 \\ 5 \end{pmatrix}$ and $\overrightarrow{OQ} = \begin{pmatrix} n \\ 3 \end{pmatrix}$, find
(a) $|\overrightarrow{OP}|$
(b) A value for n if $|\overrightarrow{OP}| = |\overrightarrow{OQ}|$.

17. Given that $\overrightarrow{OA} = \begin{pmatrix} 5 \\ 12 \end{pmatrix}$ and $\overrightarrow{OB} = \begin{pmatrix} 0 \\ m \end{pmatrix}$, find
(a) $|\overrightarrow{OA}|$
(b) A value for m if $|\overrightarrow{OA}| = |\overrightarrow{OB}|$.

18. Given that $\overrightarrow{LM} = \begin{pmatrix} -3 \\ 4 \end{pmatrix}$ and $\overrightarrow{MN} \begin{pmatrix} -15 \\ p \end{pmatrix}$, find
(a) $|\overrightarrow{LM}|$
(b) The value of p if $|\overrightarrow{MN}| = 3\,|\overrightarrow{LM}|$.

19. **a** and **b** are two vectors and $|\mathbf{a}| = 3$.
Find the value of $|\mathbf{a}+\mathbf{b}|$ when:
(a) $\mathbf{b} = 2\mathbf{a}$
(b) $\mathbf{b} = -3\mathbf{a}$
(c) **b** is perpendicular to **a** and $|\mathbf{b}| = 4$.

20. **r** and **s** are two vectors and $|\mathbf{r}| = 5$.
Find the value of $|\mathbf{r}+\mathbf{s}|$ when:
(a) $\mathbf{s} = 5\mathbf{r}$
(b) $\mathbf{s} = -2\mathbf{r}$
(c) **r** is perpendicular to **s** and $|\mathbf{s}| = 5$
(d) **s** is perpendicular to $(\mathbf{r}+\mathbf{s})$ and $|\mathbf{s}| = 3$.

Position vector

The position vector of point A is the vector $\overrightarrow{OA}$ where O is the origin.

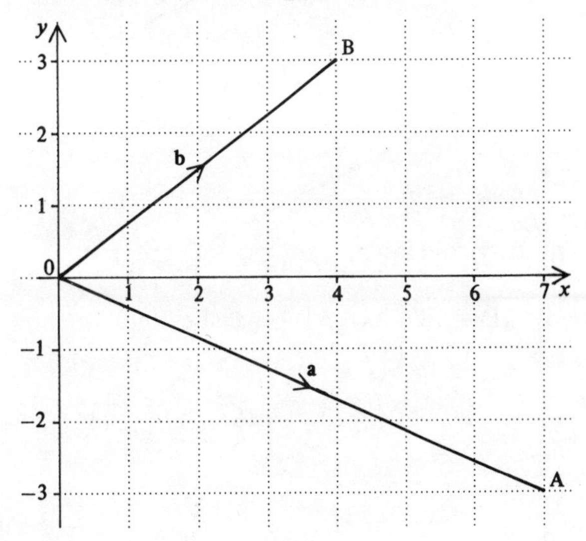

Thus in the diagram on the previous page, A has coordinates (7, −3) and the position vector of A is $\begin{pmatrix} 7 \\ -3 \end{pmatrix}$.

We write $\mathbf{a} = \begin{pmatrix} 7 \\ -3 \end{pmatrix}$.

Similarly $\mathbf{b} = \begin{pmatrix} 4 \\ 3 \end{pmatrix}$.

Example 4

Write down the position vectors of the following points.
C(3, −1), D(5, 4), E(0, 10), Z(m, n).

$$\mathbf{c} = \begin{pmatrix} 3 \\ -1 \end{pmatrix};\ \mathbf{d} = \begin{pmatrix} 5 \\ 4 \end{pmatrix};\quad \mathbf{e} = \begin{pmatrix} 0 \\ 10 \end{pmatrix};\ \mathbf{z} = \begin{pmatrix} m \\ n \end{pmatrix}$$

Example 5

Given that point A has coordinates (1, 2), $\overrightarrow{AB} = \begin{pmatrix} 1 \\ 3 \end{pmatrix}$ and $\overrightarrow{BC} = \begin{pmatrix} 4 \\ -2 \end{pmatrix}$, find

(a) the coordinates of D so that ABCD is a parallelogram.

(b) $\overrightarrow{AM}$ is M is the mid-point of BC.

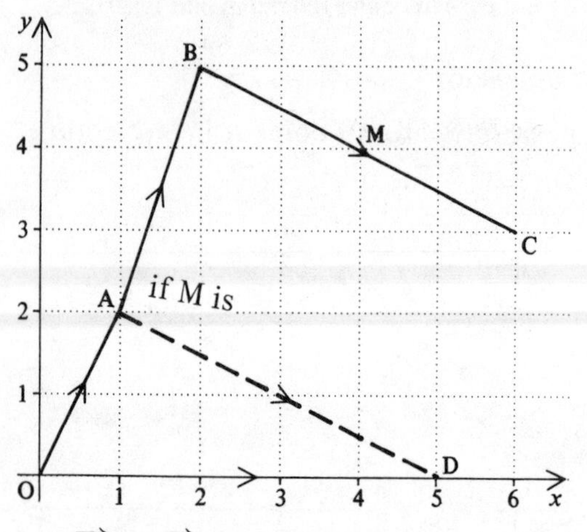

(a) $\overrightarrow{AD} = \overrightarrow{BC}$ since AD is parallel and equal in length to BC.

$$\overrightarrow{BC} = \begin{pmatrix} 4 \\ -2 \end{pmatrix},\quad \therefore\ \overrightarrow{AD} = \begin{pmatrix} 4 \\ -2 \end{pmatrix}.$$

Now $\overrightarrow{OD} = \overrightarrow{OA} + \overrightarrow{AD}$

$$\overrightarrow{OD} = \begin{pmatrix} 1 \\ 2 \end{pmatrix} + \begin{pmatrix} 4 \\ -2 \end{pmatrix} = \begin{pmatrix} 5 \\ 0 \end{pmatrix}$$

The coordinates of D are (5, 0).

(b) $\overrightarrow{AM} = \overrightarrow{AB} + \overrightarrow{BM}$

and $\overrightarrow{BM} = \frac{1}{2}\overrightarrow{BC} = \frac{1}{2}\begin{pmatrix} 4 \\ -2 \end{pmatrix} = \begin{pmatrix} 2 \\ -1 \end{pmatrix}$

$\therefore\quad \overrightarrow{AM} = \begin{pmatrix} 1 \\ 3 \end{pmatrix} + \begin{pmatrix} 2 \\ -1 \end{pmatrix} = \begin{pmatrix} 3 \\ 2 \end{pmatrix}$

Exercise 10

1. If D has coordinates (7, 2) and E has coordinates (9, 0), find the column vector for $\overrightarrow{DE}$.

2. Find the column vector $\overrightarrow{XY}$ where X and Y have coordinates (−1, 4) and (5, 2) respectively.

3. In the diagram $\overrightarrow{AB}$ represents the vector $\begin{pmatrix} 5 \\ 2 \end{pmatrix}$ and $\overrightarrow{BC}$ represents the vector $\begin{pmatrix} 0 \\ 3 \end{pmatrix}$.

(a) Copy the diagram and mark point D such that ABCD is a parallelogram.

(b) Write $\overrightarrow{AD}$ and $\overrightarrow{CA}$ as column vectors.

4. (a) On squared paper draw $\overrightarrow{AB} = \begin{pmatrix} 3 \\ -2 \end{pmatrix}$ and $\overrightarrow{BC} = \begin{pmatrix} 4 \\ 2 \end{pmatrix}$ and mark point D such that ABCD is a parallelogram.

(b) Write $\overrightarrow{AD}$ and $\overrightarrow{CA}$ as column vectors.

5. (a) On squared paper draw $\overrightarrow{XY} = \begin{pmatrix} 1 \\ 4 \end{pmatrix}$ and $\overrightarrow{YZ} = \begin{pmatrix} 4 \\ 0 \end{pmatrix}$ and mark point D such that XYZD is a parallelogram.

(b) Write $\overrightarrow{XD}$ and $\overrightarrow{ZX}$ as column vectors.

6. (a) On squared paper draw $\overrightarrow{XY} = \begin{pmatrix} -3 \\ 2 \end{pmatrix}$ and $\overrightarrow{YZ} = \begin{pmatrix} 0 \\ -5 \end{pmatrix}$ and mark point D such that XYZD is a parallelogram.

(b) Write $\overrightarrow{XD}$ and $\overrightarrow{ZX}$ as column vectors.

7. Copy the diagram below in which $\overrightarrow{OA} = \begin{pmatrix} 5 \\ 2 \end{pmatrix}$, $\overrightarrow{OB} = \begin{pmatrix} 2 \\ 5 \end{pmatrix}$.

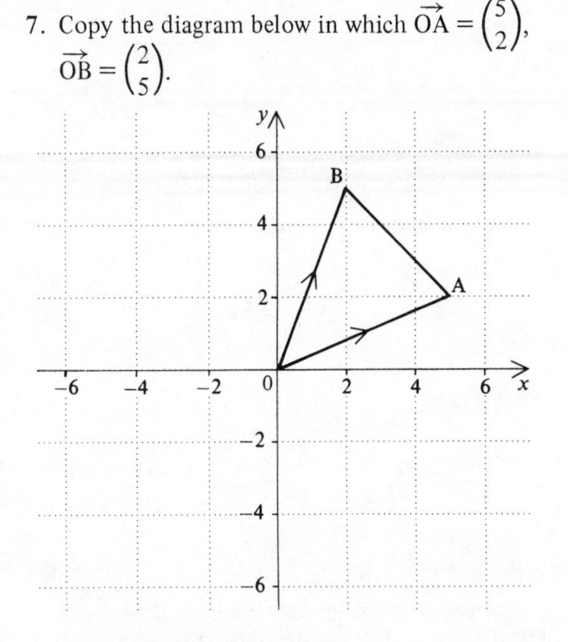

M is the mid-point of AB. Express the following as column vectors.
(a) $\overrightarrow{BA}$ (b) $\overrightarrow{BM}$
(c) $\overrightarrow{OM}$ (use $\overrightarrow{OM} = \overrightarrow{OB} + \overrightarrow{BM}$).
Hence write down the coordinates of M.

8. On a graph with origin at O, draw $\overrightarrow{OA} = \begin{pmatrix} 5 \\ -1 \end{pmatrix}$ and $\overrightarrow{OB} = \begin{pmatrix} 6 \\ -7 \end{pmatrix}$. Given that M is the mid-point of AB, express the following as column vectors.
(a) $\overrightarrow{BA}$ (b) $\overrightarrow{BM}$ (c) $\overrightarrow{OM}$
Hence write down the coordinates of M.

9. On a graph with origin at O, draw $\overrightarrow{OA} = \begin{pmatrix} -3 \\ -5 \end{pmatrix}$ and $\overrightarrow{OB} = \begin{pmatrix} -7 \\ -2 \end{pmatrix}$. Given that M is the mid-point of AB, express the following as column vectors.
(a) $\overrightarrow{BA}$ (b) $\overrightarrow{BM}$ (c) $\overrightarrow{OM}$
Hence write down the coordinates of M.

10. On a graph with origin at O, draw $\overrightarrow{OA} = \begin{pmatrix} -2 \\ 5 \end{pmatrix}$, $\overrightarrow{OB} = \begin{pmatrix} 4 \\ 2 \end{pmatrix}$ and $\overrightarrow{OC} = \begin{pmatrix} -2 \\ -4 \end{pmatrix}$.
(a) Given that M divides AB such that AM : MB = 2 : 1, express the following as column vectors
(i) $\overrightarrow{BA}$ (ii) $\overrightarrow{BM}$ (iii) $\overrightarrow{OM}$
(b) Given that N divides AC such that AN : NC = 1 : 2, express the following as column vectors
(i) $\overrightarrow{AC}$ (ii) $\overrightarrow{AN}$ (iii) $\overrightarrow{ON}$.

11. In square ABCD, side AB has column vector $\begin{pmatrix} 2 \\ 1 \end{pmatrix}$. Find two possible column vectors for $\overrightarrow{BC}$.

12. Rectangle KLMN has an area of 10 square units and $\overrightarrow{KL}$ has column vector $\begin{pmatrix} 5 \\ 0 \end{pmatrix}$. Find two possible column vectors for $\overrightarrow{LM}$.

13. In the diagram, ABCD is a trapezium in which $\overrightarrow{DC} = 2\overrightarrow{AB}$.
If $\overrightarrow{AB} = \mathbf{p}$ and $\overrightarrow{AD} = \mathbf{q}$ express in terms of **p** and and **q**

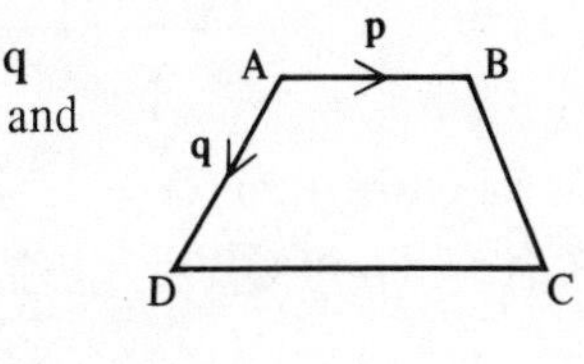

(a) $\overrightarrow{BD}$ (b) $\overrightarrow{AC}$
(c) $\overrightarrow{BC}$

14. Find the image of the vector $\begin{pmatrix} 1 \\ 3 \end{pmatrix}$ after reflection in the following lines:
(a) $y = 0$ (b) $x = 0$
(c) $y = x$ (d) $y = -x$

In questions **15** to **17**,

$\mathbf{a} = \begin{pmatrix} 3 \\ 4 \end{pmatrix}; \mathbf{b} = \begin{pmatrix} 1 \\ 4 \end{pmatrix}; \mathbf{c} = \begin{pmatrix} 4 \\ -3 \end{pmatrix};$

$\mathbf{e} = \begin{pmatrix} 5 \\ 12 \end{pmatrix}; \mathbf{f} = \begin{pmatrix} 3 \\ -2 \end{pmatrix}; \mathbf{g} = \begin{pmatrix} -4 \\ -2 \end{pmatrix}; \mathbf{h} = \begin{pmatrix} -12 \\ 5 \end{pmatrix}.$

15. If $m\mathbf{a} + n\mathbf{b} = \mathbf{e}$ (where m and n are numbers) write down two simultaneous equations in m and n and solve them.

16. If $m\mathbf{c} + n\mathbf{g} = \begin{pmatrix} -4 \\ -12 \end{pmatrix}$, find the numbers m and n.

17. If $s\mathbf{f} + t\mathbf{h} = \begin{pmatrix} 18 \\ -9 \end{pmatrix}$, find the numbers s and t.

9.5 VECTOR GEOMETRY

Example 1

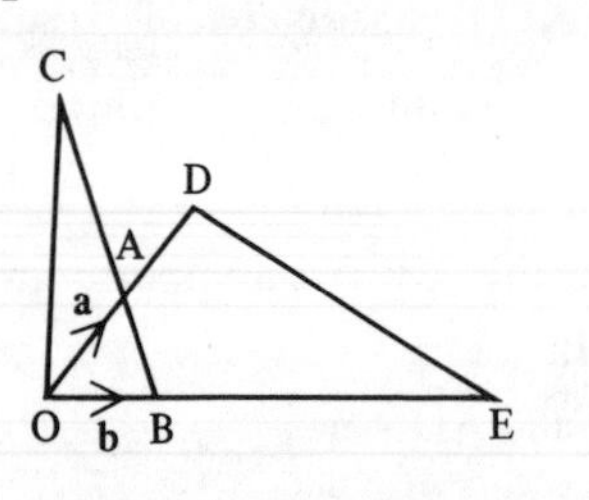

In the diagram, $\overrightarrow{OD} = 2\overrightarrow{OA}$, $\overrightarrow{OE} = 4\overrightarrow{OB}$, $\overrightarrow{OA} = \mathbf{a}$ and $\overrightarrow{OB} = \mathbf{b}$.

(a) Express $\overrightarrow{OD}$ and $\overrightarrow{OE}$ in terms of **a** and **b** respectively.
(b) Express $\overrightarrow{BA}$ in terms of **a** and **b**.
(c) Express $\overrightarrow{ED}$ in terms of **a** and **b**.
(d) Given that $\overrightarrow{BC} = 3\overrightarrow{BA}$, express $\overrightarrow{OC}$ in terms of **a** and **b**.
(e) Express $\overrightarrow{EC}$ in terms of **a** and **b**.
(f) Hence show that the points E, D and C lie on a straight line.

(a) $\overrightarrow{OD} = 2\mathbf{a}$
$\overrightarrow{OE} = 4\mathbf{b}$

(b) $\overrightarrow{BA} = -\mathbf{b} + \mathbf{a}$

(c) $\overrightarrow{ED} = -4\mathbf{b} + 2\mathbf{a}$

(d) $\overrightarrow{OC} = \overrightarrow{OB} + \overrightarrow{BC}$
$\overrightarrow{OC} = \mathbf{b} + 3(-\mathbf{b} + \mathbf{a})$
$\overrightarrow{OC} = -2\mathbf{b} + 3\mathbf{a}$

(e) $\overrightarrow{EC} = \overrightarrow{EO} + \overrightarrow{OC}$
$\overrightarrow{EC} = -4\mathbf{b} + (-2\mathbf{b} + 3\mathbf{a})$
$\overrightarrow{EC} = -6\mathbf{b} + 3\mathbf{a}$

(f) Using the results for $\overrightarrow{ED}$ and $\overrightarrow{EC}$, we see that $\overrightarrow{EC} = \frac{3}{2}\overrightarrow{ED}$.

Since $\overrightarrow{EC}$ and $\overrightarrow{ED}$ are parallel vectors which both pass through the point E, the points E, D and C must lie on a straight line.

Example 2

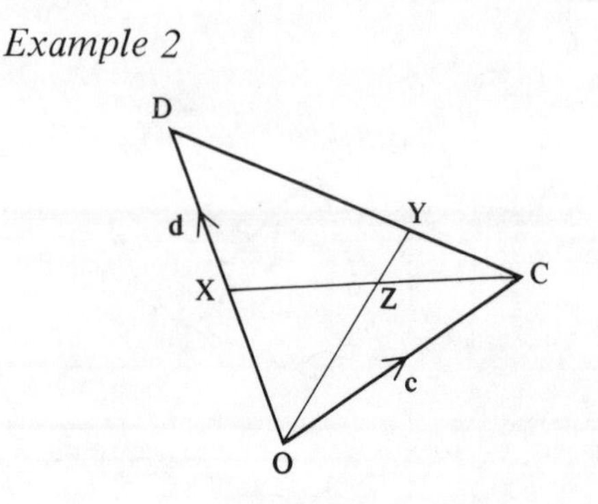

In the diagram, X is the mid-point of OD and Y lies on CD such that CY : YD = 1 : 2. $\overrightarrow{OC} = \mathbf{c}$ and $\overrightarrow{OD} = \mathbf{d}$.

(a) Express $\overrightarrow{CD}$ and $\overrightarrow{OY}$ in terms of **c** and **d**.
(b) Express $\overrightarrow{CX}$ in terms of **c** and **d**.
(c) Given that OY meets CX at Z and $\overrightarrow{CZ} = h\overrightarrow{CX}$, express OZ in terms of **c**, **d** and h.
(d) If $\overrightarrow{OZ} = k\overrightarrow{OY}$, form an equation and hence find the values of h and k.

(a) $\overrightarrow{CD} = \overrightarrow{CO} + \overrightarrow{OD} = -\mathbf{c} + \mathbf{d}$ ... [1]
$\overrightarrow{OY} = \overrightarrow{OC} + \overrightarrow{CY}$
and $\overrightarrow{CY} = \frac{1}{3}\overrightarrow{CD}$
$\therefore\ \overrightarrow{OY} = \mathbf{c} + \frac{1}{3}(-\mathbf{c} + \mathbf{d}) = \frac{2}{3}\mathbf{c} + \frac{1}{3}\mathbf{d}$ ... [2]

(b) $\overrightarrow{CX} = \overrightarrow{CO} + \overrightarrow{OX} = -\mathbf{c} + \frac{1}{2}\mathbf{d}$ ... [3]

(c) $\overrightarrow{OZ} = \overrightarrow{OC} + \overrightarrow{CZ}$
$= \mathbf{c} + h\overrightarrow{CX} = \mathbf{c} + h(-\mathbf{c} + \frac{1}{2}\mathbf{d})$
$= \mathbf{c}(1 - h) + \mathbf{d}(\frac{1}{2}h)$ ... [4]

(d) $\overrightarrow{OZ} = k\overrightarrow{OY} = k(\frac{2}{3}\mathbf{c} + \frac{1}{3}\mathbf{d})$ (from [2])
$= \mathbf{c}(\frac{2}{3}k) + \mathbf{d}(\frac{1}{3}k)$ ... [5]

equating the expressions for $\overrightarrow{OZ}$ from [4] and [5]

$\mathbf{c}(1 - h) + \mathbf{d}(\frac{1}{2}h) = \mathbf{c}(\frac{2}{3}k) + \mathbf{d}(\frac{1}{3}k)$.

For this to be true the coefficients of **c** and **d** must be equal.

$\therefore \quad 1-h = \frac{2}{3}k \qquad \ldots [6]$

$\frac{1}{2}h = \frac{1}{3}k \qquad \ldots [7]$

From [7] $h = \frac{2}{3}k$

$\therefore \quad 1-h = h$

$h = \frac{1}{2}$ and $k = \frac{3}{4}$.

Exercise 11

1. $\overrightarrow{OD} = 2\overrightarrow{OA}$,
$\overrightarrow{OE} = 3\overrightarrow{OB}$,
$\overrightarrow{OA} = \mathbf{a}$
$\overrightarrow{OB} = \mathbf{b}$.

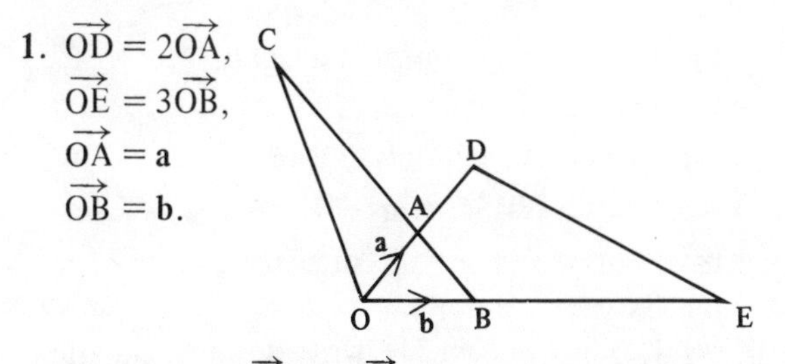

(a) Express $\overrightarrow{OD}$ and $\overrightarrow{OE}$ in terms of $\mathbf{a}$ and $\mathbf{b}$ respectively.
(b) Express $\overrightarrow{BA}$ in terms of $\mathbf{a}$ and $\mathbf{b}$.
(c) Express $\overrightarrow{ED}$ in terms of $\mathbf{a}$ and $\mathbf{b}$.
(d) Given that $\overrightarrow{BC} = 4\overrightarrow{BA}$, express OC in terms of $\mathbf{a}$ and $\mathbf{b}$.
(e) Express $\overrightarrow{EC}$ in terms of $\mathbf{a}$ and $\mathbf{b}$.
(f) Use the results for $\overrightarrow{ED}$ and $\overrightarrow{EC}$ to show that points E, D and C lie on a straight line.

2. $\overrightarrow{OY} = 2\overrightarrow{OB}$,
$\overrightarrow{OX} = \frac{5}{2}\overrightarrow{OA}$,
$\overrightarrow{OA} = \mathbf{a}$,
$\overrightarrow{OB} = \mathbf{b}$.

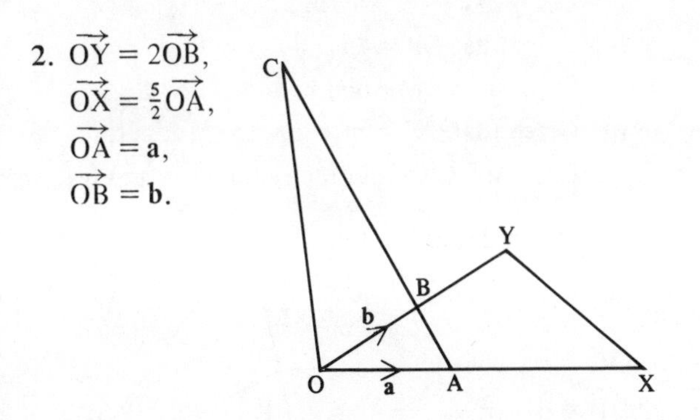

(a) Express $\overrightarrow{OY}$ and $\overrightarrow{OX}$ in terms of $\mathbf{b}$ and $\mathbf{a}$ respectively.
(b) Express $\overrightarrow{AB}$ in terms of $\mathbf{a}$ and $\mathbf{b}$.
(c) Express $\overrightarrow{XY}$ in terms of $\mathbf{a}$ and $\mathbf{b}$.
(d) Given that $\overrightarrow{AC} = 6\overrightarrow{AB}$, express $\overrightarrow{OC}$ in terms of $\mathbf{a}$ and $\mathbf{b}$.
(e) Express $\overrightarrow{XC}$ in terms of $\mathbf{a}$ and $\mathbf{b}$.
(f) Use the results for $\overrightarrow{XY}$ and $\overrightarrow{XC}$ to show that points X, Y and C lie on a straight line.

3. $\overrightarrow{OA} = \mathbf{a}$,
$\overrightarrow{OB} = \mathbf{b}$;
$\overrightarrow{AQ} = \frac{1}{2}\mathbf{a}$,
$\overrightarrow{BR} = \mathbf{b}$,
$\overrightarrow{AP} = 2\overrightarrow{BA}$.

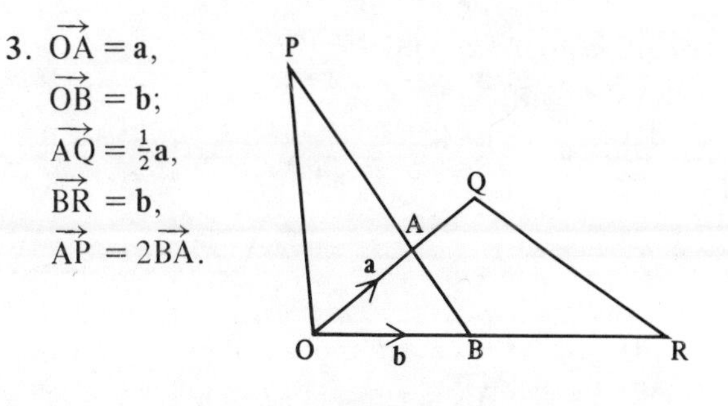

(a) Express $\overrightarrow{BA}$ and $\overrightarrow{BP}$ in terms of $\mathbf{a}$ and $\mathbf{b}$.
(b) Express $\overrightarrow{RQ}$ in terms of $\mathbf{a}$ and $\mathbf{b}$.
(c) Express $\overrightarrow{QA}$ and $\overrightarrow{QP}$ in terms of $\mathbf{a}$ and $\mathbf{b}$.
(d) Using the vectors for $\overrightarrow{RQ}$ and $\overrightarrow{QP}$, show that R, Q and P lie on a straight line.

4. In the diagram, $\overrightarrow{OA} = \mathbf{a}$ and $\overrightarrow{OB} = \mathbf{b}$, M is the mid-point of OA and P lies on AB such that $\overrightarrow{AP} = \frac{2}{3}\overrightarrow{AB}$.

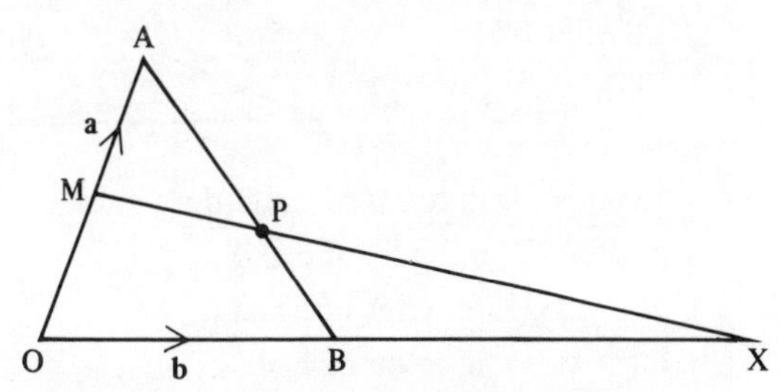

(a) Express $\overrightarrow{AB}$ and $\overrightarrow{AP}$ in terms of $\mathbf{a}$ and $\mathbf{b}$.
(b) Express $\overrightarrow{MA}$ and $\overrightarrow{MP}$ in terms of $\mathbf{a}$ and $\mathbf{b}$.
(c) If X lies on OB produced such that OB = BX, express $\overrightarrow{MX}$ in terms of $\mathbf{a}$ and $\mathbf{b}$.
(d) Show that MPX is a straight line.

5. $\overrightarrow{OP} = \mathbf{a}$,
$\overrightarrow{OA} = 3\mathbf{a}$,
$\overrightarrow{OB} = \mathbf{b}$ and
M is the mid-point of AB.

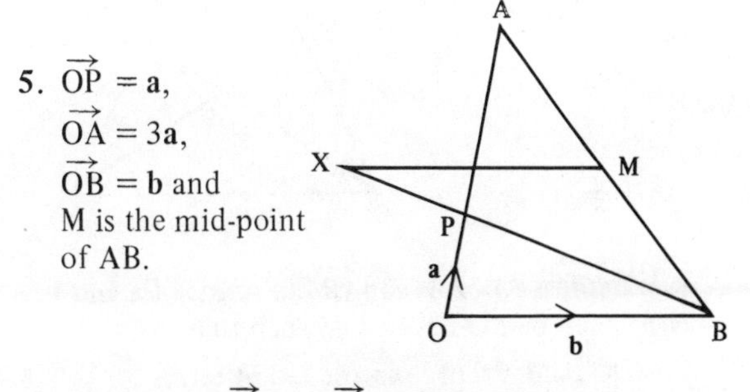

(a) Express $\overrightarrow{BP}$ and $\overrightarrow{AB}$ in terms of $\mathbf{a}$ and $\mathbf{b}$.
(b) Express $\overrightarrow{MB}$ in terms of $\mathbf{a}$ and $\mathbf{b}$.
(c) If X lies on BP produced so that $\overrightarrow{BX} = k.\overrightarrow{BP}$, express $\overrightarrow{MX}$ in terms of $\mathbf{a}$, $\mathbf{b}$ and k.
(d) Find the value of k if MX is parallel to BO.

6. AC is parallel to OB, $\overrightarrow{AX} = \frac{1}{4}\overrightarrow{AB}$;
$\overrightarrow{OA} = \mathbf{a}$,
$\overrightarrow{OB} = \mathbf{b}$,
$\overrightarrow{AC} = m\mathbf{b}$.

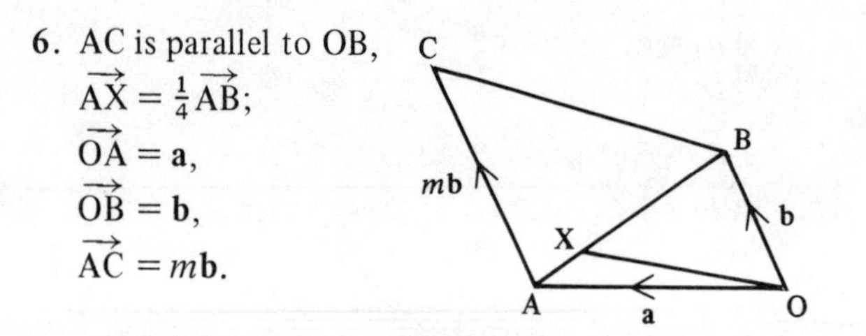

(a) Express $\overrightarrow{AB}$ in terms of **a** and **b**.
(b) Express $\overrightarrow{AX}$ in terms of **a** and **b**.
(c) Express $\overrightarrow{OX}$ in terms of **a** and **b**.
(d) Express $\overrightarrow{BC}$ in terms of **a**, **b** and **m**.
(e) Given that OX is parallel to BC, find the value of m.

7. CY is parallel to OD
$\overrightarrow{CX} = \frac{1}{5}\overrightarrow{CD}$;
$\overrightarrow{OC} = \mathbf{c}$,
$\overrightarrow{OD} = \mathbf{d}$,
$\overrightarrow{CY} = n\mathbf{d}$.

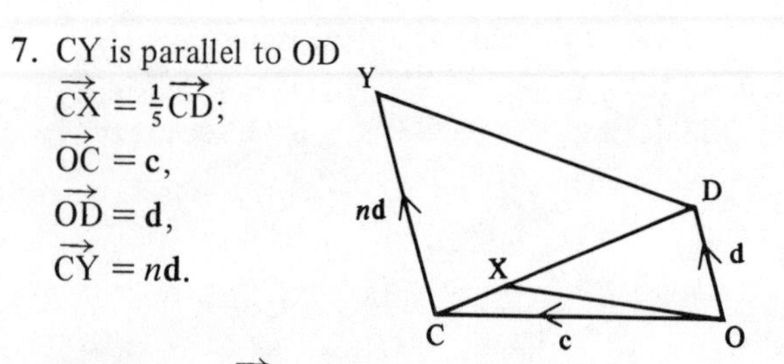

(a) Express $\overrightarrow{CD}$ in terms of **c** and **d**.
(b) Express $\overrightarrow{CX}$ in terms of **c** and **d**.
(c) Express $\overrightarrow{OX}$ in terms of **c** and **d**.
(d) Express $\overrightarrow{DY}$ in terms of **c**, **d** and n.
(e) Given that OX is parallel to DY, find the value of n.

8. M is the mid-point of AB
N is the mid-point of OB;
$\overrightarrow{OA} = \mathbf{a}$
$\overrightarrow{OB} = \mathbf{b}$.

(a) Express $\overrightarrow{AB}$, $\overrightarrow{AM}$ and $\overrightarrow{OM}$ in terms of **a** and **b**.
(b) Given that G lies on OM such that OG:GM = 2:1, express $\overrightarrow{OG}$ in terms of **a** and **b**.
(c) Express $\overrightarrow{AG}$ in terms of **a** and **b**.
(d) Express $\overrightarrow{AN}$ in terms of **a** and **b**.
(e) Show that $\overrightarrow{AG} = m\,\overrightarrow{AN}$ and find the value of m.

9. M is the mid-point of AC and N is the mid-point of OB;
$\overrightarrow{OA} = \mathbf{a}$,
$\overrightarrow{OB} = \mathbf{b}$,
$\overrightarrow{OC} = \mathbf{c}$.

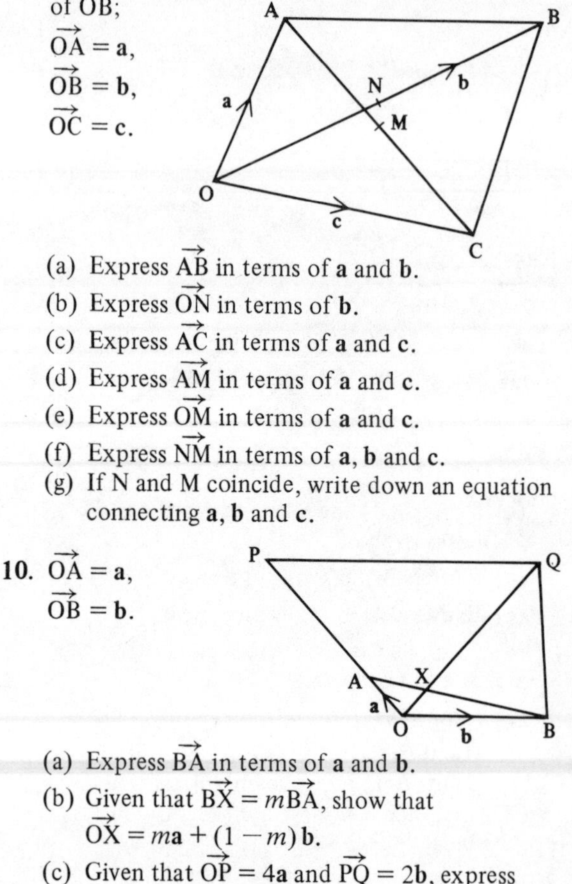

(a) Express $\overrightarrow{AB}$ in terms of **a** and **b**.
(b) Express $\overrightarrow{ON}$ in terms of **b**.
(c) Express $\overrightarrow{AC}$ in terms of **a** and **c**.
(d) Express $\overrightarrow{AM}$ in terms of **a** and **c**.
(e) Express $\overrightarrow{OM}$ in terms of **a** and **c**.
(f) Express $\overrightarrow{NM}$ in terms of **a**, **b** and **c**.
(g) If N and M coincide, write down an equation connecting **a**, **b** and **c**.

10. $\overrightarrow{OA} = \mathbf{a}$,
$\overrightarrow{OB} = \mathbf{b}$.

(a) Express $\overrightarrow{BA}$ in terms of **a** and **b**.
(b) Given that $\overrightarrow{BX} = m\overrightarrow{BA}$, show that $\overrightarrow{OX} = m\mathbf{a} + (1 - m)\mathbf{b}$.
(c) Given that $\overrightarrow{OP} = 4\mathbf{a}$ and $\overrightarrow{PQ} = 2\mathbf{b}$, express $\overrightarrow{OQ}$ in terms of **a** and **b**.
(d) Given that $\overrightarrow{OX} = n\overrightarrow{OQ}$ use the results for $\overrightarrow{OX}$ and $\overrightarrow{OQ}$ to find the values of m and n.

11. X is the mid-point of OD,
Y lies on CD such that
$\overrightarrow{CY} = \frac{1}{4}\overrightarrow{CD}$;
$\overrightarrow{OC} = \mathbf{c}$,
$\overrightarrow{OD} = \mathbf{d}$.

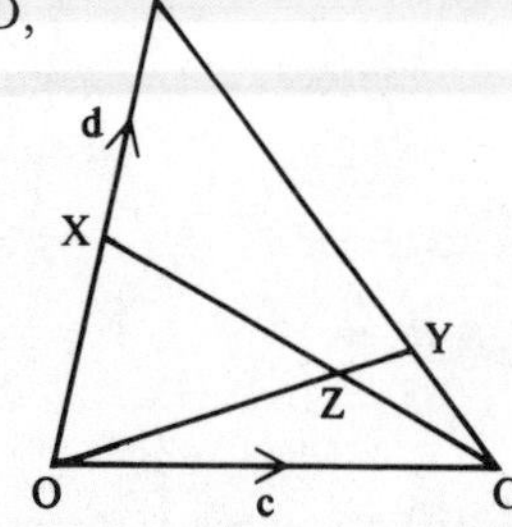

(a) Express $\overrightarrow{CD}$, $\overrightarrow{CY}$ and $\overrightarrow{OY}$ in terms of **c** and **d**.
(b) Express $\overrightarrow{CX}$ in terms of **c** and **d**.
(c) Given that $\overrightarrow{CZ} = h\overrightarrow{CX}$, express $\overrightarrow{OZ}$ in terms of **c**, **d** and h.
(d) If $\overrightarrow{OZ} = k\overrightarrow{OY}$, form an equation and hence find the values of h and k.

9.6 FUNCTIONS

The idea of a function is used in almost every branch of mathematics. The two common notations used are:
(a) $f(x) = x^2 + 4$,
(b) $f: x \to x^2 + 4$.

We may interpret (b) as follows

'function f such that x is mapped onto $x^2 + 4$'.

Numbers are mapped from a *domain* set onto a *range* set.

Example 1

The function h is defined on the domain set $\{1, 2, 3, 4\}$ by $h: x \to 3x - 1$. Find the range set.

$h(1) = 2, h(2) = 5, h(3) = 8, h(4) = 11.$

$\therefore$ The range set is $\{2, 5, 8, 11\}$.

Notice $h(1)$ gives us the result of applying function h to the number 1.

Flow diagrams

The function h in Example 1 consisted of two simpler functions as illustrated by a flow diagram.

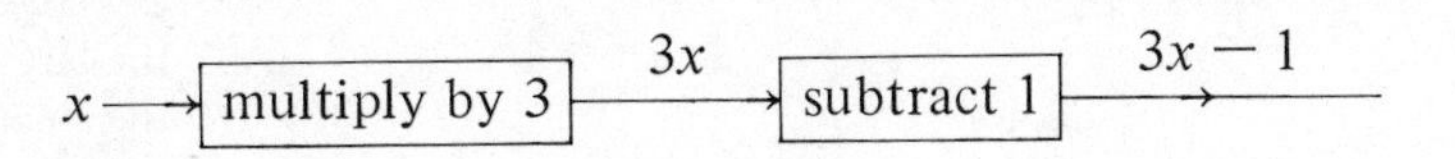

It is obviously important to 'multiply by 3' and 'subtract 1' in the correct order.

Example 2

Draw flow diagrams for the functions
(a) $f: x \to (2x + 5)^2$,
(b) $g(x) = \dfrac{5 - 7x}{3}$

(a) $x \to$ [multiply 2] $\xrightarrow{2x}$ [add 5] $\xrightarrow{(2x+5)}$ [square] $\xrightarrow{(2x+5)^2}$

(b) $x \to$ [multiply by (-7)] $\xrightarrow{-7x}$ [add 5] $\xrightarrow{5-7x}$ [divide by 3] $\xrightarrow{\frac{5-7x}{3}}$

Exercise 12

1. Given the functions $h: x \to x^2 + 1$ and $g: x \to 10x + 1$. Find
   (a) $h(2), h(-3), h(0)$,
   (b) $g(2), g(10), g(-3)$,
   (c) the range set of h for the domain $\{0, 5, 10\}$,
   (d) the range set of g for the domain $\{-1, 0, 3, 4{\cdot}5\}$.

For questions **2**, **3** and **4**, the functions f, g and h are defined as follows: $f: x \to 1 - 2x$

$$g: x \to \frac{x^3}{10}$$

$$h: x \to \frac{12}{x}$$

2. Find
   (a) $f(5), f(-5), f(\frac{1}{4})$
   (b) $g(2), g(-3), g(\frac{1}{2})$
   (c) $h(3), h(10), h(\frac{1}{3})$.
3. Find
   (a) x if $f(x) = 1$
   (b) x if $f(x) = -11$
   (c) x if $h(x) = 1$.
4. Find
   (a) y if $g(y) = 100$
   (b) z if $h(z) = 24$
   (c) w if $g(w) = 0{\cdot}8$.

For questions **5** and **6**, the functions k, l and m are defined on the domain of real numbers as follows:

$$k: x \to \frac{2x^2}{3}$$

$$l: y \to \sqrt{[(y-1)(y-2)]}$$

$$m: x \to 10 - x^2.$$

5. Find
   (a) $k(1), k(6), k(-3)$,
   (b) $l(2), l(0), l(4)$,
   (c) $m(4), m(-2), m(\frac{1}{2})$.
6. Find
   (a) x if $k(x) = 6$
   (b) x if $m(x) = 1$
   (c) y if $k(y) = 2\frac{2}{3}$
   (d) p if $m(p) = -26$.
7. The function f is defined as $f: x \to ax + b$ where a and b are constants.
   If $f(1) = 8$ and $f(4) = 17$, find the values of a and b.
8. The function g is defined as $g(x) = ax^2 + b$ where a and b are constants.
   If $g(2) = 3$ and $g(-3) = 13$, find the values of a and b.
9. Functions h and k are defined as follows: $h: x \to x^2 + 1, k: x \to ax + b$, where a and b are constants.
   If $h(0) = k(0)$ and $k(2) = 15$, find the values of a and b.

In questions **10** to **23**, draw a flow diagram for each function.

10. $f: x \to 5x + 4$
11. $f: x \to 3(x - 4)$
12. $f: x \to (2x + 7)^2$
13. $f: x \to \left(\frac{9 + 5x}{4}\right)$
14. $f: x \to \frac{4 - 3x}{5}$
15. $f: x \to 2x^2 + 1$
16. $f: x \to \frac{3x^2}{2} + 5$
17. $f: x \to \sqrt{(4x - 5)}$
18. $f: x \to 4\sqrt{(x^2 + 10)}$
19. $f: x \to (7 - 3x)^2$
20. $f: x \to 4(3x + 1)^2 + 5$
21. $f: x \to 5 - x^2$
22. $f: x \to \frac{10\sqrt{(x^2 + 1)} + 6}{4}$
23. $f: x \to \left(\frac{x^3}{4} + 1\right)^2 - 6$
24. $f(x)$ is defined as the product of the digits of x, e.g. $f(12) = 1 \times 2 = 2$.
    (a) Find (i) $f(25)$ (ii) $f(713)$
    (b) If x is an integer with three digits, find
    (i) x such that $f(x) = 1$
    (ii) the largest x such that $f(x) = 4$
    (iii) the largest x such that $f(x) = 0$
    (iv) the smallest x such that $f(x) = 2$.
25. $g(x)$ is defined as the sum of the prime factors of x, e.g. $g(12) = 2 + 3 = 5$. Find
    (a) $g(10)$ (b) $g(21)$ (c) $g(36)$
    (d) $g(99)$ (e) $g(100)$ (f) $g(1000)$
26. $h(x)$ is defined as the number of letters in the English word describing x, e.g. $h(1) = 3$. Find
    (a) $h(2)$ (b) $h(11)$ (c) $h(18)$
    (d) the largest value of x for which $h(x) = 3$.
27. If $f: x \to$ next prime number greater than x, find:
    (a) $f(7)$ (b) $f(14)$ (c) $f[f(3)]$
28. If $g: x \to 2^x + 1$, find:
    (a) $g(2)$ (b) $g(4)$ (c) $g(-1)$
    (d) the value of x if $g(x) = 9$

Composite functions

The function $f : x \to 3x + 2$ is itself a composite function, consisting of two simpler functions: 'multiply by 3' and 'add 2'.

If $f : x \to 3x + 2$ and $g : x \to x^2$ then fg is a composite function where g is performed first and then f is performed on the result of g.

The function fg may be found using a flow diagram.

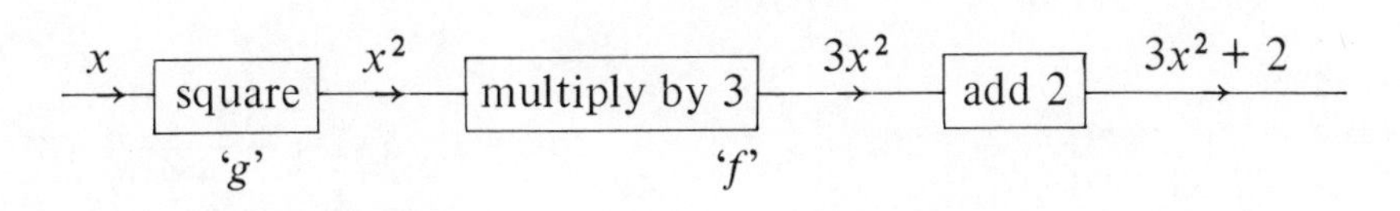

Thus $fg : x \to 3x^2 + 2$.

Inverse functions

If a function f maps a number n onto m, then the inverse function f^{-1} maps m onto n. The inverse of a given function is found using a flow diagram.

Example 3

Find the inverse of f where $f : x \to \dfrac{5x - 2}{3}$

(a) Draw a flow diagram for f

x → multiply by 5 → $5x$ → subtract 2 → $5x - 2$ → divide by 3 → $\dfrac{5x-2}{3}$

(b) Draw a new flow diagram with each operation replaced by its inverse. Start with x on the right.

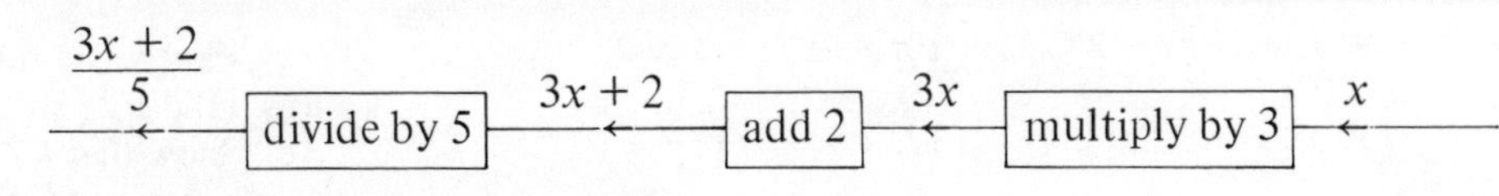

Thus the inverse of f is given by

$$f^{-1} : x \to \frac{3x + 2}{5}$$

Exercise 13

For questions **1** and **2**, the functions f, g and h are as follows: $f : x \to 4x$
$g : x \to x + 5$
$h : x \to x^2$

1. Find the following in the form '$x \to \ldots$'
(a) fg (b) gf (c) hf
(d) fh (e) gh (f) fgh
(g) hfg

2. Find
(a) x if $hg(x) = h(x)$
(b) x if $fh(x) = gh(x)$

For questions **3, 4** and **5**, the functions f, g and h are as follows: $f : x \to 2x$
$g : x \to x - 3$
$h : x \to x^2$

3. Find the following in the form '$x \to \ldots$'
(a) fg (b) gf (c) gh
(d) hf (e) ghf (f) hgf

4. Evaluate
(a) $fg(4)$ (b) $gf(7)$ (c) $gh(-3)$
(d) $fgf(2)$ (e) $ggg(10)$ (f) $hfh(-2)$

5. Find
(a) x if $f(x) = g(x)$ (b) x if $hg(x) = gh(x)$
(c) x if $gf(x) = 0$ (d) x if $fg(x) = 4$

For questions **6, 7** and **8**, the functions l, m and n are as follows: $l : x \to 2x + 1$
$m : x \to 3x - 1$
$n : x \to x^2$

6. Find the following in the form '$x \to \ldots$'
(a) lm (b) ml (c) ln
(d) nm (e) lnm (f) mln

7. Find
(a) $lm(2)$ (b) $nl(1)$ (c) $mn(-2)$
(d) $mm(2)$ (e) $nln(2)$ (f) $llm(0)$

8. Find
(a) x if $l(x) = m(x)$
(b) two values of x if $nl(x) = nm(x)$
(c) x if $ln(x) = mn(x)$

In questions **9** to **26**, find the inverse of each function in the form '$x \to \ldots$'

9. $f : x \to 5x - 2$

10. $f : x \to 5(x - 2)$

11. $f : x \to 3(2x + 4)$

12. $g : x \to \dfrac{2x + 1}{3}$

13. $f : x \to \dfrac{3(x - 1)}{4}$

14. $g : x \to 2(3x + 4) - 6$

15. $h : x \to \frac{1}{2}(4 + 5x) + 10$

16. $k : x \to -7x + 3$

17. $j : x \to \dfrac{12 - 5x}{3}$

18. $l : x \to \dfrac{4 - x}{3} + 2$

19. $m : x \to \dfrac{\left[\dfrac{(2x - 1)}{4} - 3\right]}{5}$

20. $f : x \to \dfrac{3(10 - 2x)}{7}$

21. $g : x \to \left[\dfrac{\dfrac{x}{4} + 6}{5}\right] + 7$

22. $h : x \to \dfrac{3\left(\dfrac{x}{2} - 3\right)}{4} - 10$

23. $m : x \to \dfrac{\left(9 + \dfrac{x}{3}\right)3}{7} - 2$

24. $n : x \to \dfrac{\left[\dfrac{5 - 2x}{3} + 8\right]}{9}$

25. $h : x \to \frac{1}{5}\left[\dfrac{(10 + \frac{1}{2}x)}{6} + 100\right]$

26. $f : x \to \frac{3}{4}[3(4 + \frac{1}{3}x) - 4]$

For questions **27** to **30**, the functions f, g and h are defined as follows: $f : x \to 3x$
$g : x \to x - 5$
$h : x \to 2x + 1$

27. Find in the form '$x \to \ldots$'
(a) f^{-1} (b) g^{-1} (c) h^{-1}
(d) fg (e) $(fg)^{-1}$ (f) $g^{-1}f^{-1}$

28. Find in the form '$x \to \ldots$'
(a) hf (b) hf^{-1} (c) $f^{-1}h^{-1}$
(d) hg (e) $(hg)^{-1}$ (f) $g^{-1}h^{-1}$

29. Find
(a) $g^{-1}(2)$ (b) $fg^{-1}(2)$ (c) $(gf)^{-1}(10)$
(d) $f^{-1}g^{-1}(10)$ (e) $f^{-1}f^{-1}ff(2)$

30. Find
(a) x if $h(x) = f(x)$
(b) the set of values of x for which
(i) $f(x) > g(x)$
(ii) $fg(x) > 0$

31. The function f is defined by $f : x \to \dfrac{2x + 5}{3}$
(a) Find f^{-1} in the form '$x \to \ldots$'
(b) Find $f^{-1}(3)$
(c) Show that $f^{-1}(3)$ is the solution of the equation $f(x) = 3$.

9.7 BINARY OPERATIONS

Example 1 .

Given $a * b = a + b + 1$.

(a) Evaluate (i) $1 * 3$ (ii) $(1 * 3) * 5$
(b) Find x if $x * 3 = 10$
(c) Find y if $(y * 4) * y = 12$.

(a) (i) $1 * 3 = 1 + 3 + 1 = 5$
(ii) $(1 * 3) * 5 = (5) * 5 = 5 + 5 + 1 = 11$

(b) $x * 3 = 5$
$x + 3 + 1 = 5 \quad \therefore \quad x = 1$

(c) $(y * 4) * y = 12$
$(y + 4 + 1) * y = 12$
$y + 4 + 1 + y + 1 = 12 \quad \therefore \quad y = 3$

Note: the operation $\circ$ is *commutative* if
$a \circ b = b \circ a$
the operation $\circ$ is *associative* if
$(a \circ b) \circ c = a \circ (b \circ c)$

The operation $*$ in Example 1 is both commutative and associative.

Exercise 14

1. If $a * b$ denotes $a + 3b$, evaluate
(a) $1 * 2$ (b) $2 * 1$ (c) $3 * 5$
(d) $5 * 3$ (e) $3 * (-1)$ (f) $2 * \frac{1}{3}$
2. If $x \circ y$ denotes $2x - y$, evaluate
(a) $4 \circ 1$ (b) $8 \circ 0$ (c) $3 \circ (-1)$
(d) $(-2) \circ (-4)$
(e) $(4 \circ 1) \circ 2$ (f) $(3 \circ 2) \circ 1$
3. If $m \square n$ denotes $m^2 + n^2$, evaluate
(a) $2 \square 3$ (b) $4 \square 1$ (c) $(-2) \square (-3)$
(d) $(1 \square 2) \square 3$ (e) $1 \square (2 \square 3)$
(f) $(1 \square 2) \square (2 \square 3)$

In questions **4** to **9** use the operation given to evaluate
(a) $3 * 2$ (b) $4 * 5$ (c) $2 * 1$
(d) $(3 * 2) * 4$ (e) $1 * (3 * 2)$ (f) $(2 * 1) * (3 * 2)$

4. $a * b$ denotes ab^2
5. $x * y$ denotes $3x - y$
6. $m * n$ denotes $3mn$
7. $a * b$ denotes $a^2 + b^2$
8. $x * y$ denotes $(x - y)^2$
9. $m * n$ denotes $\frac{m}{n}$
10. If $p * q$ denotes pq^2, solve the equations
(a) $x * 2 = 16$ (b) $x * 3 = 18$
(c) $x * 1 = 7$ (d) $x * -2 = 20$
(e) $3 * x = 12$ (f) $5 * x = 45$
(g) $x * x = 8$
11. If $a * b$ denotes $3a - b$, solve the equations
(a) $x * 2 = 7$ (b) $x * 1 = 5$
(c) $x * 5 = 2$ (d) $x * 4 = 10$
(e) $3 * x = 2$ (f) $4 * x = 5$
(g) $x * x = 12$
12. If $a * b$ denotes $a + 2b + 2$, solve the equations
(a) $x * 3 = 12$ (b) $x * 6 = 24$
(c) $2 * x = 14$ (d) $3 * x = 3$
(e) $x * x = 8$ (f) $x * 3x = 23$
13. $x * y$ denotes $\frac{x}{y} + 3$.
(a) Evaluate $8 * 2$ and $2 * 8$
(b) Is operation $*$ commutative?
(c) Find z if $z * 2 = 7$
(d) Find m if $m * 5 = -4$
(e) Find k if $4 * k = 5$
14. $a \square b$ denotes $3ab$.
(a) Evaluate $4 \square 5$ and $5 \square 4$
(b) Is operation $\square$ commutative?
(c) Evaluate $(1 \square 3) \square 4$ and $1 \square (3 \square 4)$
(d) Is operation $\square$ associative?
(e) Find x if $x \square 4 = 108$
(f) Find y if $y \square y = 147$
(g) Find z if $(z \square 3) \square 2 = 270$
15. $m * n$ denotes 'the smallest integer greater than $\frac{m}{n}$'.
(a) Find (i) $3 * 2$ (ii) $2 * 3$
(iii) $8 * 3$ (iv) $3 * 8$
(b) Is operation $*$ commutative?
(c) Find (i) $(3 * 2) * 3$
(ii) $(8 * 3) * (3 * 2)$
16. $x \circ y$ denotes $2x + 3y$. Solve the equations
(a) $m \circ 4 = 20$ (b) $5 \circ m = 11$
(c) $m \circ 3 = 4 \circ m$ (d) $m \circ m = 1$
Is the operation $\circ$ commutative?
17. $a * b$ denotes $\frac{a + b}{ab}$
(a) Evaluate (i) $1 * 2$ (ii) $5 * 2$
(iii) $(1 * 2) * \frac{1}{2}$
(b) Solve the equations
(i) $x * 3 = 2$ (ii) $5 * x = 1$
(c) Simplify $2x * 3x$.
18. $m \circ n$ denotes $m^2 + 2n + 3$.
Solve the equations.
(a) $3 \circ x = 22$
(b) $x \circ 5 = 22$
(c) $x \circ x = 2$
(d) $3 \circ (x \circ 1) = 24$.

9.8 OPERATION TABLES

The operation $*$ on $\{a, b, c, d\}$ is shown in the table.

$*$	a	b	c	d
a	b	d	a	c
b	d	c	b	a
c	a	b	c	d
d	c	a	d	b

(a) The *identity* element is c since when c is combined with any element the result is the same element. The identity element is easily identified by drawing 'loops' around elements inside the table which appear in the same order outside the table. The identity is at the intersection of the loops as shown.

(b) a is the inverse of d, since

$$a * d = c$$

b is 'self inverse', since $b * b = c$.

(c) The system is *closed*. The only elements in the table are those in the set $\{a, b, c, d\}$.

Exercise 15

1. The operation table for $\circ$ is shown.

   Find
   (a) the identity element
   (b) the inverse of w
   (c) the inverse of z
   (d) the inverse of x.

$\circ$	w	x	y	z
w	x	z	w	y
x	z	y	x	w
y	w	x	y	z
z	y	w	z	x

2. The operation table for $*$ is shown.

   Find
   (a) the identity element
   (b) $b * d$
   (c) $(b * d) * b$
   (d) the inverse of d
   (e) x if $a * x = c$
   (f) y if $y * c = d$.

$*$	a	b	c	d
a	a	b	c	d
b	b	a	d	c
c	c	d	a	b
d	d	c	b	a

3. The operation table for $\square$ is shown.

$\square$	A	B	C	D	E
A	C	D	E	A	B
B	D	E	A	B	C
C	E	A	B	C	D
D	A	B	C	D	E
E	B	C	D	E	A

   Find
   (a) the identity element
   (b) the inverses of A and C
   (c) $A \square C$
   (d) $(A \square C) \square D$
   (e) $(B \square E) \square (D \square A)$
   (f) X if $X \square C = A$
   (g) Y if $Y \square Y = B$
   (h) Z if $(Z \square B) \square C = A$
   Is the set $\{A, B, C, D, E\}$ closed under $\square$?

4. For the operations $\circ$, $*$ and $\square$ in questions **1**, **2** and **3**, decide whether or not the operation is commutative.

5. Construct an operation table to illustrate matrix multiplication on the set $\{\mathbf{A}, \mathbf{B}, \mathbf{C}, \mathbf{D}\}$ where

$$\mathbf{A} = \begin{pmatrix} 1 & 0 \\ 0 & 1 \end{pmatrix}, \quad \mathbf{B} = \begin{pmatrix} -1 & 0 \\ 0 & 1 \end{pmatrix}$$

$$\mathbf{C} = \begin{pmatrix} -1 & 0 \\ 0 & -1 \end{pmatrix}, \quad \mathbf{D} = \begin{pmatrix} 1 & 0 \\ 0 & -1 \end{pmatrix}$$

   Find
   (a) the identity element
   (b) $\mathbf{X}$ if $\mathbf{XA} = \mathbf{D}$ (c) $\mathbf{Y}$ if $\mathbf{CY} = \mathbf{B}$.

6. The operation $*$ is defined on the set $\{0, 1, 2, 3, 4\}$ by the relation $a * b =$ the remainder when $a \times b$ is divided by 3.
   e.g. $3 * 2 = 0$, since $3 \times 2 = 6$ and $\frac{6}{3} = 2$ remainder 0.
   (a) Construct an operation table for the operation $*$ on the set given.
   (b) Is there an identity element?
   (c) Find x if $2 * x = 1$.
   (d) Find y if $y * 1 = 2$.

7. Construct an operation table for the operation $\circ$ on the set $\{1, 2, 3, 4\}$, where $a \circ b =$ the remainder when $a \times b$ is divided by 5.
(a) What is the identity element?
Solve the equations
(b) $x \circ 3 = 1$ (c) $x \circ x = 4$
(d) $(4 \circ 2) \circ x = 1$ (e) $x \circ (3 \circ 3) = 3$
(f) $x \circ (x \circ 2) = 3$ (g) $(x \circ 3) \circ x = 3$

8. The operation table for the set $X = \{a, b, c, d, e, f\}$ under the operation $*$ is shown.

$*$	a	b	c	d	e	f
a				c		
b	a					
c		d		a		
d		b				
e		e	c			
f	d	f	b	e	a	

Complete the table, given that
(a) the operation $*$ is commutative,
(b) each element of X appears just once in each row and column,
(c) the set X is closed under $*$.
(There may be more than one solution.)

9. Repeat question 8 for the operation tables below which are subject to the same three conditions.

(a)

$*$	a	b	c	d	e	f
a		a	b			
b			d		e	
c					c	
d	c			b		e
e	f					
f		b				

(b)

$*$	1	2	3	4	5	6
1		3	6			
2			5		4	
3					1	
4					5	4
5	2					
6		6				

REVISION EXERCISE 9A

1. Given that $\mathcal{E} = \{1, 2, 3, 4, 5, 6, 7, 8\}$, $A = \{1, 3, 5\}$, $B = \{5, 6, 7\}$, list the members of the sets
(a) $A \cap B$ (b) $A \cup B$
(c) A' (d) $A' \cap B'$
(e) $A \cup B'$

2. The set P and Q are such that $n(P \cup Q) = 50$, $n(P \cap Q) = 9$ and $n(P) = 27$. Find the value of $n(Q)$.

3. Draw three diagrams similar to Figure 1 below, and shade the following.
(a) $Q \cap R'$
(b) $(P \cup Q) \cap R$
(c) $(P \cap Q) \cap R'$

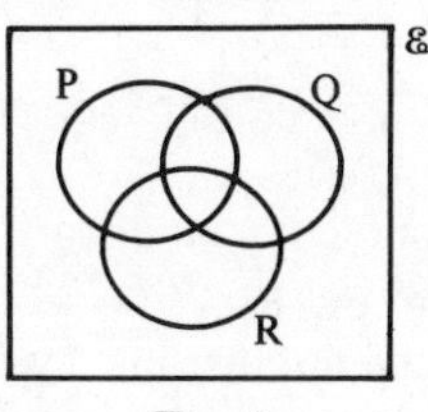

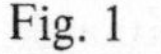
Fig. 1

4. Describe the shaded regions in Figures 2 and 3.

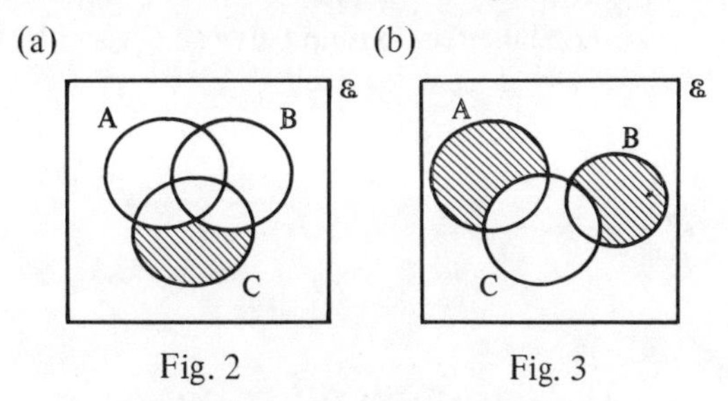

Fig. 2 Fig. 3

5. Given that $\mathcal{E} = \{\text{people on a train}\}$, $M = \{\text{males}\}$, $T = \{\text{people over 25 years old}\}$ and $S = \{\text{snooker players}\}$
(a) express in set notation:
(i) all the snooker players are over 25
(ii) some snooker players are women.
(b) express in words $T \cap M' = \emptyset$.

6. The figures in the diagram indicate the number of elements in each subset of &.

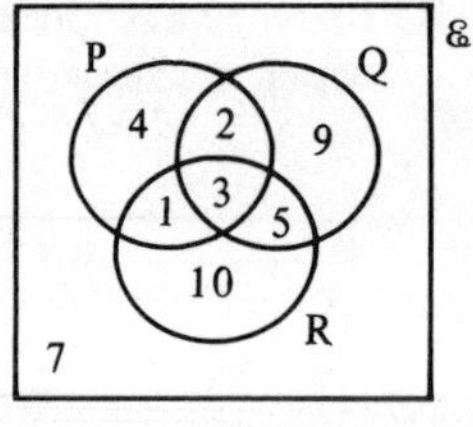

(a) Find $n(P \cap R)$
(b) Find $n(Q \cup R)'$
(c) Find $n(P' \cap Q')$.

7. In $\triangle$OPR, the mid-point of PR is M.

If $\overrightarrow{OP} = \mathbf{p}$ and $\overrightarrow{OR} = \mathbf{r}$, find in terms of **p** and **r**
(a) $\overrightarrow{PR}$ (b) $\overrightarrow{PM}$
(c) $\overrightarrow{OM}$

8. If $\mathbf{a} = \begin{pmatrix}1\\4\end{pmatrix}$ and $\mathbf{b} = \begin{pmatrix}-3\\4\end{pmatrix}$, find
(a) $|\mathbf{b}|$ (b) $|\mathbf{a}+\mathbf{b}|$ (c) $|2\mathbf{a}-\mathbf{b}|$

9. If $4\begin{pmatrix}1\\3\end{pmatrix} + 2\begin{pmatrix}1\\m\end{pmatrix} = 3\begin{pmatrix}n\\-6\end{pmatrix}$, find the values of m and n.

10. The points O, A and B have coordinates (0, 0), (5, 0) and (−1, 4) respectively. Write as column vectors.
(a) $\overrightarrow{OB}$ (b) $\overrightarrow{OA} + \overrightarrow{OB}$ (c) $\overrightarrow{OA} - \overrightarrow{OB}$
(d) $\overrightarrow{OM}$ where M is the mid-point of AB.

11. In the parallelogram OABC, M is the mid-point of AB and N is the mid-point of BC.
If $\overrightarrow{OA} = \mathbf{a}$ and $\overrightarrow{OC} = \mathbf{c}$, express in terms of **a** and **c**.
(a) $\overrightarrow{CA}$ (b) $\overrightarrow{ON}$ (c) $\overrightarrow{NM}$
Describe the relationship between CA and NM.

12. The vectors **a**, **b**, **c** are given by
$\mathbf{a} = \begin{pmatrix}1\\5\end{pmatrix}, \mathbf{b} = \begin{pmatrix}-2\\1\end{pmatrix}, \mathbf{c} = \begin{pmatrix}-1\\17\end{pmatrix}$.
Find numbers m and n so that $m\mathbf{a} + n\mathbf{b} = \mathbf{c}$.

13. Given that $\overrightarrow{OP} = \begin{pmatrix}3\\2\end{pmatrix}$, $\overrightarrow{OQ} = \begin{pmatrix}0\\4\end{pmatrix}$ and that M is the mid-point of PQ, express as column vectors
(a) $\overrightarrow{PQ}$ (b) $\overrightarrow{PM}$ (c) $\overrightarrow{OM}$.

14. Given $f: x \to 2x - 3$ and $g: x \to x^2 - 1$, find
(a) $f(-1)$ (b) $g(-1)$ (c) $fg(-1)$
(d) $gf(3)$.
Write the function ff in the form '$ff: x \to \ldots$'

15. If $f: x \to 3x + 4$ and $h: x \to \dfrac{x-2}{5}$
express f^{-1} and h^{-1} in the form '$x \to \ldots$'.
Find (a) $f^{-1}(13)$
(b) the value of z if $f(z) = 20$.

16. Given that $f(x) = x - 5$, find
(a) the value of s such that $f(s) = -2$,
(b) the values of t such that $t \times f(t) = 0$.

17. The operation $*$ is defined by $a * b = a(b + 2)$,
(a) evaluate $2 * 5$
(b) evaluate $1 * (2 * 3)$
(c) solve the equation $4 * x = 16$.

18. $a * b$ denotes $\dfrac{ab}{a+1}$
(a) evaluate $2 * 3$ (b) evaluate $\frac{1}{2} * \frac{1}{3}$
(c) solve the equation $x * 5 = 2$.

19. $a * b$ denotes $2a - 3b$
(a) evaluate $-1 * -3$
(b) solve the equation $x * 2 = 2 * x$
(c) Is the operation $*$ commutative?

20. In the operation table
(a) What is the identity element?
(b) What is the inverse of element d?
(c) Solve the equation, $x * b = b$.

$*$	a	b	c	d
a	b	d	a	c
b	d	c	b	a
c	a	b	c	d
d	c	a	d	b

EXAMINATION EXERCISE 9B

1. From a Universal set {1, 2, 3, 4, 5, 6, 7, 8, 9, 10, 11, 12} three subsets are defined:

A = {x : x is divisible by 3},
B = {x : x is a factor of 18},
C = {x : x is even and not divisible by 5}.

Draw a Venn diagram to show the Universal set and the three subsets.

Insert the numbers 1–12 in the appropriate regions of the diagram. List the members of:
(i) $A \cap B$, (ii) $B \cup C$, (iii) $A \cap (B \cup C)$, (iv) $A' \cup B'$, (v) $(A \cup B)'$. [O]

2. A universal set & includes subsets A, B and C. There are 5 members of & which are not in any of these subsets. Every member of B is a member of A but C contains members which do not belong either to A or B. Draw a Venn diagram to incorporate these features.
Given further that,
(a) there are 20 members of C not in A or B,
(b) there are 3 members common to A, B and C,
(c) there are 13 members of B,
(d) there are 27 members of A not in C,
(e) there are 10 members common to A and C,
insert appropriate numbers in the regions of your diagram. Hence calculate,
(i) $n(A)$, (ii) $n(A \cap C \cap B')$, (iii) $n(\&)$. [O]

3. (i) Draw a single Venn diagram to illustrate the relations between the following sets:

P = {parallelograms}, Q = {quadrilaterals},
R = {rectangles}, S = {squares},
Z = {quadrilaterals having *one and only one* pair of parallel sides}.

State which one of the sets P, R, S, Z, ϕ is equal to
(a) $R \cap S$, (b) $P \cap Z$.

(ii) The 89 members of the Fifth Form all belong to one or more of the Chess Club, the Debating Society and the Jazz Club. Denoting these sets by C, D and J respectively, it is known that 20 pupils belong to C only, 15 to J only and 12 to D only. Given that $n(C \cap J) = 18$, $n(C \cap D) = 20$ and $n(D \cap J) = 16$, calculate
(a) $n(C \cap J \cap D)$, (b) $n(D')$. [L]

4. The operation $*$ is defined by $p * q = p + q + pq$.
(a) Evaluate (i) $3 * 5$, (ii) $5 * -2$.
(b) Find n, when $7 * n = 23$.
(c) Find y when $(1 * -\frac{1}{2}) * y = 5$. [C 16+]

5. (a) In a survey carried out at a sports centre, men were asked about their sporting activities. Of the men questioned, 12 played rugby, 16 played squash, 13 played tennis. 8 played none of these games. 3 played both rugby and squash, 5 played both rugby and tennis. 2 men played tennis only.
Let R, S and T be the sets of rugby, squash and tennis players respectively. Let the number of men playing all three games be x. Draw a Venn diagram and show, in terms of x, the number of men in each region of the diagram in set R. Also show the number in each of the other four regions.
Find the total number of men questioned and state the possible values of x.

(b) O is the origin. OPQR is a quadrilateral in which $\mathbf{OP} = \mathbf{a}$, $\mathbf{PQ} = \mathbf{b}$ and $\mathbf{OR} = 2\mathbf{b}$. M is the mid-point of PR and S is the point such that OPSR is a parallelogram.
(i) State the type of quadrilateral OPQR.
(ii) Find **OS** in terms of **a** and **b**.
(iii) Find **PR** in terms of **a** and **b**.
(iv) Find the position vector **OM** in terms of **a** and **b**. [W]

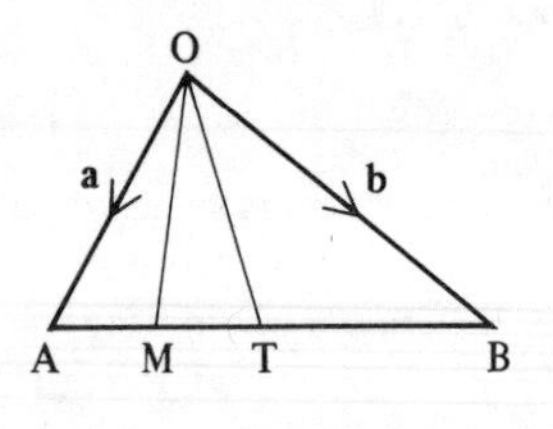

6. (a) In the diagram, T is the mid-point of AB and M is the mid-point of AT. Given that **OA** = **a** and **OB** = **b**, express as simply as possible in terms of **a** and **b**,
   (i) **AB**, (ii) **AM**. (iii) **OM**.
   (b) Two points P and Q have position vectors **p** and **q** respectively, relative to the origin O. Given that $\mathbf{p} = \begin{pmatrix} 5 \\ 3 \end{pmatrix}$ and $\mathbf{PQ} = \begin{pmatrix} -2 \\ 1 \end{pmatrix}$, find
   (i) **q**, (ii) |**PQ**|,
   (iii) the coordinates of the point R, which is such that **OR** = **QP**.
   Given also that $\mathbf{s} = \begin{pmatrix} 1 \\ 1 \end{pmatrix}$, $\mathbf{t} = \begin{pmatrix} 8 \\ 2 \end{pmatrix}$ and $l\mathbf{p} + m\mathbf{s} = \mathbf{t}$, write down two simultaneous equations in l and m, and solve them. [C]

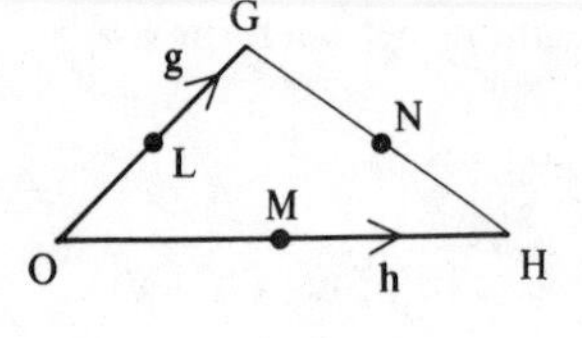

7. In the triangle OGH the midpoints of OG, OH and GH are L, M and N respectively.

   OG = **g**, **OH** = **h**.

   (i) Write down expressions, in terms of **g** and **h**, for
   (a) **OL**, (b) **GN**, (c) **ON**, (d) **OL** + **OM** + **ON**.
   (ii) (a) Use the vector equation **OG** + **GM** = **OM** to express **GM** in terms of **g** and **h**.
   (b) By a similar method, express **HL** in terms of **g** and **h**.
   (c) Hence obtain an expression for **GM** + **HL** + **ON** and simplify it.
   (iii) (a) Use your results from (ii) to express **GM** − **HL** in terms of **g** and **h**.
   (b) K is a point, not shown on the diagram, such that **OK** = **GM** − **HL**. What can you say about OK and GH? [SMP]

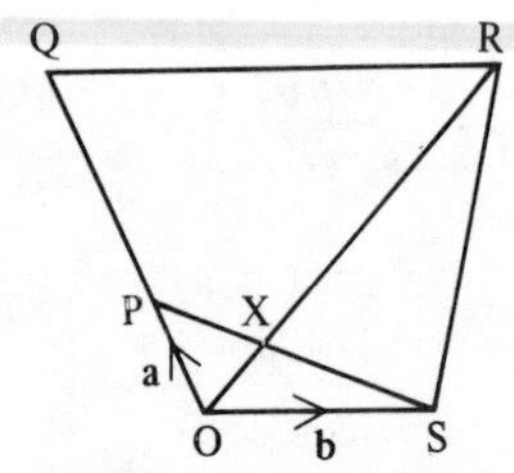

8. In the diagram **OP** = **a** and **OS** = **b**.
   (i) Express **SP** in terms of **a** and **b**.
   (ii) Given that **SX** = h**SP**, show that **OX** = h**a** + $(1-h)$**b**.
   (iii) Given that **OQ** = 3**a** and **QR** = 2**b**, write down an expression for **OR** in terms of **a** and **b**.
   (iv) Given that **OX** = k**OR** use the results of parts (ii) and (iii) to find the values of h and k.
   (v) Find the numerical value of the ratio $\dfrac{PX}{XS}$. [C]

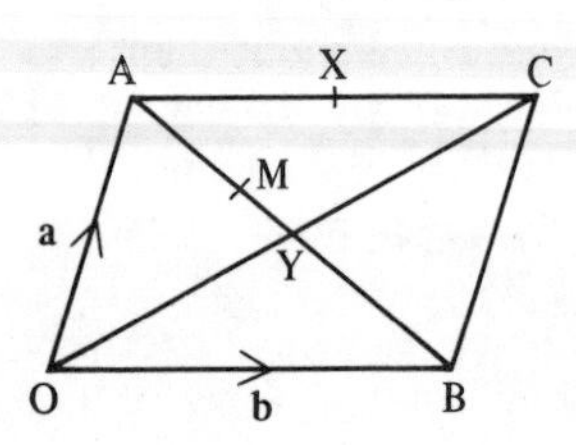

9. In the diagram **OA** = **a**, **OB** = **b**. OACB is a parallelogram, X is the mid-point of AC, and M is the point on AB such that AM = $\frac{1}{3}$AB.
   (i) Express the following vectors in terms of **a** and **b** as simply as possible: **OX**, **AB**, **AM**, **OM**.
   (ii) L is the point on OX such that OL = $\frac{2}{3}$OX. Express the vector **OL** in terms of **a** and **b**. What can you now say about L and M?
   (iii) AB meets OC at Y. Express the vectors **OY** and **AY** in terms of **a** and **b**, and hence, or otherwise show that AM : AY = OM : OX. [SMP]

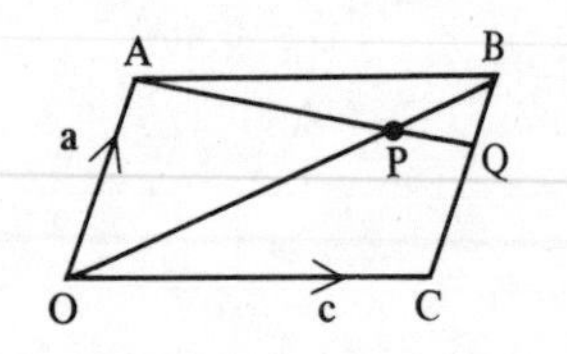

10. In the parallelogram OABC, **OP** = $\frac{3}{4}$**OB** and APQ is a straight line.

    OA = **a** and **OC** = **c**.

    (i) Find **OB**, **OP** and **AP** in terms of **a** and **c**.
    (ii) By writing **OQ** as **OA** + x**AP** express **OQ** in terms of **a**, **c** and x.
    (iii) By writing **OQ** as **OC** + y**CB** express **OQ** in terms of **a**, **c** and y.
    (iv) Find the value of x which makes the terms in **c** equal in the two expressions for **OQ**. Hence find the value of y.
    (v) Use the value of y to find $\dfrac{CQ}{QB}$. [SMP]

10 Calculus

Gottfried Leibniz (1646–1716) from Leipzig was a phenomenal child prodigy having many and varied talents which he employed throughout his life. He was a philosopher, lawyer, ecclesiastical diplomat and historian. He was also a self-taught mathematician. Although Newton had already used his own version of the calculus nearly twenty years earlier, Leibniz was the first to publish his discovery in 1684. It is clear that they had both worked independently, but this remained a source of intense rivalry between their supporters for years.

10.1 RATE OF CHANGE

When we find the gradient of a function, we are finding the *rate of change* of one quantity with respect to another quantity.

Figure 1 shows a graph connecting the height h of a ball thrown in the air and its time of flight t.

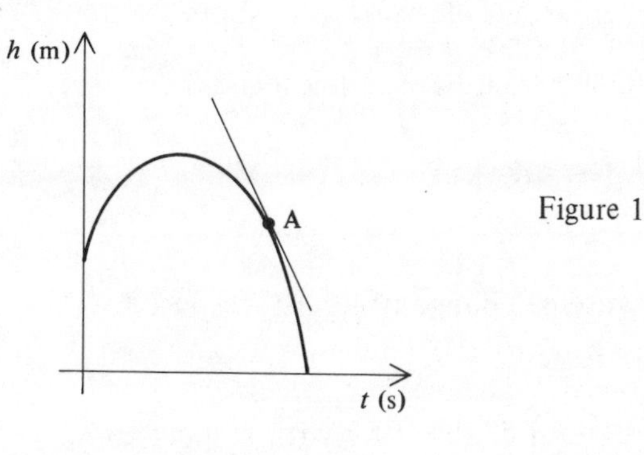

Figure 1

The gradient of the graph at A is the rate of change of h with respect to t. The units for this rate of change are metres per second.

Exercise 1

1. Draw the graph of the function $y = x^2 - 2x$ for $-2 \leqslant x \leqslant 5$ using a scale of 2 cm to 1 unit on the x-axis and 1 cm to 1 unit on the y-axis. Draw tangents to the curve at the following points and find the gradient of each
(a) at $x = 2$, (b) at $x = 4$, (c) at $x = -1$.

2. A graph connecting variables z and t is drawn.

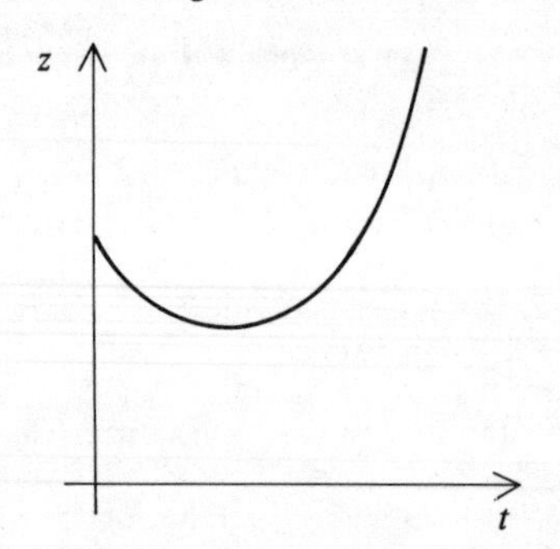

Give the units for the gradient of the curve when z and t have units as follows:
(a) z in metres, t in seconds.
(b) z in kg, t in seconds.
(c) z in cm^3, t in hours.
(d) z in cm^2, t in days.
(e) z in m/s, t in seconds.
(f) z in km^2, t in hours.

3. The height h km of a rocket is plotted as a function of the time of flight t in hours.

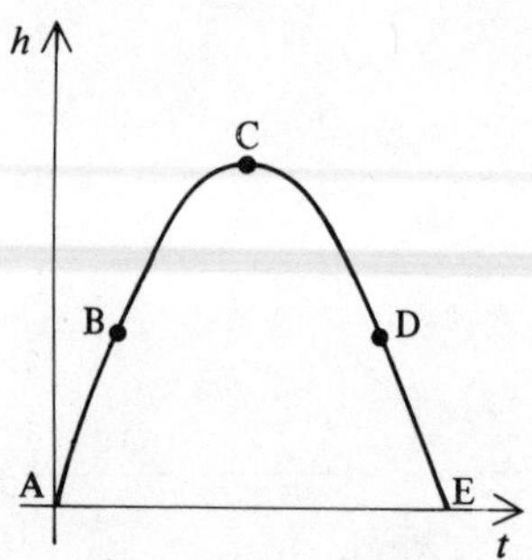

Find the point (or points) at which
(a) $h = 0$,
(b) the gradient $= 0$,
(c) the rate of change of h with respect to $t > 0$,
(d) the rate of change of h with respect to $t < 0$.
In what units is the gradient given?

4. m in kg is plotted as a function of x in metres.

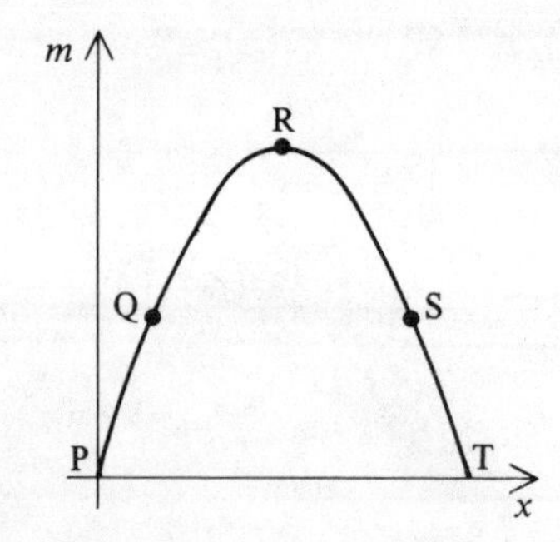

Find the point (or points) at which
(a) the rate of change of m with respect to x is zero,
(b) $m = 0$,
(c) the rate of change of m with respect to $x > 0$,
(d) the rate of change of m with respect to $x < 0$.
In what units is the gradient given?

5. x in metres is plotted as a function of t in seconds.

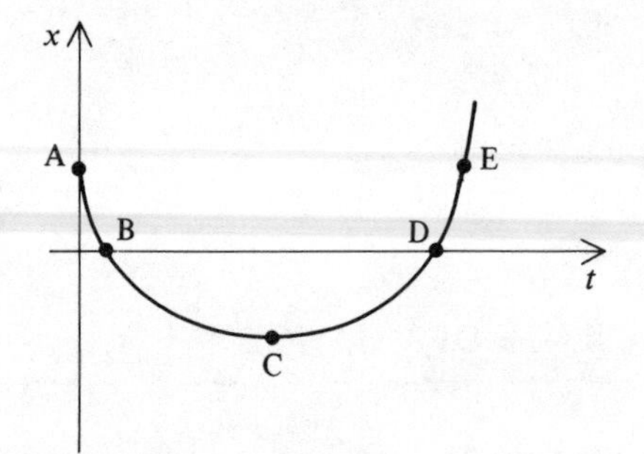

Find the point (or points) at which
(a) $x = 0$,
(b) the rate of change of x with respect to t is zero,
(c) the rate of change of x with respect to $t < 0$,
(d) the rate of change of x with respect to $t > 0$.
In what units is the gradient given?

10.2 DIFFERENTIATION

The gradient (or rate of change) of a function is calculated by differentiating.

If $y = x^n$

$$\frac{dy}{dx} = nx^{n-1}$$

The rate of change of y with respect to x is written as '$\frac{dy}{dx}$'.

The rate of change of v with respect to t would be written as '$\frac{dv}{dt}$'.

Example 1

Find $\frac{dy}{dx}$ for the following equations

(a) $y = x^3$
(b) $y = x^4 + x^1$
(c) $y = 3x^2$
(d) $y = 5x + 6$

(a) $\frac{dy}{dx} = 3x^2$

(b) $\frac{dy}{dx} = 4x^3 + 1x^0$
$= 4x^3 + 1$

(c) $\frac{dy}{dx} = 3(2x)$
$= 6x$

(d) $\frac{dy}{dx} = 5.$

Exercise 2

Find $\frac{dy}{dx}$ for the following equations.

1. $y = x^2$
2. $y = x^5$
3. $y = 5x$
4. $y = x^7$
5. $y = x^3 + x^4$
6. $y = x^2 + x^5$
7. $y = x^6 + 4x$
8. $y = x^3 - x^2$
9. $y = x^4 + x^2 + x$
10. $y = 2x^2$
11. $y = 4x^3$
12. $y = 6x^5$
13. $y = 7x + 3$
14. $y = 3x^2 + 4x$
15. $y = 10x^3 - 7x$
16. $y = 4x^{10} - 2x^9$
17. $y = x^2 + 4x + 5$
18. $y = x^2 - 3x - 7$
19. $y = 2x^2 - 6$
20. $y = x^3 + x^2 - 6x$
21. $y = 3x^4 - x^3$
22. $y = 100x - 90$
23. $y = 5 - 3x$
24. $y = 10 - x^2$
25. $y = 15 - 2x^2$
26. $y = x^3 - 17x$
27. $y = 5x^3 - 2x^2 + 4x$
28. $y = x^2 - 11x - 12$
29. $y = x^{100} - 999$
30. $y = 0{\cdot}1x^{10} - 17$
31. $y = \frac{1}{2}x^2$
32. $y = \frac{1}{4}x^8$
33. $y = \frac{2}{3}x^2 + 3x$
34. $y = \frac{3}{4}x^6$
35. $y = 0{\cdot}01x^2$
36. $y = \frac{1}{10}x^5$
37. $y = 1 + x + x^3$
38. $y = \frac{1}{2}x^2 + \frac{1}{3}x^3 + \frac{1}{4}x^4$
39. $y = 17 - x^{17}$
40. $y = -0{\cdot}4x^3$

Example 2

Find $\frac{dy}{dx}$ for the following equations.

(a) $y = (x - 1)(x + 3)$

(b) $y = \frac{1}{x^2}$

(c) $y = \frac{3}{x} - \frac{2}{x^3}$

(a) $y = (x - 1)(x + 3)$
$y = x^2 + 2x - 3$
$\frac{dy}{dx} = 2x + 2$

(b) $y = \frac{1}{x^2}$
$y = x^{-2}$
$\frac{dy}{dx} = -2x^{-3}$
$= -2 \cdot \frac{1}{x^3} = \frac{-2}{x^3}$

(c) $y = \frac{3}{x} - \frac{2}{x^3}$
$y = 3x^{-1} - 2x^{-3}$
$\frac{dy}{dx} = -3x^{-2} - 2(-3x^{-4})$
$\frac{dy}{dx} = \frac{-3}{x^2} + \frac{6}{x^4}$

Exercise 3

Find $\frac{dy}{dx}$ for the equations in questions **1** to **30**.

1. $y = (x - 2)(x + 3)$
2. $y = (x - 5)(x + 1)$
3. $y = (2x - 3)(x + 1)$
4. $y = x(3x - 16)$
5. $y = x^2(x - 5)$
6. $y = x^2(x + 11)$
7. $y = (x + 3)^2$
8. $y = (x - 4)^2$
9. $y = (2x - 3)^2$
10. $y = x^2(x - 1)^2$
11. $y = x(x + 5)^2$
12. $y = 2(x + 3)^2$
13. $y = x^{-3}$
14. $y = x^{-4}$
15. $y = x^{-2} + x^2$
16. $y = x^{-1} + x^3$
17. $y = \frac{1}{x^5}$
18. $y = \frac{1}{x^4}$
19. $y = \frac{3}{x^2}$
20. $y = \frac{4}{x^3}$
21. $y = \frac{3}{x} + \frac{2}{x^2}$
22. $y = \frac{4}{x} - \frac{3}{x^3}$
23. $y = x^3 + \frac{1}{x^3}$
24. $y = x^2 - \frac{1}{x^2}$
25. $y = x^2 + 1 + \frac{1}{x^2}$
26. $y = x^2\left(x + \frac{1}{x}\right)$
27. $y = x\left(x + \frac{1}{x}\right)$
28. $y = \frac{1}{x^2}(x^2 + x + 1)$
29. $y = \left(x + \frac{1}{x}\right)^2$
30. $y = \left(2x + \frac{1}{x}\right)^2$

Find $\frac{dz}{dt}$ for the equations in questions **31** to **44**.

31. $z = 4t^3 - 5t$
32. $z = t^{19} - t^{15}$
33. $z = 3t^4 - \frac{1}{3}t$
34. $z = 1 - \frac{1}{2}t + 4t^3$
35. $z = 6 - 3t - t^5$
36. $z = (t^2 - 1)(t + 1)$
37. $z = (2t - 3)^2$
38. $z = \frac{1}{t^3} + \frac{1}{t^4}$
39. $z = \frac{3}{t} - \frac{5}{t^2} + 4t$
40. $z = \frac{t^2 + 1}{t}$
41. $z = 4t^2 - \frac{4}{t^2} + 10$
42. $z = \frac{3t^3 + 4t}{t}$
43. $z = \frac{1}{t^2}(3t^3 + t)$
44. $z = t^{-2}(t^{-3} + t^4)$

In questions **45** to **60**, differentiate each equation with respect to x.

45. $y = x^3 + x$
46. $y = x$
47. $z = (x^2 + 1)^2$
48. $z = x^{10} + x^9 + x^8$
49. $w = 9 - \frac{5}{6}x - 6x^2$
50. $v = (x + 1)^3$
51. $t = 4x^3 - \frac{3}{4x}$
52. $a = \frac{4}{x^2} - \frac{3}{x^3}$
53. $s = 4 - \frac{3}{2x} + \frac{5}{x^3}$
54. $y = (x^2 + 1)^3$
55. $z = \frac{6x^2 + 4x + 3}{x}$
56. $p = x^{\frac{1}{2}}$
57. $v = 2x^{\frac{1}{2}}$
58. $w = x^2 - x^{\frac{1}{2}}$
59. $t = \sqrt{x}$
60. $a = (\sqrt{x})^3$

Example 3

Find the gradient of the curve $y = 3x^2 - 7x + 1$ at the point $(2, -1)$.

$$\frac{dy}{dx} = 6x - 7$$

At $x = 2$, $\frac{dy}{dx} = 12 - 7 = 5$.

$\therefore$ The gradient at the point $(2, -1)$ is 5.

Note that the y-coordinate is not required to find the gradient at the point.

Example 4

Find the coordinates of the point on the curve $y = 2x^2 - 6x + 1$ where the gradient is equal to 2.

$$y = 2x^2 - 6x + 1$$

$$\frac{dy}{dx} = 4x - 6$$

Where the gradient is 2, $\frac{dy}{dx} = 2$.

$$\therefore \quad 4x - 6 = 2$$
$$4x = 8$$
$$x = 2.$$

On the curve when $x = 2$,

$$y = 2(2)^2 - 6(2) + 1$$
$$y = -3$$

The gradient is 2 at the point $(2, -3)$.

Exercise 4

For questions **1** to **10**, find the gradient from the equation of the curve at the point indicated.

1. $y = x^2 + 7x$ at $x = 3$.
2. $y = 3x^2 - 6x + 1$ at $x = 0$.
3. $y = 5x + 9$ at $x = 10$.
4. $y = 4x^3 - 7x + 5$ at $x = 2$.
5. $y = 5x^2 - 3x + 1$ at $(1, 3)$.
6. $y = 3x^3 - 7x + 10$ at $(-1, 14)$.
7. $z = 4 - 3x - 3x^2$ at $(0, 4)$.
8. $z = \frac{1}{x} + x$ at $x = 2$.
9. $y = \frac{4}{x} - 3x + 4$ at $(1, 5)$.
10. $y = \frac{4}{x^2} + x^2$ at $(-2, 5)$.
11. Find the gradient of the curve $y = x^2 + 3x - 4$ at the two points where $y = 6$.
12. Find the gradient of the curve $y = x^2 - 5x + 7$ at the two points where $y = 21$.
13. Find the x-coordinate of the point where the gradient of the curve $y = x^2 - 3x + 5$ is equal to 1.
14. Find the x-coordinate of the point where the gradient of the curve $y = 2x^2 + 3x - 11$ is equal to -1.
15. Find the coordinates of the point on the curve $y = x^2 - 6x + 7$ where the gradient is zero.
16. Find the points on the curve $z = 2x^2 - 8x + 3$ where the gradient is (a) 0 (b) 4.

17. Find the x-coordinates of the points on the curve $y = x + \frac{1}{x}$ at which the tangent is parallel to the x-axis.
18. A(1, 5) and B(4, 11) are two points on the curve $y = x^2 - 3x + 7$. Calculate the gradient of the chord AB. The point C is on the curve between A and B where the tangent to the curve is parallel to the chord AB. Find the x-coordinate of C.
19. The equation of a curve is $y = x^3 - 2x + 10$. Calculate the acute angle between the x-axis and the tangent to the curve at the point $(-1, 11)$.
20. The equation of a curve is $y = x^2 - 4x + 9$. Calculate the acute angle between the x-axis and the tangent to the curve at the points
(a) (3, 6) (b) $(\frac{9}{4}, \frac{81}{16})$.
21. Find the x-coordinate of the point on the curve $y = \frac{x^3}{3} - 3x^2 + \frac{2}{3}$ where the tangent is parallel to the tangent drawn at the point $(1, -2)$.
22. Tangents are drawn to the curve $y = x^2 - 5x$ at the points (0, 0) and (4, −4). Calculate the acute angle between these tangents.
23. Find the coordinates of the two points on the curve $y = 4x + \frac{1}{x}$ where the gradient is zero.
24. Find the coordinates of the point on the curve $y = 3x + \frac{1}{x^2}$ where the gradient is 1.

10.3 MAXIMUM AND MINIMUM VALUES

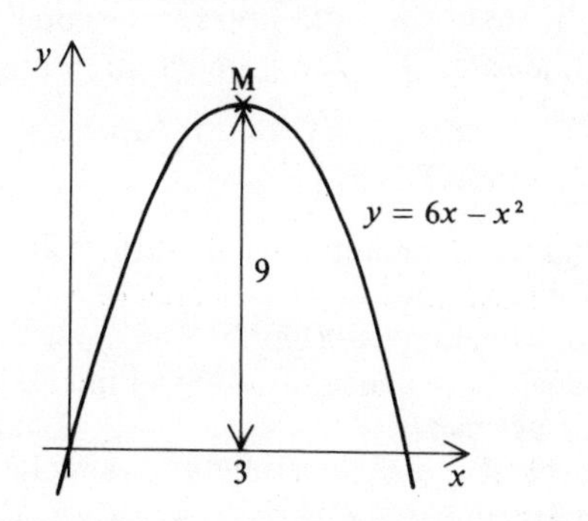

The function $y = 6x - x^2$ has a maximum value of 9 at $x = 3$.
The gradient of the curve at M is clearly zero.

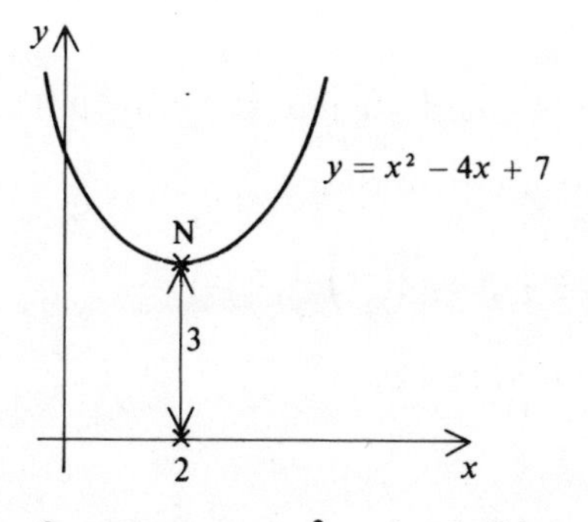

The function $y = x^2 - 4x + 7$ has a minimum value of 3 at $x = 2$.
The gradient of the curve at N is zero.

There is a maximum point at M and a minimum point at N.

Maximum and minimum points are also called *turning points*.

We locate turning points by finding where the gradient of the function is zero.

To decide whether a point is a maximum or a minimum we look at the gradient just to the left and just to the right of the point where the gradient is zero.

Example 1

Find the maximum or minimum values of the function $y = 2x^2 - 8x + 7$.

$$y = 2x^2 - 8x + 7$$
$$\frac{dy}{dx} = 4x - 8$$

For turning points, $\frac{dy}{dx} = 0$.

$$\therefore \quad 4x - 8 = 0$$
$$x = 2$$

There is only one turning point since $\frac{dy}{dx} = 0$ for only one value of x.

Take a value of x slightly less than 2, say $x = 1, \frac{dy}{dx} = -4$.

Take a value of x slightly greater than 2, say $x = 3, \frac{dy}{dx} = +4$.

So from left to right the gradient goes from
−4 to 0 to +4
i.e. \ — /

We deduce that the function has a minimum value at $x = 2$.

The minimum value is the y-value at this point.

At $x = 2, y = 2(2)^2 - 8(2) + 7$
$y = -1$

The function has a minimum value of −1 at $x = 2$.

Exercise 5

Find the turning values of the following functions and state whether the value is a maximum or minimum.

1. $y = x^2 - 6x + 5$
2. $y = 10 + 8x - x^2$
3. $y = 1 + 4x - 2x^2$
4. $y = 3x^2 + 6x - 5$
5. $y = (x - 3)(x - 5)$
6. $y = (1 - x)(3 + x)$
7. $y = 3x^2 - 3x + 1$
8. $y = 5 + 2x - 4x^2$
9. $y = x^3 - 3x^2 - 24x + 10$
10. $y = x^3 - 6x^2 + 9x + 17$
11. $y = 2x^3 + 9x^2 + 2$
12. $y = x^3 - 3x^2 + 10$
13. $y = \frac{1}{3}x^3 - \frac{1}{2}x^2 - 2x$
14. $y = x^3 - 2x^2 + x + 7$
15. $y = x + \frac{1}{x}$
16. $y = 4x + \frac{1}{x}$
17. $y = 5 - x - \frac{1}{x}$
18. $y = 1 - \frac{1}{4}x - \frac{4}{x}$
19. The height h metres, after t seconds, of a ball thrown into the air is given by $h = 10 + 40t - 5t^2$. Find the maximum height reached by the ball.
20. A rectangle has a perimeter of 24 cm. If one side has length x cm, find the length of the other side in terms of x and hence find an expression for the area of the rectangle in terms of x. Find the maximum possible area of the rectangle as x varies.
21. A rectangle has a perimeter of 60 cm. Find its maximum possible area. (Follow the method used in question **20**.)
22. A solid cylinder has a volume of 16π cm³.
(a) Express its height h in terms of its radius r.
(b) Write down an expression for the total surface area A of the cylinder, in terms of r and h.
(c) Show that $A = 2\pi\left(\frac{16}{r} + r^2\right)$.
(d) Find the minimum possible surface area.
23. Find the least possible surface area of a solid cylinder of volume 54π cm³.
24. A rectangle has an area of 64 cm². Find its least possible perimeter.
25. Find the least possible perimeter of a rectangle of area 100 m².

10.4 INTEGRATION

Integration is the reverse process of differentiation.

If $\frac{dy}{dx} = x^n$

$y = \frac{x^{n+1}}{n+1}$ + constant. [unless $n = -1$]

Example 1

Integrate to find y in terms of x

(a) $\frac{dy}{dx} = x^2 + 1$ (b) $\frac{dy}{dx} = 2x^2 + 4x - 5$

(a) $y = \frac{x^3}{3} + x + c$ (where c is a constant)

(b) $y = 2.\left(\frac{x^3}{3}\right) + 4.\left(\frac{x^2}{2}\right) - 5x + c.$

$y = \frac{2x^3}{3} + 2x^2 - 5x + c.$

Example 2

A curve passes through the point (1, 2) and has a gradient function $2x - 4$. Find the equation of the curve.

$$\frac{dy}{dx} = 2x - 4$$

$$\therefore \quad y = x^2 - 4x + c \quad (c \text{ constant})$$

The point (1, 2) satisfies the equation.

$$\therefore \quad 2 = 1^2 - 4(1) + c$$

$$\therefore \quad c = 5$$

The equation of the curve is $y = x^2 - 4x + 5$.

Exercise 6

In questions **1** to **20**, find y in terms of x given the following expressions for $\frac{dy}{dx}$.

1. x^3 **2.** x^2
3. $x^4 + x$ **4.** $x^5 + 3$
5. $x^2 - 4$ **6.** $x^6 + 3x$
7. $2x^3 - 3x^2$ **8.** $x^4 - x$
9. $\frac{x^2}{2} - 7$ **10.** $\frac{x}{3} + \frac{1}{4}$
11. $3x^4 - 6x$ **12.** $x + x^2 + x^3$
13. $1 - 7x$ **14.** $(x + 1)(x + 2)$
15. $(2x - 1)(x + 3)$ **16.** $x^2(x - 3)$
17. $x(x - 1)(x + 1)$ **18.** $(2x + 1)^2$
19. x^{-2} **20.** x^{-3}

In questions **21** to **26**, find the equation of the curve given the gradient of the curve and the coordinates of a point on the curve.

21. $\frac{dy}{dx} = 2x - 3; (0, 4)$ **22.** $\frac{dy}{dx} = 2x + 2; (2, -7)$

23. $\frac{dy}{dx} = 4x - 7; (1, -4)$ **24.** $\frac{dy}{dx} = 6x - 4; (0, -10)$

25. $\frac{dy}{dx} = 3x^2 - 2; (2, 5)$ **26.** $\frac{dy}{dx} = 6x^2, (1, 7)$

27. If $\frac{dv}{dt} = 2t + 5$ and $v = 11$ when $t = 2$, find v in terms of t.

28. If $\frac{dz}{dy} = 3y^2 - 4$ and $z = 1$ when $y = 0$, find z in terms of y.

29. If $\frac{ds}{dt} = 6t^2 - 2t$ and $s = 55$ when $t = 3$, find s when $t = 1$.

30. If $\frac{dv}{dt} = 2t - 1$ and $v = 8$ when $t = 3$, find v when $t = 2$.

Integral notation

$\int(x^2 + 1)dx$ means
'integrate $x^2 + 1$ with respect to x'.

$$\text{so} \quad \int(x^2 + 1)dx = \frac{x^3}{3} + x + c.$$

Area under a curve

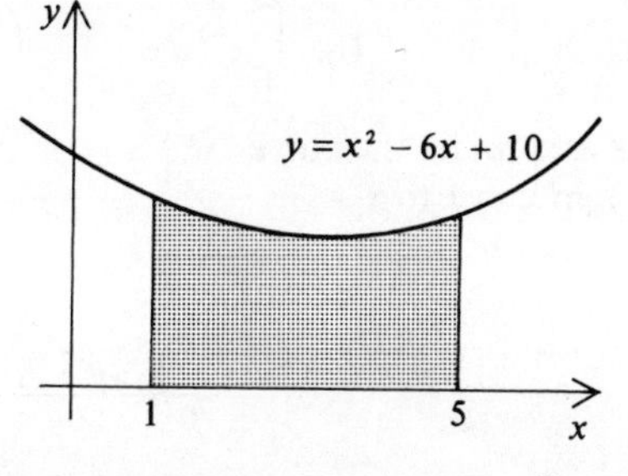

The area under the curve $y = x^2 - 6x + 10$ between $x = 1$ and $x = 5$ is given by

$$\int_1^5 y\,dx = \int_1^5 (x^2 - 6x + 10)dx$$

where 1 and 5 are *limits* of integration.

$$\text{Area} = \left[\frac{x^3}{3} - \frac{6x^2}{2} + 10x\right]_1^5$$

Note
(a) Square brackets are used conventionally.
(b) No constant of integration is needed since it disappears in the working.

$$\text{Area} = \left(\frac{5^3}{3} - 3(25) + 50\right) - \left(\frac{1^3}{3} - 3(1^2) + 10\right)$$

$$= \frac{125}{3} - 75 + 50 - \frac{1}{3} + 3 - 10$$

$$= 9\tfrac{1}{3} \text{ square units.}$$

Exercise 7

Perform the following integrals.

1. $\int (x^2 + 2)dx$
2. $\int (2x - 1)dx$
3. $\int (x^3 - 4x)dx$
4. $\int (x^2 + x + 1)dx$
5. $\int \left(\frac{x}{2} - \frac{1}{4}\right)dx$
6. $\int (\pi x + 3)dx$
7. $\int (t^2 + 1)dt$
8. $\int (3t - 5)dt$
9. $\int (z^2 - 5z)dz$
10. $\int (y - 10)dy$

Evaluate the following integrals

11. $\int_1^3 (2x - 1)\,dx$
12. $\int_0^4 (4x + 3)\,dx$
13. $\int_1^4 (x^2 - 2x + 3)\,dx$
14. $\int_{-1}^1 (3x^2 - 7)\,dx$
15. $\int_0^2 (6t^2 - 4t)\,dt$
16. $\int_{-2}^{-1} (2t + 4)\,dt$
17. $\int_{10}^{11} 2z\,dz$
18. $\int_1^2 x^{-2}dx$
19. Find the area under the curve $y = x^2$ from $x = 1$ to $x = 3$.

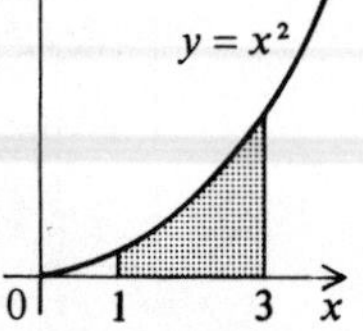

20. Find the area under the curve $y = x^2 - 4x + 5$ between $x = 0$ and $x = 3$.

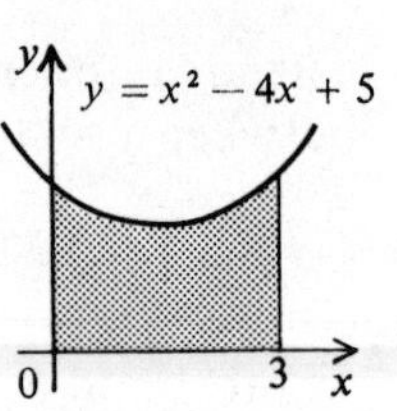

21. Find the area under the curve $y = 3x^2 - 4x + 2$ between $x = -1$ and $x = 2$.
22. Find the area under the line $v = 2t + 5$ between $t = 2$ and $t = 5$.
23. Find where the curve $y = 5x - x^2$ cuts the x-axis and hence find the area shaded.

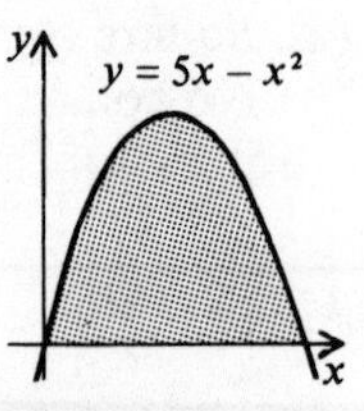

24. Find where the curve $y = (2 - x)(3 + x)$ cuts the x-axis and hence sketch the curve. Calculate the area enclosed between the curve and the x-axis.
25. Find where the curve $y = x^2 - 4x + 3$ cuts the x-axis and hence sketch the curve. Calculate the area enclosed between the curve and the x-axis.
26. Show that the line $y = 8$ cuts the curve $y = 6x - x^2$ at the points (2, 8) and (4, 8). Calculate the area enclosed by the curve and the line $y = 8$.
27. Find the points where the line $y = x$ cuts the curve $y = 4x - x^2$. Calculate the area enclosed by the curve and the line $y = x$.
28. Find where the curve $y = 4 - x^2$ cuts the x- and y-axes and sketch the curve. Find where the line $y = x + 2$ cuts the curve and hence find the area enclosed between the line and the curve.

Approximate integration

There are many functions which cannot be integrated exactly or which are very difficult to integrate. When a computer is used to evaluate a difficult integral the computer program will probably use the method shown below.

Example 3

Find an approximate value for the area under the curve $y = \frac{12}{x}$ between $x = 2$ and $x = 5$.

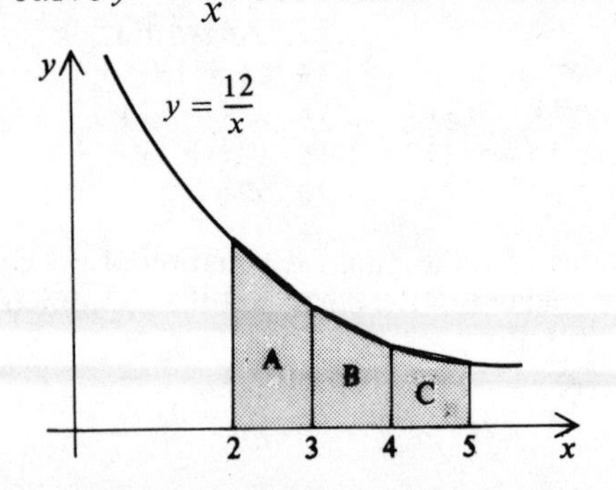

(a) Draw lines parallel to the y-axis at $x = 2, 3, 4, 5$ to form three trapeziums. The area under the curve is approximately equal to the area of the three trapeziums A, B and C.

(b) Calculate the values of y at $x = 2, 3, 4, 5$

at $x = 2, y = \frac{12}{2} = 6$
at $x = 3, y = \frac{12}{3} = 4$
at $x = 4, y = \frac{12}{4} = 3$
at $x = 5, y = \frac{12}{5} = 2{\cdot}4$

(c) Calculate the area of each of the trapeziums using the formula area $= \frac{1}{2}(a+b)h$, where a and b are the parallel sides and h is the distance in between.

Area A $= \frac{1}{2}(6+4) \times 1 = 5$ sq. units

Area B $= \frac{1}{2}(4+3) \times 1 = 3{\cdot}5$ sq. units

Area C $= \frac{1}{2}(3+2{\cdot}4) \times 1 = 2{\cdot}7$ sq. units

Total area under the curve $\approx (5+3{\cdot}5+2{\cdot}7)$

$\approx 11{\cdot}2$ sq. units

It can be shown that the actual area under the curve is $11{\cdot}0$ sq. units (to 3 S.F.). Can you see why our approximate value is a higher value?

Exercise 8

1. Find an approximate value for the area under the curve $y = \frac{8}{x}$ between $x = 2$ and $x = 5$.

Divide the area into three trapeziums as shown.

2. Find an approximate value for the area under the curve $y = \frac{12}{(x+2)}$ between $x = 0$ and $x = 4$.

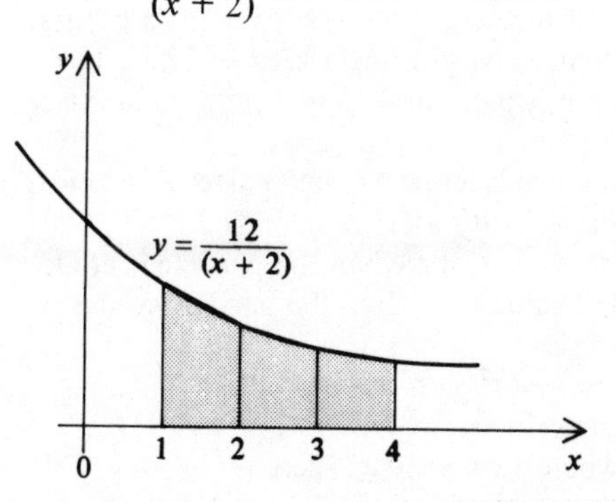

Divide the area into four trapeziums as shown.

3. Find an approximate value for the area under the curve $y = 6x - x^2$ between $x = 1$ and $x = 4$. Divide the area into three trapeziums of equal width.

4. (a) Sketch the curve $y = 16 - x^2$ for $-4 \leqslant x \leqslant 4$.
   (b) Find an approximate value for the area under the curve $y = 16 - x^2$ between $x = 0$ and $x = 3$. (Divide the area into three trapeziums of equal width.)
   (c) State whether your approximate value is greater than or less than the actual value for the area.
   (d) Confirm your answer to part (c) by evaluating the integral $\int_0^3 (16 - x^2)dx$.

5. (a) Find an approximate value for the area under the curve $y = x^2 + 2x + 5$ between $x = 0$ and $x = 4$. Divide the area into four trapeziums.

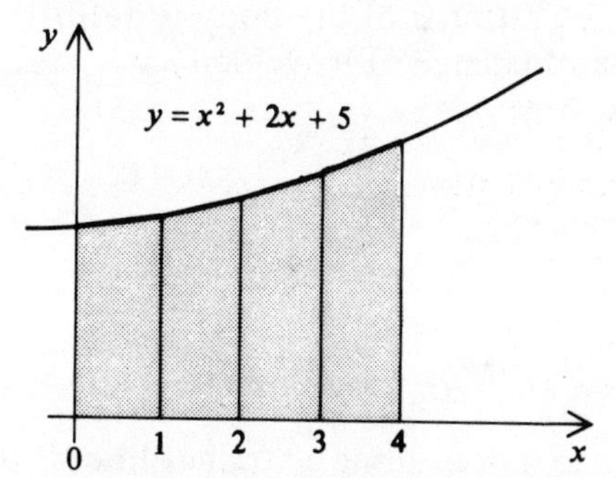

   (b) State whether your approximate value is greater than or less than the actual value for the area.
   (c) Confirm your answer to part (b) by evaluating the integral $\int_0^4 (x^2 + 2x + 5)dx$.

In questions **6** to **10** find an approximate value for the integral using the number of trapeziums indicated.

6. $\int_1^5 \frac{12}{x} dx$: four trapeziums, width 1 unit.

7. $\int_0^5 (30 - x^2)dx$: five trapeziums, width 1 unit.

8. $\int_1^{\frac{5}{2}} \frac{9}{x}.dx$: three trapeziums, width $\frac{1}{2}$ unit.

9. $\int_0^1 x^2.\,dx$: five trapeziums, width $0{\cdot}2$ unit.

10. $\int_{-4}^0 \frac{12}{(x+6)} dx$: four trapeziums, width 1 unit.

10.5 VELOCITY AND ACCELERATION

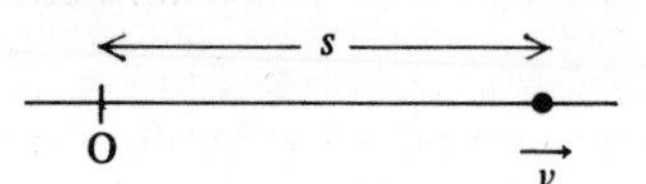

Suppose a body moves along a straight line so that its displacement (or distance) s m from a fixed point O after t seconds is given by the formula $s = t^2 + 5t + 10$ e.g. after 3 seconds it is $(3^2 + (5 \times 3) + 10)$ m from O, i.e. 34 m from O.

The velocity v of the body is defined as the rate of change of its displacement from O with respect to time.

so $v = \dfrac{ds}{dt} = 2t + 5$ m/s.

The acceleration a of the body is defined as the rate of change of its velocity with respect to time.

so $a = \dfrac{dv}{dt} = 2$ m/s^2.

Example 1

A particle moves along a straight line so that its displacement s metres from a fixed point O after t seconds is given by $s = t^3 - 4t^2 + 7$. Find
(a) its displacement from O after 1 second
(b) its velocity after 2 seconds
(c) its acceleration after 4 seconds
(d) when its acceleration is zero.

(a) When $t = 1, s = 1^3 - 4(1)^2 + 7 = 4$.
After one second the particle is 4 metres from O.

(b) $v = \dfrac{ds}{dt} = 3t^2 - 8t$

when $t = 2, v = 3\,.(2)^2 - 8(2)$
$v = -4$ m/s.

The negative sign means it is moving in the opposite direction to that in which s is increasing. If s is positive to the right, then a negative velocity means the particle is moving to the left.

(c) $a = \dfrac{dv}{dt} = 6t - 8$.
When $t = 4, a = 16$ m/s^2.

(d) When $a = 0, 6t - 8 = 0$
$t = \frac{4}{3}$.
The acceleration is zero after $1\frac{1}{3}$ seconds.

Example 2

The acceleration a m/s^2 of a body after t seconds is given by the formula $a = 2t - 4$. If $v = 10$ m/s when $t = 2$, find
(a) an expression for v in terms of t;
(b) the initial velocity.

(a) $a = \dfrac{dv}{dt} = 2t - 4$.
Integrating, $v = t^2 - 4t + c$ (c is a constant)
$v = 10$ when $t = 2$

so $10 = 2^2 - 4(2) + c$
$\therefore\ c = 14$
$\therefore\ v = t^2 - 4t + 14$.

(This is an expression for v 'in terms of t'.)

(b) 'Initially' means when $t = 0$,

so initial velocity $= 0^2 - 4(0) + 14$
$= 14$ m/s.

Exercise 9

In this exercise s, v and a are the displacement, velocity and acceleration respectively of a particle moving along a straight line after t seconds. The displacement is measured from a point O in metres.

1. If $s = t^2 - 7t + 12$, calculate
(a) the displacement from O after 2 seconds,
(b) the velocity after 2 seconds,
(c) the values of t when the particle is at O,
(d) the value of t when the particle comes to rest.
2. If $s = t^3 - 6t^2 + 9t + 1$, calculate
(a) the velocity when $t = 0$,
(b) when the velocity is zero,
(c) the acceleration after 3 seconds,
(d) the initial acceleration (i.e. when $t = 0$).

3. If $v = 2t^2 - 11t + 5$, calculate
   (a) the initial velocity,
   (b) the initial acceleration,
   (c) when the particle comes to rest,
   (d) when the acceleration is zero.
4. If $v = 3t^2 - 13t + 4$, calculate
   (a) the velocity after 5 seconds,
   (b) the acceleration after 3 seconds,
   (c) the acceleration when the particle first comes to rest.
5. If $v = 2t + 3$ and $s = 17$ when $t = 2$, find
   (a) the initial velocity,
   (b) an expression for s in terms of t,
   (c) the displacement from O after 10 seconds,
   (d) the acceleration after 3 seconds.
6. If $a = 2t - 6$ and $v = 8$ when $t = 0$, find
   (a) an expression for v in terms of t,
   (b) when the particle comes to rest,
   (c) the minimum velocity.
7. If $v = (t - 3)(t - 7)$ and $s = 10$ when $t = 0$, find
   (a) when the particle comes to rest,
   (b) an expression for s in terms of t,
   (c) the displacement from O when the particle first comes to rest,
   (d) the initial acceleration.
8. If $s = t^2 + 5t$, calculate
   (a) the distance moved in the first 3 seconds,
   (b) the distance moved in the third second,
   (c) the initial velocity.
9. If $s = t^3 + 4t^2 + 5$, calculate
   (a) the distance moved in the fourth second,
   (b) the initial velocity,
   (c) the initial acceleration.
10. If $v = 6t - t^2$ and the particle is initially at O, find
    (a) the maximum velocity,
    (b) an expression for s in terms of t,
    (c) when the particle passes through O.
11. If $v = 10 + t^2 - t^3$, calculate
    (a) the maximum acceleration,
    (b) the displacement from O after 2 seconds, given $s = 10\frac{1}{12}$ when $t = 1$.
12. If $v = 3t$ for $0 \leqslant t \leqslant 3$ and $v = 9$ for $3 < t \leqslant 10$.
    (a) sketch a velocity-time graph for the first 10 seconds of the motion,
    (b) calculate the distance travelled during the first 10 seconds.
13. A particle starts from rest at O and its acceleration is given by $a = 8 - t$. Find
    (a) when the acceleration is zero,
    (b) an expression for v in terms of t,
    (c) an expression for s in terms of t,
    (d) when the particle returns to O.
14. A slice of toast starts from rest at point O and during its motion it has an acceleration given by $a = 1 - 2t$. Find
    (a) an expression for v in terms of t,
    (b) an expression for s in terms of t,
    (c) its maximum velocity,
    (d) when the slice of toast returns to O.

10.6 SPEED-TIME GRAPHS

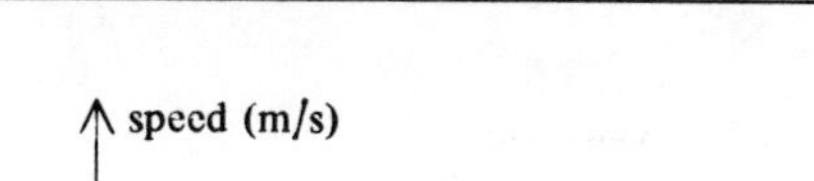
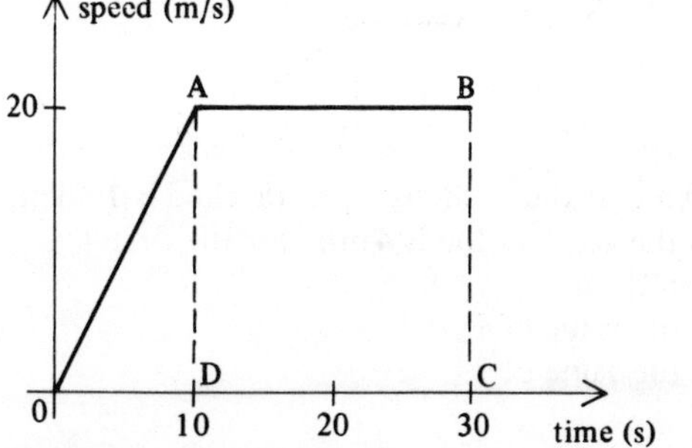

The diagram is the speed-time graph of the first 30 seconds of a car journey. Two quantities are obtained from such graphs
(a) acceleration = gradient of speed-time graph,
(b) distance travelled = area under graph.

In this example,

(a) The gradient of line OA $= \frac{20}{10} = 2$.

$\therefore$ The acceleration in the first 10 seconds is $2\ \text{m/s}^2$.

(b) The distance travelled in the first 30 seconds is given by the area of OAD plus the area of ABCD.

$$\text{Distance} = (\tfrac{1}{2} \times 10 \times 20) + (20 \times 20)$$
$$= 500\ \text{m}.$$

Exercise 10

The graphs show speed v in m/s and time t in seconds.

1. Find
   (a) the acceleration when $t = 4$,
   (b) the total distance travelled,
   (c) the average speed for the whole journey.

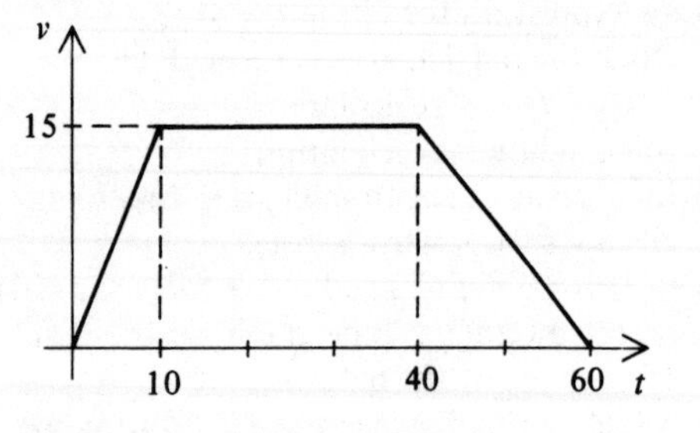

2. Find
   (a) the total distance travelled,
   (b) the average speed for the whole journey,
   (c) the distance travelled in the first 10 seconds,
   (d) the acceleration when $t = 20$.

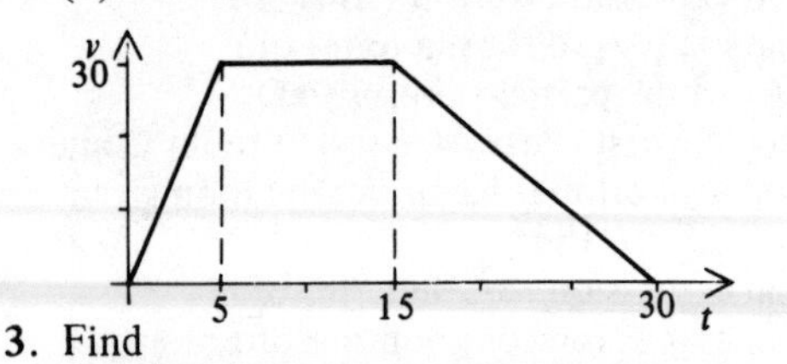

3. Find
   (a) the total distance travelled,
   (b) the distance travelled in the first 40 seconds,
   (c) the acceleration when $t = 15$.

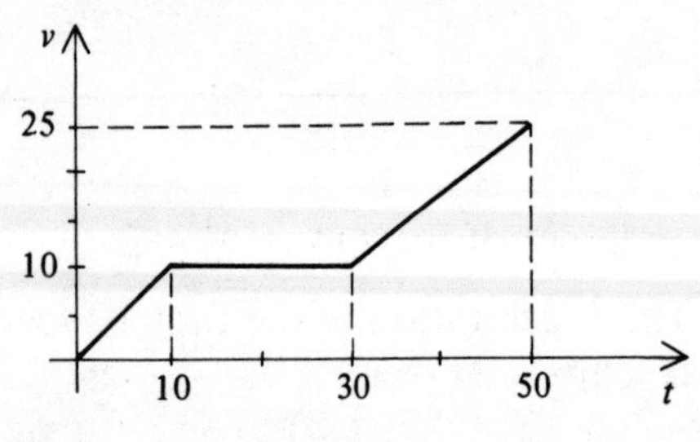

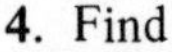

4. Find
   (a) V if the total distance travelled is 900 m,
   (b) the distance travelled in the first 60 seconds.

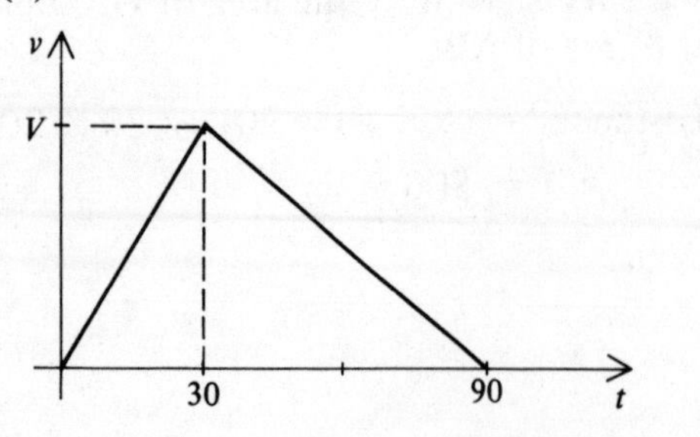

5. Find
   (a) T if the initial acceleration is $2\,\text{m/s}^2$,
   (b) the total distance travelled.
   (c) the average speed for the whole journey.

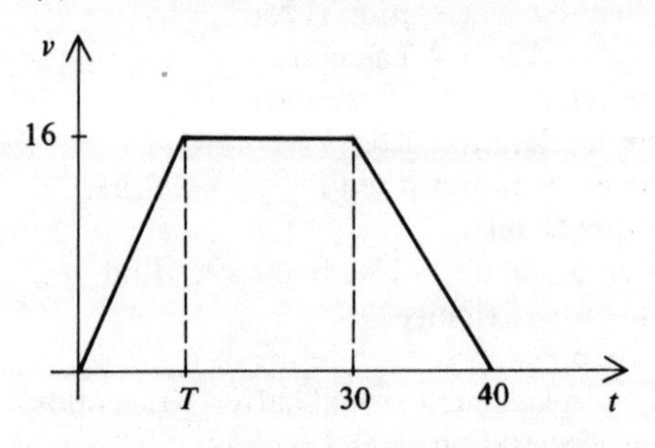

6. Given that the total distance travelled = 810 m, find
   (a) the value of V,
   (b) the rate of change of the speed when $t = 30$,
   (c) the time taken to travel the first 420 m of the journey.

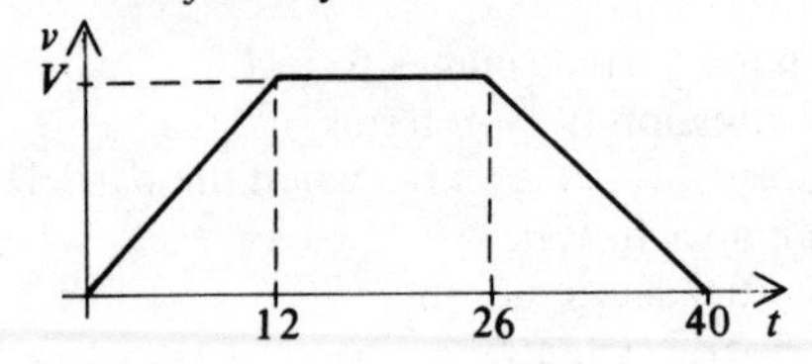

7. Given that the total distance travelled is 1·5 km, find
   (a) the value of V,
   (b) the rate of deceleration after 10 seconds.

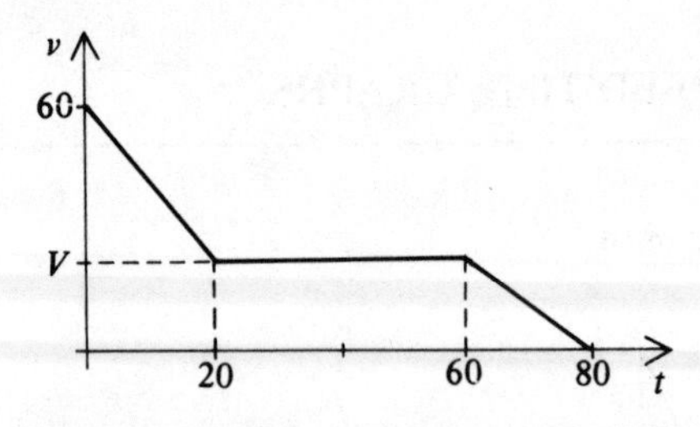

8. Given that the total distance travelled is 1·4 km, and the acceleration is $4\,\text{m/s}^2$ for the first T seconds, find
   (a) the value of V,
   (b) the value of T.

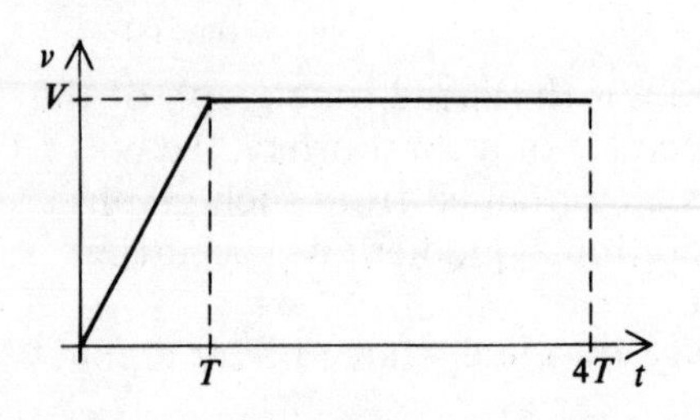

9. Given that the average speed for the whole journey is $37{\cdot}5$ m/s and that the deceleration between T and $2T$ is $2{\cdot}5$ m/s^2, find
(a) the value of V, (b) the value of T.

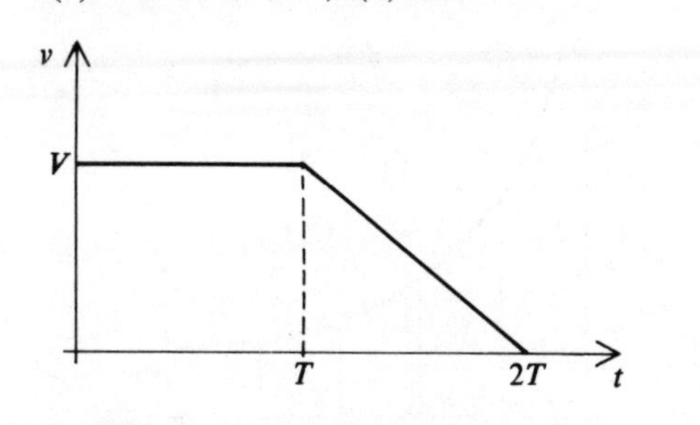

10. Given that the total distance travelled is 4 km and that the initial deceleration is 4 m/s^2, find
(a) the value of V,
(b) the value of T.

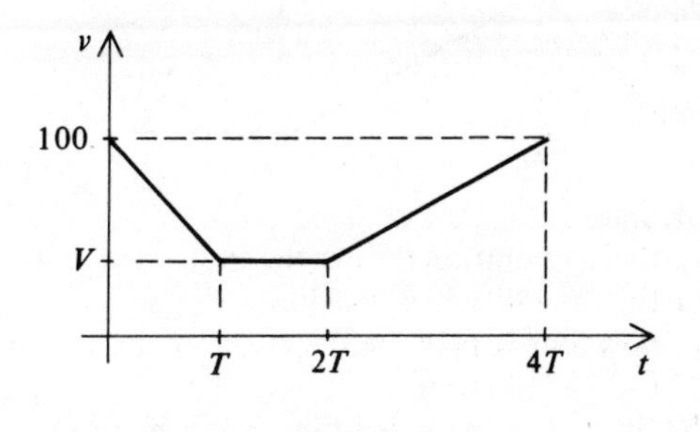

REVISION EXERCISE 10A

1. Differentiate with respect to x
(a) $y = x^3 - 7x + 10$ (b) $y = x^4 + \frac{1}{2}x^2 - 3x$
(c) $y = 3x + \dfrac{1}{x}$ (d) $y = \dfrac{5}{x} + \dfrac{x}{5}$

2. The variables y, x, v and t have units kg, m, m/s and s respectively. Write down the units for the following rates of change,
(a) $\dfrac{dy}{dt}$ (b) $\dfrac{dx}{dt}$ (c) $\dfrac{dy}{dx}$
(d) $\dfrac{dv}{dt}$ (e) $\dfrac{dx}{dy}$

3. Find the gradient of the curve $y = x^3 - 3x + 10$ at the point $(1, 8)$.

4. Find the gradient of the tangent to the curve $y = 8 + 4x - 2x^2$ at the point where the curve cuts the y-axis.

5. Calculate the angle between the x-axis and the tangent to the curve $y = 3x^2 + 5x - 12$ at the point $(-1, -14)$.

6. Find the coordinates of the point on the curve $y = x^2 - 3x + 7$ where the tangent to the curve is parallel to the line $y = 3x + 4$.

7. Determine the maximum and/or minimum values of the following functions
(a) $y = 9 + 4x - x^2$ (b) $y = x^3 - 3x + 7$
(c) $y = 3x + \dfrac{3}{x}$ (d) $y = x^4 - 5$.

8. Find the greatest possible area of a rectangle with a perimeter of 76 cm.

9. Work out the following integrals.
(a) $\int (x^2 + 7x)\,dx$ (b) $\int (4x - 6)\,dx$
(c) $\int (3t^2 - t + 1)\,dt$ (d) $\int \left(\dfrac{y^2}{2} - 2y\right) dy$

10. Evaluate
(a) $\displaystyle\int_0^2 (4x - 5)\,dx$ (b) $\displaystyle\int_{-1}^1 (6x^2 - 5x)\,dx$
(c) $\displaystyle\int_{-2}^1 \left(\frac{x}{2} + \frac{x^2}{3}\right) dx$ (d) $\displaystyle\int_2^3 \frac{1}{x^2}\,dx$

11. The gradient of a curve at the point (x, y) is $2x - 3$. The curve passes through the point $(2, 3)$. Find the equation of the curve.

12. Calculate the area under the curve $y = x^2 - 4x + 10$ between $x = 1$ and $x = 3$.

13. Find where the curve $y = 6 + x - x^2$ cuts the x-axis and sketch the curve. Calculate the area enclosed by the curve and the x-axis.

14. Find where the curve $y = 5x - x^2$ cuts the x-axis and sketch the curve. Find also where the curve meets the line $y = x$ and hence calculate the area enclosed by the curve and the line.

15. A particle moves along a straight line so that its displacement s metres from a fixed point O after t seconds is given by $s = t^3 + t^2 + 2t - 5$. Find
(a) the initial velocity,
(b) the acceleration when $t = 2$,
(c) the distance of the particle from O when it is accelerating at 8 m/s^2.

16. A particle moves along a straight line so that after t seconds its acceleration a m/s^2 is given by $a = 6t + 4$. Find
(a) an expression for its velocity in terms of t if its initial velocity is 3 m/s,
(b) an expression for its displacement s from a fixed point O in terms of t if it is 16 m from O after one second.

17. The diagram is the speed–time graph of a bus.

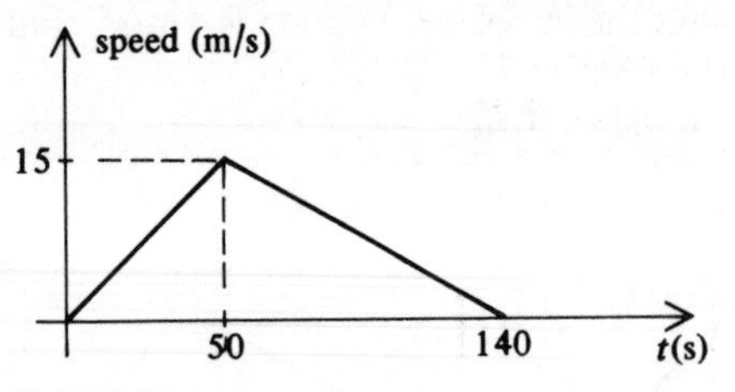

Calculate
(a) the acceleration during the first 50 seconds,
(b) the total distance travelled,
(c) how long it takes before it is moving at 12 m/s for the first time.

18. The diagram is the speed-time graph of a car.

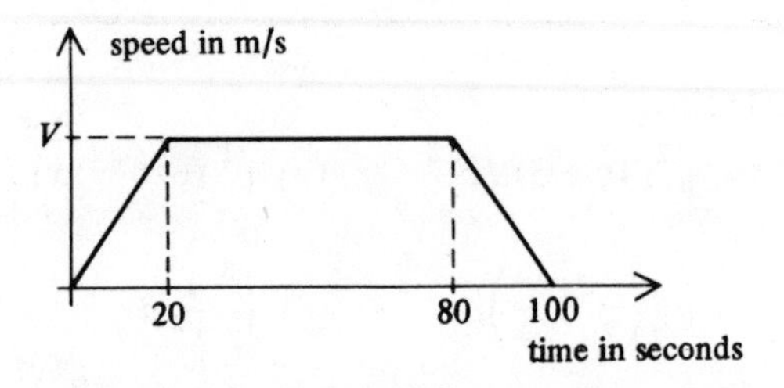

Given that the total distance travelled is 2·4 km, calculate
(a) the value of the maximum speed V,
(b) the distance travelled in the first 30 seconds of the motion.

19. (a) Find an approximate value for the area under the curve $y = \frac{12}{x^2}$ between $x = 1$ and $x = 4$.
Draw three trapeziums as shown below.

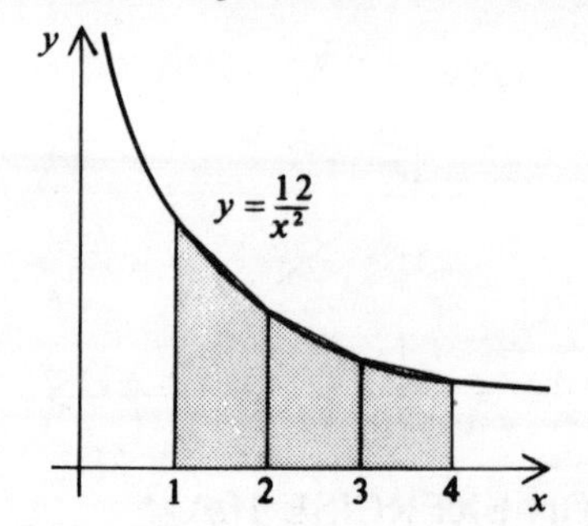

(b) State whether your approximate value is greater than or less than the actual area under the curve.
(c) Confirm your answer to part (b) by evaluating the integral $\int_1^4 12x^{-2}\,dx$.

20. Find an approximate value for the integral $\int_0^2 (5 - x^2)\,dx$, by drawing four trapeziums of width $\frac{1}{2}$ unit each.

EXAMINATION EXERCISE 10B

1. The equation of a curve is given by $y = 2x^3 - 5x^2 + 4$.
(i) Show that the curve meets the x-axis at the point where $x = 2$.
(ii) Calculate the acute angle between the tangent to the curve at the point where $x = 2$ and the x-axis.
The curve has a turning point at $x = \frac{5}{3}$. Calculate the coordinates of the other turning point on the curve and determine for which turning point y is a maximum and for which y is a minimum. Hence sketch the curve for values of x from -1 to 3. [AEB]

2. Using the methods of the calculus, find the coordinates of the maximum point and the minimum point on the curve $y = 2x^3 - 6x + 7$ and distinguish between the maximum point and the minimum point, giving reasons.
At a certain point on the curve, the x-coordinate is h and the gradient of the curve is $12h$. Prove that $h^2 - 2h - 1 = 0$. Given also that h is positive, calculate the value of h, giving your answer correct to three significant figures. [JMB]

3. (a) The diagram represents part of the curve $y = 2x - x^2$.
   (i) Write down the value of the x coordinate of P.
   (ii) Evaluate the area bounded by the curve and the x-axis.
   (b) The equation of a curve is $y = 2x^3 - 5x^2 - x$.
   Calculate the acute angle between the x-axis and the tangent to the curve at the point $(2, -6)$. [AEB]

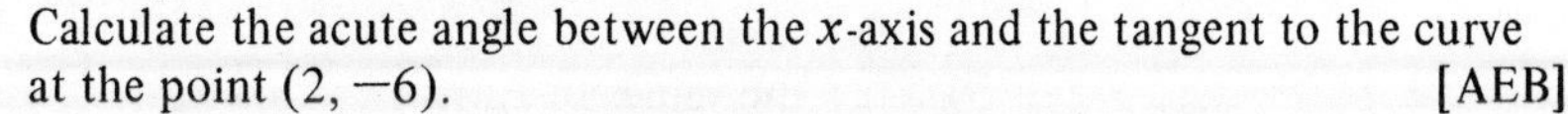

4. A ball was thrown vertically upwards and, after t seconds, its height, h metres, above the ground was given by

$$h = 33 + 4t - 5t^2.$$

   (a) Calculate the height from which the ball was thrown.
   (b) Find the speed with which it was thrown.
   (c) Find the time when the speed became zero.
   (d) Calculate the greatest height above the ground reached by the ball.
   (e) Find how many seconds elapsed from the time the ball was thrown until it reached ground level. [L]

5. A rectangular block of wood is $3x$ cm long, x cm wide and h cm high.
   (i) Find an expression for the total length of the edges of the block in terms of x and h.

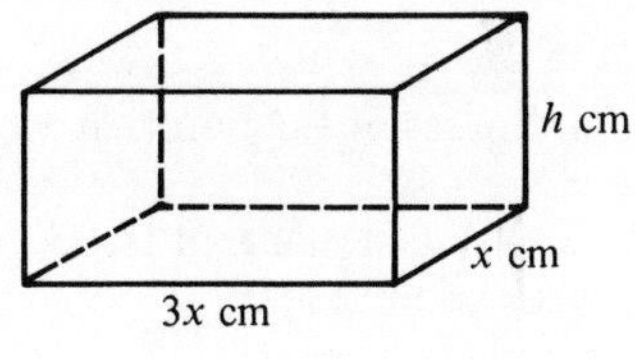

   Given that the total length of the edges is 96 cm,
   (ii) express h in terms of x,
   (iii) show that the volume, V cm^3 of the block in terms of x is given by the equation $V = 72x^2 - 12x^3$,
   (iv) find the value of x for which V is maximum, and hence calculate the maximum volume of the block. [NEA]

6. A rectangular box, made of thin sheet metal and without a lid, is of length $2x$ cm, width x cm, and height h cm. Write down an expression, in terms of x and h, for the area of sheet metal required to make the box.
   Given that this area is 600 cm^2, show that

$$h = \frac{600 - 2x^2}{6x}.$$

   Hence show that the volume, V cm^3, of the box is given by the formula

$$V = 200x - \frac{2x^3}{3}.$$

   Find $\frac{\mathrm{d}V}{\mathrm{d}x}$ and hence find the value of x for which V is a maximum.
   Calculate the volume of the largest such box which can be constructed using exactly 600 cm^2 of sheet metal. [L]

7. (a) The equation of a curve is

$$y = x^3 - 3x^2 - 9x.$$

   Show that the curve has a turning point at $(3, -27)$ and calculate the coordinates of the other turning point on the curve. Determine which of these turning points is a maximum and which is a minimum.

   (b) (i) Given that $\frac{dy}{dx} = 2x$, find y in terms of x if $y = 1$ when $x = 2$.
   (ii) Evaluate $\int_0^2 (3x - 1)(x + 1)\,dx$. [AEB]

8. A particle moves along a straight line so that, t seconds after observations are commenced, its distance, s metres, from a fixed point O in the line is given by

$$s = 8 - 6t + 2t^3.$$

Calculate
(a) the values of t when $s = 8$,
(b) the distance travelled by the particle in the first second after observations are commenced,
(c) the speed of the particle when $t = 2$,
(d) the acceleration of the particle when $t = 1$,
(e) the positive value of t for which the particle is momentarily at rest. [L]

9. In this question, the time t is measured in seconds, the distance s in metres, the velocity v in m/s and the acceleration a in m/s^2.
The formula for the acceleration of a particle moving in a straight line is $a = 7 - 2t$ and $v = 12$ when $t = 2$. Calculate
(i) the time at which $a = 3$,
(ii) the formula for v in terms of t,
(iii) the maximum velocity,
(iv) the distance travelled in the first second.
[JMB]

10. Calculate the co-ordinates of the maximum point of the curve

$$y = (1 - x)(x - 4).$$

Draw a rough sketch of the curve indicating the co-ordinates of the points at which the curve crosses the axes.
(i) Find the point, P, of the curve at which the gradient of the tangent is -3.
(ii) Find the equation of the tangent to the curve at the point $(1, 0)$.
(iii) Calculate the area contained between the curve and the x-axis. [O]

11 Statistics and probability

Blaise Pascal (1623–1662) suffered the most appalling ill-health throughout his short life. He is best known for his work with Fermat on probability. This followed correspondence with a gentleman gambler who was puzzled as to why he lost so much in betting on the basis of the appearance of a throw of dice. Pascal's work on probability became of enormous importance and showed for the first time that absolute certainty is not a necessity in mathematics and science. He also studied physics, but his last years were spent in religious meditation and illness.

11.1 DATA DISPLAY

Bar chart

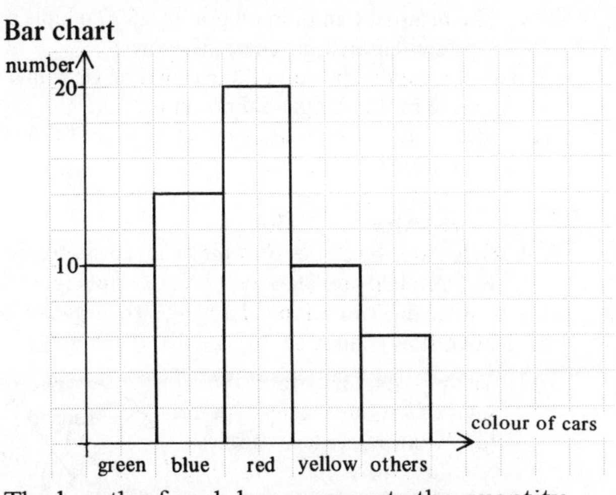

The length of each bar represents the quantity in question. The width of each bar has no significance. In the bar chart above, the number of the cars of each colour in a car park is shown.

Histograms are similar to bar charts but the *area* of each rectangle represents the quantity in question.

Pie chart

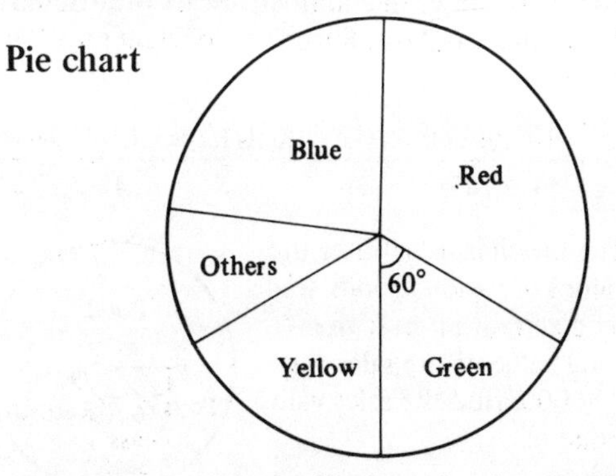

The information is displayed using sectors of a circle.

This pie-chart shows the same information as the bar chart on the left.

The angles of the sectors are calculated as follows:

Total number of cars $= 10 + 14 + 20 + 10 + 6 = 60$

Angle representing green cars $= \frac{10}{60} \times 360° = 60°$

Angle representing blue cars $= \frac{14}{60} \times 360°$, etc.

Exercise 1

1. The bar chart shows the number of children playing various games on a given day.

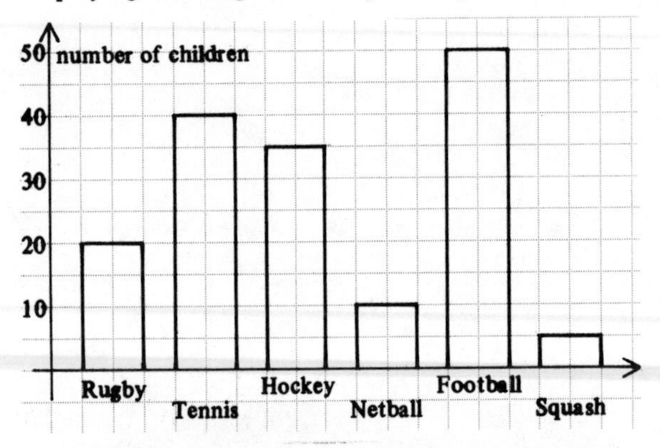

(a) Which game had the least number of players?
(b) What was the total number of children playing all the games?
(c) How many more footballers were there than tennis players?

2. The table shows the number of cars of different makes in a car park. Illustrate this data on a bar chart.

Make	Fiat	Renault	Leyland	Rolls Royce	Ford	Datsun
Number	14	23	37	5	42	18

3. The pie-chart illustrates the values of various goods sold by a certain shop. If the total value of the sales was £24 000, find the sales value value of
(a) toys
(b) grass seed
(c) records
(d) food.

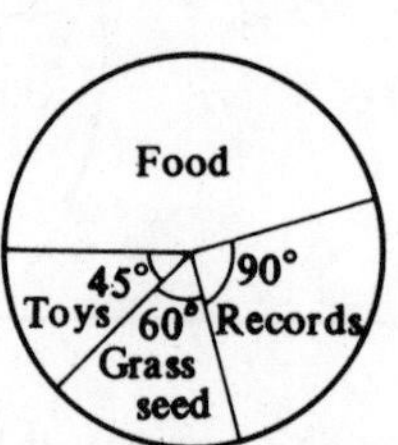

4. The table shows the colours of a random selection of 'Smarties'. Calculate the angles on a pie chart corresponding to each colour.

colour	red	green	blue	yellow	pink
number	5	7	11	4	9

5. A quantity of scrambled eggs is made using the following recipe:

ingredient	eggs	milk	butter	cheese	salt/pepper
mass	450g	20 g	39 g	90 g	1 g

Calculate the angle on a 'scrambled egg chart' corresponding to each ingredient.

6. Calculate the angles on a pie-chart corresponding to quantities A, B, C, D and E given in the tables.

(a)

quantity	A	B	C	D	E
number	3	5	3	7	0

(b)

quantity	A	B	C	D	E
mass	10 g	15 g	34 g	8 g	5 g

(c)

quantity	A	B	C	D	E
length	7	11	9	14	11

7. The weights of A, B, C are in the ratio 2:3:4. Calculate the angles representing A, B and C on a pie-chart.

8. The cooking times for meals L, M and N are in the ratio $3:7:x$. On a pie-chart, the angle corresponding to L is 60°. Find x.

9. The results of an opinion poll of 2000 people are represented on a pie chart. The angle corresponding to 'don't know' is 18°. How many people in the sample did not know?

10. The transfer fees of five footballers are as follows:

Gibson	£4·50	Crisp	£9
Campbell	£6	Raynor	£80
Hawley	50p		

Calculate the angles on a pie chart corresponding to Campbell and Hawley. (This example illustrates one of the limitations of a pie-chart: negative quantities cannot be displayed.)

11. The pie chart illustrates the sales of various makes of petrol.
(a) What percentage of sales does 'Esso' have?
(b) If 'Jet' accounts for $12\frac{1}{2}$% of total sales, calculate the angles x and y.

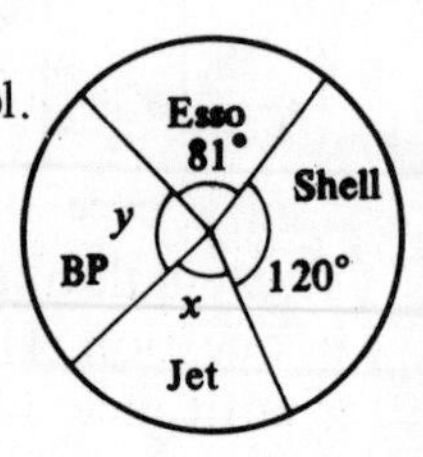

11.2 MEAN, MEDIAN AND MODE

(a) The *mean* of a series of numbers is obtained by adding the numbers and dividing the result by the number of numbers.

(b) The *median* of a series of numbers is obtained by arranging the numbers in ascending order and then choosing the number in the 'middle'. If there are *two* 'middle' numbers the median is the average (mean) of these two numbers.

(c) The *mode* of a series of numbers is simply the number which occurs most often.

Example 1

Find the mean, median and mode of the following numbers:
5, 4, 10, 3, 3, 4, 7, 4, 6, 5.

(a) Mean

$$= \frac{(5 + 4 + 10 + 3 + 3 + 4 + 7 + 4 + 6 + 5)}{10}$$

$$= \frac{51}{10} = 5{\cdot}1$$

(b) Median: arranging numbers in order of size
3, 3, 4, 4, 4, 5, 5, 6, 7, 10
↑
The median is the 'average' of 4 and 5

$\therefore$ median = 4·5.

(c) Mode = 4 (there are more 4's than any other number).

Frequency tables

A frequency table shows a number x such as a mark or a score, against the frequency f or number of times that x occurs.

The next example shows how these symbols are used in calculating the mean, the median and the mode.

The symbol Σ (or sigma) means 'the sum of'.

Example 2

The marks obtained by 100 students in a test were as follows:

mark (x)	0	1	2	3	4
frequency (f)	4	19	25	29	23

Find
(a) the mean mark (b) the median mark
(c) the modal mark

(a) Mean $= \frac{\Sigma xf}{\Sigma f}$,

where Σxf means 'the sum of the products'
i.e. Σ(number × frequency)

and Σf means 'the sum of the frequencies'

$$\text{Mean} = \frac{(0 \times 4) + (1 \times 19) + (2 \times 25) + (3 \times 29) + (4 \times 23)}{100}$$

$$= \frac{248}{100} = 2{\cdot}48$$

(b) The median mark is the number between the 50th and 51st numbers. By inspection, both the 50th and 51st numbers are 3.

$\therefore$ Median = 3 marks.

(c) The modal mark = 3.

Exercise 2

1. Find the mean, median and mode of the following sets of numbers:
(a) 3, 12, 4, 6, 8, 5, 4
(b) 7, 21, 2, 17, 3, 13, 7, 4, 9, 7, 9
(c) 12, 1, 10, 1, 9, 3, 4, 9, 7, 9
(d) 8, 0, 3, 3, 1, 7, 4, 1, 4, 4.

2. Find the mean, median and mode of the following sets of numbers:
(a) 3, 3, 5, 7, 8, 8, 8, 9, 11, 12, 12
(b) 7, 3, 4, 10, 1, 2, 1, 3, 4, 11, 10, 4
(c) −3, 4, 0, 4, −2, −5, 1, 7, 10, 5
(d) $1, \frac{1}{2}, \frac{1}{2}, \frac{3}{4}, \frac{1}{4}, 2, \frac{1}{2}, \frac{1}{4}, \frac{3}{4}$.

3. A group of 50 people were asked how many books they had read in the previous year; the results are shown in the frequency table below. Calculate the mean number of books read per person.

Number of books	0	1	2	3	4	5	6	7	8
Frequency	5	5	6	9	11	7	4	2	1

4. A number of people were asked how many coins they had in their pockets; the results are shown below. Calculate the mean number of coins per person.

Number of coins	0	1	2	3	4	5	6	7
Frequency	3	6	4	7	5	8	5	2

5. The following tables give the distribution of marks obtained by different classes in various tests. For each table, find the mean, median and mode.

(a)

mark	0	1	2	3	4	5	6
frequency	3	5	8	9	5	7	3

(b)

mark	15	16	17	18	19	20
frequency	1	3	7	1	5	3

(c)

mark	0	1	2	3	4	5	6
frequency	10	11	8	15	25	20	11

6. One hundred golfers play a certain hole and their scores are summarised below.

score	2	3	4	5	6	7	8
number of players	2	7	24	31	18	11	7

Find
(a) the mean score
(b) the median score.

7. (a) The mean of 3, 7, 8, 10 and x is 6. Find x.
(b) The mean of 3, 3, 7, 8, 10, x and x is 7. Find x.

8. The mean height of 12 men is 1·70 m, and the mean height of 8 women is 1·60 m. Find
(a) the total height of the 12 men,
(b) the total height of the 8 women,
(c) the mean height of the 20 men and women.

9. The total weight of 6 rugby players is 540 kg and the mean weight of 14 ballet dancers is 40 kg. Find the mean weight of the group of 20 rugby players and ballet dancers.

10. The number of goals scored in a series of football matches was as follows:

number of goals	1	2	3
number of matches	8	8	x

(i) If the mean number of goals is 2·04, find x.
(ii) If the modal number of goals is 3, find the smallest possible value of x.
(iii) If the median number of goals is 2, find the largest possible value of x.

11. In a survey of the number of occupants in a number of cars, the following data resulted.

number of occupants	1	2	3	4
number of cars	7	11	7	x

(i) If the mean number of occupants is $2\frac{1}{3}$, find x.
(ii) If the mode is 2, find the largest possible value of x.
(iii) If the median is 2, find the largest possible value of x.

12. The numbers 3, 5, 7, 8 and N are arranged in ascending order. If the mean of the numbers is equal to the median, find N.

13. The mean of 5 numbers is 11. The numbers are in the ratio 1:2:3:4:5. Find the smallest number.

14. The mean of a set of 7 numbers is 3·6 and the mean of a different set of 18 numbers is 5·1. Calculate the mean of the 25 numbers.

15. The number of illnesses suffered by the members of a herd of cows is as follows:

number of illnesses	0	1	2	3
number of cows	31	11	5	3

(a) Find
(i) the mean number of illnesses,
(ii) the median number of illnesses.
(b) In a different herd of 30 cows, the mean number of illnesses was 0·8. Find the mean number of illnesses for the 80 cows.

16. The median of five consecutive integers is N.
(a) Find the mean of the five numbers.
(b) Find the mean and the median of the squares of the integers.
(c) Find the difference between these values.

17. The marks obtained by the members of a class are summarised in the table.

mark	x	y	z
frequency	a	b	c

Calculate the mean mark in terms of a, b, c, x, y and z.

11.3 CUMULATIVE FREQUENCY

It is possible to record data in groups (class intervals) giving the frequency of occurrence in each group. A cumulative frequency curve (or ogive) shows the *median* at the 50th percentile of the cumulative frequency.
The value at the 25th percentile is known as the *lower quartile*, and that at the 75th percentile as the *upper quartile.*
A measure of the spread or dispersion of the data is given by the *inter-quartile range* where

inter-quartile range = upper quartile − lower quartile.

Example 1

The marks obtained by 80 students in an examination are shown below.

mark	frequency	cumulative frequency	marks represented by cumulative frequency
1–10	3	3	$\leqslant 10$
11–20	5	8	$\leqslant 20$
21–30	5	13	$\leqslant 30$
31–40	9	22	$\leqslant 40$
41–50	11	33	$\leqslant 50$
51–60	15	48	$\leqslant 60$
61–70	14	62	$\leqslant 70$
71–80	8	70	$\leqslant 80$
81–90	6	76	$\leqslant 90$
91–100	4	80	$\leqslant 100$

The table also shows the cumulative frequency.
Plot a cumulative frequency curve and hence estimate
(a) the median (b) the inter-quartile range.

The points on the graph are plotted at the upper limit of each group of marks.

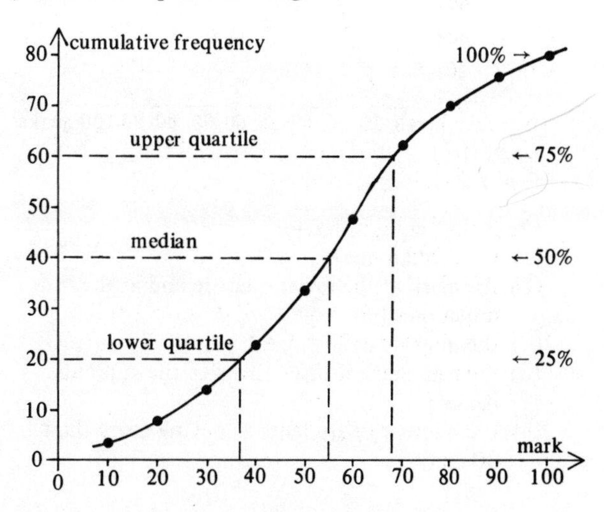

From the cumulative frequency curve

median = 55 marks
lower quartile = 37·5 marks
upper quartile = 68 marks

∴ inter-quartile range = 68 − 37·5
= 30·5 marks.

Exercise 3

1. Figure 1 shows the cumulative frequency curve for the marks of 60 students in an examination.

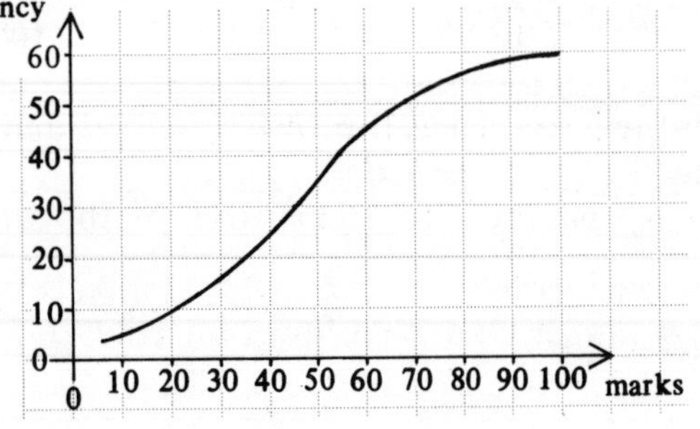

Figure 1

From the graph, estimate
(a) the median mark
(b) the mark at the lower quartile and at the upper quartile
(c) the inter-quartile range
(d) the pass mark if two-thirds of the students passed
(e) the number of students achieving less than 40 marks.

2. Figure 2 shows the cumulative frequency curve for the marks of 140 students in an examination.

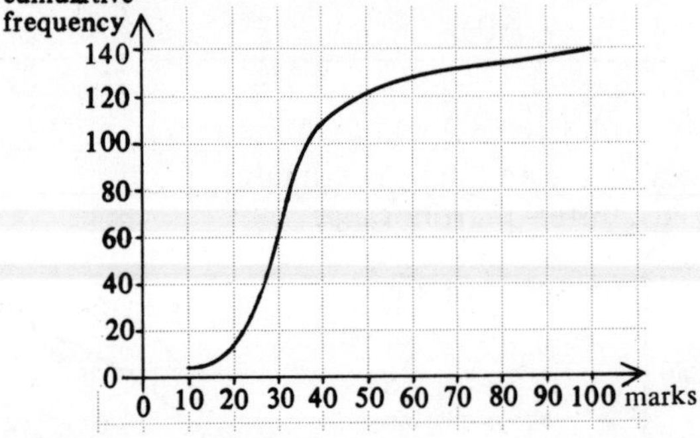

Figure 2

From the graph, estimate
(a) the median mark
(b) the mark at the lower quartile and at the upper quartile
(c) the inter-quartile range
(d) the pass mark if three-fifths of the students passed
(e) the number of students achieving more than 30 marks.

In questions **3** to **6**, draw a cumulative frequency curve, and find
(a) the median, (b) the interquartile range.

3.

mass (kg)	frequency
1-5	4
6-10	7
11-15	11
16-20	18
21-25	22
26-30	10
31-35	5
36-40	3

4.

length (cm)	frequency
41-50	6
51-60	8
61-70	14
71-80	21
81-90	26
91-100	14
101-110	7
111-120	4

5.

time (seconds)	frequency
36-45	3
46-55	7
56-65	10
66-75	18
76-85	12
86-95	6
96-105	4

6.

number of marks	frequency
1-10	0
11-20	2
21-30	4
31-40	10
41-50	17
51-60	11
61-70	3
71-80	3

7. In an experiment, 50 people were asked to guess the weight of a bunch of daffodils in grams. The guesses were as follows:

47 39 21 30 42 35 44 36 19 52
23 32 66 29 5 40 33 11 44 22
27 58 38 37 48 63 23 40 53 24
47 22 44 33 13 59 33 49 57 30
17 45 38 33 25 40 51 56 28 64

Construct a frequency table using intervals 0-9, 10-19, 20-29, etc. Hence draw a cumulative frequency curve and estimate
(a) the median weight
(b) the inter-quartile range
(c) the number of people who guessed a weight within 10 grams of the median.

8. Sketch the shape of the cumulative frequency curve which corresponds to each of the following histograms. Each histogram shows on the vertical axis the number of items (frequency f) in each of the groups or class intervals on the horizontal axis.

(a)

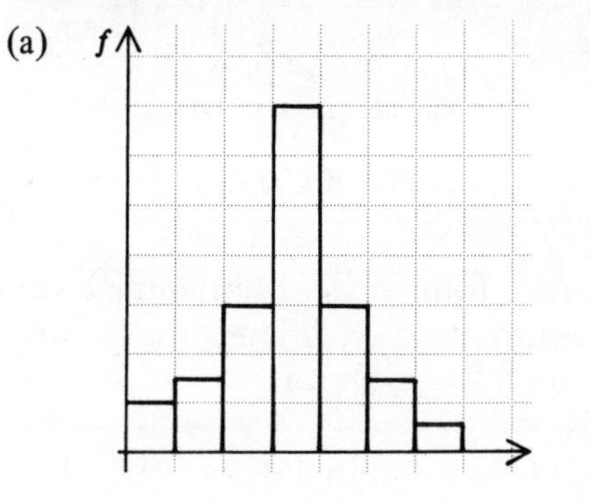

(b)

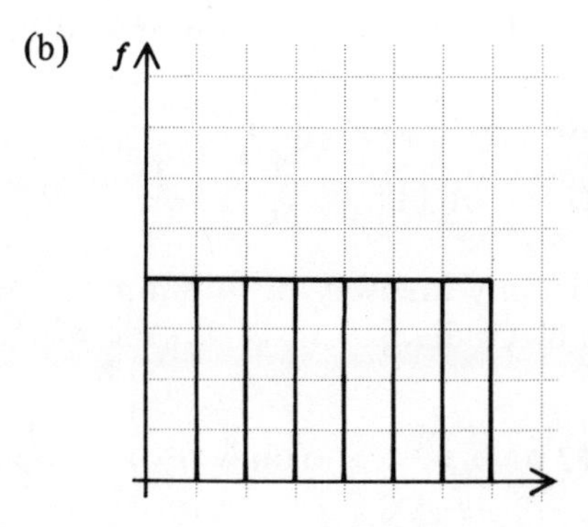

(c)

(d)

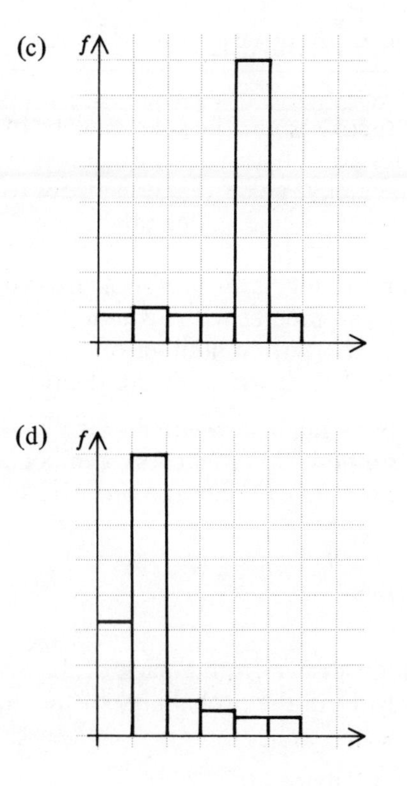

9. In a competition, 30 children had to pick up as many paper clips as possible in one minute using a pair of tweezers. The results were as follows.

3 17 8 11 26 23 18 28 33 38
12 38 22 50 5 35 39 30 31 43
27 34 9 25 39 14 27 16 33 49

Construct a frequency table using intervals 1-10, 11-20, etc. and hence draw a cumulative frequency curve.
(a) From the curve, estimate the median number of clips picked up.
(b) From the frequency table, estimate the mean of the distribution using the mid-interval values 5·5, 15·5, etc.
(c) Calculate the exact value of the mean using the original data.
(d) Why is it possible only to estimate the mean in part (b)?

11.4 SIMPLE PROBABILITY

Probability theory is not the sole concern of people interested in betting, although it is true to say that a 'lucky' poker player is likely to be a player with a sound understanding of probability.
All major airlines regularly overbook aircraft because they can predict with accuracy the probability that a certain number of passengers will fail to arrive for the flight.

Suppose a 'trial' can have n equally likely results and suppose that a 'success' can occur in s ways (from the n). Then the probability of a 'success' $= \frac{s}{n}$.

Example 1

A single card is drawn from a pack of 52 playing cards. Find the probability of the following results:

(a) the card is an ace
(b) the card is the ace of hearts
(c) the card is a spade
(d) the card is a picture card.

We will use the notation 'p (an ace)' to represent 'the probability of selecting an ace'.

(a) p (an ace) $= \frac{4}{52} = \frac{1}{13}$.

(b) p (ace of hearts) $= \frac{1}{52}$.

(c) p (a spade) $= \frac{13}{52} = \frac{1}{4}$.

(d) p (a picture card) $= \frac{12}{52} = \frac{3}{13}$.

In each case, we have counted the number of ways in which a 'success' can occur and divided by the number of possible results of a 'trial'.

Example 2

A black die and a white die are thrown at the same time. Display all the possible outcomes. Find the probability of obtaining:
(a) a total of 5,
(b) a total of 11,
(c) a 'two' on the black die and a 'six' on the white die.

It is convenient to display all the possible outcomes on a grid.

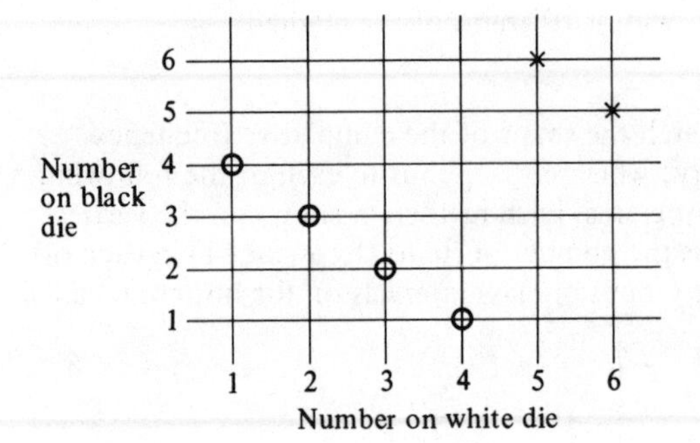

There are 36 possible outcomes, shown where the lines cross.

(a) There are four ways of obtaining a total of 5 on the two dice. They are shown circled on the diagram.

$\therefore$ Probability of obtaining a total of 5 $= \frac{4}{36}$

(b) There are two ways of obtaining a total of 11. They are shown with a cross on the diagram.

$\therefore \quad p$ (total of 11) $= \frac{2}{36} = \frac{1}{18}$

(c) There is only one way of obtaining a 'two' on the black die and a 'six' on the white die.

$\therefore \quad p$ (2 on black and 6 on white) $= \frac{1}{36}$.

Exercise 4

1. One card is drawn at random from a pack of 52 playing cards. Find the probability of drawing
   (a) a 'King',
   (b) a red card,
   (c) the seven of clubs,
   (d) either the King, Queen or Jack of diamonds.
2. A fair die is thrown once. Find the probability of obtaining
   (a) a six,
   (b) an even number,
   (c) a number greater than 3,
   (d) a three or a five.
3. A 10p and a 5p coin are tossed at the same time. List all the possible outcomes. Find the probability of obtaining
   (a) two heads, (b) a head and a tail.
4. A bag contains 6 red balls and 4 green balls.
   (a) Find the probability of selecting at random:
       (i) a red ball (ii) a green ball.
   (b) One red ball is removed from the bag. Find the new probability of selecting at random
       (i) a red ball (ii) a green ball.
5. A 'hand' of 13 cards contains the cards shown.

   A card is selected at random from the 13. Find the probability of selecting:
   (a) any card of the heart suit,
   (b) any card of the club suit,
   (c) a 'six' of any suit,
   (d) any 'picture' card [not including an ace],
   (e) the 'four' of clubs,
   (f) an 'eight' of any suit,
   (g) any 'six' or 'four'.
6. Three coins are tossed at the same time. List all the possible outcomes. Find the probability of obtaining
   (a) three heads,
   (b) two heads and one tail,
   (c) no heads,
   (d) at least one head.
7. A bag contains 10 red balls, 5 blue balls and 7 green balls. Find the probability of selecting at random:
   (a) a red ball,
   (b) a green ball,
   (c) a blue *or* a red ball,
   (d) a red *or* a green ball.
8. Cards with the numbers 2 to 101 are placed in a hat. Find the probability of selecting:
   (a) an even number,
   (b) a number less than 14,
   (c) a square number,
   (d) a prime number less than 40,
   (e) a prime number greater than 90.
9. A red die and a blue die are thrown at the same time. List all the possible outcomes in a systematic way. Find the probability of obtaining:
   (a) a total of 10,
   (b) a total of 12,
   (c) a total less than 6,
   (d) the same number of both dice,
   (e) a total more than 9.
   What is the most likely total?
10. A die is thrown; when the result has been recorded, the die is thrown a second time. Display all the possible outcomes of the two throws. Find the probability of obtaining:
    (a) a total of 4 from the two throws,
    (b) a total of 8 from the two throws,
    (c) a total between 5 and 9 inclusive from the two throws,
    (d) a number on the second throw which is double the number on the first throw,
    (e) a number on the second throw which is four times the number on the first throw.
11. Find the probability of the following:
    (a) throwing a number less than 8 on a single die,
    (b) obtaining the same number of heads and tails when five coins are tossed,
    (c) selecting a square number from the set $A = \{4, 9, 16, 25, 36, 49\}$,
    (d) selecting a prime number from the set A.
12. Four coins are tossed at the same time. List all the possible outcomes in a systematic way. Find the probability of obtaining:
    (a) two heads and two tails,
    (b) four tails,
    (c) at least one tail,
    (d) three heads and one tail.

13. A man glances at his digital watch watch and it reads 09.41.
Some time later he looks again.
Find the probability that:
(a) the number in column A is a '3',
(b) the number in column A is less than 6,
(c) the number in column B is a '3',
(d) the number in column B is a '7'.

14. A digital watch reads 17.36.
A man glances at the watch at some other time. Find the probability that:

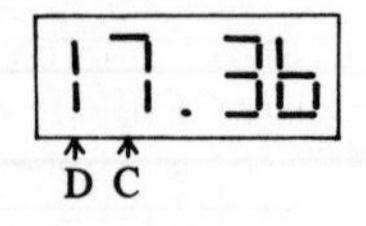

(a) the number in column C is a '7',
(b) the number in column C is a '1',
(c) the number in column D is less than 5,
(d) the number in column D is a '4'.

15. Two dice and two coins are thrown at the same time. Find the probability of obtaining:
(a) two heads and a total of 12 on the dice,
(b) a head, a tail and a total of 9 on the dice,
(c) two tails and a total of 3 on the dice.
What is the most likely outcome?

16. A red, a blue and a white die are all thrown at the same time. Display all the possible outcomes in a suitable way. Find the probability of obtaining:
(a) a total of 18 on the three dice,
(b) a total of 4 on the three dice,
(c) a total of 10 on the three dice,
(d) a total of 15 on the three dice,
(e) a total of 7 on the three dice,
(f) the same number on each dice.

11.5 EXCLUSIVE AND INDEPENDENT EVENTS

Two events are *exclusive* if they cannot occur at the same time:
e.g. Selecting an 'ace' or selecting a 'ten' from a pack of cards.
For exclusive events A and B

$$p(\text{A or B}) = p(\text{A}) + p(\text{B})$$

Two events are *independent* if the occurence of one event is unaffected by the occurrence of the other.
e.g. Obtaining a 'head' on one coin, and a tail on another coin when the coins are tossed at the same time

$$p(\text{A } \textit{and} \text{ B}) = p(\text{A}) \,.\, p(\text{B})$$

where $p(\text{A})$ = probability of A occurring etc.

This is the multiplication law.

Example 1

A ball is selected at random from a bag containing 4 red balls and 6 blue balls.
The first ball is replaced and a second ball is selected. Find the probability of selecting
(a) two red balls,
(b) one red ball and one blue ball.

(a) The selection of a red ball first is independent of the selection of a red ball second.

$\therefore$ p(red ball first *and* red ball second)
$= p(\text{red ball first}) \times p(\text{red ball second})$

$$\therefore \quad p(\text{2 red balls}) = \frac{4}{10} \times \frac{4}{10} = \frac{16}{100}$$

(b) There are two ways of choosing a red and a blue ball, either 'red then blue' or 'blue then red'.

$$\text{(i) } p(\text{red then blue}) = \frac{4}{10} \times \frac{6}{10} = \frac{24}{100}$$

$$\text{(ii) } p(\text{blue then red}) = \frac{6}{10} \times \frac{4}{10} = \frac{24}{100}$$

$$\therefore p(\text{one red and one blue}) = \frac{24}{100} + \frac{24}{100} = \frac{48}{100}$$

(adding because (i) and (ii) are exclusive).

Exercise 5

1. A card is drawn from a pack of playing cards and a die is thrown. Events A and B are as follows:
   A : 'a Jack is drawn from the pack'
   B : 'a three is thrown on the die'.
   (a) Is A independent or exclusive of B?
   (b) Write down the values of $p(\text{A}), p(\text{B})$.
   (c) Write down the value of $p(\text{A and B})$.
2. A coin is tossed and a die is thrown.
   (a) Is the result of the coin toss independent or exclusive of the throw of the die?
   (b) Write down the probability of obtaining
      (i) a 'head' on the coin,
      (ii) an odd number on the die,
      (iii) a 'head' on the coin and an odd number on the die.
3. A card is drawn from a pack of playing cards. Events X and Y are as follows:
   X : 'a King is drawn'
   Y : 'a Queen is drawn'.
   (a) Is X independent or exclusive of Y?
   (b) Write down the values of
      (i) $p(\text{X})$ (ii) $p(\text{Y})$ (iii) $p(\text{X or Y})$.
4. In an experiment, a card is drawn from a pack of playing cards and a die is thrown. Find the probability of obtaining:
   (a) A card which is an ace and a six on the die,
   (b) the king of clubs and an even number on the die,
   (c) a heart and a 'one' on the die.
5. A card is taken at random from a pack of playing cards and replaced. After shuffling, a second card is selected. Find the probability of obtaining:
   (a) two cards which are clubs,
   (b) two Kings,
   (c) two picture cards,
   (d) a heart and a club in any order,
   (e) the ace of diamonds and the ace of spades in any order.
6. A ball is selected at random from a bag containing 3 red balls, 4 black balls and 5 green balls. The first ball is replaced and a second is selected. Find the probability of obtaining:
   (a) two red balls,
   (b) two green balls,
   (c) one black and one green,
   (d) two of the same colour.
7. The letters of the word 'INDEPENDENT' are written on individual cards and the cards are put into a box. A card is selected and then replaced and then a second card is selected. Find the probability of obtaining:
   (a) the letter 'P' twice,
   (b) the letter 'E' twice,
   (c) an 'E' and an 'N' in any order.
8. Three coins are tossed and two dice are thrown at the same time. Find the probability of obtaining:
   (a) three heads and a total of 12 on the dice,
   (b) three tails and a total of 9 on the dice,
   (c) two heads, one tail and a total of 10 on the dice.
9. When a golfer plays any hole, he will take 3, 4, 5, 6, or 7 strokes with probabilities of $\frac{1}{10}, \frac{1}{5}, \frac{2}{5}, \frac{1}{5}$ and $\frac{1}{10}$ respectively. He never takes more than 7 strokes. Find the probability of the following events:
   (a) scoring 4 on each of the first three holes,
   (b) scoring 3, 4 and 5 (in that order) on the first three holes,
   (c) scoring a total of 28 for the first four holes,
   (d) scoring a total of 10 for the first three holes,
   (e) scoring a total of 20 for the first three holes.
10. A bag contains 3 red balls and 1 white ball. One ball is selected from the bag and then replaced. A second ball is selected. Events A, B, C and D are as follows:
   A : 'a red ball is selected on the first draw'
   B : 'a white ball is selected on the first draw'
   C : 'a red ball is selected on the second draw'
   D : 'a white ball is selected on the second draw'.
   (a) Is A independent or exclusive of B?
   (b) Is A independent or exclusive of C?
   (c) Is C independent or exclusive of D?
   (d) Is A independent or exclusive of D?
   (e) Is B independent or exclusive of C?
   (f) Write down the values of $p(\text{A}), p(\text{B}), p(\text{C})$, $p(\text{D})$.
   (g) Write down the values of
      (i) $p(\text{A and C})$ (ii) $p(\text{B and D})$
      (iii) $p(\text{A and D})$ (iv) $p(\text{B and C})$
      (v) $p(\text{A or B})$ (vi) $p(\text{C or D})$
   (h) Is the 'double' event (A and D) independent or exclusive of the 'double' event (B and C)?
   (i) What is the probability of obtaining one ball of each colour in any order?

11. A bag contains 2 blue discs and 3 white discs. One disc is selected from the bag and then replaced. A second disc is selected.
Events P, Q, R and S are as follows:
P : 'a blue disc is selected on the first draw'
Q : 'a white disc is selected on the first draw'
R : 'a blue disc is selected on the second draw'
S : 'a white disc is selected on the second draw'.
(a) Write down the values of $p(\text{P}), p(\text{Q}), p(\text{R}), p(\text{S})$.
(b) Write down the values of
(i) $p(\text{Q and S})$ (ii) $p(\text{P and R})$
(iii) $p(\text{P and S})$ (iv) $p(\text{Q and R})$
(c) What is the probability of obtaining one disc of each colour?
(d) What is the probability of obtaining either two blue discs or two white discs?

12. If a hedgehog crosses a certain road before 7.00 a.m., the probability of being run over is $\frac{1}{10}$. After 7.00 a.m., the corresponding probability is $\frac{3}{4}$. The probability of the hedgehog waking up early enough to cross before 7.00 a.m., is $\frac{4}{5}$.
What is the probability of the following events:
(a) the hedgehog waking up too late to reach the road before 7.00 a.m.,
(b) the hedgehog waking up early and crossing the road in safety,
(c) the hedgehog waking up late and crossing the road in safety,
(d) the hedgehog waking up early and being run over,
(e) the hedgehog crossing the road in safety.

13. A coin is biased so that it shows 'Heads' with a probability of $\frac{2}{3}$. The same coin is tossed three times. Find the probability of obtaining:
(a) two tails on the first two tosses,
(b) a head, a tail and a head (in that order),
(c) two heads and one tail (in any order).

11.6 TREE DIAGRAMS

Example 1

A bag contains 5 red balls and 3 green balls. A ball is drawn at random and then replaced. Another ball is drawn.

What is the probability that both balls are green?

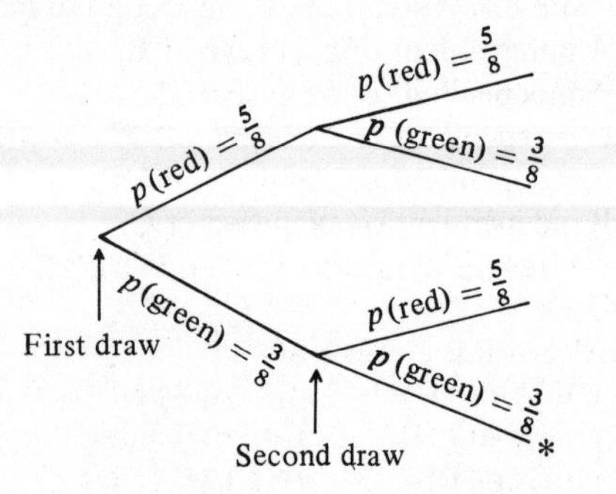

The branch marked * involves the selection of a green ball twice.
The probability of this event is obtained by simply multiplying the fractions on the two branches.

$\therefore \quad p(\text{two green balls}) = \frac{3}{8} \times \frac{3}{8} = \frac{9}{64}$

Example 2

A bag contains 5 red balls and 3 green balls. A ball is selected at random and not replaced. A second ball is then selected.
Find the probability of selecting
(a) two green balls
(b) one red ball and one green ball.

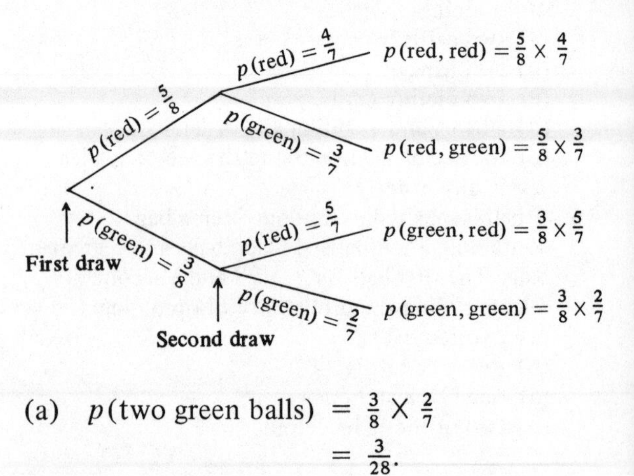

(a) $p(\text{two green balls}) = \frac{3}{8} \times \frac{2}{7} = \frac{3}{28}$.

(b) $p(\text{one red, one green}) = (\frac{5}{8} \times \frac{3}{7}) + (\frac{3}{8} \times \frac{5}{7}) = \frac{15}{28}$.

Exercise 6

1. A bag contains 10 discs, 7 are black and 3 white. A disc is selected, and then replaced. A second disc is selected. Draw a tree diagram to show all the possible outcomes. Find the probability of the following:
   (a) both discs are black,
   (b) both discs are white,
   (c) the two discs are different colours.
2. A bag contains 10 discs, 7 are green and 3 blue. A disc is selected and *not* replaced. A second disc is selected. Draw a tree diagram to show all the possible outcomes. Find the probability of the following:
   (a) both discs are green,
   (b) both discs are blue,
   (c) the two discs are different colours.
3. A bag contains 5 red balls, 3 blue balls and 2 yellow balls. A ball is drawn and not replaced. A second ball is drawn. Find the probability of drawing:
   (a) two red balls,
   (b) one blue ball and one yellow ball,
   (c) two yellow balls,
   (d) two balls of the same colour.
4. A bag contains 4 red balls, 2 green balls and 3 blue balls. A ball is drawn and not replaced. A second ball is drawn. Find the probability of drawing:
   (a) two blue balls,
   (b) two red balls,
   (c) one red ball and one blue ball,
   (d) one green ball and one red ball.
5. A six-sided die is thrown three times. Draw a tree diagram, showing at each branch the two events: 'six' and 'not six'. What is the probability of throwing a total of:
   (a) three sixes, (b) no sixes,
   (c) one six,
   (d) at least one six (use part (b)).
6. A card is drawn at random from a pack of 52 playing cards. The card is replaced and a second card is drawn. This card is replaced and a third card is drawn. What is the probability of drawing:
   (a) three hearts?
   (b) at least two hearts?
   (c) exactly one heart?
7. A bag contains 6 red marbles and 4 blue marbles. A marble is drawn at random and not replaced. Two further draws are made, again without replacement. Find the probability of drawing:
   (a) three red marbles, (b) three blue marbles,
   (c) no red marbles, (d) at least one red marble.
8. A coin is biased so that the probability of a 'head' is $\frac{3}{4}$. Find the probability that, when tossed three times, it shows:
   (a) three tails,
   (b) two heads and one tail,
   (c) one head and two tails,
   (d) no tails.
   Write down the sum of the probabilities in (a), (b), (c) and (d).
9. A teacher decides to award exam grades A, B or C by a new fairer method. Out of 20 children, three are to receive A's, five B's and the rest C's. He writes the letters A, B and C on 20 pieces of paper and invites the pupils to draw their exam result, going through the class in alphabetical order. Find the probability that:
   (a) the first three pupils all get grade 'A',
   (b) the first three pupils all get grade 'B',
   (c) the first three pupils all get different grades,
   (d) the first four pupils all get grade B.
   (Do not cancel down the fractions.)
10. The probability that an amateur golfer actually hits the ball is (regrettably for all concerned) only $\frac{1}{10}$. If four separate attempts are made, find the probability that the ball will be hit:
    (a) four times, (b) at least twice,
    (c) not at all.
11. A box contains x milk chocolates and y plain chocolates. Two chocolates are selected at random. Find, in terms of x and y, the probability of choosing:
    (a) a milk chocolate on the first choice,
    (b) two milk chocolates,
    (c) one of each sort,
    (d) two plain chocolates.
12. A pack of z cards contains x 'winning cards'. Two cards are selected at random. Find, in terms of x and z, the probability of choosing:
    (a) a 'winning' card on the first choice,
    (b) two 'winning cards' in the two selections,
    (c) exactly one 'winning' card in the pair.
13. Bag A contains 3 red balls and 3 blue balls. Bag B contains 1 red ball and 3 blue balls. A ball is taken at random from bag A and placed in bag B. A ball is then chosen from bag B. What is the probability that the ball taken from B is red?
14. On a Monday or a Thursday, Mr Gibson paints a 'masterpiece' with a probability of $\frac{1}{5}$. On any other day, the probability of producing a 'masterpiece' is $\frac{1}{100}$. In common with other great painters, Gibson never knows what day it is. Find the probability that on one day chosen at random, he will in fact paint a masterpiece.

15. In the Venn diagram, & = {pupils in a class of 15}, G = {girls}, S = {swimmers}, F = {pupils who believe in Father Christmas}.

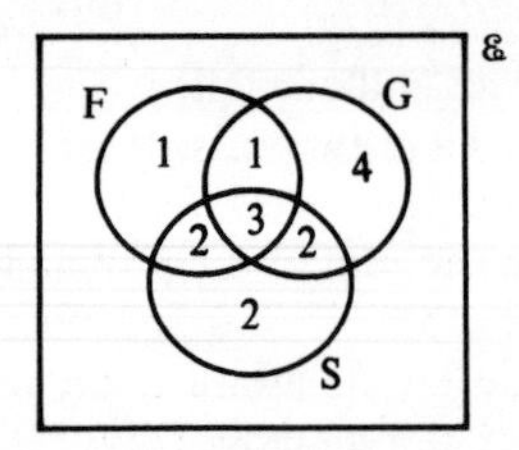

A pupil is chosen at random. Find the probability that the pupil:
(a) can swim, (b) is a girl swimmer,
(c) is a boy swimmer who believes in Father Christmas.
Two pupils are chosen at random. Find the probability that:
(d) both are boys, (e) neither can swim,
(f) both are girl swimmers who believe in Father Christmas.

16. A bag contains 3 red, 4 white and 5 green balls. Three balls are selected without replacement. Find the probability that the three balls chosen are:
(a) all red, (b) all green,
(c) one of each colour.
If the selection of the three balls was carried out 1100 times, how often would you expect to choose:
(d) three red balls? (e) one of each colour?

17. There are 1000 components in a box of which 10 are known to be defective. Two components are selected at random. What is the probability that:
(a) both are defective,
(b) neither are defective,
(c) just one is defective?
(Do *not* simplify your answers.)

18. There are 10 boys and 15 girls in a class. Two children are chosen at random. What is the probability that:
(a) both are boys (b) both are girls,
(c) one is a boy and one is a girl?

19. There are 500 ball bearings in a box of which 100 are known to be undersize. Three ball bearings are selected at random. What is the probability that:
(a) all three are undersize,
(b) none are undersize?
Give your answers as decimals correct to three significant figures.

20. There are 9 boys and 15 girls in a class. Three children are chosen at random. What is the probability that:
(a) all three are boys,
(b) all three are girls,
(c) one is a boy and two are girls?
Give your answers as fractions.

REVISION EXERCISE 11A

1. A pie chart is drawn with sectors to represent the following percentages:

   20%, 45%, 30%, 5%.

   What is the angle of the sector which represents 45%?

2. The pie chart shows the numbers of votes for candidates A, B and C in an election.
   What percentage of the votes were cast in favour of candidate C?

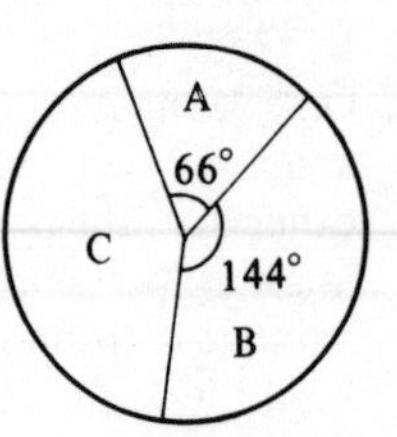

3. A pie chart is drawn showing the expenditure of a football club as follows:

Wages	£41,000
Travel	£9,000
Rates	£6,000
Miscellaneous	£4,000

   What is the angle of the sector showing the expenditure on travel?

4. The mean of four numbers is 21.
   (a) Calculate the sum of the four numbers,
   Six other numbers have a mean of 18.
   (b) Calculate the mean of the ten numbers.

5. Find
(a) the mean, (b) the median,
(c) the mode,
of the numbers 3, 1, 5, 4, 3, 8, 2, 3, 4, 1.

6.

Marks	3	4	5	6	7	8
Number of pupils	2	3	6	4	3	2

The table shows the number of pupils in a class who scored marks 3 to 8 in a test. Find
(a) the mean mark, (b) the modal mark,
(c) the median mark.

7. The mean height of 10 boys is 1·60 m and the mean height of 15 girls is 1·52 m. Find the mean height of the 25 boys and girls.

8.

Mark	3	4	5
Number of pupils	3	x	4

The table shows the number of pupils who scored marks 3, 4 or 5 in a test. Given that the mean mark is 4·1, find x.

9. When two dice are thrown simultaneously, what is the probability of obtaining the same number on both dice?

10. A bag contains 20 discs of equal size of which 12 are red, x are blue and the rest are white.
(a) If the probability of selecting a blue disc is $\frac{1}{4}$, find x.
(b) A disc is drawn and then replaced. A second disc is drawn. Find the probability that neither disc is red.

11. Three dice are thrown. What is the probability that none of them shows a 1 or a 6?

12. A coin is tossed four times. What is the probability of obtaining at least three 'heads'?

13. A bag contains 8 balls of which 2 are red and 6 are white. A ball is selected and not replaced. A second ball is selected. Find the probability of obtaining:
(a) two red balls, (b) two white balls,
(c) one ball of each colour.

14. A bag contains x green discs and 5 blue discs. A disc is selected and replaced. A second disc is drawn. Find, in terms of x, the probability of selecting:
(a) a green disc on the first draw,
(b) a green disc on the first and second draws.
Find the corresponding probabilities when the first disc is *not* replaced.

15. In a group of 20 people, 5 cannot swim. If two people are selected at random, what is the probability that neither of them can swim?

16. (a) What is the probability of winning the toss in five consecutive test matches?
(b) What is the probability of winning the toss in all the matches in the FA cup from the first round to the final (i.e. 8 matches)?

17. Mr and Mrs Stringer have three children. What is the probability that:
(a) all the children are boys,
(b) there are more girls than boys?
(Assume that a boy is as likely as a girl.)

18. The probability that it will be wet today is $\frac{1}{6}$. If it is dry today, the probability that it will be wet tomorrow is $\frac{1}{8}$. What is the probability that both today and tomorrow will be dry?

19. Two dice are thrown. What is the probability that the *product* of the numbers on top is
(a) 12, (b) 4, (c) 11?

20. The probability of snow on Christmas day is $\frac{1}{20}$. What is the probability that snow will fall on the next three Christmas days?

EXAMINATION EXERCISE 11B

1. A bag contains eight red balls and two black balls, which are otherwise identical.
(a) A ball is drawn from the bag and then replaced before a ball is drawn again. Find, as a fraction, the probability that
(i) the first ball drawn is black,
(ii) the first ball drawn is black and the second is red.
(b) If the balls are drawn from the bag one at a time but are not replaced, find the probability, as a fraction, that the first ball drawn is black and the second is red.
[C 16+]

2. In last year's mathematics examination, there were 20 000 candidates who were awarded grades as follows:

Grade A	2 780	Grade D	1 600
Grade B	4 360	Grade E	1 380
Grade C	6 320	Unclassified	3 560

(a) On graph paper, draw a bar chart to represent this information. Use 2 cm to represent 1 000 candidates, and 1 cm to represent each grade.
(b) If the information were presented on a pie chart, calculate the angle which would be used for the sector representing Grade C, giving your answer to the nearest degree.
(c) Find, as a decimal correct to three decimal places, the probability that a candidate picked at random will have
(i) grade C, (ii) grade A, B or C.
(d) If two candidates are to be picked at random, write down, but do not simplify, an expression for the probability that they will both have grade C. [L]

3. 36 girls were given a test in which the maximum mark available was 100. The table below shows the cumulative frequency of the results obtained.

Mark	10	20	30	40	50	60	70	80	90	100
Number of girls scoring this mark or less	1	4	8	16	24	29	32	34	35	36

(i) Calculate how many girls scored a mark between 61 and 70 inclusive.
(ii) Using a vertical scale of 2 cm to represent 5 girls and a horizontal scale of 1 cm to represent 10 marks, plot these values on graph paper and draw a smooth curve through your points.
(iii) Showing your method clearly, use your graph to estimate the median mark.
(iv) State the probability that a girl chosen at random will have a mark
(a) less than or equal to 50, (b) greater than 70.
(v) A second group of girls was tested and a quarter of them scored more than 70 marks. If one girl is now chosen at random from each group, find the probability that
(a) both will have scored more than 70,
(b) just one will have scored more than 70. [C]

4. (a) A survey was taken of the number of cars passing a road junction during 35 equal intervals of time. The results were recorded in the following table. For example, 4 cars passed the junction during each of 6 intervals.

Number of cars	0	1	2	3	4	5	6
Frequency	8	7	4	4	6	3	3

Find, for this distribution,
(i) the mode, (ii) the median, (iii) the mean.

(b) From a group of five children, consisting of three girls and two boys, one child is chosen at random. Write down the probability that the child chosen is a girl.
A second child is then chosen at random from the remaining four children. Given that the first child chosen is a girl, write down the probability that the second child chosen is also a girl.
On another occasion, two children are chosen at random from this same group of three girls and two boys. Calculate the probability that
(i) both are girls, (ii) both are boys, (iii) they are of different sexes. [C]

5. *In this question all probabilities should be expressed as fractions.*
   Each member of a class of 30 boys supports one and only one of three football teams; 13 boys support City, 10 support Rovers and 7 support United.
   (a) If a boy is to be chosen at random, what is the probability that he will support City?
   (b) If two boys are to be chosen at random, what is the probability that they will both support City?
   Draw a tree diagram to show all the possible outcomes of choosing two boys at random, showing the probability of each outcome.
   Hence find
   (c) the probability that the two boys chosen will support the same team,
   (d) the probability that the two boys chosen will support different teams.
   [L]

6. The probabilities that each of three marksmen, Tom, Dick and Harry, will hit a target at a single attempt are $\frac{1}{3}, \frac{1}{4}, \frac{1}{5}$ respectively.
   (i) If they all fire simultaneously at the target, find the probabilities that:
   (a) Harry misses it, (b) all three men hit it,
   (c) all three men miss it, (d) at least one man hits it.
   (ii) They now each prepare to fire once at the target in the order Tom, Dick and Harry. In this case once the target has been hit no more shots are fired. Copy and complete the tree diagram by inserting all the outcomes and one-stage probabilities in the correct places on it.

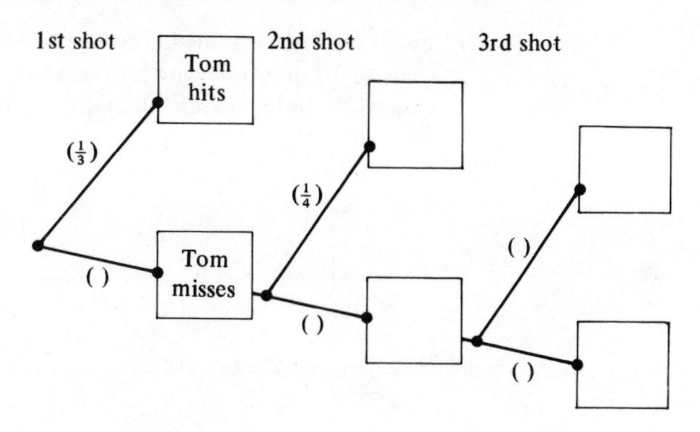

   Find the probabilities that
   (a) the target is hit by Dick, (b) the target is hit by either Tom or Dick,
   (c) the target is hit by Harry, (d) the target is hit. [SMP]

7. Six blank cards are taken and one is labelled '1', two are labelled '3', and three are labelled '5'.
   (i) From the six cards one is drawn at random. What is the probability that the number on the card is
   (a) 1, (b) 5?
   (ii) The first card is replaced and a card is again drawn at random. What is the probability that the sum of the numbers on the two cards is
   (a) 4, (b) 6?
   (iii) If, instead, the first card is not replaced and a second card is drawn at random, what is the probability in this case that the sum of the numbers on the two cards is 6?
   [In (ii) and (iii) possibility-space diagrams or tree diagrams may be helpful.]
   [SMP]

8. A group of 50 boys each threw a dart at a dartboard and each measured the distance in centimetres (to the nearest millimetre) of the tip of his dart from the centre of the board. They recorded the following results:

0·5	6·5	8·4	3·6	2·8	6·6	7·3	9·2	8·1	4·9
5·1	4·1	8·0	6·1	9·2	4·4	5·9	7·9	5·8	6·2
12·1	8·3	4·9	1·1	8·4	10·2	8·1	5·7	11·2	4·7
6·1	5·1	9·0	4·8	4·2	5·1	9·1	5·1	8·9	7·8
7·2	8·3	4·3	6·3	7·5	6·4	7·8	6·1	7·1	6·5

(i) Make a grouped frequency distribution of these results using classes of 0·0–, 2·0–, 4·0–, 6·0–, 8·0–, 10·0–, 12·0–13·9.
(ii) Draw a cumulative frequency polygon to illustrate the grouped frequency distribution.
(iii) By using your diagram, or otherwise, estimate the median and the quartiles.
(iv) Estimate the percentage of boys whose throws were less than 6·5 cm from the centre. [JMB]

9. Two normal unbiased dice are thrown. Find the probability that
(i) the total score is 7, (ii) the total score is 7 or 11.
In a game, if a player gets a total score of 7 or 11 with one throw of the pair of dice it is called a win. If a player throws the pair of dice twice calculate the probability that
(iii) he wins exactly once, (iv) he wins at least once. [AEB]

10. From a starting point O on an open plain, a man walks a sequence of stages, each of 1 km either North or East. The direction of each stage is decided by spinning a coin: 'heads' due North, 'tails' due East.

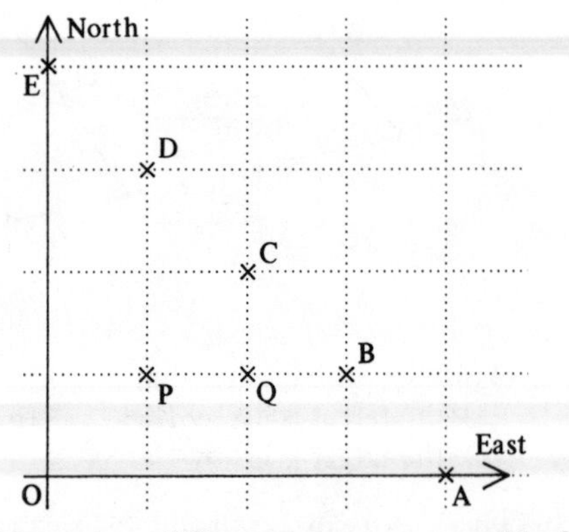

Scale: grid squares are 1 km by 1 km.

(i) How many different routes are possible from O to
(a) P, (b) Q, (c) C?
(ii) $p(\text{A})$ denotes the probability that, after 4 stages, the man will be at A, and so on. Calculate $p(\text{A}), p(\text{B}), p(\text{C}), p(\text{D})$ and $p(\text{E})$.
(iii) Calculate how far each point A, B, C, D and E is from O in a straight line.
(iv) What is the probability that the man ends up more than 3 km from O after 4 stages?
(v) When 20 men each walked 4 stages it was noticed that 2 ended at A, 5 at B, 5 at C, 4 at D and 4 at E. Find the average distance of the 20 men from O. [SMP]

Answers

PART 1

Exercise 1 *page 1*

1. 7·91 **2.** 0·214 **3.** 22·22 **4.** 112·02 **5.** 7·372 **6.** 165·1
7. 0·066 **8.** 25·84 **9.** 466·2 **10.** 10·668 **11.** 1·22 **12.** 3·18
13. 1·67 **14.** 0·008 **15.** 1·61 **16.** 2·8 **17.** 16·63 **18.** 199·24
19. 16·859 **20.** 0·856 **21.** 24·1 **22.** 25·74 **23.** 26·7 **24.** 1·296
25. 3·86 **26.** 30·647 **27.** 21·76 **28.** 0·006223 **29.** 0·001 **30.** 6 015 000
31. 1·56 **32.** 1·684 **33.** 0·0288 **34.** 6·48 **35.** 2·176 **36.** 0·0042
37. 0·02 **38.** 0·04 **39.** 0·0001 **40.** 0·027 **41.** 7·56 **42.** 1·8
43. 0·7854 **44.** 0·10098 **45.** 63·423 **46.** 7200 **47.** 360 **48.** 0·05184
49. 34000 **50.** 5·8081 **51.** 18 **52.** 5·3 **53.** 0·74 **54.** 111
55. 2·34 **56.** 44·734 **57.** 1620 **58.** 3·56 **59.** 8·8 **60.** 0·36
61. 2·6 **62.** 0·06975 **63.** 2·4 **64.** 600 **65.** 1200 **66.** 10
67. 0·00175 **68.** 37·5 **69.** 13·2 **70.** 0·944 **71.** 200 **72.** 290
73. 0·804 **74.** 0·1 **75.** 0·8 **76.** 270 000 **77.** 12·83 **78.** 0·1027
79. 0·077 **80.** 0·0024 **81.** 82 800 **82.** 0·089 **83.** 0·0009 **84.** 4
85. 0·0008 **86.** 8 **87.** 152·4 **88.** 18 **89.** 0·01 **90.** 10 million
91. 184 **92.** 3·3 **93.** 20 **94.** 2·2 **95.** 0·099 **96.** $66\frac{2}{3}$
97. 3·3 **98.** 2·7 **99.** 3 **100.** 2·15

Exercise 2 *page 2*

1. $1\frac{11}{20}$ **2.** $\frac{11}{24}$ **3.** $1\frac{1}{2}$ **4.** $\frac{5}{12}$ **5.** $\frac{4}{15}$ **6.** $\frac{1}{10}$
7. $\frac{8}{15}$ **8.** $\frac{5}{42}$ **9.** $\frac{15}{26}$ **10.** $\frac{5}{12}$ **11.** $4\frac{1}{2}$ **12.** $1\frac{2}{3}$
13. $1\frac{5}{12}$ **14.** $\frac{1}{12}$ **15.** $\frac{1}{2}$ **16.** $1\frac{1}{8}$ **17.** $\frac{55}{56}$ **18.** $\frac{41}{56}$
19. $\frac{3}{28}$ **20.** $6\frac{6}{7}$ **21.** $\frac{23}{40}$ **22.** $\frac{7}{40}$ **23.** $\frac{3}{40}$ **24.** $1\frac{7}{8}$
25. $2\frac{5}{12}$ **26.** $1\frac{1}{12}$ **27.** $1\frac{1}{6}$ **28.** $2\frac{5}{8}$ **29.** $3\frac{13}{15}$ **30.** $1\frac{7}{15}$
31. $3\frac{1}{5}$ **32.** $2\frac{2}{9}$ **33.** $6\frac{1}{10}$ **34.** $\frac{9}{10}$ **35.** $9\frac{1}{10}$ **36.** $1\frac{9}{26}$
37. $\frac{1}{9}$ **38.** $\frac{2}{3}$ **39.** $1\frac{1}{3}$ **40.** $5\frac{1}{4}$ **41.** 2 **42.** $6\frac{5}{29}$
43. $\frac{18}{25}$ **44.** $10\frac{2}{5}$ **45.** $1\frac{51}{70}$ **46.** 1
47. (a) $\frac{1}{2}, \frac{7}{12}, \frac{2}{3}$ (b) $\frac{2}{3}, \frac{3}{4}, \frac{5}{6}$ (c) $\frac{1}{3}, \frac{5}{8}, \frac{17}{24}, \frac{3}{4}$ (d) $\frac{5}{6}, \frac{8}{9}, \frac{11}{12}$
48. (a) $\frac{1}{2}$ (b) $\frac{3}{4}$ (c) $\frac{17}{24}$ (d) $\frac{7}{18}$ (e) $\frac{3}{10}$ (f) $\frac{5}{12}$

Exercise 3 *page 3*

1. 0·25 **2.** 0·4 **3.** 0·8 **4.** 0·75 **5.** 0·5 **6.** 0·375
7. 0·9 **8.** 0·625 **9.** $0{\cdot}41\dot{6}$ **10.** $0{\cdot}1\dot{6}$ **11.** $0{\cdot}\dot{6}$ **12.** $0{\cdot}8\dot{3}$
13. $0{\cdot}\dot{2}8571\dot{4}$ **14.** $0{\cdot}\dot{4}2857\dot{1}$ **15.** $0{\cdot}\dot{4}$ **16.** $0{\cdot}\dot{4}\dot{5}$ **17.** 1·2 **18.** 2·625
19. $2{\cdot}\dot{3}$ **20.** 1·7 **21.** 2·1875 **22.** $2{\cdot}\dot{2}8571\dot{4}$ **23.** $2{\cdot}\dot{8}5714\dot{2}$ **24.** 3·19

25. $\frac{1}{5}$ **26.** $\frac{7}{10}$ **27.** $\frac{1}{4}$ **28.** $\frac{9}{20}$ **29.** $\frac{9}{25}$ **30.** $\frac{13}{25}$
31. $\frac{1}{8}$ **32.** $\frac{5}{8}$ **33.** $\frac{21}{25}$ **34.** $2\frac{7}{20}$ **35.** $3\frac{19}{20}$ **36.** $1\frac{1}{20}$
37. $3\frac{1}{5}$ **38.** $\frac{27}{100}$ **39.** $\frac{7}{1000}$ **40.** $\frac{11}{100\,000}$ **41.** 0·58 **42.** 1·42
43. 0·65 **44.** 1·61 **45.** 0·07 **46.** 0·16 **47.** 3·64 **48.** 0·60
49. $\frac{4}{15}$, 0·33, $\frac{1}{3}$ **50.** $\frac{2}{7}$, 0·3, $\frac{4}{9}$ **51.** $\frac{7}{11}$, 0·705, 0·71 **52.** $\frac{5}{18}$, 0·3, $\frac{4}{13}$

Exercise 4 page 4

1. (a) 8 (b) 8·17 (c) 8·17 **2.** (a) 20 (b) 19·6 (c) 19·62
3. (a) 20 (b) 20·0 (c) 20·04 **4.** (a) 1 (b) 0·815 (c) 0·81
5. (a) 311 (b) 311 (c) 311·14 **6.** (a) 0 (b) 0·275 (c) 0·28
7. (a) 0 (b) 0·00747 (c) 0·01 **8.** (a) 16 (b) 15·6 (c) 15·62
9. (a) 900 (b) 900 (c) 900·12 **10.** (a) 4 (b) 3·56 (c) 3·56
11. (a) 5 (b) 5·45 (c) 5·45 **12.** (a) 21 (b) 21·0 (c) 20·96
13. (a) 0 (b) 0·0851 (c) 0·09 **14.** (a) 1 (b) 0·515 (c) 0·52
15. (a) 3 (b) 3·07 (c) 3·07 **16.** 5·7 **17.** 0·8 **18.** 11·2
19. 0·1 **20.** 0·0 **21.** 11·1 **22.** 5·6 **23.** 10·0 **24.** 8·0
25. 2200 **26.** 3600 **27.** 400 **28.** 41 800 **29.** 200 **30.** 55 600
31. 10 000 **32.** 500 **33.** 0
34. (i) 10·5 (ii) 4·3 **35.** (i) 13, 11 (ii) 3, 1 (iii) 0·846, 0·6
36. (a) 56 cm^2 (b) 28 cm^2 **37.** 3·30, 2·87

Exercise 5 page 4

1. 3000 **2.** 10 000 **3.** 10 **4.** 200 000 **5.** 5 **6.** 70
7. 0·1 **8.** 0·5 **9.** 100 000 **10.** 300 000 **11.** 0·001 **12.** 0·00001
13. 400 000 **14.** 1000 **15.** 0·1 **16.** 0·0001 **17.** 4 **18.** 2500
19. 50 000 **20.** 0·01

Exercise 6 page 5

1. 4×10^{3} **2.** 5×10^{2} **3.** 7×10^{4} **4.** 6×10 **5.** $2·4 \times 10^{3}$
6. $3·8 \times 10^{2}$ **7.** $4·6 \times 10^{4}$ **8.** $4·6 \times 10$ **9.** 9×10^{5} **10.** $2·56 \times 10^{3}$
11. 7×10^{-3} **12.** 4×10^{-4} **13.** $3·5 \times 10^{-3}$ **14.** $4·21 \times 10^{-1}$ **15.** $5·5 \times 10^{-5}$
16. 1×10^{-2} **17.** $5·64 \times 10^{5}$ **18.** $1·9 \times 10^{7}$ **19.** $1·23 \times 10^{-6}$ **20.** $2·5 \times 10^{7}$
21. $3·83 \times 10^{4}$ **22.** $1·96 \times 10^{-3}$ **23.** 7×10^{-4} **24.** $6·2 \times 10^{-4}$ **25.** 2×10^{-9}
26. $5·5 \times 10^{9}$ **27.** 8 **28.** 7

Exercise 7 page 6

1. 10^{7} **2.** 10^{8} **3.** 10^{5} **4.** 10^{6} **5.** 10^{-6}
6. 10^{-17} **7.** 10 **8.** 10^{2} **9.** 10^{-5} **10.** 10^{10}
11. 10^{5} **12.** 10^{-2} **13.** 10^{6} **14.** 10^{8} **15.** 10^{-6}
16. 10^{10} **17.** 10^{6} **18.** 10^{-6} **19.** 6×10^{8} **20.** 9×10^{11}
21. $8·8 \times 10^{4}$ **22.** $9·3 \times 10^{-10}$ **23.** $4·8 \times 10^{2}$ **24.** 2×10^{4} **25.** $3·1 \times 10^{10}$
26. $4·3 \times 10^{3}$ **27.** 2×10^{-11} **28.** 2×10^{3} **29.** $1·5 \times 10^{7}$ **30.** 3×10^{8}
31. $2·8 \times 10^{-2}$ **32.** 7×10^{-9} **33.** 2×10^{6} **34.** 4×10^{-6} **35.** 9×10^{-2}
36. $6·6 \times 10^{-8}$ **37.** 2×10^{-2} **38.** $3·44 \times 10^{-4}$ **39.** $5·3 \times 10^{-2}$ **40.** $1·86 \times 10^{-3}$
41. $3·5 \times 10^{-7}$ **42.** 1×10^{-16} **43.** 8×10^{9} **44.** $7·4 \times 10^{-7}$ **45.** $1·67 \times 10^{6}$
46. 9×10 **47.** 7×10^{7} **48.** $4·5 \times 10^{9}$ **49.** 3×10^{-8} **50.** 4
51. c, a, b **52.** 13 **53.** 16
54. (i) $8·75 \times 10^{2}$, $3·75 \times 10^{2}$ (ii) 10^{8}, $4·29 \times 10^{7}$ **55.** (a) 20·5 s (b) $6·3 \times 10^{91}$ years (!)

Exercise 8 page 7

1. 1:3
2. 1:6
3. 1:50
4. 1:1·6
5. 1:0·75
6. 1:0·375
7. 1:25
8. 1:8
9. 2·4:1
10. 2·5:1
11. 0·8:1
12. 0·02:1
13. £15, £25
14. £36, £84
15. 140 m, 110 m
16. £18, £27, £72
17. 15 kg, 75 kg, 90 kg
18. 46 min, 69 min, 69 min
19. £39
20. 18 kg, 36 kg, 54 kg, 72 kg
21. £400, £1000, £1000, £1600
22. 5:3
23. £200
24. 3:7
25. $\frac{1}{7}x$
26. 6
27. 12
28. £120
29. 300 g
30. 625

Exercise 9 page 7

1. £1·68
2. £84
3. 6 days
4. $2\frac{1}{2}$ litres
5. 60 km
6. 119 g
7. £68·40
8. $2\frac{1}{4}$ weeks
9. 80p
10. (a) 12; (b) 2100
11. 4
12. 5·6 days
13. £175
14. 540°
15. £1·20

Exercise 10 page 8

1. (a) Fr 218 (b) \$121·80 (c) Ptas 36 400 (d) DM 6·3 (e) Lire 5244 (f) \$1·566
2. (a) £45·87 (b) £1436·78 (c) £1·79 (d) £10·34 (e) £219·30 (f) £5·22
3. £1·80
4. Spain by £1·20
5. £20 000
6. Germany £12 400; USA £14 300; Britain £15 000; Belgium £17 200, France £17 800
7. 3·25 Swiss francs = £1
8. DM 1690
9. £2197·46
10. $62\frac{1}{2}$p

Exercise 11 page 9

1. (a) 70 m (b) 16 m (c) 3·55 m (d) 108·5 m
2. (a) 5 cm (b) 3·5 cm (c) 0·72 cm (d) 2·86 cm
3. (a) 450 000 cm (b) 4500 m (c) 4·5 km
4. 12·3 km
5. 4·71 km
6. 50 cm
7. 640 cm
8. 5·25 cm

Exercise 12 page 9

1. 40 m by 30 m; 12 cm^2; 1200 m^2
2. 1 m^2, 6 m^2
3. 0·32 km^2
4. 5 m^2
5. 150 km^2
6. 1200 hectares
7. 240 cm^2
8. 192 m^2

Exercise 13 page 10

1. (a) $\frac{3}{5}$ (b) $\frac{6}{25}$ (c) $\frac{7}{20}$ (d) $\frac{1}{50}$
2. (a) 25% (b) 10% (c) $87\frac{1}{2}$% (d) $33\frac{1}{3}$% (e) 72% (f) 31% (g) 57·5% (h) 7·5%
3. (a) 0·36 (b) 0·28 (c) 0·07 (d) 0·134 (e) 0·6 (f) 0·875 (g) 0·01 (h) $0\cdot\dot{6}$
4. (a) $1\frac{1}{5}$ (b) $2\frac{1}{2}$ (c) $\frac{99}{100}$ (d) $\frac{1}{8}$ (e) $\frac{1}{40}$ (f) $\frac{5}{8}$ (g) $\frac{1}{30}$ (h) $\frac{17}{400}$
5. (a) 70% (b) 0·7% (c) 0·07% (d) 350% (e) 225% (f) $55\frac{5}{9}$% (g) 320% (h) 900%
6. (a) 0·075 (b) 3·01 (c) 0·005 (d) 0·0125 (e) 0·009 (f) 0·09 (g) 0·375 (h) 0·096
7. (a) 45%; $\frac{1}{2}$; 0·6 (b) 4%; $\frac{6}{16}$; 0·38 (c) 11%; 0·111; $\frac{1}{9}$ (d) 0·3; 32%, $\frac{1}{3}$
   (e) ($\frac{1}{2}$ of 0·4); $(\frac{1}{2})^2$; (15% of 2) (f) 0·075%; 8×10^{-4}; 8×10^{-3}
   (g) $(0\cdot2)^2$; $\frac{1}{20}$; 6% (h) 5×10^{-3}; 1·3%; 4×10^{-2} (i) $(0\cdot3)^2$, (0·4 of 80%); $\frac{4}{12}$
   (j) 10^{-3}; 0·2%, $\frac{1}{200}$
8. (a) 85% (b) 77·5% (c) 23·75% (d) 56% (e) 10% (f) 37·5%

Exercise 14 page 11

1. (a) £15 (b) 900 kg (c) \$2·80 (d) 125
2. £32 **3.** 13·2 p **4.** 52·8 kg **5.** 200 **6.** 29 000 **7.** 500 cm
8. £6·30 **9.** 400 kg **10.** 325

Exercise 15 page 11

1. (a) 25%, profit (b) 25%, profit (c) 10%, loss (d) 20%, profit (e) 30%, profit (f) 7·5%, profit
(g) 12%, loss (h) 54%, loss
2. 28% **3.** $44\frac{4}{9}$% **4.** 46·9% **5.** 12% **6.** $5\frac{1}{3}$%
7. (a) £50 (b) £450 (c) £800 (d) £12·40
8. £500 **9.** £12 **10.** £5 **11.** 60p **12.** £220 **13.** 14·3%
14. 20% **15.** 8 : 11

Exercise 16 page 12

1. (a) £1380 (b) £412·50 (c) £3337·50 (d) £3855 (e) £4890·50 (f) £13 875
2. (a) £10000 (b) £5 200 (c) £13 150 (d) £15 965·56 (e) £16 212·50 (f) £20 471·67
3. (a) £183·20 (b) £268
4. 32% **5.** 29%

Exercise 17 page 13

1. (a) $2\frac{1}{2}$h (b) $3\frac{1}{8}$h (c) 75 s (d) 4 h
2. (a) 20 m/s (b) 30 m/s (c) $83\frac{1}{3}$m/s (d) 108 km/h (e) 79·2 km/h (f) 1·2 cm/s
(g) 90 m/s (h) 25 mph (i) 0·03 miles per second
3. (a) 75 km/h (b) 4·52 km/h (c) 7·6 m/s (d) 4×10^6 m/s (e) 2.5×10^8 m/s
(f) 200 km/h (g) 3 km/h
4. (a) 110 000 m (b) 10 000 m (c) 56 400 m (d) 4500 m (e) 50 400 m (f) 80 m
(g) 960 000 m
5. (a) 3·125 h (b) 76·8 km/h **6.** (a) 4·45 h (b) 23·6 km/h
7. 46 km/h
8. (a) 8 m/s (b) 7·6 m/s (c) 102·63 s (d) 7·79 m/s
9. 1230 km/h **10.** 3 h **11.** 100 s **12.** $1\frac{1}{2}$ minutes **13.** 600 m **14.** $53\frac{1}{3}$s
15. 5 cm/s **16.** 60 s

Exercise 18 page 15

1. 100 **2.** 9 **3.** 900 **4.** 6000 **5.** 90000 **6.** 500000
7. 1000 000 **8.** 50 **9.** 80 000 000 **10.** 0·0009 **11.** 0·8 **12.** 0·00005
13. 0·04 **14.** 6 **15.** 6 **16.** 200 **17.** 90 **18.** 30
19. 300 **20.** 0·3 **21.** 0·1 **22.** 1 **23.** 0·02 **24.** 800
25. 0·8 **26.** 200 **27.** 0·08 **28.** 31·70 **29.** 50·27 **30.** 2·002
31. 139·2 **32.** 557·0 **33.** 1274 **34.** 4984 **35.** 74·2 **36.** 45800
37. 273500 **38.** 0·5227 **39.** 0·04709 **40.** 0·001722 **41.** 0·0002938 **42.** 0·01850
43. 6.561×10^{-5} **44.** 73440 **45.** 7327 **46.** 370900 **47.** 1232 000
48. 2.601×10^7 **49.** 4.84×10^{-6} **50.** 1.369×10^5 **51.** 5.29×10^{-8} **52.** 2.89×10^{10} **53.** 1.103×10^{-8}
54. 9.860×10^{-6}

Exercise 19 page 15

1. 2·995 **2.** 5·138 **3.** 9·223 **4.** 1·084 **5.** 8·939 **6.** 25·44
7. 80·44 **8.** 14·63 **9.** 46·26 **10.** 21·74 **11.** 13·39 **12.** 75·50
13. 251·0 **14.** 0·8843 **15.** 0·6099 **16.** 0·3782 **17.** 0·2891 **18.** 0·2381
19. 0·3332 **20.** 0·09701 **21.** 0·02040 **22.** 0·09487 **23.** 800 **24.** 3162

25. 10 000 **26.** 0·003162 **27.** 90 000 **28.** 0·04 **29.** 0·2 **30.** 300
31. 0·04 **32.** 9×10^6 **33.** 0·0049 **34.** 0·007 **35.** 800 **36.** 110
37. $6{\cdot}4 \times 10^9$ **38.** 10^{-2} **39.** $\frac{1}{10}$ **40.** 10 000 **41.** $\frac{1}{50}$ **42.** 100
43. $\frac{1}{3}$ **44.** $\frac{1}{5}$ **45.** $1\frac{1}{2}$ **46.** $2\frac{1}{2}$ **47.** $2\frac{1}{3}$ **48.** 10 000
49. 10^8 **50.** $\frac{3}{8}$

Exercise 20 page 16

1. 54 **2.** 151 **3.** 70 **4.** 16 **5.** 58 **6.** 32
7. 89 **8.** 224 **9.** 190 **10.** 1036 **11.** 522 **12.** 444
13. 11 **14.** 24 **15.** 15 **16.** 23
17. (a) 1111_2; 101_4; 33_5 (b) 21_3; 21_4; 1011_2 (c) 131_4; 51_6; 34_{10} (d) 66_9; 55_{12}; 102_8
(e) 1011_3; 40_8; 100001_2
18. 10322_4 **19.** 3002_4 **20.** 2040_5 **21.** 452_6 **22.** 11001_2 **23.** 100101_2
24. 1000011_2 **25.** 101111_2 **26.** 636_8 **27.** 456_9 **28.** 66_{12} **29.** $E3_{12}$
30. 119_{12} **31.** $16E_{12}$ **32.** 273_{12} **33.** 420_{12} **34.** 1010_2 **35.** 10011_2
36. 10000_2 **37.** 100100_2 **38.** 100_2 **39.** 110_2 **40.** 111_2 **41.** 110_2
42. 335_8 **43.** 625_8 **44.** 475_8 **45.** 1311_8 **46.** 144_8 **47.** 17_8
48. 2007_8 **49.** 3513_8 **50.** 1111_2 **51.** 11110_2 **52.** $1\,000\,010_2$ **53.** $1\,000\,001_2$
54. 10101_2 **55.** (a) 7 (b) 3
56. (a) 1111_4 (b) 11111_6 (c) 1111_n (d) 1231_n (e) 572_n (f) 3024_n
(g) 10005_n
57. (i) (a) base 8 (b) 5 (ii) (a) base 7 (b) 6
58. (a) 1111_3 (b) $10\,000_3$ (c) 2222_3 (d) $k = 2$ (e) $n = 11$
59. (a) 1111_5 (b) $10\,000_5$ (c) 4444_4 (d) $k = 4$ (e) $n = 22, c = 4$
60. $x = 3, y = 1$

Exercise 21 page 17

1. 3·041 **2.** 1460 **3.** 0·03083 **4.** 47·98 **5.** 130·6 **6.** 0·4771
7. 0·3658 **8.** 37·54 **9.** 8·000 **10.** 0·6537 **11.** 0·03716 **12.** 34·31
13. 0·7195 **14.** 3·598 **15.** 0·2445 **16.** 2·043 **17.** 0·3798 **18.** 0·7683
19. −0·5407 **20.** 0·07040 **21.** 2·526 **22.** 0·09478 **23.** 0·2110 **24.** 3·123
25. 2·230 **26.** 128·8 **27.** 4·268 **28.** 3·893 **29.** 0·6290 **30.** 0·4069
31. 9·298 **32.** 0·1010 **33.** 0·3692 **34.** 1·125 **35.** 1·677 **36.** 0·9767
37. 0·8035 **38.** 0·3528 **39.** 2·423 **40.** 1·639 **41.** 0·0004659 **42.** 0·3934
43. −0·7526 **44.** 2·454 **45.** 40 000 **46.** 0·07049 **47.** 405 400 **48.** 471·3
49. 20810 **50.** $2{\cdot}218 \times 10^6$ **51.** $1{\cdot}237 \times 10^{-24}$ **52.** 3·003 **53.** 0·03581 **54.** 47·40
55. −1748 **56.** 0·01138 **57.** 1757 **58.** 0·02635 **59.** 0·1651 **60.** 5447
61. 0·006562 **62.** 0·1330 **63.** 0·4451 **64.** 0·03616 **65.** 19·43 **66.** $1{\cdot}296 \times 10^{-15}$
67. $5{\cdot}595 \times 10^{14}$ **68.** $1{\cdot}022 \times 10^{-8}$ **69.** 0·01922 **70.** 0·9613

Revision Exercise 1A page 18

1. (a) 185 (b) 150 (c) 40 (d) $\frac{11}{12}$ (e) $2\frac{4}{5}$ (f) $\frac{2}{5}$
2. 128 cm **3.** $\frac{2}{5}$ **4.** $\frac{a}{b}$ **5.** (a) 0·0547 (b) 0·055 (c) $5{\cdot}473 \times 10^{-2}$
6. 1·238 **7.** (a) 3×10^7 (b) $3{\cdot}7 \times 10^4$ (c) $2{\cdot}7 \times 10^{13}$
8. (a) £26 (b) 6 : 5 (c) 6 or −6 **9.** £75
10. (a) (i) 57·2% (ii) $87\frac{1}{2}$% (b) 40% (c) 80p **11.** 5%
12. (a) £500 (b) $37\frac{1}{2}$% **13.** £350 **14.** £150·50, £8910
15. (a) 2·4 km (b) 1 km^2 **16.** (a) 300 m (b) 60 cm (c) 150 cm^2
17. (a) 1 : 50 000 (b) 1 : 4 000 000 **18.** (a) 22% (b) 20·8% (c) £240
19. (a) (i) 7 m/s (ii) 200 m/s (iii) 5 m/s (b) (i) 144 km/h (ii) 2·16 km/h

20. (a) 0·005 m/s (b) 1·6 s (c) 172·8 km **21.** $33\frac{1}{3}$ mph
22. (a) (i) 0·05 (ii) $1\frac{1}{2}$ (b) (i) 12·2 (ii) 38·7 (iii) 0·0387
(c) (i) 138 (ii) 13·8 (iii) 0·0138
23. (a) 5%; $(0{\cdot}3)^2$; $\frac{1}{11}$; $\sqrt{0{\cdot}04}$ (b) 91×10^{-3}; $\sqrt{0{\cdot}01}$; 11%; $\frac{1}{9}$ (c) 310%; $\sqrt{90\,000}$; $21{\cdot}7^2$; $2{\cdot}96 \times 10^3$
24. (a) 11011_2 (b) 10010_2 (c) 1111_2 (d) 111_2 (e) $1\,000\,0001_2$
25. (a) 103_8 (b) 10003_8 (c) 121_8 (d) 474_8 (e) 64_8
26. (a) $n = 4$ (b) base 9 (c) 324_y
27. (a) 0·5601 (b) 3·215 (c) 0·6161 (d) 0·4743
28. (a) 0·340 (b) $4{\cdot}08 \times 10^{-6}$ (c) 64·9 (d) 0·119

Examination Exercise 1B page 20

1. (i) £14 (ii) £185 (iii) £579; £93·24; £731·64
2. (a) £7·70 (b) $33\frac{1}{3}$% (c) £8 (d) £776 (e) £127; 19·6%
3. £2246; £394 **4.** £982 600; 520 km **5.** £8800; £12 000; 36·4%
6. £1·40/h; £88·20; (i) £1·75/h (ii) 25%; £105·84 **7.** £1074; 22·9%; £4320
8. £9000; 6·2% increase; (i) $x = 20$ (ii) $x = 11{\cdot}1$
9. (a) 12 min, 18 min, 50 km/h (b) 162 g **10.** (i) 20 000 (ii) 21840; 10%
11. (i) 82 000 (ii) 25% (iii) 9900 (iv) 67·1%

PART 2

Exercise 1 page 23

1. 13 **2.** 211 **3.** −12 **4.** −31 **5.** −66 **6.** 6·1 **7.** 9·1
8. −35 **9.** 18·7 **10.** −9 **11.** −3 **12.** 3 **13.** −2 **14.** −14
15. −7 **16.** 3 **17.** 181 **18.** −2·2 **19.** 8·2 **20.** 17 **21.** 2
22. −6 **23.** −15 **24.** −14 **25.** −2 **26.** −12 **27.** −80 **28.** −13·1
29. −4·2 **30.** 12·4 **31.** −7 **32.** 8 **33.** 4 **34.** −10 **35.** 11
36. 4 **37.** −20 **38.** 8 **39.** −5 **40.** −10 **41.** −26 **42.** −21
43. 8 **44.** 1 **45.** −20·2 **46.** −50 **47.** −508 **48.** −29 **49.** 0
50. −21 **51.** −0·1 **52.** −4 **53.** 6·7 **54.** 1 **55.** −850 **56.** 4
57. 6 **58.** −4 **59.** −12 **60.** −31

Exercise 2 page 24

1. −8 **2.** 28 **3.** 12 **4.** 24 **5.** 18 **6.** −35 **7.** 49
8. −12 **9.** −2 **10.** 9 **11.** −4 **12.** 4 **13.** −4 **14.** 8
15. 70 **16.** −7 **17.** $\frac{1}{4}$ **18.** $-\frac{3}{5}$ **19.** −0·01 **20.** 0·0002 **21.** 121
22. 6 **23.** −600 **24.** −1 **25.** −20 **26.** −2·6 **27.** −700 **28.** 18
29. −1000 **30.** 640 **31.** −6 **32.** −42 **33.** −0·4 **34.** −0·4 **35.** −200
36. −35 **37.** −2 **38.** $\frac{1}{2}$ **39.** $-\frac{1}{4}$ **40.** −90

Exercise 3 page 24

1. −10 **2.** 1 **3.** 12 **4.** −28 **5.** −2 **6.** 16 **7.** −3
8. 14 **9.** −28 **10.** 4 **11.** $-\frac{1}{6}$ **12.** 9 **13.** −30 **14.** 24
15. −1 **16.** −2 **17.** −30 **18.** 7 **19.** 3 **20.** 16 **21.** 93
22. 2400 **23.** 10 **24.** 1 **25.** −4 **26.** 48 **27.** −1 **28.** 0
29. −8 **30.** 170 **31.** −3 **32.** 1 **33.** 1 **34.** 0 **35.** 15
36. 5 **37.** −2·4 **38.** −180 **39.** 5 **40.** −994 **41.** 2 **42.** −48
43. 60 **44.** −2·5 **45.** −32 **46.** 0 **47.** −0·1 **48.** −16 **49.** −4·3
50. $-\frac{1}{16}$

Exercise 4 page 25

1. 7	**2.** 13	**3.** 13	**4.** 22	**5.** 1	**6.** -1	**7.** 18
8. -4	**9.** -3	**10.** 37	**11.** 0	**12.** -4	**13.** -7	**14.** -2
15. -3	**16.** -8	**17.** -30	**18.** 16	**19.** -10	**20.** 0	**21.** 7
22. -6	**23.** -2	**24.** -7	**25.** -5	**26.** 3	**27.** 4	**28.** -8
29. -2	**30.** 2	**31.** 0	**32.** 4	**33.** -4	**34.** -3	**35.** -9
36. 4						

Exercise 5 page 25

1. 9	**2.** 27	**3.** 4	**4.** 16	**5.** 36	**6.** 18	**7.** 1
8. 6	**9.** 2	**10.** 8	**11.** -7	**12.** 15	**13.** -23	**14.** 3
15. 32	**16.** 36	**17.** 144	**18.** -8	**19.** -7	**20.** 13	**21.** 5
22. -16	**23.** 84	**24.** 17	**25.** 6	**26.** 0	**27.** -25	**28.** -5
29. 17	**30.** $-1\frac{1}{2}$	**31.** 19	**32.** 8	**33.** 19	**34.** 16	**35.** -16
36. 12	**37.** 36	**38.** -12	**39.** 2	**40.** 11	**41.** -23	**42.** -26
43. 5	**44.** 31	**45.** $4\frac{1}{2}$				

Exercise 6 page 26

1. -20	**2.** 16	**3.** -42	**4.** -4	**5.** -90	**6.** -160	**7.** -2
8. -81	**9.** 4	**10.** 22	**11.** 14	**12.** 5	**13.** 1	**14.** $\sqrt{5}$
15. 4	**16.** $-6\frac{1}{2}$	**17.** 54	**18.** 25	**19.** 4	**20.** 312	**21.** 45
22. 22	**23.** 14	**24.** -36	**25.** -7	**26.** 1	**27.** 901	**28.** -30
29. -5	**30.** $7\frac{1}{2}$	**31.** -7	**32.** $-\frac{3}{13}$	**33.** 7	**34.** -2	**35.** 0
36. $-4\frac{1}{2}$	**37.** 6	**38.** 2	**39.** 26	**40.** -9	**41.** $3\frac{1}{4}$	**42.** $-\frac{5}{6}$
43. 4	**44.** $2\frac{2}{3}$	**45.** $3\frac{1}{4}$	**46.** $-2\frac{1}{6}$	**47.** -13	**48.** 12	**49.** $1\frac{1}{3}$
50. $-\frac{5}{36}$						

Exercise 7 page 26

1. $3x + 11y$	**2.** $2a + 8b$	**3.** $3x + 2y$	**4.** $5x + 5$
5. $9 + x$	**6.** $3 - 9y$	**7.** $5x - 2y - x^2$	**8.** $2x^2 + 3x + 5$
9. $-10y$	**10.** $3a^2 + 2a$	**11.** $7 + 7a - 7a^2$	**12.** $5x$
13. $\frac{10}{a} - b$	**14.** $\frac{5}{x} - \frac{5}{y}$	**15.** $\frac{3m}{x}$	**16.** $\frac{1}{2} - \frac{2}{x}$
17. $\frac{5}{a} + 3b$	**18.** $-\frac{n}{4}$	**19.** $7x^2 - x^3$	**20.** $2x^2$
21. $x^2 + 5y^2$	**22.** $-12x^2 - 4y^2$	**23.** $5x - 11x^2$	**24.** $\frac{8}{x^2}$
25. $5x + 2$	**26.** $12x - 7$	**27.** $3x + 4$	**28.** $11 - 6x$
29. $-5x - 20$	**30.** $7x - 2x^2$	**31.** $3x^2 - 5x$	**32.** $x - 4$
33. $5x^2 + 14x$	**34.** $-4x^2 - 3x$	**35.** $5a + 8$	**36.** $a + 9$
37. $ab + 4a$	**38.** $y^2 + y$	**39.** $2x - 2$	**40.** $6x + 3$
41. $x - 4$	**42.** $7x + 5y$	**43.** $4x^2 - 11x$	**44.** $2x^2 + 14x$
45. $3y^2 - 4y + 1$	**46.** $12x + 12$	**47.** $4ab - 3a + 14b$	**48.** $2x - 4$

Exercise 8 page 27

1. $x^2 + 4x + 3$	**2.** $x^2 + 5x + 6$	**3.** $y^2 + 9y + 20$	**4.** $x^2 + x - 12$
5. $x^2 + 3x - 10$	**6.** $x^2 - 5x + 6$	**7.** $a^2 - 2a - 35$	**8.** $z^2 + 7z - 18$
9. $x^2 - 9$	**10.** $k^2 - 121$	**11.** $2x^2 - 5x - 3$	**12.** $3x^2 - 2x - 8$
13. $2y^2 - y - 3$	**14.** $49y^2 - 1$	**15.** $9x^2 - 4$	**16.** $6a^2 + 5ab + b^2$

17. $3x^2+7xy+2y^2$ 18. $6b^2+bc-c^2$ 19. $-5x^2+16xy-3y^2$ 20. $15b^2+ab-2a^2$
21. $2x^2+2x-4$ 22. $6x^2+3x-9$ 23. $24y^2+4y-8$ 24. $6x^2-10x-4$
25. $4a^2-16b^2$ 26. x^3-3x^2+2x 27. $8x^3-2x$ 28. $3y^3+3y^2-18y$
29. $x^3+x^2y+x^2z+xyz$ 30. $3za^2+3zam-6zm^2$

Exercise 9 page 27

1. $x^2+8x+16$ 2. x^2+4x+4 3. x^2-4x+4 4. $4x^2+4x+1$
5. $y^2-10y+25$ 6. $9y^2+6y+1$ 7. $x^2+2xy+y^2$ 8. $4x^2+4xy+y^2$
9. $a^2-2ab+b^2$ 10. $4a^2-12ab+9b^2$ 11. $3x^2+12x+12$ 12. $9-6x+x^2$
13. $9x^2+12x+4$ 14. $a^2-4ab+4b^2$ 15. $2x^2+6x+5$ 16. $2x^2+2x+13$
17. $5x^2+8x+5$ 18. $2y^2-14y+25$ 19. $10x-5$ 20. $-8x+8$
21. $-10y+5$ 22. $3x^2-2x-8$ 23. $2x^2+4x-4$ 24. $-x^2-18x+15$

Exercise 10 page 28

1. $2x^2+y^2+3xy+11x+7y+12$ 2. $2x^2-6y^2+xy-8x+5y+6$
3. $x^2+y^2+2xy+4x+4y+4$ 4. $a^2-b^2-2c^2+ac-3bc$
5. $2x^2+5+\frac{3}{x^2}$ 6. $x^2+1-\frac{2}{x^2}$ 7. $x^2+4+\frac{4}{x^2}$ 8. $y^2+2+\frac{1}{y^2}$
9. $4x^2+12+\frac{9}{x^2}$ 10. $x^2-2+\frac{1}{x^2}$ 11. $2x^2+3-\frac{5}{x^2}$ 12. $x^3+x+1+\frac{1}{x^2}$
13. $x^4-4x+\frac{4}{x^2}$ 14. $4x^2+\frac{4}{x}+\frac{1}{x^4}$ 15. $2x^2+3x+4+\frac{2}{x}+\frac{1}{x^2}$
16. $2x^2-3x+5+\frac{1}{x}+\frac{3}{x^2}$ 17. x^2-x+4 18. $4x^2+3x+4$
19. $-10x+5$ 20. $2x^2+2+\frac{13}{x^2}$ 21. $x^2+x+4+\frac{1}{x}+\frac{1}{x^2}$ 22. $5x^2+18+\frac{17}{x^2}$
23. $4x^2-12x+25-\frac{24}{x}+\frac{16}{x^2}$ 24. $\frac{4x}{y}$

Exercise 11 page 28

1. 8 2. 9 3. 7 4. 10 5. $\frac{1}{3}$ 6. 10 7. $1\frac{1}{2}$
8. -1 9. $-1\frac{1}{2}$ 10. $\frac{1}{3}$ 11. 35 12. 130 13. 14 14. $\frac{2}{3}$
15. $3\frac{1}{3}$ 16. $-2\frac{1}{2}$ 17. 3 18. $1\frac{1}{8}$ 19. $\frac{3}{10}$ 20. $-1\frac{1}{4}$ 21. 10
22. 27 23. 20 24. 18 25. 28 26. -15 27. $\frac{99}{100}$ 28. 0
29. 1000 30. $-\frac{1}{1000}$ 31. 1 32. -7 33. -5 34. $1\frac{1}{6}$ 35. 1
36. 2 37. -5 38. -3 39. $-1\frac{1}{2}$ 40. 2 41. 1 42. $3\frac{1}{2}$
43. 2 44. -1 45. $10\frac{2}{3}$ 46. 1·1 47. -1 48. 2 49. $2\frac{1}{2}$
50. $1\frac{1}{3}$

Exercise 12 page 29

1. $-1\frac{1}{2}$ 2. 2 3. $-\frac{2}{5}$ 4. $-\frac{1}{3}$ 5. $1\frac{2}{3}$ 6. 6 7. $-\frac{2}{5}$
8. $-3\frac{1}{5}$ 9. $\frac{1}{2}$ 10. -4 11. 18 12. 5 13. 4 14. 3
15. $2\frac{3}{4}$ 16. $-\frac{7}{22}$ 17. $\frac{1}{4}$ 18. 1 19. 4 20. -11 21. $-7\frac{1}{3}$
22. $1\frac{1}{4}$ 23. -5 24. 6 25. 3 26. 6 27. 2 28. 3
29. 4 30. 3 31. $10\frac{1}{2}$ 32. 5 33. 2 34. -1 35. -17
36. $-2\frac{9}{10}$ 37. $2\frac{10}{21}$ 38. $\frac{1}{3}$ 39. 14 40. 6

Exercise 13 page 30

1. $\frac{1}{4}$ 2. -3 3. 4 4. $-7\frac{2}{3}$ 5. -43 6. 11 7. $-\frac{1}{2}$
8. 0 9. 1 10. $-1\frac{2}{3}$ 11. $\frac{1}{4}$ 12. 0 13. $-\frac{6}{7}$ 14. $1\frac{9}{17}$
15. $1\frac{22}{23}$ 16. $\frac{2}{11}$ 17. 10, 8, 6 18. 13, 12, 5 19. 10, 8, 6 20. 13, 12, 5 21. 5, 4, 3
22. 13, 12, 5 23. 4 cm 24. 5 m 25. 4

Exercise 14 page 31

1. $\frac{1}{3}$ 2. $\frac{1}{5}$ 3. $1\frac{2}{3}$ 4. -3 5. $\frac{5}{11}$ 6. -2 7. 6
8. $3\frac{3}{4}$ 9. -7 10. $-7\frac{2}{3}$ 11. 2 12. 3 13. 4 14. -2
15. -3 16. 3 17. $1\frac{5}{7}$ 18. $4\frac{4}{5}$ 19. 10 20. 24 21. 2
22. 3 23. 5 24. -4 25. $6\frac{3}{4}$ 26. -3 27. 0 28. 3
29. 0 30. 1 31. 2 32. 3 33. 4 34. $\frac{3}{5}$ 35. $1\frac{1}{8}$
36. -1 37. 1 38. 1 39. $\frac{1}{4}$ 40. $-\frac{1}{3}$ 41. $\frac{9}{10}$ 42. 1
43. 2 44. $-\frac{1}{7}$ 45. 2 46. 3 47. 3 48. 1 49. 2
50. 1 51. -2 52. $-\frac{2}{3}$ 53. 7 54. $1\frac{1}{4}$

Exercise 15 page 32

1. 91, 92, 93 2. 21, 22, 23, 24 3. 57, 59, 61 4. 506, 508, 510 5. $12\frac{1}{2}$
6. $12\frac{1}{2}$ 7. $11\frac{2}{3}$ 8. $8\frac{1}{3}, 41\frac{2}{3}$ 9. $1\frac{1}{4}, 13\frac{3}{4}$ 10. $3\frac{1}{3}$ cm
11. 12 cm 12. 20° 13. 5 cm 14. 7 cm 15. $18\frac{1}{2}, 27\frac{1}{2}$
16. 20°, 60°, 100° 17. 45°, 60°, 75° 18. 5 19. 6, 8 20. 12, 24, 30
21. 5, 15, 8 22. $59\frac{2}{3}$ kg, $64\frac{2}{3}$ kg, $72\frac{2}{3}$ kg 23. 24, 22, 15 24. 48, 12
25. 40, 8 26. 6 27. 168·84 cm^2 28. 14 29. 45p, 31p
30. £21·50

Exercise 16 page 34

1. £3700 2. 3 3. 8 4. $1\frac{3}{7}$ m 5. 80°, 100°
6. 30°, 60°, 90°, 120°, 150°, 270° 7. 26, 58 8. 2 km 9. 8 km 10. 400 m

Exercise 17 page 34

1. $x = 2, y = 1$ 2. $x = 4, y = 2$ 3. $x = 3, y = 1$ 4. $x = -2, y = 1$
5. $x = 3, y = 2$ 6. $x = 5, y = -2$ 7. $x = 2, y = 1$ 8. $x = 5, y = 3$
9. $x = 3, y = -1$ 10. $a = 2, b = -3$ 11. $a = 5, b = \frac{1}{4}$ 12. $a = 1, b = 3$
13. $m = \frac{1}{2}, n = 4$ 14. $w = 2, x = 3$ 15. $x = 6, y = 3$ 16. $x = \frac{1}{2}, z = -3$
17. $m = 1\frac{15}{17}, n = \frac{11}{17}$ 18. $c = 1\frac{16}{23}, d = -2\frac{12}{23}$

Exercise 18 page 35

1. 1 2. -3 3. -2 4. 15 5. -12 6. -3 7. -2 8. -11
9. -21 10. 1 11. 0 12. 15 13. -10 14. 3 15. 6 16. -11
17. 2 18. 5 19. -19 20. -4 21. x 22. $-3x$ 23. $4x$ 24. $4y$
25. $9y$ 26. $3x$ 27. $-8x$ 28. $4x$ 29. $2x$ 30. $3y$ 31. $10a$ 32. $2a$
33. $-8x$ 34. $5y$ 35. $9x$ 36. k 37. $-3x$ 38. $-11y$ 39. $9x$ 40. $4k$

Exercise 19 page 36

1. $x = 2, y = 4$ 2. $x = 1, y = 4$ 3. $x = 2, y = 5$ 4. $x = 3, y = 7$
5. $x = 5, y = 2$ 6. $a = 3, b = 1$ 7. $x = 1, y = 3$ 8. $x = 1, y = 3$
9. $x = -2, y = 3$ 10. $x = 4, y = 1$ 11. $x = 1, y = 5$ 12. $x = 0, y = 2$

13. $x = \frac{5}{7}, y = 4\frac{3}{7}$ 14. $x = 1, y = 2$ 15. $x = 2, y = 3$ 16. $x = 4, y = -1$
17. $x = 3, y = 1$ 18. $x = 1, y = 2$ 19. $x = 2, y = 1$ 20. $x = -2, y = 1$
21. $x = 1, y = 2$ 22. $a = 4, b = 3$ 23. $x = -23, y = -78$ 24. $x = 3, y = \frac{1}{2}$
25. $x = 4, y = 3$ 26. $x = 5, y = -2$ 27. $x = \frac{1}{3}, y = -2$ 28. $x = 5\frac{5}{14}, y = \frac{2}{7}$
29. $x = 3, y = -1$ 30. $x = 5, y = 0{\cdot}2$

Exercise 20 page 36

1. $5\frac{1}{2}, 9\frac{1}{2}$ 2. 6, 3 or $2\frac{2}{5}, 5\frac{2}{5}$ 3. 4, 10 4. $a = 2, c = 7$
5. $m = 4, c = -3$ 6. $a = 1, b = -2$ 7. $m = 1$p, $w = 3$p 8. TV £200, video £450
9. 7, 3 10. white 2oz, brown $3\frac{1}{2}$ oz 11. 120 m, 240 m
12. 150 m, 350 m 13. 2p × 15, 5p × 25 14. 10p × 14, 50p × 7 15. 20
16. man £50, woman £70 17. current 4 m/s, kipper 10 m/s 18. $\frac{5}{7}$
19. $\frac{3}{5}$ 20. boy 10, mouse 3 21. 4, 7 22. $y = 3x - 2$
23. walks 4 m/s, runs 5 m/s 24. £1 × 15, £5 × 5 25. 36, 9
26. wind $4\frac{1}{2}$ knots, submarine $20\frac{1}{2}$ knots 27. $a = 1, b = 2, c = 5$ 28. $y = 2x^2 - 3x + 5$
29. $y = x^2 + 3x + 4$ 30. $y = x^2 + 2x - 3$

Exercise 21 page 38

1. $x(x + 5)$ 2. $x(x - 6)$ 3. $x(7 - x)$ 4. $y(y + 8)$
5. $y(2y + 3)$ 6. $2y(3y - 2)$ 7. $8x(x - 7)$ 8. $2a(8 - a)$
9. $3c(2c - 7)$ 10. $3x(5 - 3x)$ 11. $7y(8 - 3y)$ 12. $x(a + b + 2c)$
13. $x(x + y + 3z)$ 14. $y(x^2 + y^2 + z^2)$ 15. $ab(3a + 2b)$ 16. $xy(x + y)$
17. $2a(3a + 2b + c)$ 18. $m(a + 2b + m)$ 19. $2k(x + 3y + 2z)$ 20. $a(x^2 + y + 2b)$
21. $xk(x + k)$ 22. $ab(a^2 + 2b)$ 23. $bc(a - 3b)$ 24. $ae(2a - 5e)$
25. $ab(a^2 + b^2)$ 26. $x^2y(x + y)$ 27. $2xy(3y - 2x)$ 28. $3ab(b^2 - a^2)$
29. $a^2b(2a + 5b)$ 30. $ax^2(y - 2z)$ 31. $2ab(x + b + a)$ 32. $yx(a + x^2 - 2yx)$

Exercise 22 page 38

1. $(a + b)(x + y)$ 2. $(a + b)(y + z)$ 3. $(x + y)(b + c)$ 4. $(x + y)(h + k)$
5. $(x + y)(m + n)$ 6. $(a + b)(h - k)$ 7. $(a + b)(x - y)$ 8. $(m + n)(a - b)$
9. $(h + k)(s + t)$ 10. $(x + y)(s - t)$ 11. $(a - b)(x - y)$ 12. $(x - y)(s - t)$
13. $(a - x)(s - y)$ 14. $(h - b)(x - y)$ 15. $(m - n)(a - b)$ 16. $(x - z)(k - m)$
17. $(2a + b)(x + 3y)$ 18. $(2a + b)(x + y)$ 19. $(2m + n)(h - k)$ 20. $(m - n)(2h + 3k)$
21. $(2x + y)(3a + b)$ 22. $(2a - b)(x - y)$ 23. $(x^2 + y)(a + b)$ 24. $(m - n)(s + 2t^2)$

Exercise 23 page 39

1. $(x + 2)(x + 5)$ 2. $(x + 3)(x + 4)$ 3. $(x + 3)(x + 5)$ 4. $(x + 3)(x + 7)$
5. $(x + 2)(x + 6)$ 6. $(y + 5)(y + 7)$ 7. $(y + 3)(y + 8)$ 8. $(y + 5)(y + 5)$
9. $(y + 3)(y + 12)$ 10. $(a + 2)(a - 5)$ 11. $(a + 3)(a - 4)$ 12. $(z + 3)(z - 2)$
13. $(x + 5)(x - 7)$ 14. $(x + 3)(x - 8)$ 15. $(x - 2)(x - 4)$ 16. $(y - 2)(y - 3)$
17. $(x - 3)(x - 5)$ 18. $(a + 2)(a - 3)$ 19. $(a + 5)(a + 9)$ 20. $(b + 3)(b - 7)$
21. $(x - 4)(x - 4)$ 22. $(y + 1)(y + 1)$ 23. $(y - 7)(y + 4)$ 24. $(x - 5)(x + 4)$
25. $(x - 20)(x + 12)$ 26. $(x - 15)(x - 11)$ 27. $(y + 12)(y - 9)$ 28. $(x - 7)(x + 7)$
29. $(x - 3)(x + 3)$ 30. $(x - 4)(x + 4)$

Exercise 24 page 39

1. $(2x + 3)(x + 1)$ 2. $(2x + 1)(x + 3)$ 3. $(3x + 1)(x + 2)$ 4. $(2x + 3)(x + 4)$
5. $(3x + 2)(x + 2)$ 6. $(2x + 5)(x + 1)$ 7. $(3x + 1)(x - 2)$ 8. $(2x + 5)(x - 3)$
9. $(2x + 7)(x - 3)$ 10. $(3x + 4)(x - 7)$ 11. $(2x + 1)(3x + 2)$ 12. $(3x + 2)(4x + 5)$
13. $(3x - 2)(x - 3)$ 14. $(y - 2)(3y - 5)$ 15. $(4y - 3)(y - 5)$ 16. $(2y + 3)(3y - 1)$

17. $(2x-5)(3x-6)$ 18. $(5x+2)(2x+1)$ 19. $(6x-1)(x-3)$ 20. $(4x+1)(2x-3)$
21. $(6x+5)(2x-1)$ 22. $(16x+3)(x+1)$ 23. $(2a-1)(2a-1)$ 24. $(x+2)(12x-7)$
25. $(x+3)(15x-1)$ 26. $(8x+1)(6x+5)$ 27. $(16x-3)(4x+1)$ 28. $(15x-1)(8x+5)$
29. $(3x-1)(3x+1)$ 30. $(2a-3)(2a+3)$

Exercise 25 page 39

1. $(y-a)(y+a)$ 2. $(m-n)(m+n)$ 3. $(x-t)(x+t)$ 4. $(y-1)(y+1)$
5. $(x-3)(x+3)$ 6. $(a-5)(a+5)$ 7. $(x-\frac{1}{2})(x+\frac{1}{2})$ 8. $(x-\frac{1}{3})(x+\frac{1}{3})$
9. $(2x-y)(2x+y)$ 10. $(a-2b)(a+2b)$ 11. $(5x-2y)(5x+2y)$ 12. $(3x-4y)(3x+4y)$
13. $\left(x-\frac{y}{2}\right)\left(x+\frac{y}{2}\right)$ 14. $(3m-\frac{2}{3}n)(3m+\frac{2}{3}n)$ 15. $(4t-\frac{2}{5}s)(4t+\frac{2}{5}s)$ 16. $\left(2x-\frac{z}{10}\right)\left(2x+\frac{z}{10}\right)$
17. $x(x-1)(x+1)$ 18. $a(a-b)(a+b)$ 19. $x(2x-1)(2x+1)$ 20. $2x(2x-y)(2x+y)$
21. $3x(2x-y)(2x+y)$ 22. $2m(3m-2n)(3m+2n)$ 23. $5(x-\frac{1}{2})(x+\frac{1}{2})$
24. $2a(5a-3b)(5a+3b)$ 25. $3y(2x-z)(2x+z)$ 26. $4ab(3a-b)(3a+b)$ 27. $2a^3(5a-2b)(5a+2b)$
28. $9xy(2x-5y)(2x+5y)$ 29. 161 30. 404 31. 4400
32. 2421 33. 4329 34. 0·75 35. 4·8 36. −2469 37. 0·0761 38. −10 900
39. 53·6 40. 0·000 005

Exercise 26 page 40

1. $-3, -4$ 2. $-2, -5$ 3. $3, -5$ 4. $2, -3$ 5. $2, 6$
6. $-3, -7$ 7. $2, 3$ 8. $5, -1$ 9. $-7, 2$ 10. $-\frac{1}{2}, 2$
11. $\frac{2}{3}, -4$ 12. $1\frac{1}{2}, -5$ 13. $\frac{2}{3}, 1\frac{1}{2}$ 14. $\frac{1}{4}, 7$ 15. $\frac{3}{5}, -\frac{1}{2}$
16. $7, 8$ 17. $\frac{5}{6}, \frac{1}{2}$ 18. $7, -9$ 19. $-1, -1$ 20. $3, 3$
21. $-5, -5$ 22. $7, 7$ 23. $-\frac{1}{3}, \frac{1}{2}$ 24. $-1\frac{1}{4}, 2$ 25. $13, -5$
26. $-3, \frac{1}{6}$ 27. $\frac{1}{10}, -2$ 28. $1, 1$ 29. $\frac{2}{9}, -\frac{1}{4}$ 30. $-\frac{1}{4}, \frac{3}{5}$

Exercise 27 page 41

1. $0, 3$ 2. $0, -7$ 3. $0, 1$ 4. $0, \frac{1}{3}$ 5. $4, -4$
6. $7, -7$ 7. $\frac{1}{2}, -\frac{1}{2}$ 8. $\frac{2}{3}, -\frac{2}{3}$ 9. $0, -1\frac{1}{2}$ 10. $0, 1\frac{1}{2}$
11. $0, 5\frac{1}{2}$ 12. $\frac{1}{4}, -\frac{1}{4}$ 13. $\frac{1}{2}, -\frac{1}{2}$ 14. $0, \frac{5}{8}$ 15. $0, \frac{1}{12}$
16. $0, 6$ 17. $0, 11$ 18. $0, 1\frac{1}{2}$ 19. $0, 1$ 20. $0, 4$
21. $0, 3$ 22. $\frac{1}{2}, -\frac{1}{2}$ 23. $1\frac{1}{3}, -1\frac{1}{3}$ 24. $3, -3$ 25. $0, 2\frac{2}{5}$
26. $\frac{1}{3}, -\frac{1}{3}$ 27. $0, \frac{1}{4}$ 28. $0, \frac{1}{6}$ 29. $\frac{1}{4}, -\frac{1}{4}$ 30. $0, \frac{1}{5}$

Exercise 28 page 41

1. $-\frac{1}{2}, -5$ 2. $-\frac{2}{3}, -3$ 3. $-\frac{1}{2}, -\frac{2}{3}$ 4. $\frac{1}{3}, 3$ 5. $\frac{2}{5}, 1$
6. $\frac{1}{3}, 1\frac{1}{2}$ 7. −0·63, −2·37 8. −0·27, −3·73 9. 0·72, 0·28 10. 6·70, 0·30
11. 0·19, −2·69 12. 0·85, −1·18 13. 0·61, −3·28 14. $-1\frac{2}{3}, 4$ 15. $-1\frac{1}{2}, 5$
16. 3·56, −0·56 17. 0·16, −3·16 18. $-\frac{1}{2}, 2\frac{1}{3}$ 19. $-\frac{1}{3}, -8$ 20. $1\frac{2}{3}, -1$
21. 2·28, 0·22 22. −0·35, −5·65 23. $-\frac{2}{3}, \frac{1}{2}$ 24. −0·58, 2·58 25. −2·69, 0·19
26. 0·22, −1·55 27. −0·37, 5·37 28. $-\frac{5}{6}, 1\frac{3}{4}$ 29. $-\frac{7}{9}, 1\frac{1}{4}$ 30. $1\frac{2}{5}, -2\frac{1}{4}$
31. $-4, 1\frac{1}{2}$ 32. $-3, 1\frac{2}{3}$ 33. $-2, 1\frac{2}{3}$ 34. $-3\frac{1}{2}, \frac{1}{5}$ 35. $-3, \frac{4}{5}$
36. $-8\frac{1}{2}, 11$

Exercise 29 page 42

1. $-3, 2$ 2. $-3, -7$ 3. $-\frac{1}{2}, 2$ 4. $1, 4$ 5. $-1\frac{2}{3}, \frac{1}{2}$
6. −0·39, −4·28 7. −0·16, 6·16 8. 3 9. $2, -1\frac{1}{3}$ 10. $-3, -1$

11. $-2, 2\frac{1}{2}$ **12.** $-3, 7$ **13.** $-4, \frac{2}{3}$ **14.** $-1{\cdot}83, 3{\cdot}83$ **15.** $-0{\cdot}16, 3{\cdot}16$
16. $-\frac{1}{3}, 2$ **17.** $0, \frac{1}{2}$ **18.** $0, -4$ **19.** $-\frac{1}{5}, -1$ **20.** $-1\frac{4}{5}, 1$
21. $0{\cdot}66, -22{\cdot}66$ **22.** $-7, 2$ **23.** $\frac{1}{4}, 7$ **24.** $-\frac{1}{2}, \frac{3}{5}$ **25.** $0, 3\frac{1}{2}$
26. $-\frac{1}{4}, \frac{1}{4}$ **27.** $-2{\cdot}77, 1{\cdot}27$ **28.** $-\frac{2}{3}, 1$ **29.** $-\frac{1}{2}, 2$ **30.** $0, 3$
31. $-1, 1\frac{3}{5}$ **32.** $-1\frac{2}{3}, 1$ **33.** $-\frac{2}{3}, 2$ **34.** $-2{\cdot}27, 1{\cdot}77$ **35.** $-1{\cdot}4, -1$
36. $-2, 0$ **37.** $-1\frac{6}{7}, 1$ **38.** -1 **39.** $\frac{3}{5}, 4$ **40.** $\frac{3}{4}, 1\frac{2}{5}$

Exercise 30 page 43

1. 8, 11 **2.** 11, 13 **3.** 12 cm **4.** 6 cm **5.** $x = 11$
6. 10 cm × 24 cm **7.** 8 km north, 15 km east **8.** 12 eggs **9.** 13 eggs
10. 4 **11.** 2, 5 **12.** $\frac{40}{x}$ h, $\frac{40}{x-2}$ h, 10 km/h **13.** 4 km/h
14. 20 mph **15.** 5 mph **16.** 157 km **17.** $x = 2$ **18.** $x = 3$
19. $\frac{3}{4}$ **20.** 9 cm or 13 cm

Revision Exercise 2A page 44

1. (a) $-2\frac{1}{2}$ (b) $2\frac{2}{3}$ (c) $0, -5$ (d) $2, -2$ (e) $-5, 2\frac{2}{3}$
2. (a) 14 (b) 18 (c) 28
3. (a) $(2x-y)(2x+y)$ (b) $2(x+3)(x+1)$ (c) $(2-3k)(3m+2n)$ (d) $(2x+1)(x-3)$
4. (a) $x = 3, y = -2$ (b) $m = 1\frac{1}{2}, n = -3$ (c) $x = 7, y = \frac{1}{2}$ (d) $x = -1, y = -2$
5. (a) 8 (b) 140 (c) 29 (d) 42 (e) 6 (f) -6
6. (a) $2x - 21$ (b) $(1-2x)(2a-3b)$ (c) 23
7. (a) 1 (b) $10\frac{1}{2}$ (c) $0, 3\frac{1}{2}$ (d) $-3, -2$ (e) 12
8. (a) $z(z-4)(z+4)$ (b) $(x^2+1)(y^2+1)$ (c) $(2x+3)(x+4)$ **9.** $\frac{7}{8}$
10. (a) $c = 5, d = -2$ (b) $x = 2, y = -1$ (c) $x = 9, y = -14$ (d) $s = 5, t = -3$
11. (a) $\frac{1}{2}, -\frac{1}{2}$ (b) $\frac{7}{11}$ (c) 3 (d) 0, 5
12. (a) $1{\cdot}78, -0{\cdot}28$ (b) $1{\cdot}62, -0{\cdot}62$ (c) $0{\cdot}87, -1{\cdot}54$ (d) $1{\cdot}54, -4{\cdot}54$
13. (a) $x = 9$ (b) $x = 10$
14. (a) 2 (b) -3 (c) 36 (d) 0 (e) 36 (f) 4
15. speed = 5 mph **16.** 8 cm × 6·5 cm **17.** (a) $-2, 4$ (b) 16 (c) $6{\cdot}19, 0{\cdot}81$
18. $-\frac{1}{5}, 3$ **19.** 8 **20.** $x = 13$

Examination Exercise 2B page 45

1. (i) $\frac{126}{x}$ (ii) $\frac{143}{x+2}$; $x = 9$; 19
2. (i) $(3x + 2y)$ pence (ii) $\left(\frac{3x+2y}{5}\right)$ pence (iii) $4x$ pence (iv) $\frac{100(x-2y)}{(3x+2y)}$
3. (a) $x - 1$ (b) $\frac{x}{1}; \frac{1}{x-1}$ (c) 1·62 **4.** $p = 30, q = 5$; $t = 0{\cdot}63$ s or $5{\cdot}37$ s
5. (a) $m = -7$ (b) (i) 6 cm²; (ii) $2x^2 + 7x - 6$, 5·05 cm
6. $\frac{135}{x}, \frac{135}{x-15}$; 60 km/h, 45 km/h **7.** (i) 10·5 (ii) 4, 5 (iii) 2·76, 7·24
8. $x^2 + 16$; $x^2 - 16x + 73$; 2, 6; 15 cm², 13 cm² **9.** $\frac{75}{V}$ h; $\frac{75}{V+20}$ h; $2\frac{1}{2}$ h; $37\frac{1}{2}$ km/h
10. (i) $5, -1$ (ii) $a = 3, -3$; $b = 4$ or -8 respectively; $(x^2 + 2x + 3)(x^2 - 2x + 3)$ **11.** $5, 7\frac{1}{2}$
12. (i) $-1{\cdot}6, 4{\cdot}1$ (ii) $6x^2$ cm²; $(x+1) \times (x-2) \times \frac{3x}{2}$, $8x^2 - 5x - 4$; volume = 68·9 cm³
13. (a) $-1\frac{2}{7}, 2$ (b) (i) $(6x + 10)$ cm (ii) $2x^2 + 9x + 4$; 31·8 cm

PART 3

Exercise 1 page 50

1. $10{\cdot}2\ m^2$ 2. $22\ cm^2$ 3. $103\ m^2$ 4. $9\ cm^2$ 5. $31\ m^2$
6. $6000\ cm^2$ 7. $39\ cm^2$ 8. $26\ m^2$ 9. $18\ cm^2$ 10. $20\ m^2$
11. 13 m 12. 15 cm 13. 56 m 14. 4·5 cm, 12·8 cm 15. 8 m, 10 m
16. 12 cm 17. $3\frac{1}{2}$ cm, $6\frac{1}{2}$ cm 18. 2500

Exercise 2 page 51

1. $48{\cdot}3\ cm^2$ 2. $28{\cdot}4\ cm^2$ 3. $66{\cdot}4\ m^2$ 4. $3{\cdot}07\ cm^2$ 5. $18{\cdot}2\ cm^2$
6. $12{\cdot}3\ cm^2$ 7. $2{\cdot}78\ cm^2$ 8. $36{\cdot}4\ m^2$ 9. $62{\cdot}4\ m^2$ 10. $30{\cdot}4\ m^2$
11. $44{\cdot}9\ cm^2$ 12. $0{\cdot}277\ m^2$ 13. $63\ m^2$ 14. $70{\cdot}7\ m^2$ 15. $14\ m^2$
16. $65{\cdot}8\ cm^2$ 17. $18{\cdot}1\ cm^2$ 18. $8{\cdot}03\ m^2$ 19. $14\ m^2$ 20. $52{\cdot}0\ cm^2$
21. $124\ cm^2$ 22. $69{\cdot}8\ m^2$ 23. $57{\cdot}1\ cm^2$ 24. 10·7 cm 25. 50·9°
26. 4·10 m 27. 4·85 m 28. 7·23 cm

Exercise 3 page 53

1. 31·4 cm 2. 12·6 m 3. 0·314 cm 4. 314 m 5. 17·9 m
6. $3{\cdot}33 \times 10^5$ m 7. 0·0254 m 8. 9·86 km 9. $3{\cdot}14 \times 10^6$ km 10. $314\ cm^2$
11. $12{\cdot}6\ m^2$ 12. $0{\cdot}0314\ m^2$ 13. $31400\ m^2$ 14. $7{\cdot}07 \times 10^8\ m^2$ 15. $3{\cdot}14 \times 10^{-6}\ m^2$
16. $0{\cdot}196\ cm^2$ 17. $31{\cdot}0\ cm^2$ 18. $1{\cdot}26 \times 10^{-3}\ cm^2$ 19. 44 cm, $154\ cm^2$ 20. 88 cm, $616\ cm^2$
21. 220 m, $3850\ m^2$ 22. 2·2 cm, $0{\cdot}385\ cm^2$ 23. 10 cm 24. 4 m
25. 0·7 m 26. 0·1 km 27. 200 m 28. 4000 m 29. 2 cm
30. 10 m 31. 3 m 32. 3·99 cm 33. 15·5 km 34. 0·152 cm
35. 1670 m 36. 0·564 m

Exercise 4 page 54

1. (a) $c = 15{\cdot}7$ cm, $A = 19{\cdot}6\ cm^2$ (b) $r = 4{\cdot}46$ cm, $A = 62{\cdot}4\ cm^2$ (c) $r = 2{\cdot}52$ m, $c = 15{\cdot}8$ m
(d) $r = 796$ km, $A = 1{\cdot}99 \times 10^6\ km^2$ (e) $r = 0{\cdot}155$ cm, $c = 0{\cdot}971$ cm
2. $21{\cdot}5\ cm^2$ 3. (a) $40{\cdot}8\ m^2$ (b) 6 4. 30; (a) $1510\ cm^2$ (b) $509\ cm^2$
5. 5307 6. 29 7. 970 8. (a) 80 (b) 7
9. 5·39 cm ($\sqrt{29}$) 10. (a) 33·0 cm (b) $70{\cdot}8\ cm^2$ 11. (a) $98\ cm^2$ (b) $14{\cdot}0\ cm^2$
12. 1 : 3 : 5

Exercise 5 page 55

1. $4\ cm^2$ 2. 7·5 cm, $37{\cdot}5\ cm^2$ 3. 7·83 m, $7{\cdot}83\ m^2$ 4. 1·5 m, $0{\cdot}225\ m^2$
5. 6 m, $60\ m^2$ 6. 90°, $18{\cdot}75\ cm^2$ 7. 300°, $490\ cm^2$ 8. 60°, $13{\cdot}52\ cm^2$
9. 60°, 2 m 10. 90°, 15 m 11. 317°, 31·7 cm 12. 4 cm, $12\ cm^2$
13. 12 cm, $288\ cm^2$ 14. 109 m, $5450\ m^2$ 15. 5 cm, 6 cm 16. 12 m, 27 m
17. 5·48 cm, 9·86 cm 18. (a) 12 cm (b) 30° 19. (a) 3·98 cm (b) 74·9°
20. (a) 30° (b) 10·5 cm 21. (a) 18 cm (b) 40° 22. (a) 10 cm (b) 45°

Exercise 6 page 57

1. (a) 14·5 cm (b) $72{\cdot}6\ cm^2$ (c) $24{\cdot}5\ cm^2$ (d) $48{\cdot}1\ cm^2$
2. (a) 6·88 cm, 7·33 cm (b) 22·1 cm, 25·4 cm (c) 6·98 cm, $5{\cdot}13\ cm^2$ (d) 12·7 cm, $27{\cdot}7\ cm^2$
(e) 60°, 10·5 cm, $9{\cdot}03\ cm^2$ (f) 106·3°, $11{\cdot}2\ cm^2$ (g) 97·2°, 10·2 cm, $12{\cdot}7\ cm^2$
(h) 6·53 cm, $16{\cdot}2\ cm^2$ (i) 7·63 m, 8·42 m (j) 25·7 cm, 30·9 cm (k) 120°, $6140\ cm^2$
(l) 9·88 cm, 12·9 cm (m) 8·38 cm (n) 54·8 cm (o) 10·4 cm, 20·1 cm
3. 3 cm 4. 3·97 cm 5. $13{\cdot}5\ cm^2$, $404\ cm^3$
6. (a) $130\ cm^2$ (b) $184\ cm^2$ 7. $459\ cm^2$, $651\ cm^2$ 8. $19{\cdot}5\ cm^2$ 9. $0{\cdot}312r^2$

Exercise 7 page 58

1. (a) 160 cm^2 (b) 300 cm^3 (c) 255 cm^3 (d) 240 cm^3
 (e) 5400 cm^3 (f) 70700 cm^3
2. (a) 502 cm^3 (b) 760 m^3 (c) 12·5 cm^3 (d) 6·28 cm^3
3. 3·98 cm 4. 6·37 cm 5. 1·89 cm 6. 5·37 cm
7. 9·77 cm 8. 7·38 cm 9. 1270 cm 10. 4·24 litres
11. 10 cm/s 12. 1570 cm^3, 12·56 kg 13. 3:4 14. cubes by 77 cm^3
15. No 16. 1·19 cm 17. 53 times 18. 191 cm

Exercise 8 page 60

1. 20·9 cm^3 2. 523 cm^3 3. 4190 cm^3 4. 100 cm^3 5. 268 cm^3
6. 4·19x^3 cm^3 7. 0·00419 m^3 8. 3 cm^3 9. 93·3 cm^3 10. 48 cm^3
11. 92·4 cm^3 12. 262 cm^3 13. 234 cm^3 14. 414 cm^3 15. 5 m
16. 2·43 cm 17. 23·9 cm 18. 6 cm 19. 3·72 cm 20. 1930 g
21. 106 s 22. (a) 125 (b) 2744 (c) 2·7 × 10^7
23. (a) 0·36 cm (b) 0·427 cm (c) 3·6 cm 24. (a) 6·69 cm (b) 39·1 cm (c) 2·71 cm
25. $10\frac{2}{3}$ cm^3 26. 1·05 cm^3 27. 488 cm^3 28. 4·19 cm^3 29. 53·6 cm^3
30. 74·4 cm^3 31. 4·24 cm 32. 122 cm^3 33. 54 400 cm^3 34. 943 cm^3 35. 5050 cm^3
36. (a) 0·118 cm^3 (c) 8 cm^3

Exercise 9 page 62

1. (a) 36π cm^2 (b) 40π cm^2, 72π cm^2 (c) 60π cm^2 (d) $1{\cdot}4\pi$ m^2, $2{\cdot}38\pi$ m^2
 (e) 400π m^2 (f) 65π cm^2 (g) 192π mm^2 (h) $10{\cdot}2\pi$ cm^2
 (i) $0{\cdot}0004\pi$ m^2 (j) 98π cm^2, 147π cm^2
2. 1·65 cm 3. 2·12 cm 4. 3·46 cm
5. (a) 3 cm (b) 4 cm (c) 3 cm (d) 0·2 m (e) 6 cm (f) 2·5 cm (g) 6 cm
6. 303 cm^2 7. 147 cm^2 8. £1180 9. £3870 10. 94·1 cm^3
11. 44·6 cm^2 12. 675 cm^2 13. 1·62 × 10^8 years 14. 377 cm^2
15. 20 cm, 10 cm 16. 71·7 cm^2

Exercise 10 page 64

1. (a) 4450 km (b) 7790 km (c) 6340 km (d) 8340 km (e) 4180 km (f) 2960 km
2. (a) 4200 n.m. (b) 2400 n.m. (c) 1052 n.m. (d) 625 n.m. (e) 2070 n.m. (f) 690 n.m.
3. (a) 5560 n.m. (b) 7110 n.m. (c) 2330 n.m. (d) 18900 n.m. (e) 11400 n.m. (f) 6450 n.m.
4. (a) 3000 n.m. (b) 3840 n.m. (c) 1260 n.m. (d) 10200 n.m. (e) 6160 n.m. (f) 3480 n.m.
5. (a) 2440 km (b) 4000 km (c) 2500 km (d) 4720 km
6. (a) (60°N, 10°W) (b) (25°N, 17°W) (c) (16°N, 111°W) (d) (59° 30′S, 19°E)
 (e) (66° 30′N, 11° 10′W) (f) (17° 40′N, 10°W)
7. (a) (0°, 18°W) (b) (0°, 22°W)
8. (a) (60° 30′N, 10°W) (b) (51° 30′N, 10°W) (c) (33° 30′N, 10°W)
9. (a) (17° 30′S, 112°W) (b) (5°N, 112°W) 10. 750 km/h

Exercise 11 page 66

1. (a) 3185 km (b) 5460 km (c) 1300 km (d) 6270 km (e) 2580 km (f) 5740 km
2. (a) 5100 km (b) 8560 km (c) 3830 km (d) 4460 km (e) 1010 km (f) 2500 km
3. (a) 3300 n.m. (b) 2860 n.m. 4. (a) 3600 n.m. (b) 2010 n.m.
5. (a) 1170 km/h (b) 1250 km/h (c) 2360 km/h
6. (a) 1020 n.m. (b) 4440 n.m. (c) 1800 n.m. (d) 1310 n.m. (e) 479 n.m. (f) 2460 n.m.
7. (46° 30′N, 10°W) 8. (25°N, 92° 15′W) 9. (a) 5560 km (b) 8450 km
10. (a) 5960 km (b) 5600 km (c) 52·1° 11. (a) 3350 km (b) 3050 km (c) 27·7°
12. (32°N, 10°W), 600 knots 13. 6 hours $37\frac{1}{2}$ minutes after noon
14. (25°N, 22° 25′W) 15. (66°S, 85° 54′W)

Revision Exercise 3A page 67

1. (a) $14\,cm^2$ (b) $54\,cm^2$ (c) $50\,cm^2$ (d) $18\,m^2$
2. (a) $56{\cdot}5\,m$, $254\,m^2$ (b) $10{\cdot}8\,cm$ (c) $3{\cdot}99\,cm$
3. (a) $9\pi\,cm^2$ (b) 8:1
4. $3{\cdot}44\,cm^2$, $4{\cdot}56\,cm^2$
5. (a) $12{\cdot}2\,cm$ (b) $61{\cdot}1\,cm^2$
6. (a) $11{\cdot}2\,cm$ (b) $10{\cdot}3\,cm$ (c) $44{\cdot}7\,cm^2$ (d) $31{\cdot}5\,cm^2$ (e) $13{\cdot}1\,cm^2$
7. $103{\cdot}2^\circ$
8. $9{\cdot}95\,cm$
9. (a) $904\,cm^3$ (b) $5{\cdot}76\,cm$
10. $8{\cdot}06\,cm$
11. $99{\cdot}5\,cm^3$
12. $332\,cm^3$, $201\,cm^3$
13. 4 cm
14. (a) $15{\cdot}6\,cm^2$ (b) $93{\cdot}5\,cm^2$ (c) $3740\,cm^3$
15. $0{\cdot}370\,cm$
16. $104\,cm^2$
17. (a) 3910 km (b) 4280 km (c) 6920 km
18. (a) 2220 n.m. (b) 1970 n.m. (c) 2100 n.m.
19. (a) $(80^\circ N, 15^\circ E)$ (b) $(55^\circ N, 58^\circ 35' E)$
20. (a) $70^\circ W$ (b) $70^\circ N$

Examination Exercise 3B page 69

1. (a) 8 cm (b) $512\,cm^3$ (c) $384\,cm^2$ (d) (i) 8 (ii) 8 (iii) 24 (iv) 24
   (e) $1152\,cm^2$
2. (a) $\frac{a^2}{16}\,cm^2$ (b) $\frac{b}{2\pi}\,cm$, $\frac{b^2}{4\pi}\,cm^2$ (c) $\frac{c}{\pi+2}\,cm$, $\frac{\pi c^2}{2(\pi+2)^2}\,cm^2$
3. (i) $628{\cdot}4\,cm^3$ (ii) $329{\cdot}9\,cm^2$ (iii) $4{\cdot}6\,cm$
4. (ii) (a) $0{\cdot}866\,cm$ (b) $0{\cdot}433\,cm^2$ (iii) (a) $\frac{\pi}{3}\,cm^2$ (b) $1{\cdot}228\,cm^2$, $39{\cdot}1\%$
5. 840 g/s; $1050\,cm^3$; 2 min 18 s
6. (a) $270\,000\,cm^2$ (b) $415\,800\,cm^3$ (c) $3850\,cm^2$ (d) 108 cm; 83 times
7. $4\pi x^2\,cm^2$; $\frac{x^2}{2}(\pi - 2)\,cm^2$; 145 cm
8. (i) $3{\cdot}46\,mm$ (ii) $41{\cdot}6\,mm^2$ (iii) $3{\cdot}14\,mm^2$ (iv) $6920\,mm^3$; $4{\cdot}8\,cm^3$
9. 5280 km; $34^\circ N$; 48°; $58^\circ E$ or $38^\circ W$
10. (i) 5400 n.m. (ii) 16200 n.m. (iii) 3020 n.m. (iv) 9740 n.m., 61 h
11. (i) $22{\cdot}45\,cm$, $15{\cdot}2\,cm$ (ii) $10{\cdot}11\,cm^2$ (iii) 11 cm (iv) $113{\cdot}1\,cm$

PART 4

Exercise 1 page 73

1. 95°
2. 49°
3. 100°
4. 77°
5. 129°
6. 95°
7. $a = 30^\circ$
8. $e = 30^\circ, f = 60^\circ$
9. 110°
10. $x = 54^\circ$
11. $a = 40^\circ$
12. $a = 36^\circ, b = 72^\circ, c = 144^\circ, d = 108^\circ$
13. 105°
14. $a = 30^\circ, b = 120^\circ, c = 150^\circ$
15. $x = 20^\circ, y = 140^\circ$
16. $a = 120^\circ, b = 34^\circ, c = 26^\circ$
17. 117°
18. $a = 30^\circ, b = 60^\circ, c = 150^\circ, d = 120^\circ$
19. $a = 10^\circ, b = 80^\circ$
20. $e = 71^\circ, f = 21^\circ$
21. 144°
22. 70°
23. $41^\circ, 66^\circ$
24. $46^\circ, 122^\circ$
25. $a = 72^\circ, b = 108^\circ$
26. $x = 60^\circ, y = 120^\circ$
27. 110°
28. 60°
29. $128\frac{4}{7}^\circ$
30. 15
31. 12
32. 9
33. 18
34. 12

Exercise 2 page 76

1. $a = 116^\circ, b = 64^\circ, c = 64^\circ$
2. $a = 64^\circ, b = 40^\circ$
3. $x = 68^\circ$
4. $a = 40^\circ, b = 134^\circ, c = 134^\circ$
5. $m = 69^\circ, y = 65^\circ$
6. $t = 48^\circ, u = 48^\circ, v = 42^\circ$
7. $a = 118^\circ, b = 100^\circ, c = 62^\circ$
8. $a = 34^\circ, b = 76^\circ, c = 70^\circ, d = 70^\circ$
9. $72^\circ, 108^\circ$
10. 88°
11. 298°
12. 50°
13. $36^\circ, 54^\circ$
14. $36^\circ, 36^\circ, 72^\circ$
15. 90°
16. $a = 52^\circ, b = 139^\circ, c = 100^\circ, d = 100^\circ$
17. $m = 20^\circ, n = 130^\circ$
18. 160°

19. $a+b+c=360°$ **20.** $l+m-n=180°$ **21.** 144°, 120°, 60°, 36°
22. 74°, 166° **23.** 108°, 160°, 160°

Exercise 3 *page 79*

1. (a) $\frac{5}{8}$ (b) $3\frac{3}{4}$ cm **2.** (a) $\frac{4}{7}$ (b) $2\frac{6}{7}$ cm **3.** (a) $\frac{5}{6}$ (b) $\frac{5}{11}$
4. 4 cm **5.** $6\frac{2}{13}$ cm, $3\frac{11}{13}$ cm **6.** $2\frac{4}{7}$ cm, $3\frac{3}{7}$ cm **7.** 6 cm **8.** 4 cm **9.** $3\frac{1}{2}$ cm

Exercise 4 *page 80*

1. 10 cm **2.** 4·12 cm **3.** 4·24 cm **4.** 9·90 cm **5.** 8·72 cm **6.** 5·66 cm
7. 6·63 cm **8.** 5 cm **9.** 17 cm **10.** 4 cm **11.** 9·85 cm **12.** 7·07 cm
13. 3·46 m **14.** 40·3 km **15.** 13·6 cm **16.** 6·34 m **17.** 4·58 cm **18.** 84·9 km
19. 24 cm **20.** 9·80 cm **21.** 5, 4, 3; 13, 12, 5; 25, 24, 7; 41, 40, 9; 61, 60, 11
22. $x=4$ m, 20·6 m

Exercise 5 *page 82*

1. Yes, S.S.S. **2.** Yes, S.A.S. **3.** No **4.** Yes, A.A.S. **5.** No **6.** Yes, A.A.S.
7. Yes, A.A.S. **8.** No **9.** Yes, S.S.S. **10.** No

Exercise 6 *page 85*

1. (a) 1, 1 (b) 1, 1 (c) 2, 2 (d) 2, 2 (e) 4, 4 (f) 0, 2
(g) 0, 2 (h) 0, 1 (i) 1, 1 (j) 0, 2 (k) 0, 2 (l) ∞, ∞
(m) 5, 5
2. square 4, 4; rectangle 2, 2; parallelogram 0, 2; rhombus 2, 2; trapezium 0, 1; kite 1, 1; equilateral triangle 3, 3; regular hexagon 6, 6
3. 34°, 56° **4.** 35°, 35° **5.** 72°, 108°, 80° **6.** 40°, 30°, 110° **7.** 116°, 32°, 58°
8. 55°, 55° **9.** 26°, 26°, 77° **10.** 52°, 64°, 116° **11.** 70°, 40°, 110° **12.** 54°, 72°, 36°
13. 60°, 15°, 75°, 135°

Exercise 7 *page 86*

1. $a=2\frac{1}{2}$ cm, $e=3$ cm **2.** $x=6$ cm, $y=10$ cm **3.** $x=12$ cm, $y=8$ cm **4.** $m=10$ cm, $a=16\frac{2}{3}$ cm
5. $y=6$ cm **6.** $x=4$ cm, $w=1\frac{1}{2}$ cm **7.** $e=9$ cm, $f=4\frac{1}{2}$ cm **8.** $x=13\frac{1}{3}$ cm, $y=9$ cm
9. $m=6$ cm, $n=6$ cm **10.** $m=5\frac{1}{3}$ cm, $z=4\frac{4}{5}$ cm **11.** $v=5\frac{1}{3}$ cm, $w=6\frac{2}{3}$ cm **12.** Yes
13. No **14.** Yes **15.** No **16.** No **17.** 2 cm, 6 cm **18.** 16 m
19. (a) Yes (b) No (c) No (d) Yes (e) Yes (f) No (g) No (h) Yes

Exercise 8 *page 89*

1. 16 cm^2 **2.** 27 cm^2 **3.** $11\frac{1}{4}$ cm^2 **4.** $14\frac{1}{2}$ cm^2 **5.** 128 cm^2 **6.** 12 cm^2
7. 8 cm **8.** 18 cm **9.** $4\frac{1}{2}$ cm **10.** $7\frac{1}{2}$ cm **11.** $2\frac{1}{2}$ cm **12.** 6 cm
13. (a) $16\frac{2}{3}$ cm^2 (b) $10\frac{2}{3}$ cm^2 **14.** (a) 25 cm^2 (b) 21 cm^2
15. 8 cm^2 **16.** 6 cm **17.** 24 cm^2 **18.** (a) $1\frac{4}{5}$ cm (b) 3 cm (c) 3:5 (d) 9:25

Exercise 9 *page 91*

1. 480 cm^3 **2.** 540 cm^3 **3.** 160 cm^3 **4.** 4500 cm^3 **5.** 81 cm^3 **6.** 11 cm^3
7. 16 cm^3 **8.** $85\frac{1}{3}$ cm^3 **9.** 4 cm **10.** 21 cm **11.** 4·6 cm **12.** 9 cm
13. 6·6 cm **14.** $4\frac{1}{2}$ cm **15.** $168\frac{3}{4}$ cm^3 **16.** 106·3 cm^3 **17.** 12 cm
18. (a) 2:3 (b) 8:27 **19.** 8:125 **20.** ${x_1}^3:{x_2}^3$ **21.** 54 kg **22.** 240 cm^2
23. $9\frac{3}{8}$ cm^3 **24.** $2812\frac{1}{2}$ cm^2 **25.** 100 g

Exercise 10 page 94

1. $a = 27^\circ, b = 30^\circ$ 2. $c = 20^\circ, d = 45^\circ$ 3. $c = 58^\circ, d = 41^\circ, e = 30^\circ$
4. $f = 40^\circ, g = 55^\circ, h = 55^\circ$ 5. $a = 32^\circ, b = 80^\circ, c = 43^\circ$ 6. $x = 34^\circ, y = 34^\circ, z = 56^\circ$
7. 43° 8. 92° 9. 42° 10. $c = 46^\circ, d = 44^\circ$
11. $e = 49^\circ, f = 41^\circ$ 12. $g = 76^\circ, h = 52^\circ$ 13. 48° 14. 32°
15. 22° 16. $a = 36^\circ, x = 36^\circ$

Exercise 11 page 96

1. $a = 94^\circ, b = 75^\circ$ 2. $c = 101^\circ, d = 84^\circ$ 3. $x = 92^\circ, y = 116^\circ$ 4. $c = 60^\circ, d = 45^\circ$
5. 37° 6. 118° 7. $e = 36^\circ, f = 72^\circ$ 8. 35° 9. 18° 10. 90° 11. 30°
12. $22\frac{1}{2}^\circ$ 13. $n = 58^\circ, t = 64^\circ, w = 45^\circ$ 14. $a = 32^\circ, b = 40^\circ, c = 40^\circ$
15. $a = 18^\circ, c = 72^\circ$ 16. 55° 17. $e = 41^\circ, f = 41^\circ, g = 41^\circ$ 18. 8°
19. $x = 30^\circ, y = 115^\circ$ 20. $x = 80^\circ, z = 10^\circ$

Exercise 12 page 98

1. 115° 2. 33° 3. 53° 4. $f = 29^\circ, g = 55^\circ, h = 96^\circ$ 5. 35° 6. 58°
7. 30° 8. 84° 9. $e = 106^\circ, f = 74^\circ, g = 106^\circ$ 10. $x = 18^\circ, y = 54^\circ$
11. $a = 36^\circ, b = 108^\circ$ 12. $m = 24^\circ, n = 42^\circ$ 13. $x = 12^\circ, y = 12^\circ, z = 80^\circ$
14. $k = 64^\circ, m = 52^\circ, n = 38^\circ$ 15. 28° 16. 24° 17. $a = 28^\circ, b = 62^\circ, c = 34^\circ$
18. 60° 19. (a) 30° (b) 30° (c) 36° (d) 132° 20. (a) 100°
21. (a) $2z$ (b) $180 - 4z$ (c) $3z$ 22. (a) $40 + b$ (b) 25°
23. $14^\circ, 26^\circ, 140^\circ$ 24. $52^\circ, 52^\circ, 26^\circ, 116^\circ$ 26. (a) 18° (b) 108° (c) 126°
27. (a) 111° (b) 119°

Exercise 13 page 101

1. $a = 18^\circ$ 2. $x = 40^\circ, y = 65^\circ, z = 25^\circ$ 3. $c = 30^\circ, e = 15^\circ$
4. $f = 50^\circ, g = 40^\circ$ 5. $h = 70^\circ, k = 40^\circ, i = 40^\circ$ 6. $m = 108^\circ, n = 36^\circ$
7. $x = 50^\circ, y = 68^\circ$ 8. $a = 74^\circ, b = 32^\circ$ 9. $e = 36^\circ$ 10. $k = 63^\circ, m = 54^\circ$
11. $k = 50^\circ, m = 50^\circ, n = 80^\circ, p = 80^\circ$ 12. $n = 16^\circ, p = 46^\circ$ 13. (a) 24° (b) 78° (c) 48°
14. (a) p (b) $2p$ (c) $90 - 2p$ 15. $x = 70^\circ, y = 20^\circ, z = 55^\circ$
16. (b) $2a, 180 - 3a$ 19. $55^\circ, 60^\circ, 65^\circ$ 20. (a) 64° (b) $180 - 2x$

Exercise 14 page 104

1. 3 2. 4 5. 3 6. 1 7. 8 8. 4
9. (a) 3 (b) 12 (c) 3:2 (d) 4:1
10. (a) 3 (b) 6 (c) 9:1 (d) 4:9

Exercise 15 page 105

1. 93° 2. 36° 7. 7·8 cm
8. (a) 7·2 cm, 5·85 cm (b) 5·2 cm, 10·4 cm (c) 12·2 cm (d) 8·2 cm, 11·15 cm
10. 5·0 cm 13. 10·4 cm 14. 3·5 cm 15. 6·6 cm 16. 4·9 cm, 7·75 cm

Revision Exercise 4A page 109

2. 80° 3. (a) 30° (b) $22\frac{1}{2}^\circ$ (c) 12 4. $5\frac{1}{4}$ cm 5. 5 cm 6. (a) 40° (b) 100°
7. 4·12 cm 8. (i) 3 cm (ii) 5·66 cm 11. (c) $2\frac{4}{5}$ cm 12. (b) 6 cm
13. $3\frac{2}{3}$ cm, $1\frac{1}{11}$ cm 14. 6 cm 15. (c) $5\frac{1}{3}$ cm 16. 250 cm^3
17. (a) $3\frac{1}{3}$ cm (b) 1620 cm^3 18. (a) 1 m^2 (b) 1000 cm^3
19. (a) 50° (b) 128° (c) $c = 50^\circ, d = 40^\circ$ (d) $x = 10^\circ, y = 40^\circ$
20. (a) 15° (b) $109^\circ, 71^\circ, 35\frac{1}{2}^\circ, 35\frac{1}{2}^\circ$ (c) $31^\circ, 59^\circ, 31^\circ$ (d) $32^\circ, 58^\circ, 26^\circ$

21. (a) 5·29 cm (b) 2 cm (c) 7·07 cm (d) 1 cm **22.** (b) 3 cm (c) $3\frac{3}{4}$ cm^2
23. (a) 55° (b) 45° **24.** (b) 3·6 cm

Examination Exercise 4B page 113

1. $x°, x°, x°, (2x-y)°, (180-2x)°, (x-y)°$; 4·3 cm **2.** 40°, 27°, 25°, 25°, 52°
3. 33°, 33°, 44° **5.** (c) $1\frac{1}{3}$ cm^2 (d) $\frac{16}{9}$ **6.** (a) 85 cm^2 (b) 16·1 cm (d) 34°
7. (c) $\frac{1}{2}$ (d) $\frac{4}{1}$ **9.** 7:16 **10.** 8·65 cm; 23·3 cm^2
12. (i) 12 cm (ii) 4·8 cm (iii) 3:5 (iv) 9:25 (v) $xy = 24$

PART 5

Exercise 1 page 116

1. $\frac{5}{7}$ **2.** $\frac{7}{8}$ **3.** $\frac{15}{4}$ **4.** $\frac{11}{2}$ **5.** 5 **6.** $\frac{3}{4}$
7. 1 **8.** 1 **9.** $8x$ **10.** $5y$ **11.** $\frac{1}{2}$ **12.** $4x$
13. $\frac{11}{12}$ **14.** $\frac{7}{9}$ **15.** $\frac{5b}{6}$ **16.** $\frac{x}{2y}$ **17.** 2 **18.** $\frac{a}{2}$
19. $\frac{2b}{3}$ **20.** $\frac{11}{12x}$ **21.** $\frac{a}{5b}$ **22.** a **23.** $\frac{7}{8}$ **24.** $\frac{-3}{5x}$
25. $\frac{y^2}{2}$ **26.** $\frac{2y}{3a}$ **27.** $\frac{4m}{5n}$ **28.** $\frac{9xb}{5}$ **29.** $\frac{7}{11xy}$ **30.** $\frac{3y}{4x}$
31. $\frac{7m}{2a}$ **32.** $\frac{4a}{3}$ **33.** $\frac{2a}{3b}$ **34.** $\frac{7y}{6x}$ **35.** $\frac{4c^2}{3a^2}$ **36.** $\frac{a}{2}$
37. $\frac{6x}{5}$ **38.** $-\frac{4y}{x}$ **39.** $100a^3$

Exercise 2 page 117

1. $\frac{5+2x}{3}$ **2.** $\frac{3x+1}{x}$ **3.** $\frac{32}{25}$ **4.** $\frac{4+5a}{5}$ **5.** $\frac{3}{4-x}$ **6.** $\frac{b}{3+2a}$
7. $\frac{3-x}{2}$ **8.** $\frac{5x+4}{8x}$ **9.** 12 **10.** $\frac{3x+2}{4x+1}$ **11.** $\frac{2x+1}{y}$ **12.** $\frac{2-x}{x}$
13. $y+xy$ **14.** $\frac{x+2y}{3xy}$ **15.** $\frac{2x+5}{2}$ **16.** $\frac{10x}{5x-3}$ **17.** $\frac{5x+4}{3x-2}$ **18.** $\frac{3+6x}{2x}$
19. $\frac{2x-1}{2}$ **20.** $\frac{6-b}{2a}$ **21.** $\frac{2b+4a}{b}$ **22.** $2ab-3b^2$ **23.** $\frac{1-3x^2}{x}$ **24.** $\frac{1+4a}{b+b^2}$
25. $\frac{1+2x+2x^2}{x}$ **26.** $\frac{6n-3m}{2mn}$ **27.** $\frac{1-3ac}{2c}$ **28.** $2y+3x-5$
29. $\frac{3a+2}{4d-5a}$ **30.** $\frac{7-28x}{xz+x}$

Exercise 3 page 118

1. $\frac{x+2}{x-3}$ **2.** $\frac{x}{x+1}$ **3.** $\frac{x+4}{2(x-5)}$ **4.** $\frac{x+5}{x-2}$ **5.** $\frac{x+3}{x+2}$ **6.** $\frac{x+5}{x-2}$
7. $\frac{x}{2x-1}$ **8.** $\frac{2x+1}{2x+3}$ **9.** $\frac{x-1}{3x-4}$ **10.** $\frac{5-x}{3+x}$ **11.** $\frac{x-4}{2}$ **12.** $\frac{x}{x-7}$

13. $\frac{5+x}{1-x}$ 14. $\frac{x^2+1}{x^2}$ 15. $2x^2-1$ 16. $2x-1$ 17. $12x+1$ 18. $\frac{x}{x-3}$

19. $3x^2-1$ 20. $30x-2$ 21. $\frac{1-4x}{2}$ 22. $\frac{3x^2+1}{x^2+2}$ 23. $\frac{2x+7}{x-2}$ 24. $\frac{6x+2}{12x-3}$

25. -1 26. -2 27. $-\frac{1}{4}$ 28. $\frac{1-4x}{2+8x}$ 29. $\frac{4x-2}{1-4x}$ 30. $\frac{3x-2}{5+3x}$

31. $\frac{1}{2-x}$ 32. $\frac{3x+2}{x^2}$ 33. $\frac{x+2}{x}$ 34. $\frac{x}{2x+3}$ 35. x^2-1 36. $\frac{x+2}{x}$

37. $\frac{x+3}{x+4}$ 38. $\frac{x+1}{x-3}$ 39. $\frac{x+1}{2x+1}$ 40. $\frac{x+3}{x}$

Exercise 4 page 119

1. $\frac{3}{5}$ 2. $\frac{3x}{5}$ 3. $\frac{3}{x}$ 4. $\frac{4}{7}$ 5. $\frac{4x}{7}$ 6. $\frac{4}{7x}$ 7. $\frac{7}{8}$ 8. $\frac{7x}{8}$

9. $\frac{7}{8x}$ 10. $\frac{5}{6}$ 11. $\frac{5x}{6}$ 12. $\frac{5}{6x}$ 13. $\frac{23}{20}$ 14. $\frac{23x}{20}$ 15. $\frac{23}{20x}$ 16. $\frac{1}{12}$

17. $\frac{x}{12}$ 18. $\frac{1}{12x}$ 19. $\frac{1}{6}$ 20. $\frac{x}{6}$ 21. $\frac{1}{6x}$ 22. $\frac{19x}{15}$ 23. $\frac{x}{14}$ 24. $\frac{7}{12x}$

25. $\frac{23}{12}$ 26. $\frac{31x}{12}$ 27. $\frac{23}{12x}$ 28. $\frac{11}{20}$ 29. $-\frac{5}{42x}$ 30. $\frac{17}{10y}$ 31. $\frac{11}{3x}$ 32. $\frac{7}{4a}$

Exercise 5 page 120

1. $\frac{5x+2}{6}$ 2. $\frac{7x+2}{12}$ 3. $\frac{9x+13}{10}$ 4. $\frac{1-2x}{12}$

5. $\frac{2x-9}{15}$ 6. $\frac{-3x-12}{14}$ 7. $\frac{3x+1}{x(x+1)}$ 8. $\frac{7x-8}{x(x-2)}$

9. $\frac{8x+9}{(x-2)(x+3)}$ 10. $\frac{4x+11}{(x+1)(x+2)}$ 11. $\frac{-3x-17}{(x+3)(x-1)}$ 12. $\frac{11-x}{(x+1)(x-2)}$

13. $\frac{2x(2x+1)}{(x+1)(x-1)}$ 14. $\frac{x(x-7)}{(x+2)(x-1)}$ 15. $\frac{x^2+4x+6}{2(x+2)}$ 16. $\frac{x^2-6x-3}{3(x-3)}$

17. $\frac{4x-1}{x(x-1)(x+2)}$ 18. $\frac{8x-14}{(x-1)(x+2)(x-3)}$ 19. $\frac{5x+14}{(x-1)(x+2)(x+3)}$ 20. $\frac{2x-1}{x(x+2)(x-3)}$

Exercise 6 page 120

1. a 2. $\frac{10m}{3}$ 3. $\frac{2y}{3x}$ 4. $\frac{15b}{4a}$ 5. $\frac{4}{a^2}$ 6. $\frac{8}{3}$ 7. $\frac{x-1}{x+2}$ 8. $\frac{y^2}{5x}$

9. $\frac{x}{8}$ 10. $3(x+2)$ 11. $\frac{5}{2(x-1)}$ 12. $\frac{1}{2(x+3)}$ 13. $\frac{1}{x-2}$

14. $\frac{y(x-1)}{2}$ 15. $\frac{x-2}{2x}$ 16. $\frac{5x+1}{(x+1)(x-1)}$ 17. $\frac{x+18}{(x-2)(x+3)}$

18. $\frac{5(2x+7)}{x(x+5)}$ 19. $\frac{x^2-9}{x+2}$ 20. $\frac{x(1-2x)}{(x-1)(x-2)}$ 21. $\frac{6}{xy}$ 22. $x(x+1)$

23. $\frac{6c}{a}$ 24. $\frac{12y^2z}{x}$ 25. $\frac{x-3}{x(x+1)}$ 26. $\frac{7}{6x}$ 27. $\frac{3x^2+5y^2}{xy}$ 28. $\frac{11}{2(x-2)}$

29. $\frac{x^2+1}{xy}$ 30. $\frac{3x+5}{x(x+1)}$ 31. $\frac{x}{2m}$ 32. $\frac{9z}{xy}$ 33. 1 34. $\frac{ab^3}{x}$
35. $\frac{y}{x+1}$ 36. $\frac{3bc}{4}$ 37. $\frac{x}{x-1}$ 38. $\frac{z^2}{y}$

Exercise 7 page 121

1. $2\frac{1}{2}$ 2. 3 3. $\frac{B}{A}$ 4. $\frac{T}{N}$ 5. $\frac{K}{M}$ 6. $\frac{4}{y}$
7. $\frac{C}{B}$ 8. $\frac{D}{4}$ 9. $\frac{T+N}{9}$ 10. $\frac{B-R}{A}$ 11. $\frac{R+T}{C}$ 12. $\frac{N-R^2}{L}$
13. $\frac{R-S^2}{N}$ 14. 2 15. -7 16. $T-A$ 17. $S-B$ 18. $N-D$
19. $M-B$ 20. $L-D^2$ 21. $T-N^2$ 22. $N+M-L$ 23. $R-S-Z$ 24. 7
25. $A+R$ 26. $E+A$ 27. $F+B$ 28. F^2+B^2 29. $A+B+D$ 30. A^2+E
31. $L+B$ 32. $N+T$ 33. 2 34. $4\frac{1}{2}$ 35. $\frac{N-C}{A}$ 36. $\frac{L-D}{B}$
37. $\frac{F-E}{D}$ 38. $\frac{H+F}{N}$ 39. $\frac{T+Z}{Y}$ 40. $\frac{B+L}{R}$ 41. $\frac{Q-m}{V}$ 42. $\frac{n+a+m}{t}$
43. $\frac{s-t-n}{q}$ 44. $\frac{t+s^2}{n}$ 45. $\frac{c-b}{V^2}$ 46. $\frac{r+6}{n}$ 47. $\frac{s-d}{m}$ 48. $\frac{t+b}{m}$
49. $\frac{j-c}{m}$ 50. 2 51. $2\frac{2}{3}$ 52. $\frac{C-AB}{A}$ 53. $\frac{F-DE}{D}$ 54. $\frac{a-hn}{h}$
55. $\frac{q+bd}{b}$ 56. $\frac{n-rt}{r}$ 57. $\frac{b+4t}{t}$ 58. $\frac{z-St}{S}$ 59. $\frac{s+vd}{v}$ 60. $\frac{g-mn}{m}$

Exercise 8 page 122

1. 12 2. 10 3. BD 4. TB 5. RN 6. bm
7. 26 8. $BT+A$ 9. $AN+D$ 10. B^2N-Q 11. $ge+r$ 12. $4\frac{1}{2}$
13. $\frac{DC-B}{A}$ 14. $\frac{pq-m}{n}$ 15. $\frac{vS+t}{r}$ 16. $\frac{qt+m}{z}$ 17. $\frac{bc-m}{A}$ 18. $\frac{AE-D}{B}$
19. $\frac{nh+f}{e}$ 20. $\frac{qr-b}{g}$ 21. 4 22. -2 23. 2 24. $A-B$
25. $C-E$ 26. $D-H$ 27. $n-m$ 28. $q-t$ 29. $s-b$ 30. $r-v$
31. $m-t$ 32. 2 33. $\frac{T-B}{X}$ 34. $\frac{M-Q}{N}$ 35. $\frac{V-T}{M}$ 36. $\frac{N-L}{R}$
37. $\frac{v^2-r}{r}$ 38. $\frac{w-t^2}{n}$ 39. $\frac{n-2}{q}$ 40. $\frac{1}{4}$ 41. $-\frac{1}{7}$ 42. $\frac{B-DE}{A}$
43. $\frac{D-NB}{E}$ 44. $\frac{h-bx}{f}$ 45. $\frac{v^2-Cd}{h}$ 46. $\frac{NT-MB}{M}$ 47. $\frac{mB+ef}{fN}$ 48. $\frac{TM-EF}{T}$
49. $\frac{yx-zt}{y}$ 50. $\frac{k^2m-x^2}{k^2}$

Exercise 9 page 123

1. $\frac{1}{2}$ 2. $1\frac{2}{3}$ 3. $\frac{B}{C}$ 4. $\frac{T}{X}$ 5. $\frac{M}{B}$ 6. $\frac{n}{m}$
7. $\frac{v}{t}$ 8. $\frac{n}{\sin 20°}$ 9. $\frac{7}{\cos 30°}$ 10. $\frac{B}{x}$ 11. $6\frac{2}{3}$ 12. $\frac{ND}{B}$

13. $\frac{HM}{N}$ 14. $\frac{et}{b}$ 15. $\frac{vs}{m}$ 16. $\frac{mb}{t}$ 17. $1\frac{1}{2}$ 18. $3\frac{1}{3}$

19. $\frac{B-DC}{C}$ 20. $\frac{Q+TC}{T}$ 21. $\frac{V+TD}{D}$ 22. $\frac{L}{MB}$ 23. $\frac{N}{BC}$ 24. $\frac{m}{cd}$

25. $\frac{tc-b}{t}$ 26. $\frac{xy-z}{x}$ 27. 1 28. $\frac{5}{6}$ 29. $\frac{A}{C-B}$ 30. $\frac{V}{H-G}$

31. $\frac{r}{n+t}$ 32. $\frac{b}{q-d}$ 33. $\frac{m}{t+n}$ 34. $\frac{b}{d-h}$ 35. $\frac{d}{C-e}$ 36. $\frac{m}{r-e^2}$

37. $\frac{n}{b-t^2}$ 38. $\frac{d}{mn-b}$ 39. $\frac{M-Nq}{N}$ 40. $\frac{Y+Tc}{T}$ 41. $\frac{N-2MP}{2M}$ 42. $\frac{B-6Ac}{6A}$

43. $\frac{K}{(C-B)M}$ 44. $\frac{z}{y(y+z)}$ 45. $\frac{m^2}{n-p}$ 46. $\frac{q}{w-t}$

Exercise 10 page 124

1. 4 2. 24 3. 11 4. B^2-A 5. D^2-C 6. H^2+E

7. $\frac{c^2-b}{a}$ 8. a^2+m 9. $\frac{b^2+t}{g}$ 10. $b-r^2$ 11. $d-t^2$ 12. b^2+d

13. $n-c^2$ 14. $b-f^2$ 15. $c-g^2$ 16. $\frac{M-P^2}{N}$ 17. $\frac{D-B}{A}$ 18. A^4+D

19. $\pm\sqrt{g}$ 20. ± 4 21. $\pm\sqrt{B}$ 22. $\pm\sqrt{(B-A)}$ 23. $\pm\sqrt{(M+A)}$ 24. $\pm\sqrt{(b-a)}$

25. $\pm\sqrt{(C-m)}$ 26. $\pm\sqrt{(d-n)}$ 27. $\pm\sqrt{\frac{n}{m}}$ 28. $\pm\sqrt{\frac{b}{a}}$ 29. $\frac{at}{z}$ 30. $\pm\sqrt{\left(\frac{m+t}{a}\right)}$

31. $\pm\sqrt{(a-n)}$ 32. $\pm\sqrt{40}$ 33. $\pm\sqrt{(B^2+A)}$ 34. $\pm\sqrt{(x^2-y)}$ 35. $\pm\sqrt{(t^2-m)}$ 36. 8

37. $\frac{M^2-A^2B}{A^2}$ 38. $\frac{M}{N^2}$ 39. $\frac{N}{B^2}$ 40. $a-b^2$ 41. $\pm\sqrt{(a^2-t^2)}$ 42. $\pm\sqrt{(m-x^2)}$

43. $\frac{4}{\pi^2}-t$ 44. $\frac{B^2}{A^2}-1$ 45. $\pm\sqrt{\left(\frac{C^2+b}{a}\right)}$ 46. $\pm\sqrt{\left(\frac{b^2+a^2x}{a^2}\right)}$ 47. $\pm\sqrt{(x^2-b)}$

48. $\pm\sqrt{(c-b)a}$ 49. $\frac{c^2-b^2}{a}$ 50. $\pm\sqrt{\left(\frac{m}{a+b}\right)}$

Exercise 11 page 124

1. $3\frac{2}{3}$ 2. 3 3. $\frac{D-B}{2N}$ 4. $\frac{E+D}{3M}$ 5. $\frac{2b}{a-b}$ 6. $\frac{e+c}{m+n}$

7. $\frac{3}{x+k}$ 8. $\frac{C-D}{R-T}$ 9. $\frac{z+x}{a-b}$ 10. $\frac{nb-ma}{m-n}$ 11. $\frac{d+xb}{x-1}$ 12. $\frac{a-ab}{b+1}$

13. $\frac{d-c}{d+c}$ 14. $\frac{M(b-a)}{b+a}$ 15. $\frac{n^2-mn}{m+n}$ 16. $\frac{m^2+5}{2-m}$ 17. $\frac{2+n^2}{n-1}$ 18. $\frac{e-b^2}{b-a}$

19. $\frac{3x}{a+x}$ 20. $\frac{e-c}{a-d}$ 21. $\frac{d}{a-b-c}$ 22. $\frac{ab}{m+n-a}$ 23. $\frac{s-t}{b-a}$ 24. $2x$

25. $\frac{v}{3}$ 26. $\frac{a(b+c)}{b-2a}$ 27. $\frac{5x}{3}$ 28. $-\frac{4z}{5}$ 29. $\frac{mn}{p^2-m}$ 30. $\frac{mn+n}{4+m}$

Exercise 12 page 125

1. $-\left(\frac{by+c}{a}\right)$ 2. $\pm\sqrt{\left(\frac{e^2+ab}{a}\right)}$ 3. $\frac{n^2}{m^2}+m$ 4. $\frac{a-b}{1+b}$ 5. $3y$

6. $\dfrac{a}{e^2+c}$ 7. $-\left(\dfrac{a+lm}{m}\right)$ 8. $\dfrac{t^2g}{4\pi^2}$ 9. $\dfrac{4\pi^2d}{t^2}$ 10. $\pm\sqrt{\dfrac{a}{3}}$

11. $\pm\sqrt{\left(\dfrac{t^2e-ba}{b}\right)}$ 12. $\dfrac{1}{a^2-1}$ 13. $\dfrac{a+b}{x}$ 14. $\pm\sqrt{(x^4-b^2)}$ 15. $\dfrac{c-a}{b}$

16. $\dfrac{a^2-b}{a+1}$ 17. $\pm\sqrt{\left(\dfrac{G^2}{16\pi^2}-T^2\right)}$ 18. $-\left(\dfrac{ax+c}{b}\right)$ 19. $\dfrac{1+x^2}{1-x^2}$ 20. $\pm\sqrt{\left(\dfrac{a^2m}{b^2}+n\right)}$

21. $\dfrac{P-M}{E}$ 22. $\dfrac{RP-Q}{R}$ 23. $\dfrac{z-t^2}{x}$ 24. $(g-e)^2-f$ 25. $\dfrac{4np+me^2}{mn}$

Exercise 13 page 126

1. (a) $S=ke$ (b) $v=kt$ (c) $x=kz^2$ (d) $y=k\sqrt{x}$ (e) $T=k\sqrt{L}$ (f) $C=kr$ (g) $A=kr^2$ (h) $V=kr^3$
2. (a) 9 (b) $2\frac{2}{3}$ 3. (a) 35 (b) 11 4. (a) 75 (b) 4

5.

x	1	3	4	$5\frac{1}{2}$
z	4	12	16	22

6.

r	1	2	4	$1\frac{1}{2}$
V	4	32	256	$13\frac{1}{2}$

7.

h	4	9	25	$2\frac{1}{4}$
w	6	9	15	$4\frac{1}{2}$

8. (a) 18 (b) 2 9. (a) 42 (b) 4 10. 333 N/cm^2
11. 180 m; 2 s 12. 675 J; $\sqrt{\frac{4}{3}}$ cm 13. 4 cm; 49 h 14. $15\frac{5}{8}$ h 15. 9000 N; 25 m/s
16. $15^4:1$ (50625 : 1)

Exercise 14 page 127

1. (a) $x=\dfrac{k}{y}$ (b) $s=\dfrac{k}{t^2}$ (c) $t=\dfrac{k}{\sqrt{q}}$ (d) $m=\dfrac{k}{w}$ (e) $z=\dfrac{k}{t^2}$
2. (a) 1 (b) 4 3. (a) $2\frac{1}{2}$ (b) $\frac{1}{2}$ 4. (a) 36 (b) ± 4
5. (a) 1·2 (b) ± 2 6. (a) 16 (b) ± 10 7. (a) 6 (b) 16
8. (a) $\frac{1}{2}$ (b) $\frac{1}{20}$

9.

y	2	4	1	$\frac{1}{4}$
z	8	4	16	64

10.

t	2	5	20	10
v	25	4	$\frac{1}{4}$	1

11.

x	1	4	256	36
r	12	6	$\frac{3}{4}$	2

12. (a) 6 (b) 50 13. (a) 0·36 (b) 6 14. $k=100, n=3$

x	1	2	4	10
z	100	$12\frac{1}{2}$	1·5625	$\frac{1}{10}$

15. $k=12, n=2$

v	1	4	36	10000
y	12	6	2	$\frac{3}{25}$

16. 2·5 m^3; 200 N/m^2 17. 3 h; 48 men
18. 2 days; 200 days 19. 6 cm

Exercise 15 page 129

1. (a) 24 (b) 5 2. (a) 36 (b) $\frac{2}{3}$ 3. (a) 18 (b) 5
4. (a) $2\frac{1}{2}$ (b) 36 5. 20 6. $26\frac{2}{3}$; 0·4 7. $y \propto z^2$ 8. $x \propto z^4$
9. $z \propto \dfrac{1}{t^2}$ 10. $x \propto z^3$ 11. $y \propto \dfrac{1}{z^2}$ 12. 192 g

Exercise 16 page 130

1. 3^4 2. $4^2 \times 5^3$ 3. 3×7^3 4. $2^3 \times 7$ 5. 10^{-3} 6. $2^{-2} \times 3^{-3}$
7. $15^{\frac{1}{2}}$ 8. $3^{\frac{1}{3}}$ 9. $10^{\frac{1}{5}}$ 10. $5^{\frac{3}{2}}$ 11. x^7 12. y^{13}
13. z^4 14. z^{100} 15. m 16. e^{-5} 17. y^2 18. w^6
19. y 20. x^{10} 21. 1 22. w^{-5} 23. w^{-5} 24. x^7

25. a^8 26. k^3 27. 1 28. x^{29} 29. y^2 30. x^6
31. z^4 32. t^{-4} 33. $4x^6$ 34. $16y^{10}$ 35. $6x^4$ 36. $10y^5$
37. $15a^4$ 38. $8a^3$ 39. 3 40. $4y^2$ 41. $\frac{5}{2}y$ 42. $32a^4$
43. $108x^5$ 44. $4z^{-3}$ 45. $2x^{-4}$ 46. $\frac{5}{2}y^5$ 47. 1 48. $21w^{-3}$
49. $2n^4$ 50. $2x$

Exercise 17 page 130

1. 27 2. 1 3. $\frac{1}{9}$ 4. 25 5. 2 6. 4
7. 9 8. 2 9. 27 10. 3 11. $\frac{1}{3}$ 12. $\frac{1}{2}$
13. 1 14. $\frac{1}{5}$ 15. 10 16. 8 17. 32 18. 4
19. $\frac{1}{9}$ 20. $\frac{1}{8}$ 21. 18 22. 10 23. 1000 24. $\frac{1}{1000}$
25. $\frac{1}{9}$ 26. 1 27. $1\frac{1}{2}$ 28. $\frac{1}{25}$ 29. $\frac{1}{10}$ 30. $\frac{1}{4}$
31. $\frac{1}{4}$ 32. 100 000 33. 1 34. $\frac{1}{32}$ 35. 0·1 36. 0·2
37. 1·5 38. 1 39. 9 40. $1\frac{1}{2}$ 41. $\frac{3}{10}$ 42. 64
43. $\frac{1}{100}$ 44. $1\frac{2}{3}$ 45. $\frac{1}{100}$ 46. 1 47. 100 48. 6
49. 750 50. -7

Exercise 18 page 130

1. $25x^4$ 2. $49y^6$ 3. $100a^2b^2$ 4. $4x^2y^4$ 5. $2x$ 6. $\frac{1}{9y}$
7. x^2 8. $\frac{x^2}{2}$ 9. 1 10. $\frac{2}{x}$ 11. $36x^4$ 12. $25y$
13. $16x^2$ 14. $27y$ 15. 25 16. 1 17. 49 18. 1
19. $8x^6y^3$ 20. $100x^2y^6$ 21. $\frac{3x}{2}$ 22. $\frac{2}{x}$ 23. x^3y^5 24. $12x^3y^2$
25. $10y^4$ 26. $3x^3$ 27. $x^3y^2z^4$ 28. x 29. $3y$ 30. $27x^{\frac{3}{2}}$
31. $10x^3y^5$ 32. $32x^2$ 33. $\frac{5}{2}x^2$ 34. $\frac{9}{x^2}$ 35. $2a^2$ 36. $a^3b^3c^6$
37. (a) 2^5 (b) 2^7 (c) 2^6 (d) 2^0 38. (a) 3^{-3} (b) 3^{-4} (c) 3^{-1} (d) 3^{-2}
39. 16 40. $\frac{1}{4}$ 41. $\frac{1}{6}$ 42. 1 43. $16\frac{1}{8}$ 44. $\frac{3}{8}$
45. $\frac{1}{4}$ 46. $\frac{5}{256}$ 47. $1\frac{1}{16}$ 48. 0 49. $\frac{1}{4}$ 50. $\frac{1}{4}$
51. 3 52. 4 53. -1 54. -2 55. 3 56. 3
57. 1 58. $\frac{1}{5}$ 59. 0 60. -4 61. 2 62. -5
63. 1 64. $\frac{1}{18}$

Exercise 19 page 131

1. $(x-2)$ or $(x+12)$ 2. $(x+3)$ or $(x-14)$ 3. $(x+2)$ or $(2x-5)$ 4. $(x+3)$ or $(3x-7)$
5. $(x+1)$ 6. $(x-3)$ 7. $(x+2)$ 8. $(x+1)$ 9. $(x+3)$ 10. $(x-2)$
11. -9 12. -3 13. 13 14. 8 15. -2 16. -6
17. -3 18. 1 19. -3 20. 1

Exercise 20 page 132

1. (x^2+2x+1) 2. (x^2+4x+3) 3. $(2x^2+x-1)$ 4. (x^2+3x-2)
5. (x^2+1) 6. (x^3-7) 7. $-3, \frac{1}{2}, 2$ 8. $-2, 1, 3$ 9. $-2, -\frac{1}{2}, 2$ 10. $-5, \frac{1}{2}, 3$
11. $-1, 2, 3$ 12. $-3, -2, -\frac{1}{2}$ 13. $\frac{1}{2}$ 14. $-5, \frac{1}{3}, 4$ 15. 2, 3, 4 16. $-2, -5, -6, 3$
17. $2, \sqrt[3]{7}$ 18. $-3, \sqrt[3]{3}$ 19. $\frac{1}{2}, \pm\sqrt{2}$ 20. $-2, \pm 1$
21. (a) -1 (b) 0·6258 (c) 0·5961 (d) 0·2210

Exercise 21 page 133

1. (a) 2·828 (b) 4·243 (c) 6·245 (d) 2·762 (e) 10·54 (f) 20·32
2. (a) 2·080 (b) 3·037 (c) 4·121 (d) 0·9872 (e) 10·16
3. (a) 2·036 (b) 3·041 (c) 10·28 (d) 1·933
4. 4·85
5. 3·28
6. 0·076923
7. 9·472
8. 1·781
9. 3·141593

Exercise 22 page 134

1. $<$ 2. $>$ 3. $>$ 4. $=$ 5. $<$ 6. $<$ 7. $=$ 8. $>$
9. $<$ 10. $>$ 11. $<$ 12. $>$ 13. $>$ 14. $>$ 15. $=$ 16. F
17. F 18. T 19. F 20. F 21. T 22. T 23. F 24. F
25. $x > 13$ 26. $x < -1$ 27. $x < 12$ 28. $x \leqslant 2\frac{1}{2}$
29. $x > 3$ 30. $x \geqslant 8$ 31. $x < \frac{1}{4}$ 32. $x \geqslant -3$
33. $x < -8$ 34. $x < 4$ 35. $x > -9$ 36. $x < 8$
37. $x > 3$ 38. $x \geqslant 1$ 39. $x < 1$ 40. $x > 2\frac{1}{3}$

Exercise 23 page 135

1. $x > 5$ 2. $x \leqslant 3$ 3. $x > 6$ 4. $x \geqslant 4$
5. $x < 1$ 6. $x < -3$ 7. $x > 0$ 8. $x > 4$
9. $x > 2$ 10. $x < -3$ 11. $1 < x < 4$ 12. $-2 \leqslant x \leqslant 5$
13. $1 \leqslant x < 6$ 14. $0 \leqslant x < 5$ 15. $-1 \leqslant x \leqslant 7$ 16. $-2 < x < 2$
17. $-4 < x < 4$ 18. $x < -1$ 19. $x \geqslant 3$ or $x \leqslant -3$ 20. $0 < x < 4$
21. $5 \leqslant x \leqslant 9$ 22. $-1 < x < 4$ 23. $1 \leqslant x \leqslant 6$ 24. $\frac{1}{2} < x < 8$
25. $-8 < x < 2$ 26. $\{1, 2, 3, 4, 5, 6\}$ 27. $\{7, 11, 13, 17, 19\}$ 28. $\{2, 4, 6, 8, 10\}$
29. $\{4, 9, 16, 25, 36, 49\}$ 30. $\{5, 10\}$ 31. $\{-4, -3, -2, -1\}$ 32. $\{2, 3, 4, \ldots 12\}$
33. $\{1, 4, 9\}$ 34. $\{2, 3, 5, 7, 11\}$ 35. $\{2, 4, 6, \ldots 18\}$ 36. $n = 5$
37. $x = 7$ 38. $y = 5$ 39. $4 < z < 5$ 40. $4 < p < 5$
41. $\frac{1}{2}$ (or other values) 42. 1, 2, 3, ... 14 43. 19 44. $\frac{1}{2}$ (or other values)
45. 19

Exercise 24 page 136

1. $x \leqslant 3$ 2. $y \geqslant 2\frac{1}{2}$ 3. $1 \leqslant x \leqslant 6$ 4. $x < 7, y < 5$
5. $y \geqslant x$ 6. $x + y < 10$ 7. $x < 8, y > -2, y < x$
8. $x \geqslant 0, y \geqslant x - 1, x + y \leqslant 7$ 9. $y \geqslant 0, y \leqslant x + 2, x + y \leqslant 6$
32. $x + y \leqslant 11, x \geqslant 0, y \geqslant 0$ 33. $y \geqslant x,\ 2x + y \leqslant 10, x \geqslant 0$
34. $x < 5, y < 6, x + y \geqslant 5$ 35. $x < 4, y < x + 2, 4y + 2x > 8$
36. $y \geqslant \frac{1}{2}x, x + y \leqslant 9, y \leqslant 3x$ 37. $x \geqslant 0, x + 2y \geqslant 12, x + y \leqslant 12$
38. $y < 4x, x + y < 10, 3y > x$ 39. $y \geqslant 0, x + y \leqslant 12, 3x + y \geqslant 12$
40. $x \geqslant 0, x + y < 11, 2y \geqslant x + 5$

Exercise 25 page 140

1. (a) maximum value = 26 at (6, 5) (b) minimum value = 12 at (3, 3)
2. (a) maximum value = 25 at (8, 3) (b) minimum value = 9 at (7, 2)
3. (a) maximum value = 40 at (20, 0) (b) maximum value = 112 at (14, 8)
4. (3, 3), (4, 2), (4, 3), (4, 4), (5, 1), (5, 2), (6, 0)
5. (0, 6), (0, 7), (0, 8), (1, 5), (1, 6), (2, 4)
6. (3, 2), (2, 3), (2, 4), (2, 5), (1, 4), (1, 5), (0, 5), (0, 6)
7. (2, 4), (2, 5), (2, 6), (2, 7), (2, 8), (2, 9), (2, 10), (3, 6), (3, 7), (3, 8)
8. (a) (6, 7), (7, 7), (8, 6), (7, 6), (6, 8) (b) 7 defenders, 6 forwards have lowest wage bill (£190)
9. (a) 18 (b) £2·60
10. (a) 14 (b) (7, 7), £154 000 (c) (13, 3), £190 000
11. (a) 10 (b) £250, (4, 7)
12. (a) (6, 12), £1440 (b) (10, 10), £1800
13. (14, 11), £1360
14. (a) (9, 7), £44 (b) (15, 5), £45

Revision exercise 5A *page 141*

1. (a) $\frac{9x}{20}$ (b) $\frac{7}{6x}$ (c) $\frac{5x-2}{6}$ (d) $\frac{5x+23}{(x-1)(x+3)}$
2. (a) $(x-2)(x+2)$ (b) $\frac{3}{x+2}$
3. (a) $s = t(r+3)$ (b) $r = \frac{s-3t}{t}$ (c) $t = \frac{s}{r+3}$
4. (a) $z = x - 5y$ (b) $m = \frac{11}{k+3}$ (c) $z = \frac{T^2}{C^2}$
5. (a) 50 (b) 50
6. (a) 16 (b) ± 4
7. (a) (i) 3 (ii) 4 (iii) $\frac{1}{4}$ (b) (i) 4 (ii) 0
8. (a) 9, 10 (b) 2, 3, 4, 5
9. $\frac{t^2}{k^2} - 5$
10. $\frac{z+2}{z-3}$
11. (a) $\frac{3}{5}$ (b) $\frac{k(1-y)}{y}$
13. (a) $1\frac{5}{6}$ (b) 0·09
14. (a) $-1, -3, 3$ (b) $-2, -\frac{1}{2}, \frac{1}{2}$ (c) 3
15. (a) $\frac{5+a^2}{2-a}$ (b) $-\left(\frac{cz+b}{a}\right)$ (c) $\frac{a^2+1}{a^2-1}$
16. $y \geqslant 2, x + y \leqslant 6, y \leqslant 3x$
17. $x \geqslant 0, y \geqslant x - 2, x + y \leqslant 7$
19. (a) $\frac{7}{2x}$ (b) $\frac{3a+7}{a^2-4}$ (c) $\frac{x-8}{x(x+1)(x-2)}$
20. $p = \frac{10t^2}{s}$

Examination exercise 5B *page 143*

1. (a) $\frac{2t^3}{t^2-4}; \frac{4t^2}{t^2-4}; \frac{t^4}{t^2-4}$ (b) $\frac{9}{5}; \frac{9}{2}$
2. (i) 126 (ii) 4 (iii) 1, −12 (iv) $\frac{2(y-ax)}{x^2}$
3. $x^6 = 1$, 2; 1
4. (i) −2 (ii) −1 (iii) no values (iv) 3·30, −0·30 (2 D.P.)
5. (i) $3\frac{1}{2}, 8\frac{1}{2}$ (ii) −2, 5 (iii) t; $t = 3$ or $-3\frac{4}{5}$; $x = 8$ or $-12\frac{2}{5}$; $y = 10$ or $3\frac{1}{5}$
6. (a) (i) $\frac{1}{2}$ or 4 (ii) 0 or $4\frac{1}{2}$ (iii) 4·27 or 0·23 (b) $F = \frac{k}{R^2}$; 68 600; 14
7. (a) $\frac{2(x-7)}{(x-5)(3-x)}$ (b) 5·8 or 0·2 (c) cuts x-axis where $x = 5{\cdot}8, 0{\cdot}2$
8. (a) $p = -3, q = 2$ (b) $x = \frac{5ay}{2y+3}$ (c) 10
9. (i) 18·5 at (5·5, 7·5); 21·5 at (4·3, 8·6) (ii) 18 at (5, 8) or (6, 6); 21 at (5, 8)
10. (iii) 32 (iv) 16

PART 6

Exercise 1 *page 146*

1. (a) 0·2924 (b) 0·9563 (c) 0·3057
2. (a) 0·8572 (b) 0·5150 (c) 1·6643
3. (a) 0·9882 (b) 0·1530 (c) 6·460
4. (a) 0·9553 (b) 0·2957 (c) 3·2305
5. (a) 0·2974 (b) 0·9548 (c) 0·3115
6. (a) 0·8660 (b) 0·5000 (c) 1·7321
7. (a) 0·1219 (b) 0·9925 (c) 0·1228
8. (a) 0·1547 (b) 0·9880 (c) 0·1566
9. (a) 0·7083 (b) 0·7059 (c) 1·0035
10. (a) 0·0558 (b) 0·9984 (c) 0·0559
11. (a) 0·9128 (b) 0·4083 (c) 2·2355
12. (a) 0·3074 (b) 0·9516 (c) 0·3230
13. (a) 0·8965 (b) 0·4431 (c) 2·0233
14. (a) 0·5864 (b) 0·8100 (c) 0·7239

15. (a) 0·9888 (b) 0·1495 (c) 6·612
16. (a) 0·3123 (b) 0·9500 (c) 0·3288
17. (a) 0·4617 (b) 0·8870 (c) 0·5206
18. (a) 0·7563 (b) 0·6543 (c) 1·1558
19. (a) 0·3437 (b) 0·9391 (c) 0·3659
20. (a) 0·4589 (b) 0·8885 (c) 0·5165
21. (a) 0·7359 (b) 0·6771 (c) 1·0869
22. (a) 0·2882 (b) 0·9576 (c) 0·3009
23. (a) 0·9793 (b) 0·2025 (c) 4·8361
24. (a) 0·1423 (b) 0·9898 (c) 0·1438
25. (a) 0·3308 (b) 0·9437 (c) 0·3505
26. (a) 0·3934 (b) 0·9193 (c) 0·4279
27. (a) 0·9886 (b) 0·1507 (c) 6·560
28. (a) 0·8411 (b) 0·5410 (c) 1·5547
29. (a) 0·9882 (b) 0·1535 (c) 6·435
30. (a) 0·2611 (b) 0·9653 (c) 0·2704

Exercise 2 page 146

1. 46·4°, 46° 24′ **2.** 71·6°, 71° 36′ **3.** 20·8°, 20° 48′ **4.** 78·0°, 78°
5. 70·7°, 70° 42′ **6.** 0·8°, 0° 48′ **7.** 16·8°, 16° 49′ **8.** 23·5°, 23° 31′
9. 70·3°, 70° 15′ **10.** 83·3°, 83° 16′ **11.** 1·5°, 1° 31′ **12.** 7·8°, 7° 46′
13. 54·2°, 54° 12′ **14.** 69·5°, 69° 30′ **15.** 18·6°, 18° 36′ **16.** 29·3°, 29° 18′
17. 88·6°, 88° 36′ **18.** 73·9°, 73° 54′ **19.** 57·3°, 57° 19′ **20.** 23·5°, 23° 31′
21. 35·0°, 35° 1′ **22.** 14·5°, 14° 28′ **23.** 56·4°, 56° 26′ **24.** 65·9°, 65° 51′
25. 18·5°, 18° 30′ **26.** 30·6°, 30° 36′ **27.** 53·6°, 53° 36′ **28.** 63·5°, 63° 30′
29. 2·9°, 2° 54′ **30.** 52·3°, 52° 18′ **31.** 16·7°, 16° 44′ **32.** 30·5°, 30° 32′
33. 49·6°, 49° 37′ **34.** 53·0°, 53° 2′ **35.** 68·4°, 68° 25′ **36.** 30·6°, 30° 37′

Exercise 3 page 147

1. BC, AB, AC **2.** DE, FE, DF **3.** LN, LM, MN **4.** XZ, XY, YZ
5. AC, AB, BC **6.** PR, QR, PQ **7.** YZ, XZ, XY **8.** KM, KL, LM
9. AC, BC, AB **10.** XZ, XY, YZ **11.** MO, MN, NO **12.** AC, BC, AB
13. KL, KM, LM **14.** DF, DE, EF **15.** XZ, XY, YZ **16.** AC, BC, AB
17. AD, BD, AB **18.** WZ, WX, XZ **19.** PR, PQ, QR **20.** BC, AB, AC

Exercise 4 page 148

1. 4·54 **2.** 3·50 **3.** 3·71 **4.** 6·62 **5.** 8·01
6. 31·9 **7.** 45·4 **8.** 4·34 **9.** 17·1 **10.** 13·2
11. 38·1 **12.** 3·15 **13.** 516 **14.** 79·1 **15.** 5·84
16. 2·56 **17.** 18·3 **18.** 8·65 **19.** 11·9 **20.** 10·6
21. 119 **22.** 10·1 **23.** 3·36 cm **24.** 4·05 cm **25.** 4·10 m
26. 11·7 m **27.** 9·48 cm **28.** 5·74 m **29.** 9·53 cm **30.** 100 m
31. 56·7 m **32.** 16·3 cm **33.** 0·952 cm **34.** 8·27 m

Exercise 5 page 150

1. 5, 5·55 **2.** 13·1, 27·8 **3.** 34·6, 41·3 **4.** 20·4, 11·7 **5.** 94·1, 94·1
6. 15·2, 10, 6·43 **7.** 4·26 **8.** 3·50 **9.** 26·2 **10.** 8·82
11. (a) 17·4 cm (b) 11·5 cm (c) 26·5 cm **12.** (a) 6·82 cm (b) 6·01 cm (c) 7·31 cm

Exercise 6 page 151

1. 36·9° **2.** 44·4° **3.** 48·2° **4.** 60° **5.** 36·9° **6.** 50·2°
7. 29·0° **8.** 56·4° **9.** 38·9° **10.** 43·9° **11.** 41·8° **12.** 39·3°
13. 60·3° **14.** 50·5° **15.** 13·6° **16.** 34·8° **17.** 60·0° **18.** 42·0°
19. 36·9° **20.** 51·3° **21.** 19·6° **22.** 17·9° **23.** 32·5° **24.** 59·6°
25. 54·8° **26.** 46·3°

Exercise 7 page 152

1. $19{\cdot}5°$ **2.** 4·1 m **3.** (a) 26·0 km (b) 23·4 km **4.** (a) 88·6 km (b) 179·3 km
5. 4·1 m **6.** 8·6 m **7.** (a) 484 km (b) 858 km (c) 986 km, 060·6°
8. 954 km, 133° **9.** 56·3° **10.** 54·5° **11.** 71·6° **12.** 91·8°
13. 402 m **14.** 36·4° **15.** 10·3 cm **16.** 9·51 cm **17.** 71·1° **18.** 67·1 m
19. 138 m **20.** 83·2 km **21.** 60° **22.** 13·9 cm **23.** Yes
24. 11·1 m; 11·1 s; 222 m **25.** 4·4 m **26.** 3·13 m

Exercise 8 page 155

1. −0·9397 **2.** −1·732 **3.** 0·766 **4.** −0·9397 **5.** −0·8387 **6.** 0·6561
7. −0·5878 **8.** −6·314 **9.** −2·605 **10.** −0·2924 **11.** −0·1736 **12.** −0·0523
13. 0·5774 **14.** −0·342 **15.** 0·1959 **16.** −0·9336 **17.** 0·5793 **18.** −11·43
19. 0·2487 **20.** 0·5906 **21.** 3·732 **22.** −0·0175 **23.** −0·8829 **24.** −0·1219
25. −0·3019 **26.** −1·6643 **27.** 0·6873 **28.** −0·342 **29.** 0·9994 **30.** 0·9184
31. 53·1°, 126·9° **32.** 24·9°, 155·1° **33.** 60·2°, 299·8°
34. 76·7°, 283·3° **35.** 49·6°, 229·6° **36.** 75·7°, 104·3°
37. 45°, 225° **38.** 30·8°, 329·2° **39.** 64·6°, 244·6°
40. 41·5°, 318·5° **41.** 228·6°, 311·4° **42.** 128·9°, 308·9°
43. 150°, 210° **44.** 116·6°, 296·6° **45.** 252·2°, 287·8°

Exercise 9 page 156

1. 6·38 m **2.** 12·5 m **3.** 5·17 cm **4.** 40·4 cm **5.** 7·81 m, 7·10 m
6. 3·55 m, 6·68 m **7.** 8·61 cm **8.** 9·97 cm **9.** 8·52 cm **10.** 15·2 cm
11. 35·8° **12.** 42·9° **13.** 32·3° **14.** 37·8° **15.** 35·5°, 48·5° **16.** 68·8°, 80·0°
17. 64·6° **18.** 34·2° **19.** 50·6° **20.** 39·1° **21.** 39·5° **22.** 21·6°

Exercise 10 page 158

1. 6·24 **2.** 6·05 **3.** 5·47 **4.** 9·27 **5.** 10·1 **6.** 8·99
7. 5·87 **8.** 4·24 **9.** 11·9 **10.** 154 **11.** 25·2° **12.** 78·5°
13. 115·0° **14.** 111·1° **15.** 24·0° **16.** 92·5° **17.** 99·9° **18.** 38·2°
19. 137·8° **20.** 34·0° **21.** 60·2° **22.** 8·72 **23.** 1·40 **24.** 7·38

Exercise 11 page 159

1. BC = 7·83, AC = 4·77 **2.** DE = 7·99, $\hat{D}$ = 51·9° **3.** $\hat{Q}$ = 92·9°, $\hat{R}$ = 38·6°
4. 25·4°, 9·71 cm **5.** 10·0 m, 58·6° **6.** 7·33 cm, 9·57 cm
7. 45·2°, 102·6°, 32·2° **8.** 65·2° **9.** 104·5°, 46·6° **10.** 16·8° **11.** 8·84
12. 182 m **13.** 42·5° **14.** 42·8° **15.** 64·0° **16.** 11·7 m **17.** 40·4°
18. 55·8°, 33·4°

Exercise 12 page 160

1. 6·0 cm **2.** 10·8 m **3.** 35·6 km **4.** 25·2 m **5.** 38·6°, 48·5°, 92·9°
6. 40·4 m **7.** 9·8 km; 085·7° **8.** (a) 29·6 km (b) 050·5°
9. $B\hat{A}C$ = 27·7°; $A\hat{B}C$ = 111·8°; $B\hat{C}A$ = 40·5°; $A\hat{C}D$ = 63·8°; $A\hat{D}C$ = 85·2°; CD = 5·2 m
10. 378 km, 048·1° **11.** 12·7 km from M, 23·9 km from L
12. (a) 6·1 km (b) 11·5 km **13.** 9·64 m **14.** 8·6°
15. 43·9 m. They will fly towards the mouse but stop because at some point they will be less than 40 m apart.
16. (a) 10·8 m (b) 72·6° (c) 32·6° **17.** 118 m **18.** (a) 146° (b) 023·1°
19. (a) 62·2° (b) 2·33 km **20.** (a) 5·66 cm (b) 4·47 cm (c) 3·66 cm

Exercise 13 *page 161*

1. (a) 13 cm (b) 13·6 cm (c) 17·1°
2. (a) 4·04 m (b) 38·9° (c) 11·2 m (d) 19·9°
3. (a) 8·49 cm (b) 8·49 cm (c) 10·4 cm (d) 35·3° (e) 35·3°
4. (a) 10 m (b) 7·81 m (c) 9·43 m (d) 70·2°
5. (a) 14·1 cm (b) 18·7 cm (c) 69·3° (d) 29·0° (e) 41·4°
6. (a) 4·47 m (b) 7·48 m (c) 63·4° (d) 74·5° (e) 53·3°
7. 10·8 cm; 21·8°
8. (a) $h \tan 65°$ (b) $h \tan 57°$ (c) 22·7 m
9. 22·6 m
10. 55·0 m
11. 7·26 m

Exercise 14 *page 163*

1. (a) 11·3 cm, 12·4 cm (b) 23·8° (c) 32·0°
2. (a) 10 cm, 9·43 cm (b) 26·6° (c) 32·5° (d) 32·0°
3. (a) 29·5° (b) 38·7° (c) 29·5°
4. (a) 58·0° (b) 66·1°
5. (a) 57·5° (b) 61·0° (c) 61·0°
6. (a) 16·0° (b) 19·3° (c) 19·3°
7. 43·3°
8. (a) 8·16 cm (b) 54·7° (c) 70·5°
9. (a) 8·87 cm (b) 62·5° (c) 75·4°
10. (a) 8·31 cm (b) 67·4° (c) 78·2°

Revision exercise 6A *page 165*

1. (a) 45·6° (b) 58·0° (c) 3·89 cm (d) 33·8 m
2. (a) 1·75 (b) 60° 19′
3. (a) 12·7 cm (b) 5·92 cm (c) 36·1°
4. 5·39 cm
5. (a) 220° (b) 295°
6. 0·335 m
7. (a) 6·61 cm (b) 12·8 cm (c) 5·67 cm
8. (a) −0·574 (b) −1·37 (c) −0·891 (d) 0·894 (e) −1·73 (f) 0·245
9. (a) −0·7, −5·6 (b) 117°, 243°
10. 73·4°
11. 8·76 m, 9·99 m
12. 0·539
13. (a) 86·9 cm (b) 53·6 cm (c) 133 cm
14. 26·4°
15. 52·4 m
16. 45·2 km, 33·6 km
17. 4·12 cm, 9·93 cm
18. (a) 14·1 cm (b) 35·3° (c) 35·3°
19. (a) 11·3 cm (b) 8·25 cm (c) 55·6°
20. (a) 6·63 cm (b) 41·8°

Examination exercise 6B *page 166*

1. (i) 6·08 cm, 1 cm (ii) 165·8° (iii) 15·1 cm^2
2. (i) 68° 26′ (ii) 111° 34′ (iii) 3·81 km^2 (iv) 2·77 km
3. (i) 41·4° (ii) 1058 m
4. (i) 36·9° (ii) 5·66 cm (iii) 46·7° (iv) 64 cm^3
5. (i) 5·84 cm (ii) 42·4° (iii) 9·73 cm (iv) 24·4 cm^2
6. (i) 115 m, 48·3 m (ii) 93·4 m (iii) 4870 m^2
7. BH = 97·8 m; 94·0 m; 34·2 m; 4·2 m
8. (i) 10 km (ii) 126·8° (iii) 1940 m (iv) 9·2°
9. $101\frac{1}{2}°$, $164\frac{1}{2}°$; 8·36 km
10. (i) 66·4° (ii) 4·84 cm (iii) 148, 84·3°

PART 7

Exercise 1 *page 170*

1. (1, 3)
2. (0, −2), (−1, 8)
3. (−3, 8)
4. (−1, 12)
5. (0, 3), (4, 30)
6. $(\frac{1}{2}, 3)$
7. (0, 2)
8. (0, 7)
9. (−2, 9)
10. (−1, 180)
11. (4, 6)
12. (−1, 4)
13. (−4, −2), $(\frac{1}{2}, 6)$
14. (3, 10), (−3, 10)
15. (−2, 2)

16. $y = x - 3$ 17. $y = 2x$ 18. $y = \frac{x}{3}$ 19. $x + y = 10$ 20. $y = x + 6$ 21. $y = 2x + 1$
22. $y = 3x + 4$ 23. $x + y = 12$ 24. $y = 2x - 5$ 25. $y = 10 - 2x$ 26. $v = 6 - 3t$
27. $v = \frac{1}{2}(10 - 3t)$ 28. $y = \frac{1}{2}x$ 29. (a) $y = \frac{1}{2}x + 1$ (b) $x + y = 4$
30. (a) $y = 2x - 1$ (b) $y = -\frac{1}{2}x + 2$ 31. (a) $y = \frac{1}{2}x - 2$ (b) $x + y = 3$

Exercise 2 page 172

11. (0, 0), (1, 4), (1·6, 1·6) 12. (0, 1), $(2\frac{1}{4}, 1)$, $(4\frac{1}{2}, 10)$
13. (−2, −6), (1·25, 3·75), (4·5, 0·5) 14. (−1·5, 1·5), (0·67, 8), (3·5, 8), (3·5, −3·5)
15. (4, −2), (0·33, 5·33), (−2·28, −5·14) 16. (−2, 3), (0·6, 8·2), (2·5, 2·5), (1·33, 1·33)
17. (a) £560 (b) 2400 miles 18. (a) 1·53 kg (b) 7·1 h
19. (a) £188 (b) 158 km/h (c) £210 20. (a) £4315 (b) 26 000 miles (c) £380

Exercise 3 page 173

1. $1\frac{1}{2}$ 2. 2 3. 3 4. $1\frac{1}{2}$ 5. $\frac{1}{2}$ 6. $-\frac{1}{6}$ 7. −7 8. −1
9. 4 10. −4 11. 5 12. $-1\frac{3}{7}$ 13. 6 14. 0 15. 0 16. infinite
17. infinite 18. −8 19. $5\frac{1}{3}$ 20. 0 21. $\frac{b-d}{a-c}$ 22. $\frac{n+b}{m-a}$ 23. $\frac{2f}{a}$ 24. −4
25. 0 26. $-\frac{6d}{c}$ 27. (a) $-1\frac{1}{5}$ (b) $\frac{1}{10}$ (c) $\frac{4}{5}$
28. (a) infinite (b) $-\frac{3}{10}$ (c) $\frac{3}{10}$ 29. $3\frac{1}{2}$
30. (a) $\frac{n+4}{2m-3}$ (b) $n = -4$ (c) $m = 1\frac{1}{2}$

Exercise 4 page 174

1. 1, 3 2. 1, −2 3. 2, 1 4. 2, −5 5. 3, 4 6. $\frac{1}{2}$, 6
7. 3, −2 8. 2, 0 9. $\frac{1}{4}$, −4 10. −1, 3 11. −2. 6 12. −1, 2
13. −2, 3 14. −3, −4 15. $\frac{1}{2}$, 3 16. $-\frac{1}{3}$, 3 17. 4, −5 18. $1\frac{1}{2}$, −4
19. 10, 0 20. 0, 4 21. $y = 3x + 7$ 22. $y = 2x - 9$ 23. $y = -x + 5$ 24. $y = 2x - 1$
25. $y = 3x + 5$ 26. $y = -x + 7$ 27. $y = \frac{1}{2}x - 3$ 28. $y = 2x - 3$ 29. $y = 3x - 11$ 30. $y = -x + 5$
31. $y = \frac{1}{3}x - 4$ 32. (a) $y = x + 3$ (b) $y = 6 - 2x$
33. (a) $y = \frac{x}{4}$ (b) $7y = 2x + 28$ (c) $y = 10 - x$ 34. (a) $y = 3x$ (b) $y = 12 - x$ (c) $y = \frac{1}{2}x$
35. $m = 3, c = 5$ 36. $m = 2, c = 3$ 37. $n = -2{\cdot}5, k = 15$ 38. $m = 0{\cdot}2, c = 3$

Exercise 6 page 178

1. (a) 4 (b) 8 (c) 10·6 2. (a) 3 (b) −5 (c) 1·5
3. (a) 7·25 (b) −2 (c) −0·8, 3·8 15. (a) 0·75 (b) 1·23
16. (a) 3·13 (b) 3·35 17. (a) −2·45 (b) 1·4
18. (a) 5 (b) 10·1 (c) −1·25 20. (a) 245 (b) 41 (c) $25 < x < 67$
21. (a) 10·7 cm^2 (b) 1·7 cm × 5·3 cm (c) 12·25 cm^2 (d) 3·5 cm × 3·5 cm (e) square
22. (a) 2·5 s (b) 31·3 m (c) $2 < t < 3$ 23. (a) 108 m/s (b) 1·4 s (c) $2{\cdot}3 < t < 3{\cdot}6$
25. 3·3 30. (a) 1 (b) 2·7 31. (a) 0, 2·9 (b) −0·65, 1·35, 5·3

Exercise 7 page 181

1. (a) 0·4, 2·4 (b) −0·8, 3·8 (c) −1, 3 (d) −0·4, 2·4
2. 1, 3 3. 2·9 4. 0·3, 3·7
5. (a) $y = 3$ (b) $y = -2$ (c) $y = x + 4$ (d) $y = x$ (e) $y = 6$

6. (a) $y = 6$ (b) $y = 0$ (c) $y = 4$ (d) $y = 2x$ (e) $y = 2x + 4$
7. (a) $y = -4$ (b) $y = 2x$ (c) $y = x - 2$ (d) $y = -3$ (e) $y = 2$
8. (a) $y = 5$ (b) $y = 2x$ (c) $y = 0{\cdot}2$ (d) $y = 3 - x$ (e) $y = 3$
9. (a) $y = 0$ (b) $y = -2\frac{1}{2}$ (c) $y = -8x$ (d) $y = -3$ (e) $y = -5\frac{1}{2}x$
10. (a) $-1{\cdot}65, 3{\cdot}65$ (b) $-1{\cdot}3, 2{\cdot}3$ (c) $-1{\cdot}45, 3{\cdot}45$
11. (a) $1{\cdot}7, 5{\cdot}3$ (b) $0{\cdot}2, 4{\cdot}8$
12. (a) $-3{\cdot}3, 0{\cdot}3$ (b) $-4{\cdot}6, -0{\cdot}4$
13. (a) $-2{\cdot}35, 0{\cdot}85$ (b) $-2{\cdot}8, 1{\cdot}8$
14. (a) (i) $-0{\cdot}4, 2{\cdot}4$ (ii) $-0{\cdot}5, 2$ (b) $-1{\cdot}25 < x < 2{\cdot}75$
15. (a) $3{\cdot}35$ (b) $2{\cdot}4, 7{\cdot}6$ (c) $4{\cdot}25$
16. (a) $\pm 3{\cdot}74$ (c) $\pm 2{\cdot}83$
17. (a) $1{\cdot}75$ (b) $0, \pm 1{\cdot}4$
18. (a) $1{\cdot}6 < x < 7{\cdot}4$ (b) $6{\cdot}9$
19. (a) $40°, 140°$ (b) $30°, 150°$
20. (a) (i) $16°, 111°$ (ii) $153°$ (b) $2{\cdot}24$ (c) $63°$
21. (a) (i) $48°, 205°$ (ii) $37°, 217°$ (b) $90° < x < 164°$
22. (a) $2{\cdot}6$ (b) $0{\cdot}45$ (c) $0{\cdot}64; 5{\cdot}66$
23. (a) $-1{\cdot}62, 0{\cdot}62$ (b) $-\frac{1}{2}, 1$
24. (a) $4, -0{\cdot}77$ (b) $0{\cdot}6 < x < 2{\cdot}8$
25. (a) $-0{\cdot}65, 1{\cdot}3, 5{\cdot}3$ (b) $-0{\cdot}8, 1{\cdot}5$

Revision exercise 7A *page 183*

1. (a) $(0, 4)$ (b) $(7, 7)$ (c) $(4, 5{\cdot}5), (-1, -12)$ (d) $(2, 8)$ (e) $(-1, 4), (2, 4)$
2. (a) $y = x - 7$ (b) $y = 2x + 5$ (c) $y = -2x + 10$ (d) $y = \dfrac{x+1}{2}$
3. (a) 2 (b) 1 (c) $-3\frac{1}{2}$ (d) 0 (e) 10
4. (a) $2, -7$ (b) $-4, 5$ (c) $\frac{1}{2}, 4$ (d) $-\frac{1}{2}, 5$ (e) $-2, 12$ (f) $-\frac{2}{3}, 8$
5. A: $y = 6$; B: $y = \frac{1}{2}x - 3$; C: $y = 10 - x$; D: $y = 3x$
6. A: $4y = 3x - 16$; B: $2y = x - 8$; C: $2y + x = 8$; D: $4y + 3x = 16$
7. (a) $y = 2x - 3$ (b) $y = 3x + 4$ (c) $y = 10 - x$ (d) $y = 7$
8. (a) A$(0, 8)$, B$(8, 0)$ (b) -1 (c) $y = 8 - x$
9. (a) A$(0, -8)$, B$(4, 0)$ (b) 2 (c) $y = 2x - 8$
10. (a) OM: $y = 2\frac{1}{2}x$; ON: $y = \frac{1}{2}x$; MN: $2y = 12 - x$ (b) 12 sq. units
11. $y = 2x - 7$
12. 25 sq. units
13. -3
14. 219
16. (a) $y = 3x$ (b) $y = 0$ (c) $y = 11 - x$ (d) $y = 5x$
17. (a) $y = x + 3$ (b) $y = 0$ (c) $y = 3$ (d) $y = 5$
18. (a) $1{\cdot}56, -2{\cdot}56$ (b) $\pm 2{\cdot}24$ (c) $\pm 2{\cdot}65$
19. (a) $0{\cdot}84, 4{\cdot}15$ (b) $0{\cdot}65 < x < 3{\cdot}85$ (c) $3{\cdot}3$
20. (a) $9{\cdot}2$ (b) $0{\cdot}6$ (c) $1{\cdot}4$ (d) $1{\cdot}65$

Examination exercise 7B *page 185*

1. $x^2 + x - 4 = 0$; $1{\cdot}56$
2. (iii) $3x^2 - 6x - 20 = 0$; $3{\cdot}75$ or $-1{\cdot}75$ (iv) -5
3. (i) $8x^2$ m^2 (ii) $(20 - 3x)$ m (iii) $(400 - 120x + 9x^2)$ m^2; $x = 3{\cdot}53$
4. (i) 2000 (ii) 270 (iii) $1{\cdot}6 \leqslant x \leqslant 2{\cdot}4$. 380 plates
5. (a) $-0{\cdot}4$ (c) $1{\cdot}22, 5{\cdot}64$ $\quad \dfrac{4}{x} + \dfrac{x}{4} = -\dfrac{x}{3} + 4$
6. (i) $1{\cdot}27 < x < 4{\cdot}73$ (ii) $1{\cdot}1 < x < 5{\cdot}8$
7. (a) $2{\cdot}62, 0{\cdot}30$ (b) $1{\cdot}76, 4{\cdot}64$
8. (i) $1{\cdot}62, -0{\cdot}62$ (ii) $3{\cdot}2, -2{\cdot}2$ (iii) $3{\cdot}45, -1{\cdot}45$
9. (i) $x < 1{\cdot}4$ (ii) $0{\cdot}27 < x < 3{\cdot}73$ (iii) $0{\cdot}27, 3{\cdot}73$
10. (ii) (a) 135 (b) 78 mm (iii) $2y + 5x = 575$

PART 8

Exercise 1 page 190

1. $\begin{pmatrix} 2 & 4 \\ 4 & 2 \end{pmatrix}$ **2.** $\begin{pmatrix} 1 & 6 & -1 \\ 7 & -10 & 6 \end{pmatrix}$ **3.** — **4.** $\begin{pmatrix} -4 & 2 \\ 0 & 0 \end{pmatrix}$

5. $(8 \quad 10)$ **6.** $\begin{pmatrix} 0 & 15 \\ 3 & -6 \end{pmatrix}$ **7.** $\begin{pmatrix} 0 & -3 \\ 1 & 0 \\ -9 & -5 \end{pmatrix}$ **8.** $\begin{pmatrix} 4 & 3 \\ 7 & 6 \end{pmatrix}$ **9.** —

10. $\begin{pmatrix} 6 & 7 \\ 5 & 0 \end{pmatrix}$ **11.** $\begin{pmatrix} -1 & 9 \\ -2 & -1 \\ 25 & 10 \end{pmatrix}$ **12.** $\begin{pmatrix} 1 & -5\frac{1}{2} \\ \frac{1}{2} & 4 \end{pmatrix}$ **13.** $\begin{pmatrix} -1 & 12 \\ 4 & 7 \end{pmatrix}$ **14.** $\begin{pmatrix} 15 & 20 \\ -4 & -9 \end{pmatrix}$

15. $\begin{pmatrix} 5 & -10 \\ 2 & 7 \end{pmatrix}$ **16.** $\begin{pmatrix} 3 & 14 \\ -2 & 9 \end{pmatrix}$ **17.** $\begin{pmatrix} 12 \\ 13 \end{pmatrix}$ **18.** $\begin{pmatrix} 5 \\ 13 \end{pmatrix}$ **19.** $\begin{pmatrix} 14 & 1 \\ -32 & -13 \end{pmatrix}$

20. $\begin{pmatrix} 8 & -27 \\ 23 & -2 \end{pmatrix}$ **21.** $\begin{pmatrix} 8 & -27 \\ 23 & -2 \end{pmatrix}$ **22.** — **23.** — **24.** $\begin{pmatrix} 16 & 20 \\ 4 & 5 \\ 12 & 15 \end{pmatrix}$

25. $\begin{pmatrix} 30 & 40 \\ -8 & -18 \end{pmatrix}$ **26.** $\begin{pmatrix} 7 \\ 36 \end{pmatrix}$ **27.** $\begin{pmatrix} 12 & 15 \\ 4 & 5 \end{pmatrix}$ **28.** (17) **29.** $\begin{pmatrix} -18 & 14 & -6 \\ 26 & -26 & 8 \end{pmatrix}$

30. $\begin{pmatrix} 1 & -6 \\ 18 & 13 \end{pmatrix}$ **31.** $\begin{pmatrix} -107 & -84 \\ 252 & 61 \end{pmatrix}$ **32.** — **33.** $\begin{pmatrix} -9 & 13 & -17 \\ 3 & -4 & 5 \\ 0 & -7 & 14 \end{pmatrix}$

34. $\begin{pmatrix} 59 \\ -21 \end{pmatrix}$ **35.** $\begin{pmatrix} 1 & 5 & 1 \\ 3 & -11 & 0 \\ 22 & -20 & 7 \end{pmatrix}$ **36.** $\begin{pmatrix} 45 & -140 \\ -28 & 101 \end{pmatrix}$

37. $x = 6, y = 3, z = 0$ **38.** $x = 4, y = 5, z = 7, w = -5, v = 0$
39. $a = 4, b = 9, c = 15, d = 2$ **40.** $x = 1, y = 4$ **41.** $m = 5, n = -\frac{1}{3}$
42. $p = 3, q = -1$ **43.** $x = 1, y = 2, z = -1, w = -2$ **44.** $y = 2, z = -1, x = 1, w = -2$
45. $a = -3, e = 4, k = 2$ **46.** $m = 3, n = 5, p = 3, q = 3$ **47.** $x = 2\frac{2}{3}$
48. $k = \pm 1$ **49.** (a) $k = 2$ (b) $m = 4$ **50.** (a) $n = 3$ (b) $q = 9$

Exercise 2 page 192

1. $\begin{pmatrix} 1 & -1 \\ -3 & 4 \end{pmatrix}$ **2.** $\begin{pmatrix} 5 & -2 \\ -2 & 1 \end{pmatrix}$ **3.** $\frac{1}{2}\begin{pmatrix} 2 & -4 \\ -1 & 3 \end{pmatrix}$ **4.** $\frac{1}{3}\begin{pmatrix} 1 & -2 \\ -1 & 5 \end{pmatrix}$ **5.** $\frac{1}{2}\begin{pmatrix} 2 & 2 \\ 1 & 2 \end{pmatrix}$

6. $\frac{1}{5}\begin{pmatrix} 2 & 3 \\ 1 & 4 \end{pmatrix}$ **7.** $\frac{1}{8}\begin{pmatrix} 3 & -1 \\ 2 & 2 \end{pmatrix}$ **8.** $\frac{1}{6}\begin{pmatrix} 4 & 3 \\ -2 & 0 \end{pmatrix}$ **9.** $\frac{1}{5}\begin{pmatrix} -3 & 2 \\ -1 & -1 \end{pmatrix}$ **10.** no inverse

11. $\frac{1}{14}\begin{pmatrix} 4 & 2 \\ -1 & 3 \end{pmatrix}$ **12.** $-\frac{1}{5}\begin{pmatrix} 1 & -1 \\ -2 & -3 \end{pmatrix}$ **13.** $-\frac{1}{5}\begin{pmatrix} -4 & 3 \\ -1 & 2 \end{pmatrix}$ **14.** $\frac{1}{7}\begin{pmatrix} 1 & 0 \\ 5 & 7 \end{pmatrix}$ **15.** $-\frac{1}{6}\begin{pmatrix} -4 & -1 \\ 2 & 2 \end{pmatrix}$

16. $\frac{1}{2}\begin{pmatrix} 3 & -4 \\ -1 & 2 \end{pmatrix}$ **17.** $-\frac{1}{2}\begin{pmatrix} 1 & 0 \\ -3 & -2 \end{pmatrix}$ **18.** $\begin{pmatrix} -1 & 3 \\ 0 & -3 \end{pmatrix}$ **19.** $\begin{pmatrix} 3 & 2 \\ 1 & 3 \end{pmatrix}$ **20.** $\begin{pmatrix} 2 & -3 \\ 2 & 4 \end{pmatrix}$

21. $\frac{2}{3}\begin{pmatrix} 1 & -1 \\ 2 & 1 \end{pmatrix}$ **22.** (a) $\begin{pmatrix} 5 & -11 \\ -1 & 3 \end{pmatrix}$ (b) $\frac{1}{2}\begin{pmatrix} 1 & 3 \\ 0 & 2 \end{pmatrix}$ (c) $\frac{1}{2}\begin{pmatrix} 3 & 1 \\ 1 & 1 \end{pmatrix}$ **23.** $\begin{pmatrix} 1 & -3 \\ 4 & 0 \end{pmatrix}$

24. $\begin{pmatrix} 4 \\ 3 \end{pmatrix}$ **25.** 4; 1; $1\frac{1}{2}$ **26.** $x = -2$ **27.** (a) 42 (b) 14

Exercise 3 page 193

1. $x = 2, y = 3$ 2. $x = 4, y = -1$ 3. $x = -4, y = 3$ 4. $x = \frac{1}{2}, y = -5$ 5. $x = -3, y = 0$
6. $x = \frac{2}{3}, y = \frac{1}{4}$ 7. $a = 4, b = -5$ 8. $a = -3, b = 7$ 9. $m = -3, n = \frac{1}{2}$ 10. $m = 0, n = -5$
11. $p = 2, q = -7$ 12. $p = 0{\cdot}2, q = 0{\cdot}7$ 13. $s = 1, t = 4$ 14. $\begin{pmatrix} -2 \\ 3 \end{pmatrix}$ 15. No
16. Only (b) has a solution. 17. $x = 1, y = 2, z = 3$
18. (a) $x = 2, y = 3, z = -2$ (b) $x = 8, y = 12, z = 4$

Exercise 5 page 195

1. (b) A′(8, 4) B′(5, 6) C′(4, 9) D′(9, 8) E′(3, 0) (c) A″(2, 6) B″(5, 4) C″(6, 1) D″(1, 2) E″(7, 10)
(d) A* (4, 2) B* (6, 5) C* (9, 6) D* (8, 1) E* (0. 7)
2. (a) A′(−2, 4) B′(−6, 8) C′(−9, 0) (b) A′(2, −4) B′(6, −8) C′(9, 0)
(c) A′(2, 4) B′(−2, 8) C′(−5, 0) (d) A′(2, −8) B′(6, −12) C′(9, −4)
(e) A′(4, 2) B′(8, 6) C′(0, 9)
3. (a) $x = 0$ (b) $x = 3$ (c) $y = x$
4. (a) A′(2, 4) B′(6, 8) C′(−2, 11) (b) A′(−4, −2) B′(−8, −6) C′(0, −9)
(c) A′(0, 2) B′(−4, −2) C′(4, −5)
5. (a) A′(5, 0) B′(11, −3) C′(12, 7) (b) A′(2, 5) B′(−4, 8) C′(−5, −2)
6. (a) $x + y = 2$ (b) $y = x + 3$

Exercise 7 page 197

1. (a) A′(3, −1) B′(6, −1) C′(6, −3) (b) D′(3, −3) E′(3, −6) F′(1, −6)
(c) P′(−7, −4) Q′(−5, −4) R′(−5, −1)
2. (a) A′(−4, 2) B′(−10, 6) C′(−4, 12) (b) A″(−4, −2) B″(2, −6) C″(−4, −12)
(c) A*(2, −4) B*(−2, −10) C*(−8, −4)
3. (a) P′(−1, −3) Q′(−5, −8) R′(2, −10) (b) P″(5, −9) Q″(10, −13) R″(12, −6)
(c) P*(1, 5) Q*(5, 10) R*(−2, 12)
4. (a) A′(1, 2) B′(−10, 2) C′(2, 7) (b) A″(−9, −4) B″(2, −4) C″(−10, −9)
(c) A*(13, −4) B*(2, −4) C*(14, 1)
5. (a) A′(−2·5, 3·7) B′(−5·7, 10·2) C′(2·5, 12·4) (b) A″(−5·1, 2·1) B″(−12·3, 2·6) C″(−10·1, 10·8)
(c) A*(0, 3·2) B*(7·1, 4·6) C*(7·1, −3·9)
6. (a) +90°, centre (0, 0) (b) 180°, centre (2, 1) (c) −90°, centre (2, 0)
(d) 180°, centre $(3\frac{1}{2}, 2\frac{1}{2})$ (e) −90°, centre (−2, 0) (f) +90°, centre (0, −4)

Exercise 8 page 198

1. $\begin{pmatrix} 6 \\ 2 \end{pmatrix}$ 2. $\begin{pmatrix} 5 \\ -3 \end{pmatrix}$ 3. $\begin{pmatrix} -4 \\ 4 \end{pmatrix}$ 4. $\begin{pmatrix} -5 \\ -4 \end{pmatrix}$ 5. $\begin{pmatrix} 0 \\ -5 \end{pmatrix}$
6. (a) $\begin{pmatrix} 7 \\ 3 \end{pmatrix}$ (b) $\begin{pmatrix} 0 \\ -9 \end{pmatrix}$ (c) $\begin{pmatrix} 9 \\ 10 \end{pmatrix}$ (d) $\begin{pmatrix} -10 \\ 3 \end{pmatrix}$ (e) $\begin{pmatrix} -1 \\ 13 \end{pmatrix}$ (f) $\begin{pmatrix} 10 \\ 0 \end{pmatrix}$
(g) $\begin{pmatrix} -9 \\ -4 \end{pmatrix}$ (h) $\begin{pmatrix} -10 \\ 0 \end{pmatrix}$
7. (5, 2) 8. (5, 6) 9. (8, −5) 10. (0, 6) 11. (4, −7)
12. (−3, 4) 13. (−3, −5) 14. (−1, −8) 15. (6, −1) 16. (5, 2)
17. (5, 2) 18. (0, 6) 19. (8, 6) 20. (0, 6) 21. (−3, 2)
22. (8, −5) 23. (−3, −5)

Exercise 9 page 200

7. (4, 8), (8, 4), (10, 10) 8. (3, 6), (7, 2), (9, 8) 9. (1, 1), (10, 4), (4, 7)
10. (1, 4), (7, 8), (11, 2) 11. (0, 2), (6, 2), (6, 6), (0, 6) 12. (0, 0); +2

13. (2, 1); +3
14. (7, 6); +2
15. $(5\frac{1}{2}, 6)$; +3
16. (11, 9); $+\frac{1}{2}$
17. (4, 6); $+1\frac{1}{2}$
18. $(\frac{1}{2}, 1)$ $(6\frac{1}{2}, 1)$ $(\frac{1}{2}, 5)$
19. (6, 6) (12, 6) (6, 10)
20. (3, 3) (4, 3) (4, 5) (3, 5)
21. (1, 1) (2, 1) (2, 3) (1, 3)
22. (4, 9) (4, 5) (4, 1) (0, 1) (0, 5)
23. (3, 7) (1, 7) (3, 2)
24. (10, 7) (6, 7) (6, 5) (10, 5)
25. (12, 9) (3, 9) (3, 3) (12, 3)
26. (6, 5) $(3\frac{1}{2}, 5)$ $(3\frac{1}{2}, 3)$ (6, 3)
27. (5, 4); −2
28. (4, 3); −2
29. (4, 6); $-\frac{1}{3}$
30. (6, 6); −1

Exercise 10 *page 203*

1. (2, −3)
2. (5, −1)
3. (6, 4)
4. (4, −6)
5. (0, 0)
6. (−6, 4)
7. (3, −2)
8. (3, 2)
9. (−2, 3)
10. (0, 0)
11. (−3, 2)
12. (−3, −2)
13. (−3, −2)
14. (−6, −4)
15. (6, 0)
16. (−2, 3)
17. (0, 4)
18. (6, 8)
19. (3, 2)
20. (0, 0)

Exercise 11 *page 203*

1. $T_6, T_5; \mathbf{R_b}$
2. $T_3, T_1; \mathbf{M_2}$
3. $T_4, T_2; \mathbf{M_4}$
4. $T_2, T_8; \mathbf{M_4}$
5. $T_7, T_5; \mathbf{M_4}$
6. $T_1, T_6; \mathbf{M_3}$
7. $T_4, T_3; \mathbf{M_3}$
8. $T_3, T_6; \mathbf{M_2}$
9. $T_8, T_1; \mathbf{R_c}$
10. $T_8, T_2; \mathbf{R_a}$
11. $T_8, T_5; \mathbf{M_2}$
12. $T_8, T_1; \mathbf{R_b}$
13. $T_4, T_1; \mathbf{R_b}$
14. $T_3, T_8; \mathbf{I}$
15. $T_4, T_3; \mathbf{M_4}$
16. $T_6, T_7, T_1; \mathbf{R_b}$
17. $T_6, T_2, T_7; \mathbf{R_a}$
18. $T_7, T_2, T_4; \mathbf{M_4}$
19. $T_5, T_4, T_8; \mathbf{M_2}$
20. $T_7, T_2, T_3; \mathbf{M_1}$
21. $T_2, T_8, T_7; \mathbf{I}$
22. $T_5, T_8, T_6; \mathbf{I}$
23. $T_4, T_6, T_5; \mathbf{M_4}$
24. $T_5, T_3, T_8; \mathbf{R_a}$
25. No

Exercise 12 *page 204*

1. (a) (−4, 4) (b) (2, −2) (c) (0, 0) (d) (0, 4) (e) (0, 0)
2. (a) (−2, 5) (b) (−4, 0) (c) (2, −2) (d) (1, −1)
3. A^{-1} : reflection in $x = 2$
   B^{-1} : B
   C^{-1} : translation $\begin{pmatrix} 6 \\ -2 \end{pmatrix}$
   D^{-1} : D
   E^{-1} : E
   F^{-1} : translation $\begin{pmatrix} -4 \\ -3 \end{pmatrix}$
   G^{-1} : 90° rotation anticlockwise, centre (0, 0)
   H^{-1} : enlargement, scale factor 2, centre (0, 0)
4. (a) (4, 0) (b) (−6, −1) (c) (−2, −2) (d) (2, −2) (e) (6, 2)
5. (a) (1, −6) (b) (4, −2) (c) (2, 7) (d) (4, −6) (e) (2, −4)
6. (a) reflection in y-axis
   (b) rotation 180°, centre (−2, 2)
   (c) rotation −90°, centre (2, 2)
7. (a) rotation +90°, centre (0, 0) (b) translation $\begin{pmatrix} -2 \\ 5 \end{pmatrix}$
   (c) rotation +90°, centre (2, −4) (d) rotation +90°, centre $(-\frac{1}{2}, 3\frac{1}{2})$
8. (a) (−1, −2) (b) (8, 2) (c) (4, −6) (d) (0, −2)
9. (a) rotation +90°, centre (2, 2) (b) enlargement, scale factor $\frac{1}{2}$, centre (8, 6)
   (c) rotation −90°, centre $(-\frac{1}{2}, -3\frac{1}{2})$

Exercise 13 *page 205*

2. (a) enlargement : scale factor 2, centre (0, 0) (b) enlargement : scale factor $-\frac{1}{2}$, centre (0, 0)
   (c) reflection in $y = -x$ (d) enlargement : scale factor −2, centre (0, 0)
3. A : reflection in x-axis B : reflection in y-axis C : reflection in $y = x$
   D : rotation, −90°, centre (0, 0) E : reflection in $y = -x$ F : rotation, 180°, centre (0, 0)
   G : rotation, +90°, centre (0, 0) H : identity (no change)
4. (a) ratio = 4 : 1 (b) ratio = 4 : 1
5. (a) $\begin{pmatrix} 0 & -1 \\ 1 & 0 \end{pmatrix}$ (b) $\begin{pmatrix} -1 & 0 \\ 0 & -1 \end{pmatrix}$ (c) $\begin{pmatrix} 0{\cdot}866 & -0{\cdot}5 \\ 0{\cdot}5 & 0{\cdot}866 \end{pmatrix}$ (d) $\begin{pmatrix} 0 & 1 \\ -1 & 0 \end{pmatrix}$
   (e) $\begin{pmatrix} 0{\cdot}5 & -0{\cdot}866 \\ 0{\cdot}866 & 0{\cdot}5 \end{pmatrix}$ (f) $\begin{pmatrix} -0{\cdot}866 & -0{\cdot}5 \\ 0{\cdot}5 & -0{\cdot}866 \end{pmatrix}$ (g) $\begin{pmatrix} 0{\cdot}707 & -0{\cdot}707 \\ 0{\cdot}707 & 0{\cdot}707 \end{pmatrix}$ (h) $\begin{pmatrix} 0{\cdot}6 & -0{\cdot}8 \\ 0{\cdot}8 & 0{\cdot}6 \end{pmatrix}$

6. (a) $+90°$ (b) $36{\cdot}9°$ (c) $-60°$ (d) $-53{\cdot}1°$ (e) $-45°$
7. rotation $+45°$; enlargement scale factor $\sqrt{2}$
8. rotation $-26{\cdot}6°$; enlargement scale factor $\sqrt{5}$
9. (d) rotation $-90°$, centre $(0, 0)$
10. $y = 2x$
11. $y = 3x$
12. (c) $OB = \sqrt{20}$, $OB' = 3\sqrt{20}$ (d) $36{\cdot}9°$ (e) rotation $36{\cdot}9°$; enlargement scale factor 3

Exercise 14 *page 208*

1. $x = 0$
2. $y = 0$
3. $y = 0$
4. $x = 0$
5. stretch : parallel to x-axis, scale factor 2
6. stretch : parallel to x-axis, scale factor 3
7. stretch : parallel to y-axis, scale factor 2
8. stretch : parallel to x-axis, scale factor $1\frac{1}{2}$
9. shear : x-axis invariant
10. stretch : parallel to x-axis, scale factor -2
11. stretch : parallel to y-axis, scale factor 3
12. stretch : parallel to x-axis, scale factor $\frac{1}{2}$
13. rotation : $+90°$, centre $(0, 0)$
14. reflection in y-axis
15. reflection in $y = -x$
16. reflection in $y = x$
17. enlargement : scale factor 2, centre $(0, 0)$
18. enlargement : scale factor $\frac{1}{2}$, centre $(0, 0)$
19. stretch : parallel to x-axis, scale factor 3
20. shear : x-axis invariant
21. shear : y-axis invariant
22. stretch : parallel to y-axis, scale factor 2
23. enlargement : scale factor -2, centre $(0, 0)$
24. enlargement : scale factor $-\frac{1}{2}$, centre $(0, 0)$
25. invariant line $y = -x$
26. invariant line $y = -x$
28. two-way stretch : scale factor 2 parallel to x-axis, scale factor 4 parallel to y-axis
29. invariant line : x-axis; $\begin{pmatrix} 1 & 1 \\ 0 & 1 \end{pmatrix}$
30. invariant line : y-axis; $\begin{pmatrix} 1 & 0 \\ 2 & 1 \end{pmatrix}$
31. $\begin{pmatrix} 0 & -1 \\ 1 & 0 \end{pmatrix}$
32. $\begin{pmatrix} 0 & 1 \\ 1 & 0 \end{pmatrix}$
33. $\begin{pmatrix} -1 & 0 \\ 0 & 1 \end{pmatrix}$
34. $\begin{pmatrix} -1 & 0 \\ 0 & -1 \end{pmatrix}$
35. $\begin{pmatrix} 3 & 0 \\ 0 & 3 \end{pmatrix}$
36. $\begin{pmatrix} 0 & -1 \\ -1 & 0 \end{pmatrix}$
37. $\begin{pmatrix} -2 & 0 \\ 0 & -2 \end{pmatrix}$
38. $\begin{pmatrix} 1 & 0 \\ 0 & 4 \end{pmatrix}$
39. $\begin{pmatrix} 1 & 0 \\ 0 & -1 \end{pmatrix}$
40. $\begin{pmatrix} 0 & 1 \\ -1 & 0 \end{pmatrix}$
41. $\begin{pmatrix} \frac{1}{2} & 0 \\ 0 & \frac{1}{2} \end{pmatrix}$
42. $\begin{pmatrix} 3 & 0 \\ 0 & 1 \end{pmatrix}$

Exercise 15 *page 211*

1. (a) reflection in $y = x - 1$ (b) reflection in $y = 1$
   (c) rotation $-90°$, centre $(2, -2)$ (d) enlargement, scale factor 3, centre $(2, -1)$
2. $\mathbf{BA} \equiv \mathbf{A}$ then $\mathbf{B}$
3. $\mathbf{BA} \equiv \mathbf{A}$ then $\mathbf{B}$
4. $\mathbf{ZYX} \equiv \mathbf{X}$ then $\mathbf{Y}$ then $\mathbf{Z}$
5. $\mathbf{RQP} \equiv \mathbf{P}$ then $\mathbf{Q}$ then $\mathbf{R}$
7. (a) $(14, 3)$ (b) $m = 3, n = \frac{1}{2}$ (c) $h = 1, k = -2$
8. (a) 4 (b) 2 (c) 1 (d) 2 (e) 1 (f) 1
9. 1
10. object area = 4 sq. units; image area = 8 sq. units
11. (a) 16 sq. units (b) 16 sq. units (c) 36 sq. units
12. 2 sq. units
13. (a) $(-2, 2)$ (c) $\begin{pmatrix} 2 \\ -2 \end{pmatrix}$ (d) $\begin{pmatrix} 2 & 0 \\ 0 & 2 \end{pmatrix}\begin{pmatrix} x \\ y \end{pmatrix} + \begin{pmatrix} 2 \\ -2 \end{pmatrix}$
14. (a) $x = 2$ (c) $\begin{pmatrix} 4 \\ 0 \end{pmatrix}$ (d) $\begin{pmatrix} -1 & 0 \\ 0 & 1 \end{pmatrix}\begin{pmatrix} x \\ y \end{pmatrix} + \begin{pmatrix} 4 \\ 0 \end{pmatrix}$
15. (a) $\begin{pmatrix} 2 & 0 \\ 0 & 2 \end{pmatrix}\begin{pmatrix} x \\ y \end{pmatrix} + \begin{pmatrix} -1 \\ -3 \end{pmatrix}$ (b) $\begin{pmatrix} 2 & 0 \\ 0 & 2 \end{pmatrix}\begin{pmatrix} x \\ y \end{pmatrix} + \begin{pmatrix} -\frac{1}{2} \\ -1 \end{pmatrix}$ (c) $\begin{pmatrix} 0 & 1 \\ 1 & 0 \end{pmatrix}\begin{pmatrix} x \\ y \end{pmatrix} + \begin{pmatrix} -3 \\ 3 \end{pmatrix}$
   (d) $\begin{pmatrix} -1 & 0 \\ 0 & -1 \end{pmatrix}\begin{pmatrix} x \\ y \end{pmatrix} + \begin{pmatrix} 3 \\ 5 \end{pmatrix}$ (e) $\begin{pmatrix} 1 & 0 \\ 0 & -1 \end{pmatrix}\begin{pmatrix} x \\ y \end{pmatrix} + \begin{pmatrix} 0 \\ 2 \end{pmatrix}$ (f) $\begin{pmatrix} 0 & 1 \\ -1 & 0 \end{pmatrix}\begin{pmatrix} x \\ y \end{pmatrix} + \begin{pmatrix} 4 \\ 0 \end{pmatrix}$

Revision exercise 8A *page 212*

1. (a) $\begin{pmatrix} 6 & 4 \\ 2 & 8 \end{pmatrix}$ (b) $\begin{pmatrix} 4 & -1 \\ 1 & 2 \end{pmatrix}$ (c) $\begin{pmatrix} 1\frac{1}{2} & 1 \\ \frac{1}{2} & 2 \end{pmatrix}$ (d) $\begin{pmatrix} -3 & 13 \\ -1 & 11 \end{pmatrix}$ (e) $\begin{pmatrix} 1 & 3 \\ 0 & 4 \end{pmatrix}$

2. (a) $\begin{pmatrix} -9 & -1 \\ 5 & 1\frac{1}{3} \end{pmatrix}$ (b) $\begin{pmatrix} 12 & 6 \\ 4 & 2 \end{pmatrix}$ (c) $\begin{pmatrix} 9 & -2 \\ 2 & -7 \end{pmatrix}$

3. (a) (14); $\begin{pmatrix} -1 & -5 \\ 3 & 15 \end{pmatrix}$ (b) $\mathbf{X} = \begin{pmatrix} 1 & 3 \\ 2 & 4 \end{pmatrix}$

4. (a) $\frac{1}{13}\begin{pmatrix} 5 & 1 \\ -3 & 2 \end{pmatrix}$ (b) $x = 3, y = -2$

5. $x = 3$; $-\frac{1}{9}\begin{pmatrix} -1 & -2 \\ -3 & 3 \end{pmatrix}$

6. $h = 4, k = -4$

7. (a) $a = \pm 4$ (b) $a = \pm 3$

8. (a) reflection in y-axis (b) reflection in $y = x$ (c) rotation $-90°$, centre (0, 0) (d) reflection in $y = -x$ (e) rotation 180°, centre (0, 0) (f) rotation $-90°$, centre (0, 0)

9. (a) (4, −1) (b) (4, 1) (c) (−3, 2)

10. (a) (−1, −3) (b) (−1, 3) (c) (6, 2) (d) (−3, 1) (e) (−2, 6) (f) (0, 2)

11. (a) (−1, 2) (b) (1, −2) (c) (10, −2) (d) (6, −2) (e) (−10, 2) (f) (12, 2)

12. (a) A′(−3, −1) B′(1, −1) C′(−3, −7) (b) A′(2, −2) B′(6, −2) C′(2, −8) (c) A′(1, 1) B′(2, 1) C′($1, -\frac{1}{2}$) (d) A′(4, 2) B′(3, 2) C′($4, 3\frac{1}{2}$) (e) A′(−2, 2) B′(−6, 2) C′(−2, 8)

13. (a) A′(−1, −2) B′(−7, −2) C′(−1, 2) (b) A′(2, −2) B′(8, −2) C′(2, −6) (c) A′(2, 1) B′($\frac{1}{2}$, 1) C′(2, 2) (d) A′(1, 7) B′(4, 7) C′(1, 5)

14. (a) (−2, 5) (b) (−4, −3) (c) rotation +90°, centre (0, 0)

15. (a) rotation +90°, centre (0, 0) (b) reflection in x-axis (c) rotation 180°, centre (0, 0) (d) rotation −90°, centre (0, 0) (e) reflection in $y = -x$

16. (a) reflection in $y = x$ (b) reflection in y-axis (c) enlargement, scale factor 3, centre (0, 0) (d) shear, x-axis invariant (e) shear, y-axis invariant (f) stretch, parallel to x-axis, scale factor 3

17. (a) $\begin{pmatrix} -1 & 0 \\ 0 & -1 \end{pmatrix}$ (b) $\begin{pmatrix} 1 & 0 \\ 0 & -1 \end{pmatrix}$ (c) $\begin{pmatrix} 4 & 0 \\ 0 & 4 \end{pmatrix}$ (d) $\begin{pmatrix} 0 & -1 \\ -1 & 0 \end{pmatrix}$ (e) $\begin{pmatrix} 0 & 1 \\ -1 & 0 \end{pmatrix}$

18. (a) enlargement, scale factor 2, centre (0, 0); translation $\begin{pmatrix} 5 \\ -2 \end{pmatrix}$ (b) (11, −4) (c) (3, −1) (d) (1, 3)

19. (a) $\begin{pmatrix} 0 & 1 \\ 1 & 0 \end{pmatrix}$ (b) $\begin{pmatrix} -1 & 0 \\ 0 & 1 \end{pmatrix}$ (c) $\begin{pmatrix} 0 & 1 \\ -1 & 0 \end{pmatrix}$ (d) $\begin{pmatrix} 0 & -1 \\ 1 & 0 \end{pmatrix}$ $\mathbf{AB} \equiv$ rotation −90°, centre (0, 0) $\mathbf{BA} \equiv$ rotation +90°, centre (0, 0)

20. (a) reflection in $x = \frac{1}{2}$ (b) reflection in $y = -x$ (c) rotation, 180°, centre (1, 1)

Examination Exercise 8B page 214

1. **RS** : rotation, 180°, centre ($4\frac{1}{2}, -2\frac{1}{2}$); **SR** : rotation, 180°, centre ($-\frac{1}{2}, 3\frac{1}{2}$); **M** : translation $\begin{pmatrix} 10 \\ -12 \end{pmatrix}$

2. (i) rotation, 180°, centre (0, 0) (ii) reflection in $y = -x$ (iii) translation $\begin{pmatrix} -3 \\ -3 \end{pmatrix}$ (iv) rotation, −90°, centre ($-1\frac{1}{2}, 1\frac{1}{2}$) $\mathbf{T} = \begin{pmatrix} -1 & 0 \\ 0 & -1 \end{pmatrix}$; $\mathbf{U} = \begin{pmatrix} 0 & -1 \\ -1 & 0 \end{pmatrix}$

3. (ii) scale factor = $\sqrt{2}$ (1·4 to 2 S.F.) (iii) (a) 5 cm², 10 cm² (b) 2 (c) $\sqrt{2}$ (iv) enlargement, scale factor 2, centre (0, 0)

4. (ii) enlargement, scale factor 3, centre (0, 0) (v) reflection in $y = -x$

5. (ii) OA′ = 12 cm, AOA′ = 36·9° (iii) 36·9°; scale factor 3 (iv) 144 cm²; 17·0 cm

6. (a) $\begin{pmatrix} -2 & 3 \\ 3 & -4 \end{pmatrix}$, $\begin{pmatrix} 7 & 8 \\ -9 & -10 \end{pmatrix}$ (b) **M** : reflection in x-axis; **R** : rotation, +90°, centre (0, 0);

$\mathbf{RM} = \begin{pmatrix} 0 & 1 \\ 1 & 0 \end{pmatrix}$ reflection in $y = x$;

$\mathbf{MR} = \begin{pmatrix} 0 & -1 \\ -1 & 0 \end{pmatrix}$ reflection in $y = -x$.

7. (ii) $P_2(-6, 8)$, $Q_2(14, -2)$, $R_2(15, 5)$; $4y = 3x$ (iii) 73·7°

8. (i) 90° (ii) $\begin{pmatrix} 0 & -1 \\ 1 & 0 \end{pmatrix}$ (iii) $\begin{pmatrix} 8 \\ -4 \end{pmatrix}$ (iv) (6, 2) (vi) $x + y + 5 = 0$

9. (i) $T_2 : (0, 0), (-4, 0), (0, -8)$. $T_3 : (0, 0), (0, -8), (16, 0)$
(ii) enlargement, scale factor 2, centre (0, 0) (iv) $n = 4$

10. (a) (i) reflection in $y = x$, translation $\begin{pmatrix} -5 \\ 5 \end{pmatrix}$ (ii) (−4, 3) (iii) any points on $y = x + 5$
(iv) reflection in $y = x + 5$ (b) $y = 3x$

PART 9

Exercise 1 *page 218*

1. (a) {5, 7} (b) {1, 2, 3, 4, 5, 6, 7, 8, 9, 11, 13} (c) 5 (d) 11
(e) true (f) true (g) false (h) true

2. (a) {2, 3, 5, 7} (b) {1, 2, 3, . . . 9} (c) 4 (d) ∅ (e) false
(f) true (g) false (h) true

3. (a) {2, 4, 6, 8, 10} (b) {16, 18, 20} (c) ∅ (d) 15 (e) 11
(f) 21 (g) false (h) false (i) true (j) true

4. (a) {1, 3, 4, 5} (b) {1, 5} (c) 1 (d) {1, 5} (e) {1, 3, 5, 10} (f) 4
(g) true (h) false (i) true

5. (a) 4 (b) 3 (c) {b, d} (d) {a, b, c, d, e} (e) 5 (f) 2

6. (a) 2 (b) 4 (c) {1, 2, 4, 6, 7, 8, 9} (d) {7, 9} (e) {1, 2, 4}
(f) {1, 2, 4, 7, 9} (g) {1, 2, 4, 6, 8} (h) {6, 7, 8, 9} (i) {1, 2, 4, 7, 9}

7. (a) 4 (b) 5 (c) 8 (d) {1} (e) {1, 2, 4, 8} (f) {2, 4, 6, 7, 8}
(g) {6, 7} (h) 4 (i) {1, 2, 4, 6, 7, 8}

8. (a) {a, b, c, h} (b) 5 (c) 3 (d) {b, c, h} (e) 8 (f) {a, b, c, d, e, h}
(g) true (h) false (i) true

9.

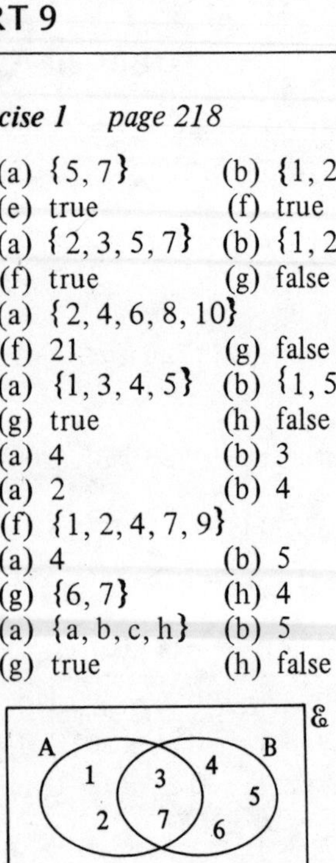

10. 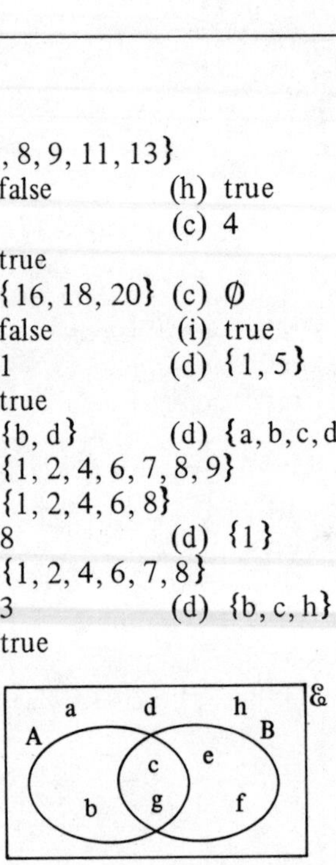 (a) {a, b, d, h} (b) {b, c, e, f, g}
(c) {a, d, h} (d) 6

11. (a) {4, 5} (b) {5} (c) {1, 2, 4, 5, 6, 7, 9} (d) {3, 4} (e) {2, 3}
(f) {1, 8, 9, 10} (g) 1 (h) 8

12. (a) {5, 9} (b) {5, 9} (c) {1, 2, 3, 4, 5, 6, 8, 9, 10, 12} (d) {7, 8, 11} (e) {6, 8}
(f) {1, 10} (g) 1 (h) 4

Exercise 2 *page 220*

7. (a) $A \cup B$ (b) $A' \cap B$ (c) $(A \cup B)'$ (d) $Y' \cap X$ (e) $A \cap B \cap C$ (f) $A \cap B$
(g) $(A \cap B) \cup C$ (h) $L' \cap (M \cap N)$ (i) $(X \cap Y) \cup (Y \cap Z)$ (j) $(A \cup C)' \cap B$
(k) $(A \cap B) \cup (B \cap C) \cup (A \cap C)$ (l) $(A \cap B)'$

8. (a) (i) 12 (ii) 4 (iii) 15 (iv) 17 (v) 10 (vi) 3
(vii) 8 (viii) 7 (ix) 11 (x) 19 (xi) 19 (xii) 8
(xiii) 7 (xiv) 6 (xv) 11
(b) (i) 9 (ii) 2 (iii) 12 (iv) 13 (v) 8 (vi) 1
(vii) 7 (viii) 4 (ix) 9 (x) 17 (xi) 15 (xii) 6
(xiii) 4 (xiv) 5 (xv) 8

9. (a) A (b) A (c) $P \cup Q$ (d) $X \cap Y$ (e) A (f) $A \cap B$
(g) $A' \cap B$ (h) $A \cup B$ (i) A

Exercise 3 *page 222*

1. 22 **2.** 11 **3.** 19 **4.** (a) 4; 0 (b) 14; 10
5. (a) 6; 0 (b) 16; 10 **6.** (a) 8, 2 (b) 16; 10
7. (a) 9; 3 (b) 21; 15 **8.** (a) 9 (b) 5
9. {squares} **10.** $I \cap E$ = {equilateral triangles}; $I \cup E$ = {isosceles triangles}
11. **12.** **13.** 25 **14.**

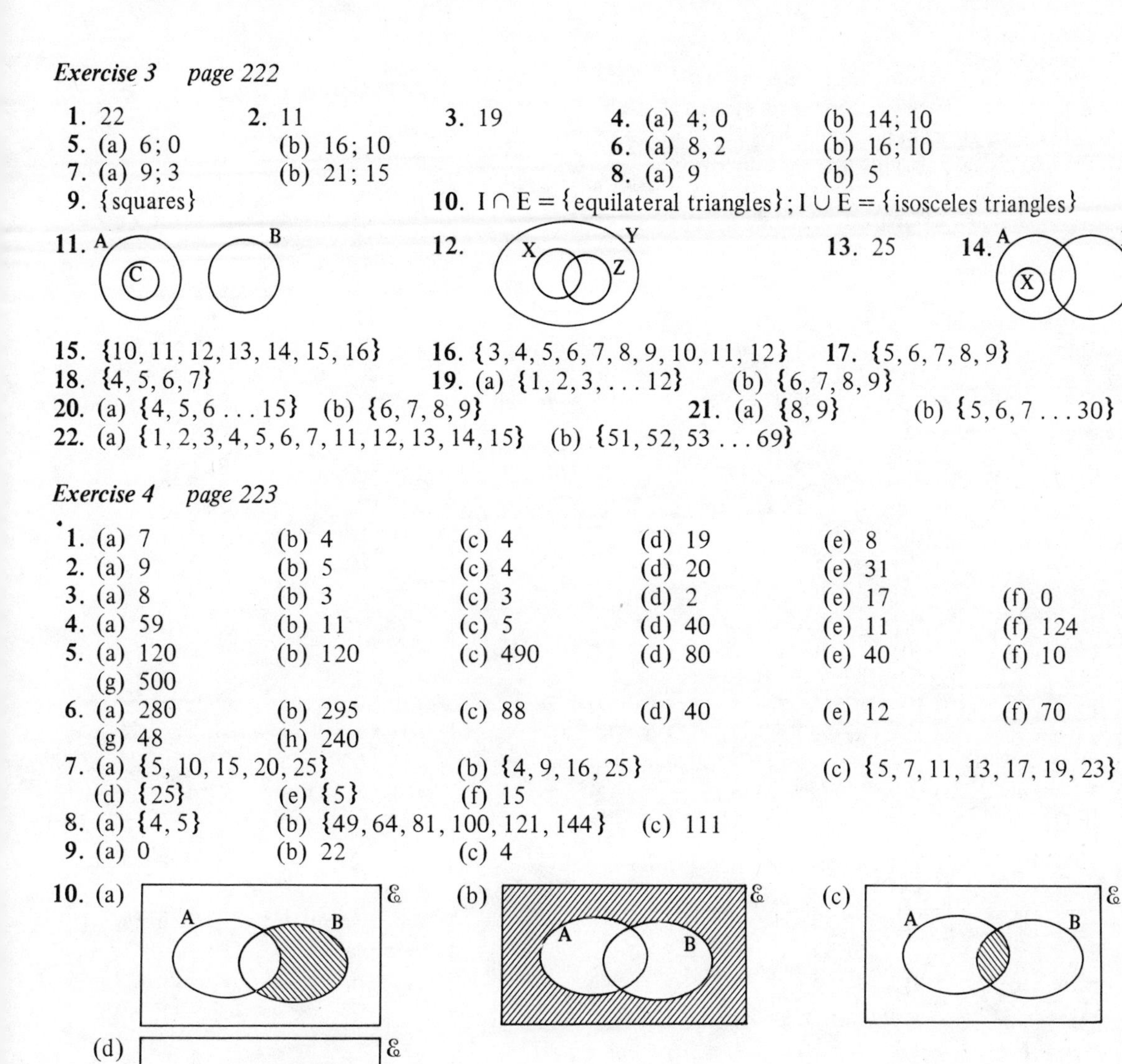

15. {10, 11, 12, 13, 14, 15, 16} **16.** {3, 4, 5, 6, 7, 8, 9, 10, 11, 12} **17.** {5, 6, 7, 8, 9}
18. {4, 5, 6, 7} **19.** (a) {1, 2, 3, . . . 12} (b) {6, 7, 8, 9}
20. (a) {4, 5, 6 . . . 15} (b) {6, 7, 8, 9} **21.** (a) {8, 9} (b) {5, 6, 7 . . . 30}
22. (a) {1, 2, 3, 4, 5, 6, 7, 11, 12, 13, 14, 15} (b) {51, 52, 53 . . . 69}

Exercise 4 *page 223*

1. (a) 7 (b) 4 (c) 4 (d) 19 (e) 8
2. (a) 9 (b) 5 (c) 4 (d) 20 (e) 31
3. (a) 8 (b) 3 (c) 3 (d) 2 (e) 17 (f) 0
4. (a) 59 (b) 11 (c) 5 (d) 40 (e) 11 (f) 124
5. (a) 120 (b) 120 (c) 490 (d) 80 (e) 40 (f) 10
(g) 500
6. (a) 280 (b) 295 (c) 88 (d) 40 (e) 12 (f) 70
(g) 48 (h) 240
7. (a) {5, 10, 15, 20, 25} (b) {4, 9, 16, 25} (c) {5, 7, 11, 13, 17, 19, 23}
(d) {25} (e) {5} (f) 15
8. (a) {4, 5} (b) {49, 64, 81, 100, 121, 144} (c) 111
9. (a) 0 (b) 22 (c) 4

10. (a) (b) (c)

(d)

Exercise 5 *page 225*

1. (a) $10 - x$ (b) $13 - x$ (d) 5 **2.** 9 **3.** 14 **4.** 3
5. 36 **6.** 3 **7.** 11 **8.** 5 **9.** 28 **10.** $x = 6$; 26
11. 34 **12.** (a) 12 (b) $\frac{1}{2}$ **13.** $x = 10$; 30 **14.** 2
15. (a) All good footballers are Scottish men. (b) No good footballers are Scottish men.
(c) There are some good footballers who are Scottish men.
16. (a) The football players are all boys. (b) The hockey players are all girls.
(c) There are some people who play both football and hockey.
(d) There are no boys who play hockey. (e) $B \cap F = \emptyset$ (f) $H \cup F = \mathcal{E}$
17. (a) $S \cap T = \emptyset$ (b) $F \subset T$ (c) $S \cap F \neq \emptyset$
(d) All living creatures are either spiders, animals that fly or animals which taste nice.
(e) Animals which taste nice are all spiders.

18. (a) All tigers who believe in fairies also believe in Eskimos.
    (b) All tigers who believe in fairies or Eskimos are in hospital.
    (c) There are no tigers in hospital who believe in Eskimos. (d) $H \subset T$ (e) $T \cap X \neq \emptyset$
19. (a) There are no good bridge players called Peter.
    (b) All school teachers are either called Peter, are good bridge players or are women.
    (c) There are some women teachers called Peter. (d) $W \cap B = \emptyset$ (e) $B \subset (W \cap P)$

Exercise 6 page 227

1. d	2. 2c	3. 3c	4. 3d	5. 5d	6. 3c
7. −2d	8. −2c	9. −3c	10. −c	11. c + d	12. c + 2d
13. 2c + d	14. 3c + d	15. 2c + 2d	16. 2c + 3d	17. 2c − d	18. 3c − d
19. −c + 2d	20. −c + 3d	21. −c + d	22. −c − 2d	23. −2c − 2d	24. −3c − 6d
25. −2c + 3d	26. c + 6d				
27. $\overrightarrow{QI}$	28. $\overrightarrow{QU}$	29. $\overrightarrow{QH}$	30. $\overrightarrow{QB}$	31. $\overrightarrow{QF}$	32. $\overrightarrow{QJ}$
33. $\overrightarrow{QZ}$	34. $\overrightarrow{QL}$	35. $\overrightarrow{QE}$	36. $\overrightarrow{QX}$	37. $\overrightarrow{QW}$	38. $\overrightarrow{QK}$

	(a)	(b)	(c)	(d)	(e)
39.	−a	a + b	2a − b	−a + b	
40.	a + b	a − 2b	−a + b	−a − b	
41.	−a − b	3a − b	2a − b	−2a + b	
42.	a − 2b	a − b	2a	−2a + 3b	
43.	−2a + 2b	2a − b	3a − b	−3a + b	
44.	2a − c	2a − c	3a	a + b + c	−3a − b
45.	b − c	2b + 2c	a + 2b + 2c	−a − b	−a − b + c
46.	a + c	−a + c	a + b + c	b − c	−a + 2c

Exercise 7 page 229

1. (a) a (b) −a + b (c) 2b (d) −2a (e) −2a + 2b
   (f) −a + b (g) a + b (h) b (i) −b + 2a (j) −2b + a
2. (a) a (b) −a + b (c) 3b (d) −2a (e) −2a + 3b
   (f) $-a + \frac{3}{2}b$ (g) $a + \frac{3}{2}b$ (h) $\frac{3}{2}b$ (i) −b + 2a (j) −3b + a
3. (a) 2a (b) −a + b (c) 2b (d) −3a (e) −3a + 2b
   (f) $-\frac{3}{2}a + b$ (g) $\frac{3}{2}a + b$ (h) $\frac{1}{2}a + b$ (i) −b + 3a (j) −2b + a
4. (a) $\frac{1}{2}a$ (b) −a + b (c) 4b (d) $-\frac{3}{2}a$ (e) $-\frac{3}{2}a + 4b$
   (f) $-a + \frac{8}{3}b$ (g) $\frac{1}{2}a + \frac{8}{3}b$ (h) $-\frac{1}{2}a + \frac{8}{3}b$ (i) $\frac{3}{2}a - b$ (j) a − 4b
5. (a) 5a (b) b − a (c) $\frac{3}{2}b$ (d) −6a (e) $\frac{3}{2}b - 6a$
   (f) b − 4a (g) 2a + b (h) a + b (i) 6a − b (j) $a - \frac{3}{2}b$
6. (a) 4a (b) b − a (c) 3b (d) −5a (e) 3b − 5a
   (f) $\frac{3}{4}b - \frac{5}{4}a$ (g) $\frac{15}{4}a + \frac{3}{4}b$ (h) $\frac{11}{4}a + \frac{3}{4}b$ (i) 5a − b (j) a − 3b
7. $\frac{1}{2}s - \frac{1}{2}t$ 8. $\frac{1}{3}a + \frac{2}{3}b$ 9. a + c − b 10. 2m + 2n
11. (a) b − a (b) b − a (c) 2b − 2a (d) b − 2a (e) b − 2a
    (f) 2b − 3a
12. (a) y − z (b) $\frac{1}{2}y - \frac{1}{2}z$ (c) $\frac{1}{2}y + \frac{1}{2}z$ (d) $-x + \frac{1}{2}y + \frac{1}{2}z$ (e) $-\frac{2}{3}x + \frac{1}{3}y + \frac{1}{3}z$
    (f) $\frac{1}{3}x + \frac{1}{3}y + \frac{1}{3}z$

Exercise 8 page 232

13. $\begin{pmatrix}-11\\9\end{pmatrix}$ 14. $\begin{pmatrix}-1\\-4\end{pmatrix}$ 15. $\begin{pmatrix}4\\8\end{pmatrix}$ 16. $\begin{pmatrix}4\\3\end{pmatrix}$ 17. $\begin{pmatrix}8\\-7\end{pmatrix}$ 18. $\begin{pmatrix}18\\-1\end{pmatrix}$

19. $\begin{pmatrix}7\\-4\end{pmatrix}$ 20. $\begin{pmatrix}-18\\-16\end{pmatrix}$ 21. $\begin{pmatrix}13\\19\end{pmatrix}$ 22. $\begin{pmatrix}8\\5\end{pmatrix}$ 23. $\begin{pmatrix}10\\-13\end{pmatrix}$ 24. $\begin{pmatrix}17\\35\end{pmatrix}$

25. $\begin{pmatrix}4\\8\end{pmatrix}$ 26. $\begin{pmatrix}4\\3\end{pmatrix}$ 27. $\begin{pmatrix}-1\\1\end{pmatrix}$ 28. $\begin{pmatrix}-16\\3\end{pmatrix}$ 29. $\begin{pmatrix}-2\frac{1}{2}\\-3\end{pmatrix}$ 30. $\begin{pmatrix}-3\frac{1}{2}\\\frac{1}{2}\end{pmatrix}$
31. $\begin{pmatrix}-3\\-7\end{pmatrix}$ 32. $\begin{pmatrix}-2\\12\end{pmatrix}$ 33. $\begin{pmatrix}4\\8\end{pmatrix}$ 34. $\begin{pmatrix}-5\frac{1}{2}\\4\frac{1}{2}\end{pmatrix}$ 35. $\begin{pmatrix}0\\0\end{pmatrix}$ 36. $\begin{pmatrix}-2\frac{1}{2}\\-6\end{pmatrix}$
37. (b) **l** and **s**; **n** and **r**; **p** and **t**; **m** and **q**
38. (a) true (b) true (c) true (d) true (e) false (f) false

Exercise 9 page 233

1. 5 2. $\sqrt{17}$ 3. 13 4. 3 5. 5 6. $\sqrt{45}$
7. $\sqrt{74}$ 8. $\sqrt{208}$ 9. 10 10. $\sqrt{89}$
11. (a) $\sqrt{320}$ (b) no 12. (a) $\sqrt{148}$ (b) no 13. $\sqrt{29}$ 14. $\sqrt{26}$
15. $\sqrt{5}$ 16. (a) 5 (b) $n = \pm 4$ 17. (a) 13 (b) $m = \pm 13$
18. (a) 5 (b) $p = 0$ 19. (a) 9 (b) 6 (c) 5
20. (a) 30 (b) 5 (c) $\sqrt{50}$ (d) 4

Exercise 10 page 234

1. $\begin{pmatrix}2\\-2\end{pmatrix}$ 2. $\begin{pmatrix}6\\-2\end{pmatrix}$ 3. (b) $\begin{pmatrix}0\\3\end{pmatrix}$; $\begin{pmatrix}-5\\-5\end{pmatrix}$ 4. (b) $\begin{pmatrix}4\\2\end{pmatrix}$; $\begin{pmatrix}-7\\0\end{pmatrix}$
5. (b) $\begin{pmatrix}4\\0\end{pmatrix}$; $\begin{pmatrix}-5\\-4\end{pmatrix}$ 6. (b) $\begin{pmatrix}0\\-5\end{pmatrix}$; $\begin{pmatrix}3\\3\end{pmatrix}$
7. (a) $\begin{pmatrix}3\\-3\end{pmatrix}$ (b) $\begin{pmatrix}1\frac{1}{2}\\-1\frac{1}{2}\end{pmatrix}$ (c) $\begin{pmatrix}3\frac{1}{2}\\3\frac{1}{2}\end{pmatrix}$; $M(3\frac{1}{2}, 3\frac{1}{2})$
8. (a) $\begin{pmatrix}-1\\6\end{pmatrix}$ (b) $\begin{pmatrix}-\frac{1}{2}\\3\end{pmatrix}$ (c) $\begin{pmatrix}5\frac{1}{2}\\-4\end{pmatrix}$; $M(5\frac{1}{2}, -4)$
9. (a) $\begin{pmatrix}4\\-3\end{pmatrix}$ (b) $\begin{pmatrix}2\\-1\frac{1}{2}\end{pmatrix}$ (c) $\begin{pmatrix}-5\\-3\frac{1}{2}\end{pmatrix}$; $M(-5, -3\frac{1}{2})$
10. (a) (i) $\begin{pmatrix}-6\\3\end{pmatrix}$ (ii) $\begin{pmatrix}-2\\1\end{pmatrix}$ (iii) $\begin{pmatrix}2\\3\end{pmatrix}$ (b) (i) $\begin{pmatrix}0\\-9\end{pmatrix}$ (ii) $\begin{pmatrix}0\\-3\end{pmatrix}$ (iii) $\begin{pmatrix}-2\\2\end{pmatrix}$
11. $\begin{pmatrix}1\\-2\end{pmatrix}$ or $\begin{pmatrix}-1\\2\end{pmatrix}$ 12. $\begin{pmatrix}0\\2\end{pmatrix}$ or $\begin{pmatrix}0\\-2\end{pmatrix}$
13. (a) $\mathbf{q} - \mathbf{p}$ (b) $\mathbf{q} + 2\mathbf{p}$ (c) $\mathbf{p} + \mathbf{q}$
14. (a) $\begin{pmatrix}1\\-3\end{pmatrix}$ (b) $\begin{pmatrix}-1\\3\end{pmatrix}$ (c) $\begin{pmatrix}3\\1\end{pmatrix}$ (d) $\begin{pmatrix}-3\\-1\end{pmatrix}$ 15. $m = 1, n = 2$
16. $m = 2, n = 3$ 17. $s = 2, t = -1$

Exercise 11 page 237

1. (a) $2\mathbf{a}$; $3\mathbf{b}$ (b) $-\mathbf{b} + \mathbf{a}$ (c) $-3\mathbf{b} + 2\mathbf{a}$ (d) $4\mathbf{a} - 3\mathbf{b}$ (e) $4\mathbf{a} - 6\mathbf{b}$ (f) $\overrightarrow{EC} = 2\overrightarrow{ED}$
2. (a) $2\mathbf{b}$; $\frac{5}{2}\mathbf{a}$ (b) $-\mathbf{a} + \mathbf{b}$ (c) $-\frac{5}{2}\mathbf{a} + 2\mathbf{b}$ (d) $-5\mathbf{a} + 6\mathbf{b}$ (e) $-\frac{15}{2}\mathbf{a} + 6\mathbf{b}$ (f) $\overrightarrow{XC} = 3\overrightarrow{XY}$
3. (a) $-\mathbf{b} + \mathbf{a}$; $-3\mathbf{b} + 3\mathbf{a}$ (b) $-2\mathbf{b} + \frac{3}{2}\mathbf{a}$ (c) $-\frac{1}{2}\mathbf{a}$; $-2\mathbf{b} + \frac{3}{2}\mathbf{a}$
4. (a) $-\mathbf{a} + \mathbf{b}$; $-\frac{2}{3}\mathbf{a} + \frac{2}{3}\mathbf{b}$ (b) $\frac{1}{2}\mathbf{a}$; $-\frac{1}{6}\mathbf{a} + \frac{2}{3}\mathbf{b}$ (c) $-\frac{1}{2}\mathbf{a} + 2\mathbf{b}$ (d) $\overrightarrow{MX} = 3\overrightarrow{MP}$
5. (a) $-\mathbf{b} + \mathbf{a}$; $-3\mathbf{a} + \mathbf{b}$ (b) $-\frac{3}{2}\mathbf{a} + \frac{1}{2}\mathbf{b}$ (c) $(k - \frac{3}{2})\mathbf{a} + (\frac{1}{2} - k)\mathbf{b}$ (d) $k = \frac{3}{2}$
6. (a) $-\mathbf{a} + \mathbf{b}$ (b) $-\frac{1}{4}\mathbf{a} + \frac{1}{4}\mathbf{b}$ (c) $\frac{3}{4}\mathbf{a} + \frac{1}{4}\mathbf{b}$ (d) $\mathbf{a} + (m - 1)\mathbf{b}$ (e) $m = \frac{4}{3}$
7. (a) $-\mathbf{c} + \mathbf{d}$ (b) $-\frac{1}{5}\mathbf{c} + \frac{1}{5}\mathbf{d}$ (c) $\frac{4}{5}\mathbf{c} + \frac{1}{5}\mathbf{d}$ (d) $\mathbf{c} + (n - 1)\mathbf{d}$ (e) $n = \frac{5}{4}$
8. (a) $-\mathbf{a} + \mathbf{b}$; $-\frac{1}{2}\mathbf{a} + \frac{1}{2}\mathbf{b}$; $\frac{1}{2}\mathbf{a} + \frac{1}{2}\mathbf{b}$ (b) $\frac{1}{3}\mathbf{a} + \frac{1}{3}\mathbf{b}$ (c) $-\frac{2}{3}\mathbf{a} + \frac{1}{3}\mathbf{b}$ (d) $-\mathbf{a} + \frac{1}{2}\mathbf{b}$ (e) $m = \frac{2}{3}$

9. (a) $-\mathbf{a}+\mathbf{b}$ (b) $\frac{1}{2}\mathbf{b}$ (c) $-\mathbf{a}+\mathbf{c}$ (d) $-\frac{1}{2}\mathbf{a}+\frac{1}{2}\mathbf{c}$ (e) $\frac{1}{2}\mathbf{a}+\frac{1}{2}\mathbf{c}$ (f) $-\frac{1}{2}\mathbf{b}+\frac{1}{2}\mathbf{a}+\frac{1}{2}\mathbf{c}$
(g) $\mathbf{a}+\mathbf{c}=\mathbf{b}$

10. (a) $-\mathbf{b}+\mathbf{a}$ (b) $m\mathbf{a}+(1-m)\mathbf{b}$ (c) $4\mathbf{a}+2\mathbf{b}$ (d) $n=\frac{1}{6}, m=\frac{2}{3}$

11. (a) $-\mathbf{c}+\mathbf{d}$; $-\frac{1}{4}\mathbf{c}+\frac{1}{4}\mathbf{d}$; $\frac{3}{4}\mathbf{c}+\frac{1}{4}\mathbf{d}$
(b) $-\mathbf{c}+\frac{1}{2}\mathbf{d}$ (c) $(1-h)\mathbf{c}+\frac{h}{2}\mathbf{d}$ (d) $(1-h)\mathbf{c}+\frac{h}{2}\mathbf{d}=k.\frac{3}{4}\mathbf{c}+\frac{k}{4}\mathbf{d}$; $h=\frac{2}{5}, k=\frac{4}{5}$

Exercise 12 *page 240*

1. (a) 5, 10, 1 (b) 21, 101, −29 (c) {1, 26, 101} (d) {−9, 1, 31, 46}
2. (a) $-9, 11, \frac{1}{2}$ (b) $0{\cdot}8, -2{\cdot}7, \frac{1}{80}$ (c) $4, 1{\cdot}2, {\cdot}36$
3. (a) 0 (b) 6 (c) 12
4. (a) 10 (b) $\frac{1}{2}$ (c) 2
5. (a) $\frac{2}{3}, 24, 6$ (b) $0, \sqrt{2}, \sqrt{6}$ (c) $-6, 6, 9\frac{3}{4}$
6. (a) ± 3 (b) ± 3 (c) ± 2 (d) ± 6
7. $a=3, b=5$ **8.** $a=2, b=-5$ **9.** $a=7, b=1$ **10.** ×5, +4 **11.** −4, ×3
12. ×2, +7, square **13.** ×5, +9, ÷4 **14.** ×−3, +4, ÷5 **15.** square, ×2, +1 **16.** square, ×$\frac{3}{2}$, +5
17. ×4, −5, √ **18.** square, +10, √, ×4 **19.** ×−3, +7, square
20. ×3, +1, square, ×4, +5 **21.** square, ×−1, +5
22. square, +1, √, ×10, +6, ÷4 **23.** cube, ÷4, +1, square, −6
24. (a) 10, 21 (b) 111, 411, 990, 112
25. (a) 7 (b) 10 (c) 5 (d) 14 (e) 7
(f) 7
26. (a) 3 (b) 6 (c) 8 (d) 10
27. (a) 11 (b) 17 (c) 7
28. (a) 5 (b) 17 (c) $1\frac{1}{2}$ (d) 3

Exercise 13 *page 242*

1. (a) $x \to 4(x+5)$ (b) $x \to 4x+5$ (c) $x \to (4x)^2$ (d) $x \to 4x^2$ (e) $x \to x^2+5$
(f) $x \to 4(x^2+5)$ (g) $x \to [4(x+5)]^2$
2. (a) $-2{\cdot}5$ (b) $\pm\sqrt{\frac{5}{3}}$
3. (a) $x \to 2(x-3)$ (b) $x \to 2x-3$ (c) $x \to x^2-3$ (d) $x \to (2x)^2$ (e) $x \to (2x)^2-3$
(f) $x \to (2x-3)^2$
4. (a) 2 (b) 11 (c) 6 (d) 2 (e) 1 (f) 64
5. (a) −3 (b) 2 (c) $1\frac{1}{2}$ (d) 5
6. (a) $x \to 2(3x-1)+1$ (b) $x \to 3(2x+1)-1$ (c) $x \to 2x^2+1$
(d) $x \to (3x-1)^2$ (e) $x \to 2(3x-1)^2+1$ (f) $x \to 3(2x^2+1)-1$
7. (a) 11 (b) 9 (c) 11 (d) 14 (e) 81 (f) −1
8. (a) 2 (b) 0, 2 (c) $\pm\sqrt{2}$

9. $x \to \frac{x+2}{5}$ **10.** $x \to \frac{x}{5}+2$ **11.** $x \to \frac{x}{6}-2$ **12.** $x \to \frac{3x-1}{2}$ **13.** $x \to \frac{4x}{3}+1$

14. $x \to \frac{(x+6)/2-4}{3}$ **15.** $x \to \frac{2(x-10)-4}{5}$ **16.** $x \to \frac{x-3}{-7}$

17. $x \to \frac{3x-12}{-5}$ **18.** $x \to \frac{3(x-2)-4}{-1}$ **19.** $x \to \frac{4(5x+3)+1}{2}$

20. $x \to \frac{(7x/3)-10}{-2}$ **21.** $x \to 4\,[5(x-7)-6]$

22. $x \to 2\left[\frac{4(x+10)}{3}+3\right]$ **23.** $x \to 3\left[\frac{7(x+2)}{3}-9\right]$

24. $x \to \frac{3(9x-8)-5}{-2}$ **25.** $x \to 2[6(5x-100)-10]$ **26.** $x \to 3\left[\left(\frac{4x}{3}+4\right) \div 3-4\right]$

27. (a) $x \to \frac{x}{3}$ (b) $x \to x + 5$ (c) $x \to \frac{x-1}{2}$ (d) $x \to 3(x-5)$ (e) $x \to \frac{x}{3} + 5$ (f) $x \to \frac{x}{3} + 5$
28. (a) $x \to 6x + 1$ (b) $x \to \frac{2x}{3} + 1$ (c) $x \to \frac{x-1}{6}$ (d) $x \to 2x - 9$ (e) $x \to \frac{x+9}{2}$ (f) $x \to \frac{x+9}{2}$
29. (a) 7 (b) 21 (c) 5 (d) 5 (e) 2
30. (a) 1 (b) $x > -2\frac{1}{2}$; $x > 5$
31. (a) $x \to \frac{3x-5}{2}$ (b) 2

Exercise 14 page 243

1.	(a) 7	(b) 5	(c) 18	(d) 14	(e) 0	(f) 3	
2.	(a) 7	(b) 16	(c) 7	(d) 0	(e) 12	(f) 7	
3.	(a) 13	(b) 17	(c) 13	(d) 34	(e) 170	(f) 194	
4.	(a) 12	(b) 100	(c) 2	(d) 192	(e) 144	(f) 288	
5.	(a) 7	(b) 7	(c) 5	(d) 17	(e) -4	(f) 8	
6.	(a) 18	(b) 60	(c) 6	(d) 216	(e) 54	(f) 324	
7.	(a) 13	(b) 41	(c) 5	(d) 185	(e) 170	(f) 194	
8.	(a) 1	(b) 1	(c) 1	(d) 9	(e) 0	(f) 0	
9.	(a) $1\frac{1}{2}$	(b) $\frac{4}{5}$	(c) 2	(d) $\frac{3}{8}$	(e) $\frac{2}{3}$	(f) $1\frac{1}{3}$	
10.	(a) 4	(b) 2	(c) 7	(d) 5	(e) ± 2	(f) ± 3	(g) 2
11.	(a) 3	(b) 2	(c) $2\frac{1}{3}$	(d) $4\frac{2}{3}$	(e) 7	(f) 7	(g) 6
12.	(a) 4	(b) 10	(c) 5	(d) -1	(e) 2	(f) 3	
13.	(a) 7; $3\frac{1}{4}$	(b) No	(c) 8	(d) -35	(e) 2		
14.	(a) 60; 60	(b) Yes	(c) 108; 108	(d) Yes	(e) 9	(f) ± 7	(g) 5
15.	(a) (i) 2	(ii) 1	(iii) 3	(iv) 1	(b) No	(c) (i) 1	(ii) 2
16.	(a) 4	(b) $\frac{1}{3}$	(c) 1	(d) $\frac{1}{5}$			
17.	(a) (i) $1\frac{1}{2}$	(ii) $\frac{7}{10}$	(iii) $2\frac{2}{3}$	(b) (i) $\frac{3}{5}$	(ii) $1\frac{1}{4}$	(c) $\frac{5}{6x}$	
18.	(a) 5	(b) ± 3	(c) -1	(d) ± 1			

Exercise 15 page 244

1. (a) y (b) z (c) w (d) x
2. (a) a (b) c (c) d (d) d (e) c (f) b
3. (a) D (b) B, E (c) E (d) E (e) E (f) B (g) C
   (h) D; Yes
4. All three operations are commutative.
5.

×	A	B	C	D
A	A	B	C	D
B	B	A	D	C
C	C	D	A	B
D	D	C	B	A

(a) **A** (b) **D** (c) **D**

6. (a)

*	0	1	2	3	4
0	0	0	0	0	0
1	0	1	2	0	1
2	0	2	1	0	2
3	0	0	0	0	0
4	0	1	2	0	1

(b) No (c) 2 (d) 2

7.

∘	1	2	3	4
1	1	2	3	4
2	2	4	1	3
3	3	1	4	2
4	4	3	2	1

(a) 1 (b) 2 (c) 2 or 3 (d) 2 (e) 2
(f) 2 or 3 (g) 1 or 4

8.

*	*a*	*b*	*c*	*d*	*e*	*f*
a	*f*	*a*	*e*	*c*	*b*	*d*
b	*a*	*c*	*d*	*b*	*e*	*f*
c	*e*	*d*	*f*	*a*	*c*	*b*
d	*c*	*b*	*a*	*d*	*f*	*e*
e	*b*	*e*	*c*	*f*	*d*	*a*
f	*d*	*f*	*b*	*e*	*a*	*c*

9. (a)

*	*a*	*b*	*c*	*d*	*e*	*f*
a	*e*	*a*	*b*	*c*	*f*	*d*
b	*a*	*c*	*d*	*f*	*e*	*b*
c	*b*	*d*	*e*	*a*	*c*	*f*
d	*c*	*f*	*a*	*b*	*d*	*e*
e	*f*	*e*	*c*	*d*	*b*	*a*
f	*d*	*b*	*f*	*e*	*a*	*c*

(b)

*	1	2	3	4	5	6
1	4	3	6	1	2	5
2	3	1	5	2	4	6
3	6	5	4	3	1	2
4	1	2	3	6	5	4
5	2	4	1	5	6	3
6	5	6	2	4	3	1

Revision Exercise 9A *page 245*

1. (a) $\{5\}$ (b) $\{1, 3, 5, 6, 7\}$ (c) $\{2, 4, 6, 7, 8\}$ (d) $\{2, 4, 8\}$ (e) $\{1, 2, 3, 4, 5, 8\}$

2. 32 **3.** (a) (b) 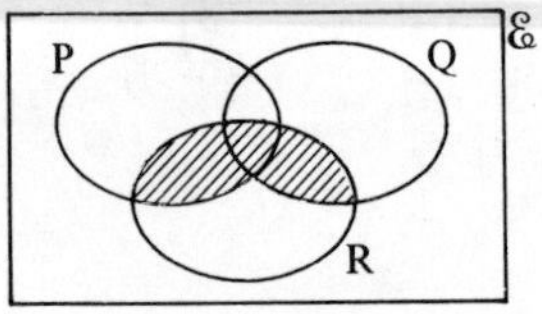 (c)

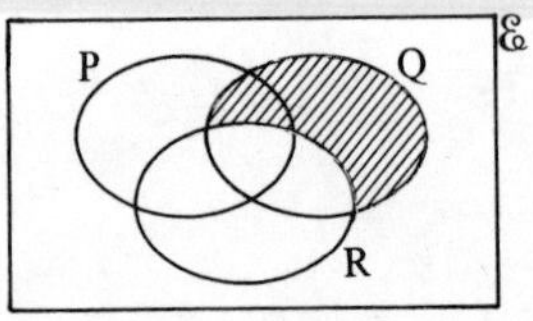

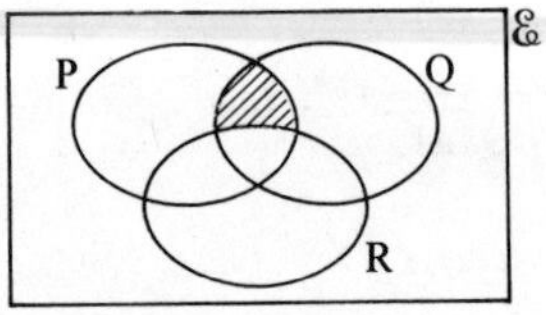

4. (a) $(A \cup B)' \cap C$ (b) $(A \cup B) \cap C'$

5. (a) (i) $S \subset T$ (ii) $S \cap M' \neq \emptyset$ (b) There are no women on the train over 25 years old.

6. (a) 4 (b) 11 (c) 17 **7.** (a) $\mathbf{r} - \mathbf{p}$ (b) $\frac{1}{2}\mathbf{r} - \frac{1}{2}\mathbf{p}$ (c) $\frac{1}{2}\mathbf{r} + \frac{1}{2}\mathbf{p}$

8. (a) 5 (b) $\sqrt{68}$ (c) $\sqrt{41}$ **9.** $n = 2, m = -15$

10. (a) $\begin{pmatrix} -1 \\ 4 \end{pmatrix}$ (b) $\begin{pmatrix} 4 \\ 4 \end{pmatrix}$ (c) $\begin{pmatrix} 6 \\ -4 \end{pmatrix}$ (d) $\begin{pmatrix} 2 \\ 2 \end{pmatrix}$

11. (a) $\mathbf{a} - \mathbf{c}$ (b) $\frac{1}{2}\mathbf{a} + \mathbf{c}$ (c) $\frac{1}{2}\mathbf{a} - \frac{1}{2}\mathbf{c}$ CA is parallel to NM and CA = 2NM.

12. $m = 3, n = 2$ **13.** (a) $\begin{pmatrix} -3 \\ 2 \end{pmatrix}$ (b) $\begin{pmatrix} -1\frac{1}{2} \\ 1 \end{pmatrix}$ (c) $\begin{pmatrix} 1\frac{1}{2} \\ 3 \end{pmatrix}$

14. (a) -5 (b) 0 (c) -3 (d) 8; $ff: x \rightarrow 4x - 9$

15. $f^{-1}: x \rightarrow \frac{(x-4)}{3}$; $h^{-1}: x \rightarrow 5x + 2$ (a) 3 (b) $5\frac{1}{3}$

16. (a) 3 (b) 0, 5 **17.** (a) 14 (b) 12 (c) 2

18. (a) 2 (b) $\frac{1}{9}$ (c) $\frac{2}{3}$ **19.** (a) 7 (b) 2 (c) No

20. (a) c (b) a (c) c

Examination Exercise 9B page 246

1. (i) {3, 6, 9} (ii) {1, 2, 3, 4, 6, 8, 9, 12} (iii) {3, 6, 9, 12} (iv) {1, 2, 4, 5, 7, 8, 10, 11, 12}
   (v) {4, 5, 7, 8, 10, 11}
2. (i) 37 (ii) 7 (iii) 62
3. (i) (a) S (b) ∅ (c) 5 (ii) (a) 6 (b) 47
4. (a) (i) 23 (ii) −7 (b) 2
5. (a) Total number of men = 35 $x = 0, 1, 2, 3.$
   (b) (i) trapezium (ii) **a** + 2**b** (iii) 2**b** − **a** (iv) $\frac{1}{2}$**a** + **b**
6. (a) (i) **b** − **a** (ii) $\frac{1}{4}$(**b** − **a**) (iii) $\frac{1}{4}$(3**a** + **b**)
   (b) (i) $\begin{pmatrix}3\\4\end{pmatrix}$ (ii) 2·24 (iii) (2, −1) (iv) $l = 3, m = -7$
7. (i) (a) $\frac{1}{2}$**g** (b) $\frac{1}{2}$(**h** − **g**) (c) $\frac{1}{2}$(**g** + **h**) (d) **g** + **h**
   (ii) (a) $\frac{1}{2}$**h** − **g** (b) $\frac{1}{2}$**g** − **h** (c) **0**
   (iii) (a) $\frac{3}{2}$(**h** − **g**) (b) OK is parallel to GH and OK = $\frac{3}{2}$GH
8. (i) **a** − **b** (iii) 3**a** + 2**b** (iv) $h = \frac{3}{5}, k = \frac{1}{5}$ (v) $\frac{2}{3}$
9. (i) **OX** = **a** + $\frac{1}{2}$**b**; **AB** = **b** − **a**; **AM** = $\frac{1}{3}$(**b** − **a**); **OM** = $\frac{1}{3}$(2**a** + **b**)
   (ii) **OL** = $\frac{1}{3}$(2**a** + **b**); L and M coincide. (iii) **OY** = $\frac{1}{2}$(**a** + **b**); **AY** = $\frac{1}{2}$(**b** − **a**)
10. (i) **OB** = **a** + **c**; **OP** = $\frac{3}{4}$(**a** + **c**); **AP** = $\frac{1}{4}$(3**c** − **a**)
    (ii) **OQ** = **a**$\left(1 - \frac{x}{4}\right)$ + **c**$\left(\frac{3x}{4}\right)$ (iii) **OQ** = y**a** + **c** (iv) $x = \frac{4}{3}$ $y = \frac{2}{3}$ (v) 2

PART 10

Exercise 1 page 249

1. (a) 2 (b) 6 (c) −4
2. (a) m/s (b) kg/s (c) cm³/h (d) cm²/day (e) m/s/s (f) km²/h
3. (a) A, E (b) C (c) A, B (d) D, E (e) km/h
4. (a) R (b) P, T (c) P, Q (d) S, T (e) kg/m
5. (a) B, D (b) C (c) A, B (d) D, E (e) m/s

Exercise 2 page 251

1. $2x$ 2. $5x^4$ 3. 5 4. $7x^6$ 5. $3x^2 + 4x^3$
6. $2x + 5x^4$ 7. $6x^5 + 4$ 8. $3x^2 - 2x$ 9. $4x^3 + 2x + 1$ 10. $4x$
11. $12x^2$ 12. $30x^4$ 13. 7 14. $6x + 4$ 15. $30x^2 - 7$
16. $40x^9 - 18x^8$ 17. $2x + 4$ 18. $2x - 3$ 19. $4x$ 20. $3x^2 + 2x - 6$
21. $12x^3 - 3x^2$ 22. 100 23. −3 24. $-2x$ 25. $-4x$
26. $3x^2 - 17$ 27. $15x^2 - 4x + 4$ 28. $2x - 11$ 29. $100x^{99}$ 30. x^9
31. x 32. $2x^7$ 33. $\frac{4}{3}x + 3$ 34. $\frac{9}{2}x^5$ 35. $0{\cdot}02x$
36. $\frac{1}{2}x^4$ 37. $1 + 3x^2$ 38. $x + x^2 + x^3$ 39. $-17x^{16}$ 40. $-1{\cdot}2x^2$

Exercise 3 page 251

1. $2x + 1$ 2. $2x - 4$ 3. $4x - 1$ 4. $6x - 16$ 5. $3x^2 - 10x$
6. $3x^2 + 22x$ 7. $2x + 6$ 8. $2x - 8$ 9. $8x - 12$ 10. $4x^3 - 6x^2 + 2x$
11. $3x^2 + 20x + 25$ 12. $4x + 12$ 13. $-3x^{-4}$ 14. $-4x^{-5}$ 15. $-2x^{-3} + 2x$
16. $-x^{-2} + 3x^2$ 17. $-\frac{5}{x^6}$ 18. $-\frac{4}{x^5}$ 19. $-\frac{6}{x^3}$ 20. $-\frac{12}{x^4}$

21. $-\frac{3}{x^2}-\frac{4}{x^3}$ **22.** $-\frac{4}{x^2}+\frac{9}{x^4}$ **23.** $3x^2-\frac{3}{x^4}$ **24.** $2x+\frac{2}{x^3}$ **25.** $2x-\frac{2}{x^3}$

26. $3x^2+1$ **27.** $2x$ **28.** $-\frac{1}{x^2}-\frac{2}{x^3}$ **29.** $2x-\frac{2}{x^3}$ **30.** $8x-\frac{2}{x^3}$

31. $12t^2-5$ **32.** $19t^{18}-15t^{14}$ **33.** $12t^3-\frac{1}{3}$ **34.** $-\frac{1}{2}+12t^2$ **35.** $-3-5t^4$

36. $3t^2+2t-1$ **37.** $8t-12$ **38.** $-\frac{3}{t^4}-\frac{4}{t^5}$ **39.** $-\frac{3}{t^2}+\frac{10}{t^3}+4$ **40.** $1-\frac{1}{t^2}$

41. $8t+\frac{8}{t^3}$ **42.** $6t$ **43.** $3-\frac{1}{t^2}$ **44.** $-5t^{-6}+2t$ **45.** $3x^2+1$

46. 1 **47.** $4x^3+4x$ **48.** $10x^9+9x^8+8x^7$ **49.** $-\frac{5}{6}-12x$ **50.** $3x^2+6x+3$

51. $12x^2+\frac{3}{4x^2}$ **52.** $-\frac{8}{x^3}+\frac{9}{x^4}$ **53.** $\frac{3}{2x^2}-\frac{15}{x^4}$ **54.** $6x^5+12x^3+6x$ **55.** $6-\frac{3}{x^2}$

56. $\frac{1}{2}x^{-\frac{1}{2}}$ **57.** $x^{-\frac{1}{2}}$ **58.** $2x-\frac{1}{2}x^{-\frac{1}{2}}$ **59.** $\frac{1}{2}x^{-\frac{1}{2}}$ **60.** $\frac{3}{2}\sqrt{x}$

Exercise 4 *page 252*

1. 13 **2.** −6 **3.** 5 **4.** 41 **5.** 7
6. 2 **7.** −3 **8.** $\frac{3}{4}$ **9.** −7 **10.** −3
11. −7, 7 **12.** 9, −9 **13.** 2 **14.** −1 **15.** (3, −2)
16. (a) (2, −5) (b) (3, −3) **17.** −1, 1 **18.** 2; $\frac{5}{2}$ **19.** 45°
20. (a) 63·4° (b) 26·6° **21.** 5 **22.** 29·7° **23.** $(\frac{1}{2}, 4), (-\frac{1}{2}, -4)$
24. (1, 4)

Exercise 5 *page 254*

1. −4; min. **2.** 26; max. **3.** 3; max. **4.** −8; min. **5.** −1; min.
6. 4; max. **7.** $\frac{1}{4}$; min. **8.** $5\frac{1}{4}$; max. **9.** −70, min.; −40, max.
10. 21, max.; 17, min. **11.** 2, min.; 29, max. **12.** 10, max.; 6, min.
13. $-3\frac{1}{3}$, min.; $1\frac{1}{6}$, max. **14.** $7\frac{4}{27}$, max.; 7, min. **15.** 2, min.; −2, max.
16. 4, min.; −4, max. **17.** 3, max.; 7, min. **18.** −1, max.; 3, min.

19. 90 m **20.** 36 cm² **21.** 225 cm² **22.** (a) $h=\frac{16}{r^2}$ (b) 24π cm²
23. 54π cm² **24.** 32 cm **25.** 40 m

Exercise 6 *page 255*

1. $\frac{x^4}{4}+c$ **2.** $\frac{x^3}{3}+c$ **3.** $\frac{x^5}{5}+\frac{x^2}{2}+c$ **4.** $\frac{x^6}{6}+3x+c$

5. $\frac{x^3}{3}-4x+c$ **6.** $\frac{x^7}{7}+\frac{3x^2}{2}+c$ **7.** $\frac{x^4}{2}-x^3+c$ **8.** $\frac{x^5}{5}-\frac{x^2}{2}+c$

9. $\frac{x^3}{6}-7x+c$ **10.** $\frac{x^2}{6}+\frac{1}{4}x+c$ **11.** $\frac{3}{5}x^5-3x^2+c$ **12.** $\frac{x^2}{2}+\frac{x^3}{3}+\frac{x^4}{4}+c$

13. $x-\frac{7}{2}x^2+c$ **14.** $\frac{x^3}{3}+\frac{3}{2}x^2+2x+c$ **15.** $\frac{2}{3}x^3+\frac{5}{2}x^2-3x+c$ **16.** $\frac{x^4}{4}-x^3+c$

17. $\frac{x^4}{4}-\frac{x^2}{2}+c$ **18.** $\frac{4}{3}x^3+2x^2+x+c$ **19.** $-x^{-1}+c$ **20.** $-\frac{1}{2}x^{-2}+c$

21. $y=x^2-3x+4$ **22.** $y=x^2+2x-15$ **23.** $y=2x^2-7x+1$ **24.** $y=3x^2-4x-10$
25. $y=x^3-2x+1$ **26.** $y=2x^3+5$ **27.** $v=t^2+5t-3$ **28.** $z=y^3-4y+1$
29. 11 **30.** 4

Exercise 7 page 256

1. $\frac{x^3}{3} + 2x + c$
2. $x^2 - x + c$
3. $\frac{x^4}{4} - 2x^2 + c$
4. $\frac{x^3}{3} + \frac{x^2}{2} + x + c$
5. $\frac{x^2}{4} - \frac{x}{4} + c$
6. $\pi\frac{x^2}{2} + 3x + c$
7. $\frac{t^3}{3} + t + c$
8. $\frac{3t^2}{2} - 5t + c$
9. $\frac{z^3}{3} - \frac{5z^2}{2} + c$
10. $\frac{y^2}{2} - 10y + c$
11. 6
12. 44
13. 15
14. −12
15. 8
16. 1
17. 21
18. $\frac{1}{2}$
19. $8\frac{2}{3}$ square units
20. 6
21. 9
22. 36
23. $20\frac{5}{6}$
24. $20\frac{5}{6}$
25. $1\frac{1}{3}$
26. $1\frac{1}{3}$
27. $4\frac{1}{2}$
28. $4\frac{1}{2}$

Exercise 8 page 257

1. 7·47
2. 13·4
3. 23·5
4. (b) 38·5 (c) less (d) 39
5. (a) 58 (b) greater (c) $57\frac{1}{3}$
6. 20·2
7. 107·5
8. 8·4
9. 0·34
10. 13·4

Exercise 9 page 258

1. (a) 2 m (b) −3 m/s (c) 3 s, 4 s (d) $3\frac{1}{2}$ s
2. (a) 9 m/s (b) 1 s, 3 s (c) 6 m/s² (d) −12 m/s²
3. (a) 5 m/s (b) −11 m/s² (c) $\frac{1}{2}$ s, 5 s (d) $2\frac{3}{4}$ s
4. (a) 14 m/s (b) 5 m/s² (c) −11 m/s²
5. (a) 3 m/s (b) $s = t^2 + 3t + 7$ (c) 137 m (d) 2 m/s²
6. (a) $v = t^2 - 6t + 8$ (b) 2 s, 4 s (c) −1 m/s
7. (a) 3 s, 7 s (b) $s = \frac{t^3}{3} - 5t^2 + 21t + 10$ (c) 37 m (d) −10 m/s²
8. (a) 24 m (b) 10 m (c) 5 m/s
9. (a) 65 m (b) 0 m/s (c) 8 m/s²
10. (a) 9 m/s (b) $s = 3t^2 - \frac{t^3}{3}$ (c) 0 s, 9 s
11. (a) $\frac{1}{3}$ m/s² (b) $18\frac{2}{3}$ m
12. (b) $76\frac{1}{2}$ m
13. (a) 8 s (b) $v = 8t - \frac{t^2}{2}$ (c) $s = 4t^2 - \frac{t^3}{6}$ (d) 24 s
14. (a) $v = t - t^2$ (b) $s = \frac{t^2}{2} - \frac{t^3}{3}$ (c) $\frac{1}{4}$ m/s (d) $1\frac{1}{2}$ s

Exercise 10 page 260

1. (a) $1\frac{1}{2}$ m/s² (b) 675 m (c) $11\frac{1}{4}$ m/s
2. (a) 600 m (b) 20 m/s (c) 225 m (d) −2 m/s²
3. (a) 600 m (b) $387\frac{1}{2}$ m (c) 0 m/s²
4. (a) 20 m/s (b) 750 m
5. (a) 8 s (b) 496 m (c) 12·4 m/s
6. (a) 30 m/s (b) $-2\frac{1}{7}$ m/s² (c) 20 s
7. (a) 15 m/s (b) $2\frac{1}{4}$ m/s²
8. (a) 40 m/s (b) 10 s
9. (a) 50 m/s (b) 20 s
10. (a) 20 m/s (b) 20 s

Revision Exercise 10A page 261

1. (a) $3x^2 - 7$ (b) $4x^3 + x - 3$ (c) $3 - \frac{1}{x^2}$ (d) $-\frac{5}{x^2} + \frac{1}{5}$
2. (a) kg/s (b) m/s (c) kg/m (d) m/s² (e) m/kg

3. 0
4. 4
5. 135°
6. (3, 7)
7. (a) 13, max. (b) 5, min.; 9, max. (c) 6, min.; −6, max. (d) −5, min.
8. 361 cm²
9. (a) $\frac{x^3}{3}+\frac{7x^2}{2}+c$ (b) $2x^2-6x+c$ (c) $t^3-\frac{t^2}{2}+t+c$ (d) $\frac{y^3}{6}-y^2+c$
10. (a) −2 (b) 4 (c) $\frac{1}{4}$ (d) $\frac{1}{6}$
11. $y=x^2-3x+5$
12. $12\frac{2}{3}$ square units
13. (−2, 0), (3, 0); $20\frac{5}{6}$ square units
14. (0, 0), (5, 0); (4, 4), $10\frac{2}{3}$ square units
15. (a) 2 m/s (b) 14 m/s² (c) 1 m
16. (a) $v=3t^2+4t+3$ (b) $s=t^3+2t^2+3t+10$
17. (a) 0·3 m/s² (b) 1050 m (c) 40 s
18. (a) 30 m/s (b) 600 m
19. (a) 10·7 square units (b) greater (c) 9 square units
20. $7\frac{1}{4}$ square units

Examination Exercise 10B page 262

1. (ii) 76°; (0, 4); max. at $x=0$, min. at $x=\frac{5}{3}$
2. max. (−1, 11); min. (1, 3); $h=2{\cdot}41$
3. (a) (i) 2 (ii) $1\frac{1}{3}$ square units (b) 71·6°
4. (a) 33 m (b) 4 m/s (c) 0·4 s (d) 33·8 m (e) 3 s
5. (i) $(4h+16x)$ cm (ii) $h=24-4x$ (iv) $x=4$; 384 cm³
6. area $=2x^2+6xh$; max. at $x=10$; $1333\frac{1}{3}$ cm³
7. (a) (−1, 5), max., (3, −27), min. (b) (i) $y=x^2-3$ (ii) 10
8. (a) 0 s, $\pm\sqrt{3}$ s (b) 4 m (c) 18 m/s (d) 12 m/s² (e) 1 s
9. (i) 2 s (ii) $v=2+7t-t^2$ (iii) 14·25 m/s (iv) $5\frac{1}{6}$ m
10. $(2\frac{1}{2}, 2\frac{1}{4})$; (i) (4, 0) (ii) $y=3x-3$ (iii) $4\frac{1}{2}$ square units

PART 11

Exercise 1 page 266

1. (a) Squash (b) 160 (c) 10
3. (a) £3000 (b) £4000 (c) £6000 (d) £11 000
4. red 50°; green 70°; blue 110°; yellow 40°; pink 90°
5. eggs 270°; milk 12°; butter 23·4°; cheese 54°; salt/pepper 0·6°
6. (a) A 60°; B 100°; C 60°; D 140°; E 0° (b) A 50°; B 75°; C 170°; D 40°; E 25°
   (c) A 48·5°; B 76·2°; C 62·3°; D 96·9°; E 76·2°
7. 80°, 120°, 160°
8. $x=8$
9. 100
10. 21·6°, 1·8°
11. (a) 22·5% (b) $x=45°, y=114°$

Exercise 2 page 267

1. (a) mean = 6; median = 5; mode = 4. (b) mean = 9; median = 7; mode = 7.
   (c) mean = 6·5; median = 8; mode = 9. (d) mean = 3·5; median = 3·5; mode = 4.
2. (a) mean = 7·82; median = 8; mode = 8. (b) mean = 5; median = 4; mode = 4.
   (c) mean = 2.1; median = 2.5; mode = 4. (d) mean = $\frac{13}{18}$; median = $\frac{1}{2}$; mode = $\frac{1}{2}$.
3. 3·38
4. 3·475
5. (a) mean = 3·025; median = 3; mode = 3. (b) mean = 17·75; median = 17; mode = 17.
   (c) mean = 3·38; median = 4; mode = 4.
6. (a) 5·17 (b) 5
7. (a) 2 (b) 9
8. (a) 20·4 m (b) 12·8 m (c) 1·66 m
9. 55 kg
10. (i) 9 (ii) 9 (iii) 15
11. (i) 5 (ii) 10 (iii) 10
12. 12
13. $3\frac{2}{3}$
14. 4·68

15. (a) (i) 0·6 (ii) 0 (b) 0·675
16. (a) N (b) mean $= N^2 + 2$; median $= N^2$ (c) 2
17. $\dfrac{ax + by + cz}{a + b + c}$

Exercise 3 page 270

1. (a) 45 (b) 28, 60 (c) 32 (d) 35 (e) 25
2. (a) 31 (b) 26, 38 (c) 12 (d) 29 (e) 80
3. (a) 20 kg (b) 10·5 kg
4. (a) 80·5 cm (b) 22 cm
5. (a) 71 s (b) 20 s
6. (a) 45 (b) 14
7. (a) 36·5 g (b) 20 g (c) 15
9. (a) 26 (b) 25·8 (c) 26·1

Exercise 4 page 273

1. (a) $\frac{1}{13}$ (b) $\frac{1}{2}$ (c) $\frac{1}{52}$ (d) $\frac{3}{52}$
2. (a) $\frac{1}{6}$ (b) $\frac{1}{2}$ (c) $\frac{1}{2}$ (d) $\frac{1}{3}$
3. (a) $\frac{1}{4}$ (b) $\frac{1}{2}$
4. (a) (i) $\frac{3}{5}$ (ii) $\frac{2}{5}$ (b) (i) $\frac{5}{9}$ (ii) $\frac{4}{9}$
5. (a) $\frac{5}{13}$ (b) $\frac{4}{13}$ (c) $\frac{2}{13}$ (d) $\frac{3}{13}$ (e) $\frac{1}{13}$ (f) 0 (g) $\frac{3}{13}$
6. (a) $\frac{1}{8}$ (b) $\frac{3}{8}$ (c) $\frac{1}{8}$ (d) $\frac{7}{8}$
7. (a) $\frac{5}{11}$ (b) $\frac{7}{22}$ (c) $\frac{15}{22}$ (d) $\frac{17}{22}$
8. (a) $\frac{1}{2}$ (b) $\frac{3}{25}$ (c) $\frac{9}{100}$ (d) $\frac{3}{25}$ (e) $\frac{1}{50}$
9. (a) $\frac{1}{12}$ (b) $\frac{1}{36}$ (c) $\frac{5}{18}$ (d) $\frac{1}{6}$ (e) $\frac{1}{6}$; most likely total = 7.
10. (a) $\frac{1}{12}$ (b) $\frac{5}{36}$ (c) $\frac{2}{3}$ (d) $\frac{1}{12}$ (e) $\frac{1}{36}$
11. (a) 1 (b) 0 (c) 1 (d) 0
12. (a) $\frac{3}{8}$ (b) $\frac{1}{16}$ (c) $\frac{15}{16}$ (d) $\frac{1}{4}$
13. (a) $\frac{1}{10}$ (b) $\frac{3}{5}$ (c) $\frac{1}{6}$ (d) 0
14. (a) $\frac{1}{12}$ (b) $\frac{1}{8}$ (c) 1 (d) 0
15. (a) $\frac{1}{144}$ (b) $\frac{1}{18}$ (c) $\frac{1}{72}$; head, tail and total of 7.
16. (a) $\frac{1}{216}$ (b) $\frac{1}{72}$ (c) $\frac{1}{8}$ (d) $\frac{5}{108}$ (e) $\frac{5}{72}$ (f) $\frac{1}{36}$

Exercise 5 page 275

1. (a) independent (b) $\frac{1}{13}, \frac{1}{6}$ (c) $\frac{1}{78}$
2. (a) independent (b) (i) $\frac{1}{2}$ (ii) $\frac{1}{2}$ (iii) $\frac{1}{4}$
3. (a) exclusive (b) (i) $\frac{1}{13}$ (ii) $\frac{1}{13}$ (iii) $\frac{2}{13}$
4. (a) $\frac{1}{78}$ (b) $\frac{1}{104}$ (c) $\frac{1}{24}$
5. (a) $\frac{1}{16}$ (b) $\frac{1}{169}$ (c) $\frac{9}{169}$ (d) $\frac{1}{8}$ (e) $\dfrac{2}{52^2}\left(=\dfrac{1}{1352}\right)$
6. (a) $\frac{1}{16}$ (b) $\frac{25}{144}$ (c) $\frac{5}{18}$ (d) $\frac{25}{72}$
7. (a) $\frac{1}{121}$ (b) $\frac{9}{121}$ (c) $\frac{18}{121}$
8. (a) $\frac{1}{288}$ (b) $\frac{1}{72}$ (c) $\frac{1}{32}$
9. (a) $\frac{1}{125}$ (b) $\frac{1}{125}$ (c) $\frac{1}{10000}$ (d) $\frac{3}{500}$ (e) $\frac{3}{500}$
10. (a) exclusive (b) independent (c) exclusive (d) independent
    (e) independent (f) $\frac{3}{4}, \frac{1}{4}, \frac{3}{4}, \frac{1}{4}$ (g) (i) $\frac{9}{16}$ (ii) $\frac{1}{16}$ (iii) $\frac{3}{16}$ (iv) $\frac{3}{16}$ (v) 1 (vi) 1
    (h) exclusive (i) $\frac{3}{8}$
11. (a) $\frac{2}{5}, \frac{3}{5}, \frac{2}{5}, \frac{3}{5}$ (b) (i) $\frac{9}{25}$ (ii) $\frac{4}{25}$ (iii) $\frac{6}{25}$ (iv) $\frac{6}{25}$ (c) $\frac{12}{25}$ (d) $\frac{13}{25}$
12. (a) $\frac{1}{5}$ (b) $\frac{18}{25}$ (c) $\frac{1}{20}$ (d) $\frac{2}{25}$ (e) $\frac{77}{100}$
13. (a) $\frac{1}{9}$ (b) $\frac{4}{27}$ (c) $\frac{4}{9}$

Exercise 6 page 277

1. (a) $\frac{49}{100}$ (b) $\frac{9}{100}$ (c) $\frac{21}{50}$
2. (a) $\frac{7}{15}$ (b) $\frac{1}{15}$ (c) $\frac{7}{15}$
3. (a) $\frac{2}{9}$ (b) $\frac{2}{15}$ (c) $\frac{1}{45}$ (d) $\frac{14}{45}$
4. (a) $\frac{1}{12}$ (b) $\frac{1}{6}$ (c) $\frac{1}{3}$ (d) $\frac{2}{9}$
5. (a) $\frac{1}{216}$ (b) $\frac{125}{216}$ (c) $\frac{25}{72}$ (d) $\frac{91}{216}$
6. (a) $\frac{1}{64}$ (b) $\frac{5}{32}$ (c) $\frac{27}{64}$
7. (a) $\frac{1}{6}$ (b) $\frac{1}{30}$ (c) $\frac{1}{30}$ (d) $\frac{29}{30}$
8. (a) $\frac{1}{64}$ (b) $\frac{27}{64}$ (c) $\frac{9}{64}$ (d) $\frac{27}{64}$; Sum = 1.
9. (a) $\frac{3}{20} \times \frac{2}{19} \times \frac{1}{18} (= \frac{1}{1140})$ (b) $\frac{1}{4} \times \frac{4}{19} \times \frac{1}{6} (= \frac{1}{114})$ (c) $(\frac{3}{20} \times \frac{5}{19} \times \frac{12}{18}) \times 6$ (d) $\frac{5}{20} \times \frac{4}{19} \times \frac{3}{18} \times \frac{2}{17}$
10. (a) $\frac{1}{10000}$ (b) $\frac{523}{10000}$ (c) $\frac{9^4}{10^4}$
11. (a) $\frac{x}{x+y}$ (b) $\frac{x(x-1)}{(x+y)(x+y-1)}$ (c) $\frac{2xy}{(x+y)(x+y-1)}$ (d) $\frac{y(y-1)}{(x+y)(x+y-1)}$
12. (a) $\frac{x}{z}$ (b) $\frac{x(x-1)}{z(z-1)}$ (c) $\frac{2x(z-x)}{z(z-1)}$
13. $\frac{3}{10}$
14. $\frac{9}{140}$
15. (a) $\frac{3}{5}$ (b) $\frac{1}{3}$ (c) $\frac{2}{15}$ (d) $\frac{2}{21}$ (e) $\frac{1}{7}$ (f) $\frac{1}{35}$
16. (a) $\frac{1}{220}$ (b) $\frac{1}{22}$ (c) $\frac{3}{11}$ (d) 5 (e) 300
17. (a) $\frac{10 \times 9}{1000 \times 999}$ (b) $\frac{990 \times 989}{1000 \times 999}$ (c) $\frac{2 \times 10 \times 990}{1000 \times 999}$
18. (a) $\frac{3}{20}$ (b) $\frac{7}{20}$ (c) $\frac{1}{2}$
19. (a) 0·00781 (b) 0·511
20. (a) $\frac{21}{506}$ (b) $\frac{455}{2024}$ (c) $\frac{945}{2024}$

Revision Exercise 11A page 278

1. 162°
2. 41·7%
3. 54°
4. (a) 84 (b) 19·2
5. (a) 3·4 (b) 3 (c) 3
6. (a) 5·45 (b) 5 (c) 5
7. 1·552 m
8. 3
9. $\frac{1}{6}$
10. $\frac{4}{25}$
11. $\frac{8}{27}$
12. $\frac{5}{16}$
13. (a) $\frac{1}{28}$ (b) $\frac{15}{28}$ (c) $\frac{3}{7}$
14. (a) $\frac{x}{x+5}$ (b) $\left(\frac{x}{x+5}\right)^2$; $\frac{x}{x+5}$, $\frac{x(x-1)}{(x+5)(x+4)}$
15. $\frac{1}{19}$
16. (a) $\frac{1}{32}$ (b) $\frac{1}{256}$
17. (a) $\frac{1}{8}$ (b) $\frac{1}{2}$
18. $\frac{35}{48}$
19. (a) $\frac{1}{9}$ (b) $\frac{1}{12}$ (c) 0
20. $\frac{1}{20^3}$

Examination Exercise 11B page 279

1. (a) (i) $\frac{1}{5}$ (ii) $\frac{4}{25}$ (b) $\frac{8}{45}$
2. (b) 114° (c) (i) 0·316 (ii) 0·673 (d) $\frac{6320}{20\,000} \times \frac{6319}{19\,999}$
3. (i) 3 (iii) 43 (iv) (a) $\frac{2}{3}$ (b) $\frac{1}{9}$ (v) (a) $\frac{1}{36}$ (b) $\frac{11}{36}$
4. (a) (i) 0 (ii) 2 (iii) 2·4 (b) $\frac{3}{5}$; $\frac{1}{2}$ (i) $\frac{3}{10}$, (ii) $\frac{1}{10}$, (iii) $\frac{3}{5}$
5. (a) $\frac{13}{30}$ (b) $\frac{26}{145}$ (c) $\frac{48}{145}$ (d) $\frac{97}{145}$
6. (i) (a) $\frac{4}{5}$ (b) $\frac{1}{60}$ (c) $\frac{2}{5}$ (d) $\frac{3}{5}$ (ii) (a) $\frac{1}{6}$ (b) $\frac{1}{2}$ (c) $\frac{1}{10}$ (d) $\frac{3}{5}$
7. (i) (a) $\frac{1}{6}$ (b) $\frac{1}{2}$ (ii) (a) $\frac{1}{9}$ (b) $\frac{5}{18}$ (iii) $\frac{4}{15}$
8. (iii) median 6·7 cm, lower quartile 5·2 cm, upper quartile 8·4 cm. (iv) 46%
9. (i) $\frac{1}{6}$ (ii) $\frac{2}{9}$ (iii) $\frac{28}{81}$ (iv) $\frac{32}{81}$
10. (i) (a) 2 (b) 3 (c) 6
    (ii) $\frac{1}{16}, \frac{1}{4}, \frac{3}{8}, \frac{1}{4}, \frac{1}{16}$ (iii) 4 km, 3·16 km, 2·83 km, 3·16 km, 4 km. (iv) $\frac{5}{8}$ (v) 3·33 km